Boundary Value Problems on Time Scales, Volume I

Advances in Applied Mathematics
Series Editors:
Daniel Zwillinger, H. T. Banks

Advanced Engineering Mathematics with MATLAB, 4th Edition
Dean G. Duffy

Quadratic Programming with Computer Programs
Michael J. Best

Introduction to Radar Analysis
Bassem R. Mahafza

CRC Standard Mathematical Tables and Formulas, 33rd Edition
Edited by Daniel Zwillinger

The Second-Order Adjoint Sensitivity Analysis Methodology
Dan Gabriel Cacuci

Operations Research
A Practical Introduction, 2nd Edition
Michael Carter, Camille C. Price, Ghaith Rabadi

Handbook of Mellin Transforms
Yu. A. Brychkov, O. I. Marichev, N. V. Savischenko

Advanced Mathematical Modeling with Technology
William P. Fox, Robert E. Burks

Introduction to Quantum Control and Dynamics
Domenico D'Alessandro

Handbook of Radar Signal Analysis
Bassem R. Mahafza, Scott C. Winton, Atef Z. Elsherbeni

Separation of Variables and Exact Solutions to Nonlinear PDEs
Andrei D. Polyanin, Alexei I. Zhurov

Boundary Value Problems on Time Scales, Volume I
Svetlin G. Georgiev, Khaled Zennir

Boundary Value Problems on Time Scales, Volume II
Svetlin G. Georgiev, Khaled Zennir

Observability and Mathematics
Fluid Mechanics, Solutions of Navier-Stokes Equations, and Modeling
Boris Khots

Handbook of Differential Equations, 4th Edition
Daniel Zwillinger, Vladimir Dobrushkin

Experimental Statistics and Data Analysis for Mechanical and Aerospace Engineers
James Middleton

Advanced Engineering Mathematics with MATLAB
Dean G. Duffy

Handbook of Fractional Calculus for Engineering and Science
Harendra Singh, H. M. Srivastava, Juan J. Nieto

Advanced Engineering Mathematics
A Second Course with MatLab
Dean G. Duffy

https://www.routledge.com/Advances-in-Applied-Mathematics/book-series/CRCADVAPPMTH?pd=published,forthcoming&pg=1&pp=12&so=pub&view=list

Boundary Value Problems on Time Scales, Volume I

Svetlin G. Georgiev
Khaled Zennir

CRC Press
Taylor & Francis Group
Boca Raton London New York

CRC Press is an imprint of the
Taylor & Francis Group, an **informa** business
A CHAPMAN & HALL BOOK

First edition published 2022
by CRC Press
6000 Broken Sound Parkway NW, Suite 300, Boca Raton, FL 33487-2742

and by CRC Press
2 Park Square, Milton Park, Abingdon, Oxon, OX14 4RN

CRC Press is an imprint of Taylor & Francis Group, LLC

ISBN: 978-1-032-00291-0 (hbk)
ISBN: 978-1-032-00293-4 (pbk)
ISBN: 978-1-003-17355-7 (ebk)

DOI: 10.1201/9781003173557

Typeset in Nimbus font
by KnowledgeWorks Global Ltd.

Contents

Preface

Time scale theory was first initiated by Stefan Hilger in 1988 in his PhD thesis to unify both approaches of dynamic modelling: difference and differential equations. Similar ideas have been used before and go back to the introduction of the Riemann-Stieltjes integral which unifies sums and integrals. Many results to differential equations carry over easily to corresponding results for difference equations, while other results seem to be totally different in nature. Because of these reasons, the theory of dynamic equations is an active area of research. The time scale calculus can be applied to any fields in which dynamic processes are described by discrete or continuous time models. So, the calculus of time scales has various applications involving non continuous domains such as certain bug populations, phytoremediation of metals, wound healing, maximization problems in economics and traffic problems. Boundary value problems (BVPs) on time scales have been extensively investigated in simulating processes and the phenomena subject to short-time perturbations during their evolution.

This book is devoted to the qualitative theory of boundary value problems on time scales. It summarizes the most recent contributions in this area. The book is intended for senior undergraduate students and beginning graduate students of engineering and science courses. The book contains two volumes. Volume I is devoted to boundary value problems for first and second order dynamic equations on time scales. Volume II investigates boundary value problems for third, fourth and higher order dynamic equations on time scales. The chapters in the book are pedagogically organized. Each chapter concludes with a section with practical problems.

There are seven chapters in Volume I of this book. Chapter 1 investigates some classes of first order boundary value problems with nonlinear boundary conditions. Results are obtained for existence of at least one, at least two and at least three solutions of the considered BVPs. Some classes of periodic boundary value problems are studied. Chapter 2 is devoted to some boundary value problems for first order impulsive dynamic equations. Criteria are given under which the investigated boundary value problems have at least one, at least two and at least three solutions. Chapter 3 gives a definition for the Cauchy function and investigates some of its properties. In the chapter the Green functions are constructed for general two point BVPs, Dirichlet BVPs, focal BVPs and periodic BVPs, as well as for some three points BVPs. In Chapter 4 the Sturm-Liouville theory on time scales is introduced. The first, second and third eigenvalue problems are investigated. Chapter 5 is devoted to some classes of second order boundary value problems on time scales. Results are deducted for existence of one, two and three positive solutions for the considered boundary value problems. In Chapter 6 some nonlinear second order eigenvalue problems are considered. Existence results are deducted and proved for iterative eigenvalue problems. Chapter 7 explores the main classes of nonlinear second order boundary value problems for impulsive dynamic equations.

This book is addressed to a wide audience of specialists such as mathematicians, physicists, engineers and biologists. It can be used as a textbook at the graduate level and as a reference book for several disciplines.

Paris, July 2020 *Svetlin G. Georgiev and Khaled Zennir*

Chapter 1
Boundary Value Problems for Nonlinear First Order Dynamic Equations

Suppose that $\mathbb{T}$ is a time scale with forward jump operator and delta differentiation operator σ and Δ, respectively. Let also, $a, b \in \mathbb{T}$, $a < b$. Throughout this chapter, for $y, z \in \mathbb{R}^n$, with $\langle y, z\rangle$ we will denote the usual inner product and $||z||$ denotes the Euclidean norm of z on $\mathbb{R}^n$.

1.1 Existence of Solutions for First Order Dynamic Equations with Linear Boundary Conditions

In this section we will investigate the first order dynamic equation

$$x^{\Delta} = f(t, x), \quad t \in [a, b], \tag{1.1}$$

subject to the boundary conditions

$$Mx(a) + Rx(\sigma(b)) = 0, \tag{1.2}$$

where

(A1) $f : [a, b] \times \mathbb{R}^n \to \mathbb{R}^n$ is continuous,
(A2) M and R are given constants so that $M + R \neq 0$.

We will start our investigations with the following useful lemma.

Lemma 1.1. *Suppose* (A1) *and* (A2). *Then the boundary value problem (BVP)* (1.1), (1.2) *is equivalent to the integral equation*

$$x(t) = \int_a^t f(s, x(s))\Delta s - \frac{R}{M+R}\int_a^{\sigma(b)} f(s, x(s))\Delta s, \quad t \in [a, b]. \tag{1.3}$$

Proof. 1. Suppose that $x : [a, \sigma(b)] \to \mathbb{R}^n$ satisfies the BVP (1.1), (1.2). Then we integrate (1.1) from a to t and we get

DOI: 10.1201/9781003173557-1

$$x(t) = x(a) + \int_a^t f(s,x(s))\Delta s, \quad t \in [a,\sigma(b)].$$

Hence,

$$x(\sigma(b)) = x(a) + \int_a^{\sigma(b)} f(s,x(s))\Delta s$$

and using (1.2), we arrive at

$$\begin{aligned} 0 &= Mx(a) + Rx(\sigma(b)) \\ &= Mx(a) + R\left(x(a) + \int_a^{\sigma(b)} f(s,x(s))\Delta s\right) \\ &= (M+R)x(a) + R\int_a^{\sigma(b)} f(s,x(s))\Delta s, \end{aligned}$$

whereupon,

$$x(a) = -\frac{R}{M+R}\int_a^{\sigma(b)} f(s,x(s))\Delta s.$$

Therefore,

$$x(t) = \int_a^t f(s,x(s))\Delta s - \frac{R}{M+R}\int_a^{\sigma(b)} f(s,x(s))\Delta s, \quad t \in [a,b],$$

i.e., x satisfies (1.3).

2. Suppose that x satisfies (1.3) on $[a,\sigma(b)]$. Then

$$x^{\Delta}(x) = f(t,x(t)), \quad t \in [a,b],$$

and

$$\begin{aligned} x(a) &= -\frac{R}{M+R}\int_a^{\sigma(b)} f(s,x(s))\Delta s, \\ x(\sigma(b)) &= \int_a^{\sigma(b)} f(s,x(s))\Delta s - \frac{R}{M+R}\int_a^{\sigma(b)} f(s,x(s))\Delta s \\ &= \frac{M}{M+R}\int_a^{\sigma(b)} f(s,x(s))\Delta s. \end{aligned}$$

From here,

$$\begin{aligned} Mx(a) + Rx(\sigma(b)) &= -\frac{MR}{M+R}\int_a^{\sigma(b)} f(s,x(s))\Delta s + \frac{MR}{M+R}\int_a^{\sigma(b)} f(s,x(s))\Delta s \\ &= 0. \end{aligned}$$

Thus, x satisfies (1.1) and (1.2). This completes the proof.

For our next result, we need the following fixed point theorem.

Theorem 1.1. *[23][Schaefer's Fixed Point Theorem] Assume that S is a normed linear space, the operator $F : S \to S$ is compact. Define*

$$H(F) = \{x \in S : x = \mu F x, \quad \mu \in (0,1)\}.$$

Then, either

1. *the set $H(F)$ is unbounded, or*
2. *F has a fixed point in S.*

Theorem 1.2. *Suppose (A1) and (A2). Let, also, there exist non-negative constants α and K so that*

$$\|f(t,q)\| \le 2\alpha\langle q, f(t,q)\rangle + K, \quad q \in \mathbb{R}^n, \quad t \in [a,\sigma(b)], \tag{1.4}$$

and $R \neq 0$, $\left|\frac{M}{R}\right| \le 1$. Then the BVP (1.1), (1.2) has at least one solution $x \in \mathscr{C}([a,\sigma(b)])$.

Proof. Set

$$P = \left(1 + \left|\frac{R}{M+R}\right|\right) K(\sigma(b) - a) + 1$$

and

$$B_P = \left\{x \in \mathscr{C}([a,\sigma(b)]) : \max_{t\in[a,\sigma(b)]} \|x(t)\| < P\right\}.$$

For $x \in \mathscr{C}([a,\sigma(b)])$, define the operator $T : \mathscr{C}([a,\sigma(b)]) \to \mathscr{C}([a,\sigma(b)])$, as follows

$$Tx(t) = \int_a^t f(s,x(s))\Delta s - \frac{R}{M+R}\int_a^{\sigma(b)} f(s,x(s))\Delta s, \quad t \in [a,\sigma(b)].$$

Denote

$$r(t) = \|x(t)\|^2, \quad t \in [a,\sigma(b)].$$

Then

$$\begin{aligned}
r^\Delta(t) &= \langle x^\Delta(t), x^\sigma(t)\rangle + \langle x(t), x^\Delta(t)\rangle \\
&= \langle x^\Delta(t), x^\sigma(t) + x(t)\rangle \\
&= \langle x^\Delta(t), x(t) + \mu(t)x^\Delta(t) + x(t)\rangle
\end{aligned}$$

$$= 2\langle x(t), x^{\Delta}(t)\rangle + \mu(t)\langle x^{\Delta}(t), x^{\Delta}(t)\rangle$$

$$= 2\langle x(t), x^{\Delta}(t)\rangle + \mu(t)\|x^{\Delta}(t)\|^2, \quad t \in [a, \sigma(b)].$$

Suppose that x satisfies the BVP

$$x^{\Delta} = \lambda f(t,x), \quad t \in [a, \sigma(b)],$$

$$Mx(a) + Rx(\sigma(b)) = 0.$$

Then

$$r^{\Delta}(t) = 2\langle x(t), \lambda f(t,x(t))\rangle + \mu(t)\|\lambda f(t,x(t))\|^2$$

$$\geq 2\langle x(t), \lambda f(t,x(t))\rangle, \quad t \in [a, \sigma(b)].$$

By (1.4), we find

$$\|\lambda f(t,x(t))\| \leq 2\alpha\langle x(t), \lambda f(t,x(t))\rangle + \lambda K$$

$$\leq \alpha r^{\Delta}(t) + K, \quad t \in [a, \sigma(b)].$$

Also,

$$\begin{aligned} r(\sigma(b)) &= \|x(\sigma(b))\| \\ &= \frac{1}{|R|}\|-Rx(\sigma(b))\| \\ &= \frac{1}{|R|}\|Mx(a)\| \\ &= \left|\frac{M}{R}\right| \|x(a)\| \\ &= \left|\frac{M}{R}\right| r(a) \\ &\leq r(a), \end{aligned}$$

and

$$\begin{aligned} \|x(\sigma(b))\| &= \left|\frac{M}{R}\right| \|x(a)\| \\ &\leq \|x(a)\|. \end{aligned}$$

Take $x \in \overline{B}_P$ arbitrarily. Then

$$\begin{aligned}
\|\lambda Tx(t)\| &= \| -\frac{R}{M+R}\int_a^{\sigma(b)} \lambda f(s,x(s))\Delta s + \int_a^t \lambda f(s,x(s))\Delta s\| \\
&\leq \| -\frac{R}{M+R}\int_a^{\sigma(b)} \lambda f(s,x(s))\Delta s\| + \|\int_a^t \lambda f(s,x(s))\Delta s\| \\
&\leq \left|\frac{R}{M+R}\right|\int_a^{\sigma(b)} \|\lambda f(s,x(s))\|\Delta s + \int_a^t \|\lambda f(s,x(s))\|\Delta s \\
&= \left(1+\left|\frac{R}{M+R}\right|\right)\int_a^{\sigma(b)} \|\lambda f(s,x(s))\|\Delta s \\
&\leq \left(1+\left|\frac{R}{M+R}\right|\right)\int_a^{\sigma(b)} \left(\alpha r^{\Delta}(s)+K\right)\Delta s \\
&= \left(1+\left|\frac{R}{M+R}\right|\right)(\alpha(r(\sigma(b))-r(a))+K(\sigma(b)-a)) \\
&\leq \left(1+\left|\frac{R}{M+R}\right|\right)K(\sigma(b)-a) \\
&< P, \quad t \in [a,\sigma(b)].
\end{aligned}$$

Thus, $T : \overline{B}_P \to \mathscr{C}([a,\sigma(b)])$ satisfies

$$x \neq \lambda Tx, \quad x \in \partial B_P, \quad \lambda \in [0,1].$$

By the Arzela-Ascoli theorem, it follows that $T : \overline{B}_P \to \mathscr{C}([a,\sigma(b)])$ is compact. From here and from Schaefer's fixed point theorem it follows that the operator T has a fixed point in $\overline{B}_P$. This completes the proof.

Example 1.1. Let $\mathbb{T} = \mathbb{Z}$ and $n = 1$. Consider the BVP

$$\begin{aligned}
x^{\Delta} &= \frac{1}{1+x^2}, \quad t \in [0,10], \\
x(0)+3x(11) &= 0.
\end{aligned} \tag{1.5}$$

Here $\sigma(t) = t+1, t \in \mathbb{T}$,

$$f(t,x) = \frac{1}{1+x^2}, \quad t \in [0,11], \quad x \in \mathbb{R}, \quad M = 1, \quad R = 3.$$

Take $\alpha = \frac{1}{2}$, $K = 2$. Then

$$|f(t,x)| \leq 2\alpha x f(t,x) + K \quad \Longleftrightarrow$$

$$\begin{aligned}
f(t,x) \le x f(t,x) + 2 \quad &\Longleftrightarrow \\
(1-x) f(t,x) \le 2 \quad &\Longleftrightarrow \\
\frac{1-x}{1+x^2} \le 2 \quad &\Longleftrightarrow \\
1 - x \le 2 + 2x^2 \quad &\Longleftrightarrow \\
0 \le 1 + x + 2x^2, \quad x \in \mathbb{R}, \quad t \in [0,11]&.
\end{aligned}$$

Next,

$$\left|\frac{M}{R}\right| = \frac{M}{R} = \frac{1}{3} \le 1.$$

Hence and applying Theorem 1.2, it follows that the BVP (1.5) has at least one solution $x \in \mathscr{C}([0,11])$.

Exercise 1.1. Let $\mathbb{T} = 2^{\mathbb{N}_0}$ and $n = 1$. Prove that the BVP

$$x^{\Delta} = \frac{x^2}{(1+t+t^2)(1+x^4)(1+x^6)}, \quad t \in [1,64],$$

$$10x(1) + 21x(128) = 0,$$

has at least one solution $x \in \mathscr{C}([1,128])$.

1.2 Existence of Solutions of First Order Dynamic Equations with Nonlinear Functional Boundary Value Conditions

In this section, we study the existence of solutions of the following first order dynamic equation with nonlinear functional boundary value conditions

$$\begin{aligned}
x^{\Delta}(t) &= f(t, x^{\sigma}(t)) \quad \text{for} \quad \Delta-\text{a.e.} \quad t \in [a,b], \\
B(x(a), x) &= 0,
\end{aligned} \tag{1.6}$$

where

(A3) $B \in \mathscr{C}(\mathbb{R} \times \mathscr{C}(\mathbb{T}))$, $f : [a,b] \times \mathbb{R} \to \mathbb{R}$ is L^1 Δ-Carathéodory function such that

1. for any $r \in \mathbb{R}$, $f(\cdot, r)$ is Δ-measurable in $[a,b]$,
2. $f(t,\cdot) \in \mathscr{C}(\mathbb{R})$ for $\Delta-\text{a.e.}$ $t \in [a,b]$,
3. for any $p > 0$ there exists $m_p \in L^1([a,b])$ such that

$$|f(t,x)| \le m_p(t) \quad \text{for} \quad \Delta-\text{a.e.} \quad t \in [a,b]$$

and for all $|x| \le p$.

Definition 1.1. A function $x : \mathbb{T} \to \mathbb{R}$ is said to be a solution of the BVP (1.6) if

1.

$$x \in AC([a,b]) = \left\{ x \in \mathscr{C}(\mathbb{T}) : x^{\Delta} \in L^1([a,b]), \;\; x(t) = x(a) + \int_a^t x^{\Delta}(s)\Delta s, \;\; t \in \mathbb{T} \right\},$$

2. $B(x(a),x) = 0$,
3. it satisfies the dynamic equation (1.6) for $\Delta-$a.e. $t \in [a,b]$.

For $y,z \in \mathscr{C}(\mathbb{T})$ such that $y(t) \le z(t)$, $t \in \mathbb{T}$, denote

$$[y,z] = \{x \in \mathscr{C}(\mathbb{T}), \quad y(t) \le x(t) \le z(t), \quad t \in \mathbb{T}\}.$$

Definition 1.2. We say that $\alpha, \beta \in AC([a,b])$ is a pair of coupled lower and upper solutions of the dynamic equation (1.6), if

1. $\alpha \le \beta$ in $\mathbb{T}$,
2. $\alpha^{\Delta}(t) - f(t,\alpha^{\sigma}(t)) \le 0 \le \beta^{\Delta}(t) - f(t,\beta^{\sigma}(t))$ for $\Delta-$a.e. $t \in [a,b]$,
3. $B(\alpha(a),x) \le 0 \le B(\beta(a),x)$ for any $x \in [\alpha,\beta]$.

In addition, in this section we suppose

(A4) There exist $\alpha, \beta \in AC([a,b])$, a pair of lower and upper solutions of (1.6).

Consider the following modified problem

$$\begin{aligned} x^{\Delta}(t) &= F(t,x^{\sigma}(t)) \quad \text{for} \quad \Delta-\text{a.e.} \quad t \in [a,b], \\ x(a) &= p(a,x(a) - B(x(a),x)), \end{aligned} \tag{1.7}$$

where

$$F(t,x) = f(t,p(\sigma(t),x)) - \frac{r - p(\sigma(t),r)}{1+|r|},$$

$$p(t,r) = \max\{\alpha(t), \min\{r, \beta(t)\}\}, \quad r \in \mathbb{R}, \quad t \in [a,b].$$

Before our existence result, we need the following useful lemma.

Lemma 1.2. *Suppose* (A3) *and* (A4). *Then any solution of the dynamic equation* (1.7) *belong to the section* $[\alpha,\beta]$ *and are solutions of the dynamic equation* (1.6).

Proof. Let $x : \mathbb{T} \to \mathbb{R}$ be a solution of the dynamic equation (1.7). Set $z = x - \beta$. Suppose that it is false $z \le 0$ in $\mathbb{T}$. Since

$$z(a) = x(a) - \beta(a)$$

$$= p(a,x(a)-B(x(a),x))-\beta(a)$$

$$= \max\{\alpha(a),\min\{x(a)-B(x(a),x),\beta(a)\}\}-\beta(a)$$

$$\leq \beta(a)-\beta(a)$$

$$= 0.$$

Then there exists a $c \in \mathbb{T}\backslash\{a\}$ such that

$$z(c) = \max\{z(t) : t \in \mathbb{T}\} > 0.$$

1. Suppose that $\rho(c) = c$. Then there exists a $\delta > 0$ such that $z(t) > 0$ for any $t \in [\rho(c-\delta),c]$. Note that there exists $z^{\Delta}(\rho(t))$ for Δ-a.e. $t \in [c-\delta,c]$ and

$$\begin{aligned}
z^{\Delta}(\rho(t)) &= x^{\Delta}(\rho(t)) - \beta^{\Delta}(\rho(t)) \\
&= x^{\Delta}(t) - \beta^{\Delta}(t) \\
&\leq x^{\Delta}(t) - f(t,\beta(t)) \\
&= F(t,x(t)) - f(t,\beta(t)) \\
&= f(t,p(t,x(t))) - \frac{x(t)-p(t,x(t))}{1+|x(t)|} \\
&\quad -f(t,\beta(t)) \\
&= f(t,\beta(t)) - \frac{x(t)-\beta(t)}{1+|x(t)|} - f(t,\beta(t)) \\
&= -\frac{x(t)-\beta(t)}{1+|x(t)|} \\
&< 0, \quad t \in [c-\delta,c].
\end{aligned}$$

Since $z \in AC([a,b])$, for any $t \in [\rho(c-\delta),c]$, we have

$$\begin{aligned}
z(c)-z(t) &= \int_t^c z^{\Delta}(s)\Delta s \\
&< 0.
\end{aligned}$$

This is a contradiction with the definition of c.

2. Suppose that $\rho(c) < c$. Then

$$z^{\Delta}(\rho(c)) = \frac{z(c)-z(\rho(c))}{c-\rho(c)} \qquad (1.8)$$

$$> 0$$

and

$$x^{\Delta}(\rho(c)) = F(\rho(c),x(c))$$

$$= f(\rho(c),\beta(c)) - \frac{x(c)-\beta(c)}{1+|x(c)|}$$

$$\leq f(\rho(c),\beta(c))$$

$$\leq \beta^{\Delta}(\rho(c)).$$

By the last inequality, we get

$$z^{\Delta}(\rho(c)) \leq 0,$$

which contradicts with (1.8).

Therefore, $u \leq \beta$ in $\mathbb{T}$. As in the above, one can prove that $u \geq \alpha$ in $\mathbb{T}$. If

$$x(a) - B(x(a),x) < \alpha(a),$$

then

$$x(a) = p(a,x(a)-B(x(a),x))$$

$$= \max\{\alpha(a), \min\{x(a)-B(x(a),x), \beta(a)\}\}$$

$$= \alpha(a)$$

and then

$$B(\alpha(a),x) > 0.$$

The last inequality contradicts with the definition of α. Consequently,

$$x(a) - B(x(a),x) \geq \alpha(a).$$

As in the above, one can prove that

$$x(a) - B(x(a),x) \leq \beta(a).$$

Thus, any solution x of equation (1.7) is a solution of equation (1.6) and belongs to $[\alpha,\beta]$. This completes the proof.

Suppose that $\mathscr{C}(\mathbb{T})$ is endowed with the supremum norm. Set

$$M_F(t) = \int_a^t m_F(s)\Delta s, \quad t \in \mathbb{T},$$

$$K_F = \|M_F\|,$$

$$K_P = \sup\{|p(t,r)| : (t,r) \in \mathbb{T}\times\mathbb{R}\},$$

$$K = K_P + K_F,$$

$$D = \{x \in \mathscr{C}(\mathbb{T}) : \|x\| \le K\}.$$

Define the operator $T : \mathscr{C}(\mathbb{T}) \to \mathscr{C}(\mathbb{T})$ as follows

$$Tx(t) = p(a,x(a) - B(x(a),x)) + \int_a^t F(s,x^\sigma(s))\Delta s, \quad t \in \mathbb{T}.$$

Note that any fixed point $x \in \mathscr{C}(\mathbb{T})$ of the operator T is a solution to the BVP (1.7). For our next result, we need the following fixed point theorem.

Theorem 1.3. *[23][Schauder Fixed Point Theorem] Any compact convex nonempty subset Y of a normed space has the fixed point property.*

Theorem 1.4. *Suppose* (A3) *and* (A4). *Then problem* (1.6) *has at least one solution in* $[\alpha,\beta]$.

Proof. We have that $T : \mathscr{C}(\mathbb{T}) \to \mathscr{C}(\mathbb{T})$. Take $x \in D$ arbitrarily. Then

$$\begin{aligned}
|Tx(t)| &= \left| p(a,x(a) - B(x(a),x)) + \int_a^t F(s,x^\sigma(s))\Delta s \right| \\
&\le |p(a,x(a) - B(x(a),x))| + \left| \int_a^t F(s,x^\sigma(s))\Delta s \right| \\
&\le K_P + \int_a^t |F(s,x^\sigma(s))|\Delta s \\
&\le K_P + \int_a^t m_F(s)\Delta s \\
&= K_P + M_F(t)
\end{aligned}$$

$$\leq K_P + \|M_F\|$$

$$\leq K_P + K_F$$

$$= K, \quad t \in \mathbb{T}.$$

Hence,

$$\|Tx\| \leq K.$$

Therefore, $T : D \to D$. Note that D is a nonempty, bounded, closed and convex subset of $\mathscr{C}(\mathbb{T})$. Hence and by the Schauder fixed point theorem, we conclude that T has a fixed point in D, which is a solution to the BVP (1.7). By Lemma 1.2, we conclude that the BVP (1.6) has a solution in D. This completes the proof.

Now, we consider a particular case of the BVP (1.6). Namely, consider the BVP

$$\begin{aligned} x^{\Delta}(t) &= f(t, x^{\sigma}(t)) \quad \text{for} \quad \Delta - \text{a.e.} \quad t \in [a,b], \\ x(a) &= g(x^{\sigma}(b)), \end{aligned} \tag{1.9}$$

where g is a continuous and increasing function. By Theorem 1.4, we get the following result.

Theorem 1.5. *Assume that* $f : [a,b] \times \mathbb{R} \to \mathbb{R}$ *is a* L^1 Δ*-Carathéodory function and there exists a pair of coupled lower and upper solutions* $\alpha, \beta \in AC([a,b])$ *of the BVP* (1.9)*, i.e.,*

$$\alpha^{\Delta}(t) \leq f(t, \alpha^{\sigma}(t) \quad \textit{for} \quad \Delta - \textit{a.e.} \quad t \in [a,b],$$

$$\alpha(a) \leq g(\beta^{\sigma}(b)),$$

$$\beta^{\Delta}(t) \geq f(t, \beta^{\sigma}(t)) \quad \textit{for} \quad \Delta - \textit{a.e.} \quad t \in [a,b],$$

$$\beta(a) \geq g(\alpha^{\sigma}(b))$$

and $\alpha \leq \beta$ *in* $\mathbb{T}$*. Let, also, g be a continuous and non-increasing function in* $\mathbb{T}$*. Then the BVP* (1.9) *has at least one solution in* $[\alpha, \beta]$*.*

Now, suppose that

(A5) there exists a function $h \in L^1([a,b])$ such that

$$1 - \mu(t)h(t) > 0, \quad t \in [a,b],$$

and if

$$\alpha^\sigma(t) \le r \le r' \le \beta^\sigma(t), \quad t \in [a,b],$$

then

$$f(t,r') - f(t,r) \le h(t)(r'-r)$$

$$\le 0, \quad t \in [a,b].$$

Theorem 1.6. *Suppose that all conditions of Theorem 1.5 hold and (A5) holds. Then the BVP (1.9) has exactly one solution in $[\alpha,\beta]$.*

Proof. By Theorem 1.5, it follows that the BVP (1.9) has at least one solution in $[\alpha,\beta]$. Suppose that the BVP (1.9) has two solutions $\bar{x}$ and x in $[\alpha,\beta]$ such that $\bar{x} \not\equiv x$. We will consider the following two cases.

1. Let $x(a) \ne \bar{x}(a)$. Without loss of generality, suppose that $x(a) > \bar{x}(a)$. Then

$$g(x^\sigma(b)) = x(a)$$

$$> \bar{x}(a)$$

$$= g(\bar{x}^\sigma(b)).$$

Since g is non-increasing, by the last inequalities we conclude that

$$x^\sigma(b) \le \bar{x}^\sigma(b).$$

Hence, there exists a $t_0 \in [a,b]$ such that

$$\bar{x}(t_0) \le x(t_0) \quad \text{and} \quad x(t) \le \bar{x}(t)$$

for any $t \in [\sigma(t_0), \sigma(b)]$. By (A5), we get

$$(\bar{x}-x)^\Delta(t) = f(t,\bar{x}^\sigma(t)) - f(t,x^\sigma(t))$$

$$\le h(t)(\bar{x}^\sigma(t) - x^\sigma(t)) \quad \text{for} \quad \Delta-\text{a.e.} \quad t \in [t_0,b].$$

Therefore,

$$r(t) = (\bar{x}-x)^\Delta(t) - h(t)(\bar{x}^\sigma(t) - x^\sigma(t))$$

$$\le 0 \quad \text{for} \quad \Delta-\text{a.e.} \quad t \in [t_0,b].$$

Hence,

$$(\overline{x}-x)(t) = e_{\ominus(-h)}(t,t_0)(\overline{x}-x)(t_0) + \int_{t_0}^{t} e_{\ominus(-h)}(t,s)r(s)\Delta s, \quad t \in [t_0,\sigma(b)]. \tag{1.10}$$

By $(A5)$, we get

$$\begin{aligned}1+\mu(t)\,(\ominus(-h)(t)) &= 1+\frac{\mu(t)h(t)}{1-\mu(t)h(t)}\\ &= \frac{1}{1-\mu(t)h(t)}\\ &> 0, \quad t \in [a,b].\end{aligned}$$

Therefore,

$$e_{\ominus(-h)}(t,t_0) > 0, \quad t \in [t_0,\sigma(b)]. \tag{1.11}$$

Hence and (1.10), we obtain

$$\overline{x}(t) \le x(t), \quad t \in [t_0,\sigma(b)].$$

In particular, by the last inequality, we find

$$\overline{x}^{\sigma}(b) \le x^{\sigma}(b),$$

which is a contradiction.

2. Let $x(a) = \overline{x}(a)$. Since $x \not\equiv \overline{x}$, there exists a $t_1 \in \mathbb{T}\backslash\{a\}$ such that $x(t_1) \neq \overline{x}(t_1)$. Without loss of generality, assume that $x(t_1) < \overline{x}(t_1)$. Then there exists a $t_2 \in (a,t_1]$ so that]

$$x(t) \le \overline{x}(t), \quad t \in (t_2,t_1],$$

and

$$x(t_2) = \overline{x}(t_2) \tag{1.12}$$

or

$$x(t_2) < \overline{x}(t_2) \quad \text{and} \quad x(\rho(t_2)) > \overline{x}(\rho(t_2)). \tag{1.13}$$

Set $\overline{t} = t_2$ in the case when we have (1.12) and $\overline{t} = \rho(t_2)$ in the case when we have (1.13). Then

$$\begin{aligned}(\overline{x}-x)^{\Delta}(t) &= \overline{x}^{\Delta}(t) - x^{\Delta}(t)\\ &= f(t,\overline{x}^{\sigma}(t)) - f(t,x^{\sigma}(t))\\ &\le h(t)(\overline{x}-x)^{\sigma}(t)\\ &\le 0) \quad \text{for} \quad \Delta-\text{a.e.} \quad t \in [\overline{t},t_1].\end{aligned}$$

Thus, we get the following initial value problem (IVP)

$$r_1(t) = (\overline{x}-x)^{\Delta}(t) - h(t)(\overline{x}-x)^{\sigma}(t)$$

$$\leq 0 \quad \text{for} \quad \Delta-\text{a.e.} \quad t \in [\overline{t}, t_1],$$

$$(\overline{x}-x)(\overline{t}) = 0.$$

Hence and applying (1.11), we arrive at

$$(\overline{x}-x)(t) = \int_{\overline{t}}^{t} e_{\ominus(-h)}(t,s) r_1(s) \Delta s$$

$$\leq 0, \quad t \in [\overline{t}, t_1].$$

By the last inequality, in particular we get

$$\overline{x}(t_1) \leq x(t_1),$$

which is a contradiction. This completes the proof.

In addition, below we suppose

(A6) there exists $M > 0$ such that for any $t \in [a,b]$ and

$$\alpha^{\sigma}(t) \leq r \leq r' \leq \beta^{\sigma}(t),$$

we have

$$f(t,r') - f(t,r) \geq -M(r'-r).$$

(A7) there exists $N > 0$ such that for any $t \in [a,b]$ and

$$\alpha^{\sigma}(t) \leq r \leq r' \leq \beta^{\sigma}(t),$$

we have

$$g(r') - g(r) \geq -N(r'-r)$$

and

$$N e_{\ominus(-h)}(\sigma(b), a) < 1. \tag{1.14}$$

Theorem 1.7. *Suppose that all conditions of Theorem 1.6 hold and (A6) and (A7) hold. Then there exists two monotone sequences $\{y_n\}_{n\in\mathbb{N}}$ and $\{z_n\}_{n\in\mathbb{N}}$ such that*

$$\alpha = y_0 \leq y_n \leq z_n \leq z_0 \leq \beta \quad \textit{in} \quad \mathbb{T}$$

for any $n \in \mathbb{N}$, which converge in $\mathscr{C}(\mathbb{T})$ to the unique solution of the BVP (1.9).

Proof. For any $\eta, \xi \in [\alpha, \beta]$ consider the BVPs

$$\begin{aligned} x^{\Delta}(t) &= f(t,\eta^{\sigma}(t)) + M(\eta^{\sigma}(t) - x^{\sigma}(t)) \quad \text{for} \quad \Delta - \text{a.e.} \quad t \in [a,b], \\ x(a) &= g(\xi^{\sigma}(b)), \end{aligned} \tag{1.15}$$

$$\begin{aligned} x^{\Delta}(t) &= f(t,\alpha^{\sigma}(t)) + M(\alpha^{\sigma}(t) - x^{\sigma}(t)) \quad \text{for} \quad \Delta - \text{a.e.} \quad t \in [a,b], \\ x(a) &= g(\beta^{\sigma}(b)), \end{aligned} \tag{1.16}$$

$$\begin{aligned} x^{\Delta}(t) &= f(t,\beta^{\sigma}(t)) + M(\beta^{\sigma}(t) - x^{\sigma}(t)) \quad \text{for} \quad \Delta - \text{a.e.} \quad t \in [a,b], \\ x(a) &= g(\alpha^{\sigma}(b)). \end{aligned} \tag{1.17}$$

Define the function

$$G_{\eta}(t,r) = f(t,\eta^{\sigma}(t)) + M(\eta^{\sigma}(t) - r), \quad (t,r) \in [a,b] \times \mathbb{R}.$$

We have that G_{η} is a L^1 Δ-Carathéodory function. Now, for any $\eta, \xi \in [\alpha, \beta]$, using $(A6)$ and the monotonicity of the function g, we get

$$\begin{aligned} \alpha^{\Delta}(t) &\leq f(t,\alpha^{\sigma}(t)) \\ &\leq f(t,\alpha^{\sigma}(t)) + M(\eta^{\sigma}(t) - \alpha^{\sigma}(t)) \\ &= G_{\eta}(t,\alpha^{\sigma}(t)) \quad \text{for} \quad \Delta - \text{a.e.} \quad t \in [a,b], \\ \alpha(a) &\leq g(\beta^{\sigma}(b)) \\ &\leq g(\alpha^{\sigma}(b)), \\ \beta^{\Delta}(t) &\geq f(t,\beta^{\sigma}(t)) \\ &\geq f(t,\beta^{\sigma}(t)) + M(\eta^{\sigma}(t) - \beta^{\sigma}(t)) \\ &= G_{\eta}(t,\beta^{\sigma}(t)) \quad \text{for} \quad \Delta - \text{a.e.} \quad t \in [a,b], \\ \beta(a) &\geq g(\alpha^{\sigma}(b)) \\ &\geq g(\beta^{\sigma}(b)). \end{aligned}$$

By Theorem 1.6, it follows that (1.16) and (1.17) have unique solutions in $[\alpha, \beta]$ which we will denote by y_1 and z_1, respectively. Set

$$p_1 = y_1 - z_1 \quad \text{in} \quad \mathbb{T}.$$

By $(A6)$, we get

$$\begin{aligned}
r(t) &= p_1^\Delta(t) + M p_1^\sigma(t) \\
&= y_1^\Delta(t) - z_1^\Delta(t) + M(y_1^\sigma(t) - z_1^\sigma(t)) \\
&= f(t, \alpha^\sigma(t)) + M(\alpha^\sigma(t) - y_1^\sigma(t)) \\
&\quad - f(t, \beta^\sigma(t)) - M(\beta^\sigma(t) - z_1^\sigma(t)) \\
&\quad + M(y_1^\sigma(t) - z_1^\sigma(t)) \\
&= f(t, \alpha^\sigma(t)) - f(t, \beta^\sigma(t)) + M(\alpha^\sigma(t) - \beta^\sigma(t)) \\
&\leq M(\alpha^\sigma(t) - \beta^\sigma(t)) - M(\alpha^\sigma(t) - \beta^\sigma(t)) \\
&= 0 \quad \text{for} \quad \Delta - \text{a.e.} \quad t \in [a,b],
\end{aligned}$$

and

$$\begin{aligned}
p_1(a) &= y_1(a) - z_1(a) \\
&= g(\beta^\sigma(b)) - g(\alpha^\sigma(b)) \\
&\leq 0.
\end{aligned}$$

Then

$$p_1(t) = e_{\ominus M}(t,a) p_1(a) + \int_a^t e_{\ominus M}(t, \sigma(s)) r(s) \Delta s, \quad t \in \mathbb{T}.$$

Note that

$$\begin{aligned}
1 + \mu(t)(\ominus M)(t) &= 1 - \frac{M\mu(t)}{1 + M\mu(t)} \\
&= \frac{1}{1 + M\mu(t)} \\
&> 0, \quad t \in [a,b].
\end{aligned}$$

Then $e_{\ominus M}(t,s) > 0$, $t \in \mathbb{T}$, $s \in [a,b]$. Thus,

$$p_1(t) \leq 0, \quad t \in \mathbb{T},$$

and

$$y_1(t) \leq z_1(t), \quad t \in \mathbb{T}.$$

Therefore, y_1 and z_1 are a pair of coupled lower and upper solutions of the BVP (1.9). Now, we consider the BVPs

$$\begin{aligned} x^{\Delta}(t) &= f(t, y_1^{\sigma}(t)) + M(y_1^{\sigma}(t) - x^{\sigma}(t)) \quad \text{for} \quad \Delta-\text{a.e.} \quad t \in [a,b], \\ x(a) &= g(z_1^{\sigma}(b)), \end{aligned} \tag{1.18}$$

$$\begin{aligned} x^{\Delta}(t) &= f(t, z_1^{\sigma}(t)) + M(z_1^{\sigma}(t) - x^{\sigma}(t)) \quad \text{for} \quad \Delta-\text{a.e.} \quad t \in [a,b], \\ x(a) &= g(y_1^{\sigma}(b)). \end{aligned} \tag{1.19}$$

By Theorem 1.6, we have that the BVP (1.18) has a unique solution $y_2 \in [y_1, z_1]$ and the BVP (1.19) has a unique solution $z_2 \in [y_1, z_1]$. As in the above, we get

$$y_2 \leq z_2 \quad \text{in} \quad [a,b].$$

Continuing this process, we get sequences $\{y_n\}_{n\in\mathbb{N}}$ and $\{z_n\}_{n\in\mathbb{N}}$ such that y_n is the unique solution to the BVP

$$\begin{aligned} y_n^{\Delta}(t) &= f(t, y_{n-1}^{\sigma}(t)) + M(y_{n-1}^{\sigma}(t) - y_n^{\sigma}(t)) \quad \text{for} \quad \Delta-\text{a.e.} \quad t \in [a,b], \\ y_n(a) &= g(z_{n-1}^{\sigma}(b)), \end{aligned} \tag{1.20}$$

and z_n is the unique solution to the BVP

$$\begin{aligned} z_n^{\Delta}(t) &= f(t, z_{n-1}^{\sigma}(t)) + M(z_{n-1}^{\sigma}(t) - z_n^{\sigma}(t)) \quad \text{for} \quad \Delta-\text{a.e.} \quad t \in [a,b], \\ z_n(a) &= g(y_{n-1}^{\sigma}(b)). \end{aligned} \tag{1.21}$$

For these sequences, we have

$$\alpha = y_0 \leq y_n \leq y_{n+1} \leq z_{n+1} \leq z_n \leq z_0 = \beta \quad \text{in} \quad \mathbb{T}.$$

Hence, the limits

$$y(t) = \lim_{n\to\infty} y_n(t), \quad z(t) = \lim_{n\to\infty} z_n(t), \quad t \in [a,b],$$

exist for any $t \in [a,b]$, and $y_n \to y$, $z_n \to z$ in $\mathscr{C}(\mathbb{T})$, as $n \to \infty$, where

$$\begin{aligned} y^{\Delta}(t) &= f(t, y^{\sigma}(t)) \quad \text{for} \quad \Delta-\text{a.e.} \quad t \in [a,b], \\ y(a) &= g(z^{\sigma}(b)), \end{aligned}$$

and

$$z^{\Delta}(t) = f(t, z^{\sigma}(t)) \quad \text{for} \quad \Delta - \text{a.e.} \quad t \in [a,b],$$

$$z(a) = g(y^{\sigma}(b)),$$

and

$$\alpha \le y \le z \le \beta \quad \text{in} \quad \mathbb{T}.$$

Let

$$p = z - y.$$

Then, using $(A5)$, we find

$$\begin{aligned} p^{\Delta}(t) &= z^{\Delta}(t) - y^{\Delta}(t) \\ &= f(t, z^{\sigma}(t)) - f(t, y^{\sigma}(t)) \\ &\le h(t)(z^{\sigma}(t) - y^{\sigma}(t)) \\ &= h(t)p^{\sigma}(t) \quad \text{for} \quad \Delta - \text{a.e.} \quad t \in [a,b]. \end{aligned}$$

Hence and by the Gronwall inequality, we find

$$p(t) \le e_{\ominus(-h)}(t,a)p(a), \quad t \in \mathbb{T}.$$

By $(A7)$, we obtain

$$\begin{aligned} 0 &\le p(a) \\ &= g(z^{\sigma}(b)) - g(y^{\sigma}(b)) \\ &\le N(z^{\sigma}(b) - y^{\sigma}(b)) \\ &= Np^{\sigma}(b). \end{aligned}$$

Therefore,

$$\begin{aligned} 0 &\le p(a) \\ &\le Np^{\sigma}(b) \\ &\le Ne_{\ominus(-h)}(\sigma(b),a)p(a). \end{aligned}$$

Now, applying (1.14), we conclude that $p(a) = 0$ and $p \equiv 0$ in $\mathbb{T}$. Thus, $y \equiv z$ in $\mathbb{T}$ and y and z are solutions to the BVP (1.9). Now, by the uniqueness of the solution of the BVP (1.9), we get the desired result. This completes the proof.

1.3 Positive Solutions for First Order Dynamic Equations with Nonlinear Functional Boundary Value Conditions

In this section we will investigate the following BVP

$$\begin{aligned} x^{\Delta}(t) &= -p(t)x(\sigma(t)) + f(t,x(t)), \quad t \in [a,b], \\ x(a) &= g(x(\sigma(b))), \end{aligned} \tag{1.22}$$

where

(A8) $f : [a,b] \times [0,\infty) \to [0,\infty)$ is a continuous function, $\int_a^{\sigma(b)} f(s,a)\Delta s > 0$, and there exists a constant $R > 0$ such that

$$f(t,u) \le f(t,v) \le \frac{R}{2(\sigma(b)-a)}, \quad t \in [a,b], \quad 0 \le u \le v \le R.$$

(A9) $g : [0,\infty) \to [0,\infty)$ is a continuous function,

$$g(u) \le g(v) \le \frac{R}{2}, \quad 0 \le u \le v \le R,$$

where R is the same constant as in $(A8)$.

(A10) $p : [a,b] \to (0,\infty)$, $p \in \mathscr{C}_{rd}([a,b])$.

Suppose that $\mathscr{C}([a,\sigma(b)])$ is equipped with the maximum norm. For $x \in \mathscr{C}([a,\sigma(b)])$ define the operator

$$\Phi x(t) = \frac{1}{e_p(t,a)} \left(\int_a^t e_p(s,a) f(s,x(s))\Delta s + g(x(\sigma(b))) \right), \quad t \in [a,\sigma(b)].$$

Lemma 1.3. *Suppose* $(A8)$-$(A10)$. *Then any fixed point* $x \in \mathscr{C}([a,\sigma(b)])$ *of the operator* Φ *is a solution to the BVP* (1.22).

Proof. Let $x \in \mathscr{C}([a,\sigma(b)])$ is a fixed point of the operator Φ. Then

$$x(t) = \frac{1}{e_p(t,a)} \left(\int_a^t e_p(s,a) f(s,x(s))\Delta s + g(x(\sigma(b))) \right), \quad t \in [a,\sigma(b)].$$

Then

$$x(a) = g(x(\sigma(b)))$$

and

$$
\begin{aligned}
x^{\Delta}(t) &= \left(\frac{1}{e_p(\cdot,a)}\right)^{\Delta}(t)\left(\int_a^{\sigma(t)} e_p(s,a)f(s,x(s))\Delta s + g(x(\sigma(b)))\right)\\
&\quad + \frac{1}{e_p(t,a)}\left(e_p(t,a)f(t,x(t))\right)\\
&= -\frac{p(t)e_p(t,a)}{e_p(t,a)e_p(\sigma(t),a)}\left(\int_a^{\sigma(t)} e_p(s,a)f(s,x(s))\Delta s + g(x(\sigma(b)))\right)\\
&\quad + f(t,x(t))\\
&= -\frac{p(t)}{e_p(\sigma(t),a)}\left(\int_a^{\sigma(t)} e_p(s,a)f(s,x(s))\Delta s + g(x(\sigma(b)))\right)\\
&\quad + f(t,x(t))\\
&= -p(t)x(\sigma(t)) + f(t,x(t)), \quad t\in[a,b].
\end{aligned}
$$

Thus, x is a solution to the BVP (1.22). This completes the proof.

Let

$$
K = \{x \in \mathscr{C}([a,\sigma(b)]) : x(t) \geq 0, \quad t\in[a,\sigma(b)]\}.
$$

Note that K is a normal cone in $\mathscr{C}([a,\sigma(b)])$.

Lemma 1.4. *Suppose* $(A8)$-$(A10)$. *Then the operator* $\Phi : [0,R] \to K$ *is a continuous operator.*

Proof. Let $\{x_n\}_{n\in\mathbb{N}} \subseteq [0,R]$, $x\in[0,R]$ and $\lim_{n\to\infty} x_n = x$. Then

$$
0 \leq x_n(t) \leq R,
$$

$$
0 \leq x(t) \leq R, \quad t\in[a,b], \quad n\in\mathbb{N}.
$$

Take $\varepsilon > 0$ arbitrarily. Note that f is uniformly continuous on $[a,b]\times[0,R]$. Then there exists a $\delta_1 = \delta_1(\varepsilon) > 0$ such that if $u_1,u_2\in[0,R]$ with $|u_1-u_2| < \delta_1$,

$$
|f(s,u_1) - f(s,u_2)| < \frac{\varepsilon}{2(\sigma(b)-a)e_p(\sigma(b),a)}, \quad s\in[a,b].
$$

Now, using that g is a continuous function at $\sigma(b)$, there exists a $\delta_2 = \delta_2(\varepsilon) > 0$ such that for any $u\in(0,\infty)$ with $|u - x(\sigma(b))| < \delta_2$,

$$
|g(u) - g(x(\sigma(b)))| < \frac{\varepsilon}{2}.
$$

Let $\delta = \max\{\delta_1,\delta_2\}$. From $\lim_{n\to\infty} x_n = x$, there exists an $N = N(\varepsilon) > 0$ such that

$$|x_n(s)-x(s)|<\delta, \quad s\in[a,\sigma(b)], \quad n>N.$$

Hence,

$$\begin{aligned}
|\Phi x_n(t)-\Phi x(t)| &= \left|\frac{1}{e_p(t,a)}\left(\int_a^t e_p(s,a)f(s,x_n(s))\Delta s-g(x_n(\sigma(b)))\right)\right.\\
&\quad \left.-\frac{1}{e_p(t,a)}\left(\int_a^t e_p(s,a)f(s,x(s))\Delta s-g(x(\sigma(b)))\right)\right|\\
&= \left|\frac{1}{e_p(t,a)}\int_a^t e_p(s,a)(f(s,x_n(s))-f(s,x(s)))\Delta s\right.\\
&\quad \left.-(g(x_n(\sigma(b)))-g(x(\sigma(b))))\right|\\
&\le \left|\frac{1}{e_p(t,a)}\int_a^t e_p(s,a)(f(s,x_n(s))-f(s,x(s)))\Delta s\right|\\
&\quad +|g(x_n(\sigma(b)))-g(x(\sigma(b)))|\\
&\le \frac{1}{e_p(t,a)}\int_a^t e_p(s,a)|f(s,x_n(s))-f(s,x(s))|\Delta s+\frac{\varepsilon}{2}\\
&\le \int_a^{\sigma(b)} e_p(s,a)|f(s,x_n(s))-f(s,x(s))|\Delta s+\frac{\varepsilon}{2}\\
&\le e_p(\sigma(b),a)\int_a^{\sigma(b)} |f(s,x_n(s))-f(s,x(s))|\Delta s+\frac{\varepsilon}{2}\\
&< e_p(\sigma(b),a)\frac{\varepsilon}{2(\sigma(b)-a)e_p(\sigma(b),a)}(\sigma(b)-a)+\frac{\varepsilon}{2}\\
&= \frac{\varepsilon}{2}+\frac{\varepsilon}{2}\\
&= \varepsilon, \quad n>N, \quad t\in[a,b].
\end{aligned}$$

Therefore, $\lim_{n\to\infty}\Phi x_n=\Phi x$ and $\Phi:[0,R]\to K$ is a continuous operator. This completes the proof.

Lemma 1.5. *Suppose* (A8)-(A10). *Then* $\Phi:[0,R]\to K$ *is a compact operator.*

Proof. Let $X\subseteq[0,R]$ be a bounded set. Then there exists a constant $L>0$ such that

$$0\le x(s)\le L, \quad s\in[a,\sigma(b)],$$

for any $x\in X$. Define

$$Q = \max_{(s,u)\in[a,b]\times[0,L]} f(s,u),$$

$$M = \max_{u\in[0,L]} g(u).$$

Take $x \in X$ arbitrarily. Then

$$\begin{aligned}
\Phi x(t) &= \frac{1}{e_p(t,a)}\left(\int_a^t e_p(s,a)f(s,x(s))\Delta s + g(x(\sigma(b)))\right)\\
&\le \int_a^t e_p(s,a)f(s,x(s))\Delta s + g(x(\sigma(b)))\\
&\le \int_a^{\sigma(b)} e_p(s,a)f(s,x(s))\Delta s + M\\
&\le e_p(\sigma(b),a)Q(\sigma(b)-a)+M, \quad t\in[a,\sigma(b)].
\end{aligned}$$

Thus, $\Phi(X)$ is uniformly bounded. Now, we take $t_1, t_2 \in [a,\sigma(b)]$, $t_1 < t_2$. Then

$$\begin{aligned}
|\Phi x(t_2) - \Phi x(t_1)| &= \left|\frac{1}{e_p(t_2,a)}\left(\int_a^{t_2} e_p(s,a)f(s,x(s))\Delta s + g(x(\sigma(b)))\right)\right.\\
&\quad \left. -\frac{1}{e_p(t_1,a)}\left(\int_a^{t_1} e_p(s,a)f(s,x(s))\Delta s + g(x(\sigma(b)))\right)\right|\\
&= \left|\frac{1}{e_p(t_2,a)}\int_a^{t_1} e_p(s,a)f(s,x(s))\Delta s\right.\\
&\quad +\frac{1}{e_p(t_2,a)}\int_{t_1}^{t_2} e_p(s,a)f(s,x(s))\Delta s\\
&\quad -\frac{1}{e_p(t_1,a)}\int_a^{t_1} e_p(s,a)f(s,x(s))\Delta s\\
&\quad \left. +g(x(\sigma(b)))\left(\frac{1}{e_p(t_2,a)}-\frac{1}{e_p(t_1,a)}\right)\right|\\
&= \left|\left(\frac{1}{e_p(t_2,a)}-\frac{1}{e_p(t_1,a)}\right)\int_a^{t_1} e_p(s,a)f(s,x(s))\Delta s\right.\\
&\quad +\frac{1}{e_p(t_2,a)}\int_{t_1}^{t_2} e_p(s,a)f(s,x(s))\Delta s\\
&\quad \left. +g(x(\sigma(b)))\left(\frac{1}{e_p(t_2,a)}-\frac{1}{e_p(t_1,a)}\right)\right|
\end{aligned}$$

$$\leq \left| \frac{1}{e_p(t_2,a)} - \frac{1}{e_p(t_1,a)} \right| \left(\int_a^{t_1} e_p(s,a) f(s,x(s)) \Delta s + g(x(\sigma(b))) \right)$$

$$+ \frac{1}{e_p(t_2,a)} \int_{t_1}^{t_2} e_p(s,a) f(s,x(s)) \Delta s$$

$$\leq \left(\frac{1}{e_p(t_1,a)} - \frac{1}{e_p(t_2,a)} \right) (Q e_p(\sigma(b),a) + M)(\sigma(b) - a)$$

$$+ \int_{t_1}^{t_2} e_p(s,a) f(s,x(s)) \Delta s$$

$$\leq \left(\frac{1}{e_p(t_1,a)} - \frac{1}{e_p(t_2,a)} \right) (Q e_p(\sigma(b),a) + M)(\sigma(b) - a)$$

$$+ Q e_p(\sigma(b),a)(t_2 - t_1).$$

Hence, we conclude that $\Phi(X)$ is equi-continuous. Now, applying the Arzela-Ascoli theorem, we conclude that $\Phi : [0,R] \to K$ is a compact operator. This completes the proof.

Lemma 1.6. *Suppose* (A8)-(A10). *Then* $\Phi : [0,R] \to K$ *is a monotone operator.*

Proof. Let $x,y \in [0,R]$ and $x(t) \leq y(t)$, $t \in [a,\sigma(b)]$. Then

$$\begin{aligned} \Phi x(t) &= \frac{1}{e_p(t,a)} \left(\int_a^t e_p(s,a) f(s,x(s)) \Delta s + g(x(\sigma(b))) \right) \\ &\leq \frac{1}{e_p(t,a)} \left(\int_a^t e_p(s,a) f(s,y(s)) \Delta s + g(y(\sigma(b))) \right) \\ &= \Phi y(t), \quad t \in [a,\sigma(b)]. \end{aligned}$$

This completes the proof.

Lemma 1.7. *Suppose* (A8)-(A10). *Then* $\Phi 0 \geq 0$.

Proof. We have

$$\begin{aligned} \Phi 0(t) &= \frac{1}{e_p(t,a)} \left(\int_a^t e_p(s,a) f(s,0) \Delta s + g(0) \right) \\ &\geq 0, \quad t \in [a,\sigma(b)]. \end{aligned}$$

This completes the proof.

Lemma 1.8. *Suppose* (A8)-(A10). *Then* $\Phi R \leq R$.

Proof. We have

$$\begin{aligned}
\Phi R(t) &= \frac{1}{e_p(t,a)}\left(\int_a^t e_p(s,a)f(s,R)\Delta s + g(R)\right) \\
&\leq \int_a^t \frac{e_p(s,a)}{e_p(t,a)} f(s,R)\Delta s + \frac{g(R)}{e_p(t,a)} \\
&\leq \int_a^t f(s,R)\Delta s + g(R) \\
&\leq \int_a^{\sigma(b)} f(s,R)\Delta s + \frac{R}{2} \\
&\leq \frac{R}{2(\sigma(b)-a)}(\sigma(b)-a) + \frac{R}{2} \\
&= \frac{R}{2} + \frac{R}{2} \\
&= R, \quad t \in [a,\sigma(b)].
\end{aligned}$$

This completes the proof.

Theorem 1.8. *Suppose* (A8)-(A10). *Then the BVP* (1.22) *has positive solutions.*

Proof. Let $v_0 = 0$ and $w_0 = R$. Construct the sequences $\{v_n\}_{n\in\mathbb{N}}$ and $\{w_n\}_{n\in\mathbb{N}}$ as follows

$$v_n = \Phi v_{n-1} \quad \text{and} \quad w_n = \Phi w_{n-1}.$$

We have, using Lemma 1.7,

$$\begin{aligned}
0 &= v_0 \\
&\leq \Phi v_0 \\
&= v_1.
\end{aligned}$$

Now, applying Lemma 1.6, we get

$$\begin{aligned}
v_1 &= \Phi v_0 \\
&\leq \Phi v_1 \\
&= v_2.
\end{aligned}$$

Assume that $v_n \leq v_{n+1}$ for some $n \in \mathbb{N}$. Then, applying Lemma 1.6, we get

$$v_{n+1} = \Phi v_n$$

$$\leq \Phi v_{n+1}$$

$$= v_{n+2}.$$

Thus,

$$v_0 \leq v_1 \leq \ldots \leq v_n \leq v_{n+1} \leq \ldots.$$

Next, by Lemma 1.8, we have

$$\begin{aligned} R &= w_0 \\ &\geq \Phi R \\ &= \Phi w_0 \\ &= w_1. \end{aligned}$$

Now, we apply Lemma 1.6 and we find

$$\begin{aligned} w_1 &= \Phi w_0 \\ &\geq \Phi w_1 \\ &= w_2. \end{aligned}$$

Assume that $w_n \geq w_{n+1}$ for some $n \in \mathbb{N}$. Then

$$\begin{aligned} w_{n+2} &= \Phi w_{n+1} \\ &\leq \Phi w_n \\ &= w_{n+1}. \end{aligned}$$

Consequently,

$$\ldots \leq w_n \leq \ldots \leq w_2 \leq w_1 \leq w_0.$$

Next, $v_0 \leq w_0$. Now, applying Lemma 1.6, we find

$$\begin{aligned} v_1 &= \Phi v_0 \\ &\leq \Phi w_0 \end{aligned}$$

$$= w_1.$$

Assume that $v_n \leq w_n$ for some $n \in \mathbb{N}$. Then

$$\begin{aligned} v_{n+1} &= \Phi v_n \\ &\leq \Phi w_n \\ &= w_{n+1}. \end{aligned}$$

Therefore,

$$v_0 \leq v_1 \leq \ldots \leq v_n \leq \ldots \leq w_n \leq w_{n-1} \leq \ldots \leq w_1 \leq w_0$$

and the sequences $\{v_n\}_{n\in\mathbb{N}}$ and $\{w_n\}_{n\in\mathbb{N}}$ are convergent sequences. Let

$$v = \lim_{n\to\infty} v_n, \quad w = \lim_{n\to\infty} w_n.$$

Since $\Phi : [v_0, w_0] \to K$ is a continuous operator, we find

$$\begin{aligned} v &= \lim_{n\to\infty} v_{n+1} \\ &= \lim_{n\to\infty} \Phi v_n \\ &= \Phi v \end{aligned}$$

and

$$\begin{aligned} w &= \lim_{n\to\infty} w_{n+1} \\ &= \lim_{n\to\infty} \Phi w_n \\ &= \Phi w. \end{aligned}$$

Hence and by Lemma 1.3, we conclude that v and w are solutions of the BVP (1.22). Next, by the proof of Lemma 1.7, we find

$$\begin{aligned} \Phi v_0(\sigma(b)) &= \frac{1}{e_p(\sigma(b), a)} \int_a^{\sigma(b)} e_p(s,0) f(s,0) \Delta s \\ &> 0 \end{aligned}$$

and

$$\begin{aligned}0 &< \Phi v_0(\sigma(b))\\&\le \Phi v(\sigma(b))\\&= v(\sigma(b))\\&\le w(\sigma(b)).\end{aligned}$$

Thus, v and w are positive solutions of the BVP (1.22). This completes the proof.

Example 1.2. Let $\mathbb{T} = [0,1]\bigcup[2,3]$, where $[0,1]$ and $[2,3]$ are real-valued intervals. Consider the BVP

$$\begin{aligned}x^{\Delta}(t) &= -(1+t)x(\sigma(t)) + \frac{t\sqrt{x(t)+1}}{9}, \quad t \in [0,3],\\x(0) &= \frac{(x(3))^2}{12}.\end{aligned} \tag{1.23}$$

Here

$$\begin{aligned}p(t) &= 1+t,\\f(t,x) &= \frac{t\sqrt{x+3}}{9},\\g(x) &= \frac{x^2}{12}, \quad t \in [0,3], \quad x \in \mathbb{R}.\end{aligned}$$

Take $R = 6$. Note that g is a non-decreasing function, and f is a non-decreasing function with respect to the second argument. Moreover, if $0 \le x \le 6$, then

$$\begin{aligned}f(t,x) &\le \frac{3\sqrt{6+3}}{9}\\&= 1\\&= \frac{R}{2(\sigma(3)-0)}, \quad t \in [0,3],\end{aligned}$$

and

$$\begin{aligned}g(x) &\le \frac{36}{12}\\&= 3\end{aligned}$$

$$= \frac{R}{2}.$$

Consequently, $(A8)$-$(A10)$ hold and the BVP (1.23) has positive solutions.

Exercise 1.2. Let $\mathbb{T} = \mathbb{N}_0$. Prove that the BVP

$$x^{\Delta}(t) = -(1+t+t^2)x(t+1) + \frac{(2+t)\sqrt{1+(x(t))^2}}{100}, \quad t \in [0,7],$$

$$x(0) = \frac{x(8)}{100},$$

has positive solutions.

1.4 Existence of Solutions for First Order Periodic Boundary Value Problems

In this section, we investigate the periodic boundary value problem (PBVP)

$$\begin{aligned} x^{\Delta} &= -p(t)x + g(t,x), \quad t \in [a,b], \\ x(a) &= x(\sigma(b)), \end{aligned} \tag{1.24}$$

where

(A11) $g : [a,b] \times \mathbb{R}^n \to \mathbb{R}^n$ is a continuous function,
(A12) $p : [a,b] \to \mathbb{R}$ is a continuous function such that $-p \in \mathscr{R}$ and

$$e_{-p}(\sigma(b),a) \neq 1.$$

Lemma 1.9. *Suppose* (A11) *and* (A12). *Then the PBVP* (1.24) *is equivalent to the integral equation*

$$x(t) = e_{-p}(t,a)\left(\frac{e_{-p}(\sigma(b),a)}{1-e_{-p}(\sigma(b),a)}\int_a^{\sigma(b)} \frac{g(s,x(s))}{e_{-p}(\sigma(s),a)}\Delta s + \int_a^t \frac{g(s,x(s))}{e_{-p}(\sigma(s),a)}\Delta s\right), \tag{1.25}$$

$t \in [a,b]$.

Proof. 1. Suppose that x is a solution to the PBVP (1.24). Then

$$\begin{aligned} \left(\frac{x}{e_{-p}(\cdot,a)}\right)^{\Delta}(t) &= \frac{x^{\Delta}(t)e_{-p}(t,a) + p(t)e_{-p}(t,a)x(t)}{e_{-p}(t,a)e_{-p}(\sigma(t),a)} \\ &= \frac{x^{\Delta}(t) + p(t)x(t)}{e_{-p}(\sigma(t),a)} \end{aligned}$$

$$= \frac{g(t,x(t))}{e_{-p}(\sigma(t),a)}, \quad t \in [a,b].$$

We integrate the last equality from a to t and we find

$$\frac{x(t)}{e_{-p}(t,a)} - x(a) = \int_a^t \frac{g(s,x(s))}{e_{-p}(\sigma(s),a)} \Delta s, \quad t \in [a,b],$$

or

$$\frac{x(t)}{e_{-p}(t,a)} = x(a) + \int_a^t \frac{g(s,x(s))}{e_{-p}(\sigma(s),a)} \Delta s, \quad t \in [a,b],$$

or

$$x(t) = e_{-p}(t,a) \left(x(a) + \int_a^t \frac{g(s,x(s))}{e_{-p}(\sigma(s),a)} \Delta s \right), \quad t \in [a,b]. \tag{1.26}$$

Hence,

$$x(\sigma(b)) = x(a)$$

$$= e_{-p}(\sigma(b),a) \left(x(a) + \int_a^{\sigma(b)} \frac{g(s,x(s))}{e_{-p}(\sigma(s),a)} \Delta s \right),$$

whereupon,

$$(1 - e_{-p}(\sigma(b),a))x(a) = e_{-p}(\sigma(b),a) \int_a^{\sigma(b)} \frac{g(s,x(s))}{e_{-p}(\sigma(s),a)} \Delta s$$

and

$$x(a) = \frac{e_{-p}(\sigma(b),a)}{1 - e_{-p}(\sigma(b),a)} \int_a^{\sigma(b)} \frac{g(s,x(s))}{e_{-p}(\sigma(s),a)} \Delta s.$$

From here and from (1.26), we arrive at (1.25).

2. Now, suppose that x is a solution to the integral equation (1.25). Then x is Δ-differentiable and

$$x^\Delta(t) = -p(t)e_{-p}(t,a)\Bigg(\frac{e_{-p}(\sigma(b),a)}{1 - e_{-p}(\sigma(b),a)} \int_a^{\sigma(b)} \frac{g(s,x(s))}{e_{-p}(\sigma(s),a)} \Delta s$$

$$+ \int_a^t \frac{g(s,x(s))}{e_{-p}(\sigma(s),a)} \Delta s \Bigg)$$

$$+ e_{-p}(\sigma(t),a) \frac{g(t,x(t))}{e_{-p}(\sigma(t),a)}$$

$$= -p(t)x(t) + g(t,x(t)), \quad t \in [a,b].$$

Moreover,

$$x(a) = \frac{e_{-p}(\sigma(b),a)}{1-e_{-p}(\sigma(b),a)} \int_a^{\sigma(b)} \frac{g(s,x(s))}{e_{-p}(\sigma(s),a)} \Delta s$$

and

$$\begin{aligned} x(\sigma(b)) &= e_{-p}(\sigma(b),a)\left(\frac{e_{-p}(\sigma(b),a)}{1-e_{-p}(\sigma(b),a)} \int_a^{\sigma(b)} \frac{g(s,x(s))}{e_{-p}(\sigma(s),a)} \Delta s \right. \\ &\quad \left. + \int_a^{\sigma(b)} \frac{g(s,x(s))}{e_{-p}(\sigma(s),a)} \Delta s \right) \\ &= \frac{e_{-p}(\sigma(b),a)}{1-e_{-p}(\sigma(b),a)} \int_a^{\sigma(b)} \frac{g(s,x(s))}{e_{-p}(\sigma(s),a)} \Delta s \\ &= x(a). \end{aligned}$$

Thus, x satisfies (1.24). This completes the proof.

Theorem 1.9. *Suppose* (A11) *and* (A12) *and there exist non-negative constants* α *and* K *so that*

$$\frac{\|\lambda g(t,q)\|}{|e_{-p}(\sigma(t),a)|} \leq 2\alpha\langle q, \lambda g(t,q) - p(t)q\rangle + K \tag{1.27}$$

for any $(t,q,\lambda) \in [a,b] \times \mathbb{R}^n \times [0,1]$. *Then the PBVP* (1.24) *has at least one solution.*

Proof. Suppose that $\mathscr{C}([a,\sigma(b)])$ is provided with the maximum norm. For $x \in \mathscr{C}([a,\sigma(b)])$ define the operator

$$Tx(t) = e_{-p}(t,a)\left(\frac{e_{-p}(\sigma(b),a)}{1-e_{-p}(\sigma(b),a)} \int_a^{\sigma(b)} \frac{g(s,x(s))}{e_{-p}(\sigma(s),a)} \Delta s + \int_a^t \frac{g(s,x(s))}{e_{-p}(\sigma(s),a)} \Delta s \right),$$

$t \in [a,b]$. Let

$$P_1 = \max_{t\in[a,\sigma(b)]} \left(|e_{-p}(t,a)| \left(\frac{e_{-p}(\sigma(b),a)}{|1-e_{-p}(\sigma(b),a)|} + 1 \right) \right) K(\sigma(b)-a) + 1,$$

$$B_{P_1} = \left\{ x \in \mathscr{C}([a,\sigma(b)]), x : [a,\sigma(b)] \to \mathbb{R}^n, \max_{t\in[a,\sigma(b)]} \|x(t)\| < P_1 \right\}.$$

Suppose that x is a solution of equation

$$\lambda Tx = x, \quad \lambda \in [0,1].$$

Consider the functions

$$r(t) = \|x(t)\|^2,$$

$$H_1(t) = |e_{-p}(t,a)| \left(\frac{e_{-p}(\sigma(b),a)}{|1-e_{-p}(\sigma(b),a)|} + 1 \right), \quad t \in [a,\sigma(b)].$$

Then

$$
\begin{aligned}
r^{\Delta}(t) &= \langle x^{\Delta}(t), x(t)\rangle + \langle x^{\sigma}(t), x^{\Delta}(t)\rangle \\
&= \langle x^{\Delta}(t), x(t)\rangle + \langle x^{\Delta}(t), x(t) + \mu(t)x^{\Delta}(t)\rangle \\
&= 2\langle x^{\Delta}(t), x(t)\rangle + \mu(t)\langle x^{\Delta}(t), x^{\Delta}(t)\rangle \\
&= 2\langle x(t), \lambda g(t,x(t)) - p(t)x(t)\rangle \\
&\quad + \mu(t)\langle \lambda g(t,x(t)) - p(t)x(t), \lambda g(t,x(t)) - p(t)x(t)\rangle \\
&\geq 2\langle x(t), \lambda g(t,x(t)) - p(t)x(t)\rangle, \quad t \in [a,\sigma(b)].
\end{aligned}
$$

Consequently,

$$
\begin{aligned}
\|Tx\| &= \left\| e_{-p}(t,a)\left(\frac{e_{-p}(\sigma(b),a)}{1-e_{-p}(\sigma(b),a)} \int_a^{\sigma(b)} \frac{g(s,x(s))}{e_{-p}(\sigma(s),a)} \Delta s \right.\right. \\
&\quad \left.\left. + \int_a^t \frac{g(s,x(s))}{e_{-p}(\sigma(s),a)} \Delta s \right) \right\| \\
&\leq |e_{-p}(t,a)| \left(\frac{e_{-p}(\sigma(b),a)}{|1-e_{-p}(\sigma(b),a)|} \int_a^{\sigma(b)} \frac{\|\lambda g(s,x(s))\|}{|e_{-p}(\sigma(s),a)|} \Delta s \right. \\
&\quad \left. + \int_a^t \frac{\|\lambda g(s,x(s))\|}{|e_{-p}(\sigma(s),a)|} \Delta s \right) \\
&\leq |e_{-p}(t,a)| \left(\frac{e_{-p}(\sigma(b),a)}{|1-e_{-p}(\sigma(b),a)|} + 1 \right) \int_a^{\sigma(b)} \frac{\|\lambda g(s,x(s))\|}{|e_{-p}(\sigma(s),a)|} \Delta s \\
&\leq H_1(t) \int_a^{\sigma(b)} \frac{\|\lambda g(s,x(s))\|}{|e_{-p}(\sigma(s),a)|} \Delta s \\
&\leq H_1(t) \int_a^{\sigma(b)} (2\alpha\langle x(s), \lambda g(s,x(s)) - p(s)x(s)\rangle + K) \Delta s \\
&\leq H_1(t) \int_a^{\sigma(b)} \left(\alpha r^{\Delta}(s) + K \right) \Delta s \\
&= H_1(t) \left(\alpha \left(\|x(\sigma(b))\|^2 - \|x(a)\|^2 \right) + K(\sigma(b)-a) \right) \\
&= K(\sigma(b)-a) H_1(t)
\end{aligned}
$$

$$< P_1, \quad t \in [a, \sigma(b)].$$

Therefore,

$$\lambda Tx \neq x, \quad x \in \partial B_{P_1}, \quad \lambda \in [0,1].$$

Now, applying the Schaefer theorem, Theorem 1.3, we conclude that the operator T has a fixed point in $\mathscr{C}([a,\sigma(b)]$. By Lemma 1.9, we obtain that the PBVP (1.24) has at least one solution in $\mathscr{C}([a,\sigma(b)])$. This completes the proof.

Corollary 1.1. *Suppose* (A11) *and* (A12) *and* $p < 0$ *on* $[a,b]$. *If there exist non-negative constants* α *and* K *so that*

$$\frac{\|g(t,q)\|}{|e_{-p}(\sigma(t),a)|} \leq 2\alpha\langle q, g(t,q)\rangle + K, \tag{1.28}$$

$(t,q) \in [a,b] \times \mathbb{R}^n$, *then the PBVP* (1.24) *has at least one solution.*

Proof. Let $\lambda \in [0,1]$. Then, multiplying (1.28) by λ and using that $p < 0$ on $[a,b]$, we get

$$\begin{aligned}
\frac{\lambda\|g(t,q)\|}{|e_{-p}(\sigma(t),a)|} &\leq 2\alpha\lambda\langle q, g(t,q)\rangle + K\lambda \\
&\leq 2\alpha\langle q, \lambda g(t,q)\rangle + K \\
&\leq 2\alpha\langle q, \lambda g(t,q) - p(t)q\rangle + K, \quad t \in [a,b].
\end{aligned}$$

Thus, (1.27) holds. Hence and by Theorem 1.9, we conclude that the PBVP (1.24) has at least one solution. This completes the proof.

Corollary 1.2. *Suppose* (A11), (A12), *and* $g : [a,b] \times \mathbb{R}^n \to \mathbb{R}^n$ *is bounded. Then the PBVP* (1.24) *has at least one solution.*

Proof. Since g is bounded on $[a,b] \times \mathbb{R}^n$, then there exists a constant $L > 0$ so that

$$\sup_{(t,q)\in[a,b]\times\mathbb{R}^n} \|g(t,q)\| \leq L.$$

Let, also,

$$K = \max_{t\in[a,b]} \frac{L}{|e_{-p}(\sigma(t),a)|}.$$

Then

$$\begin{aligned}
\frac{\|\lambda g(t,q)\|}{|e_{-p}(\sigma(t),a)|} &\leq \frac{\|g(t,q)\|}{|e_{-p}(\sigma(t),a)|} \\
&\leq \frac{L}{|e_{-p}(\sigma(t),a)|}
\end{aligned}$$

$$\leq K, \quad t \in [a,b].$$

Thus, (1.27) holds for $\alpha = 0$. Hence and by Theorem 1.9, it follows that the PBVP (1.24) has at least one solution. This completes the proof.

Corollary 1.3. *Let $n = 1$. Suppose (A11), (A12), and there exist non-negative constants α and K so that*

$$\frac{|\lambda g(t,q)|}{|e_{-p}(\sigma(t),a)|} \leq 2\alpha(q\lambda g(t,q) - p(t)q^2) + K \tag{1.29}$$

for any $(t,q,\lambda) \in [a,b] \times \mathbb{R} \times [0,1]$. Then the PBVP (1.24) has at least one solution.

Example 1.3. Let $n = 1$ and $\mathbb{T} = 2^{\mathbb{N}_0}$. Consider the PBVP

$$\begin{aligned} x^{\Delta} &= x + \frac{1}{1+x^{100}}, \quad t \in [1,16], \\ x(1) &= x(32). \end{aligned} \tag{1.30}$$

Here

$$\sigma(t) = 2t,$$

$$p(t) = -1,$$

$$g(t,x) = \frac{1}{1+x^{100}}, \quad t \in [1,16], \quad x \in \mathbb{R}.$$

We have

$$g(t,x) \leq 1, \quad t \in [1,16], \quad x \in \mathbb{R},$$

and

$$1 + \mu(t)(-p(t)) = 1 + t$$

$$> 0, \quad t \in [1,16],$$

i.e., $-p \in \mathscr{R}_+$. Then we apply (1.29) for $\alpha = 0$ and $K = 1$ and we conclude that the PBVP (1.29) has at least one solution.

Exercise 1.3. Let $n = 1$ and $\mathbb{T} = 3^{\mathbb{N}_0}$. Prove that the PBVP

$$x^{\Delta} = 2x + \frac{1}{1+x^{20}+x640}, \quad t \in [1,81],$$

$$x(1) = x(243),$$

has at least one solution.

Remark 1.1. Consider the PBVP

$$\begin{aligned} x^{\Delta} &= f(t,x), \quad t \in [a,b], \\ x(a) &= x(\sigma(b)). \end{aligned} \tag{1.31}$$

We can reduce (1.31) to (1.24) if we set $g(t,x) = f(t,x) - x$ and consider the PBVP

$$\begin{aligned} x^{\Delta} &= x + g(t,x), \quad t \in [a,b], \\ x(a) &= x(\sigma(b)). \end{aligned}$$

Exercise 1.4. Find conditions for the function f so that the PBVP (1.31) has at least one solution.

Now, we will investigate the following PBVP

$$\begin{aligned} x^{\Delta} &= -p(t)x^{\sigma} + g(t,x^{\sigma}), \quad t \in [a,b], \\ x(a) &= x(\sigma(b)), \end{aligned} \tag{1.32}$$

where

(A13) $g : [a,b] \times \mathbb{R}^n \to \mathbb{R}^n$ is a continuous function.
(A14) $p : [a,b] \to \mathbb{R}$ is a continuous function such that $p \in \mathscr{R}$ and

$$e_p(\sigma(b),a) \neq 1.$$

Lemma 1.10. *Suppose* (A13) *and* (A14). *Then the PBVP* (1.32) *is equivalent to the integral equation*

$$\begin{aligned} x(t) = \frac{1}{e_p(t,a)} \Bigg(&\frac{1}{e_p(\sigma(b),a) - 1} \int_a^{\sigma(b)} g(s,x^{\sigma}(s)) e_p(s,a) \Delta s \\ &+ \int_a^t e_p(s,a) g(s,x^{\sigma}(s)) \Delta s \Bigg), \quad t \in [a,\sigma(b)]. \end{aligned} \tag{1.33}$$

Proof. 1. Let x be a solution to the PBVP (1.32). Then

$$\begin{aligned} (e_p(\cdot,a)x)^{\Delta}(t) &= p(t)e_p(t,a)x^{\sigma}(t) + e_p(t,a)x^{\Delta}(t) \\ &= p(t)e_p(t,a)x^{\sigma}(t) - p(t)e_p(t,a)x^{\sigma}(t) \\ &\quad + e_p(t,a)g(t,x^{\sigma}(t)) \end{aligned}$$

$$= e_p(t,a)g(t,x^\sigma(t)), \quad t \in [a,b].$$

We integrate the last equality from a to t and we find

$$e_p(t,a)x(t) - x(a) = \int_a^t e_p(s,a)g(s,x^\sigma(s))\Delta s, \quad t \in [a,b],$$

or

$$e_p(t,a)x(t) = x(a) + \int_a^t e_p(s,a)_g(s,x^\sigma(s))\Delta s, \quad t \in [a,b],$$

or

$$x(t) = \frac{x(a)}{e_p(t,a)} + \frac{1}{e_p(t,a)}\int_a^t e_p(s,a)g(s,x^\sigma(s))\Delta s, \quad t \in [a,b]. \tag{1.34}$$

Hence,

$$\begin{aligned} x(a) &= x(\sigma(b)) \\ &= \frac{x(a)}{e_p(\sigma(b),a)} + \frac{1}{e_p(\sigma(b),a)}\int_a^{\sigma(b)} e_p(s,a)g(s,x^\sigma(s))\Delta s, \end{aligned}$$

whereupon,

$$\frac{e_p(\sigma(b),a)-1}{e_p(\sigma(b),a)}x(a) = \frac{1}{e_p(\sigma(b),a)}\int_a^{\sigma(b)} e_p(s,a)g(s,x^\sigma(s))\Delta s$$

and

$$x(a) = \frac{1}{e_p(\sigma(b),a)-1}\int_a^{\sigma(b)} e_p(s,a)g(s,x^\sigma(s))\Delta s.$$

We put $x(a)$ in (1.34) and we get (1.33).

2. Let x be a solution to the integral equation (1.33). Then x is Δ-differentiable and

$$\begin{aligned} x^\Delta(t) &= \left(\frac{1}{e_p(\cdot,a)}\right)^\Delta(t)\left(\frac{1}{e_p(\sigma(b),a)-1}\int_a^{\sigma(b)} g(s,x^\sigma(s))e_p(s,a)\Delta s\right. \\ &\quad \left. + \int_a^{\sigma(t)} e_p(s,a)g(s,x^\sigma(s))\Delta s\right) \\ &\quad + \frac{1}{e_p(t,a)}g(t,x^\sigma(t))e_p(t,a) \\ &= -\frac{p(t)}{e_p(\sigma(t),a)}\left(\frac{1}{e_p(\sigma(b),a)-1}\int_a^{\sigma(b)} g(s,x^\sigma(s))e_p(s,a)\Delta s\right. \\ &\quad \left. + \int_a^{\sigma(t)} e_p(s,a)g(s,x^\sigma(s))\Delta s\right) + g(t,x^\sigma(t)) \end{aligned}$$

$$= -p(t)x^{\sigma}(t) + g(t, x^{\sigma}(t)), \quad t \in [a, b].$$

Moreover,

$$x(a) = \frac{1}{e_p(\sigma(b), a) - 1} \int_a^{\sigma(b)} g(s, x^{\sigma}(s)) e_p(s, a) \Delta s$$

and

$$\begin{aligned} x(\sigma(b)) &= \frac{1}{e_p(\sigma(b), a)} \left(\frac{1}{e_p(\sigma(b), a) - 1} \int_a^{\sigma(b)} g(s, x^{\sigma}(s)) e_p(s, a) \Delta s \right. \\ &\quad \left. + \int_a^{\sigma(b)} e_p(s, a) g(s, x^{\sigma}(s)) \Delta s \right) \\ &= \frac{e_p(\sigma(b), a)}{e_p(\sigma(b), a)(e_p(\sigma(b), a) - 1)} \int_a^{\sigma(b)} e_p(s, a) g(s, x^{\sigma}(s)) \Delta s \\ &= \frac{1}{e_p(\sigma(b), a) - 1} \int_a^{\sigma(b)} e_p(s, a) g(s, x^{\sigma}(s)) \Delta s \\ &= x(a), \end{aligned}$$

i.e., x satisfies the PBVP (1.32). This completes the proof.

Theorem 1.10. *Suppose* (A13) *and* (A14) *and let there exist non-positive constants* α *and* K *such that*

$$\|\lambda g(t, q) e_p(t, q)\| \leq -2\alpha \langle q, \lambda g(t, q) - p(t) q \rangle + K$$

for any $(t, q, \lambda) \in [a, b] \times \mathbb{R}^n \times [0, 1]$. *Then the PBVP* (1.32) *has at least one solution.*

Proof. Let $\mathscr{C}([a, \sigma(b)])$ be endowed with the maximum norm. For $x \in \mathscr{C}([a, \sigma(b)])$, define the operator

$$\begin{aligned} Tx(t) &= \frac{1}{e_p(t, a)} \left(\frac{1}{e_p(\sigma(b), a) - 1} \int_a^{\sigma(b)} g(s, x^{\sigma}(s)) e_p(s, a) \Delta s \right. \\ &\quad \left. + \int_a^t e_p(s, a) g(s, x^{\sigma}(s)) \Delta s \right), \quad t \in [a, \sigma(b)]. \end{aligned}$$

Next, define

$$P_2 = \max_{t \in [a, \sigma(b)]} \left(\frac{1}{|e_p(t, a)|} \left(\frac{1}{|1 - e_p(\sigma(b), a)|} + 1 \right) \right) K(\sigma(b) - a) + 1,$$

$$B_{P_2} = \left\{ x \in \mathscr{C}([a, \sigma(b)]) : x : [a, \sigma(b)] \to \mathbb{R}^n, \max_{t \in [a, \sigma(b)]} \|x(t)\| < P_2 \right\},$$

$$r(t) = \|x(t)\|^2,$$

$$H_2(t) = \frac{1}{|e_p(t,a)|}\left(\frac{1}{|e_p(\sigma(b),a)-1|}+1\right), \quad t \in [a,\sigma(b)].$$

Assume that $x \in \mathscr{C}([a,\sigma(b)])$ is a solution of equation

$$\lambda Tx = x, \quad \lambda \in [0,1].$$

Then

$$\begin{aligned}
-\alpha r^{\Delta}(t) &= -\alpha\langle x^{\Delta}(t), x(t)\rangle - \alpha\langle x^{\sigma}(t), x^{\Delta}(t)\rangle \\
&= -\alpha\langle x^{\Delta}(t), x(t)\rangle - \alpha\langle x^{\Delta}(t), x(t)\rangle \\
&\quad -\alpha\mu(t)\langle x^{\Delta}(t), x^{\Delta}(t)\rangle \\
&= -\alpha\left(2\langle x^{\Delta}(t), x(t)\rangle + \mu(t)\|x^{\Delta}(t)\|^2\right) \\
&\geq -2\alpha\langle x^{\Delta}(t), x^{\Delta}(t)\rangle \\
&= -2\alpha\langle x(t), g(t,x^{\sigma}(t)) - p(t)x^{\sigma}(t)\rangle, \quad t \in [a,b].
\end{aligned}$$

Then

$$\begin{aligned}
\|\lambda Tx(t)\| &= \left\| \frac{1}{e_p(t,a)}\left(\frac{1}{e_p(\sigma(b),a)-1}\int_a^{\sigma(b)} \lambda g(s,x^{\sigma}(s))e_p(s,a)\Delta s \right.\right. \\
&\quad \left.\left. + \int_a^t \lambda e_p(s,a)g(s,x^{\sigma}(s))\Delta s\right)\right\| \\
&\leq \frac{1}{|e_p(t,a)||e_p(\sigma(b),a)-1|}\int_a^{\sigma(b)} \|\lambda g(s,x^{\sigma}(s))e_p(s,a)\|\Delta s \\
&\quad + \frac{1}{|e_p(t,a)|}\int_a^t \|\lambda g(s,x^{\sigma}(s))\||e_p(s,a)|\Delta s \\
&= \frac{1}{|e_p(t,a)|}\left(\frac{1}{|e_p(\sigma(b),a)-1|}\int_a^{\sigma(b)} \|\lambda g(s,x^{\sigma}(s))e_p(s,a)\|\Delta s \right. \\
&\quad \left. + \int_a^{\sigma(b)} \|\lambda g(s,x^{\sigma}(s))e_p(s,a)\|\Delta s\right)
\end{aligned}$$

$$
\begin{aligned}
&= \frac{1}{|e_p(t,a)|}\left(\frac{1}{|e_p(\sigma(b),a)-1|}+1\right)\int_a^{\sigma(b)} \|\lambda g(s,x^\sigma(s))e_p(s,a)\|\Delta s \\
&= H_2(t)\int_a^{\sigma(b)} \|\lambda g(s,x^\sigma(s))e_p(s,a)\|\Delta s \\
&\le H_2(t)\int_a^{\sigma(b)} (-2\lambda\langle x^\sigma(s),\lambda g(s,x^\sigma(s))-p(s)x^\sigma(s)\rangle + K)\Delta s \\
&\le H_2(t)\int_a^{\sigma(b)} \left(-\alpha r^\Delta(s)+K\right)\Delta s \\
&= H_2(t)\left(-\alpha\|x(\sigma(b))\|^2+\alpha\|x(a)\|^2+K(\sigma(b)-a)\right) \\
&= K(\sigma(b)-a)H_2(t) \\
&< P_2, \quad t\in[a,\sigma(b)].
\end{aligned}
$$

Thus,

$$\lambda Tx \neq x, \quad x\in\partial B_{P_2}, \quad \lambda\in[0,1].$$

Hence and by the Schaefer fixed point Theorem 1.3, we conclude that the operator T has a fixed point in $\mathscr{C}([a,\sigma(b)])$. Now, applying Lemma 1.10, we obtain that the PBVP has at least one solution in $\mathscr{C}([a,\sigma(b)])$. This completes the proof.

Exercise 1.5. Let $n=1$. Write an analogue Theorem 1.10.

1.5 Multiplicity of Positive Solutions to Nonlinear First Order Periodic Boundary Value Problems

In this section we will investigate the following nonlinear first order PBVP

$$
\begin{aligned}
x^\Delta &= -p(t)x^\sigma + f(x), \quad t\in[a,b], \\
x(a) &= x(\sigma(b)),
\end{aligned}
\tag{1.35}
$$

where

(A15) $f:[0,\infty)\to[0,\infty)$ is continuous.
(A16) $p:[a,b]\to[0,\infty)$ is continuous.

Set

$$A=\frac{1}{e_p(\sigma(b),a)-1}, \quad \delta=\frac{A}{1+A}.$$

Since $p>0$, we have $p\in\mathscr{R}_+$, $A>0$ and $0<\delta<1$.

Lemma 1.11. *Suppose* (A15) *and* (A16). *Then the PBVP* (1.35) *is equivalent to the integral equation*

$$x(t) = \frac{1}{e_p(t,a)}\left(A\int_a^{\sigma(b)} f(x(s))e_p(s,a)\Delta s\right.$$
$$\left.+\int_a^t e_p(s,a)f(x(s))\Delta s\right), \quad t\in[a,\sigma(b)]. \tag{1.36}$$

Proof. 1. Let x be a solution to the PBVP (1.32). Then

$$\begin{aligned}(e_p(\cdot,a)x)^{\Delta}(t) &= p(t)e_p(t,a)x^{\sigma}(t)+e_p(t,a)x^{\Delta}(t)\\ &= p(t)e_p(t,a)x^{\sigma}(t)-p(t)e_p(t,a)x^{\sigma}(t)\\ &\quad +e_p(t,a)f(x(t))\\ &= e_p(t,a)f(x(t)), \quad t\in[a,b].\end{aligned}$$

We integrate the last equality from a to t and we find

$$e_p(t,a)x(t)-x(a) = \int_a^t e_p(s,a)f(x(s))\Delta s, \quad t\in[a,b],$$

or

$$e_p(t,a)x(t) = x(a)+\int_a^t e_p(s,a)_f(x(s))\Delta s, \quad t\in[a,b],$$

or

$$x(t) = \frac{x(a)}{e_p(t,a)}+\frac{1}{e_p(t,a)}\int_a^t e_p(s,a)f(x(s))\Delta s, \quad t\in[a,b]. \tag{1.37}$$

Hence,

$$\begin{aligned}x(a) &= x(\sigma(b))\\ &= \frac{x(a)}{e_p(\sigma(b),a)}+\frac{1}{e_p(\sigma(b),a)}\int_a^{\sigma(b)} e_p(s,a)f(x(s))\Delta s,\end{aligned}$$

whereupon,

$$\frac{e_p(\sigma(b),a)-1}{e_p(\sigma(b),a)}x(a) = \frac{1}{e_p(\sigma(b),a)}\int_a^{\sigma(b)} e_p(s,a)f(x(s))\Delta s$$

and

$$x(a) = \frac{1}{e_p(\sigma(b),a)-1}\int_a^{\sigma(b)} e_p(s,a)f(x(s))\Delta s.$$

We put $x(a)$ in (1.37) and we get (1.36).

2. Let x is a solution to the integral equation (1.36). Then x is Δ-differentiable and

$$\begin{aligned}
x^{\Delta}(t) &= \left(\frac{1}{e_p(\cdot,a)}\right)^{\Delta}(t)\left(\frac{1}{e_p(\sigma(b),a)-1}\int_a^{\sigma(b)} f(x(s))e_p(s,a)\Delta s\right.\\
&\quad\left.+\int_a^{\sigma(t)} e_p(s,a)f(x(s))\Delta s\right)\\
&\quad+\frac{1}{e_p(t,a)}f(x(t))e_p(t,a)\\
&= -\frac{p(t)}{e_p(\sigma(t),a)}\left(\frac{1}{e_p(\sigma(b),a)-1}\int_a^{\sigma(b)} f(x(s))e_p(s,a)\Delta s\right.\\
&\quad\left.+\int_a^{\sigma(t)} e_p(s,a)f(x(s))\Delta s\right)+f(x(t))\\
&= -p(t)x^{\sigma}(t)+f(x(t)), \quad t\in[a,b].
\end{aligned}$$

Moreover,

$$x(a)=\frac{1}{e_p(\sigma(b),a)-1}\int_a^{\sigma(b)} f(x(s))e_p(s,a)\Delta s$$

and

$$\begin{aligned}
x(\sigma(b)) &= \frac{1}{e_p(\sigma(b),a)}\left(\frac{1}{e_p(\sigma(b),a)-1}\int_a^{\sigma(b)} f(x(s))e_p(s,a)\Delta s\right.\\
&\quad\left.+\int_a^{\sigma(b)} e_p(s,a)f(x(s))\Delta s\right)\\
&= \frac{e_p(\sigma(b),a)}{e_p(\sigma(b),a)(e_p(\sigma(b),a)-1)}\int_a^{\sigma(b)} e_p(s,a)f(x(s))\Delta s\\
&= \frac{1}{e_p(\sigma(b),a)(e_p(\sigma(b),a)-1)}\int_a^{\sigma(b)} e_p(s,a)f(x(s))\Delta s\\
&= x(a),
\end{aligned}$$

i.e., x satisfies the PBVP (1.35). This completes the proof.

Suppose that $\mathscr{C}([a,\sigma(b)])$ is endowed with the maximum norm. Define

$$P=\{x\in\mathscr{C}([a,\sigma(b)]) : x(t)\geq\delta\|x\|, \quad t\in[a,\sigma(b)]\}.$$

Define the operator $\Phi : P\to E$ as follows

$$\Phi x(t) = \frac{1}{e_p(t,a)}\left(A\int_a^{\sigma(b)} f(x(s))e_p(s,a)\Delta s \right.$$
$$\left. + \int_a^t e_p(s,a)f(x(s))\Delta s\right), \quad t \in [a,\sigma(b)],$$

By Lemma 1.11, it follows that any fixed points of the operator Φ are solutions to the PBVP (1.35).

Lemma 1.12. *Suppose* $(A15)$ *and* $(A16)$. *Then* $\Phi : P \to P$.

Proof. Let $x \in P$ be arbitrarily chosen. Then, by $(A15)$ and $(A16)$, we get

$$\begin{aligned}
0 &\le \Phi x(t) \\
&= \frac{1}{e_p(t,a)}\left(A\int_a^{\sigma(b)} f(x(s))e_p(s,a)\Delta s \right. \\
&\quad \left. + \int_a^t e_p(s,a)f(x(s))\Delta s\right) \\
&\le \int_a^t e_p(s,a)f(x(s))\Delta s + A\int_a^{\sigma(b)} e_p(s,a)f(x(s))\Delta s \\
&\le \int_a^{\sigma(b)} e_p(s,a)f(x(s))\Delta s + A\int_a^{\sigma(b)} e_p(s,a)f(x(s))\Delta s \\
&= (1+A)\int_a^{\sigma(b)} e_p(s,a)f(x(s))\Delta s, \quad t \in [a,\sigma(b)].
\end{aligned}$$

Then

$$\|\Phi x\| \le (1+A)\int_a^{\sigma(b)} e_p(s,a)f(x(s))\Delta s.$$

Hence,

$$\begin{aligned}
\Phi x(t) &= \frac{1}{e_p(t,a)}\left(A\int_a^{\sigma(b)} f(x(s))e_p(s,a)\Delta s \right. \\
&\quad \left. + \int_a^t e_p(s,a)f(x(s))\Delta s\right) \\
&\ge \frac{A}{e_p(t,a)}\int_a^{\sigma(b)} e_p(s,a)f(x(s))\Delta s \\
&\ge \frac{A}{e_p(\sigma(b),a)}\int_a^{\sigma(b)} e_p(s,a)f(x(s))\Delta s \\
&= \frac{A^2}{1+A}\int_a^{\sigma(b)} e_p(s,a)f(x(s))\Delta s
\end{aligned}$$

$$\geq \left(\frac{A}{1+A}\right)^2 \|\Phi x\|$$

$$= \delta\|\Phi x\|, \quad t \in [a, \sigma(b)].$$

This completes the proof.

Lemma 1.13. *Suppose* (A15) *and* (A16). *Then* $\Phi : P \to P$ *is continuous.*

Proof. Let $\{x_n\}_{n\in\mathbb{N}}$ be a sequence of elements of P such that $x_n \to x$, $x \in P$, in $\mathscr{C}([a, \sigma(b)])$. Take $\varepsilon > 0$ arbitrarily. Then there exists an $N_1 \in \mathbb{N}$ so that

$$|x_n(s) - x(s)| < \varepsilon, \quad s \in [a, \sigma(b)], \quad n > N_1.$$

Since $f : [0, \infty) \to [0, \infty)$ is continuous, there exists an $N \in \mathbb{N}$ such that

$$|f(x_n(s)) - f(x(s))| < \frac{\varepsilon}{e_p(\sigma(b), a)(1+A)(\sigma(b) - a)}, \quad s \in [a, \sigma(b)], \quad n > N.$$

Hence,

$$\begin{aligned}
|\Phi x_n(t) - \Phi x(t)| &= \left| \frac{1}{e_p(t,a)} \left(A \int_a^{\sigma(b)} f(x_n(s)) e_p(s,a) \Delta s \right.\right. \\
&\quad \left. + \int_a^t e_p(s,a) f(x_n(s)) \Delta s \right) \\
&\quad - \frac{1}{e_p(t,a)} \left(A \int_a^{\sigma(b)} f(x(s)) e_p(s,a) \Delta s \right. \\
&\quad \left.\left. + \int_a^t e_p(s,a) f(x(s)) \Delta s \right) \right| \\
&= \left| \frac{1}{e_p(t,a)} \left(A \int_a^{\sigma(b)} (f(x_n(s)) - f(x(s))) e_p(s,a) \Delta s \right.\right. \\
&\quad \left.\left. + \int_a^t e_p(s,a) (f(x_n(s)) - f(x(s))) \Delta s \right) \right| \\
&\leq \frac{1}{e_p(t,a)} \left(\left| A \int_a^{\sigma(b)} (f(x_n(s)) - f(x(s))) e_p(s,a) \Delta s \right| \right. \\
&\quad \left. + \left| \int_a^t e_p(s,a) (f(x_n(s)) - f(x(s))) \Delta s \right| \right) \\
&\leq \int_a^{\sigma(b)} e_p(s,a) |f(x_n(s)) - f(x(s))| \Delta s
\end{aligned}$$

$$
\begin{aligned}
&+A\int_a^{\sigma(b)} e_p(s,a)|f(x_n(s))-f(x(s))|\Delta s\\
&=(1+A)\int_a^{\sigma(b)} e_p(s,a)|f(x_n(s))-f(x(s))|\Delta s\\
&\le (1+A)e_p(\sigma(b),a)\int_a^{\sigma(b)} |f(x_n(s))-f(x(s))|\Delta s\\
&< (1+A)e_p(\sigma(b),a)\frac{\varepsilon}{e_p(\sigma(b),a)(1+A)(\sigma(b)-a)}(\sigma(b)-a)\\
&=\varepsilon, \quad t\in[a,\sigma(b)], \quad n>N.
\end{aligned}
$$

Hence,

$$
\|\Phi x_n-\Phi x\|<\varepsilon, \quad n>N.
$$

Therefore, $\Phi : P \to P$ is a continuous operator. This completes the proof.

Lemma 1.14. *Suppose* (A15) *and* (A16). *Then* $\Phi : P \to P$ *is completely continuous.*

Proof. Let $X \subset P$ be a bounded set. Then there exists a constant $L>0$ so that

$$
0\le x(t)\le L, \quad t\in[a,\sigma(b)].
$$

Since $f:[0,\infty)\to[0,\infty)$ is a continuous function, there exists a constant $M>0$ such that

$$
0\le f(u)\le M, \quad 0\le u\le L.
$$

Take $x\in X$ arbitrarily. Then

$$
\begin{aligned}
0\le\ &\Phi x(t)\\
&=\frac{1}{e_p(t,a)}\left(A\int_a^{\sigma(b)} f(x(s))e_p(s,a)\Delta s\right.\\
&\quad\left.+\int_a^t e_p(s,a)f(x(s))\Delta s\right)\\
&\le \int_a^t e_p(s,a)f(x(s))\Delta s+A\int_a^{\sigma(b)} e_p(s,a)f(x(s))\Delta s\\
&\le \int_a^{\sigma(b)} e_p(s,a)f(x(s))\Delta s+A\int_a^{\sigma(b)} e_p(s,a)f(x(s))\Delta s\\
&=(1+A)\int_a^{\sigma(b)} e_p(s,a)f(x(s))\Delta s\\
&\le (1+A)(\sigma(b)-a)Me_p(\sigma(b),a), \quad t\in[a,\sigma(b)].
\end{aligned}
$$

Therefore, $\Phi(X)$ is uniformly bounded. Now, take $t_1, t_2 \in [a, \sigma(b)]$, $t_1 < t_2$. Then

$$\begin{aligned}
|\Phi x(t_2) - \Phi x(t_1)| &= \Bigg| \frac{1}{e_p(t_1,a)} \left(A \int_a^{\sigma(b)} f(x(s)) e_p(s,a) \Delta s \right. \\
&\quad \left. + \int_a^{t_1} e_p(s,a) f(x(s)) \Delta s \right) \\
&\quad - \frac{1}{e_p(t_2,a)} \left(A \int_a^{\sigma(b)} f(x(s)) e_p(s,a) \Delta s \right. \\
&\quad \left. + \int_a^{t_2} e_p(s,a) f(x(s)) \Delta s \right) \Bigg| \\
&= \Bigg| \frac{1}{e_p(t_1,a)} \int_a^{t_1} e_p(s,a) f(x(s)) \Delta s \\
&\quad - \frac{1}{e_p(t_2,a)} \int_a^{t_1} e_p(s,a) f(x(s)) \Delta s \\
&\quad - \frac{1}{e_p(t_2,a)} \int_{t_1}^{t_2} e_p(s,a) f(x(s)) \Delta s \\
&\quad + A \left(\frac{1}{e_p(t_1,a)} - \frac{1}{e_p(t_2,a)} \right) \int_a^{\sigma(b)} e_p(s,a) f(x(s)) \Delta s \Bigg| \\
&= \Bigg| \left(\frac{1}{e_p(t_1,a)} - \frac{1}{e_p(t_2,a)} \right) \int_a^{t_1} e_p(s,a) f(x(s)) \Delta s \\
&\quad - \frac{1}{e_p(t_2,a)} \int_{t_1}^{t_2} e_p(s,a) f(x(s)) \Delta s \\
&\quad + A \left(\frac{1}{e_p(t_1,a)} - \frac{1}{e_p(t_2,a)} \right) \int_a^{\sigma(b)} e_p(s,a) f(x(s)) \Delta s \Bigg| \\
&\leq \left| \frac{1}{e_p(t_1,a)} - \frac{1}{e_p(t_2,a)} \right| \int_a^{t_1} e_p(s,a) f(x(s)) \Delta s \\
&\quad + \frac{1}{e_p(t_2,a)} \int_{t_1}^{t_2} e_p(s,a) f(x(s)) \Delta s \\
&\quad + A \left| \frac{1}{e_p(t_1,a)} - \frac{1}{e_p(t_2,a)} \right| \int_a^{\sigma(b)} e_p(s,a) f(x(s)) \Delta s \\
&\leq (1+A) \left| \frac{1}{e_p(t_1,a)} - \frac{1}{e_p(t_2,a)} \right| \int_a^{\sigma(b)} e_p(s,a) f(x(s)) \Delta s
\end{aligned}$$

$$+\frac{1}{e_p(t_2,a)}\int_{t_1}^{t_2} e_p(s,a)f(x(s))\Delta s$$

$$\leq (1+A)\left|\frac{1}{e_p(t_1,a)}-\frac{1}{e_p(t_2,a)}\right| Me_p(\sigma(b),a)(\sigma(b)-a)$$

$$+Me_p(\sigma(b),a)(\sigma(b)-a)(t_2-t_1).$$

Thus, $\Phi(X)$ is equi-continuous. Hence and by the Arzela-Ascoli theorem, we conclude that $\Phi : P \to P$ is completely continuous. This completes the proof.

Define

$$f_0 = \lim_{x\to 0+}\frac{f(x)}{x} \quad \text{and} \quad f_\infty = \lim_{x\to\infty}\frac{f(x)}{x}.$$

For our next result, we need the following fixed point theorem.

Theorem 1.11. *[28][Guo-Krasnosel'skii Fixed Point Theorem] Let X be a Banach space with a cone K. Let Ω_1, Ω_2 be bounded open subsets of X with*

$$0 \in \Omega_1 \subset \overline{\Omega_1} \subset \Omega_2,$$

$\Phi : K\bigcap(\overline{\Omega_2}\backslash\Omega_1) \to K$ is a completely continuous operator such that either

1.
$$\|\Phi x\| \leq \|x\|, \quad x\in K\bigcap\partial\Omega_1, \quad \|\Phi x\| \geq \|x\|, \quad x\in K\bigcap\partial\Omega_2,$$

or

2.
$$\|\Phi x\| \geq \|x\|, \quad x\in K\bigcap\partial\Omega_1, \quad \|\Phi x\| \leq \|x\|, \quad x\in K\bigcap\partial\Omega_2.$$

Then Φ has at least one fixed point in $K\bigcap(\overline{\Omega_2}\backslash\Omega_1)$.

Theorem 1.12. *Assume* (A15) *and* (A16). *Let, also,*

$$f_0 = 0, \quad f_\infty = \infty.$$

Then the PBVP (1.35) *has at least one positive solution.*

Proof. Take $\varepsilon > 0$ and $\rho > 0$ so that

$$\varepsilon(1+A)(\sigma(b)-a)e_p(\sigma(b),a) \leq 1$$

and

$$\rho A\delta(\sigma(b)-a) \geq 1.$$

Since $f_0 = 0$, we can choose $H_1 > 0$ so that

$$f(x) \leq \varepsilon x, \quad x\in[0,H_1].$$

Because $f_\infty = \infty$, we can choose $\hat{H}_2 > 0$ so that

$$f(x) \geq \rho x, \quad x \in [\hat{H}_2, \infty).$$

Let

$$H_2 = \max\left\{2H_1, \frac{\hat{H}_2}{\delta}\right\}.$$

Define

$$\Omega_i = \{x \in \mathscr{C}([a, \sigma(b)]) : \|x\| < H_i\}, \quad i = 1, 2.$$

Take $x_1 \in P \bigcap \partial\Omega_1$ arbitrarily. Then

$$\begin{aligned}
0 &\leq \Phi x_1(t) \\
&= \frac{1}{e_p(t_1, a)}\left(A \int_a^{\sigma(b)} f(x(s)) e_p(s, a) \Delta s \right. \\
&\quad \left. + \int_a^t e_p(s, a) f(x_1(s)) \Delta s\right) \\
&\leq \int_a^t e_p(s, a) f(x_1(s)) \Delta s + A \int_a^{\sigma(b)} e_p(s, a) f(x_1(s)) \Delta s \\
&\leq \int_a^{\sigma(b)} e_p(s, a) f(x_1(s)) \Delta s + A \int_a^{\sigma(b)} e_p(s, a) f(x_1(s)) \delta s \\
&= (1 + A) \int_a^{\sigma(b)} e_p(s, a) f(x_1(s)) \Delta s \\
&\leq \varepsilon(1 + A) \int_a^{\sigma(b)} e_p(s, a) x_1(s) \Delta s \\
&\leq \varepsilon(1 + A) e_p(\sigma(b), a)(\sigma(b) - a)\|x_1\| \\
&\leq \|x_1\|, \quad t \in [a, \sigma(b)].
\end{aligned}$$

Thus,

$$\|\Phi x_1\| \leq \|x_1\|, \quad x_1 \in P \bigcap \partial\Omega_1.$$

Take $x_2 \in P \bigcap \partial\Omega_2$. Then

$$\begin{aligned}
x_2(t) &\geq \delta\|x_2\| \\
&= \delta H_2
\end{aligned}$$

$$\geq \hat{H}_2, \quad t \in [a, \sigma(b)].$$

Hence,

$$\begin{aligned}
\Phi x_2(a) &\geq A \int_a^{\sigma(b)} e_p(s,a) f(x_2(s)) \Delta s \\
&\geq \rho A \int_a^{\sigma(b)} e_p(s,a) x_2(s) \Delta s \\
&\geq \rho A \delta \int_a^{\sigma(b)} e_p(s,a) \|x_2\| \Delta s \\
&\geq \rho A \delta (\sigma(b) - a) \|x_2\| \\
&\geq \|x_2\|.
\end{aligned}$$

Therefore,

$$\|\Phi x_2\| \geq \|x_2\|, \quad x_2 \in P \bigcap \partial \Omega_2.$$

By the Guo-Krasnosel'skii theorem, it follows that the operator Φ has a fixed point in $P \bigcap (\overline{\Omega}_2 \backslash \Omega_1)$. This completes the proof.

Example 1.4. Consider the PBVP

$$x^{\Delta} = -x^{\sigma} + x^2(1 + x^2), \quad t \in [a, b],$$

$$x(a) = x(\sigma(b)).$$

Here

$$p(t) = 1, \quad t \in [a, \sigma(b)],$$

$$f(x) = x^2(1 + x^2), \quad x \in [0, \infty).$$

Then

$$\begin{aligned}
f_0 &= \lim_{x \to 0+} \frac{f(x)}{x} \\
&= \lim_{x \to 0+} \frac{x^2(1 + x^2)}{x} \\
&= \lim_{x \to 0+} \left(x(1 + x^2)\right) \\
&= 0
\end{aligned}$$

and

$$\begin{aligned} f_\infty &= \lim_{x\to\infty} \frac{f(x)}{x} \\ &= \lim_{x\to\infty} \frac{x^2(1+x^2)}{x} \\ &= \lim_{x\to\infty} \left(x(1+x^2)\right) \\ &= \infty. \end{aligned}$$

Now, applying Theorem 1.12, we conclude that the considered PBVP has at least one positive solution.

Exercise 1.6. Let $\mathbb{T} = \mathbb{N}_0$. Prove that the PBVP

$$x^\Delta = -(1+t^2)x^\sigma + \frac{x^8(1+x^{16})}{1+x+x^2}, \quad t \in [1,10],$$

$$x(1) = x(11),$$

has at least one positive solution.

Theorem 1.13. *Let* (A15) *and* (A16) *hold and*

$$f_0 = \infty, \quad f_\infty = 0.$$

Then the PBVP (1.35) *has at least one positive solution.*

Proof. Take the constants $M > 0$ and $\lambda > 0$ so that

$$MA\delta(\sigma(b) - a) \geq 1$$

and

$$\lambda(1+A)(\sigma(b) - 1)e_p(\sigma(b), a) \leq 1.$$

Since $f_0 = \infty$, there exists $H_3 > 0$ so that

$$f(x) \geq Mx, \quad x \in [0, H_3].$$

Because $f_\infty = 0$, there exists $\hat{H}_4 > 0$ such that

$$f(x) \leq \lambda x, \quad x \in [\hat{H}_4, \infty).$$

Define

$$\Omega_3 = \{x \in \mathscr{C}([a, \sigma(b)]) : \|x\| < H_3\}.$$

Take $x_3 \in P \bigcap \partial\Omega_3$ arbitrarily. Then

$$\begin{aligned}\Phi x_3(a) &= A\int_a^{\sigma(b)} e_p(s,a)f(x_3(s))\Delta s\\ &\geq AM\int_a^{\sigma(b)} e_p(s,a)x_3(s)\Delta s\\ &\geq AM\delta\|x\|\int_a^{\sigma(b)} e_p(s,a)\Delta s\\ &\geq AM\delta(\sigma(b)-a)\|x\|\\ &\geq \|x\|.\end{aligned}$$

So,

$$\|\Phi x\| \geq \|x\| \quad \text{for} \quad x\in P\bigcap\partial\Omega_3.$$

Now, we consider the following cases for the function f.

1. Let f is bounded. Then there exists a constant $N>0$ so that

$$f(x)\leq N, \quad x\in[0,\infty).$$

Set

$$H_4 = \max\{2H_3, (1+A)N(\sigma(b)-a)e_p(\sigma(b),a)\}$$

and

$$\Omega_4 = \{x\in\mathscr{C}([a,\sigma(b)]) : \|x\| < H_4\}.$$

Take $x_4\in P\bigcap\partial\Omega_4$ arbitrarily. Then

$$\begin{aligned}\Phi x_4(t) &= \frac{1}{e_p(t,a)}\left(\int_a^t e_p(s,a)f(x_4(s))\Delta s\right.\\ &\quad \left.+A\int_a^{\sigma(b)} e_p(s,a)f(x_4(s))\Delta s\right)\\ &\leq N(1+A)\int_a^{\sigma(b)} e_p(s,a)\Delta s\\ &\leq N(1+A)(\sigma(b)-a)e_p(\sigma(b),a)\\ &\leq H_4\\ &= \|x_4\|, \quad t\in[a,\sigma(b)].\end{aligned}$$

So,

$$\|\Phi x_4\| \leq \|x_4\|, \quad x_4\in P\bigcap\partial\Omega_4.$$

2. Let f be unbounded. Then there exists $H_4 > \max\{2H_3, \hat{H}_4\}$ such that

$$f(x) \leq f(H_4), \quad x \in [0, H_4].$$

Then, for $x_4 \in P \bigcap \partial\Omega_4$, we arrive at

$$\begin{aligned}
\Phi x_4(t) &= \frac{1}{e_p(t,a)}\left(\int_a^t e_p(s,a)f(x_4(s))\Delta s\right.\\
&\quad \left.+A\int_a^{\sigma(b)} e_p(s,a)f(x_4(s))\Delta s\right)\\
&\leq f(H_4)(1+A)\int_a^{\sigma(b)} e_p(s,a)\Delta s\\
&\leq f(H_4)(1+A)e_p(\sigma(b),a)(\sigma(b)-a)\\
&\leq \lambda H_4(1+A)e_p(\sigma(b),a)(\sigma(b)-a)\\
&\leq H_4\\
&= \|x_4\|, \quad t \in [a,\sigma(b)].
\end{aligned}$$

Thus,

$$\|\Phi x_4\| \leq \|x_4\|, \quad x_4 \in P \bigcap \partial\Omega_4.$$

Now, applying the Guo-Krasnosel'skii theorem, Theorem 1.11, we conclude that the PBVP (1.35) has at least one positive solution. This completes the proof.

Example 1.5. Consider the PBVP

$$\begin{aligned}
x^{\Delta} &= -4x^{\sigma} + \frac{1}{1+x^2}, \quad t \in [a,\sigma(b)],\\
x(a) &= x(\sigma(b)).
\end{aligned}$$

Here

$$p(t) = 4, \quad t \in [a,\sigma(b)],$$

$$f(x) = \frac{1}{1+x^2}, \quad x \in [0,\infty).$$

Then

$$f_0 = \lim_{x \to 0+} \frac{f(x)}{x}$$

$$= \lim_{x\to 0+} \frac{1}{x(1+x^2)}$$

$$= \infty$$

and

$$f_\infty = \lim_{x\to\infty} \frac{f(x)}{x}$$

$$= \lim_{x\to\infty} \frac{1}{x(1+x^2)}$$

$$= 0.$$

Hence and by Theorem 1.13, we conclude that the considered PBVP has at least one positive solution.

Exercise 1.7. Let $\mathbb{T} = 2\mathbb{Z}$. Prove that the PBVP

$$x^\Delta = -t^2 x^\sigma + \frac{1}{1+x^4+x^8}, \quad t \in [-2,10],$$

$$x(-2) = x(12),$$

has at least one positive solution.

For $0 < a < b$ and α non-negative continuous concave functional, define

$$P_a = \{x \in P : \|x\| < a\},$$

$$P(\alpha,a,b) = \{x \in P : a \leq \alpha(x), \|x\| \leq b\}.$$

Next, denote

$$D = (1+A)(\sigma(b)-a)e_p(\sigma(b),a),$$

$$C = \frac{A(\sigma(b)-a)}{e_p(\sigma(b),a)}.$$

For the next our result, we need of the following fixed point theorem.

Theorem 1.14. *[36][Leggett-Williams Fixed Point Theorem] Let $\Phi : \overline{P}_c \to \overline{P}_c$ be a completely continuous operator and α be a non-negative continuous concave functional on P such that $\alpha(x) \leq \|x\|$ for any $x \in \overline{P}_c$. Suppose that there exist $0 < d < a < b \leq c$ such that*

1. $\{x \in P(\alpha,a,b) : \alpha(x) > a\} \neq \emptyset$ and $\alpha(\Phi x) > a$, $x \in P(\alpha,a,b)$,

2. $\|\Phi x\| < d$ *for* $\|x\| \leq d$*,*
3. $\alpha(\Phi x) > a$ *for* $x \in P(\alpha, a, c)$ *with* $\|\Phi x\| > b$*.*

Then Φ *has at least three fixed points* $x_1, x_2, x_3 \in \overline{P}_c$ *satisfying*

$$\|x_1\| < d, \quad a < \alpha(x_2), \quad \|x_3\| > d \quad \text{and} \quad \alpha(x_3) < a.$$

Theorem 1.15. *Assume* (A15), (A16) *and there exist numbers* d*,* a_1 *and* c *with* $0 < d < a_1 < \dfrac{a_1}{\delta} < c$ *such that*

$$f(x) < \frac{d}{D}, \quad x \in [a, d], \tag{1.38}$$

$$f(x) > \frac{a_1}{C}, \quad x \in \left[a_1, \frac{a_1}{\delta}\right], \tag{1.39}$$

and

$$f(x) < \frac{c}{D}, \quad x \in [a_1, c]. \tag{1.40}$$

Then the PBVP (1.35) *has at least three positive solutions.*

Proof. For $x \in P$, define

$$\alpha(x) = \min_{t \in [a, \sigma(b)]} x(t).$$

We have that α is a non-negative continuous and concave functional on P with

$$\alpha(x) \leq \|x\|, \quad x \in P.$$

Assume that there exists a positive number r such that

$$f(x) < \frac{r}{D}, \quad x \in [0, r].$$

Then, for $x \in \overline{P}_r$ and $t \in [a, \sigma(b)]$, we get

$$\begin{aligned}
\Phi x(t) &= \frac{1}{e_p(t,a)}\left(\int_a^t e_p(s,a) f(x(s))\Delta s \right.\\
&\quad \left. + A \int_a^{\sigma(b)} e_p(s,a) f(x(s))\Delta s\right)\\
&\leq (1+A)\int_a^{\sigma(b)} e_p(s,a) f(x(s))\Delta s\\
&\leq \frac{r}{D}(1+A)\int_a^{\sigma(b)} e_p(s,a)\Delta s\\
&\leq \frac{r}{D}(1+A) e_p(\sigma(b),a)(\sigma(b)-a)
\end{aligned}$$

$$= r.$$

Thus, $\Phi x \in P_r$ and $\Phi : \overline{P}_r \to P_r$. Hence, we conclude that if (1.38) and (1.40) hold, then $\Phi : \overline{P}_d \to P_d$ and $\Phi : \overline{P}_c \to P_c$, respectively. Now, we will show

$$\left\{x \in P\left(\alpha, a_1, \frac{a_1}{\delta}\right) : \alpha(x) > a_1\right\} \neq \emptyset. \tag{1.41}$$

In fact, we have

$$\frac{(1+\delta)a_1}{2\delta} = \alpha\left(\frac{(1+\delta)a_1}{2\delta}\right)$$

$$> a_1,$$

i.e.,

$$\frac{(1+\delta)a_1}{2\delta} \in \left\{x \in P\left(\alpha, a_1, \frac{a_1}{\delta}\right) : \alpha(x) > a_1\right\}$$

and (1.41) holds. Moreover, for $x \in P\left(\alpha, a_1, \frac{a_1}{\delta}\right)$, we have

$$\begin{aligned}
\frac{a_1}{\delta} &\geq \|x\| \\
&\geq x(t) \\
&\geq \min_{t \in [a,\sigma(b)]} x(t) \\
&= \alpha(x) \\
&\geq a_1, \quad t \in [a, \sigma(b)].
\end{aligned}$$

Then, using (1.39), for $x \in P\left(\alpha, a_1, \frac{a_1}{\delta}\right)$, we get

$$\begin{aligned}
\alpha(\Phi x) &= \min_{t \in [a,\sigma(b)]} \frac{1}{e_p(t,a)}\left(\int_a^t e_p(s,a) f(x(s))\Delta s\right. \\
&\quad \left. + A\int_a^{\sigma(b)} e_p(s,a) f(x(s))\Delta s\right) \\
&\geq \frac{A}{e_p(\sigma(b),a)} \int_a^{\sigma(b)} e_p(s,a) f(x(s))\Delta s
\end{aligned}$$

$$
\begin{aligned}
&\geq \frac{A}{e_p(\sigma(b),a)} \int_a^{\sigma(b)} f(x(s))\Delta s \\
&\geq \frac{A(\sigma(b)-a)}{Ce_p(\sigma(b),a)} a_1 \\
&= a_1.
\end{aligned}
$$

Therefore, all conditions of the Leggett-Williams theorem are satisfied. Consequently, the PBVP (1.35) has at least three positive solutions u, v and w such that

$$\|u\| < \delta, \quad a_1 < \min_{t\in[a,\sigma(b)]} v(t), \quad \|w\| > d$$

and

$$\min_{t\in[a,\sigma(b)]} w(t) < a_1.$$

This completes the proof.

Corollary 1.4. *Assume* (A15), (A16) *and* $m \in \mathbb{N}$. *Suppose that there exist numbers* d_j, $j \in \{1,\ldots,m\}$, *and* a_h, $h \in \{1,\ldots,m-1\}$, *such that*

$$0 < d_1 < a_1 < \frac{a_1}{\delta} < d_2 < a_2 < \frac{a_2}{\delta} < \ldots < d_{m-1} < a_{m-1} < \frac{a_{m-1}}{\delta}$$

such that

$$f(x) < \frac{d_j}{D}, \quad x \in [0, d_j], \quad j \in \{1,\ldots,m\},$$

and

$$f(x) > \frac{a_h}{C}, \quad x \in \left[a_h, \frac{a_h}{\delta}\right], \quad h \in \{1,\ldots,m-1\}.$$

Then the PBVP (1.35) *has at least* $2m-1$ *positive solutions.*

Proof. For $m=1$, by Theorem 1.13, it follows that the PBVP (1.35) has at least one positive solution. For $m=2$, by Theorem 1.15, it follows that the PBVP (1.35) has at least three positive solutions. By induction, for arbitrarily chosen $m \in \mathbb{N}$, we get the desired result. This completes the proof.

Example 1.6. Let $\mathbb{T} = [0,1]\bigcup[2,3]$, where $[0,1]$ and $[2,3]$ are the real valued intervals. Consider the PBVP

$$
\begin{aligned}
x^\Delta &= -x^\sigma + f(x), \quad t \in [0,3], \\
x(0) &= x(3),
\end{aligned}
\tag{1.42}
$$

where

$$f(x) = \begin{cases} x^2 + \dfrac{(2e^2-1)^2}{1152e^8}, \quad x \in \left[0, \dfrac{8e^6(2e^2-1)}{3}\right], \\ \dfrac{2e^2-1}{24e^4}x + \dfrac{(8192e^{20} - 128e^{10} + 1)(2e^2-1)}{1152e^8}, \\ x \in \left[\dfrac{8e^6(2e^2-1)}{3}, \infty\right). \end{cases}$$

Here

$$p(t) = 1, \quad t \in \mathbb{T},$$

$$a = 0,$$

$$b = 3.$$

We have $\sigma(b) = 3$ and

$$\begin{aligned} e_p(\sigma(b), a) &= e_1(3,0) \\ &= e^{1+\log(1+1)+1} \\ &= e^{2+\log 2} \\ &= 2e^2, \\ A &= \frac{1}{e_p(\sigma(b),a)-1} \\ &= \frac{1}{2e^2-1}, \\ 1+A &= 1 + \frac{1}{2e^2-1} \\ &= \frac{2e^2}{2e^2-1}, \\ \frac{A}{1+A} &= \frac{1}{2e^2}, \\ \delta &= \frac{1}{4e^4}, \end{aligned}$$

$$
\begin{aligned}
C &= \frac{A\sigma(b)}{e_p(\sigma(b),a)} \\
&= \frac{\frac{1}{2e^2-1}\sigma(3)}{e_1(3,0)} \\
&= \frac{\frac{3}{e^2-1}}{2e^2} \\
&= \frac{3}{2e^2(2e^2-1)}, \\
D &= (1+A)\sigma(b)e_p(\sigma(b),a) \\
&= 3\left(\frac{2e^2}{2e^2-1}\right)(2e^2) \\
&= \frac{12e^4}{2e^2-1}.
\end{aligned}
$$

Take

$$
\begin{aligned}
d &= \frac{2e^2-1}{24e^4}, \\
a_1 &= \frac{2e^2(2e^2-1)}{3}, \\
c &> \frac{8192e^{20}-128e^{10}+1}{e^8}.
\end{aligned}
$$

Then

$$
\frac{a_1}{\delta} = \frac{8e^6(2e^2-1)}{3}
$$

and

$$
0 < d < a_1 < \frac{a_1}{\delta} < c.
$$

Next,

$$
\begin{aligned}
\frac{d}{D} &= \frac{\frac{2e^2-1}{24e^4}}{\frac{12e^4}{2e^2-1}} \\
&= \frac{(2e^2-1)^2}{288e^8},
\end{aligned}
$$

$$\frac{a_1}{C} = \frac{\frac{2e^2(2e^2-1)}{3}}{\frac{3}{2e^2(2e^2-1)}}$$

$$= \frac{4e^4(2e^2-1)^2}{9},$$

$$\frac{c}{D} = \frac{c(2e^2-1)}{12e^4}.$$

Then, for $x \in \left[0, \frac{2e^2-1}{24e^4}\right]$, we have

$$\begin{aligned} f(x) &= x^2 + \frac{(2e^2-1)^2}{1152e^8} \\ &\leq \frac{(2e^2-1)^2}{576e^8} + \frac{(2e^2-1)^2}{1152e^8} \\ &= \frac{3(2e^2-1)^2}{1152e^8} \\ &= \frac{(2e^2-1)^2}{384e^8} \\ &< \frac{d}{D} \\ &< \frac{c}{D}. \end{aligned}$$

Next, for $x \in \left[\frac{2e^2(2e^2-1)}{3}, \frac{8e^6(2e^2-1)}{3}\right]$, we have

$$\begin{aligned} f(x) &= x^2 + \frac{(2e^2-1)^2}{1152e^8} \\ &\geq \frac{4e^4(2e^2-1)^2}{9} + \frac{(2e^2-1)^2}{1152e^8} \\ &> \frac{a_1}{C}. \end{aligned}$$

Now, for $x \in \left[\frac{8e^6(2e^2-1)}{3}, c\right]$, we get

$$f(x) = \frac{2e^2-1}{24e^4}x + \frac{\left(8192e^{20} - 128e^{10} + 1\right)(2e^2-1)}{1152e^8}$$

$$
\begin{aligned}
&< c\frac{2e^2-1}{24e^4} + \frac{\left(8192e^{20}-128e^{10}+1\right)(2e^2-1)}{1152e^8} \\
&< c\frac{2e^2-1}{24e^4} + c\frac{2e^2-1}{24e^4} \\
&= \frac{c(2e^2-1)}{12e^4} \\
&= \frac{c}{D}.
\end{aligned}
$$

Thus, all conditions of Theorem 1.15 are satisfied. Therefore, the PBVP (1.42) has at least three positive solutions.

Theorem 1.16. *Assume* (A15), (A16) *and*

$$
f_0 = \infty, \quad f_\infty = \infty.
$$

Let, also, there exists a positive constant R such that

$$
f(x) < \frac{R}{(1+A)e_p(\sigma(b),a)(\sigma(b)-a)} \quad \textit{for} \quad 0 < x \le R.
$$

Then the PBVP (1.35) *has at least two positive solutions.*

Proof. Let

$$
\Omega_R = \{x \in P : \|x\| < R\}.
$$

Then, for $x \in P\bigcap \partial\Omega_R$, we get

$$
\begin{aligned}
\Phi x(t) &= \frac{1}{e_p(t,a)}\left(\int_a^t e_p(s,a)f(x(s))\Delta s \right.\\
&\quad \left. +A\int_a^{\sigma(b)} e_p(s,a)f(x(s))\Delta s\right) \\
&\le (1+A)\int_a^{\sigma(b)} e_p(s,a)f(x(s))\Delta s \\
&\le (1+A)e_p(\sigma(b),a)\int_a^{\sigma(b)} f(x(s))\Delta s \\
&< (1+A)e_p(\sigma(b),a)\frac{R}{(1+A)e_p(\sigma(b),a)(\sigma(b)-a)}(\sigma(b)-a) \\
&= R, \quad t \in [a,\sigma(b)].
\end{aligned}
$$

Thus,

$$\|\Phi x\| \le \|x\|, \quad x \in P \bigcap \partial \Omega_R.$$

Since $f_0 = \infty$, for

$$M \ge \frac{e_p(\sigma(b),a)}{A\delta(\sigma(b)-a)},$$

there exists a positive constant $R_1 < \delta R$ such that

$$f(x) \ge Mx, \quad x \in [0, R_1].$$

Let

$$\Omega_{R_1} = \{x \in P : \|x\| < R_1\}.$$

Then, for $x \in P \bigcap \partial \Omega_{R_1}$, we have

$$x(t) \ge \delta \|x\|, \quad t \in [a, \sigma(b)],$$

and

$$\begin{aligned}
\Phi x(t) &= \frac{1}{e_p(t,a)} \left(\int_a^t e_p(s,a) f(x(s)) \Delta s \right. \\
&\quad \left. + A \int_a^{\sigma(b)} e_p(s,a) f(x(s)) \Delta s \right) \\
&\ge \frac{A}{e_p(\sigma(b),a)} \int_a^{\sigma(b)} e_p(s,a) f(x(s)) \Delta s \\
&\ge \frac{AM}{e_p(\sigma(b),a)} \int_a^{\sigma(b)} e_p(s,a) x(s) \Delta s \\
&\ge \frac{AM\delta}{e_p(\sigma(b),a)} \|x\| (\sigma(b) - a) \\
&\ge \|x\|, \quad t \in [a, \sigma(b)],
\end{aligned}$$

whereupon,

$$\|\Phi x\| \ge \|x\|.$$

Because $f_\infty = \infty$, for

$$N \ge \frac{e_p(\sigma(b),a)}{A\delta(\sigma(b)-a)},$$

there exists a positive constant $R_2 > \dfrac{R}{\delta}$ such that

$$f(x) \ge Nx, \quad x \in [\delta R_2, \infty).$$

Let

$$\Omega_{R_2} = \{x \in P : \|x\| < R_2\}.$$

Then, for $x \in P\bigcap \partial\Omega_{R_2}$, we obtain

$$x(t) \geq \delta\|x\|, \quad t \in [a, \sigma(b)],$$

and

$$\begin{aligned}
\Phi x(t) &= \frac{1}{e_p(t,a)}\left(\int_a^t e_p(s,a)f(x(s))\Delta s\right. \\
&\quad \left. + A\int_a^{\sigma(b)} e_p(s,a)f(x(s))\Delta s\right) \\
&\geq \frac{A}{e_p(\sigma(b),a)}\int_a^{\sigma(b)} e_p(s,a)f(x(s))\Delta s \\
&\geq \frac{A}{e_p(\sigma(b),a)}\int_a^{\sigma(b)} f(x(s))\Delta s \\
&\geq \frac{AN}{e_p(\sigma(b),a)}\int_a^{\sigma(b)} x(s)\Delta s \\
&\geq \frac{AN\delta}{e_p(\sigma(b),a)}(\sigma(b)-a)\|x\| \\
&\geq \|x\|, \quad t \in [a, \sigma(b)],
\end{aligned}$$

whereupon,

$$\|\Phi x\| \geq \|x\|.$$

Now, applying the Guo-Krasnosel'skii fixed point Theorem 1.11, we get that Φ has a fixed point in $P\bigcap(\overline{\Omega}_R \backslash \Omega_{R_1})$ and Φ has a fixed point in $P\bigcap(\overline{\Omega}_{R_2} \backslash \Omega_R)$. This completes the proof.

Theorem 1.17. *Assume* (A15), (A16) *and*

$$f_0 = 0, \quad f_\infty = 0.$$

Let, also, there exists a positive constant r such that

$$f(x) \geq \frac{re_p(\sigma(b),a)}{A(\sigma(b)-a)}, \quad x \in [0, r].$$

Then the PBVP (1.35) *has at least two positive solutions.*

Proof. Let

$$\Omega_r = \{x \in P : \|x\| < r\}.$$

Then, for $x \in P \bigcap \partial \Omega_r$, we have

$$\begin{aligned}
\Phi x(t) &= \frac{1}{e_p(t,a)}\Big(\int_a^t e_p(s,a)f(x(s))\Delta s \\
&\quad + A\int_a^{\sigma(b)} e_p(s,a)f(x(s))\Delta s\Big) \\
&\geq \frac{A}{e_p(t,a)}\int_a^{\sigma(b)} e_p(s,a)f(x(s))\Delta s \\
&\geq \frac{A}{e_p(\sigma(b),a)}\int_a^{\sigma(b)} f(x(s))\Delta s \\
&\geq \frac{A}{e_p(\sigma(b),a)}\left(\frac{re_p(\sigma(b),a)}{A(\sigma(b)-a)}\right)(\sigma(b)-a) \\
&= r \\
&= \|x\|, \quad t \in [a,\sigma(b)],
\end{aligned}$$

and

$$\|\Phi x\| \geq \|x\|.$$

Since $f_0 = 0$, for

$$\varepsilon_1 \leq \frac{1}{(1+A)e_p(\sigma(b),a)(\sigma(b)-a)},$$

there exists a positive constant $r_1 < \delta r$ such that

$$f(x) \leq \varepsilon_1 x, \quad x \in (0,r_1].$$

Let

$$\Omega_{r_1} = \{x \in P : \|x\| < r_1\}.$$

Then, for $x \in P \bigcap \partial \Omega_{r_1}$, we find

$$\begin{aligned}
\Phi x(t) &= \frac{1}{e_p(t,a)}\Big(\int_a^t e_p(s,a)f(x(s))\Delta s \\
&\quad + A\int_a^{\sigma(b)} e_p(s,a)f(x(s))\Delta s\Big) \\
&\leq (1+A)\int_a^{\sigma(b)} e_p(s,a)f(x(s))\Delta s \\
&\leq \varepsilon_1(1+A)e_p(\sigma(b),a)\int_a^{\sigma(b)} x(s)\Delta s
\end{aligned}$$

$$\leq \varepsilon_1(1+A)e_p(\sigma(b),a)(\sigma(b)-a)\|x\|$$

$$\leq \|x\|, \quad t \in [a,\sigma(b)],$$

and

$$\|\Phi x\| \leq \|x\|.$$

Since $f_\infty = 0$, for

$$\varepsilon_2 \leq \frac{1}{(1+A)e_p(\sigma(b),a)(\sigma(b)-a)},$$

there exists a positive constant $r_2 > \dfrac{r}{\delta}$ such that

$$f(x) \leq \varepsilon_2 x, \quad x \in (0,r_2].$$

Let

$$\Omega_{r_2} = \{x \in P : \|x\| < r_2\}.$$

Then, for $x \in P\bigcap \partial\Omega_{r_2}$, we obtain

$$\begin{aligned}
\Phi x(t) &= \frac{1}{e_p(t,a)}\left(\int_a^t e_p(s,a)f(x(s))\Delta s \right.\\
&\quad \left. +A\int_a^{\sigma(b)} e_p(s,a)f(x(s))\Delta s\right)\\
&\leq (1+A)\int_a^{\sigma(b)} e_p(s,a)f(x(s))\Delta s\\
&\leq \varepsilon_2(1+A)e_p(\sigma(b),a)\int_a^{\sigma(b)} x(s)\Delta s\\
&\leq \varepsilon_2(1+A)e_p(\sigma(b),a)(\sigma(b)-a)\|x\|\\
&\leq \|x\|, \quad t \in [a,\sigma(b)],
\end{aligned}$$

and

$$\|\Phi x\| \leq \|x\|.$$

Then, by the Guo-Krasnosel'skii fixed point theorem, Theorem 1.11, we get that Φ has a fixed point in $P\bigcap(\overline{\Omega}_R\backslash\Omega_{r_1})$ and Φ has a fixed point in $P\bigcap(\overline{\Omega}_{r_2}\backslash\Omega_R)$. This completes the proof.

1.6 Positive Solutions for Nonlinear First Order m-Point Boundary Value Problems

Suppose that $m \in \mathbb{N}$, $m \geq 2$, and

$$a = t_1 < t_2 < \ldots < t_m = b$$

and consider the m-point BVP

$$\begin{aligned} x^{\Delta} &= -px^{\sigma} + f(t, x^{\sigma}), \quad t \in [t_1, t_m], \\ x(t_1) &= \sum_{k=2}^{m-1} \alpha_k x(t_k) + \alpha_1 x(\sigma(t_m)), \end{aligned} \tag{1.43}$$

where

(A17) $f : [t_1, \sigma(t_m)] \times [0, \infty) \to [0, \infty)$ is continuous,
(A18) $p \in \mathscr{C}_{rd}([t_1, \sigma(t_m)])$, $p \in \mathscr{R}$,
(A19) $\alpha_1, \ldots, \alpha_{m-1} > 0$ and

$$1 - \sum_{k=2}^{m-1} \frac{\alpha_k}{e_p(t_k, t_1)} - \frac{\alpha_1}{e_p(\sigma(t_m), t_1)} > 0.$$

Set

$$\Gamma = \frac{1}{1 - \sum_{k=2}^{m-1} \frac{\alpha_k}{e_p(t_k, t_1)} - \frac{\alpha_1}{e_p(\sigma(t_m), t_1)}}$$

and consider the BVP

$$\begin{aligned} x^{\Delta} &= -px^{\sigma} + h(t), \quad t \in [t_1, t_m], \\ x(t_1) &= \sum_{k=2}^{m-1} \alpha_k x(t_k) + \alpha_1 x(\sigma(t_m)), \end{aligned} \tag{1.44}$$

where

(A20) $h \in \mathscr{C}_{rd}([t_1, t_m])$.

We will start our investigations in this section with the following useful lemma.

Lemma 1.15. *Suppose* (A19) *and* (A20). *Then the BVP* (1.44) *is equivalent to the following integral equation*

$$
\begin{aligned}
x(t) &= \frac{1}{e_p(t,t_1)}\Bigg(\Gamma\Bigg(\frac{\alpha_1}{e_p(\sigma(t_m),t_1)}\int_{t_1}^{\sigma(t_m)} e_p(s,t_1)h(s)\Delta s \\
&\quad + \sum_{k=2}^{m-1}\frac{\alpha_k}{e_p(t_k,t_1)}\int_{t_1}^{t_k} e_p(s,t_1)h(s)\Delta s\Bigg) \qquad (1.45)\\
&\quad + \int_{t_1}^{t} e_p(s,t_1)h(s)\Delta s\Bigg), \quad t\in[t_1,\sigma(t_m)].
\end{aligned}
$$

Proof. 1. Let x be a solution to the BVP (1.44). Then

$$
x(t) = \frac{1}{e_p(t,t_1)}\left(x(t_1) + \int_{t_1}^{t} e_p(s,t_1)h(s)\Delta s\right), \quad t\in[t_1,t_m]. \tag{1.46}
$$

Then

$$
\begin{aligned}
x(t_k) &= \frac{1}{e_p(t_k,t_1)}\left(x(t_1) + \int_{t_1}^{t_k} e_p(s,t_1)h(s)\Delta s\right),\\
\alpha_k x(t_k) &= \frac{\alpha_k}{e_p(t_k,t_1)}\left(x(t_1) + \int_{t_1}^{t_k} e_p(s,t_1)h(s)\Delta s\right), \quad k\in\{2,\ldots,m-1\},\\
x(\sigma(t_m)) &= \frac{1}{e_p(\sigma(t_m),t_1)}\left(x(t_1) + \int_{t_1}^{\sigma(t_m)} e_p(s,t_1)h(s)\Delta s\right),\\
\alpha_1 x(\sigma(t_m)) &= \frac{\alpha_1}{e_p(\sigma(t_m),t_1)}\left(x(t_1) + \int_{t_1}^{\sigma(t_m)} e_p(s,t_1)h(s)\Delta s\right).
\end{aligned}
$$

Hence,

$$
\begin{aligned}
x(t_1) &= \sum_{k=2}^{m-1}\alpha_k x(t_k) + \alpha_1 x(\sigma(t_m))\\
&= \sum_{k=2}^{m-1}\frac{\alpha_k}{e_p(t_k,t_1)}\left(x(t_1) + \int_{t_1}^{t_k} e_p(s,t_1)h(s)\Delta s\right)\\
&\quad + \frac{\alpha_1}{e_p(\sigma(t_m),t_1)}\left(x(t_1) + \int_{t_1}^{\sigma(t_m)} e_p(s,t_1)h(s)\Delta s\right)\\
&= \left(\sum_{k=2}^{m-1}\frac{\alpha_k}{e_p(t_k,t_1)} + \frac{\alpha_1}{e_p(\sigma(t_m),t_1)}\right)x(t_1)\\
&\quad + \sum_{k=2}^{m-1}\frac{\alpha_k}{e_p(t_k,t_1)}\int_{t_1}^{t_k} e_p(s,t_1)h(s)\Delta s\\
&\quad + \frac{\alpha_1}{e_p(\sigma(t_m),t_1)}\int_{t_1}^{\sigma(t_m)} e_p(s,t_1)h(s)\Delta s,
\end{aligned}
$$

whereupon,

$$\left(1-\sum_{k=2}^{m-1}\frac{\alpha_k}{e_p(t_k,t_1)}-\frac{\alpha_1}{e_p(\sigma(t_m),t_1)}\right)x(t_1)$$
$$=\sum_{k=2}^{m-1}\frac{\alpha_k}{e_p(t_k,t_1)}\int_{t_1}^{t_k}e_p(s,t_1)h(s)\Delta s$$
$$+\frac{\alpha_1}{e_p(\sigma(t_m),t_1)}\int_{t_1}^{\sigma(t_m)}e_p(s,t_1)h(s)\Delta s$$

and

$$x(t_1)=\Gamma\left(\sum_{k=2}^{m-1}\frac{\alpha_k}{e_p(t_k,t_1)}\int_{t_1}^{t_k}e_p(s,t_1)h(s)\Delta s\right.$$
$$\left.+\frac{\alpha_1}{e_p(\sigma(t_m),t_1)}\int_{t_1}^{\sigma(t_m)}e_p(s,t_1)h(s)\Delta s\right),$$

which we put in (1.46) and we get (1.45).

2. Let x be a solution to the integral equation (1.45). We differentiate (1.45) and we find

$$x^{\Delta}(t)=\left(\frac{1}{e_p(\cdot,t_1)}\right)^{\Delta}(t)\left(\Gamma\left(\frac{\alpha_1}{e_p(\sigma(t_m),t_1)}\int_{t_1}^{\sigma(t_m)}e_p(s,t_1)h(s)\Delta s\right.\right.$$
$$\left.+\sum_{k=2}^{m-1}\frac{\alpha_k}{e_p(t_k,t_1)}\int_{t_1}^{t_k}e_p(s,t_1)h(s)\Delta s\right)$$
$$\left.+\int_{t_1}^{\sigma(t)}e_p(s,t_1)h(s)\Delta s\right)$$
$$+\frac{1}{e_p(t,t_1)}e_p(t,t_1)h(t)$$
$$=-\frac{p(t)}{e_p(\sigma(t),t_1)}\left(\Gamma\left(\frac{\alpha_1}{e_p(\sigma(t_m),t_1)}\int_{t_1}^{\sigma(t_m)}e_p(s,t_1)h(s)\Delta s\right.\right.$$
$$\left.+\sum_{k=2}^{m-1}\frac{\alpha_k}{e_p(t_k,t_1)}\int_{t_1}^{t_k}e_p(s,t_1)h(s)\Delta s\right)$$
$$\left.+\int_{t_1}^{\sigma(t)}e_p(s,t_1)h(s)\Delta s\right)+h(t)$$
$$=-p(t)x^{\sigma}(t)+h(t),\quad t\in[t_1,t_m].$$

Thus, x satisfies the first equation of (1.44). Now, we put $t=t_1$, $t=t_2$ and so on $t=t_m$ in (1.45) and we find

$$\begin{aligned} x(t_1) &= \Gamma\Bigg(\frac{\alpha_1}{e_p(\sigma(t_m),t_1)}\int_{t_1}^{\sigma(t_m)} e_p(s,t_1)h(s)\Delta s \\ &\quad + \sum_{k=2}^{m-1}\frac{\alpha_k}{e_p(t_k,t_1)}\int_{t_1}^{t_k} e_p(s,t_1)h(s)\Delta s\Bigg), \\ x(t_j) &= \frac{1}{e_p(t_j,t_1)}\Bigg(\Gamma\Bigg(\frac{\alpha_1}{e_p(\sigma(t_m),t_1)}\int_{t_1}^{\sigma(t_m)} e_p(s,t_1)h(s)\Delta s \\ &\quad + \sum_{k=2}^{m-1}\frac{\alpha_k}{e_p(t_k,t_1)}\int_{t_1}^{t_k} e_p(s,t_1)h(s)\Delta s\Bigg) \\ &\quad + \int_{t_1}^{t_j} e_p(s,t_1)h(s)\Delta s\Bigg) \\ &= \frac{1}{e_p(t_j,t_1)}x(t_1) + \frac{1}{e_p(t_j,t_1)}\int_{t_1}^{t_j} e_p(s,t_1)h(s)\Delta s, \\ x(\sigma(t_m)) &= \frac{1}{e_p(\sigma(t_m),t_1)}\Bigg(\Gamma\Bigg(\frac{\alpha_1}{e_p(\sigma(t_m),t_1)}\int_{t_1}^{\sigma(t_m)} e_p(s,t_1)h(s)\Delta s \\ &\quad + \sum_{k=2}^{m-1}\frac{\alpha_k}{e_p(t_k,t_1)}\int_{t_1}^{t_k} e_p(s,t_1)h(s)\Delta s\Bigg) \\ &\quad + \int_{t_1}^{\sigma(t_m)} e_p(s,t_1)h(s)\Delta s\Bigg) \\ &= \frac{1}{e_p(\sigma(t_m),t_1)}x(t_1) + \frac{1}{e_p(\sigma(t_m),t_1)}\int_{t_1}^{\sigma(t_m)} e_p(s,t_1)h(s)\Delta s. \end{aligned}$$

Hence,

$$\begin{aligned} &\sum_{k=2}^{m-1}\alpha_k x(t_k) + \alpha_1 x(\sigma(t_m)) \\ &= \sum_{j=2}^{m-1}\alpha_j\left(\frac{1}{e_p(t_j,t_1)}x(t_1) + \frac{1}{e_p(t_j,t_1)}\int_{t_1}^{t_j} e_p(s,t_1)h(s)\Delta s\right) \\ &\quad + \frac{\alpha_1}{e_p(\sigma(t_m),t_1)}x(t_1) + \frac{\alpha_1}{e_p(\sigma(t_m),t_1)}\int_{t_1}^{\sigma(t_m)} e_p(s,t_1)h(s)\Delta s \\ &= \left(\sum_{k=2}^{m-1}\frac{\alpha_k}{e_p(t_k,t_1)} + \frac{\alpha_1}{e_p(\sigma(t_m),t_1)}\right)x(t_1) \end{aligned}$$

$$
\begin{aligned}
&+\sum_{k=2}^{m-1}\frac{\alpha_k}{e_p(t_k,t_1)}\int_{t_1}^{t_k}e_p(s,t_1)h(s)\Delta s\\
&+\frac{\alpha_1}{e_p(\sigma(t_m),t_1)}\int_{t_1}^{\sigma(t_m)}e_p(s,t_1)h(s)\Delta s\\
&=-\left(1-\sum_{k=2}^{m-1}\frac{\alpha_k}{e_p(t_k,t_1)}-\frac{\alpha_1}{e_p(\sigma(t_m),t_1)}\right)x(t_1)+x(t_1)\\
&+\sum_{k=2}^{m-1}\frac{\alpha_k}{e_p(t_k,t_1)}\int_{t_1}^{t_k}e_p(s,t_1)h(s)\Delta s\\
&+\frac{\alpha_1}{e_p(\sigma(t_m),t_1)}\int_{t_1}^{\sigma(t_m)}e_p(s,t_1)h(s)\Delta s\\
&=-\sum_{k=2}^{m-1}\frac{\alpha_k}{e_p(t_k,t_1)}\int_{t_1}^{t_k}e_p(s,t_1)h(s)\Delta s\\
&-\frac{\alpha_1}{e_p(\sigma(t_m),t_1)}\int_{t_1}^{\sigma(t_m)}e_p(s,t_1)h(s)\Delta s\\
&+\sum_{k=2}^{m-1}\frac{\alpha_k}{e_p(t_k,t_1)}\int_{t_1}^{t_k}e_p(s,t_1)h(s)\Delta s\\
&+\frac{\alpha_1}{e_p(\sigma(t_m),t_1)}\int_{t_1}^{\sigma(t_m)}e_p(s,t_1)h(s)\Delta s+x(t_1)\\
&=x(t_1),
\end{aligned}
$$

i.e., x satisfies the second equation of (1.44). This completes the proof.

Define

$$
G_j(t,s)=\begin{cases}\dfrac{e_p(s,t_1)}{e_p(t,t_1)}\left(\Gamma\left(\displaystyle\sum_{k=2}^{m-1}\frac{\alpha_k}{e_p(t_k,t_1)}+\frac{\alpha_1}{e_p(\sigma(t_m),t_1)}\right)+1\right), & \sigma(s)\le r,\\[2ex] \Gamma\dfrac{e_p(s,t_1)}{e_p(t,t_1)}\left(\displaystyle\sum_{k=2}^{m-1}\frac{\alpha_k}{e_p(t_k,t_1)}+\frac{\alpha_1}{e_p(\sigma(t_m),t_1)}\right), & t\le s,\end{cases}
$$

$j\in\{1,\ldots,m-1\}$, and

$$G(t,s) = \begin{cases} G_1(t,s) & t_1 \le s \le \sigma(s) \le t_2 \\ G_2(t,s) & t_2 \le s \le \sigma(s) \le t_3 \\ \vdots & \\ G_{m-2}(t,s) & t_{m-2} \le s \le \sigma(s) \le t_{m-1} \\ G_{m-1}(t,s) & t_{m-1} \le s \le t_m. \end{cases}$$

Then, using Lemma 1.15, the BVP (1.44) is equivalent to the integral equation

$$x(t) = \int_{t_1}^{\sigma(t_m)} G(t,s)h(s)\Delta s, \quad t \in [t_1, \sigma(t_m)].$$

Lemma 1.16. *Assume* $(A18)$ *and* $(A19)$. *Then*

1. $G(t,s) \ge 0$, $(t,s) \in [t_1, \sigma(t_m)] \times [t_1, t_m]$.
2. $m \le G(t,s) \le M$, *where*

$$m = \Gamma \frac{\alpha_1}{(e_p(\sigma(t_m), t_1))^2},$$

$$M = \Gamma \sum_{k=1}^{m-2} \alpha_k + e_p(t_{m-1}, t_1).$$

3. $G(t,s) \ge \dfrac{m}{M} \sup_{(t,s)\in[t_1,\sigma(t_m)]\times[t_1,t_m]} G(t,s)$.

Proof. 1. By $(A18)$ and $(A19)$, we have $G > 0$. Hence,

$$G_j(t,s) \ge 0, \quad (t,s) \in [t_1, \sigma(t_m)] \times [t_1, t_m],$$

and then

$$G(t,s) \ge 0, \quad (t,s) \in [t_1, \sigma(t_m)] \times [t_1, t_m].$$

2. For any $j \in \{1, \ldots, m-1\}$, we have

$$G_j(t,s) \le \begin{cases} \Gamma \sum_{k=1}^{m-1} \alpha_k + e_p(t_{m-1}, t_1) & \sigma(s) \le t \\ \Gamma \sum_{k=1}^{m-1} \alpha_k + e_p(t_{m-1}, t_1) & t \le s \end{cases}$$

$$\le M, \quad (t,s) \in [t_1, \sigma(t_m)] \times [t_1, t_m],$$

and

$$G_j(t,s) \geq \begin{cases} \gamma \dfrac{\alpha_1}{(e_p(\sigma(t_m),t_1))^2} & \sigma(s) \leq t \\ \Gamma \dfrac{\alpha_1}{(e_p(\sigma(t_m),t_1))^2} & t \leq s \end{cases}$$

$$\geq m, \quad (t,s) \in [t_1,\sigma(t_m)] \times [t_1,t_m].$$

Therefore,

$$m \leq G(t,s) \leq M, \quad (t,s) \in [t_1,\sigma(t_m)] \times [t_1,t_m].$$

3. By (2), we have

$$M \geq G(t,s)$$

for any $(t,s) \in [t_1,\sigma(t_m)] \times [t_1,t_m]$. Then

$$M \geq \sup_{(t,s)\in[t_1,\sigma(t_m)]\times[t_1,t_m]} G(t,s)$$

and

$$\begin{aligned} G(t,s) &\geq m \\ &= \frac{m}{M} M \\ &\geq \frac{m}{M} \sup_{(t,s)\in[t_1,\sigma(t_m)]\times[t_1,t_m]} G(t,s), \end{aligned}$$

$(t,s) \in [t_1,\sigma(t_m)] \times [t_1,t_m]$. This completes the proof.

Suppose that $\mathscr{C}([t_1,\sigma(t_m)])$ is endowed with the maximum norm. Let

$$P = \{x \in \mathscr{C}([t_1,\sigma(t_m)]) : x(t) \geq \frac{m}{M}\|x\|, \quad t \in [t_1,\sigma(t_m)]\}.$$

For $x \in \mathscr{C}([t_1,\sigma(t_m)])$, define the operator

$$\Phi x(t) = \int_{t_1}^{\sigma(t_m)} G(t,s) f(s,x(\sigma(s)))\Delta s, \quad t \in [t_1,\sigma(t_m)].$$

Lemma 1.17. *Suppose* (A17)-(A19). *Then* $\Phi : P \to P$.

Proof. Let $x \in P$ be arbitrarily chosen. By Lemma 1.16, we have

$$G(t,s) \geq 0, \quad (t,s) \in [t_1,\sigma(t_m)] \times [t_1,t_m].$$

Hence and $(A17)$, we conclude that

$$\Phi x(t) \geq 0, \quad t \in [t_1, \sigma(t_m)].$$

Next, by Lemma 1.16, we have

$$m \leq G(t,s) \leq M, (t,s) \in [t_1, \sigma(t_m)] \times [t_1, t_m].$$

Then

$$\begin{aligned}
\Phi x(t) &= \int_{t_1}^{\sigma(t_m)} G(t,s) f(s, x(\sigma(s)))\Delta s \\
&\geq m \int_{t_1}^{\sigma(t_m)} f(s, x(\sigma(s)))\Delta s \\
&= \frac{m}{M} \int_{t_1}^{\sigma(t_m)} M f(s, x(\sigma(s)))\Delta s \\
&\geq \frac{m}{M} \int_{t_1}^{\sigma(t_m)} \left(\sup_{(t,s)\in[t_1,\sigma(t_m)]\times[t_1,t_m]} G(t,s) \right) f(s, x(\sigma(s)))\Delta s \\
&\geq \frac{m}{M} \int_{t_1}^{\sigma(t_m)} \max_{t\in[t_1,\sigma(t_m)]} G(t,s) f(s, x(\sigma(s)))\Delta s \\
&\geq \frac{m}{M} \max_{t\in[t_1,\sigma(t_m)]} \int_{t_1}^{\sigma(t_m)} G(t,s) f(s, x(\sigma(s)))\Delta s \\
&= \frac{m}{M} \|\Phi x\|, \quad x \in [t_1, \sigma(t_m)].
\end{aligned}$$

Thus, $\Phi : P \to P$. This completes the proof.

Lemma 1.18. *Suppose* (A17)-(A19). *Then the operator* $\Phi : P \to P$ *is a continuous operator.*

Proof. Let $\{x_n\}_{n\in\mathbb{N}} \subseteq P$, $x \in P$ and $x_n \to x$, as $n \to \infty$, in $\mathscr{C}([t_1, \sigma(t_m)])$. Take $\varepsilon > 0$ arbitrarily. Then there exists an $N_1 \in \mathbb{N}$ so that

$$|x_n(t) - x(t)| < \varepsilon, \quad t \in [t_1, \sigma(t_m)], \quad n > N_1.$$

Since $f : [t_1, \sigma(t_m)] \times [0, \infty) \to [0, \infty)$ is a continuous function, then there exists an $N \in \mathbb{N}$ so that

$$|f(s, x_n(\sigma(s))) - f(s, x(\sigma(s)))| < \frac{\varepsilon}{M(\sigma(t_m) - \sigma(t_1))}, \quad n \in \mathbb{N}, \quad s \in [t_1, \sigma(t_m)].$$

Hence,

$$|\Phi x_n(t) - \Phi x(t)| = \left| \int_{t_1}^{\sigma(t_m)} G(t,s) f(s, x_n(\sigma(s)))\Delta s \right.$$

$$
\begin{aligned}
&- \int_{t_1}^{\sigma(t_m)} G(t,s) f(s, x(\sigma(s)))\Delta s \Big| \\
&= \left| \int_{t_1}^{\sigma(t_m)} G(t,s)(f(s,x_n(\sigma(s))) - f(s,x(\sigma(s))))\Delta s \right| \\
&\leq \int_{t_1}^{\sigma(t_m)} G(t,s)|f(s,x_n(\sigma(s))) - f(s,x(\sigma(s)))|\Delta s \\
&< \frac{\varepsilon}{M(\sigma(t_m)-t_1)} \int_{t_1}^{\sigma(t_m)} G(t,s)\Delta s \\
&\leq \frac{\varepsilon}{M(\sigma(t_m)-t_1)} M(\sigma(t_m)-t_1) \\
&= \varepsilon, \quad t \in [t_1, \sigma(t_m)], \quad n > N.
\end{aligned}
$$

Then

$$\|\Phi x_n - \Phi x\| < \varepsilon, \quad n > N.$$

Therefore, $\Phi : P \to P$ is a continuous operator. This completes the proof.

Lemma 1.19. *Assume* (A17)-(A19). *Then the operator $\Phi : P \to P$ is a completely continuous operator.*

Proof. Let $X \subset P$ is a bounded set. Then there exists a positive constant L such that

$$0 \leq x(t) \leq L, \quad t \in [t_1, \sigma(t_m)].$$

Because $f : [t_1, \sigma(t_m)] \times [0, L] \to [0, \infty)$ is a continuous function, then there exists a positive constant Q such that

$$f(s,u) \leq Q, \quad (s,u) \in [t_1, \sigma(t_m)] \times [0, L].$$

Take $x \in X$ arbitrarily. Then

$$
\begin{aligned}
0 &\leq \Phi x(t) \\
&= \int_{t_1}^{\sigma(t_m)} G(t,s) f(s, x(\sigma(s)))\Delta s \\
&\leq Q \int_{t_1}^{\sigma(t_m)} G(t,s)\Delta s \\
&\leq MQ(\sigma(t_m)-t_1), \quad t \in [t_1, \sigma(t_m)],
\end{aligned}
$$

whereupon,

$$\|\Phi x\| \leq MQ(\sigma(t_m)-t_1).$$

Therefore, $\Phi(X)$ is uniformly bounded. Let now, $t^1, t^2 \in [t_1, \sigma(t_m)]$, $t^1 < t^2$. Then

$$\begin{aligned}
|\Phi x(t^1) - \Phi x(t^2)| &= \left| \int_{t_1}^{\sigma(t_m)} G(t^1,s) f(s,x(\sigma(s)))\Delta s \right. \\
&\quad \left. - \int_{t_1}^{\sigma(t_m)} G(t^2,s) f(s,x(\sigma(s)))\Delta s \right| \\
&= \left| \int_{t_1}^{\sigma(t_m)} (G(t^1,s) - G(t^2,s)) f(s,x(\sigma(s)))\Delta s \right| \\
&\leq \int_{t_1}^{\sigma(t_m)} |G(t^1,s) - G(t^2,s)| f(s,x(\sigma(s)))\Delta s \\
&\leq Q \int_{t_1}^{\sigma(t_m)} |G(t^1,s) - G(t^2,s)| \Delta s.
\end{aligned}$$

Thus, $\Phi(X)$ is equi-continuous. Hence and applying the Arzela-Ascoli theorem, we conclude that the operator $\Phi : P \to P$ is a completely continuous operator. This completes the proof.

Theorem 1.18. *Assume* (A17)-(A19). *Let there exist positive numbers r and R satisfying $0 < r < R < \infty$ such that for any $t \in [t_1, \sigma(t_m)]$*

$$f(t,x) \leq \frac{x}{M\sigma(t_m)}, \quad x \in [0,r],$$

$$f(t,x) \geq \frac{Mx}{m^2(\sigma(t_m) - t_1)}, \quad x \in [R,\infty).$$

Then the BVP (1.43) *has at least one positive solution.*

Proof. Define

$$\Omega_1 = \{x \in \mathscr{C}([t_1, \sigma(t_m)]) : \|x\| < r\},$$

$$\Omega_2 = \{x \in \mathscr{C}([t_1, \sigma(t_m)]) : \|x\| < \frac{RM}{m}\}.$$

We have

$$\overline{\Omega}_1 \subset \Omega_2.$$

For $x \in P \bigcap \partial\Omega_1$, we obtain

$$\begin{aligned}
\Phi x(t) &= \int_{t_1}^{\sigma(t_m)} G(t,s) f(s,x(\sigma(s)))\Delta s \\
&\leq M \int_{t_1}^{\sigma(t_m)} f(s,x(\sigma(s)))\Delta s
\end{aligned}$$

$$
\begin{aligned}
&\leq \frac{M}{M\sigma(t_m)} \int_{t_1}^{\sigma(t_m)} x(\sigma(s))\Delta s \\
&\leq \frac{1}{\sigma(t_m)} \|x\| \sigma(t_m) \\
&= \|x\|, \quad t \in [t_1, \sigma(t_m)].
\end{aligned}
$$

Therefore,

$$\|\Phi x\| \leq \|x\|, \quad x \in P \bigcap \partial \Omega_1.$$

For $x \in P \bigcap \partial \Omega_2$, we obtain

$$
\begin{aligned}
\Phi x(t) &= \int_{t_1}^{\sigma(t_m)} G(t,s) f(s, x(\sigma(s)))\Delta s \\
&\geq m \int_{t_1}^{\sigma(t_m)} f(s, x(\sigma(s)))\Delta s \\
&\geq \frac{mM}{m^2(\sigma(t_m) - t_1)} \int_{t_1}^{\sigma(t_m)} x(\sigma(s))\Delta s \\
&\geq \frac{M}{m(\sigma(t_m) - t_1)} \left(\frac{m}{M} \int_{t_1}^{\sigma(t_m)} \|x\| \Delta s \right) \\
&= \|x\|, \quad t \in [t_1, \sigma(t_m)].
\end{aligned}
$$

Therefore,

$$\|\Phi x\| \geq \|x\|, \quad x \in P \bigcap \partial \Omega_2.$$

Hence and by the Guo-Krasnosel'skii Theorem 1.11, it follows that the operator Φ has at least one fixed point in $P \bigcap (\overline{\Omega}_2 \backslash \Omega_1)$. Therefore, the BVP (1.43) has at least one positive solution for which

$$r \leq x(t) \leq \frac{RM}{m}, \quad t \in [t_1, \sigma(t_m)].$$

This completes the proof.

Theorem 1.19. *Assume* (A17)-(A19). *Let there exist positive numbers r and R satisfying $0 < r < R < \infty$ such that for any $t \in [t_1, \sigma(t_m)]$*

$$
\begin{aligned}
f(t,x) &\leq \frac{Mx}{m^2 \sigma(t_m)}, \quad x \in [0, r], \\
f(t,x) &\geq \frac{x}{M(\sigma(t_m) - t_1)}, \quad x \in [R, \infty).
\end{aligned}
$$

Then the BVP (1.43) *has at least one positive solution.*

Proof. Let Ω_1 and Ω_2 be as in the proof of Theorem 1.18. For $x \in P\bigcap\partial\Omega_1$, we obtain

$$\begin{aligned}\Phi x(t) &= \int_{t_1}^{\sigma(t_m)} G(t,s)f(s,x(\sigma(s)))\Delta s\\ &\geq m\int_{t_1}^{\sigma(t_m)} f(s,x(\sigma(s)))\Delta s\\ &\geq \frac{mM}{m^2(\sigma(t_m)-t_1)}\int_{t_1}^{\sigma(t_m)} x(\sigma(s))\Delta s\\ &\geq \frac{M}{m(\sigma(t_m)-t_1)}\left(\frac{m}{M}\int_{t_1}^{\sigma(t_m)} \|x\|\Delta s\right)\\ &= \|x\|, \quad t\in[t_1,\sigma(t_m)].\end{aligned}$$

Therefore,

$$\|\Phi x\| \geq \|x\|, \quad x\in P\bigcap\partial\Omega_1.$$

For $x \in P\bigcap\partial\Omega_2$, we obtain

$$\begin{aligned}\Phi x(t) &= \int_{t_1}^{\sigma(t_m)} G(t,s)f(s,x(\sigma(s)))\Delta s\\ &\leq M\int_{t_1}^{\sigma(t_m)} f(s,x(\sigma(s)))\Delta s\\ &\leq \frac{M}{M\sigma(t_m)}\int_{t_1}^{\sigma(t_m)} x(\sigma(s))\Delta s\\ &\leq \frac{1}{\sigma(t_m)}\|x\|\sigma(t_m)\\ &= \|x\|, \quad t\in[t_1,\sigma(t_m)].\end{aligned}$$

Therefore,

$$\|\Phi x\| \leq \|x\|, \quad x\in P\bigcap\partial\Omega_2.$$

Hence and by the Guo-Krasnosel'skii Theorem 1.11, it follows that the operator Φ has at least one fixed point in $P\bigcap(\overline{\Omega}_2\backslash\Omega_1)$. Therefore, the BVP (1.43) has at least one positive solution for which

$$r \leq x(t) \leq \frac{RM}{m}, \quad t\in[t_1,\sigma(t_m)].$$

This completes the proof.

Theorem 1.20. *Assume* (A17)-(A19) *and there exist positive constants R, R_1, R_2 satisfying*

$$R_1 < \frac{m}{M}R < R < \frac{M}{m}R < R_2$$

such that for any $t \in [t_1, \sigma(t_m)]$ we have

$$f(t,x) \leq \frac{R}{M\sigma(t_m)}, \quad x \in (0,R],$$

$$f(t,x) \geq \frac{Mx}{m^2(\sigma(t_m)-t_1)}, \quad x \in (0,R_1]\bigcup\left[\frac{m}{M}R_2,\infty\right).$$

Then the BVP (1.43) *has at least two positive solutions.*

Proof. Let

$$\Omega_R = \{x \in P : \|x\| < R\},$$

$$\Omega_{R_1} = \{x \in P : \|x\| < R_1\},$$

$$\Omega_{R_2} = \{x \in P : \|x\| < R_2\}.$$

For $x \in P\bigcap\partial\Omega_R$, we get

$$\begin{aligned}
\Phi x(t) &= \int_{t_1}^{\sigma(t_m)} G(t,s)f(s,x(\sigma(s)))\Delta s \\
&\leq M\int_{t_1}^{\sigma(t_m)} f(s,x(\sigma(s)))\Delta s \\
&\leq M\frac{R}{M\sigma(t_m)}\sigma(t_m) \\
&= R \\
&= \|x\|, \quad t \in [t_1, \sigma(t_m)].
\end{aligned}$$

Therefore,

$$\|\Phi x\| \leq \|x\|, \quad x \in P\bigcap\partial\Omega_R.$$

For $x \in P\bigcap\Omega_{R_1}$, we get

$$\begin{aligned}
\Phi x(t) &= \int_{t_1}^{\sigma(t_m)} G(t,s) f(s,x(\sigma(s)))\Delta s \\
&\geq m \int_{t_1}^{\sigma(t_m)} f(s,x(\sigma(s)))\Delta s \\
&\geq m \frac{M}{m^2(\sigma(t_m)-t_1)} \int_{t_1}^{\sigma(t_m)} x(\sigma(s))\Delta s \\
&\geq \frac{M}{m(\sigma(t_m)-t_1)} \left(\frac{m}{M} \int_{t_1}^{\sigma(t_m)} \|x\| \Delta s \right) \\
&= \frac{1}{\sigma(t_m)-t_1} (\sigma(t_m)-t_1)\|x\| \\
&= \|x\|, \quad t \in [t_1, \sigma(t_m)].
\end{aligned} \tag{1.47}$$

Therefore,

$$\|\Phi x\| \geq \|x\|, \quad x \in P \bigcap \partial\Omega_{R_1}.$$

For $x \in P \bigcap \partial\Omega_{R_2}$, we find

$$\begin{aligned}
\|x\| &= R_2, \\
x(t) &\geq \delta \|x\| \\
&\geq \delta R_2, \quad t \in [t_1, \sigma(t_m)],
\end{aligned}$$

and using (1.47), we get

$$\|\Phi x\| \geq \|x\|, \quad x \in P \bigcap \partial\Omega_{R_2}.$$

Hence and by the Guo-Krasnosel'skii fixed point Theorem 1.11, we conclude that the operator Φ has at least two fixed points. This completes the proof.

Theorem 1.21. *Assume* (A17)-(A19) *and there exist positive constants* r_1, r *and* r_2 *satisfying*

$$r_1 < \frac{m}{M} r < r < \frac{M}{m} r < r_2$$

such that for any $t \in [t_1, \sigma(t_m)]$ *we have*

$$\begin{aligned}
f(t,x) &\geq \frac{r}{m(\sigma(t_m)-t_1)}, \quad x \in (0,r], \\
f(t,x) &\leq \frac{x}{M\sigma(t_m)}, \quad x \in (0,r_1] \bigcup \left[\frac{m}{M} r_2, \infty \right).
\end{aligned}$$

Then the BVP (1.43) *has at least two positive solutions.*

Proof. Define

$$\Omega_r = \{x \in P : \|x\| < r\},$$

$$\Omega_{r_1} = \{x \in P : \|x\| < r_1\},$$

$$\Omega_{r_2} = \{x \in P : \|x\| < r_2\}.$$

For $x \in P \bigcap \partial\Omega_r$, we get

$$\begin{aligned}
\Phi x(t) &= \int_{t_1}^{\sigma(t_m)} G(t,s) f(s, x(\sigma(s)))\Delta s \\
&\geq m \int_{t_1}^{\sigma(t_m)} f(s, x(\sigma(s)))\Delta s \\
&\geq m \frac{r}{m(\sigma(t_m) - t_1)} (\sigma(t_m) - t_1) \\
&= r \\
&= \|x\|, \quad t \in [t_1, \sigma(t_m)].
\end{aligned}$$

Thus,

$$\|\Phi x\| \geq \|x\|, \quad x \in P \bigcap \partial\Omega_r.$$

For $x \in P \bigcap \partial\Omega_{r_1}$, we get

$$\begin{aligned}
\Phi x(t) &= \int_{t_1}^{\sigma(t_m)} G(t,s) f(s, x(\sigma(s)))\Delta s \\
&\leq M \int_{t_1}^{\sigma(t_m)} f(s, x(\sigma(s)))\Delta s \\
&\leq \frac{1}{\sigma(t_m)} \int_{t_1}^{\sigma(t_m)} x(\sigma(s))\Delta s \qquad (1.48) \\
&\leq \frac{1}{\sigma(t_m)} \sigma(t_m) \|x\| \\
&= \|x\|, \quad t \in [t_1, \sigma(t_m)].
\end{aligned}$$

Therefore,

$$\|\Phi x\| \leq \|x\|, \quad x \in P\bigcap \partial\Omega_{r_1}.$$

For $x \in P\bigcap \partial\Omega_{r_2}$, we have

$$x(t) \geq \delta\|x\|$$

$$= \delta r_2, \quad t \in [t_1, \sigma(t_m)],$$

and using (1.48), we get

$$\|\Phi x\| \leq \|x\|, \quad x \in P\bigcap \partial\Omega_{r_2}.$$

Hence and by the Guo-Krasnosel'skii fixed point Theorem 1.11, we conclude that the operator Φ has at least two fixed points. This completes the proof.

For the next our result, we need of the following fixed point theorem.

Theorem 1.22. *[5][Avery-Henderson Fixed Point Theorem] Let P be a cone in a real Banach space E. Set*

$$P(\phi, r) = \{u \in P : \phi(u) < r\}.$$

If η and ϕ are increasing, non-negative continuous functionals on P, let θ be a non-negative continuous functional on P with $\theta(0) = 0$ such that, for some positive constants r and M,

$$\phi(u) \leq \theta(u) \leq \eta(u), \quad \|u\| \leq M\phi(u),$$

for all $u \in \overline{P(\phi, r)}$. Suppose that there exist positive numbers $p < q < r$ such that

$$\theta(\lambda u) \leq \lambda\theta(u), \quad \lambda \in [0,1], \quad u \in \partial P(\theta, q).$$

If $A : \overline{P(\phi, r)} \to P$ is a completely continuous operator satisfying

1. *$\phi(Au) > r$ for all $u \in \partial P(\phi, r)$,*
2. *$\theta(Au) < q$ for all $u \in \partial P(\theta, q)$,*
3. *$P(\eta, p) \neq \emptyset$ and $\eta(Au) > p$ for all $u \in \partial P(\eta, p)$,*

then A has at least two fixed-points u_1 and u_2 such that

$$p < \eta(u_1) \quad \text{with} \quad \theta(u_1) < q$$

and

$$q < \theta(u_2) \quad \text{with} \quad \phi(u_2) < r.$$

Theorem 1.23. *Suppose* (A17)-(A19) *and there exist positive numbers h, q, r satisfying*

$$0 < h < q < r$$

such that

$$f(t,x) > \frac{r}{m(\sigma(t_m) - t_{m-1})}, \quad t \in [t_{m-1}, \sigma(t_m)], \quad x \in \left[r, \frac{rM}{m}\right],$$

$$f(t,x) \le \frac{q}{M\sigma(t_m)}, \quad t \in [t_1, \sigma(t_m)], \quad x \left[0, \frac{qM}{m}\right],$$

$$f(t,x) > \frac{h}{m(\sigma(t_m) - t_{m-1})}, \quad t \in [t_{m-1}, \sigma(t_m)], \quad x \in \left[\frac{hm}{M}, h\right].$$

Then the BVP (1.43) *has at least two positive solutions* x_1 *and* x_2 *such that*

$$\begin{aligned} h &< \sup_{t \in [t_1, \sigma(t_m)]} x_1(t), \qquad \sup_{t \in [t_{m-1}, \sigma(t_m)]} x_1(t) < q, \\ q &< \sup_{t \in [t_{m-1}, \sigma(t_m)]} x_2(t), \qquad \inf_{t \in [t_{m-1}, \sigma(t_m)]} x_2(t) < r. \end{aligned} \tag{1.49}$$

Proof. For $x \in P$, define

$$\phi(x) = \inf_{t \in [t_{m-1}, \sigma(t_m)]} x(t),$$

$$\theta(x) = \sup_{t \in [t_{m-1}, \sigma(t_m)]} x(t),$$

$$\eta(x) = \sup_{t \in [t_1, \sigma(t_m)]} x(t).$$

For each $x \in P$ we have

$$\phi(x) \le \theta(x) \le \eta(x)$$

and

$$\inf_{t \in [t_1, \sigma(t_m)]} x(t) \ge \frac{m}{M} \|x\|,$$

and

$$\begin{aligned} \phi(x) &= \inf_{t \in [t_{m-1}, \sigma(t_m)]} x(t) \\ &\ge \inf_{t \in [t_1, \sigma(t_m)]} x(t) \\ &\ge \frac{m}{M} \|x\| \end{aligned}$$

or

$$\|x\| \leq \frac{M}{m}\phi(x).$$

Next, $\theta(0) = 0$ and for any $x \in P$, $\lambda \in [0,1]$, we obtain

$$\begin{aligned}\theta(\lambda x) &= \sup_{t\in[t_{m-1},\sigma(t_m)]} (\lambda x(t))\\ &= \lambda \sup_{t\in[t_{m-1},\sigma(t_m)]} x(t)\\ &= \lambda\theta(x).\end{aligned}$$

Let

$$P(\phi,r) = \{u \in P : \phi(u) < r\}.$$

1. Suppose that $x \in \partial P(\phi,r)$. Then

$$\begin{aligned}\inf_{t\in[t_{m-1},\sigma(t_m)]} x(t) &\leq \|x\|\\ &\leq r\frac{M}{m}, \quad t \in [t_{m-1},\sigma(t_m)].\end{aligned}$$

   Hence,

$$\begin{aligned}\phi(\Phi x) &= \inf_{t\in[t_{m-1},\sigma(t_m)]} \Phi x(t)\\ &= \inf_{t\in[t_{m-1},\sigma(t_m)]} \int_{t_1}^{\sigma(t_m)} G(t,s)f(s,x(\sigma(s)))\Delta s\\ &= \int_{t_1}^{\sigma(t_m)} \min_{t\in[t_{m-1},\sigma(t_m)]} G(t,s)f(s,x(\sigma(s)))\Delta s\\ &\geq m\int_{t_{m-1}}^{\sigma(t_m)} f(s,x(\sigma(s)))\Delta s\\ &> m\left(\frac{r}{m(\sigma(t_m)-t_{m-1})}\right)(\sigma(t_m)-t_{m-1})\\ &= r.\end{aligned}$$

2. Let $x \in \partial P(\theta,q)$. Then

$$\begin{aligned}0 &\leq x(t)\\ &\leq \|x\|\end{aligned}$$

$$\leq \frac{M}{m}\phi(x)$$

$$\leq \frac{M}{m}\theta(x)$$

$$= q\frac{M}{m}.$$

Thus,

$$\begin{aligned}\theta(\Phi x) &= \sup_{t\in[t_{m-1},\sigma(t_m)]} \Phi x(t)\\ &= \sup_{t\in[t_{m-1},\sigma(t_m)]} \int_{t_1}^{\sigma(t_m)} G(t,s)f(s,x(\sigma(s)))\Delta s\\ &= \int_{t_1}^{\sigma(t_m)} \max_{t\in[t_{m-1},\sigma(t_m)]} G(t,s)f(s,x(\sigma(s)))\Delta s\\ &\leq M\int_{t_1}^{\sigma(t_m)} f(s,x(\sigma(s)))\Delta s\\ &< M\frac{q}{M\sigma(t_m)}(\sigma(t_m)-t_1)\\ &< q.\end{aligned}$$

3. Since $0\in P$ and $h>0$, we have that

$$P(\eta,h)\neq\emptyset.$$

Let $x\in\partial P(\eta,h)$. Then

$$\begin{aligned}\frac{M}{M}h &\leq \phi(x)\\ &\leq x(t)\\ &\leq \|x\|\\ &= h, \quad t\in[t_{m-1},\sigma(t_m)].\end{aligned}$$

Hence,

$$\eta(\Phi x) = \sup_{t\in[t_1,\sigma(t_m)]} \Phi x(t)$$

$$\geq \int_{t_1}^{\sigma(t_m)} G(t,s)f(s,x(\sigma(s)))\Delta s$$

$$\geq m\int_{t_{m-1}}^{\sigma(t_m)} f(s,x(\sigma(s)))\Delta s$$

$$> m\left(\frac{h}{m(\sigma(t)m)-t_{m-1}}\right)(\sigma(t_m)-t_{m-1})$$

$$= h.$$

Hence and by the Avery-Henderson theorem, we conclude that the operator Φ has at least two positive solutions x_1 and x_2 that satisfy (1.49). This completes the proof.

Theorem 1.24. *Assume* (A17)-(A19) *and there exist positive numbers h, q and r satisfying*

$$0 < h < q < \frac{qM}{m} \leq r$$

such that for any $t \in [t_1, \sigma(t_m)]$ we have

$$f(t,x) \leq \frac{r}{M\sigma(t_m)}, \quad x \in [0,r],$$

$$f(t,x) > \frac{q}{m(\sigma(t_m)-t_1)}, \quad x \in \left[q, \frac{qM}{m}\right],$$

$$f(t,x) < \frac{h}{M\sigma(t_m)}, \quad x \in [0,h].$$

Then the BVP (1.43) *has at least three positive solutions x_1, x_2 and x_3 such that*

$$\begin{aligned} &\sup_{t\in[t_1,\sigma(t_m)]} x_1(t) < h, \quad q < \inf_{t\in[t_1,\sigma(t_m)]} x_2(t), \\ &h < \sup_{t\in[t_1,\sigma(t_m)]} x_3(t), \quad \inf_{t\in[t_1,\sigma(t_m)]} x_3(t) < q. \end{aligned} \tag{1.50}$$

Proof. Define the non-negative continuous concave functional $\psi : P \to [0,\infty)$ as follows

$$\psi(x) = \inf_{t\in[t_1,\sigma(t_m)]} x(t).$$

For any $x \in P$, we have

$$\psi(x) \leq \|x\|.$$

1. If $x \in \overline{P}_r$, then

$$0 \leq x \leq r$$

and

$$f(t,x) < \frac{r}{M\sigma(t_m)}, \quad t \in [t_1, \sigma(t_m)],$$

and

$$\begin{aligned}\|\Phi x\| &= \sup_{t\in[t_1,\sigma(t_m)]}\int_{t_1}^{\sigma(t_m)} G(t,s)f(s,x(\sigma(s)))\Delta s\\ &\le M\int_{t_1}^{\sigma(t_m)} f(s,x(\sigma(s)))\Delta s\\ &< M\left(\frac{r}{M\sigma(t_m)}\right)(\sigma(t_m)-t_1)\\ &< r.\end{aligned}$$

Therefore, $\Phi : \overline{P}_r \to \overline{P}_r$.

2. Let $x \in \overline{P}_h$. Then

$$0 \le x \le h$$

and

$$f(t,x) \le \frac{h}{M\sigma(t_m)}, \quad t \in [t_1,\sigma(t_m)],$$

and

$$\begin{aligned}\|\Phi x\| &= \sup_{t\in[t_1,\sigma(t_m)]}\int_{t_1}^{\sigma(t_m)} G(t,s)f(s,x(\sigma(s)))\Delta s\\ &\le M\int_{t_1}^{\sigma(t_m)} f(s,x(\sigma(s)))\Delta s\\ &\le M\left(\frac{h}{M\sigma(t_m)}\right)(\sigma(t_m)-t_1)\\ &< h.\end{aligned}$$

Consequently, $\Phi : \overline{P}_h \to \overline{P}_h$.

3. Note that

$$\frac{qM}{m} \in P\left(\psi,q,\frac{qM}{m}\right).$$

Therefore,

$$\left\{y \in P\left(\psi,q,\frac{qM}{m}\right),\psi(y) > q\right\} \ne \emptyset.$$

4. Let $x \in P\left(\psi,q,\dfrac{qM}{m}\right)$. Then

$$\begin{aligned}q &\le \inf_{t\in[t_1,\sigma(t_m)]} x(t)\\ &\le \|x\|\end{aligned}$$

$$\leq \frac{qM}{m}, \quad t \in [t_1, \sigma(t_m)].$$

Hence,

$$\begin{aligned}
\psi(\Phi x) &= \int_{t_1}^{\sigma(t_m)} \inf_{t\in[t_1,\sigma(t_m)]} G(t,s) f(s,x(\sigma(s)))\Delta s \\
&\geq m \int_{t_1}^{\sigma(t_m)} f(s,x(\sigma(s)))\Delta s \\
&> \frac{q}{\sigma(t_m)-t_1}(\sigma(t_m)-t_1) \\
&= q.
\end{aligned}$$

5. Let $x \in P(\psi, q, r)$ with $\|\Phi x\| > \dfrac{qM}{m}$. Then

$$\begin{aligned}
\psi(\Phi x) &= \inf_{t\in[t_1,\sigma(t_m)]} \Phi x(t) \\
&\geq \frac{m}{M}\|\Phi x\| \\
&> q.
\end{aligned}$$

Hence and by the Leggett-Williams Theorem 1.14, we conclude that the BVP (1.43) has at least three positive solutions x_1, x_2 and x_3 that satisfy (1.50). This completes the proof.

Exercise 1.8. Let $\mathbb{T} = \mathbb{Z}$. Prove that the BVP

$$\begin{aligned}
x^\Delta &= -x^\sigma + \frac{x+1}{x^4+5}, \quad t \in [0,5], \\
x(0) &= 2x(1) + 3x(2) + x(6),
\end{aligned}$$

has at least two positive solutions.

1.7 Existence of Solutions for First Order p-Laplacian Boundary Value Problems

In this section we will investigate the BVP

$$\phi_p\left(u^\Delta(t)\right) = h(t) f(u^\sigma(t)), \quad t \in [a,b], \tag{1.51}$$

$$u(a) = B_0(u^{\Delta}(b)), \tag{1.52}$$

where

(A21) $B_0 : \mathbb{R} \to \mathbb{R}$ is a continuous function such that there exist positive constants B_1 and B_2 for which

$$0 \le B_1 \le B_2$$

and

$$B_1 x \le B_0(x) \le B_2 x, \quad x \ge 0,$$

(A22) $\phi_p(x) = |x|^{p-2}x$, $p > 2$, $\dfrac{1}{p} + \dfrac{1}{q} = 1$,
(A23) $f : [0,\infty) \to [0,\infty)$ is a continuous function,
(A24) $h : [a,b] \to [0,\infty)$, $h \in \mathscr{C}_{rd}([a,b])$, $h(t)f(u^{\sigma}(t))$ is non-increasing in t for non-decreasing functions u, hf is not identically zero.

Introduce the notations

$$f_0 = \lim_{x \to 0+} \frac{f(x)}{\phi_p(x)},$$

$$f_{\infty} = \lim_{x \to \infty} \frac{f(x)}{\phi_p(x)}.$$

Lemma 1.20. *Suppose* (A21)-(A24). *Then for any* $u \in \mathscr{C}_{rd}([a,\sigma(b)])$ *the BVP* (1.51), (1.52) *is equivalent to the integral equation*

$$u(t) = B_0\left(\phi_q\left(h(b)f(u^{\sigma}(b))\right)\right) + \int_a^t \phi_q\left(h(s)f(u^{\sigma}(s))\right)\Delta s, \quad t \in [a,\sigma(b)]. \tag{1.53}$$

Proof. 1. Let $u \in \mathscr{C}_{rd}([a,\sigma(b)])$ be a solution to the BVP (1.51), (1.52). Then

$$u^{\Delta}(t) = \phi_q\left(h(t)f(u^{\sigma}(t))\right), \quad t \in [a,b]. \tag{1.54}$$

Hence,

$$u^{\Delta}(b) = \phi_q\left(h(b)f(u^{\sigma}(b))\right)$$

and

$$u(a) = B_0(u^{\Delta}(b))$$

$$= B_0\left(\phi_q\left(h(b)f(u^{\sigma}(b))\right)\right).$$

We integrate equation (1.54) from a to t and we find

$$u(t) = u(a) + \int_a^t \phi_q\left(h(\tau)f(u^{\sigma}(\tau))\right)\Delta\tau$$

$$= B_0\left(\phi_q\left(h(b)f(u^{\sigma}(b))\right)\right)$$

$$+\int_a^t B_0\left(\phi_q\left(h(\tau)f(u^\sigma(\tau))\right)\right)\Delta\tau, \quad t\in[a,\sigma(b)],$$

i.e., u satisfies equation (1.53).

2. Now, suppose that $u\in\mathscr{C}([a,\sigma(b)])$ satisfies (1.53). Then

$$u^\Delta(t)=\phi_q\left(h(t)f(u^\sigma(t))\right), \quad t\in[a,b],$$

whereupon,

$$u^\Delta(b)=\phi_q\left(h(b)f(u^\sigma(b))\right)$$

and

$$\begin{aligned}\phi_p\left(u^\Delta(t)\right) &= h(t)f(u^\sigma(t)), \quad t\in[a,b],\\ B_0\left(u^\Delta(b)\right) &= B_0\left(\phi_q\left(h(b)f(u^\sigma(b))\right)\right)\\ &= u(a).\end{aligned}$$

Therefore, u satisfies (1.51) and (1.52). This completes the proof.

Suppose that $\mathscr{C}_{rd}([a,\sigma(b)])$ is endowed with the maximum norm. Define the cone

$$P=\{u\in\mathscr{C}_{rd}([a,\sigma(b)]): u(t),u^\Delta(t)\geq 0, \quad t\in[a,b],$$

$$u^\Delta \quad \text{is} \quad \text{non-increasing}\}.$$

Lemma 1.21. *For any $u\in P$ and $t\in[a,\sigma(b)]$, we have*

$$u(t)\geq\frac{t-a}{\sigma(b)-a}\|u\|.$$

Proof. We have

$$\begin{aligned}u(t)-u(a) &= \int_a^t u^\Delta(\tau)\Delta\tau\\ &\geq (t-a)u^\Delta(t)\end{aligned}$$

and

$$\begin{aligned}u^\sigma(b)-u(t) &= \int_t^{\sigma(b)} u^\Delta(\tau)\Delta\tau\\ &\leq (\sigma(b)-t)u^\Delta(t).\end{aligned}$$

Hence,

$$u(a) \leq u(t) - (t-a)u^{\Delta}(t),$$

$$u^{\sigma}(b) \leq u(t) + (\sigma(b)-t)u^{\Delta}(t)$$

and

$$\begin{aligned}
\frac{(\sigma(b)-t)u(a)+(t-a)u^{\sigma}(b)}{\sigma(b)-a} &= \frac{\sigma(b)-t}{\sigma(b)-a}u(a) + \frac{t-a}{\sigma(b)-a}u^{\sigma}(b) \\
&\leq \frac{\sigma(b)-t}{\sigma(b)-a}\Big(u(t)-(t-a)u^{\Delta}(t)\Big) \\
&\quad + \frac{t-a}{\sigma(b)-a}\Big(u(t)+(\sigma(b)-t)u^{\Delta}(t)\Big) \\
&= \frac{\sigma(b)-t+t-a}{\sigma(b)-a}u(t) \\
&= u(t), \quad t \in [a,\sigma(b)],
\end{aligned}$$

and

$$\begin{aligned}
u(t) &\geq \frac{t-a}{\sigma(b)-a}u^{\sigma}(b) \\
&= \frac{t-a}{\sigma(b)-a}\|u\|, \quad t \in [a,\sigma(b)].
\end{aligned}$$

This completes the proof.

For $u \in \mathscr{C}_{rd}([a,\sigma(b)])$, define the operator

$$\begin{aligned}
Lu(t) &= B_0\left(\phi_q\left(h(b)f(u^{\sigma}(b))\right)\right) \\
&\quad + \int_a^t \phi_q\left(h(s)f(u^{\sigma}(s))\right)\Delta s, \quad t \in [a,\sigma(b)].
\end{aligned}$$

Lemma 1.22. *Assume* (A21)-(A24). *Then* $L: P \to P$.

Proof. Fix $u \in P$ arbitrarily. We have

$$Lu(t) \geq 0, \quad t \in [a,\sigma(b)],$$

and

$$(Lu)^{\Delta}(t) = \phi_q\left(h(t)f(u^{\sigma}(t))\right)$$

$$\geq 0, \quad t \in [a, \sigma(b)].$$

Now, take $t_1, t_2 \in [a, \sigma(b)]$ arbitrarily so that $t_1 < t_2$. Then

$$u(t_2) > u(t_1),$$

$$u^\sigma(t_2) > u^\sigma(t_1),$$

$$h(t_2) f(u^\sigma(t_2)) < h(t_1) f(u^\sigma(t_2))$$

and

$$\phi_q\left(h(t_2) f(u^\sigma(t_2))\right) \leq \phi_q\left(h(t_1) f(u^\sigma(t_1))\right),$$

i.e., $(Lu)^\Delta$ is non-increasing. Therefore, $Lu \in P$. Because $u \in P$ was arbitrarily chosen, we conclude that $L : P \to P$. This completes the proof.

Lemma 1.23. *Assume* (A21)-(A24). *Then* $L : P \to P$ *is continuous.*

Proof. Take $\varepsilon > 0$ arbitrarily. Let $\{u_n\}_{n\in\mathbb{N}} \subset P$, $u \in P$ be such that

$$u_n \to u \quad \text{in} \quad \mathscr{C}_{rd}([a, \sigma(b)]), \quad \text{as} \quad n \to \infty.$$

There exists a natural number N so that

$$\left|B_0\left(\phi_q\left(h(b) f(u_n^\sigma(b))\right)\right) - B_0\left(\phi_q\left(h(b) f(u^\sigma(b))\right)\right)\right| < \frac{\varepsilon}{2}$$

and

$$\left|\phi_q\left(h(s) f(u_n^\sigma(s))\right) - \phi_q\left(h(s) f(u^\sigma(s))\right)\right| < \frac{\varepsilon}{2(\sigma(b) - a)}, \quad s \in [a, \sigma(b)],$$

for any $n > N$. Hence, for any $n > N$, we get

$$\begin{aligned}
|Lu_n(t) - Lu(t)| = \Big| & B_0\left(\phi_q\left(h(b) f(u_n^\sigma(b))\right)\right) \\
& + \int_a^t \phi_q\left(h(s) f(u_n^\sigma(s))\right) \Delta s \\
& - B_0\left(\phi_q\left(h(b) f(u^\sigma(b))\right)\right) \\
& - \int_a^t \phi_q\left(h(s) f(u^\sigma(s))\right) \Delta s \Big| \\
= \Big| & \Big(B_0\left(\phi_q\left(h(b) f(u_n^\sigma(b))\right)\right) - B_0\left(\phi_q\left(h(b) f(u^\sigma(b))\right)\right) \Big)
\end{aligned}$$

$$
\begin{aligned}
&+\int_a^t \Big(\phi_q\left(h(s)f(u_n^\sigma(s))\right)-\phi_q\left(h(s)f(u^\sigma(s))\right)\Big)\Delta s\Big| \\
&\le \Big|\Big(B_0\left(\phi_q\left(h(b)f(u_n^\sigma(b))\right)\right)-B_0\left(\phi_q\left(h(b)f(u^\sigma(b))\right)\right)\Big)\Big| \\
&\quad+\Big|\int_a^t \Big(\phi_q\left(h(s)f(u_n^\sigma(s))\right)-\phi_q\left(h(s)f(u^\sigma(s))\right)\Big)\Delta s\Big| \\
&< \frac{\varepsilon}{2} \\
&\quad+\int_a^t \Big|\phi_q\left(h(s)f(u_n^\sigma(s))\right)-\phi_q\left(h(s)f(u^\sigma(s))\right)\Big|\Delta s \\
&< \frac{\varepsilon}{2}+\frac{\varepsilon}{2(\sigma(b)-a)}(\sigma(b)-a) \\
&= \frac{\varepsilon}{2}+\frac{\varepsilon}{2} \\
&= \varepsilon, t\in[a,\sigma(b)].
\end{aligned}
$$

From here, we conclude that $L : P \to P$ is a continuous operator. This completes the proof.

Lemma 1.24. *Assume* (A21)-(A24). *Then* $L : P \to P$ *is a completely continuous operator.*

Proof. Let $X \subset P$ be a bounded set. Then there exists a constant $A > 0$ such that

$$
0 \le u(t) \le A, \quad t \in [a, \sigma(b)],
$$

for any $u \in X$. Since ϕ_q, h, f are continuous functions, there exist a constant $Q > 0$ so that

$$
\begin{aligned}
&\left|B_0\left(\phi_q\left(h(t)f(u^\sigma(t))\right)\right)\right| \le Q, \\
&\left|\phi_q\left(h(t)f(u^\sigma(t))\right)\right| \le Q, \quad t \in [a, \sigma(b)].
\end{aligned}
$$

Then

$$
\begin{aligned}
|Lu(t)| &= \Big|B_0\left(\phi_q\left(h(b)f(u^\sigma(b))\right)\right) \\
&\quad+\int_a^t \phi_q\left(h(s)f(u^\sigma(s))\right)\Delta s\Big| \\
&\le \Big|B_0\left(\phi_q\left(h(b)f(u^\sigma(b))\right)\right)\Big|
\end{aligned}
$$

$$
\begin{aligned}
&+\left|\int_a^t \phi_q\left(h(s)f(u^\sigma(s))\right)\Delta s\right| \\
&\le Q+\int_a^t \left|\phi_q\left(h(s)f(u^\sigma(s))\right)\right|\Delta s \\
&\le Q+(\sigma(b)-a)Q \\
&=(1+\sigma(b)-a)Q, \quad t\in[a,\sigma(b)],
\end{aligned}
$$

for any $u\in X$. Thus, $L(X)$ is uniformly bounded. Now, we take $t_1,t_2\in[a,\sigma(b)]$ arbitrarily so that $t_1<t_2$. Then, for $u\in X$, we have

$$
\begin{aligned}
|Lu(t_1)-Lu(t_2)| &= \Big| B_0\left(\phi_q\left(h(b)f(u^\sigma(b))\right)\right) \\
&\quad +\int_a^{t_1} \phi_q\left(h(s)f(u^\sigma(s))\right)\Delta s \\
&\quad -B_0\left(\phi_q\left(h(b)f(u^\sigma(b))\right)\right) \\
&\quad -\int_a^{t_2} \phi_q\left(h(s)f(u^\sigma(s))\right)\Delta s\Big| \\
&=\left|-\int_{t_1}^{t_2} \phi_q\left(h(s)f(u^\sigma(s))\right)\Delta s\right| \\
&\le \int_{t_1}^{t_2} \left|\phi_q\left(h(s)f(u^\sigma(s))\right)\right|\Delta s \\
&\le Q(t_2-t_1).
\end{aligned}
$$

Therefore, $L(X)$ is equi-continuous. From here and from the Arzela-Ascoli theorem, we conclude that $L:P\to P$ is a completely continuous operator. This completes the proof.

Theorem 1.25. *Assume* (A21)-(A24) *and*

$$f_0=\infty, \quad f_\infty=0.$$

Then the BVP (1.51), (1.52) *has at least one positive solution.*

Proof. Let

$$\gamma=\min\left\{t\in\mathbb{T}:\frac{a+\sigma(b)}{2}\le t<b\right\}$$

such that $(\gamma,b)\neq\emptyset$. Take c so that

$$c\left(B_1\phi_q(h(b))+\frac{\gamma-a}{\sigma(b)-a}\int_\gamma^{\sigma(b)}\phi_q(h(s))\Delta s\right)\geq 1.$$

Since $f_0=\infty$, there exists a constant $k>0$ such that

$$f(x)\geq\phi_p(x)\phi_p(x)$$

$$=\phi_p(cx),\quad 0<x\leq k.$$

Set

$$S_k=\{u\in\mathscr{C}_{rd}([a,\sigma(b)]):\|u\|<k\}.$$

Take $u\in P\bigcap\partial S_k$. Then

$$\|u\|=u(\sigma(b))$$

$$=k$$

and

$$\|Lu\|=L(u(\sigma(b)))$$

$$=B_0\left(\phi_q\left(h(b)f(u^\sigma(b))\right)\right)$$

$$+\int_a^{\sigma(b)}\phi_q\left(h(s)f(u^\sigma(s))\right)\Delta s$$

$$\geq B_1\phi_q\left(h(b)\phi_p(cu(\sigma(b)))\right)$$

$$+\int_\gamma^{\sigma(b)}\phi_q\left(h(s)\phi_p(cu^\sigma(s))\right)\Delta s$$

$$=B_1cu^\sigma(b)\phi_q(h(b))$$

$$+c\int_\gamma^{\sigma(b)}\phi_q\left(h(s)\phi_p(cu^\sigma(s))\right)\Delta s$$

$$=B_1cu^\sigma(b)\phi_q(h(b))$$

$$+c\int_\gamma^{\sigma(b)}u^\sigma(s)\phi_q(h(s))\Delta s$$

$$\geq B_1cu^\sigma(b)\phi_q(h(b))+\frac{c(\gamma-a)}{\sigma(b)-a}\|u\|\int_\gamma^{\sigma(b)}\phi_q(h(s))\Delta s$$

$$= \left(B_1 c\phi_q(h(b)) + \frac{c(\gamma - a)}{\sigma(b) - a} \int_\gamma^{\sigma(b)} \phi_q(h(s))\Delta s \right) \|u\|$$

$$\geq \|u\|.$$

Thus,

$$\|Lu\| \geq \|u\|, \quad u \in P \bigcap \partial S_k.$$

Now, we take

$$d \left(B_2\phi_q(h(b)) + \int_a^{\sigma(b)} \phi_q(h(s))\Delta s \right) \leq 1.$$

Since $f_\infty = 0$, there exists an $m_1 > 0$ so that

$$f(u) \leq \phi_p(d)\phi_p(u)$$

$$= \phi_p(du), \quad u \geq m_1.$$

We will consider the following two cases.

1. Let f is bounded. Then there exists a constant $M > 0$ so that

$$f(u) \leq \phi_p(M)$$

for any $u \in [0, \infty)$. Set

$$m = \max \left\{ 2k, M \left(B_2\phi_q(h(b)) + \int_a^{\sigma(b)} \phi_q(h(s))\Delta s \right) \right\}$$

and

$$S_m = \{u \in \mathscr{C}_{rd}([a, \sigma(b)]) : \|u\| < m\}.$$

Take $u \in P \bigcap S_m$. Then

$$\|Lu\| = Lu(\sigma(b))$$

$$= B_0 \left(\phi_q \left(h(b) f(u^\sigma(b)) \right) \right)$$

$$+ \int_a^{\sigma(b)} \phi_q \left(h(s) f(u^\sigma(s)) \right) \Delta s$$

$$\leq B_2\phi_q \left(h(b)\phi_p(M) \right) + \int_a^{\sigma(b)} \phi_q \left(h(s)\phi_p(M) \right) \Delta s$$

$$= MB_2\phi_q(h(b)) + M \int_a^{\sigma(b)} \phi_q(h(s))\Delta s$$

$$
\begin{aligned}
&< m \\
&= \|u\|.
\end{aligned}
$$

Thus,

$$\|Lu\| \le \|u\|, \quad u \in P\bigcap \partial S_m.$$

2. Let f is unbounded. Take

$$m \ge \max\left\{2k, \frac{\sigma(b)-a}{\gamma-a} m_1\right\}$$

so that

$$f(u) \le f(m), \quad 0 \le u \le m.$$

For $u \in P\bigcap S_m$, we get

$$
\begin{aligned}
\|Lu\| &= Lu(\sigma(b)) \\
&= B_0\left(\phi_q\left(h(b) f(u^\sigma(b))\right)\right) \\
&\quad + \int_a^{\sigma(b)} \phi_q\left(h(s) f(u^\sigma(s))\right) \Delta s \\
&\le B_2 \phi_q\left(h(b)\phi_p(m)\right) + \int_a^{\sigma(b)} \phi_q\left(h(s)\phi_p(m)\right) \Delta s \\
&= m B_2 \phi_q(h(b)) + m \int_a^{\sigma(b)} \phi_q(h(s)) \Delta s \\
&< m \\
&= \|u\|.
\end{aligned}
$$

Therefore,

$$\|Lu\| \le \|u\|, \quad u \in P\bigcap \partial S_m.$$

Hence and by the Guo-Krasnosel'skii Theorem 1.11, we conclude that the operator L has at least one fixed point in $P\bigcap\left(\overline{S}_m \backslash S_k\right)$ and

$$k \le \|u\| \le m.$$

This completes the proof.

Theorem 1.26. *Assume* (A21)-(A24) *and*

$$f_0 = 0, \quad f_\infty = \infty.$$

Then the BVP (1.51), (1.52) *has at least one positive solution.*

Proof. Let γ be the same as in the proof of Theorem 1.25. Take $\varepsilon > 0$ and $\rho > 0$ so that

$$\varepsilon\left(B_2\phi_q(h(b)) + \int_a^{\sigma(b)} \phi_q(h(s))\Delta s\right) \le 1,$$

$$\rho\left(B_1\phi_q(h(b)) + \frac{\gamma - a}{\sigma(b) - a}\int_a^{\sigma(b)} \phi_q(h(s))\Delta s\right) \ge 1.$$

Since $f_0 = 0$, we can choose $H_1 > 0$ so that

$$\begin{aligned} f(x) &\le \phi_p(\varepsilon)\phi_p(x) \\ &= \phi_p(\varepsilon x), \quad x \in [0, H_1]. \end{aligned}$$

Because $f_\infty = \infty$, we can choose $H_2 > 0$ so that

$$\begin{aligned} f(x) &\ge \phi_p(\rho)\phi_p(x) \\ &= \phi_p(\rho x), \quad x \in [H_2, \infty). \end{aligned}$$

Define

$$\Omega_i = \{x \in \mathscr{C}_{rd}([a, \sigma(b)]) : \|x\| < H_i\}, \quad i = 1, 2.$$

Take $x \in P \bigcap \partial\Omega_1$. Then

$$\begin{aligned} \|Lu\| &= Lu(\sigma(b)) \\ &= B_0\left(\phi_q\left(h(b)f(u^\sigma(b))\right)\right) \\ &\quad + \int_a^{\sigma(b)} \phi_q\left(h(s)f(u^\sigma(s))\right)\Delta s \\ &\le B_2\phi_q\left(h(b)f(u^\sigma(b))\right) \\ &\quad + \int_a^{\sigma(b)} \phi_q\left(h(s)f(u^\sigma(s))\right)\Delta s \\ &\le B_2\phi_q\left(h(b)\phi_p(\varepsilon u^\sigma(b))\right) \end{aligned}$$

$$
\begin{aligned}
&+\int_a^{\sigma(b)} \phi_q\left(h(s)\phi_p(\varepsilon u^\sigma(s))\right)\Delta s\\
&= B_2\varepsilon u^\sigma(b)\phi_q(h(b))\\
&\quad +\varepsilon\int_a^{\sigma(b)} \phi_q(h(s))u^\sigma(s)\Delta s\\
&\le B_2\varepsilon u^\sigma(b)\phi_q(h(b))+\varepsilon u^\sigma(b)\int_a^{\sigma(b)}\phi_q(h(s))\Delta s\\
&= \varepsilon\left(B_2\phi_q(h(b))+\int_a^{\sigma(b)}\phi_q(h(s))\Delta s\right)u^\sigma(b)\\
&\le \|u\|.
\end{aligned}
$$

Thus,

$$
\|Lu\|\le\|u\|,\quad u\in P\bigcap\partial\Omega_1.
$$

Let $u\in P\bigcap\partial\Omega_2$. Then

$$
\begin{aligned}
\|Lu\| &= Lu(\sigma(b))\\
&= B_0\left(\phi_q\left(h(b)f(u^\sigma(b))\right)\right)\\
&\quad +\int_a^{\sigma(b)}\phi_q\left(h(s)f(u^\sigma(s))\right)\Delta s\\
&\ge B_1\phi_q\left(h(b)f(u^\sigma(b))\right)\\
&\quad +\int_a^{\sigma(b)}\phi_q(h(s)f(u^\sigma(s)))\Delta s\\
&= B_1\phi_q(h(b))\rho u^\sigma(b)+\rho\int_a^{\sigma(b)}\phi_q(h(s))u^\sigma(s)\Delta s\\
&\ge B_1\phi_q(h(b))\rho\|u\|+\rho\int_\gamma^{\sigma(b)}\phi_q(h(s))u^\sigma(s)\Delta s\\
&\ge B_1\phi_q(h(b))\rho\|u\|+\rho\frac{\gamma-a}{\sigma(b)-a}\|u\|\int_\gamma^{\sigma(b)}\phi_q(h(s))\Delta s\\
&= \rho\left(B_1\phi_q(h(b))+\frac{\gamma-a}{\sigma(b)-a}\int_\gamma^{\sigma(b)}\phi_q(h(s))\Delta s\right)\|u\|
\end{aligned}
$$

$$\geq \|u\|.$$

Therefore,

$$\|Lu\| \geq \|u\|, \quad u \in P\bigcap \partial\Omega_2.$$

Hence and by the Guo-Krasnosel'skii Theorem 1.11, it follows that the operator L has a fixed point in $P\bigcap \left(\overline{\Omega}_2 \backslash \Omega_1\right)$. This completes the proof.

Theorem 1.27. *Assume* (A21)-(A24) *and*

$$f_0 = f_\infty = 0.$$

Let, also, there exists an $r > 0$ *so that*

$$r\left(B_1\phi_q(h(b)) + \frac{\gamma - a}{\sigma(b) - a}\int_\gamma^{\sigma(b)} \phi_q(h(s))\Delta s\right) \geq 1$$

and

$$f(x) \geq \phi_p(rx), \quad x \in [0, r],$$

where γ *is the same as in Theorem 1.25. Then the BVP* (1.51), (1.52) *has at least two positive solutions.*

Proof. Let

$$\Omega_r = \{x \in \mathscr{C}_{rd}([a, \sigma(b)]) : \|x\| < r\}.$$

Then, for $u \in P\bigcap \partial\Omega_r$, we have

$$\begin{aligned}
\|Lu\| &= L(u(\sigma(b)))\\
&= B_0\left(\phi_q\left(h(b)f(u^\sigma(b))\right)\right)\\
&\quad + \int_a^{\sigma(b)} \phi_q\left(h(s)f(u^\sigma(s))\right)\Delta s\\
&\geq B_1\phi_q\left(h(b)f(u^\sigma(b))\right) + \int_a^{\sigma(b)} \phi_q\left(h(s)f(u^\sigma(s))\right)\Delta s\\
&\geq B_1\phi_q\left(h(b)\phi_p(ru^\sigma(b))\right) + \int_a^{\sigma(b)} \phi_q\left(h(s)\phi_p(ru^\sigma(s))\right)\Delta s\\
&\geq B_1 ru^\sigma(b)\phi_q(h(b)) + r\int_a^{\sigma(b)} \phi_q(h(s))u^\sigma(s)\Delta s\\
&\geq B_1 ru^\sigma(b)\phi_q(h(b)) + r\int_\gamma^{\sigma(b)} \phi_q(h(s))u^\sigma(s)\Delta s\\
&\geq B_1 ru^\sigma(b)\phi_q(h(b)) + r\frac{\gamma - a}{\sigma(b) - a}\|u\|\int_\gamma^{\sigma(b)} \phi_q(h(s))\Delta s
\end{aligned}$$

$$= r\left(B_1\phi_q(h(b)) + \frac{\gamma - a}{\sigma(b) - a}\int_\gamma^{\sigma(b)} \phi_q(h(s))\Delta s\right)\|u\|$$

$$\geq \|u\|.$$

Therefore,

$$\|Lu\| \geq \|u\|, \quad u \in P\bigcap \partial\Omega_r.$$

Take $\varepsilon > 0$ so that

$$\varepsilon\left(B_2\phi_q(h(b)) + \int_a^{\sigma(b)} \phi_q(h(s))\Delta s\right) \leq 1.$$

Since $f_0 = 0$, there exists a positive constant $r_1 < r$ so that

$$f(x) \leq \phi_p(\varepsilon x), \quad x \in [0, r_1].$$

Define

$$\Omega_{r_1} = \{u \in \mathscr{C}_{rd}([a, \sigma(b)]) : \|u\| < r_1\}.$$

Then, for $u \in P\bigcap \partial\Omega_{r_1}$, we get

$$\begin{aligned}
\|Lu\| &= Lu(\sigma(b)) \\
&= B_0\left(\phi_q\left(h(b)f(u^\sigma(b))\right)\right) \\
&\quad + \int_a^{\sigma(b)} \phi_q\left(h(s)f(u^\sigma(s))\right)\Delta s \\
&\leq B_2\phi_q(h(b)f(u^\sigma(b))) + \int_a^{\sigma(b)} \phi_q(h(s)f(u^\sigma(s)))\Delta s \\
&\leq B_2\phi_q(h(b)\phi_p(\varepsilon u^\sigma(b))) + \int_a^{\sigma(b)} \phi_q(h(s)\phi_p(\varepsilon u^\sigma(b)))\Delta s \\
&\leq \varepsilon B_2 u^\sigma(b)\phi_q(h(b)) + \varepsilon u^\sigma(b)\int_a^{\sigma(b)} \phi_q(h(s))\Delta s \\
&= \varepsilon\left(B_2\phi_q(h(b)) + \int_a^{\sigma(b)} \phi_q(h(s))\Delta s\right)\|u\| \\
&\leq \|u\|
\end{aligned} \tag{1.55}$$

and

$$\|Lu\| \leq \|u\|, \quad u \in P\bigcap \partial\Omega_{r_1}.$$

Since $f_\infty = 0$, there exists a constant $r_2 > r$ so that

$$f(x) \leq \phi_p(\varepsilon x), \quad x \in [0, r_2].$$

Define

$$\Omega_{r_2} = \{x \in \mathscr{C}_{rd}([a, \sigma(b)]) : \|x\| < r_2\}.$$

Then, using (1.55), we get

$$\|Lu\| \leq \|u\|, \quad u \in P\bigcap \partial\Omega_{r_2}.$$

Now, applying the Guo-Krasnosel'skii fixed point Theorem 1.11, we conclude that the operator L has one fixed point in $P\bigcap(\overline{\Omega}_r \backslash \Omega_{r_1})$ and one fixed point in $P\bigcap(\overline{\Omega}_{r_2} \backslash \Omega_r)$. This completes the proof.

Theorem 1.28. *Assume* (A21)-(A24) *and*

$$f_0 = f_\infty = \infty.$$

Let, also, there exists an $r > 0$ *so that*

$$r\left(B_2\phi_q(h(b)) + \int_a^{\sigma(b)} \phi_q(h(s))\Delta s\right) \leq 1$$

and

$$f(x) \leq \phi_p(rx), \quad x \in [0, r],$$

where γ *is the same as in Theorem 1.25. Then the BVP* (1.51), (1.52) *has at least two positive solutions.*

Proof. Let

$$\Omega_r = \{x \in \mathscr{C}_{rd}([a, \sigma(b)]) : \|x\| < r\}.$$

Then, for $u \in P\bigcap \partial\Omega_r$, we have

$$\begin{aligned}
\|Lu\| &= Lu(\sigma(b))\\
&= B_0\left(\phi_q\left(h(b)f(u^\sigma(b))\right)\right)\\
&\quad + \int_a^{\sigma(b)} \phi_q\left(h(s)f(u^\sigma(s))\right)\Delta s\\
&\le B_2\phi_q(h(b)f(u^\sigma(b))) + \int_a^{\sigma(b)} \phi_q(h(s)f(u^\sigma(s)))\Delta s\\
&\le B_2\phi_q(h(b)\phi_p(ru^\sigma(b))) + \int_a^{\sigma(b)} \phi_q(h(s)\phi_p(ru^\sigma(b)))\Delta s\\
&\le rB_2u^\sigma(b)\phi_q(h(b)) + ru^\sigma(b)\int_a^{\sigma(b)} \phi_q(h(s))\Delta s\\
&= r\left(B_2\phi_q(h(b)) + \int_a^{\sigma(b)} \phi_q(h(s))\Delta s\right)\|u\|\\
&\le \|u\|.
\end{aligned}$$

Therefore,

$$\|Lu\| \le \|u\|, \quad u \in P\bigcap \partial\Omega_r.$$

Take $\varepsilon > 0$ so that

$$\varepsilon\left(B_2\phi_q(h(b)) + \int_a^{\sigma(b)} \phi_q(h(s))\Delta s\right) \le 1.$$

Since $f_0 = \infty$, there exists a positive constant $r_1 < r$ so that

$$f(x) \ge \phi_p(\varepsilon x), \quad x \in [0, r_1].$$

Define

$$\Omega_{r_1} = \{u \in \mathscr{C}_{rd}([a, \sigma(b)]) : \|u\| < r_1\}.$$

Then, for $u \in P\bigcap \partial\Omega_{r_1}$, we get

$$
\begin{aligned}
\|Lu\| &= L(u(\sigma(b))) \\
&= B_0\left(\phi_q\left(h(b)f(u^\sigma(b))\right)\right) \\
&\quad + \int_a^{\sigma(b)} \phi_q\left(h(s)f(u^\sigma(s))\right)\Delta s \\
&\geq B_1\phi_q\left(h(b)f(u^\sigma(b))\right) + \int_a^{\sigma(b)} \phi_q\left(h(s)f(u^\sigma(s))\right)\Delta s \\
&\geq B_1\phi_q\left(h(b)\phi_p(\varepsilon u^\sigma(b))\right) + \int_a^{\sigma(b)} \phi_q\left(h(s)\phi_p(\varepsilon u^\sigma(s))\right)\Delta s \\
&\geq B_1\varepsilon u^\sigma(b)\phi_q(h(b)) + \varepsilon\int_a^{\sigma(b)} \phi_q(h(s))u^\sigma(s)\Delta s \\
&\geq B_1\varepsilon u^\sigma(b)\phi_q(h(b)) + \varepsilon\int_\gamma^{\sigma(b)} \phi_q(h(s))u^\sigma(s)\Delta s \\
&\geq B_1\varepsilon u^\sigma(b)\phi_q(h(b)) + r\frac{\gamma-a}{\sigma(b)-a}\|u\|\int_\gamma^{\sigma(b)} \phi_q(h(s))\Delta s \\
&= \varepsilon\left(B_1\phi_q(h(b)) + \frac{\gamma-a}{\sigma(b)-a}\int_\gamma^{\sigma(b)} \phi_q(h(s))\Delta s\right)\|u\| \\
&\geq \|u\|
\end{aligned}
\tag{1.56}
$$

and

$$\|Lu\| \geq \|u\|, \quad u \in P\bigcap \partial\Omega_{r_1}.$$

Since $f_\infty = \infty$, there exists a constant $r_2 > r$ so that

$$f(x) \geq \phi_p(\varepsilon x), \quad x \in [0, r_2].$$

Define

$$\Omega_{r_2} = \{x \in \mathscr{C}_{rd}([a, \sigma(b)]) : \|x\| < r_2\}.$$

Then, using (1.56), we get

$$\|Lu\| \geq \|u\|, \quad u \in P\bigcap \partial\Omega_{r_2}.$$

Now, applying the Guo-Krasnosel'skii fixed point Theorem 1.11, we conclude that the operator L has one fixed point in $P\bigcap(\overline{\Omega}_r \backslash \Omega_{r_1})$ and one fixed point in $P\bigcap(\overline{\Omega}_{r_2} \backslash \Omega_r)$. This completes the proof.

1.8 Advanced Practical Problems

Problem 1.1. Let $\mathbb{T} = \mathbb{N}_0^2$ and $n = 1$. Prove that the BVP

$$x^{\Delta} = \frac{x^3}{(1+t^2+t^4)(1+x^6)(1+x^8)}, \quad t \in [1,64],$$

$$x(1) + 3x(81) = 0,$$

has at least one solution $x \in \mathscr{C}([1,81])$.

Problem 1.2. Let $\mathbb{T} = 2^{\mathbb{N}_0}$. Prove that the BVP

$$x^{\Delta}(t) = -t^4 x(2t) + \frac{(1+t^2)\sqrt{1+(x(t))^4}}{1000}, \quad t \in [1,16],$$

$$x(1) = \frac{(x(32))^5}{10000},$$

has positive solutions.

Problem 1.3. Let $n = 2$ and $\mathbb{T} = 2^{\mathbb{N}_0}$. Prove that the PBVP

$$x^{\Delta} = -3x + \frac{1}{1+y^{40}}$$

$$y^{\Delta} = -y + \frac{1}{1+x+x^2}, \quad t \in [1,32],$$

$$x(1) = x(64),$$

$$y(1) = y(64),$$

has at least one solution.

Problem 1.4. Let $n = 1$. Prove that the PBVP

$$x^{\Delta} = tx^3 + 1, \quad t \in [a,b],$$

$$x(a) = -x(\sigma(b))$$

has at least one solution.

Problem 1.5. Let $g(t,x) = g_1(t)g_2(x^{\sigma})$. Write an analogue of Theorem 1.28.

Problem 1.6. Let $g(t,x) = g_1(t) + g_2(x^{\sigma})$. Write an analogue of Theorem 1.28.

Problem 1.7. Let $\mathbb{T} = 2^{\mathbb{N}_0}$. Prove that the PBVP

$$x^{\Delta} = -2x^{\sigma} + \frac{x^{10}(1+x^{10}+x^{20})}{1+x^4+x^8}, \quad t \in [1,16],$$

$$x(1) = x(32),$$

has at least one positive solution.

Problem 1.8. Let $\mathbb{T} = \mathbb{N}_0^2$. Prove that the PBVP

$$x^{\Delta} = -\frac{1}{1+t+t^2}x^{\sigma} + \frac{\sqrt{x^3}}{1+x^4+x^8}, \quad t \in [0,81],$$

$$x(0) = x(100),$$

has at least one positive solution.

Problem 1.9. Let $\mathbb{T} = \mathbb{N}_0$. Prove that the BVP

$$x^{\Delta} = -x^{\sigma} + \frac{x+1}{x^8+x^4+5}, \quad t \in [0,7],$$

$$x(0) = 2x(1) + 3x(2) + x(3) + x(5) + x(8),$$

has at least two positive solutions.

Problem 1.10. Let $\mathbb{T} = \mathbb{N}_0^2$. Prove that the BVP

$$x^{\Delta} = -(1+t)x^{\sigma} + \frac{1}{x^4+x^2+10}, \quad t \in [0,25],$$

$$x(0) = x(1) + x(4) + x(9) + x(36),$$

has at least three positive solutions.

1.9 Notes and References

In this chapter we investigated some classes of first order boundary value problems with nonlinear boundary conditions. Results are obtained for existence of at least one, at least two and at least three solutions of the considered BVPs. Some classes of periodic boundary value problems are studied too. The material in this chapter can be found in [1], [10], [11], [14], [50], [51], [52], [53], [54], [57], [62], [66], [67], and references therein.

Chapter 2
Boundary Value Problems for First Order Impulsive Dynamic Equations

2.1 Lower and Upper Solutions for First Order Impulsive Dynamic Equations

Suppose that $\mathbb{T}$ is a time scale with forward jump operator, backward jump operator and delta differentiation operator σ, ρ and Δ, respectively. Let, also, $0, t_1, T \in \mathbb{T}$, $0 < t_1 < T$, t_1 be right-dense. Set $J = [0, T]$ and consider the impulsive dynamic equation with a single impulse.

$$\begin{aligned} y^{\nabla}(t) &= g(t, y(t)), \quad t \in J \backslash \{t_1\}, \\ y(t_1^+) &= I(y(t_1)), \\ B(y(0), y(T)) &= 0. \end{aligned} \tag{2.1}$$

Let

$$J_1 = [0, t_1], \quad J_2 = (t_1, T].$$

Let, also, y_i, $i \in \{1, 2\}$, be the restriction of $y : J \to \mathbb{R}$ on J_i, $i \in \{1, 2\}$. Define

$$\mathscr{C}(J_1) = \{y : J_1 \to \mathbb{R}, \quad y \in \mathscr{C}(J_1)\},$$

$$\mathscr{C}(J_2) = \{y : J_2 \to \mathbb{R}, \quad y \in \mathscr{C}(J_2), \quad \exists y(t_1^+)\},$$

$$A = \{y : J \to \mathbb{R}, \quad y_1 \in \mathscr{C}(J_1), \quad y_2 \in \mathscr{C}(J_2)\}.$$

For $y \in A$, define

$$\|y\| = \sup\{|y(t)| : t \in J\}.$$

Then $(A, \|\cdot\|)$ is a Banach space. For $u, v \in A$, define

$$[u, v] = \{w \in A : u(t) \leq w(t) \leq v(t), \quad t \in J\}.$$

DOI: 10.1201/9781003173557-2

Definition 2.1. We say that $y : \mathbb{T} \to \mathbb{R}$ is a solution of (2.1) if

1. $y \in A$,
2.
$$y(t) = y(0) + \int_0^t g(s, y(s))\nabla s, \quad t \in J_1,$$
$$y(t) = I(y(t_1)) + \int_{t_1}^t g(s, y(s))\nabla s, \quad t \in J_2,$$
3. $B(y(0), y(T)) = 0$.

Definition 2.2. We say that $\alpha : J \to \mathbb{R}$ is a lower solution of equation (2.1) if

1. $\alpha \in A$,
2.
$$\alpha(b) - \alpha(a) \le \int_a^b g(s, \alpha(s))\nabla s$$
for $a \le b$, $a, b \in J_1$ or $a, b \in J_2$,
3. $\alpha(t_1^+) \le I(\alpha(t_1))$,
4. $B(\alpha(0), \alpha(T)) \le 0$.

Definition 2.3. We say that $\beta : J \to \mathbb{R}$ is an upper solution of equation (2.1) if

1. $\beta \in A$,
2.
$$\beta(b) - \beta(a) \le \int_a^b g(s, \beta(s))\nabla s$$
for $a \le b$, $a, b \in J_1$ or $a, b \in J_2$,
3. $\beta(t_1^+) \ge I(\beta(t_1))$,
4. $B(\beta(0), \beta(T)) \ge 0$.

Let
$$p(t,x) = \max\{\alpha(t), \quad \min\{x, \beta(t)\}\}.$$

Suppose

(B1) $g : J \times \mathbb{R} \to \mathbb{R}$ is continuous,
(B2) there is an $h : J \to [0, \infty)$ such that
$$\int_0^T h(s)\nabla s < \infty,$$
$$|g(t, p(t,x))| \le h(t), \quad (t,x) \in J \times \mathbb{R},$$

(B3) $I : \mathbb{R} \to \mathbb{R}$ is continuous and non-decreasing,
(B4) $B : \mathbb{R} \times \mathbb{R} \to \mathbb{R}$ is continuous and for each $x \in [\alpha(0), \beta(0)]$, $B(x, \cdot)$ is non-increasing.

Define an operator $G : A \to A$ as follows.

$$Gy(t) = p(0, \overline{y}(0)) + \int_0^t g(s, p(s, y(s)))\nabla s, \quad t \in J_1, \tag{2.2}$$

$$Gy(t) = I(p(t_1, y(t_1))) + \int_{t_1}^t g(s, p(s, y(s)))\nabla s, \quad t \in J_2, \tag{2.3}$$

where

$$\overline{y}(0) = y(0) - B(y(0), y(T)).$$

Lemma 2.1. *Assume* $(B1)$-$(B4)$ *and* $y \in A$ *satisfies*

$$y(t) = p(0, \overline{y}(0)) + \int_0^t g(s, p(s, y(s)))\nabla s, \quad t \in J_1, \tag{2.4}$$

$$y(t) = I(p(t_1, y(t_1))) + \int_{t_1}^t g(s, p(s, y(s)))\nabla s, \quad t \in J_2. \tag{2.5}$$

Then

$$y(t) \in [\alpha(t), \beta(t)], \quad t \in J.$$

Proof. Take $t = 0$. From (2.4), we get

$$\begin{aligned} y(0) &= p(0, \overline{y}(0)) \\ &= \max\{\alpha(0), \quad \min\{\overline{y}(0), \beta(0)\}\} \\ &= \max\{\alpha(0), \quad \min\{y(0) - B(y(0), y(T)), \beta(0)\}\}. \end{aligned}$$

If $y(0) = \alpha(0)$, we get $y(0) \in [\alpha(0), \beta(0)]$. If

$$y(0) = \min\{y(0) - B(y(0), y(T)), \quad \beta(0)\},$$

then

$$\alpha(0) \le y(0) \le \beta(0).$$

Suppose that there is a $t_2 \in (0, t_1]$ such that

$$\alpha(t_2) > y(t_2).$$

Since $\alpha(0) \le y(0)$, there is a $t_3 \in [0, t_2)$ so that

$$\alpha(t) \le y(t), \quad t \in [0, t_3],$$

and

$$\alpha(t) > y(t), \quad t \in (t_3, t_2].$$

Then

$$p(t, y(t)) = \max\{\alpha(t), \quad \min\{y(t), \beta(t)\}\}$$

$$= \alpha(t), \quad t \in (t_3, t_2],$$

and

$$g(t, p(t, y(t))) = g(t, \alpha(t)), \quad t \in (t_3, t_2].$$

Hence, for $t \in (t_3, t_2]$, we obtain

$$\int_0^t g(s, p(s, y(s)))\nabla s = y(t) - p(0, \bar{y}(0))$$

$$= y(t_3) - p(0, \bar{y}(0)) + y(t) - y(t_3)$$

$$= \int_0^{t_3} g(s, p(s, y(s)))\nabla s + y(t) - y(t_3),$$

and

$$y(t) - y(t_3) = \int_0^t g(s, p(s, y(s)))\nabla s - \int_0^{t_3} g(s, p(s, y(s)))\nabla s$$

$$= \int_{t_3}^t g(s, p(s, y(s)))\nabla s$$

$$= \int_{t_3}^t g(s, \alpha(s))\nabla s$$

$$\geq \alpha(t) - \alpha(t_3).$$

Since $y(t_3) \geq \alpha(t_3)$, we get $y(t) > \alpha(t)$, $t \in (t_3, t_2]$. This is a contradiction. Therefore, $y(t) \geq \alpha(t)$, $t \in [0, t_1]$. Similarly, $y(t) \leq \beta(t)$, $t \in [0, t_1]$. In particular, we have

$$\alpha(t_1) \leq y(t_1) \leq \beta(t_1).$$

Hence,

$$\alpha(t_1^+) \leq I(\alpha(t_1))$$

$$\leq I(y(t_1))$$

$$\leq I(\beta(t_1))$$

$$\le \beta(t_1^+).$$

Since

$$I(y(t_1)) = I(p(t_1, y(t_1)))$$

$$= y(t_1^+),$$

we obtain

$$\alpha(t_1^+) \le y(t_1^+) \le \beta(t_1^+).$$

Proceed as above, we arrive at

$$\alpha(t) \le y(t) \le \beta(t), \quad t \in J_2.$$

This completes the proof.

Remark 2.1. Suppose that all conditions of Lemma 2.1 hold. Then

$$\begin{aligned} y(t) &= y(0) + \int_0^t g(s, y(s))\nabla s, \quad t \in J_1, \\ y(t) &= I(y(t_1)) + \int_{t_1}^t g(s, y(s))\nabla s, \quad t \in J_2. \end{aligned} \tag{2.6}$$

Lemma 2.2. *Assume* $(B1)$-$(B4)$ *and* $y \in A$ *satisfies* (2.4), (2.5). *Then*

$$\bar{y}(0) \in [\alpha(0), \beta(0)].$$

Proof. Suppose that

$$\alpha(0) > \bar{y}(0).$$

Thus,

$$y(0) = p(0, \bar{y}(0))$$

$$= \max\{\alpha(0), \min\{\bar{y}(0), \beta(0)\}\}$$

$$= \max\{\alpha(0), \bar{y}(0)\}$$

$$= \alpha(0).$$

Hence,

$$B(y(0), y(T)) > 0$$

and

$$B(\alpha(0),\alpha(T)) = B(y(0),\alpha(T))$$

$$\geq B(y(0),y(T))$$

$$> 0,$$

which is a contradiction. Therefore, $\alpha(0) \leq \overline{y}(0)$. Now, assume that $\overline{y}(0) > \beta(0)$. Then

$$y(0) - B(y(0),y(T)) = \overline{y}(0)$$

$$> \beta(0)$$

and

$$y(0) = p(0,\overline{y}(0))$$

$$= \max\{\alpha(0),\min\{\overline{y}(0),\beta(0)\}\}$$

$$= \max\{\alpha(0),\beta(0)\}$$

$$= \beta(0),$$

and

$$B(y(0),y(T)) < 0.$$

Hence,

$$0 > B(y(0),y(T))$$

$$= B(\beta(0),y(T))$$

$$\geq B(\beta(0),\beta(T)),$$

which is a contradiction. Consequently, $\overline{y}(0) \leq \beta(0)$. This completes the proof.

Lemma 2.3. *Assume* $(B1)$-$(B4)$ *and* $y \in A$ *satisfies* (2.4), (2.5). *If* y *is a fixed point of the operator* G*, then* y *is a solution of* (2.1).

Proof. By Lemma 2.1, we get

$$y(t) \in [\alpha(t),\beta(t)], \quad t \in J.$$

By Remark 2.1, it follows that y satisfies (2.6). By Lemma 2.2, we obtain

$$\bar{y}(0) \in [\alpha(0), \beta(0)]$$

and

$$\begin{aligned} y(0) &= p(0, \bar{y}(0)) \\ &= \max\{\alpha(0), \min\{\bar{y}(0), \beta(0)\}\} \\ &= \max\{\alpha(0), \bar{y}(0)\} \\ &= \bar{y}(0) \\ &= y(0) - B(y(0), y(T)), \end{aligned}$$

whereupon,

$$B(y(0), y(T)) = 0.$$

Thus, y is a solution of equation (2.1). This completes the proof.

Define

$$K = \|\alpha\| + \|\beta\|$$

and $w : J \to \mathbb{R}$ as follows

$$w(t) = K + \int_0^t h(s)\nabla s.$$

Let

$$S = \Big\{ y \in A : |y(0)| \leq K, \quad |y(t_1^+)| \leq K, \quad |y(b) - y(a)| \leq w(b) - w(a)$$

$$\text{on} \quad 0 \leq a \leq b \leq t_1 \quad \text{or} \quad t_1 < a \leq b \leq T, \quad a, b \in J \Big\}.$$

Exercise 2.1. Prove that S is a convex and compact subset of $(A, \|\cdot\|)$.

Lemma 2.4. *Let* $(B1)$-$(B4)$ *hold. Then* $G(S) \subseteq S$.

Proof. Take $y \in S$ arbitrarily. Then

$$\begin{aligned} |Gy(0)| &= |p(0, \bar{y}(0))| \\ &= |\max\{\alpha(0), \min\{\bar{y}(0), \beta(0)\}\}| \\ &\leq \max\{|\alpha(0)|, |\beta(0)|\} \end{aligned}$$

$$\leq K.$$

Note that

$$\begin{aligned}\beta(t_1) &\geq p(t_1,y(t_1))\\ &= \max\{\alpha(t_1),\min\{y(t_1),\beta(t_1)\}\}\\ &\geq \alpha(t_1).\end{aligned}$$

Hence,

$$\begin{aligned}\alpha(t_1^+) &\leq I(\alpha(t_1))\\ &\leq I(p(t_1,y(t_1)))\\ &\leq I(\beta(t_1))\\ &\leq \beta(t_1^+)\end{aligned}$$

and

$$\begin{aligned}|I(p(t_1,y(t_1)))| &\leq \max\{|\alpha(t_1^+)|,|\beta(t_1^+)|\}\\ &\leq K.\end{aligned}$$

Let $t \to t_1$, $t > t_1$, in (2.3) and we get

$$\begin{aligned}|Gy(t_1^+)| &= |I(p(t_1,y(t_1)))|\\ &\leq K.\end{aligned}$$

Let $a,b \in J$, $0 \leq a \leq b \leq t_1$. Then

$$\begin{aligned}|Gy(b)-Gy(a)| &= \Big|p(0,\bar{y}(0)) + \int_0^b g(s,p(s,y(s)))\nabla s\\ &\quad -p(0,\bar{y}(0)) - \int_0^a g(s,p(s,y(s)))\nabla s\Big|\\ &= \left|\int_a^b g(s,p(s,y(s)))\nabla s\right|\end{aligned}$$

$$
\begin{aligned}
&\le \int_a^b |g(s,p(s,y(s)))|\nabla s \\
&\le \int_a^b h(s)\nabla s \\
&= w(b)-w(a).
\end{aligned}
$$

Now, take $t_1 < a \le b \le T$. Then

$$
\begin{aligned}
|Gy(b)-Gy(a)| &= \left| I(p(t_1),y(t_1)) + \int_{t_1}^b g(s,p(s,y(s)))\nabla s \right.\\
&\quad \left. -I(p(t_1),y(t_1)) - \int_{t_1}^a g(s,p(s,y(s)))\nabla s \right| \\
&= \left| \int_a^b g(s,p(s,y(s)))\nabla s \right| \\
&\le \int_a^b |g(s,p(s,y(s)))|\nabla s \\
&\le \int_a^b h(s)\nabla s \\
&= w(b)-w(a).
\end{aligned}
$$

Thus, $G : S \to S$. This completes the proof.

Lemma 2.5. *Let* $(B1)$-$(B4)$ *hold. Then* $G : S \to S$ *is continuous.*

Proof. Let $\{y_n\}_{n\in\mathbb{N}} \subseteq S$ and $y_n \to y$, as $n \to \infty$, $y \in S$, in the space $(A, \|\cdot\|)$. Note that $y_n \to y$, as $n \to \infty$, uniformly on compact subsets of J. Then, for $t \in J_1$, we get

$$
\begin{aligned}
|Gy(t)-Gy_n(t)| &= \Big| Gy(t) - p(0,\overline{y}(0)) - (Gy_n(t) - p(0,\overline{y}_n(0))) \\
&\quad + p(0,\overline{y}(0)) - p(0,\overline{y}_n(0)) \Big| \\
&= \Big| \int_0^t g(s,p(s,y(s)))\nabla s - \int_0^t g(s,p(s,y_n(s)))\nabla s \\
&\quad + p(0,\overline{y}(0)) - p(0,\overline{y}_n(0)) \Big| \\
&= \Big| \int_0^t (g(s,p(s,y(s))) - g(s,p(s,y_n(s))))\nabla s
\end{aligned}
$$

$$
\begin{aligned}
&\left. +p(0,\overline{y}(0)) - p(0,\overline{y}_n(0)) \right| \\
&\leq \left| \int_0^t (g(s,p(s,y(s))) - g(s,p(s,y_n(s))))\nabla s \right| \\
&\quad +|p(0,\overline{y}(0)) - p(0,\overline{y}_n(0))| \\
&\leq \int_0^t \left| g(s,p(s,y(s))) - g(s,p(s,y_n(s))) \right| \nabla s \\
&\quad +|p(0,\overline{y}(0)) - p(0,\overline{y}_n(0))| \\
&\to 0, \quad \text{as} \quad n \to \infty.
\end{aligned}
$$

For $t \in J_2$, we obtain

$$
\begin{aligned}
|Gy(t) - Gy_n(t)| &= \Big| Gy(t) - I(p(t_1,y(t_1))) - (Gy_n(t) - I(p(t_1,y_n(t_1)))) \\
&\quad \left. +I(p(t_1,y(t_1))) - I(p(t_1,y_n(t_1))) \right| \\
&= \left| \int_{t_1}^t g(s,p(s,y(s)))\nabla s - \int_{t_1}^t g(s,p(s,y_n(s)))\nabla s \right. \\
&\quad \left. +I(p(t_1,y(t_1))) - I(p(t_1,y_n(t_1))) \right| \\
&= \left| \int_{t_1}^t (g(s,p(s,y(s))) - g(s,p(s,y_n(s))))\nabla s \right. \\
&\quad \left. +I(p(t_1,y(t_1))) - I(p(t_1,y_n(t_1))) \right| \\
&\leq \left| \int_{t_1}^t (g(s,p(s,y(s))) - g(s,p(s,y_n(s))))\nabla s \right| \\
&\quad + \left| I(p(t_1,y(t_1))) - I(p(t_1,y_n(t_1))) \right| \\
&\leq \int_{t_1}^t \left| g(s,p(s,y(s))) - g(s,p(s,y_n(s))) \right| \nabla s \\
&\quad + \left| I(p(t_1,y(t_1))) - I(p(t_1,y_n(t_1))) \right|
\end{aligned}
$$

$$\to 0, \quad \text{as} \quad n \to \infty.$$

This completes the proof.

Lemma 2.6. *Suppose* $(B1)$*-*$(B4)$*. Then* G *has a fixed point in* S*.*

Proof. By Lemma 2.4, it follows that $G : S \to S$. From Lemma 2.5, we have that $G : S \to S$ is continuous. Hence and by Schauder's fixed point theorem, we conclude that G has a fixed point in S. This completes the proof.

Theorem 2.1. *Suppose* $(B1)$*-*$(B4)$*. Then equation* (2.1) *has at least one solution* $y \in S$ *such that* $y(t) \in [\alpha(t), \beta(t)]$, $t \in J$.

Proof. By Lemma 2.6, we have that the operator G has a fixed point $y \in S$. By Lemma 2.3, we obtain that y solves (2.1) and $y(t) \in [\alpha(t), \beta(t)]$, $t \in J$. This completes the proof.

Example 2.1. Let $\mathbb{T} = [0,1] \bigcup [2,3]$, where $[0,1]$ and $[2,3]$ are the real-valued intervals. Consider the following impulsive dynamic equation

$$y^{\nabla}(t) = t^2 + (y(t))^2, \quad t \in [0,1] \bigcup (2,3],$$

$$y(2^+) = y(2) + 1,$$

$$B(y(0), y(3)) = 0.$$

Note that $\alpha \equiv 0$ is a lower solution for the considered equation. Let β be an upper solution of the considered equation. To construct β on $[0,1]$, consider the IVP

$$\beta' = 1 + \beta^2, \quad t \in [0,1],$$

$$\beta(0) = 1.$$

We have

$$\beta(t) = \tan t, \quad t \in [0,1].$$

Moreover,

$$\begin{aligned} \beta(2) &= \beta(0) + \int_0^2 \beta^{\nabla}(s) \nabla s \\ &= \beta(0) + \int_0^1 \beta^{\nabla}(s) \nabla s + \beta^{\nabla}(2) \\ &= \beta(0) + \int_0^1 \beta'(s) ds + \frac{\beta(2) - \beta(1)}{2 - 1} \end{aligned}$$

$$= \beta(0) + \beta(1) - \beta(0) + \beta(2) - \beta(1)$$

$$= \beta(2).$$

Thus, $\beta(2)$ is arbitrary. Take $\beta(2) = 1$. We have

$$\beta(2^+) = I(\beta(2))$$

$$= \beta(2) + 1$$

$$= 1 + 1$$

$$= 2.$$

To construct β on $[2,3]$, consider the IVP

$$\beta' = 9 + \beta^2, \quad t \in [2,3],$$

$$\beta(2) = 2.$$

For its solution we have

$$\beta(t) = 3\tan\left(\tan^{-1}\left(\frac{2}{3}\right) + 3t - 6\right), \quad t \in [2,3].$$

By Theorem 2.1, it follows that the considered equation has a solution y such that

$$0 \le y(t) \le \beta(t), \quad t \in \mathbb{T}.$$

Exercise 2.2. Let $\mathbb{T} = [0,2] \bigcup [3,4]$, where $[0,2]$ and $[3,4]$ are the real-valued intervals. Investigate for existence of solutions the following impulsive dynamic equation

$$y^{\nabla}(t) = t^4 + 2(y(t)), \quad t \in [0,2] \bigcup (3,4],$$

$$y(3^+) = y(3) - 1,$$

$$B(y(0), y(4)) = 0.$$

2.2 Periodic Boundary Value Problems for First Order Linear Dynamic Equations with Impulses

Let $\mathbb{T}$ be a time scale with forward jump operator and delta differentiation operator σ and Δ, respectively. Let, also, $0, T \in \mathbb{T}$, $T > 0$, $J = [0, T]$. Assume that

$$0 = t_0 < t_1 < \ldots < t_m < t_{m+1} = T,$$

t_k, $k \in \{1, \ldots, m\}$, are right-dense. Set $J_0 = J \backslash \{t_k\}_{k=1}^m$. Consider the following linear impulsive dynamic equation

$$\begin{aligned} x^\Delta(t) + p(t)x^\sigma(t) &= q(t), \quad t \in J_0, \\ x(t_k^+) &= x(t_k^-) + I_k(x(t_k)), \end{aligned} \tag{2.7}$$

with periodic boundary condition

$$x(0) = x(\sigma(T)), \tag{2.8}$$

(B5) $p, q \in \mathscr{C}([0, \sigma(T)])$, p is regressive on $[0, \sigma(T)]$,
(B6) $I_k \in \mathscr{C}(\mathbb{R})$, $k \in \{1, \ldots, m\}$.

Define

$$PC([0,\sigma(T)]) = \Big\{ x : [0,\sigma(T)] \to \mathbb{R}, \quad x \quad \text{is} \quad \text{continuous}$$

$$\text{on} \quad J_0, \quad \text{ld-continuous} \quad \text{at} \quad t_k,$$

$$\exists \lim_{t \to t_k + 0} x(t), \quad k \in \{1, \ldots, m\} \Big\}.$$

$PC([0, \sigma(T)])$ is a Banach space with the norm

$$\|x\| = \max_{j \in \{0, \ldots, m\}} \sup_{t \in [t_j, t_{j+1}]} |x(t)|.$$

Now, we define another Banach space

$$PC^\Delta([0,\sigma(T)]) = \Big\{ x : x \in PC([0,\sigma(T)]), \quad \exists x^\Delta \quad \text{and}$$

$$\text{it} \quad \text{is} \quad \text{continuous} \quad \text{on} \quad (t_k, t_{k+1}),$$

$$\exists \lim_{t \to t_k + 0} x^\Delta(t) \quad \text{and} \quad \lim_{t \to t_k -} x^\Delta(t) \quad \text{if}$$

$$t_k \quad \text{is} \quad \text{left-scattered}, \quad k \in \{1, \dots, m\}\Big\}$$

with the norm

$$\|x\|_\Delta = \max\left\{ \|x\|, \sup_{t \in J_0} \left| x^\Delta(t) \right| \right\}.$$

Definition 2.4. A solution of problem (2.7), (2.8) means a function $x \in PC^\Delta$ $([0, \sigma(T)])$ such that (2.7) and (2.8) are satisfied.

Lemma 2.7. *Suppose* $e_p(\sigma(T), 0) \neq 1$. *If* $x \in PC^\Delta([0, \sigma(T)])$ *is a solution of problem* (2.7), (2.8), *then*

$$x(t) = \int_0^{\sigma(T)} g(t,s)q(s)\Delta s + \sum_{k=1}^m g(t,t_j)I_j(x(t_j)), \quad t \in [0, \sigma(T)], \tag{2.9}$$

where

$$g(t,s) = \frac{1}{1 - e_{\ominus p}(\sigma(T), 0)} \begin{cases} e_{\ominus p}(t,s) & 0 \le s < t \le \sigma(T) \\ e_{\ominus p}(\sigma(T), 0)e_{\ominus p}(t,s) & 0 \le t \le s \le \sigma(T), \end{cases} \tag{2.10}$$

$t, s \in [0, \sigma(T)]$.

Proof. Let

$$y(t) = e_p(t, 0)x(t), \quad t \in [0, \sigma(T)].$$

Then

$$\begin{aligned} y^\Delta(t) &= p(t)e_p(t,0)x^\sigma(t) + e_p(t,0)x^\Delta(t) \\ &= e_p(t,0)\left(p(t)x^\sigma(t) + x^\Delta(t)\right) \\ &= q(t)e_p(t,0), \quad t \in J_0, \\ y(\sigma(T)) &= e_p(\sigma(T),0)x(\sigma(T)) \\ &= e_p(\sigma(T),0)x(0) \\ &= e_p(\sigma(T),0)y(0), \\ y(t_j^+) &= e_p(t_j^+,0)x(t_j^+) \end{aligned}$$

$$= e_p(t_j,0)(x(t_j)+I_j(x(t_j)))$$

$$= e_p(t_j,0)x(t_j)+e_p(t_j,0)I_j(x(t_j))$$

$$= e_p(t_j,0)e_p(0,t_j)y(t_j)+e_p(t_j,0)I_j(x(t_j))$$

$$= y(t_j)+e_p(t_j,0)I_j(x(t_j)), \quad j\in\{1,\ldots,m\},$$

i.e., y satisfies the impulsive periodic boundary value problem

$$\begin{aligned} y^{\Delta}(t) &= q(t)e_p(t,0), \quad t\in J_0, \\ y(t_j^+) &= y(t_j)+e_p(t_j,0)I_j(x(t_j)), \quad j\in\{1,\ldots,m\}, \end{aligned} \tag{2.11}$$

$$y(\sigma(T)) = e_p(\sigma(T),0)y(0). \tag{2.12}$$

By the first equation of (2.11), we derive

$$y(t) = y(0)+\int_0^t e_p(s,0)q(s)\Delta s, \quad t\in[0,t_1],$$

and

$$y(t) = y(t_j^+)+\int_{t_j}^t e_p(s,0)q(s)\Delta s, \quad t\in(t_j,t_{j+1}], \quad j\in\{1,\ldots,m\}.$$

Thus,

$$\begin{aligned} y(t_{j+1}) &= y(t_j^+)+\int_{t_j}^{t_{j+1}} e_p(s,0)q(s)\Delta s \\ &= \int_{t_j}^{t_{j+1}} e_p(s,0)q(s)\Delta s + y(t_j)+e_p(t_j,0)I_j(x(t_j)) \\ &= \int_{t_j}^{t_{j+1}} e_p(s,0)q(s)\Delta s + \int_{t_{j-1}}^{t_j} e_p(s,0)q(s)\Delta s \\ &\quad + e_p(t_{j-1},0)I_{j-1}(x(t_{j-1}))+e_p(t_j,0)I_j(x(t_j)) \\ &= \int_{t_{j-1}}^{t_{j+1}} e_p(s,0)q(s)\Delta s + e_p(t_{j-1},0)I_{j-1}(x(t_{j-1}))+e_p(t_j,0)I_j(x(t_j)) \\ &= \cdots \\ &= \int_0^{t_{j+1}} e_p(s,0)q(s)\Delta s + \sum_{k=1}^{j} e_p(t_k,0)I_k(x(t_k))+y(0). \end{aligned}$$

Hence,

$$
\begin{aligned}
y(t) &= y(t_j^+) + \int_{t_j}^{t} e_p(s,0)q(s)\Delta s \\
&= y(t_j) + \int_{t_j}^{t} e_p(s,0)q(s)\Delta s + e_p(t_j,0)I_j(x(t_j)) \\
&= \int_0^{t_j} e_p(s,0)q(s)\Delta s + \sum_{k=1}^{j-1} e_p(t_k,0)I_k(x(t_k)) + y(0) \\
&\quad + \int_{t_j}^{t} e_p(s,0)q(s)\Delta s + e_p(t_j,0)I_j(x(t_j)) \\
&= y(0) + \int_0^t e_p(s,0)q(s)\Delta s + \sum_{k=1}^{j} e_p(t_j,0)I_j(x(t_j)), \ \ t\in(t_j,t_{j+1}], \ \ j\in\{1,\dots,m\}.
\end{aligned}
$$

Therefore,

$$
y(t) = y(0) + \int_0^t e_p(s,0)q(s)\Delta s + \sum_{j:t_j<t} e_p(t_j,0)I_j(x(t_j)), \tag{2.13}
$$

$t\in[0,\sigma(T)]$. Hence, using the periodic boundary value condition (2.12), we get

$$
\begin{aligned}
e_p(\sigma(T),0)y(0) &= y(\sigma(T)) \\
&= y(0) + \int_0^{\sigma(T)} e_p(s,0)q(s)\Delta s \\
&\quad + \sum_{j:t_j<\sigma(T)} e_p(t_j,0)I_j(x(t_j)) \\
&= y(0) + \int_0^{\sigma(T)} e_p(s,0)q(s)\Delta s \\
&\quad + \sum_{j=1}^{m} e_p(t_j,0)I_j(x(t_j)),
\end{aligned}
$$

whereupon,

$$
(e_p(\sigma(T),0)-1)y(0) = \int_0^{\sigma(T)} e_p(s,0)q(s)\Delta s + \sum_{j=1}^{m} e_p(t_j,0)I_j(x(t_j))
$$

or

$$
y(0) = \frac{1}{e_p(\sigma(T),0)-1}\left(\int_0^{\sigma(T)} e_p(s,0)q(s)\Delta s + \sum_{j=1}^{m} e_p(t_j,0)I_j(x(t_j))\right).
$$

From the last expression and from (2.13), we obtain

$$e_p(t,0)x(t) = \frac{1}{e_p(\sigma(T),0)-1}\left(\int_0^{\sigma(T)} e_p(s,0)q(s)\Delta s + \sum_{j=1}^{m} e_p(t_j,0)I_j(x(t_j))\right)$$
$$+\int_0^t e_p(s,0)q(s)\Delta s + \sum_{j:t_j<t} e_p(t_j,0)I_j(x(t_j)),$$

or

$$\begin{aligned}
x(t) &= \frac{e_{\ominus p}(t,0)}{e_p(\sigma(T),0)-1}\left(\int_0^{\sigma(T)} e_p(s,0)q(s)\Delta s + \sum_{j=1}^{m} e_p(t_j,0)I_j(x(t_j))\right)\\
&\quad + e_{\ominus p}(t,0)\left(\int_0^t e_p(s,0)q(s)\Delta s + \sum_{j:t_j<t} e_p(t_j,0)I_j(x(t_j))\right)\\
&= \frac{e_{\ominus p}(t,0)}{e_p(\sigma(T),0)-1}\Bigg(\int_0^{\sigma(T)} e_p(s,0)q(s)\Delta s + \sum_{j=1}^{m} e_p(t_j,0)I_j(x(t_j))\\
&\quad + (e_p(\sigma(T),0)-1)\int_0^t e_p(s,0)q(s)\Delta s\\
&\quad + (e_p(\sigma(T),0)-1)\sum_{j:t_j<t} e_p(t_j,0)I_j(x(t_j))\Bigg)\\
&= \frac{e_{\ominus p}(t,0)}{e_p(\sigma(T),0)-1}\Bigg(e_p(\sigma(T),0)\int_0^t e_p(s,0)q(s)\Delta s + \int_t^{\sigma(T)} e_p(s,0)q(s)\Delta s\\
&\quad + e_p(\sigma(T),0)\sum_{j:t_j<t} e_p(t_j,0)I_j(x(t_j))\\
&\quad + \sum_{j:t_j\in[t,T)} e_p(t_j,0)I_j(x(t_j))\Bigg)\\
&= \frac{1}{1-e_{\ominus p}(\sigma(T),0)}\Bigg(\int_0^t e_{\ominus p}(t,0)e_{\ominus p}(0,s)q(s)\Delta s\\
&\quad + \int_t^{\sigma(T)} e_{\ominus p}(\sigma(T),0)e_{\ominus p}(t,0)e_{\ominus p}(0,s)q(s)\Delta s\\
&\quad + \sum_{j:t_j<t} e_{\ominus p}(t,0)e_{\ominus p}(0,t_j)I_j(x(t_j))\\
&\quad + \sum_{j:t_j\in[t,T)} e_{\ominus p}(\sigma(T),0)e_{\ominus p}(t,0)e_{\ominus p}(0,t_j)I_j(x(t_j))\Bigg)
\end{aligned}$$

$$
\begin{aligned}
&= \frac{1}{1-e_{\ominus p}(\sigma(T),0)}\bigg(\int_0^t e_{\ominus p}(t,s)q(s)\Delta s + \int_t^{\sigma(T)} e_{\ominus p}(\sigma(T),0)e_{\ominus p}(t,s)q(s)\Delta s \\
&\quad + \sum_{j:t_j<t} e_{\ominus p}(t,t_j)I_j(x(t_j)) \\
&\quad + \sum_{j:t_j\in[t,T)} e_{\ominus p}(\sigma(T),0)e_{\ominus p}(t,t_j)I_j(x(t_j))\bigg) \\
&= \int_0^{\sigma(T)} g(t,s)q(s)\Delta s + \sum_{j=1}^{m} g(t,t_j)I_j(x(t_j)),
\end{aligned}
$$

where g is defined by (2.10). This completes the proof.

Lemma 2.8. *If $x \in PC([0,\sigma(T)])$ and x, q, I_j, $j \in \{1,\dots,m\}$, satisfy* (2.9)*, then $x \in PC^{\Delta}([0,\sigma(T)])$ is a solution of problem* (2.7)*,* (2.8)*.*

Proof. We have

$$
\begin{aligned}
x^{\Delta}(t) &= \left(\int_0^{\sigma(T)} g(t,s)q(s)\Delta s\right)_t^{\Delta} + \sum_{j=1}^{m} g^{\Delta}(t,t_j)I_j(x(t_j)) \\
&= \left(\int_0^t g(t,s)q(s)\Delta s + \int_t^{\sigma(T)} g(t,s)q(s)\Delta s\right)_t^{\Delta} \\
&\quad + \sum_{j=1}^{m} g^{\Delta}(t,t_j)I_j(x(t_j)) \\
&= \frac{e_{\ominus p}(\sigma(t),t)}{1-e_{\ominus p}(\sigma(T),0)}q(t) + \int_0^t \frac{\ominus p(t)e_{\ominus p}(t,s)}{1-e_{\ominus p}(\sigma(T),0)}q(s)\Delta s \\
&\quad - \frac{e_{\ominus p}(\sigma(T),0)e_{\ominus p}(\sigma(t),t)}{1-e_{\ominus p}(\sigma(T),0)}q(t) \\
&\quad + \int_t^{\sigma(T)} \frac{\ominus p(t)e_{\ominus p}(\sigma(T),0)e_{\ominus p}(t,s)}{1-e_{\ominus p}(\sigma(T),0)}q(s)\Delta s \\
&\quad + \sum_{j=1}^{m} g^{\Delta}(t,t_j)I_j(x(t_j)) \\
&= e_{\ominus p}(\sigma(t),t)q(t) + \int_0^t \frac{\ominus p(t)e_{\ominus p}(t,s)}{1-e_{\ominus p}(\sigma(T),0)}q(s)\Delta s \\
&\quad + \int_t^{\sigma(T)} \frac{\ominus p(t)e_{\ominus p}(\sigma(T),0)e_{\ominus p}(t,s)}{1-e_{\ominus p}(\sigma(T),0)}q(s)\Delta s
\end{aligned}
$$

$$
\begin{aligned}
&+\sum_{j=1}^{m} g^{\Delta}(t,t_j)I_j(x(t_j))\\
&= (1+\mu(t)\ominus p(t))\,q(t)+\int_0^t \frac{\ominus p(t)e_{\ominus p}(t,s)}{1-e_{\ominus p}(\sigma(T),0)}q(s)\Delta s\\
&+\int_t^{\sigma(T)} \frac{\ominus p(t)e_{\ominus p}(\sigma(T),0)e_{\ominus p}(t,s)}{1-e_{\ominus p}(\sigma(T),0)}q(s)\Delta s\\
&+\sum_{j=1}^{m} g^{\Delta}(t,t_j)I_j(x(t_j)),
\end{aligned}
$$

$$
\begin{aligned}
x^{\sigma}(t) &= \left(\int_0^t g(t,s)q(s)\Delta s+\int_t^{\sigma(T)} g(t,s)q(s)\Delta s\right)^{\sigma}\\
&+\sum_{j=1}^{m} g(\sigma(t),t_j)I_j(x(t_j))\\
&= \left(\int_0^t \frac{e_{\ominus p}(t,s)}{1-e_{\ominus p}(\sigma(T),0)}q(s)\Delta s\right.\\
&\left.+\int_t^{\sigma(T)} \frac{e_{\ominus p}(\sigma(T),0)e_{\ominus p}(t,s)}{1-e_{\ominus p}(\sigma(T),0)}q(s)\Delta s\right)^{\sigma}\\
&+\sum_{j=1}^{m} g(\sigma(t),t_j)I_j(x(t_j))\\
&= \int_0^{\sigma(t)} \frac{e_{\ominus p}(\sigma(t),s)}{1-e_{\ominus p}(\sigma(T),0)}q(s)\Delta s\\
&+\int_{\sigma(t)}^{\sigma(T)} \frac{e_{\ominus p}(\sigma(T),0)e_{\ominus p}(\sigma(t),s)}{1-e_{\ominus p}(\sigma(T),0)}q(s)\Delta s\\
&+\sum_{j=1}^{m} g(\sigma(t),t_j)I_j(x(t_j))\\
&= \int_0^t \frac{e_{\ominus p}(\sigma(t),s)}{1-e_{\ominus p}(\sigma(T),0)}q(s)\Delta s+\int_t^{\sigma(t)} \frac{e_{\ominus p}(\sigma(t),s)}{1-e_{\ominus p}(\sigma(T),0)}q(s)\Delta s\\
&-\int_t^{\sigma(t)} \frac{e_{\ominus p}(\sigma(T),0)e_{\ominus p}(\sigma(t),s)}{1-e_{\ominus p}(\sigma(T),0)}q(s)\Delta s\\
&+\sum_{j=1}^{m} g(\sigma(t),t_j)I_j(x(t_j))\\
&+\int_t^{\sigma(T)} \frac{e_{\ominus p}(\sigma(T),0)e_{\ominus p}(\sigma(t),s)}{1-e_{\ominus p}(\sigma(T),0)}q(s)\Delta s
\end{aligned}
$$

$$
\begin{aligned}
&= \int_0^t \frac{(1+\mu(t)\ominus p(t))e_{\ominus p}(t,s)}{1-e_{\ominus p}(\sigma(T),0)}q(s)\Delta s \\
&\quad + \frac{\mu(t)e_{\ominus p}(\sigma(t),t)}{1-e_{\ominus p}(\sigma(T),0)}q(t) \\
&\quad - \frac{\mu(t)e_{\ominus p}(\sigma(T),0)e_{\ominus p}(\sigma(t),t)}{1-e_{\ominus p}(\sigma(T),0)}q(t) \\
&\quad + \sum_{j=1}^{m} g(\sigma(t),t_j)I_j(x(t_j)) \\
&\quad + \int_t^{\sigma(T)} \frac{e_{\ominus p}(\sigma(T),0)e_{\ominus p}(\sigma(t),s)}{1-e_{\ominus p}(\sigma(T),0)}q(s)\Delta s \\
&= \mu(t)(1+\mu(t)\ominus p(t))q(t) \\
&\quad + \int_0^t \frac{(1+\mu(t)\ominus p(t))e_{\ominus p}(t,s)}{1-e_{\ominus p}(\sigma(T),0)}q(s)\Delta s \\
&\quad + \int_t^{\sigma(T)} \frac{e_{\ominus p}(\sigma(T),0)e_{\ominus p}(\sigma(t),s)}{1-e_{\ominus p}(\sigma(T),0)}q(s)\Delta s \\
&\quad + \sum_{j=1}^{m} g(\sigma(t),t_j)I_j(x(t_j)),
\end{aligned}
$$

$$
\begin{aligned}
p(t)x^{\sigma}(t) &= p(t)\mu(t)(1+\mu(t)\ominus p(t))q(t) \\
&\quad + \int_0^t \frac{p(t)(1+\mu(t)\ominus p(t))e_{\ominus p}(t,s)}{1-e_{\ominus p}(\sigma(T),0)}q(s)\Delta s \\
&\quad + \int_t^{\sigma(T)} \frac{p(t)e_{\ominus p}(\sigma(T),0)e_{\ominus p}(\sigma(t),s)}{1-e_{\ominus p}(\sigma(T),0)}q(s)\Delta s \\
&\quad + \sum_{j=1}^{m} p(t)g(\sigma(t),t_j)I_j(x(t_j)),
\end{aligned}
$$

$$
\begin{aligned}
x^{\Delta}(t)+p(t)x^{\sigma}(t) &= (1+\mu(t)\ominus p(t))\,q(t) + \int_0^t \frac{\ominus p(t)e_{\ominus p}(t,s)}{1-e_{\ominus p}(\sigma(T),0)}q(s)\Delta s \\
&\quad + \int_t^{\sigma(T)} \frac{\ominus p(t)e_{\ominus p}(\sigma(T),0)e_{\ominus p}(t,s)}{1-e_{\ominus p}(\sigma(T),0)}q(s)\Delta s \\
&\quad + \sum_{j=1}^{m} g^{\Delta}(t,t_j)I_j(x(t_j))
\end{aligned}
$$

$$
\begin{aligned}
&+p(t)\mu(t)(1+\mu(t)\ominus p(t))q(t)\\
&+\int_0^t \frac{p(t)(1+\mu(t)\ominus p(t))e_{\ominus p}(t,s)}{1-e_{\ominus p}(\sigma(T),0)}q(s)\Delta s\\
&+\int_t^{\sigma(T)} \frac{p(t)e_{\ominus p}(\sigma(T),0)e_{\ominus p}(\sigma(t),s)}{1-e_{\ominus p}(\sigma(T),0)}q(s)\Delta s\\
&+\sum_{j=1}^{m} p(t)g(\sigma(t),t_j)I_j(x(t_j))\\
=\ & q(t)+\sum_{j=1}^{m}\Big(g^{\Delta}(t,t_j)+p(t)g(\sigma(t),t_j)\Big)I_j(x(t_j))\\
=\ & q(t)+\sum_{j=1}^{m}\Big(g^{\Delta}(t,t_j)+p(t)g(t,t_j)+\mu(t)p(t)g^{\Delta}(t,t_j)\Big)I_j(x(t_j))\\
=\ & q(t)+\sum_{j=1}^{m}\Big((1+\mu(t)p(t))g^{\Delta}(t,t_j)+p(t)g(t,t_j)\Big)\\
&\times I_j(x(t_j)),\quad t\in[0,\sigma(T)].
\end{aligned}
$$

If $0\le t\le t_j$, then

$$
\begin{aligned}
(1+\mu(t)p(t))g^{\Delta}(t,t_j)+p(t)g(t,t_j) &= \frac{(1+\mu(t)p(t))\ominus p(t)e_{\ominus p}(\sigma(T),0)e_{\ominus p}(t,t_j)}{1-e_{\ominus p}(\sigma(T),0)}\\
&\quad+\frac{p(t)e_{\ominus p}(\sigma(T),0)e_{\ominus p}(t,t_j)}{1-e_{\ominus p}(\sigma(T),0)}\\
&= 0.
\end{aligned}
$$

If $0\le t_j<t\le\sigma(T)$, then

$$
\begin{aligned}
(1+\mu(t)p(t))g^{\Delta}(t,t_j)+p(t)g(t,t_j) &= \frac{(1+\mu(t)p(t))\ominus p(t)e_{\ominus p}(t,t_j)}{1-e_{\ominus p}(\sigma(T),0)}\\
&\quad+\frac{p(t)e_{\ominus p}(t,t_j)}{1-e_{\ominus p}(\sigma(T),0)}\\
&= 0.
\end{aligned}
$$

Therefore,

$$
x^{\Delta}(t)+p(t)x^{\sigma}(t)=q(t),\quad t\in J_0.
$$

Next,

$$
\begin{aligned}
x(t_k^+) &= \int_0^{\sigma(T)} g(t_k^+,s)q(s)\Delta s + \sum_{j=1}^{m} g(t_k^+,t_j)I_j(x(t_j)) \\
&= \int_0^{\sigma(T)} g(t_k^-,s)q(s)\Delta s + \sum_{j=1}^{m} g(t_k^-,t_j)I_j(x(t_j)) \\
&\quad + \left(g(t_k^+,t_k) - g(t_k^-,t_k)\right) I_k(x(t_k)) \\
&= x(t_k^-) + I_k(x(t_k)), \quad k \in \{1,\ldots,m\},
\end{aligned}
$$

and

$$
\begin{aligned}
x(0) &= \int_0^{\sigma(T)} g(0,s)q(s)\Delta s + \sum_{j=1}^{m} g(0,t_j)I_j(x(t_j)) \\
&= \int_0^{\sigma(T)} g(\sigma(T),s)q(s)\Delta s + \sum_{j=1}^{m} g(\sigma(T),t_j)I_j(x(t_j)) \\
&= x(\sigma(T)).
\end{aligned}
$$

This completes the proof.

For p,q,I_k, $k \in \{1,\ldots,m\}$, which satisfy $(B5)$ and $(B6)$, define the operator $\Phi : PC([0,\sigma(T)]) \to PC([0,\sigma(T)])$ by

$$
\Phi x(t) = \int_0^{\sigma(T)} g(t,s)q(s)\Delta s + \sum_{j=1}^{m} g(t,t_j)I_j(x(t_j)), \quad t \in [0,\sigma(T)],
$$

where the function g is defined by (2.10). Note that by Lemma 2.7 and Lemma 2.8, it follows that every fixed point $x \in PC([0,\sigma(T)])$ of the operator Φ is a solution of problem (2.7), (2.8). Let

$$
M = \max_{j\in\{1,\ldots,m\}} \left\{ \sup_{t\in[0,\sigma(T)]} |g(t,t_j)| \right\}.
$$

Theorem 2.2. *Let p,q satisfy $(B5)$, I_k, $k \in \{1,\ldots,m\}$, satisfy $(B6)$ and*

$$
|I_k(x) - I_k(y)| \le l_k|x-y|, \quad x,y \in \mathbb{R}, \quad M\sum_{k=1}^{m} l_k < 1.
$$

Then problem (2.7), (2.8) has a unique solution in $PC([0,\sigma(T)])$.

Proof. For $x,y \in PC([0,\sigma(T)])$, we have

$$
\begin{aligned}
|\Phi x(t)-\Phi y(t)| &= \left|\int_0^{\sigma(T)} g(t,s)q(s)\Delta s+\sum_{j=1}^{m} g(t,t_j)I_j(x(t_j))\right.\\
&\quad \left.-\int_0^{\sigma(T)} g(t,s)q(s)\Delta s-\sum_{j=1}^{m} g(t,t_j)I_j(y(t_j))\right|\\
&= \left|\sum_{j=1}^{m} g(t,t_j)(I_j(x(t_j))-I_j(y(t_j)))\right|\\
&\le \sum_{j=1}^{m} |g(t,t_j)||I_j(x(t_j))-I_j(y(t_j))|\\
&\le M\sum_{j=1}^{m} l_j|x(t_j)-y(t_j)|\\
&\le M\sum_{j=1}^{m} l_j\|x-y\|,
\end{aligned}
$$

whereupon,

$$
\|\Phi x-\Phi y\| \le M\sum_{j=1}^{m} l_j\|x-y\|.
$$

Therefore, $\Phi : PC([0,\sigma(T)]) \to PC([0,\sigma(T)])$ is a contraction and it has a unique fixed point in $PC([0,\sigma(T)])$. This completes the proof.

Exercise 2.3. Let $\mathbb{T}=[0,1]\bigcup[2,3]\bigcup[4,5]$, where $[0,1]$, $[2,3]$ and $[4,5]$ are the real-valued intervals. Consider the periodic boundary value problem

$$
\begin{aligned}
x^{\Delta}(t) &= -x^{\sigma}(t)+\frac{1}{t^2+1}, \quad t\in\mathbb{T}\backslash\{2,4\},\\
x(2+) &= \frac{1}{1+(x(2))^2},\\
x(3+) &= \frac{2}{3+(x(3))^4},\\
x(0) &= x(5).
\end{aligned}
$$

Find constants l_1 and l_2 as in Theorem 2.2 so that the considered PBVP has a unique solution.

2.3 Periodic Boundary Value Problems for First Order Nonlinear Dynamic Equations with Impulses-I

Let $\mathbb{T}$, $\{t_k\}_{k=0}^{m+1}$, $PC([0,\sigma(T)])$, $PC^{\Delta}([0,\sigma(T)])$, J, J_0 and M be as in Section 2.2. Consider the following first order nonlinear impulsive dynamic equation

$$\begin{aligned} x^{\Delta}(t)+p(t)x(\sigma(t)) &= f(t,x(\sigma(t))), \quad t\in J_0,\\ x(t_k^+) &= x(t_k)+I_k(x(t_k)), \quad k\in\{1,\dots,m\}, \end{aligned} \tag{2.14}$$

$$x(0)=x(\sigma(T)), \tag{2.15}$$

where

(B7) $p\in\mathscr{C}([0,\sigma(T)])$, p is regressive on $[0,\sigma(T)]$,
(B8) $I_k\in\mathscr{C}(\mathbb{R})$, $k\in\{1,\dots,m\}$,
(B9) $f\in\mathscr{C}(J\times\mathbb{R})$.

By Lemma 2.7 and Lemma 2.8, it follows that $x\in PC^{\Delta}([0,\sigma(T)])$ is a solution of problem (2.14), (2.15) if and only if it satisfies the integral equation

$$x(t)=\int_0^{\sigma(T)} g(t,s)f(s,x(\sigma(s)))\Delta s+\sum_{j=1}^{m} g(t,t_j)I_j(x(t_j)), \quad t\in[0,\sigma(T)],$$

where the function g is defined by (2.10). Define the operator $\Psi: PC([0,\sigma(T)])\to PC([0,\sigma(T)])$ by

$$\Psi x(t)=\int_0^{\sigma(T)} g(t,s)f(s,x(\sigma(s)))\Delta s+\sum_{j=1}^{m} g(t,t_j)I_j(x(t_j)), \quad t\in[0,\sigma(T)].$$

Theorem 2.3. *Let p, I_k, $k\in\{1,\dots,m\}$, f satisfy (B7), (B8), (B9), respectively, and there exist non-negative constants α, K, β_j and γ_j, $j\in\{1,\dots,m\}$, such that*

$$|f(t,x)|\le\alpha(f(t,x)-p(t)x)+K, \quad t\in J, \quad x\in\mathbb{R},$$

$$|I_j(x)|\le\beta_j|x|+\gamma_j, \quad x\in\mathbb{R},$$

$$M\sum_{j=1}^{m}\beta_j<1.$$

Then problem (2.7), (2.8) has at least one solution in $PC([0,\sigma(T)])$.

Proof. 1. $\Psi: PC([0,\sigma(T)])\to PC([0,\sigma(T)])$ is continuous. Let

$$\{x_n\}_{n\in\mathbb{N}}\subset PC([0,\sigma(T)]),$$

$x\in PC([0,\sigma(T)])$, be such that $x_n\to x$ in $PC([0,\sigma(T)])$, as $n\to\infty$. Then

$$\begin{aligned}
|\Psi x_n(t)-\Psi x(t)| &= \Bigg| \int_0^{\sigma(T)} g(t,s)f(s,x_n(\sigma(s)))\Delta s + \sum_{j=1}^m g(t,t_j)I_j(x_n(t_j)) \\
&\quad - \int_0^{\sigma(T)} g(t,s)f(s,x(\sigma(s)))\Delta s - \sum_{j=1}^m g(t,t_j)I_j(x(t_j)) \Bigg| \\
&= \Bigg| \int_0^{\sigma(T)} g(t,s)(f(s,x_n(\sigma(s)))-f(s,x(\sigma(s))))\Delta s \\
&\quad + \sum_{j=1}^m g(t,t_j)(I_j(x(t_j))-I_j(x(t_j))) \Bigg| \\
&\le \int_0^{\sigma(T)} |g(t,s)||f(s,x_n(\sigma(s)))-f(s,x(\sigma(s)))|\Delta s \\
&\quad + \sum_{j=1}^m |g(t,t_j)||I_j(x_n(t_j))-I_j(x(t_j))| \\
&\le M\Bigg(\int_0^{\sigma(T)} |f(s,x_n(\sigma(s)))-f(s,x(\sigma(s)))|\Delta s \\
&\quad + \sum_{j=1}^m |I_j(x_n(t_j))-I_j(x(t_j))| \Bigg), \quad t\in[0,\sigma(T)].
\end{aligned}$$

Hence and by Lebesgue's dominated convergence theorem, it follows that

$$\|\Psi x_n - \Psi x\| \to 0, \quad \text{as} \quad n\to\infty.$$

2. Ψ maps a bounded set on $PC([0,\sigma(T)])$ into a bounded set on $PC([0,\sigma(T)])$. Let $D\subset PC([0,\sigma(T)])$ be a bounded set such that for any $x\in D$ we have

$$\|x\|\le L$$

for some positive constant L. Note that there exist positive constants F_1, Q_j, $j\in\{1,\ldots,m\}$, such that

$$|f(t,x(t))|\le F_1, \quad |I_j(x(t_j))|\le Q_j, \quad t\in[0,\sigma(T)], \quad x\in D.$$

Hence,

$$\begin{aligned}
|\Psi x(t)| &= \Bigg| \int_0^{\sigma(T)} g(t,s)f(s,x(\sigma(s)))\Delta s + \sum_{j=1}^m g(t,t_j)I_j(x(t_j)) \Bigg| \\
&\le \int_0^{\sigma(T)} |g(t,s)||f(s,x(\sigma(s)))|\Delta s + \sum_{j=1}^m |g(t,t_j)||I_j(x(t_j))|
\end{aligned}$$

$$\leq MF_1\sigma(T)+M\sum_{j=1}^{m}Q_j, \quad t\in[0,\sigma(T)].$$

3. Let $t^1,t^2\in[0,\sigma(T)]$. Then

$$\begin{aligned}|\Psi x(t^1)-\Psi x(t^2)| &= \left|\int_0^{\sigma(T)} g(t^1,s)f(s,x(\sigma(s)))\Delta s+\sum_{j=1}^{m} g(t^1,t_j)I_j(x(t_j))\right.\\ &\quad \left.-\int_0^{\sigma(T)} g(t^2,s)f(s,x(\sigma(s)))\Delta s-\sum_{j=1}^{m} g(t^2,t_j)I_j(x(t_j))\right|\\ &= \left|\int_0^{\sigma(T)}(g(t^1,s)-g(t^2,s))f(s,x(\sigma(s)))\Delta s\right.\\ &\quad \left.+\sum_{j=1}^{m}(g(t^1,t_j)-g(t^2,t_j)I_j(x(t_j))\right|\\ &\leq \int_0^{\sigma(T)}|g(t^1,s)-g(t^2,s)||f(s,x(\sigma(s)))|\Delta s\\ &\quad +\sum_{j=1}^{m}|g(t^1,t_j)-g(t^2,t_j)||I_j(x(t_j))|\\ &\to 0, \quad \text{as} \quad t^1\to t^2.\end{aligned}$$

From here and from the previous two points, using the Arzela-Ascoli theorem, we conclude that $\Psi : PC([0,\sigma(T)])\to PC([0,\sigma(T)])$ is a compact operator.

4. Now, we consider equation

$$x(t)=\lambda\Psi x(t), \quad t\in[0,\sigma(T)], \quad \lambda\in(0,1),$$

where $x\in PC([0,\sigma(T)])$. We have

$$\begin{aligned}|x(t)| &= \left|\lambda\int_0^{\sigma(T)} g(t,s)f(s,x(\sigma(s)))\Delta s+\lambda\sum_{j=1}^{m} g(t,t_j)I_j(x(t_j))\right|\\ &\leq \lambda\int_0^{\sigma(T)}|g(t,s)||f(s,x(\sigma(s)))|\Delta s+\lambda\sum_{j=1}^{m}|g(t,t_j)||I_j(x(t_j))|\\ &\leq \lambda M\int_0^{\sigma(T)}|f(s,x(\sigma(s)))|\Delta s+\lambda M\sum_{j=1}^{m}|I_j(x(t_j))|\\ &\leq \alpha\lambda M\int_0^{\sigma(T)}(f(s,x(\sigma(s)))-p(s)x(\sigma(s))))\Delta s+K\lambda M\sigma(T)\end{aligned}$$

$$+\lambda M \sum_{j=1}^{m} (\beta_j |x(t_j)| + \gamma_j)$$

$$= \alpha\lambda M \int_0^{\sigma(T)} x^{\Delta}(s)\Delta s + K\lambda M \sigma(T)$$

$$+\lambda M \sum_{j=1}^{m} (\beta_j \|x\| + \gamma_j)$$

$$= K\lambda M \sigma(T) + \lambda M \sum_{j=1}^{m} (\beta_j \|x\| + \gamma_j), \quad t \in [0, \sigma(T)].$$

Hence,

$$\|x\| \leq K\lambda M \sigma(T) + \lambda M \sum_{j=1}^{m} \beta_j \|x\| + \lambda M \sum_{j=1}^{m} \gamma_j$$

and

$$\|x\| \leq \frac{K\lambda M \sigma(T) + \lambda M \sum\limits_{j=1}^{m} \gamma_j}{1 - \lambda M \sum\limits_{j=1}^{m} \beta_j}.$$

From here and from Schaefer's fixed point Theorem 1.3, it follows that the operator Ψ has a fixed point in $PC([0, \sigma(T)])$. This completes the proof.

Exercise 2.4. Suppose that p, I_j, $j \in \{1, \ldots, m\}$, f satisfy $(B7)$, $(B8)$, $(B9)$, respectively, and there exist non-negative constants α, K, β_j and γ_j, $j \in \{1, \ldots, m\}$, such that

$$|f(t,x)| \leq \alpha|x| + K, \quad t \in J, \quad x \in \mathbb{R},$$

$$|I_j(x)| \leq \beta_j |x| + \gamma_j, \quad x \in \mathbb{R}, \quad j \in \{1, \ldots, m\},$$

$$M\left(\alpha\sigma(T) + \sum_{j=1}^{m} \beta_j\right) < 1.$$

Prove that problem (2.14), (2.15) has at least one solution in $PC([0, \sigma(T)])$.

Now suppose that

(B10) $p : [0, T] \to (0, \infty)$ is right-dense continuous,

$t_{m+1} = \sigma(T)$, $J_1 = [0, t_1]$.

Lemma 2.9. *Suppose* $(B10)$. *For the function* g, *defined by* (2.10), *we have the following estimate*

$$A = \frac{1}{e_p(\sigma(T), 0) - 1} \leq g(t,s) \leq \frac{e_p(\sigma(T), 0)}{e_p(\sigma(T), 0) - 1} = B, \quad t, s \in [0, \sigma(T)].$$

Proof. 1. Let $0 \le s < t \le \sigma(T)$. Then

$$\begin{aligned}
e_{\ominus p}(t,s) &= e^{\int_s^t \frac{1}{\mu(\tau)} \log(1+\mu(\tau)\ominus p(\tau))\Delta\tau} \\
&= e^{\int_s^t \frac{1}{\mu(\tau)} \log\left(1-\frac{\mu(\tau)p(\tau)}{1+\mu(\tau)p(\tau)}\right)\Delta\tau} \\
&= e^{\int_s^t \frac{1}{\mu(\tau)} \log \frac{1}{1+\mu(\tau)p(\tau)}\Delta\tau} \\
&= e^{-\int_s^t \frac{1}{\mu(\tau)} \log(1+\mu(\tau)p(\tau))\Delta\tau} \\
&= \frac{1}{e^{\int_s^t \frac{1}{\mu(\tau)} \log(1+\mu(\tau)p(\tau))\Delta\tau}} \\
&\ge \frac{1}{e^{\int_0^{\sigma(T)} \frac{1}{\mu(\tau)} \log(1+\mu(\tau)p(\tau))\Delta\tau}} \\
&= \frac{1}{e_p(\sigma(T),0)}, \\
g(t,s) &= \frac{e_{\ominus p}(t,s)}{1-e_{\ominus p}(\sigma(T),0)} \\
&\ge \frac{1}{1-e_{\ominus p}(\sigma(T),0)}\frac{1}{e_p(\sigma(T),0)} \\
&= \frac{1}{e_p(\sigma(T),0)-1} \\
&= A, \\
e_{\ominus p}(t,s) &= e^{-\int_s^t \frac{1}{\mu(\tau)} \log(1+\mu(\tau)p(\tau))\Delta\tau} \\
&\le 1, \\
g(t,s) &= \frac{e_{\ominus p}(t,s)}{1-e_{\ominus p}(\sigma(T),0)} \\
&\le \frac{1}{1-e_{\ominus p}(\sigma(T),0)} \\
&= \frac{e_p(\sigma(T),0)}{e_p(\sigma(T),0)-1}
\end{aligned}$$

$$= B.$$

2. Let $0 \le t \le s \le \sigma(T)$. Then

$$\begin{aligned}
e_{\ominus p}(t,s) &= e^{-\int_s^t \frac{1}{\mu(\tau)} \log(1+\mu(\tau)p(\tau))\Delta\tau} \\
&= e^{\int_t^s \frac{1}{\mu(\tau)} \log(1+\mu(\tau)p(\tau))\Delta\tau} \\
&\geq 1, \\
g(t,s) &= \frac{e_{\ominus p}(\sigma(T),0)e_{\ominus p}(t,s)}{1-e_{\ominus p}(\sigma(T),0)} \\
&\geq \frac{e_{\ominus p}(\sigma(T),0)}{1-e_{\ominus p}(\sigma(T),0)} \\
&= \frac{1}{e_p(\sigma(T),0)-1} \\
&= A, \\
e_{\ominus p}(t,s) &= e^{\int_t^s \frac{1}{\mu(\tau)} \log(1+\mu(\tau)p(\tau))\Delta\tau} \\
&\leq e^{\int_t^{\sigma(T)} \frac{1}{\mu(\tau)} \log(1+\mu(\tau)p(\tau))\Delta\tau} \\
&\leq e^{\int_0^{\sigma(T)} \frac{1}{\mu(\tau)} \log(1+\mu(\tau)p(\tau))\Delta\tau} \\
&= e_p(\sigma(T),0), \\
g(t,s) &= \frac{e_{\ominus p}(\sigma(T),0)e_{\ominus p}(t,s)}{1-e_{\ominus p}(\sigma(T),0)} \\
&\leq \frac{e_{\ominus p}(t,s)}{e_p(\sigma(T),0)-1} \\
&\leq \frac{e_p(\sigma(T),0)}{e_p(\sigma(T),0)-1} \\
&= B.
\end{aligned}$$

This completes the proof.

Lemma 2.10. *Let p satisfy (B8). We have*

$$\int_0^{\sigma(T)} g(t,s)p(s)\Delta s = 1.$$

Proof. We have

$$\begin{aligned}
\int_0^{\sigma(T)} g(t,s)p(s)\Delta s &= \int_0^t g(t,s)p(s)\Delta s + \int_t^{\sigma(T)} g(t,s)p(s)\Delta s \\
&= \frac{1}{1-e_{\ominus p}(\sigma(T),0)}\left(\int_0^t e_p(s,t)p(s)\Delta s\right. \\
&\quad \left. + e_{\ominus p}(\sigma(T),0)\int_t^{\sigma(T)} e_p(s,t)p(s)\Delta s\right) \\
&= \frac{1}{1-e_{\ominus p}(\sigma(T),0)}\Big(e_p(t,t)-e_p(0,t) \\
&\quad + e_p(0,\sigma(T))(e_p(\sigma(T),t)-e_p(t,t))\Big) \\
&= \frac{1}{1-e_{\ominus p}(\sigma(T),0)}(1-e_p(0,t)+e_p(0,t)-e_p(0,\sigma(T))) \\
&= \frac{1-e_{\ominus p}(\sigma(T),0)}{1-e_{\ominus p}(\sigma(T),0)} \\
&= 1.
\end{aligned}$$

This completes the proof.

Denote

$$\delta = \frac{A}{B},$$

$$f^0 = \limsup_{t\to 0+} \max_{t\in[0,T]} \frac{f(t,x)}{p(t)x}, \quad f_0 = \liminf_{t\to 0+} \min_{t\in[0,T]} \frac{f(t,x)}{p(t)x},$$

$$f^\infty = \limsup_{t\to \infty} \max_{t\in[0,T]} \frac{f(t,x)}{p(t)x}, \quad f_\infty = \liminf_{t\to \infty} \min_{t\in[0,T]} \frac{f(t,x)}{p(t)x},$$

$$f^u = \sup_{t\in[\delta u,u]} \max_{t\in[0,T]} \frac{f(t,x)}{p(t)x}, \quad f_u = \inf_{t\in[\delta u,u]} \min_{t\in[0,T]} \frac{f(t,x)}{p(t)x},$$

$$I^0(k) = \limsup_{x\to 0+} \frac{I_k(x)}{x}, \quad I_0(k) = \liminf_{x\to 0+} \frac{I_k(x)}{x},$$

$$I^{\infty}(k) = \limsup_{x\to\infty} \frac{I_k(x)}{x}, \quad I_{\infty}(k) = \liminf_{x\to\infty} \frac{I_k(x)}{x},$$

$$I^{u}(k) = \sup_{x\in[\delta u,u]} \frac{I_k(x)}{x}, \quad I_{u}(k) = \inf_{x\in[\delta u,u]} \frac{I_k(x)}{x}, \quad k\in\{1,\dots,m\}.$$

For our next results we will use the following fixed point theorem.

Theorem 2.4. *Let X be a Banach space and K is a cone in X. Assume that Ω_1, Ω_2 are open subsets of X with $0\in\Omega_1$, $\overline{\Omega_1}\subset\Omega_2$. Let*

$$\Phi : K\bigcap(\overline{\Omega_2}\backslash\Omega_1)\to K$$

be a continuous and completely continuous operator such that

1. $\|\Phi x\|\le\|x\|$, $x\in K\bigcap\partial\Omega_1$,
2. *there exists an $e\in K\backslash\{0\}$ such that $x\ne\Phi x+\lambda e$ for $x\in K\bigcap\partial\Omega_2$ and $\lambda>0$.*

Then Φ has a fixed point in $K\bigcap(\overline{\Omega_2}\backslash\Omega_1)$.

Define

$$X=\{x : x\in PC([0,\sigma(T)]), \quad x(0)=x(\sigma(T))\}$$

with the norm

$$\|x\| = \sup_{t\in[0,\sigma(T)]} |x(t)|.$$

Note that X is a Banach space. Let

$$K=\{x\in X : x(t)\ge\delta\|x\|, \quad t\in[0,\sigma(T)]\}.$$

Theorem 2.5. *Suppose* (*B*8)-(*B*10) *and there exist* $0<\alpha<\beta$ *such that*

$$f(t,x)\ge 0, \quad I_k(x)\ge 0, \quad t\in[0,\sigma(T)], \quad x\in[\delta\alpha,\beta],$$

$$f_{\beta}+A\sum_{k=1}^{m} I_{\beta}(k)\ge 1, \quad f^{\alpha}+B\sum_{k=1}^{m} I^{\alpha}(k)\le 1. \tag{2.16}$$

Then problem (2.14), (2.15) *has at least one positive solution.*

Proof. Define the operator $\Psi : K\to X$ as follows

$$\Psi x(t) = \int_0^{\sigma(T)} g(t,s)f(s,x(\sigma(s)))\Delta s+\sum_{j=1}^{m} g(t,t_j)I_j(x(t_j)), \quad t\in[0,\sigma(T)].$$

As in the proof of Theorem 2.3, we have that $\Psi : K\to X$ is continuous and completely continuous. Define the sets

$$\Omega_1 = \{x\in X : \|x\|<\alpha\},$$

$$\Omega_2 = \{x \in X : \|x\| < \beta\}.$$

1. We will prove that $\Psi : K\bigcap\left(\overline{\Omega_2}\backslash\Omega_1\right) \to K$. For $x \in K\bigcap\left(\overline{\Omega_2}\backslash\Omega_1\right)$, we have

$$\delta\alpha \le x(t) \le \beta, \quad t \in [0,\sigma(T)],$$

and, using Lemma 2.10, we get

$$\begin{aligned}\Psi x(t) &= \int_0^{\sigma(T)} g(t,s)f(s,x(\sigma(s)))\Delta s + \sum_{k=1}^{m} g(t,t_k)I_k(x(t_k))\\ &\le B\left(\int_0^{\sigma(T)} f(s,x(\sigma(s)))\Delta s + \sum_{k=1}^{m} I_k(x(t_k))\right), \quad t \in [0,\sigma(T)],\\ \|\Psi x\| &\le B\left(\int_0^{\sigma(T)} f(s,x(\sigma(s)))\Delta s + \sum_{k=1}^{m} I_k(x(t_k))\right),\end{aligned}$$

and

$$\begin{aligned}\Psi x(t) &\ge A\left(\int_0^{\sigma(T)} f(s,x(\sigma(s)))\Delta s + \sum_{k=1}^{m} I_k(x(t_k))\right)\\ &= \frac{A}{B}B\left(\int_0^{\sigma(T)} f(s,x(\sigma(s)))\Delta s + \sum_{k=1}^{m} I_k(x(t_k))\right)\\ &\ge \frac{A}{B}\|\Psi x\|, \quad t \in [0,\sigma(T)].\end{aligned}$$

Thus, $\Psi : K\bigcap\left(\overline{\Omega_2}\backslash\Omega_1\right) \to K$.
2. We will prove

$$\|\Psi x\| \le \|x\|, \quad x \in K\bigcap\partial\Omega_1.$$

We have

$$f(t,x) \le f^\alpha p(t)x, \;\; I_k(x) \le I^\alpha(k)x, \;\; k \in \{1,\ldots,m\}, \;\; x \in [\delta\alpha,\alpha], \;\; t \in [0,\sigma(T)].$$

Hence and by the second inequality of (2.16), we obtain

$$f^\alpha \le 1 - B\sum_{k=1}^{m} I^\alpha(k),$$

$$\begin{aligned}f(t,x) &\le f^\alpha p(t)x\\ &\le \left(1 - B\sum_{k=1}^{m} I^\alpha(k)\right)p(t)x, \quad t \in [0,\sigma(T)], \quad x \in [\delta\alpha,\alpha].\end{aligned}$$

For $x \in K\bigcap \Omega_1$, we get

$$\begin{aligned}\delta\alpha &= \delta\|x\| \\ &\leq x(t) \\ &\leq \alpha, \quad t \in [0,\sigma(T)],\end{aligned}$$

and, using Lemma 2.10, we find

$$\begin{aligned}\Psi x(t) &= \int_0^{\sigma(T)} g(t,s)f(s,x(\sigma(s)))\Delta s + \sum_{k=1}^m g(t,t_k)I_k(x(t_k)) \\ &\leq \left(1 - B\sum_{k=1}^m I^\alpha(k)\right)\int_0^{\sigma(T)} g(t,s)p(s)x(\sigma(s))\Delta s + B\sum_{k=1}^m I^\alpha(k)x(t_k) \\ &\leq \left(\left(1 - B\sum_{k=1}^m I^\alpha(k)\right)\int_0^{\sigma(T)} g(t,s)p(s)\Delta s + B\sum_{k=1}^m I^\alpha(k)\right)\|x\| \\ &= \left(1 - B\sum_{k=1}^m I^\alpha(k) + B\sum_{k=1}^m I^\alpha(k)\right)\|x\| \\ &= \|x\|, \quad t \in [0,\sigma(T)],\end{aligned}$$

whereupon,

$$\|\Psi x\| \leq \|x\|.$$

3. Let $e = 1$. Then $e \in K\backslash\{0\}$. Assume that there exist $x_0 \in K\bigcap\partial\Omega_2$ and a $\lambda_0 > 0$ such that

$$x_0 = \Psi x_0 + \lambda_0 e.$$

Since $x_0 \in K\bigcap\partial\Omega_2$, we obtain

$$\begin{aligned}\delta\beta &= \delta\|x_0\| \\ &\leq \|x_0\| \\ &\leq \beta.\end{aligned}$$

Also,

$$f(t,x) \geq f_\beta p(t)x, \quad t \in [0,\sigma(T)], \quad x \in [\delta\beta,\beta],$$

and

$$I_k(x) \geq I_\beta(k), \quad x \in [\delta\beta,\beta], \quad k \in \{1,\ldots,m\}.$$

By the first inequality of (2.16), we get

$$f(t,x) \geq \left(1 - A\sum_{k=1}^{m} I_\beta(k)\right) p(t)x, \quad t \in [0,\sigma(T)], \quad x \in [\delta\beta,\beta].$$

Let

$$\mu = \min_{t\in[0,\sigma(T)]} x_0(t).$$

Then, using Lemma 2.10, we obtain

$$\begin{aligned}
x_0(t) &= \Psi x_0(t) + \lambda_0 e \\
&= \int_0^{\sigma(T)} g(t,s) f(s,x_0(\sigma(s)))\Delta s + \sum_{k=1}^{m} g(t,t_k) I_k(x_0(t_k)) + \lambda_0 \\
&\geq \left(1 - A\sum_{k=1}^{m} I_\beta(k)\right) \int_0^{\sigma(T)} g(t,s)p(s)x_0(\sigma(s))\Delta s \\
&\quad + A\sum_{k=1}^{m} I_\beta(k) x_0(t_k) + \lambda_0 \\
&\geq \mu\left(1 - A\sum_{k=1}^{m} I_\beta(k)\right) \int_0^{\sigma(T)} g(t,s)p(s)\Delta s \\
&\quad + \mu A\sum_{k=1}^{m} I_\beta(k) + \lambda_0 \\
&= \mu\left(1 - A\sum_{k=1}^{m} I_\beta(k)\right) + \mu A\sum_{k=1}^{m} I_\beta(k) + \lambda_0 \\
&= \mu + \lambda_0, \quad t \in [0,\sigma(T)].
\end{aligned}$$

Hence,

$$\mu \geq \mu + \lambda_0,$$

which is a contradiction.

From here and from Theorem 2.4, it follows that problem (2.14), (2.15) has at least one positive solution. This completes the proof.

Corollary 2.1. *Suppose* $(B8)$-$(B10)$ *and*

$$f_\infty + A\sum_{k=1}^{m} I_\infty(k) \geq 1, \quad f^0 + B\sum_{k=1}^{m} I^0(k) \leq 1.$$

Then problem (2.14), (2.15) *has at least one positive solution.*

Proof. By the definitions of f^0 and I^0, it follows that there exists an $\alpha > 0$ such that

$$I^0(k) \geq \limsup_{x\in[\delta\alpha,\alpha]} \frac{I_k(x)}{x}, \quad f^0 \geq \limsup_{x\in[\delta\alpha,\alpha]} \max_{t\in[0,T]} \frac{f(t,x)}{p(t)x}.$$

Hence,

$$I^\alpha(k) \leq I^0(k), \quad f^0 \geq f^\alpha.$$

Next, by the definitions of f_∞ and I_∞, it follows that there exists an $\beta > \alpha$ such that

$$I_\infty(k) \leq \liminf_{x\in[\delta\beta,\beta]} \frac{I_k(x)}{x}, \quad f_\infty \leq \liminf_{x\in[\delta\beta,\beta]} \min_{t\in[0,T]} \frac{f(t,x)}{p(t)x}.$$

From here,

$$I_\beta(k) \geq I_\infty(k), \quad f_\beta \geq f_\infty.$$

Then

$$f_\beta + A\sum_{k=1}^{m} I_\beta(k) \geq 1, \quad f^\alpha + B\sum_{k=1}^{m} I^\alpha(k) \leq 1.$$

Now, using Theorem 2.5, it follows that problem (2.14), (2.15) has at least one positive solution. This completes the proof.

Theorem 2.6. *Suppose* $(B8)$-$(B10)$ *and there exist* $0 < \alpha < \beta$ *such that*

$$f(t,x) \geq 0, \quad I_k(x) \geq 0, \quad t \in [0,\sigma(T)], \quad x \in [\delta\alpha,\beta],$$

$$f^\beta + B\sum_{k=1}^{m} I^\beta(k) \leq 1, \quad f_\alpha + A\sum_{k=1}^{m} I_\alpha(k) \geq 1. \tag{2.17}$$

Then problem (2.14), (2.15) *has at least one positive solution.*

Proof. Let the operator $\Psi : K \to X$ and the sets Ω_1, Ω_2 be as in Theorem 2.5. As in the proof of Theorem 2.3, we have that $\Psi : K \to X$ is continuous and completely continuous.

1. As in the proof of Theorem 2.5, we have $\Psi : K\bigcap\left(\overline{\Omega_2}\backslash\Omega_1\right) \to K$.
2. We will prove

$$\|\Psi x\| \leq \|x\|, \quad x \in K\bigcap\partial\Omega_2.$$

We have

$$f(t,x) \leq f^\beta p(t)x, \;\; I_k(x) \leq I^\beta(k)x, \;\; k \in \{1,\ldots,m\}, \;\; x \in [\delta\beta,\beta], \;\; t \in [0,\sigma(T)].$$

Hence and by the second inequality of (2.17), we obtain

$$f^\beta \leq 1 - B\sum_{k=1}^{m} I^\beta(k),$$

$$f(t,x) \leq f^\beta p(t)x$$

$$\leq \left(1 - B\sum_{k=1}^{m} I^{\beta}(k)\right) p(t)x, \quad t \in [0, \sigma(T)], \quad x \in [\delta\beta, \beta].$$

For $x \in K \bigcap \Omega_2$, we get

$$\begin{aligned}\delta\beta &= \delta\|x\|\\ &\leq x(t)\\ &\leq \beta, \quad t \in [0, \sigma(T)],\end{aligned}$$

and, using Lemma 2.10, we find

$$\begin{aligned}\Psi x(t) &= \int_0^{\sigma(T)} g(t,s) f(s, x(\sigma(s)))\Delta s + \sum_{k=1}^{m} g(t,t_k) I_k(x(t_k))\\ &\leq \left(1 - B\sum_{k=1}^{m} I^{\beta}(k)\right) \int_0^{\sigma(T)} g(t,s)p(s)x(\sigma(s))\Delta s + B\sum_{k=1}^{m} I^{\beta}(k)x(t_k)\\ &\leq \left(\left(1 - B\sum_{k=1}^{m} I^{\beta}(k)\right) \int_0^{\sigma(T)} g(t,s)p(s)\Delta s + B\sum_{k=1}^{m} I^{\beta}(k)\right)\|x\|\\ &= \left(1 - B\sum_{k=1}^{m} I^{\beta}(k) + B\sum_{k=1}^{m} I^{\beta}(k)\right)\|x\|\\ &= \|x\|, \quad t \in [0, \sigma(T)],\end{aligned}$$

whereupon,

$$\|\Psi x\| \leq \|x\|.$$

3. Let $e = 1$. Then $e \in K\backslash\{0\}$. Assume that there exist $x_0 \in K \bigcap \partial\Omega_1$ and a $\lambda_0 > 0$ such that

$$x_0 = \Psi x_0 + \lambda_0 e.$$

Since $x_0 \in K \bigcap \partial\Omega_1$, we obtain

$$\begin{aligned}\delta\alpha &= \delta\|x_0\|\\ &\leq \|x_0\|\\ &\leq \alpha.\end{aligned}$$

Also,

$$f(t,x) \geq f_\alpha p(t)x, \quad t \in [0,\sigma(T)], \quad x \in [\delta\alpha,\alpha],$$

and

$$I_k(x) \geq I_\alpha(k), \quad x \in [\delta\alpha,\alpha], \quad k \in \{1,\ldots,m\}.$$

By the first inequality of (2.17), we get

$$f(t,x) \geq \left(1 - A\sum_{k=1}^{m} I_\alpha(k)\right) p(t)x, \quad t \in [0,\sigma(T)], \quad x \in [\delta\alpha,\alpha].$$

Let

$$\mu = \min_{t\in[0,\sigma(T)]} x_0(t).$$

Then, using Lemma 2.10, we obtain

$$\begin{aligned}
x_0(t) &= \Psi x_0(t) + \lambda_0 e \\
&= \int_0^{\sigma(T)} g(t,s) f(s,x_0(\sigma(s)))\Delta s + \sum_{k=1}^{m} g(t,t_k) I_k(x_0(t_k)) + \lambda_0 \\
&\geq \left(1 - A\sum_{k=1}^{m} I_\alpha(k)\right) \int_0^{\sigma(T)} g(t,s)p(s)x(\sigma(s))\Delta s \\
&\quad + A\sum_{k=1}^{m} I_\alpha(k)x(t_k) + \lambda_0 \\
&\geq \mu\left(1 - A\sum_{k=1}^{m} I_\alpha(k)\right) \int_0^{\sigma(T)} g(t,s)p(s)\Delta s \\
&\quad + \mu A\sum_{k=1}^{m} I_\alpha(k) + \lambda_0 \\
&= \mu\left(1 - A\sum_{k=1}^{m} I_\alpha(k)\right) + \mu A\sum_{k=1}^{m} I_\alpha(k) + \lambda_0 \\
&= \mu + \lambda_0, \quad t \in [0,\sigma(T)].
\end{aligned}$$

Hence,

$$\mu \geq \mu + \lambda_0.$$

This is a contradiction.

From here and from Theorem 2.4, it follows that problem (2.14), (2.15) has at least one positive solution. This completes the proof.

Corollary 2.2. *Suppose* $(B8)$-$(B10)$ *and*

$$f(t,x) \geq 0, \quad I_k(x) \geq 0, \quad t \in [0,\sigma(T)], \quad x \in [\delta\alpha,\beta],$$

$$f_0 + A\sum_{k=1}^{m} I_0(k) \geq 1, \quad f^\infty + B\sum_{k=1}^{m} I^\infty(k) \leq 1.$$

Then problem (2.14), (2.15) *has at least one positive solution.*

Proof. By the definitions of f_0 and I_0, it follows that there exists $\alpha > 0$ such that

$$I_0(k) \leq \liminf_{x\in[\delta\alpha,\alpha]} \frac{I_k(x)}{x}, \quad f_0 \geq \liminf_{x\in[\delta\alpha,\alpha]} \min_{t\in[0,T]} \frac{f(t,x)}{p(t)x}.$$

Hence,

$$I_\alpha(k) \geq I_0(k), \quad f_0 \leq f_\alpha.$$

Next, by the definitions of f^∞ and I^∞, it follows that there exists $\beta > \alpha$ such that

$$I^\infty(k) \geq \limsup_{x\in[\delta\beta,\beta]} \frac{I_k(x)}{x}, \quad f^\infty \geq \limsup_{x\in[\delta\beta,\beta]} \max_{t\in[0,T]} \frac{f(t,x)}{p(t)x}.$$

From here,

$$I^\beta(k) \leq I^\infty(k), \quad f^\infty \geq f^\beta.$$

Then

$$f_\alpha + A\sum_{k=1}^{m} I_\alpha(k) \geq 1, \quad f^\beta + B\sum_{k=1}^{m} I^\beta(k) \leq 1.$$

Now, using Theorem 2.6, it follows that problem (2.14), (2.15) has at least one positive solution. This completes the proof.

Example 2.2. Consider the following problem

$$x^\Delta(t) + p(t)x(\sigma(t)) = (x(\sigma(t)))^{\frac{1}{3}} + (x(\sigma(t)))^{\frac{1}{5}}, \quad t \in J_0,$$

$$x(t_k^+) = x(t_k^-) + (x(t_k^-))^{\frac{1}{5}} + (x(t_k^-))^{\frac{1}{7}}, \quad k \in \{1,\ldots,m\},$$

$$x(0) = x(\sigma(T)),$$

where p satisfies $(B8)$. Here

$$f(t,x(t)) = (x(t))^{\frac{1}{3}} + (x(t))^{\frac{1}{5}},$$

$$I_k(x(t_k^-)) = (x(t_k^-))^{\frac{1}{5}} + (x(t_k^-))^{\frac{1}{7}}, \quad k \in \{1,\ldots,m\}.$$

We have

$$f_0 = \liminf_{x\to 0+} \min_{t\in[0,T]} \frac{f(t,x)}{p(t)x}$$

$$
\begin{aligned}
&= \liminf_{x \to 0+} \min_{t \in [0,T]} \frac{x^{\frac{1}{3}} + x^{\frac{1}{5}}}{p(t)x} \\
&= \infty,
\end{aligned}
$$

$$
\begin{aligned}
I_0(k) &= \liminf_{x \to 0+} \frac{I_k(x)}{x} \\
&= \liminf_{x \to 0+} \frac{x^{\frac{1}{5}} + x^{\frac{1}{7}}}{x} \\
&= \infty, \quad k \in \{1, \ldots, m\},
\end{aligned}
$$

$$
\begin{aligned}
f^{\infty} &= \limsup_{x \to \infty} \max_{t \in [0,T]} \frac{f(t,x)}{p(t)x} \\
&= \limsup_{x \to \infty} \max_{t \in [0,T]} \frac{x^{\frac{1}{3}} + x^{\frac{1}{5}}}{p(t)x} \\
&= 0,
\end{aligned}
$$

$$
\begin{aligned}
I^{\infty}(k) &= \limsup_{x \to \infty} \frac{I_k(x)}{x} \\
&= \limsup_{x \to \infty} \frac{x^{\frac{1}{3}} + x^{\frac{1}{7}}}{x} \\
&= 0, \quad k \in \{1, \ldots, m\}.
\end{aligned}
$$

Hence,

$$
\begin{aligned}
f_0 + A \sum_{k=1}^{m} I_0(k) &= \infty \\
&\geq 1,
\end{aligned}
$$

$$
\begin{aligned}
f^{\infty} + B \sum_{k=1}^{m} I^{\infty}(k) &= 0 \\
&\leq 1.
\end{aligned}
$$

From here and from Corollary 2.2, it follows that the considered problem has at least one positive solution.

Exercise 2.5. Prove that problem

$$x^{\Delta}(t)+p(t)x(\sigma(t)) = 2(x(\sigma(t)))^{\frac{1}{2}}+3(x(\sigma(t)))^{\frac{1}{4}}, \quad t\in J_0,$$

$$x(t_k^+) = x(t_k^-)+(x(t_k^-))^{\frac{1}{3}}+4(x(t_k^-))^{\frac{1}{9}}, \quad k\in\{1,\ldots,m\},$$

$$x(0) = x(\sigma(T)),$$

where p satisfies $(B8)$, has at least one positive solution.

Theorem 2.7. *Suppose* $(B8)$-$(B10)$ *and there exist* $0<\alpha<\rho<\beta$ *such that*

$$f(t,x)\geq 0, \quad I_k(x)\geq 0, \quad t\in[0,\sigma(T)], \quad x\in[\delta\alpha,\beta],$$

$$f^{\alpha}+B\sum_{k=1}^{m} I^{\alpha}(k)\leq 1, \quad f_{\rho}+A\sum_{k=1}^{m} I_{\rho}(k)\geq 1,$$

$$f^{\beta}+B\sum_{k=1}^{m} I^{\beta}(k)\leq 1.$$

Then problem (2.14), (2.15) *has at least two positive solutions.*

Proof. Let Ψ, Ω_1, Ω_2 be as in the proof of Theorem 2.5. Define

$$\Omega_3=\{x\in X: \|x\|<\rho\}.$$

By Theorem 2.5, it follows that problem (2.14), (2.15) has at least one positive solution in $\overline{\Omega_3}\backslash\Omega_1$. By Theorem 2.6, it follows that problem (2.14), (2.15) has at least one positive solution in $\overline{\Omega_2}\backslash\Omega_3$. This completes the proof.

Theorem 2.8. *Suppose* $(B8)$-$(B10)$ *and there exist* $0<\alpha<\rho<\beta$ *such that*

$$f(t,x)\geq 0, \quad I_k(x)\geq 0, \quad t\in[0,\sigma(T)], \quad x\in[\delta\alpha,\beta],$$

$$f^{\alpha}+A\sum_{k=1}^{m} I^{\alpha}(k)\geq 1, \quad f_{\rho}+B\sum_{k=1}^{m} I_{\rho}(k)\leq 1,$$

$$f^{\beta}+A\sum_{k=1}^{m} I^{\beta}(k)\geq 1.$$

Then problem (2.14), (2.15) *has at least two positive solutions.*

Proof. Let Ψ, Ω_1, Ω_2 and Ω_3 be as in the proof of Theorem 2.7. By Theorem 2.6, it follows that problem (2.14), (2.15) has at least one positive solution in $\overline{\Omega_3}\backslash\Omega_1$. By Theorem 2.5, it follows that problem (2.14), (2.15) has at least one positive solution in $\overline{\Omega_2}\backslash\Omega_3$. This completes the proof.

Theorem 2.9. *Suppose* $(B8)$-$(B10)$ *and there exist positive constants*

$$0 < \alpha_1 < \beta_1 < \ldots < \alpha_r < \beta_r$$

such that

$$f(t,x) \geq 0, \quad I_k(x) \geq 0, \quad t \in [0,T], \quad x \in [\delta\alpha_1, \beta_r].$$

Then problem (2.14), (2.15) *has at least r positive solutions* x_j, $j \in \{1,\ldots,r\}$, *satisfying*

$$\alpha_j \leq \|x_j\| \leq \beta_j, \quad j \in \{1,\ldots,r\},$$

if one of the following conditions hold.

1.

$$f_{\alpha_j} + A\sum_{k=1}^{m} I_{\alpha_j}(k) \geq 1, \quad f^{\beta_j} + B\sum_{k=1}^{m} I^{\beta_j}(k) \leq 1, \quad j \in \{1,\ldots,r\}.$$

2.

$$f^{\alpha_j} + B\sum_{k=1}^{m} I^{\alpha_j}(k) \leq 1, \quad f_{\beta_j} + A\sum_{k=1}^{m} I_{\beta_j}(k) \geq 1, \quad j \in \{1,\ldots,r\}.$$

Proof. Let Ψ be as in the proof of Theorem 2.5. Define

$$\Omega_{\alpha_j} = \{x \in X : \|x\| < \alpha_j\},$$

$$\Omega_{\beta_j} = \{x \in X : \|x\| < \beta_j\}, \quad j \in \{1,\ldots,r\}.$$

1. By Theorem 2.6, it follows that problem (2.14), (2.15) has at least one positive solution $x_1 \in K\bigcap(\overline{\Omega_{\beta_1}}\backslash\Omega_{\alpha_1})$. By Theorem 2.5, it follows that problem (2.14), (2.15) has at least one positive solution $x_2 \in K\bigcap(\overline{\Omega_{\alpha_2}}\backslash\Omega_{\beta_1})$. And so on.
2. By Theorem 2.5, it follows that problem (2.14), (2.15) has at least one positive solution $x_1 \in K\bigcap(\overline{\Omega_{\beta_1}}\backslash\Omega_{\alpha_1})$. By Theorem 2.6, it follows that problem (2.14), (2.15) has at least one positive solution $x_2 \in K\bigcap(\overline{\Omega_{\alpha_2}}\backslash\Omega_{\beta_1})$. And so on. This completes the proof.

Suppose $(B10)$ and

(B11) $f \in \mathscr{C}(J_0 \times [0,\infty))$, $f \geq 0$ on $J_0 \times [0,\infty)$,
(B12) $I_k \in \mathscr{C}([0,\infty))$, $I_k \geq 0$ on $[0,\infty)$, $k \in \{1,\ldots,m\}$.

Denote

$$\max f_0 = \lim_{x\to 0}\max_{t\in[0,T]} \frac{f(t,x)}{x}, \quad \min f_0 = \lim_{x\to 0}\min_{t\in[0,T]} \frac{f(t,x)}{x},$$

$$\max f_\infty = \lim_{x\to\infty}\max_{t\in[0,T]} \frac{f(t,x)}{x}, \quad \min f_\infty = \lim_{x\to\infty}\min_{t\in[0,T]} \frac{f(t,x)}{x}.$$

Theorem 2.10. *Suppose* $(B10)$-$(B12)$ *and*

$$\max f_0 = 0, \quad \min f_\infty = \infty, \quad \lim_{x\to 0} \frac{I_k(x)}{x} = 0, \quad k \in \{1,\ldots,m\}.$$

Then problem (2.14), (2.15) *has at least one positive solution.*

Proof. Let Ψ be as in the proof of Theorem 2.5. As in the proof of Theorem 2.3, we have that $\Psi : X \to X$ is completely continuous. Also, as in the proof of Theorem 2.5, it follows that $\Psi : K \to K$. Since

$$\max f_0 = 0, \quad \text{and} \quad \lim_{x\to 0} \frac{I_k(x)}{x} = 0,$$

for $\varepsilon < (B(\sigma(T)+m))^{-1}$, there exists a positive constant r such that

$$f(t,x) \le \varepsilon x, \quad I_k(x) \le \varepsilon x, \quad k \in \{1,\dots,m\}, \quad 0 < x \le r.$$

Define

$$\Omega_r = \{x \in X : \|x\| < r\}.$$

Then, for $x \in \Omega_r$, we get

$$\begin{aligned}
\Psi x(t) &= \int_0^{\sigma(T)} g(t,s) f(s,x(\sigma(s)))\Delta s + \sum_{k=1}^{m} g(t,t_k) I_k(x(t_k)) \\
&\le B\left(\int_0^{\sigma(T)} f(s,x(\sigma(s)))\Delta s + \sum_{k=1}^{m} I_k(x(t_k))\right) \\
&\le B\varepsilon\left(\int_0^{\sigma(T)} x(\sigma(s))\Delta s + \sum_{k=1}^{m} x(t_k)\right) \\
&\le B\varepsilon(\sigma(T)+m)\|x\|, \quad t \in [0,\sigma(T)],
\end{aligned}$$

whereupon,

$$\begin{aligned}
\|\Psi x\| &\le B\varepsilon(\sigma(T)+m)\|x\| \\
&\le B\frac{\sigma(T)+m}{B(\sigma(T)+m)}\|x\| \\
&= \|x\|.
\end{aligned}$$

Since $\min f_\infty = \infty$, for $N > (A\sigma(T)\delta)^{-1}$, we get that there exists a constant $R > \dfrac{r}{\delta}$ such that

$$f(t,x) \ge Nx \quad \text{for} \quad x \ge \delta R.$$

Define

$$\Omega_R = \{x \in X : \|x\| < R\}.$$

Then, for $x \in K \bigcap \partial\Omega_R$, we have

$$
\begin{aligned}
x(t) &\geq \delta\|x\| \\
&= \delta R, \quad t \in [0,\sigma(T)],
\end{aligned}
$$

and

$$
\begin{aligned}
\Psi x(t) &= \int_0^{\sigma(T)} g(t,s)f(s,x(\sigma(s)))\Delta s + \sum_{k=1}^{m} g(t,t_k)I_k(x(t_k)) \\
&\geq A\left(\int_0^{\sigma(T)} f(s,x(\sigma(s)))\Delta s + \sum_{k=1}^{m} I_k(x(t_k))\right) \\
&\geq A\int_0^{\sigma(T)} f(s,x(\sigma(s)))\Delta s \\
&\geq AN\int_0^{\sigma(T)} x(\sigma(s))\Delta s \\
&\geq AN\delta R\sigma(T) \\
&\geq R \\
&= \|x\|, \quad t \in [0,\sigma(T)],
\end{aligned}
$$

whereupon,

$$\|\Psi x\| \geq \|x\|.$$

Hence and applying the Guo-Krasnosel'skii fixed point Theorem 1.11, it follows that problem (2.14), (2.15) has at least one positive solution. This completes the proof.

Theorem 2.11. *Suppose* $(B10)$-$(B12)$ *and*

$$\max f_\infty = 0, \quad \min f_0 = \infty, \quad \lim_{x\to\infty} \frac{I_k(x)}{x} = 0, \quad k \in \{1,\ldots,m\}.$$

Then problem (2.14), (2.15) *has at least one positive solution.*

Proof. Let Ψ be as in the proof of Theorem 2.5. As in the proof of Theorem 2.3, we have that $\Psi : X \to X$ is completely continuous. Also, as in the proof of Theorem 2.5, it follows that $\Psi : K \to K$. Since

$$\max f_\infty = 0 \quad \text{and} \quad \lim_{x\to\infty} \frac{I_k(x)}{x} = 0,$$

for $\varepsilon \leq (B(\sigma(T)+m))^{-1}$, there exists a positive constant r such that

$$f(t,x) \leq \varepsilon x, \quad I_k(x) \leq \varepsilon x, \quad k \in \{1,\ldots,m\}, \quad x \geq \delta r.$$

Let

$$\Omega_r = \{x \in X : \|x\| < r\}.$$

Then, for $x \in K \bigcap \partial\Omega_r$, we have $\|x\| = r$ and

$$\begin{aligned}
\Psi x(t) &= \int_0^{\sigma(T)} g(t,s) f(s,x(\sigma(s)))\Delta s + \sum_{k=1}^{m} g(t,t_k) I_k(x(t_k)) \\
&\leq B\left(\int_0^{\sigma(T)} f(s,x(\sigma(s)))\Delta s + \sum_{k=1}^{m} I_k(x(t_k))\right) \\
&\leq B\varepsilon\left(\int_0^{\sigma(T)} x(\sigma(s))\Delta s + \sum_{k=1}^{m} x(t_k)\right) \\
&\leq B\varepsilon(\sigma(T)+m)\|x\| \\
&\leq \|x\|, \quad t \in [0,\sigma(T)],
\end{aligned}$$

whereupon,

$$\|\Psi x\| \leq \|x\|.$$

Because $\min f_0 = \infty$, for $N > (A\sigma(T)\delta)^{-1}$, there exists a constant $R < \delta r$ such that

$$f(t,x) \geq Nx \quad \text{for} \quad 0 < x \leq R.$$

Define

$$\Omega_R = \{x \in X : \|x\| < R\}.$$

Then, for $x \in K \bigcap \partial\Omega_R$, we have

$$\begin{aligned}
x(t) &\geq \delta\|x\| \\
&= \delta R, \quad t \in [0,\sigma(T)],
\end{aligned}$$

and

$$\begin{aligned}
\Psi x(t) &= \int_0^{\sigma(T)} g(t,s) f(s,x(\sigma(s)))\Delta s + \sum_{k=1}^{m} g(t,t_k) I_k(x(t_k)) \\
&\geq \int_0^{\sigma(T)} g(t,s) f(s,x(\sigma(s)))\Delta s \\
&\geq AN\int_0^{\sigma(T)} x(\sigma(s))\Delta s
\end{aligned}$$

$$\geq AN\sigma(T)\|x\|$$

$$\geq \|x\|, \quad t \in [0, \sigma(T)],$$

whereupon,

$$\|\Psi x\| \geq \|x\|.$$

Hence and applying the Guo-Krasnosel'skii fixed point Theorem 1.11, it follows that problem (2.14), (2.15) has at least one positive solution. This completes the proof.

Theorem 2.12. *Suppose* $(B10)$-$(B12)$ *and*

$$\min f_0 = \infty,$$

there exists a positive constant R *such that*

$$f(t,x) \leq \frac{R}{2B\sigma(T)}, \quad 0 < x \leq R, \quad t \in [0, \sigma(T)],$$

and

$$I_k(x) \leq \frac{|x|}{2mB}, \quad k \in \{1, \ldots, m\}, \quad x \in \mathbb{R}.$$

Then problem (2.14), (2.15) *has at least one positive solution.*

Proof. Let Ψ be as in the proof of Theorem 2.5. As in the proof of Theorem 2.3, we have that $\Psi : X \to X$ is completely continuous. Also, as in the proof of Theorem 2.5, it follows that $\Psi : K \to K$. Since $\min f_0 = \infty$, for $N > (A\sigma(T)\delta)^{-1}$, there exists a positive constant $r < \dfrac{R}{2BN\sigma(T)}$ such that

$$f(t,x) \geq Nx, \quad 0 < x \leq r, \quad t \in [0, \sigma(T)].$$

Define

$$\Omega_r = \{x \in X : \|x\| < r\}.$$

Then, for any $x \in K \bigcap \partial\Omega_r$, we have

$$x(t) \geq \delta\|x\|$$

$$= \delta r, \quad t \in [0, \sigma(T)],$$

and

$$\Psi x(t) = \int_0^{\sigma(T)} g(t,s) f(s, x(\sigma(s)))\Delta s + \sum_{k=1}^{m} g(t,t_k) I_k(x(t_k))$$

$$\geq \int_0^{\sigma(T)} g(t,s) f(s, x(\sigma(s)))\Delta s$$

$$
\begin{aligned}
&\geq A \int_0^{\sigma(T)} f(s,x(\sigma(s)))\Delta s \\
&\geq AN \int_0^{\sigma(T)} x(\sigma(s))\Delta s \\
&\geq AN\delta r \sigma(T) \\
&\geq r \\
&= \|x\|, \quad t \in [0,\sigma(T)],
\end{aligned}
$$

whereupon,

$$\|\Psi x\| \geq \|x\|.$$

Let

$$\Omega_R = \{x \in X : \|x\| < R\}.$$

Then, for $x \in K \bigcap \partial \Omega_R$, we obtain

$$
\begin{aligned}
\Psi x(t) &= \int_0^{\sigma(T)} g(t,s) f(s,x(\sigma(s)))\Delta s + \sum_{k=1}^{m} g(t,t_k) I_k(x(t_k)) \\
&\leq B\left(\int_0^{\sigma(T)} f(s,x(\sigma(s)))\Delta s + \sum_{k=1}^{m} I_k(x(t_k))\right) \\
&\leq B\left(\frac{R}{2B\sigma(T)}\sigma(T) + \frac{1}{2mB}\sum_{k=1}^{m} x(t_k)\right) \qquad (2.18) \\
&\leq B\left(\frac{R}{2B} + \frac{mR}{2mB}\right) \\
&= R \\
&= \|x\|, \quad t \in [0,\sigma(T)],
\end{aligned}
$$

whereupon,

$$\|\Psi x\| \leq \|x\|.$$

Hence and applying the Guo-Krasnosel'skii fixed point Theorem 1.11, it follows that problem (2.14), (2.15) has at least one positive solution. This completes the proof.

Theorem 2.13. *Suppose* $(B10)$-$(B12)$ *and*

$$\min f_\infty = \infty,$$

and there exists a positive constant R such that

$$f(t,s) \le \frac{R}{2B\sigma(T)}, \quad 0 < x \le R, \quad t \in [0, \sigma(T)],$$

$$I_k(x) \le \frac{|x|}{2mB}, \quad k \in \{1, \ldots, m\}, \quad x \in \mathbb{R}.$$

Then problem (2.14), (2.15) *has at least one positive solution.*

Proof. Let Ψ be as in the proof of Theorem 2.5. As in the proof of Theorem 2.3, we have that $\Psi : X \to X$ is completely continuous.Also, as in the proof of Theorem 2.5, it follows that $\Psi : K \to K$. Since $\min f_\infty = \infty$, for $M^* \ge (A\sigma(T)\delta)^{-1}$, there exists a positive constant $R^* > \frac{R}{\delta}$ such that

$$f(t,x) \ge M^* x, \quad x \ge \delta R^*, \quad t \in [0, \sigma(T)].$$

Let

$$\Omega_{R^*} = \{x \in X : \|x\| < R^*\}.$$

Then, for any $x \in K \bigcap \partial \Omega_{R^*}$, we have

$$x(t) \ge \delta \|x\|$$

$$= \delta R^*, \quad t \in [0, \sigma(T)],$$

and

$$\Psi x(t) = \int_0^{\sigma(T)} g(t,s) f(s, x(\sigma(s)))\Delta s + \sum_{k=1}^m g(t,t_k) I_k(x(t_k))$$

$$\ge \int_0^{\sigma(T)} g(t,s) f(s, x(\sigma(s)))\Delta s$$

$$\ge A \int_0^{\sigma(T)} f(s, x(\sigma(s)))\Delta s$$

$$\ge AM^* \int_0^{\sigma(T)} x(\sigma(s))\Delta s$$

$$\ge AM^* \delta R^* \sigma(T)$$

$$\ge R^*$$

$$= \|x\|, \quad t \in [0, \sigma(T)],$$

whereupon,

$$\|\Psi x\| \geq \|x\|.$$

Hence and applying the Guo-Krasnosel'skii fixed point Theorem 1.11, it follows that problem (2.14), (2.15) has at least one positive solution. This completes the proof.

2.4 Periodic Boundary Value Problems for First Order Nonlinear Dynamic Equations with Impulses-II

Definition 2.5. We say that a time scale $\mathbb{T}$ is periodic if there exists a $p > 0$ such that if $t \in \mathbb{T}$, then $t \pm p \in \mathbb{T}$. The smallest p is called the period of the time scale.

Example 2.3. Let $\mathbb{T} = h\mathbb{Z}$, where $h > 0$. Then $\mathbb{T}$ is a periodic time scale with period h.

Remark 2.2. If $\mathbb{T}$ is a periodic time scale with period p, then

$$\sigma(t \pm np) = \sigma(t) \pm np, \quad t \in \mathbb{T}.$$

Hence,

$$\begin{aligned}
\mu(t \pm np) &= \sigma(t \pm np) - (t \pm np) \\
&= \sigma(t) \pm np - t \mp np \\
&= \sigma(t) - t \\
&= \mu(t), \quad t \in \mathbb{T}.
\end{aligned}$$

Definition 2.6. Let $\mathbb{T} \neq \mathbb{R}$ be a periodic time scale with period p. We say that a function $f : \mathbb{T} \to \mathbb{R}$ is periodic with period T if there exists an $n \in \mathbb{N}$ such that $T = np$ and $f(t \pm T) = f(t)$ for any $t \in \mathbb{T}$ and T is the smallest number such that $f(t \pm T) = f(t)$.

Below, we suppose that $\mathbb{T}$ is a periodic time scale with period p such that $0, t_j \in \mathbb{T}$, $j \in \{1, \ldots, m\}$, $T = lp$ for some $l \in \mathbb{N}$, $0 < t_j < t_{j+1}$ and each t_j is dense in $\mathbb{T}$, $j \in \{1, \ldots, m\}$. In this section we will investigate the following BVP

$$
\begin{aligned}
y^{\Delta}(t)+a(t)y^{\sigma}(t) &= f(t,y(t)), \quad t\in\mathbb{T},\\
y(t_j^{+})-y(t_j^{-}) &= I(t_j,y(t_j)), \quad j\in\{1,\dots,m\},\\
y(0) &= y(T),
\end{aligned}
\tag{2.19}
$$

where

(B13) $a\in\mathscr{R}$, $a(t+T)=a(t)$, $t\in\mathbb{T}$, $\eta=e_{\ominus a}(T,0)<1$,
(B14) $f\in\mathscr{C}(\mathbb{T}\times\mathbb{R})$, $f(t+T,y(t+T))=f(t,y(t))$, $t\in\mathbb{T}$.

Define $t_{m+1}=T$, $J_0=[0,t_1]$, $J_j=(t_j,t_{j+1}]$, $j\in\{1,\dots,m\}$,

$$
PC_p=\Big\{y:\mathbb{T}\to\mathbb{R},\quad y(t\pm T)=y(t),\quad y\in\mathscr{C}(J_j),
$$

$$
\exists y(t_j^{\pm}),\quad y(t_j^{-})=y(t_j),\quad j\in\{1,\dots,m\}\Big\},
$$

$$
PC_p^1=\{y:\mathbb{T}\to\mathbb{R},\quad y(t\pm T)=y(t),\quad y\in\mathscr{C}^1(J_j),\quad j\in\{1,\dots,m\}\}.
$$

The set PC_p is a Banach space endowed with the norm

$$
\|y\|=\max_{j\in\{1,\dots,m\}}\{\sum_{t\in J_j}|y(t)|\}.
$$

Lemma 2.11. *Let $h\in PC_p$. Then the function $y\in PC_p^1$ is a solution of the BVP*

$$
\begin{aligned}
y^{\Delta}(t)+a(t)y^{\sigma}(t) &= h(t), \quad t\in\mathbb{T},\\
y(t_j^{+})-y(t_j^{-}) &= I(t_j,y(t_j)), \quad j\in\{1,\dots,m\},\\
y(0) &= y(T)
\end{aligned}
\tag{2.20}
$$

if and only if $y\in PC_p$ and y satisfies the integral equation

$$
y(t)=\int_0^T G(t,s)h(s)\Delta s+\sum_{j=1}^{m}G(t,t_j)I(t_j,y(t_j)),
\tag{2.21}
$$

where

$$
G(t,s)=\frac{1}{1-\eta}\begin{cases} e_{\ominus a}(t,s) & 0\le s\le t\le T\\ \eta e_{\ominus a}(t,s) & 0\le t<s\le T.\end{cases}
$$

Proof. Note that $y\in PC_p^1$ is a solution of the impulsive dynamic equation

$$\begin{aligned} y^{\Delta}(t)+a(t)y^{\sigma}(t) &= h(t), \quad t\in\mathbb{T},\\ y(t_j^+)-y(t_j^-) &= I(t_j,y(t_j)), \quad j\in\{1,\dots,m\}, \end{aligned} \tag{2.22}$$

if and only if $y\in PC_p$ and it satisfies the integral equation

$$y(t)=\int_0^t e_{\ominus a}(t,s)h(s)\Delta s+e_{\ominus a}(t,0)y(0)+\sum_{\{j:t_j<t\}} e_{\ominus a}(t,t_j)I(t_j,y(t_j)), \tag{2.23}$$

$t\in[0,T]$.

1. Let $y\in PC_p^1$ be a solution of the BVP (2.20). Then y satisfies (2.22) and hence, y satisfies (2.23). From here,

$$\begin{aligned} y(0) &= y(T)\\ &= \int_0^T e_{\ominus a}(T,s)h(s)\Delta s+\eta y(0)+\sum_{j=1}^m e_{\ominus a}(T,t_j)I(t_j,y(t_j)), \end{aligned}$$

whereupon,

$$(1-\eta)y(0)=\int_0^T e_{\ominus a}(T,s)h(s)\Delta s+\sum_{j=1}^m e_{\ominus a}(T,t_j)I(t_j,y(t_j))$$

or

$$y(0)=\frac{1}{1-\eta}\left(\int_0^T e_{\ominus a}(T,s)h(s)\Delta s+\sum_{j=1}^m e_{\ominus a}(T,t_j)I(t_j,y(t_j))\right).$$

From the last expression and from (2.23), we get

$$\begin{aligned} y(t) &= \int_0^t e_{\ominus a}(t,s)h(s)\Delta s\\ &\quad+\frac{e_{\ominus a}(t,0)}{1-\eta}\left(\int_0^T e_{\ominus a}(T,s)h(s)\Delta s+\sum_{j=1}^m e_{\ominus a}(T,t_j)I(t_j,y(t_j))\right)\\ &\quad+\sum_{\{j:t_j<t\}} e_{\ominus a}(t,t_j)I(t_j,y(t_j))\\ &= \int_0^t e_{\ominus a}(t,s)h(s)\Delta s\\ &\quad+\frac{e_{\ominus a}(t,0)}{1-\eta}\left(\int_0^t e_{\ominus a}(T,s)h(s)\Delta s+\sum_{\{j:t_j<t\}} e_{\ominus a}(T,t_j)I(t_j,y(t_j))\right. \end{aligned}$$

$$
\begin{aligned}
&+\int_t^T e_{\ominus a}(T,s)h(s)\Delta s+\sum_{\{j:t_j\geq t\}} e_{\ominus a}(T,t_j)I(t_j,y(t_j))\Big)\\
&+\sum_{\{j:t_j<t\}} e_{\ominus a}(t,t_j)I(t_j,y(t_j))\\
=&\int_0^t e_{\ominus a}(t,s)h(s)\Delta s+\frac{\eta}{1-\eta}\int_0^t e_{\ominus a}(t,s)h(s)\Delta s\\
&+\frac{\eta}{1-\eta}\sum_{\{j:t_j<t\}} e_{\ominus a}(t,t_j)I(t_j,y(t_j))\\
&+\frac{e_{\ominus a}(t,0)}{1-\eta}\int_t^T e_{\ominus a}(T,s)h(s)\Delta s+\frac{e_{\ominus a}(t,0)}{1-\eta}\sum_{\{j:t_j\geq t\}} e_{\ominus a}(T,t_j)I(t_j,y(t_j))\\
&+\sum_{\{j:t_j<t\}} e_{\ominus a}(t,t_j)I(t_j,y(t_j))\\
=&\frac{1}{1-\eta}\int_0^t e_{\ominus a}(t,s)h(s)\Delta s+\frac{1}{1-\eta}\sum_{\{j:t_j<t\}} e_{\ominus a}(t,t_j)I(t_j,y(t_j))\\
&+\frac{\eta}{1-\eta}\int_t^T e_{\ominus a}(t,s)h(s)\Delta s+\frac{\eta}{1-\eta}\sum_{\{j:t_j\geq t\}} e_{\ominus a}(t,t_j)I(t_j,y(t_j))\\
=&\int_0^T G(t,s)h(s)\Delta s+\sum_{j=1}^m G(t,t_j)I(t_j,y(t_j)),\quad t\in[0,T],
\end{aligned}
$$

i.e., we get (2.21).

2. Let $y\in PC_p$ satisfies (2.21). Then $y\in PC_p^1$ and y satisfies (2.22). Next,

$$
\begin{aligned}
y(t)=&\int_0^t G(t,s)h(s)\Delta s+\int_t^T G(t,s)h(s)\Delta s\\
&+\sum_{\{j:t_j<t\}} G(t,t_j)I(t_j,y(t_j))+\sum_{\{j:t_j\geq t\}} G(t,t_j)I(t_j,y(t_j))\\
=&\frac{1}{1-\eta}\int_0^t e_{\ominus a}(t,s)h(s)\Delta s+\frac{\eta}{1-\eta}\int_t^T e_{\ominus a}(t,s)h(s)\Delta s\\
&+\frac{1}{1-\eta}\sum_{\{j:t_j\leq t\}} e_{\ominus a}(t,t_j)I(t_j,y(t_j))\\
&+\frac{\eta}{1-\eta}\sum_{\{j:t_j>t\}} e_{\ominus a}(t,t_j)I(t_j,y(t_j)),\quad t\in[0,T],
\end{aligned}
$$

whereupon,

$$
\begin{aligned}
y(T) &= \frac{1}{1-\eta}\int_0^T e_{\ominus a}(T,s)h(s)\Delta s + \frac{1}{1-\eta}\sum_{j=1}^m e_{\ominus a}(T,t_j)I(t_j,y(t_j)),\\
y(0) &= \frac{\eta}{1-\eta}\int_0^T e_{\ominus a}(0,s)h(s)\Delta s + \frac{\eta}{1-\eta}\sum_{j=1}^m e_{\ominus a}(0,t_j)I(t_j,y(t_j))\\
&= \frac{1}{1-\eta}\int_0^T e_{\ominus a}(T,s)h(s)\Delta s + \frac{1}{1-\eta}\sum_{j=1}^m e_{\ominus a}(T,t_j)I(t_j,y(t_j))\\
&= y(T).
\end{aligned}
$$

This completes the proof.

Let A be a positive constant such that

$$\sup_{s,t\in[0,T]} G(t,s) \le A.$$

We have

$$0 \le G(t,s) \le A, \quad t,s \in [0,T].$$

For $y \in PC_p$, define the operator

$$\Phi y(t) = \int_0^T G(t,s)f(s,y(s))\Delta s + \sum_{j=1}^m G(t,t_j)I(t_j,y(t_j)), \quad t \in [0,T].$$

Theorem 2.14. *Let* $(B13)$ *and* $(B14)$ *hold. Suppose that there exist positive constants* L, E_j, $j \in \{1,\dots,m\}$, *such that*

$$|f(t,y)-f(t,x)| \le L|y-x|, \quad t\in\mathbb{T}, \quad x,y\in\mathbb{R},$$

$$|I(t_j,y(t_j)) - I(t_j,x(t_j))| \le E_j|y(t_j)-x(t_j)|, \quad y,x\in PC_p, \quad j\in\{1,\dots,m\},$$

and

$$A(LT + \sum_{j=1}^m E_j) < 1.$$

Then problem (2.19) *has a unique solution in* PC_p.

Proof. Note that $\Phi : PC_p \to PC_p$. For $x,y \in PC_p$, we get

$$
\begin{aligned}
|\Phi y(t) - \Phi x(t)| = \Bigg| &\int_0^T G(t,s)f(s,y(s))\Delta s + \sum_{j=1}^m G(t,t_j)I(t_j,y(t_j))\\
&- \int_0^T G(t,s)f(s,x(s))\Delta s - \sum_{j=1}^m G(t,t_j)I(t_j,x(t_j))\Bigg|
\end{aligned}
$$

$$
\begin{aligned}
&= \left| \int_0^T G(t,s)(f(s,y(s)) - f(s,x(s)))\Delta s \right. \\
&\quad \left. + \sum_{j=1}^{m} G(t,t_j)(I(t_j,y(t_j)) - I(t_j,x(t_j))) \right| \\
&\leq \left| \int_0^T G(t,s)(f(s,y(s)) - f(s,x(s)))\Delta s \right| \\
&\quad + \left| \sum_{j=1}^{m} G(t,t_j)(I(t_j,y(t_j)) - I(t_j,x(t_j))) \right| \\
&\leq \int_0^T G(t,s)|f(s,y(s)) - f(s,x(s))|\Delta s \\
&\quad + \sum_{j=1}^{m} G(t,t_j)|I(t_j,y(t_j)) - I(t_j,x(t_j))| \\
&\leq A\Bigg(\int_0^T |f(s,y(s)) - f(s,x(s))|\Delta s \\
&\quad + \sum_{j=1}^{m} |I(t_j,y(t_j)) - I(t_j,x(t_j))| \Bigg) \\
&\leq A\left(L\int_0^T |y(s) - x(s)|\Delta s + \sum_{j=1}^{m} E_j|y(t_j) - x(t_j)| \right) \\
&\leq A\left(LT + \sum_{j=1}^{m} E_j \right)\|y - x\|, \quad t \in [0,T],
\end{aligned}
$$

whereupon,

$$
\begin{aligned}
\|\Phi y - \Phi x\| &\leq A\left(LT + \sum_{j=1}^{m} E_j \right)\|y - x\| \\
&< \|y - x\|.
\end{aligned}
$$

Hence and by the contraction mapping principle, it follows that the operator Φ has a unique fixed point in PC_p. This completes the proof.

Theorem 2.15. *Let* $(B13)$ *and* $(B14)$ *hold. Assume that there exist functions* g_1, g_2, g_3, $g_4 : PC_p \to PC_p$ *with*

$$
\alpha_1 = \int_0^T g_1(s)\Delta s < \infty,
$$

$$\beta_1 = \int_0^T g_2(s)\Delta s < \infty,$$

$$\alpha_2 = \sum_{j=1}^{m} g_3(t_j) < \infty,$$

$$\beta_2 = \sum_{j=1}^{m} g_4(t_j) < \infty,$$

such that

$$|f(t,y)| \leq g_1(t) + g_2(t)|y|, \quad t \in \mathbb{T}, \quad y \in \mathbb{R},$$

$$|I(t,y)| \leq g_3(t) + g_4(t)|y|, \quad t \in \mathbb{T}, \quad y \in \mathbb{R}.$$

Suppose that

$$A(\beta_1 + \beta_2) < 1.$$

Then problem (2.19) *has at least one solution in* PC_p.

Proof. Define

$$H(\Phi) = \{y \in PC_p : y(t) = \mu \Phi y(t), \quad t \in \mathbb{T}, \quad \mu \in (0,1)\}.$$

For $y \in H(\Phi)$, we have

$$\begin{aligned}
|y(t)| &= \left| \mu \int_0^T G(t,s) f(s,y(s))\Delta s + \mu \sum_{j=1}^{m} G(t,t_j) I(t_j, y(t_j)) \right| \\
&\leq \mu \int_0^T G(t,s)|f(s,y(s))|\Delta s + \mu \sum_{j=1}^{m} G(t,t_j)|I(t_j,y(t_j))| \\
&\leq \mu \Bigg(\int_0^T G(t,s) g_1(s)\Delta s + \int_0^T G(t,s) g_2(s)|y(s)|\Delta s \\
&\quad + \sum_{j=1}^{m} G(t,t_j) g_3(t_j) + \sum_{j=1}^{m} G(t,t_j) g_4(t_j)|y(t_j)| \Bigg) \\
&\leq \mu A(\alpha_1 + \beta_1\|y\| + \alpha_2 + \beta_2\|y\|) \\
&= \mu A(\alpha_1 + \alpha_2) + \mu A(\beta_1 + \beta_2)\|y\|, \quad t \in [0,T],
\end{aligned}$$

whereupon,

$$\|y\| \leq \mu A(\alpha_1 + \alpha_2) + \mu A(\beta_1 + \beta_2)\|y\|$$

and

$$(1 - \mu A(\beta_1 + \beta_2))\|y\| \leq \mu A(\alpha_1 + \alpha_2),$$

or

$$\|y\| \leq \frac{\mu A(\alpha_1 + \alpha_2)}{1 - \mu A(\beta_1 + \beta_2)}.$$

Therefore, $H(\Phi)$ is bounded. Hence and by Schaefer's fixed point Theorem 1.3, it follows that problem (2.19) has at least one fixed point in PC_p. This completes the proof.

Theorem 2.16. *Let* $(B13)$ *and* $(B14)$ *hold and*

$$\lim_{|y|\to\infty} \frac{f(t,y)}{y} = 0 \quad \textit{uniformly},$$

$$\lim_{|y|\to\infty} \frac{I(t,y)}{y} = 0 \quad \textit{uniformly}.$$

Then problem (2.19) *has at least one solution in* PC_p.

Proof. Suppose that the set

$$H(\Phi) = \{y \in PC_p : y(t) = \mu \Phi y(t), \quad t \in \mathbb{T}, \quad \mu \in (0,1)\}$$

is unbounded. Then there exist sequences $\{y_k\}_{k\in\mathbb{N}}$ and $\{\mu_k\}_{k\in\mathbb{N}}$ with $\|y_k\| > k$ and $\mu_k \in (0,1)$, such that

$$y_k(t) = \mu_k \int_0^T G(t,s) f(s, y_k(s)) \Delta s + \mu_k \sum_{j=1}^m G(t,t_j) I(t_j, y_k(t_j)), \quad t \in [0,T], \quad k \in \mathbb{N}.$$

Define

$$v_k(t) = \frac{y_k(t)}{\|y_k\|}, \quad k \in \mathbb{N}, \quad t \in [0,T].$$

Then $\|v_k\| = 1$, $k \in \mathbb{N}$, and

$$\begin{aligned} v_k(t) &= \mu_k \int_0^T G(t,s) \frac{f(s,y_k(s))}{\|y_k\|} \Delta s + \mu_k \sum_{j=1}^m G(t,t_j) \frac{I(t_j, y_k(t_j))}{\|y_k\|} \\ &\to 0, \quad \text{as} \quad k \to \infty, \quad t \in [0,T]. \end{aligned}$$

This contradicts with $\|v_k\| = 1$, $k \in \mathbb{N}$. Therefore, $H(\Phi)$ is bounded. From here and from Schaefer's fixed point Theorem 1.3, it follows that problem (2.19) has at least one solution in PC_p. This completes the proof.

Exercise 2.6. Assume $(B13)$, $(B14)$ and f and I are bounded. Prove that problem (2.19) has at least one solution in PC_p.

2.5 Extremal Solutions of Periodic Boundary Value Problems for First Order Impulsive Integro-Dynamic Equations of Mixed Type

Suppose that $\mathbb{T}$ is a time scale with forward jump operator and delta differentiation operator σ and Δ, respectively. Let $0, t_j, T \in \mathbb{T}$, $j \in \{1, \ldots, m\}$. In this section we will investigate the following PBVP

$$\begin{aligned} y^{\Delta}(t) &= f(t, y(t), Ty(t), Sy(t)), \quad t \in J \backslash \{t_j\}_{j=1}^{m}, \\ y(t_j^+) &= y(t_j^-) + I_j(y(t_j^-)), \quad j \in \{1, \ldots, m\}, \\ y(0) &= y(T), \end{aligned} \tag{2.24}$$

where

$$0 < t_1 < \ldots < t_m < t_{m+1} = T,$$

$J = [0, T]$, t_j, $j \in \{1, \ldots, m\}$, are right-dense,

$$Ty(t) = \int_0^t k(t,s)y(s)\Delta s,$$

$$Sy(t) = \int_0^t h(t,s)y(s)\Delta s, \quad t \in J,$$

(B15) $k \in \mathscr{C}(D)$, $k \geq 0$ on D,

$$D = \{(t,s) \in J \times J : t \geq s\}, \quad k_0 = \max_{(t,s) \in D} k(t,s),$$

(B16) $h \in \mathscr{C}(J \times J)$, $h \geq 0$ on $J \times J$, $h_0 = \max_{(t,s) \in J \times J} h(t,s)$,

(B17) $f \in \mathscr{C}(J \times \mathbb{R} \times \mathbb{R} \times \mathbb{R})$,

(B18) $I_j \in \mathscr{C}(\mathbb{R})$, $j \in \{1, \ldots, m\}$.

We will start with the following useful impulsive dynamic inequalities.

Theorem 2.17. *Assume that the sequence* $\{t_k\}_{k \in \mathbb{N}}$ *satisfies*

$$0 \leq t_0 < t_1 < t_2 < \ldots < t_k < \ldots, \quad \lim_{k \to \infty} t_k = \infty,$$

$m \in PC^1(\mathbb{T})$ *is right-dense continuous at* t_k, $k \in \mathbb{N}$,

$$\inf_{t \in J}(\mu(t)p(t)) > -1,$$

and

$$\begin{aligned} m^{\Delta}(t) &\leq p(t)m(t) + q(t), \quad t \in J \backslash \{t_k\}_{k \in \mathbb{N}}, \\ m(t_k^+) &\leq d_k m(t_k^-) + b_k, \quad k \in \mathbb{N}, \end{aligned} \tag{2.25}$$

where $p, q \in \mathscr{C}(\mathbb{T})$, $d_k \geq 1$ *and* b_k *are real constants,* $k \in \mathbb{N}$. *Then*

$$m(t) \leq m(t_0) \prod_{t_0<t_k<t} d_k e_p(t,t_0) + \sum_{t_0<t_k<t} \left(\prod_{t_k<t_j<t} d_j e_p(t,t_k) \right) b_k + \int_{t_0}^{t} \prod_{s<t_k<t} d_k e_p(t,\sigma(s)) q(s) \Delta s \tag{2.26}$$

for any $t \geq t_0$.

Proof. For $t \in [t_0, t_1]$, by the Gronwall inequality, we get

$$m(t) \leq m(t_0) e_p(t,t_0) + \int_{t_0}^{t} e_p(t,\sigma(s)) q(s) \Delta s.$$

Therefore, the inequality (2.26) is true for any $t \in [t_0, t_1]$. Now, assume that the inequality (2.26) holds for $t \in [t_0, t_n]$ for some $n \in \mathbb{N}$. Let $t \in (t_n, t_{n+1}]$. Then, using the Gronwall inequality, we get

$$m(t) \leq m(t_n) e_p(t,t_n) + \int_{t_n}^{t} e_p(t,\sigma(s)) q(s) \Delta s.$$

Next, by (2.26), using (2.25), we have

$$\begin{aligned}
m(t) &\leq (d_n m(t_n^-) + b_n) e_p(t,t_n) + \int_{t_n}^{t} e_p(t,\sigma(s)) q(s) \Delta s \\
&= d_n e_p(t,t_n) m(t_n^-) \\
&\quad + b_n e_p(t,t_n) + \int_{t_n}^{t} e_p(t,\sigma(s)) q(s) \Delta s \\
&\leq d_n e_p(t,t_n) \Bigg(m(t_0) \prod_{t_0<t_k<t_n} d_k e_p(t_n,t_0) \\
&\quad + \sum_{t_0<t_k<t_n} \left(\prod_{t_k<t_j<t_n} d_j e_p(t,t_k) \right) b_k \\
&\quad + \int_{t_0}^{t_n} \prod_{s<t_k<t_n} d_k e_p(t_n,\sigma(s)) q(s) \Delta s \Bigg) \\
&\quad + b_n e_p(t,t_n) + \int_{t_n}^{t} e_p(t,\sigma(s)) q(s) \Delta s \\
&\leq m(t_0) \prod_{t_0<t_k<t} d_k e_p(t,t_0)
\end{aligned}$$

$$+ \sum_{t_0<t_k<t} \left(\prod_{t_k<t_j<t} d_j e_p(t,t_k) \right) b_k$$

$$+ \int_{t_0}^{t} \prod_{s<t_k<t} d_k e_p(t,\sigma(s)) q(s) \Delta s.$$

Then (2.26) is true for $t \in [t_0, t_{n+1}]$. By induction, we obtain that the inequality (2.26) is true for any $t \geq t_0$. This completes the proof.

As in the above, one can prove the following impulsive dynamic inequality.

Theorem 2.18. *Assume that the sequence $\{t_k\}_{k\in\mathbb{N}}$ satisfies*

$$0 \leq t_0 < t_1 < t_2 < \ldots < t_k < \ldots, \quad \lim_{k\to\infty} t_k = \infty,$$

$m \in PC^1(\mathbb{T})$ is right-dense continuous at t_k, $k \in \mathbb{N}$,

$$\inf_{t\in J}(\mu(t)p(t)) > -1,$$

and

$$m^\Delta(t) \geq p(t)m(t) + q(t), \quad t \in J \backslash \{t_k\}_{k\in\mathbb{N}},$$

$$m(t_k^+) \geq d_k m(t_k^-) + b_k, \quad k \in \mathbb{N},$$

where $p, q \in \mathscr{C}(\mathbb{T})$, $d_k \in (0,1)$ and b_k are real constants, $k \in \mathbb{N}$. Then

$$m(t) \geq m(t_0) \prod_{t_0<t_k<t} d_k e_p(t,t_0) + \sum_{t_0<t_k<t} \left(\prod_{t_k<t_j<t} d_j e_p(t,t_k) \right) b_k$$

$$+ \int_{t_0}^{t} \prod_{s<t_k<t} d_k e_p(t,\sigma(s)) q(s) \Delta s$$

for any $t \geq t_0$.

Let

$$\Omega = PC(J) \bigcap \mathscr{C}^1(J \backslash \{t_j\}_{j=1}^m).$$

Theorem 2.19. *Let $t_0 = 0$ and $y \in \Omega$ satisfies*

$$y^\Delta(t) \geq -a(t)y(t) - b(t)Ty(t) - c(t)Sy(t), \quad t \in J \backslash \{t_j\}_{j=1}^m,$$

$$y(t_j^+) \geq y(t_j) - L_j y(t_j), \quad j \in \{1, \ldots, m\},$$

$$y(0) \geq y(T),$$

where $a, b, c \in \mathscr{C}(J)$, $a, b, c \geq 0$ on J, a is not identically vanishing, $\sup_{t\in J}(\mu(t)a(t)) < 1$, $0 \leq L_j \leq 1$, $j \in \{1, \ldots, m\}$. If

$$(Bk_0 + Ch_0)e_{\ominus(-a)}(T,0) \leq \frac{\left(\prod_{0<t_k<T}(1-L_k)\right)^2}{\int_0^T \prod_{s<t_k<T}(1-L_k)\Delta s}, \tag{2.27}$$

with

$$B = \sup_{t\in J}\left(b(t)\int_0^t e_{\ominus(-a)}(\sigma(t),s)\Delta s\right),$$

$$C = \sup_{t\in J}\left(c(t)\int_0^t e_{\ominus(-a)}(\sigma(t),s)\Delta s\right),$$

then $y(t) \geq 0$, $t \in J$.

Proof. Note that

$$\ominus(-a)(t) = \frac{a(t)}{1-\mu(t)a(t)}, \quad t \in J,$$

and

$$e_{\ominus(-a)}(t,0) = e^{\int_0^t \frac{1}{\mu(\tau)}\log\frac{1}{1-\mu(\tau)a(\tau)}\Delta\tau}$$

$$\geq 1, \quad t \in J.$$

Let

$$p(t) = y(t)e_{\ominus(-a)}(t,0), \quad t \in J.$$

Then $p \in \Omega$ and

$$\begin{aligned}
p^\Delta(t) &= y^\Delta(t)e_{\ominus(-a)}(\sigma(t),0) \\
&\quad + y(t)\left(\ominus(-a)\right)(t)e_{\ominus(-a)}(t,0) \\
&= y^\Delta(t)e_{\ominus(-a)}(\sigma(t),0) \\
&\quad + \frac{y(t)a(t)}{1-\mu(t)a(t)}e_{\ominus(-a)}(t,0) \\
&= y^\Delta(t)e_{\ominus(-a)}(\sigma(t),0) + y(t)a(t)e_{\ominus(-a)}(\sigma(t),0) \\
&= \left(y^\Delta(t) + a(t)y(t)\right)e_{\ominus(-a)}(\sigma(t),0) \\
&\geq -b(t)e_{\ominus(-a)}(\sigma(t),0)\int_0^t k(t,s)y(s)\Delta s
\end{aligned}$$

$$
\begin{aligned}
&-c(t)e_{\ominus(-a)}(\sigma(t),0)\int_0^t h(t,s)y(s)\Delta s\\
&= -b(t)e_{\ominus(-a)}(\sigma(t),0)\int_0^t k(t,s)e_{\ominus(-a)}(0,s)p(s)\Delta s\\
&\quad -c(t)e_{\ominus(-a)}(\sigma(t),0)\int_0^t h(t,s)e_{\ominus(-a)}(0,s)p(s)\Delta s\\
&= -b(t)\int_0^t k(t,s)e_{\ominus(-a)}(\sigma(t),s)p(s)\Delta s\\
&\quad -c(t)\int_0^t h(t,s)e_{\ominus(-a)}(\sigma(t),s)p(s)\Delta s, \quad t \in J\backslash\{t_j\}_{j=1}^m,
\end{aligned}
$$

$$
\begin{aligned}
p(t_j^+)-p(t_j) &= y(t_j^+)e_{\ominus(-a)}(t_j^+,0) - y(t_j)e_{\ominus(-a)}(t_j,0)\\
&= \left(y(t_j^+)-y(t_j)\right)e_{\ominus(-a)}(t_j,0)\\
&\geq -L_j y(t_j)e_{\ominus(-a)}(t_j,0)\\
&= -L_j p(t_j), \quad j \in \{1,\ldots,m\},\\
p(0) &= y(0)\\
&\geq y(T)e_{\ominus(-a)}(T,0)e_{-a}(T,0)\\
&= e_{-a}(T,0)p(T).
\end{aligned}
$$

Thus, p satisfies the following impulsive dynamic inequality

$$
\begin{aligned}
p^{\Delta}(t) \quad &\geq -b(t)\int_0^t k(t,s)e_{\ominus(-a)}(\sigma(t),s)p(s)\Delta s\\
&\quad -c(t)\int_0^t h(t,s)e_{\ominus(-a)}(\sigma(t),s)p(s)\Delta s, \quad t \in J\backslash\{t_j\}_{j=1}^m,\\
p(t_j^+)-p(t_j) &\geq -L_j p(t_j), \quad j \in \{1,\ldots,m\},\\
p(0) \quad &\geq e_{-a}(T,0)p(T).
\end{aligned}
\tag{2.28}
$$

Assume that

$$
p(t) \geq 0, \quad t \in J,
$$

is not true. Then we have the following two cases.

1. There exists a $t_1^* \in J$ such that $p(t_1^*) < 0$ and $p(t) \leq 0$, $t \in J$. Hence,

$$p^{\Delta}(t) \geq 0, \quad t \in J \backslash \{t_j\}_{j=1}^{m},$$

$$p(t_j^+) - p(t_j) \geq 0, \quad j \in \{1, \ldots, m\}.$$

Therefore, p is non-decreasing on J. Thus,

$$\begin{aligned} p(0) &\leq p(t_1^*) \\ &< 0, \\ p(0) &\leq p(T) \\ &\leq 0. \end{aligned}$$

Hence,

$$\begin{aligned} p(0) &\leq p(T) \\ &\leq e_{\ominus(-a)}(T, 0) p(0), \end{aligned}$$

or

$$\left(1 - e_{\ominus(-a)}(T, 0)\right) > 0,$$

which is a contradiction.

2. There exist $t_1^*, t_2^* \in J$ such that

$$p(t_1^*) < 0 \quad \text{and} \quad p(t_2^*) > 0.$$

Let

$$\lambda = \sup_{t \in J} p(t).$$

Then $\lambda > 0$ and there exists some $j \in \{1, \ldots, m\}$ such that $t_j < t_0^* \leq t_{j+1}$ and

$$p(t_0^*) = \lambda \quad \text{or} \quad p(t_j^+) = \lambda.$$

Without loss of generality, assume that $p(t_0^*) = \lambda$. By (2.28), we get

$$\begin{aligned} p^{\Delta}(t) &\geq -\lambda k_0 b(t) \int_0^t e_{\ominus(-a)}(\sigma(t), s) \Delta s - \lambda h_0 c(t) \int_0^t e_{\ominus(-a)}(\sigma(t), s) \Delta s \\ &\geq -\lambda \left(B k_0 + C h_0\right), \quad t \in J \backslash \{t_k\}_{k=1}^{m}. \end{aligned}$$

For $t \in [t_0^*, T]$, $k \in \{j+1, \ldots, m\}$, we have

$$p^{\Delta}(t) \geq -\lambda(Bk_0 + Ch_0), \quad t \neq t_k,$$

$$p(t_k^+) \geq (1 - L_k)p(t_k).$$

Hence and by Theorem 2.18, it follows that

$$p(t) \geq p(t_0^*) \prod_{t_0^* < t_k < t} (1 - L_k) + \int_{t_0^*}^{t} \prod_{s < t_k < t} (1 - L_k)(-\lambda(Bk_0 + Ch_0))\,\Delta s, \tag{2.29}$$

$t \in [t_0^*, T]$. In particular,

$$p(T) \geq \lambda \prod_{t_0^* < t_k < T} (1 - L_k) - \lambda\,(Bk_0 + Ch_0) \int_{t_0^*}^{T} \prod_{s < t_k < T} (1 - L_k)\Delta s. \tag{2.30}$$

If $p(T) < 0$, we get

$$0 > \lambda \prod_{t_0^* < t_k < T} (1 - L_k) - \lambda\,(Bk_0 + Ch_0) \int_{t_0^*}^{T} \prod_{s < t_k < T} (1 - L_k)\Delta s$$

or

$$\lambda \prod_{t_0^* < t_k < T} (1 - L_k) < \lambda\,(Bk_0 + Ch_0) \int_{t_0^*}^{T} \prod_{s < t_k < T} (1 - L_k)\Delta s,$$

or

$$\begin{aligned} Bk_0 + Ch_0 &> \frac{\prod_{t_0^* < t_k < T} (1 - L_k)}{\int_{t_0^*}^{T} \prod_{s < t_k < T} (1 - L_k)\Delta s} \\ &\geq \frac{\prod_{0 < t_k < T} (1 - L_k)}{\int_{0}^{T} \prod_{s < t_k < T} (1 - L_k)\Delta s}, \end{aligned}$$

whereupon,

$$\begin{aligned} (Bk_0 + Ch_0)e_{\ominus(-a)}(T, 0) &\geq Bk_0 + Ch_0 \\ &> \frac{\prod_{0 < t_k < T} (1 - L_k)}{\int_{0}^{T} \prod_{s < t_k < T} (1 - L_k)\Delta s}. \end{aligned}$$

This is a contradiction. Therefore, $p(T) \geq 0$ and from here,

$$p(0) \geq p(T)e_{-a}(T,0)$$

$$\geq 0.$$

Consequently, $0 < t_1^* < T$. Let $t_r < t_1^* \leq t_{r+1}$ for some $r \in \{1,\ldots,m\}$. Assume that $t_0^* < t_1^*$. So, $j \geq r$. By (2.29), we obtain

$$\begin{aligned} 0 &> p(t_1^*) \\ &\geq \lambda \prod_{t_0^* < t_k < t_1^*} (1-L_k) + \int_{t_0^*}^{t_1^*} \prod_{s<t_k<t_1^*} (1-L_k)(-\lambda(Bk_0+Ch_0))\Delta s, \end{aligned}$$

whereupon,

$$\lambda(Bk_0+Ch_0)\int_{t_0^*}^{t_1^*} \prod_{s<t_k<t_1^*}(1-L_k)\Delta s \geq \lambda \prod_{t_0^*<t_k<t_1^*}(1-L_k),$$

or

$$\begin{aligned} Bk_0+Ch_0 &> \frac{\prod_{t_0^*<t_k<t_1^*}(1-L_k)}{\int_{t_0^*}^{t_1^*}\prod_{s<t_k<t_1^*}(1-L_k)\Delta s} \\ &\geq \frac{\prod_{0<t_k<T}(1-L_k)}{\int_0^T \prod_{s<t_k<T}(1-L_k)\Delta s}, \end{aligned}$$

or

$$\begin{aligned} (Bk_0+Ch_0)e_{\ominus(-a)}(T,0) &\geq Bk_0+Ch_0 \\ &\geq \frac{\prod_{0<t_k<T}(1-L_k)}{\int_0^T \prod_{s<t_k<T}(1-L_k)\Delta s}. \end{aligned}$$

This is a contradiction. Now, assume that $t_1^* < t_0^*$. So, $r \leq j$. Thus,

$$p^\Delta(t) \geq -\lambda(Bk_0+Ch_0), \quad t \in J\backslash\{t_j\}_{j=1}^m,$$

$$p(t_j^+) \geq (1-L_j)p(t_j), \quad j \in \{1,\ldots,m\}.$$

Hence and by Theorem 2.18, it follows that

$$p(t) \geq p(0) \prod_{0<t_k<t} (1-L_k) + \int_0^t \prod_{s<t_k<t} (1-L_k)(-\lambda(Bk_0+Ch_0))\,\Delta s,$$

$t \in J$. From here,

$$\begin{aligned} 0 &> p(t_1^*) \\ &\geq p(0) \prod_{0<t_k<t} (1-L_k) + \int_0^t \prod_{s<t_k<t} (1-L_k)(-\lambda(Bk_0+Ch_0))\,\Delta s, \end{aligned}$$

which implies

$$p(0) \prod_{0<t_k<t} (1-L_k) < \lambda(Bk_0+Ch_0) \int_0^{t_1^*} \prod_{s<t_k<t_1^*} (1-L_k)\Delta s$$

and

$$e_{-a}(T,0)p(T) \prod_{0<t_k<t} (1-L_k) < \lambda(Bk_0+Ch_0) \int_0^{t_1^*} \prod_{s<t_k<t_1^*} (1-L_k)\Delta s.$$

Hence and by (2.30), we obtain

$$\begin{aligned} \lambda(Bk_0+Ch_0) \int_0^{t_1^*} \prod_{s<t_k<t_1^*} (1-L_k)\Delta s > e_{-a}(T,0) &\prod_{0<t_k<t_1^*} (1-L_k) \\ &\times \Bigg(\lambda \prod_{t_0^*<t_k<T} (1-L_k) \\ &-\lambda(Bk_0+Ch_0) \int_{t_0^*}^T \prod_{s<t_k<T} (1-L_k)\Delta s\Bigg) \end{aligned}$$

or

$$\begin{aligned} &\prod_{0<t_k<t_1^*} (1-L_k)\Bigg(\lambda \prod_{t_0^*<t_k<T} (1-L_k) - \lambda(Bk_0+Ch_0) \\ &\quad \times \int_{t_0^*}^T \prod_{s<t_k<T} (1-L_k)\Delta s\Bigg) \\ &\leq \lambda(Bk_0+Ch_0)e_{\ominus(-a)}(T,0) \int_0^{t_1^*} \prod_{s<t_k<t_1^*} (1-L_k)\Delta s, \end{aligned}$$

or

$$\prod_{0<t_k<t_1^*} (1-L_k)\Bigg(\prod_{t_0^*<t_k<T} (1-L_k) - (Bk_0+Ch_0) \int_{t_0^*}^T \prod_{s<t_k<T} (1-L_k)\Delta s\Bigg)$$

$$\leq (Bk_0 + Ch_0) e_{\ominus(-a)}(T,0) \int_0^{t_1^*} \prod_{s<t_k<t_1^*} (1-L_k)\Delta s.$$

From the last inequality, we get

$$\begin{aligned}
\left(\prod_{0<t_k<T}(1-L_k)\right)^2 &= \left(\prod_{0<t_k<T}(1-L_k)\right)\left(\prod_{0<t_k<T}(1-L_k)\right) \\
&\leq \prod_{0<t_k<t_1^*}(1-L_k) \prod_{t_0^*<t_k<T}(1-L_k) \prod_{0<t_k<T}(1-L_k) \\
&\leq (Bk_0 + Ch_0) \prod_{0<t_k<t_1^*}(1-L_k) \prod_{0<t_k<T}(1-L_k) \\
&\quad \times \int_{t_0^*}^{T} \prod_{s<t_k<T}(1-L_k)\Delta s \\
&\quad + (Bk_0 + Ch_0) e_{\ominus(-a)}(T,0) \prod_{0<t_k<T}(1-L_k) \\
&\quad \times \int_0^{t_1^*} \prod_{s<t_k<t_1^*}(1-L_k)\Delta s \\
&< (Bk_0 + Ch_0) e_{\ominus(-a)}(T,0) \int_0^T \prod_{s<t_k<T}(1-L_k)\Delta s.
\end{aligned}$$

This is a contradiction.

Therefore, $p(t) \geq 0$, $t \in J$. Hence, $y(t) \geq 0$, $t \in J$. This completes the proof.

Suppose

(B19) $\delta \in PC(J)$, $\eta \in \Omega$,
(B20) $a, b, c \in \mathscr{C}(J)$, $a, b, c \geq 0$ on J, a is not identically vanishing,
(B21) $0 \leq L_j < 1$, $j \in \{1, \ldots, m\}$,
(B22) $\sup_{t \in J}(\mu(t)a(t)) < 1$.

Consider the following linear PBVP

$$\begin{aligned}
y^\Delta(t) + a(t)y(t) &= -b(t)Ty(t) - c(t)Sy(t) + \delta(t), \quad t \in J \backslash \{t_j\}_{j=1}^m, \\
y(t_j^+) - y(t_j) &= -L_j y(t_j) + I_j(\eta(t_j)) + L_j \eta(t_j), \quad j \in \{1, \ldots, m\}, \qquad (2.31) \\
y(0) &= y(T).
\end{aligned}$$

Lemma 2.12. *Suppose* (*B*18)-(*B*19). *Then* $y \in \Omega$ *is a solution of the PBVP* (2.31) *if and only if* $y \in PC(J)$ *and it satisfies the integral equation*

$$
\begin{aligned}
y(t) = &\int_0^T G(t,s)\,(\delta(s) - b(s)Ty(s) - c(s)Sy(s))\,\Delta s \\
&+ \sum_{\{j:0<t_j<T\}} G(t,t_j)e_{-a}(\sigma(t_j),t_j)\,(-L_j y(t_j) + I_j(\eta(t_j)) + L_j\eta(t_j)),
\end{aligned}
\tag{2.32}
$$

$t \in J$, where

$$
G(t,s) = \frac{1}{1 - e_{-a}(T,0)} \begin{cases} e_{-a}(t,\sigma(s)) & 0 \le s < t \le T \\ e_{-a}(T,0)e_{-a}(t,\sigma(s)) & 0 \le t \le s \le T. \end{cases}
$$

Proof. Note that $y \in \Omega$ is a solution of the impulsive dynamic equation

$$
\begin{aligned}
y^\Delta(t) + a(t)y(t) &= -b(t)Ty(t) - c(t)Sy(t) + \delta(t), \quad t \in J\backslash\{t_j\}_{j=1}^m, \\
y(t_j^+) - y(t_j) &= -L_j y(t_j) + I_j(\eta(t_j)) + L_j\eta(t_j), \quad j \in \{1,\ldots,m\},
\end{aligned}
\tag{2.33}
$$

if and only if $y \in PC(J)$ and it satisfies the integral equation

$$
\begin{aligned}
y(t) = &\int_0^t e_{-a}(t,\sigma(s))\,(-b(s)Ty(s) - c(s)Sy(s) + \delta(s))\,\Delta s \\
&+ e_{-a}(t,0)y(0) \\
&+ \sum_{\{j:0<t_j<t\}} e_{-a}(t,t_j)\,(-L_j y(t_j) + I_j(\eta(t_j)) + L_j\eta(t_j)), \quad t \in J.
\end{aligned}
\tag{2.34}
$$

1. Let $y \in \Omega$ satisfy (2.31). Then y satisfies (2.34). Hence,

$$
\begin{aligned}
y(0) &= y(T) \\
&= \int_0^T e_{-a}(T,\sigma(s))\,(-b(s)Ty(s) - c(s)Sy(s) + \delta(s))\,\Delta s \\
&\quad + e_{-a}(T,0)y(0) \\
&\quad + \sum_{\{j:0<t_j<T\}} e_{-a}(T,t_j)\,(-L_j y(t_j) + I_j(\eta(t_j)) + L_j\eta(t_j)),
\end{aligned}
$$

whereupon,

$$
\begin{aligned}
(1 - e_{-a}(T,0))y(0) = &\int_0^T e_{-a}(T,\sigma(s))\,(-b(s)Ty(s) - c(s)Sy(s) + \delta(s))\,\Delta s \\
&+ \sum_{\{j:0<t_j<T\}} e_{-a}(T,t_j)\,(-L_j y(t_j) + I_j(\eta(t_j)) + L_j\eta(t_j))
\end{aligned}
$$

and

$$\begin{aligned}
y(0) &= \frac{1}{1-e_{-a}(T,0)}\int_0^T e_{-a}(T,\sigma(s))\left(-b(s)Ty(s)-c(s)Sy(s)+\delta(s)\right)\Delta s \\
&\quad+\frac{1}{1-e_{-a}(T,0)}\sum_{\{j:0<t_j<T\}} e_{-a}(T,t_j)\left(-L_jy(t_j)+I_j(\eta(t_j))+L_j\eta(t_j)\right).
\end{aligned}$$

Then

$$\begin{aligned}
y(t) &= \int_0^t e_{-a}(t,\sigma(s))\left(-b(s)Ty(s)-c(s)Sy(s)+\delta(s)\right)\Delta s \\
&\quad+\frac{e_{-a}(t,0)}{1-e_{-a}(T,0)}\int_0^t e_{-a}(T,\sigma(s))\left(-b(s)Ty(s)-c(s)Sy(s)+\delta(s)\right)\Delta s \\
&\quad+\frac{e_{-a}(t,0)}{1-e_{-a}(T,0)}\int_t^T e_{-a}(T,\sigma(s))\left(-b(s)Ty(s)-c(s)Sy(s)+\delta(s)\right)\Delta s \\
&\quad+\frac{e_{-a}(t,0)}{1-e_{-a}(T,0)}\sum_{\{j:0<t_j<t\}} e_{-a}(T,t_j)\left(-L_jy(t_j)+I_j(\eta(t_j))+L_j\eta(t_j)\right) \\
&\quad+\frac{e_{-a}(t,0)}{1-e_{-a}(T,0)}\sum_{\{j:t\le t_j<T\}} e_{-a}(t,t_j)\left(-L_jy(t_j)+I_j(\eta(t_j))+L_j\eta(t_j)\right) \\
&\quad+\sum_{\{j:0<t_j<t\}} e_{-a}(t,t_j)\left(-L_jy(t_j)+I_j(\eta(t_j))+L_j\eta(t_j)\right) \\
&= \frac{1}{1-e_{-a}(T,0)}\int_0^t e_{-a}(t,\sigma(s))\left(-b(s)Ty(s)-c(s)Sy(s)+\delta(s)\right)\Delta s \\
&\quad+\frac{e_{-a}(T,0)}{1-e_{-a}(T,0)}\int_t^T e_{-a}(t,\sigma(s))\left(-b(s)Ty(s)-c(s)Sy(s)+\delta(s)\right)\Delta s \\
&\quad+\frac{1}{1-e_{-a}(T,0)}\sum_{\{j:0<t_j<t\}} e_{-a}(t,t_j)\left(-L_jy(t_j)+I_j(\eta(t_j))+L_j\eta(t_j)\right) \\
&\quad+\frac{e_{-a}(T,0)}{1-e_{-a}(T,0)}\sum_{\{j:t\le t_j<T\}} e_{-a}(t,t_j)\left(-L_jy(t_j)+I_j(\eta(t_j))+L_j\eta(t_j)\right) \\
&= \int_0^T G(t,s)\left(-b(s)Ty(s)-c(s)Sy(s)+\delta(s)\right)\Delta s \\
&\quad+\sum_{\{j:0<t_j<T\}} G(t,t_j)e_{-a}(\sigma(t_j),t_j)\left(-L_jy(t_j)+I_j(\eta(t_j))+L_j\eta(t_j)\right),
\end{aligned}$$

$t \in J$, i.e., y satisfies (2.32).

2. Let $y \in PC(J)$ satisfy (2.32). Then $y \in \Omega$ and y satisfies (2.33). We have

$$\begin{aligned} y(0) &= \frac{e_{-a}(T,0)}{1-e_{-a}(T,0)} \int_0^T e_{-a}(0,\sigma(s))\,(-b(s)Ty(s)-c(s)Sy(s)+\delta(s))\,\Delta s \\ &+ \frac{e_{-a}(T,0)}{1-e_{-a}(T,0)} \sum_{\{j:0<t_j<T\}} e_{-a}(0,t_j)\,(-L_jy(t_j)+I_j(\eta(t_j))+L_j\eta(t_j)), \\ y(T) &= \frac{1}{1-e_{-a}(T,0)} \int_0^T e_{-a}(T,\sigma(s))\,(-b(s)Ty(s)-c(s)Sy(s)+\delta(s))\,\Delta s \\ &+ \frac{1}{1-e_{-a}(T,0)} \sum_{\{j:0<t_j<T\}} e_{-a}(T,\sigma(s))\,(-b(s)Ty(s)-c(s)Sy(s)+\delta(s))\,\Delta s \\ &= y(0). \end{aligned}$$

Then y satisfies (2.31). This completes the proof.

Let $(B22)$ holds and

$$\sup_{t,s\in J} G(t,s) = A.$$

Then, if $(B22)$ holds, we have

$$0 \le G(t,s) \le A, \quad t,s \in J.$$

Theorem 2.20. *Assume* $(B15)$, $(B16)$, $(B18)-(B22)$ *and*

$$A\left(\int_0^T (k_0 sb(s)+h_0 sc(s))\,\Delta s + \sum_{j=1}^m L_j\right) < 1. \tag{2.35}$$

Then the PBVP (2.31) *has a unique solution in* Ω.

Proof. For $y \in \Omega$ define the operator

$$\begin{aligned} \Phi y(t) &= \int_0^T G(t,s)\,(\delta(s)-b(s)Ty(s)-c(s)Sy(s))\,\Delta s \\ &+ \sum_{\{j:0<t_j<t\}} G(t,t_j)e_{-a}(\sigma(t_j),t_j)\,(-L_jy(t_j)+I_j(\eta(t_j))+L_j\eta(t_j)), \end{aligned}$$

$t \in J$. Note that $\Phi : \Omega \to \Omega$. For $x,y \in \Omega$, we get

$$\begin{aligned} |\Phi y(t)-\Phi x(t)| &= \Bigg| \int_0^T G(t,s)\,(\delta(s)-b(s)Ty(s)-c(s)Sy(s))\,\Delta s \\ &+ \sum_{\{j:0<t_j<t\}} G(t,t_j)e_{-a}(\sigma(t_j),t_j)\,(-L_jy(t_j)+I_j(\eta(t_j))+L_j\eta(t_j)) \end{aligned}$$

$$
\begin{aligned}
& - \int_0^T G(t,s)\,(\delta(s) - b(s)Tx(s) - c(s)Sx(s))\,\Delta s \\
& - \sum_{\{j:0<t_j<t\}} G(t,t_j)e_{-a}(\sigma(t_j),t_j)\,(-L_jx(t_j) + I_j(\eta(t_j)) + L_j\eta(t_j))\Bigg| \\
= & \left| \int_0^T G(t,s)\,(-b(s)T(y-x)(s) - c(s)S(y-x)(s))\,\Delta s \right. \\
& \left. + \sum_{\{j:0<t_j<t\}} G(t,t_j)e_{-a}(\sigma(t_j),t_j)\,(-L_j(y-x)(t_j)) \right| \\
\le & \left| \int_0^T G(t,s)\,(-b(s)T(y-x)(s) - c(s)S(y-x)(s))\,\Delta s \right| \\
& + \left| \sum_{\{j:0<t_j<t\}} G(t,t_j)e_{-a}(\sigma(t_j),t_j)\,(-L_j(y-x)(t_j)) \right| \\
\le & \int_0^T G(t,s)\,|-b(s)T(y-x)(s) - c(s)S(y-x)(s)|\,\Delta s \\
& + \sum_{\{j:0<t_j<t\}} G(t,t_j)e_{-a}(\sigma(t_j),t_j)\,|-L_j(y-x)(t_j)| \\
\le & A\left(\int_0^T (k_0sb(s) + h_0sc(s))\Delta s + \sum_{j=1}^m L_j \right) \|y-x\|, \quad t \in J,
\end{aligned}
$$

whereupon,

$$
\begin{aligned}
\|\Phi y - \Phi x\| & \le A\left(\int_0^T (k_0sb(s) + h_0sc(s))\Delta s + \sum_{j=1}^m L_j \right) \|y-x\| \\
& < \|y-x\|.
\end{aligned}
$$

Hence and by the contraction mapping principle, we conclude that the operator Φ has a unique fixed point in Ω. This completes the proof.

Theorem 2.21. *Assume that the following conditions hold.*

1. There exist functions $u_0, v_0 \in \Omega$, $u_0(t) \le v_0(t)$, $t \in J$, such that

$$
\begin{aligned}
u_0^{\Delta}(t) &\leq f(t, u_0(t), Tu_0(t), Su_0(t)), \quad t \in J\backslash\{t_j\}_{j=1}^m, \\
u_0(t_j^+) - u_0(t_j^-) &\leq I_j(u_0(t_j)), \quad j \in \{1, \ldots, m\}, \\
u_0(0) &\leq u_0(T), \\
v_0^{\Delta}(t) &\geq f(t, v_0(t), Tv_0(t), Sv_0(t)), \quad t \in J\backslash\{t_j\}_{j=1}^m, \\
v_0(t_j^+) - v_0(t_j^-) &\geq I_j(v_0(t_j)), \quad j \in \{1, \ldots, m\}, \\
v_0(0) &\geq v_0(T).
\end{aligned}
\tag{2.36}
$$

2. *The function* $f \in \mathscr{C}(J \times \mathbb{R} \times \mathbb{R})$ *satisfies*

$$
f(t, u_2, v_2, w_2) - f(t, u_1, v_1, w_1) \geq -a(t)(u_2 - u_1) - b(t)(v_2 - v_1) - c(t)(w_2 - w_1),
\tag{2.37}
$$

whenever

$$
\begin{aligned}
u_0(t) &\leq u_1 \leq u_2 \leq v_0(t), \\
Tu_0(t) &\leq v_1 \leq v_2 \leq Tv_0(t), \\
Su_0(t) &\leq w_1 \leq w_2 \leq Sv_0(t), \quad t \in J,
\end{aligned}
$$

where a, b, c *satisfy* (*B*20)-(*B*22).

3. *The functions* $I_j \in \mathscr{C}(\mathbb{R})$, $j \in \{1, \ldots, m\}$, *satisfy*

$$
I_j(x) - I_j(y) \geq -L_j(x - y),
\tag{2.38}
$$

whenever

$$
\begin{aligned}
u_0(t_j) &\leq y \leq x \leq v_0(t_j), \\
0 &\leq L_j < 1, \quad j \in \{1, \ldots, m\}.
\end{aligned}
$$

Furthermore, assume (2.27) *and* (2.35). *Then the PBVP* (2.24) *has minimal and maximal solutions* u^* *and* v^*, *respectively. Moreover, there exist monotone iteration sequences*

$$
\{u_n(t)\}_{n \in \mathbb{N}}, \quad \{v_n(t)\}_{n \in \mathbb{N}} \subset [u_0, v_0]
$$

such that

$$
\begin{aligned}
&u_n(t) \to u^*(t), \\
&v_n(t) \to v^*(t), \quad \text{as} \quad n \to \infty \quad \text{uniformly} \quad \text{on} \quad t \in J,
\end{aligned}
$$

where $\{u_n(t)\}_{n\in\mathbb{N}}$ *and* $\{v_n(t)\}_{n\in\mathbb{N}}$ *satisfy*

$$
\begin{aligned}
u_n^{\Delta}(t) &= f(t,u_{n-1}(t),Tu_{n-1}(t),Su_{n-1}(t))-a(t)(u_n-u_{n-1})(t)\\
&\quad -b(t)T(u_n-u_{n-1})(t)-c(t)S(u_n-u_{n-1})(t), \quad t\in J\backslash\{t_j\}_{j=1}^{m},\\
u_n(t_j^+) &= u_n(t_j)-L_ju_n(t_j)+I_j(u_{n-1}(t_j))+L_ju_{n-1}(t_j), \quad j\in\{1,\ldots,m\},\\
u_n(0) &= u_n(T),\\
v_n^{\Delta}(t) &= f(t,v_{n-1}(t),Tv_{n-1}(t),Sv_{n-1}(t))-a(t)(v_n-v_{n-1})(t)\\
&\quad -b(t)T(v_n-v_{n-1})(t)-c(t)S(v_n-v_{n-1})(t), \quad t\in J\backslash\{t_j\}_{j=1}^{m},\\
v_n(t_j^+) &= v_n(t_j)-L_jv_n(t_j)+I_j(v_{n-1}(t_j))+L_jv_{n-1}(t_j), \quad j\in\{1,\ldots,m\},\\
v_n(0) &= v_n(T), \quad n\in\mathbb{N},
\end{aligned}
\tag{2.39}
$$

$$u_0\le u_1\le\ldots\le u_n\le\ldots\le u^*\le v^*\le\ldots\le v_n\le\ldots\le v_1\le v_0.$$

Proof. Note that for any $u_{n-1},v_{n-1}\in\Omega$, using Theorem 2.20, it follows that (2.39) has unique solution $u_n,v_n\in\Omega$, respectively. We have

$$
\begin{aligned}
(u_1-u_0)^{\Delta}(t) &= u_1^{\Delta}(t)-u_0^{\Delta}(t)\\
&\ge f(t,u_0(t),Tu_0(t),Su_0(t))-a(t)(u_1-u_0)(t)\\
&\quad -b(t)T(u_1-u_0)(t)-c(t)S(u_1-u_0)(t)\\
&\quad -f(t,u_0(t),Tu_0(t),Su_0(t))\\
&= -a(t)(u_1-u_0)(t)-b(t)T(u_1-u_0)(t)\\
&\quad -c(t)S(u_1-u_0)(t), \quad t\in J\backslash\{t_j\}_{j=1}^{m},\\
(u_1-u_0)(t_j^+)-(u_1-u_0)(t_j) &= u_1(t_j^+)-u_1(t_j)-(u_0(t_j^+)-u_0(t_j))\\
&\ge -L_ju_1(t_j)+I_j(u_0(t_j))+L_ju_0(t_j)-I_j(u_0(t_j))\\
&= -L_j(u_1-u_0)(t_j), \quad j\in\{1,\ldots,m\},
\end{aligned}
$$

$$
\begin{aligned}
(u_1-u_0)(0) &= u_1(0)-u_0(0)\\
&\geq u_1(T)-u_0(T)\\
&= (u_1-u_0)(T),\\
(v_1-v_0)^{\Delta}(t) &= v_1^{\Delta}(t)-v_0^{\Delta}(t)\\
&\leq f(t,v_0(t),Tv_0(t),Sv_0(t))-a(t)(v_1-v_0)(t)\\
&\quad -b(t)T(v_1-v_0)(t)-c(t)S(v_1-v_0)(t)\\
&\quad -f(t,v_0(t),Tv_0(t),Sv_0(t))\\
&= -a(t)(v_1-v_0)(t)-b(t)T(v_1-v_0)(t)\\
&\quad -c(t)S(v_1-v_0)(t), \quad t\in J\backslash\{t_j\}_{j=1}^m,\\
(v_1-v_0)(t_j^+)-(v_1-v_0)(t_j) &= v_1(t_j^+)-v_1(t_j)-(v_0(t_j^+)-v_0(t_j))\\
&\leq -L_jv_1(t_j)+I_j(v_0(t_j))+L_jv_0(t_j)-I_j(v_0(t_j))\\
&= -L_j(v_1-v_0)(t_j), \quad j\in\{1,\ldots,m\},\\
(v_1-v_0)(0) &= v_1(0)-v_0(0)\\
&\leq v_1(T)-v_0(T)\\
&= (v_1-v_0)(T),\\
(v_1-u_1)^{\Delta}(t) &= v_1^{\Delta}(t)-u_1^{\Delta}(t)\\
&= f(t,v_0(t),Tv_0(t),Sv_0(t))-a(t)(v_1-v_0)(t)\\
&\quad -b(t)T(v_1-v_0)(t)-c(t)S(v_1-v_0)(t)
\end{aligned}
$$

$$
\begin{aligned}
&-f(t,u_0(t),Tu_0(t),Su_0(t))-a(t)(u_1-u_0)(t)\\
&-b(t)T(u_1-u_0)(t)-c(t)S(u_1-u_0)(t)\\
\geq\ &-a(t)(v_0-u_0)(t)-b(t)T(v_0-u_0)(t)\\
&-c(t)S(v_0-u_0)(t)-a(t)(v_1-v_0)(t)\\
&-b(t)T(v_1-v_0)(t)-c(t)S(v_1-v_0)(t)\\
&-a(t)(u_1-u_0)(t)-b(t)T(u_1-u_0)(t)\\
&-c(t)S(u_1-u_0)(t)\\
=\ &-a(t)(v_1-u_1)(t)-b(t)T(v_1-u_1)(t)\\
&-c(t)S(v_1-u_1)(t), \quad t\in J\backslash\{t_j\}_{j=1}^m,
\end{aligned}
$$

$$
\begin{aligned}
(v_1-u_1)(t_j^+)-(v_1-u_1)(t_j) &= (v_1(t_j^+)-v_1(t_j))-(u_1(t_j^+)-u_1(t_j))\\
&\leq -L_jv_1(t_j)+I_j(v_0(t_j))+L_ju_0(t_j)\\
&\quad +L_ju_1(t_j)-I_j(v_0(t_j))-L_jv_0(t_j)\\
&= -L_j(v_1-v_0)(t_j), \quad j\in\{1,\ldots,m\},
\end{aligned}
$$

$$
(v_1-u_1)(0)\geq(v_1-u_1)(T).
$$

Hence and by Theorem 2.19, it follows that

$$
u_0\leq u_1\leq v_1\leq v_0 \quad \text{on} \quad J.
$$

Assume that

$$
u_{n-2}\leq u_{n-1}\leq v_{n-1}\leq v_{n-2} \quad \text{on} \quad J
$$

for some $n\in\mathbb{N}$, $n\geq 2$. We will prove that

$$
u_{n-1}\leq u_n\leq v_n\leq v_{n-1} \quad \text{on} \quad J. \tag{2.40}
$$

Indeed, we have

$$
\begin{aligned}
(u_n - u_{n-1})^{\Delta}(t) &= u_n^{\Delta}(t) - u_{n-1}^{\Delta}(t) \\
&= f(t, u_{n-1}(t), Tu_{n-1}(t), Su_{n-1}(t)) \\
&\quad - a(t)(u_n - u_{n-1})(t) \\
&\quad - b(t)T(u_n - u_{n-1})(t) - c(t)S(u_n - u_{n-1})(t) \\
&\quad - f(t, u_{n-2}(t), Tu_{n-2}(t), Su_{n-2}(t)) \\
&\quad + a(t)(u_{n-1} - u_{n-2})(t) \\
&\quad + b(t)T(u_{n-1} - u_{n-2})(t) \\
&\quad + c(t)S(u_{n-1} - u_{n-2})(t) \\
&\geq -a(t)(u_n - u_{n-1})(t) - b(t)T(u_n - u_{n-1})(t) \\
&\quad - c(t)S(u_n - u_{n-1})(t), \quad t \in J \backslash \{t_j\}_{j=1}^{m}, \\
(u_n - u_{n-1})(t_j^+) - (u_n - u_{n-1})(t_j) &= u_n(t_j^+) - u_n(t_j) - (u_{n-1}(t_j^+) - u_{n-1}(t_j)) \\
&= -L_j u_n(t_j) + I_j(u_{n-1}(t_j)) + L_j u_{n-1}(t_j) \\
&\quad - I_j(u_{n-2}(t_j)) \\
&\quad - L_j u_{n-2}(t_j) + L_j u_{n-1}(t_j) \\
&= -L_j(u_n - u_{n-1})(t_j) + I_j(u_{n-1} - u_{n-2})(t_j) \\
&\quad + L_j(u_{n-1} - u_{n-2})(t_j) \\
&\geq -L_j(u_n - u_{n-1})(t_j), \quad j \in \{1, \ldots, m\}, \\
(u_n - u_{n-1})(0) &= u_n(0) - u_{n-1}(0)
\end{aligned}
$$

$$\begin{aligned}
&= u_n(T) - u_{n-1}(T) \\
&= (u_n - u_{n-1})(T), \\
(v_n - v_{n-1})^{\Delta}(t) &= v_n^{\Delta}(t) - v_{n-1}^{\Delta}(t) \\
&= f(t, v_{n-1}(t), Tv_{n-1}(t), Sv_{n-1}(t)) \\
&\quad -a(t)(v_n - v_{n-1})(t) \\
&\quad -b(t)T(v_n - v_{n-1})(t) - c(t)S(v_n - v_{n-1})(t) \\
&\quad -f(t, v_{n-2}(t), Tv_{n-2}(t), Sv_{n-2}(t)) \\
&\quad +a(t)(v_{n-1} - v_{n-2})(t) \\
&\quad +b(t)T(v_{n-1} - v_{n-2})(t) \\
&\quad +c(t)S(v_{n-1} - v_{n-2})(t) \\
&\geq -a(t)(v_n - v_{n-1})(t) - b(t)T(v_n - v_{n-1})(t) \\
&\quad -c(t)S(v_n - v_{n-1})(t), \quad t \in J \backslash \{t_j\}_{j=1}^{m}, \\
(v_n - v_{n-1})(t_j^+) - (v_n - v_{n-1})(t_j) &= v_n(t_j^+) - v_n(t_j) - (v_{n-1}(t_j^+) - v_{n-1}(t_j)) \\
&= -L_j v_n(t_j) + I_j(v_{n-1}(t_j)) \\
&\quad +L_j v_{n-1}(t_j) - I_j(v_{n-2}(t_j)) \\
&\quad +L_j v_{n-1}(t_j) - L_j v_{n-2}(t_j) \\
&\leq -L_j(v_n - v_{n-1})(t_j), \quad j \in \{1, \ldots, m\}, \\
(v_n - v_{n-1})(0) &= v_n(0) - v_{n-1}(0)
\end{aligned}$$

$$
\begin{aligned}
&= v_n(T) - v_{n-1}(T)\\
&= (v_n - v_{n-1})(T),\\
(v_n - u_n)^{\Delta}(t) &= v_n^{\Delta}(t) - u_n^{\Delta}(t)\\
&= f(t, v_{n-1}(t), Tv_{n-1}(t), Sv_{n-1}(t))\\
&\quad -a(t)(v_n - v_{n-1})(t)\\
&\quad -b(t)T(v_n - v_{n-1})(t) - c(t)S(v_n - v_{n-1})(t)\\
&\quad -f(t, u_{n-1}(t), Tu_{n-1}(t), Su_{n-1}(t))\\
&\quad +a(t)(u_n - u_{n-1})(t)\\
&\quad +b(t)T(u_n - u_{n-1})(t) + c(t)S(u_n - u_{n-1})(t)\\
&\geq -a(t)(v_{n-1} - u_{n-1})(t) - b(t)T(v_{n-1} - u_{n-1})(t)\\
&\quad -c(t)S(v_{n-1} - u_{n-1})(t) - a(t)(v_n - v_{n-1})(t)\\
&\quad -b(t)T(v_n - v_{n-1})(t) - c(t)S(v_n - v_{n-1})(t)\\
&\quad +a(t)(u_n - u_{n-1})(t) - b(t)T(u_n - u_{n-1})(t)\\
&\quad +c(t)S(u_n - u_{n-1})(t)\\
&= -a(t)(v_n - u_n)(t) - b(t)T(v_n - u_n)(t)\\
&\quad -c(t)S(v_n - u_n)(t), \quad t \in J\backslash\{t_j\}_{j=1}^{m},\\
(v_n - u_n)(t_j^+) - (v_n - u_n)(t_j) &= (v_n(t_j^+) - v_n(t_j)) - (u_n(t_j^+) - u_n(t_j))\\
&= -L_j v_n(t_j) + I_j(v_{n-1}(t_j)) + L_j v_{n-1}(t_j)
\end{aligned}
$$

$$+L_j u_n(t_j) - I_j(u_{n-1}(t_j)) - L_j u_{n-1}(t_j)$$

$$\geq -L_j(v_n - u_n)(t_j), \quad j \in \{1, \ldots, m\},$$

$$(v_n - u_n)(0) = (v_n - u_n)(T).$$

From here and from Theorem 2.19, we conclude that (2.40) holds for any $n \in \mathbb{N}$. By Lemma 2.12, it follows that

$$\begin{aligned} u_n(t) = & \int_0^t G(t,s)\Big(f(s, u_{n-1}(s), Tu_{n-1}(s), Su_{n-1}(s)) + a(s)u_{n-1}(s) \\ & -b(s)T(u_n - u_{n-1})(s) - c(s)S(u_n - u_{n-1})(s) \Big)\Delta s \\ & + \sum_{\{j:0<t_j<T\}} G(t,t_j)e_{-a}(\sigma(t_j),t_j)\left(-L_j u_n(t_j) + I_j(u_{n-1}(t_j)) + L_j u_{n-1}(t_j)\right), \\ v_n(t) = & \int_0^t G(t,s)\Big(f(s, v_{n-1}(s), Tv_{n-1}(s), Sv_{n-1}(s)) + a(s)v_{n-1}(s) \\ & -b(s)T(v_n - v_{n-1})(s) - c(s)S(v_n - v_{n-1})(s) \Big)\Delta s \\ & + \sum_{\{j:0<t_j<T\}} G(t,t_j)e_{-a}(\sigma(t_j),t_j)\left(-L_j v_n(t_j) + I_j(v_{n-1}(t_j)) + L_j v_{n-1}(t_j)\right), \end{aligned} \tag{2.41}$$

$t \in J$, $n \in \mathbb{N}$. Note that the sequences $\{u_n^{\Delta}(t)\}_{n\in\mathbb{N}}$ and $\{v_n^{\Delta}(t)\}_{n\in\mathbb{N}}$ are uniformly bounded on J. Then there exist $u^*, v^* \in \Omega$ such that

$$u_n(t) \to u^*(t), \quad v_n(t) \to v^*(t), \quad \text{as} \quad n \to \infty,$$

uniformly on J. Hence and by (2.41), we obtain

$$\begin{aligned} u^*(t) = & \int_0^T G(t,s)\left(f(s,u^*(s),Tu^*(s),Su^*(s)) + a(s)u^*(s)\right)\Delta s \\ & + \sum_{\{j:0<t_j<T\}} G(t,t_j)e_{-a}(\sigma(t_j),t_j)I_j(u^*(t_j)), \\ v^*(t) = & \int_0^T G(t,s)\left(f(s,v^*(s),Tv^*(s),Sv^*(s)) + a(s)v^*(s)\right)\Delta s \\ & + \sum_{\{j:0<t_j<T\}} G(t,t_j)e_{-a}(\sigma(t_j),t_j)I_j(v^*(t_j)), \quad t \in J. \end{aligned}$$

Let $w \in [u_0, v_0]$ be a solution of the PBVP (2.24), i.e.,

$$w^{\Delta}(t) = f(t, w(t), Tw(t), Sw(t)), \quad t \in J \backslash \{t_j\}_{j=1}^{m},$$

$$w(t_j^+) - w(t_j) = I_j(w(t_j)), \quad j \in \{1, \ldots, m\},$$

$$w(0) = w(T).$$

We have

$$\begin{aligned}
(w-u_1)^{\Delta}(t) &= w^{\Delta}(t) - u_1^{\Delta}(t) \\
&\geq f(t, w(t), Tw(t), Sw(t)) - f(t, u_0(t), Tu_0(t), Su_0(t)) \\
&\geq -a(t)(w-u_0)(t) - b(t)T(w-u_0)(t) \\
&\quad -c(t)S(w-u_0)(t), \\
(w-u_1)(t_j^+) - (w-u_1)(t_j) &= w(t_j^+) - w(t_j) - (u_1(t_j^+) - u_1(t_j)) \\
&= I_j(w(t_j)) + L_j u_1(t_j) - I_j(u_0(t_j)) - L_j u_0(t_j) \\
&\geq -L_j(w(t_j) - u_0(t_j)) + L_j(u_1(t_j) - u_0(t_j)) \\
&= -L_j(w(t_j) - u_1(t_j)), \\
(w-u_1)(0) &= w(0) - u_1(0) \\
&= (w-u_1)(T), \\
(w-v_1)^{\Delta}(t) &= w^{\Delta}(t) - v_1^{\Delta}(t) \\
&= f(t, w(t), Tw(t), Sw(t)) \\
&\quad -f(t, v_0(t), Tv_0(t), Sv_0(t)) \\
&\quad +a(t)(v_1 - v_0)(t) + b(t)T(v_1 - v_0)(t)
\end{aligned}$$

$$+c(t)S(v_1 - v_0)(t)$$

$$\leq f(t, w(t), Tw(t), Sw(t)) - f(t, v_0(t), Tv_0(t), Sv_0(t))$$

$$\leq -a(t)(w - v_0)(t) - b(t)T(w - v_0)(t) - c(t)S(w - v_0)(t),$$

$$(w - v_1)(t_j^+) - (w - v_1)(t_j) = w(t_j^+) - w(t_j) - (v_1(t_j^+) - v_1(t_j))$$

$$= I_j(w(t_j)) + L_j v_1(t_j) - I_j(v_0(t_j)) - L_j v_0(t_j)$$

$$\geq -L_j(w(t_j) - v_0(t_j)) + L_j(v_1(t_j) - v_0(t_j))$$

$$= -L_j(w(t_j) - v_1(t_j)),$$

$$(w - v_1)(0) = w(0) - v_1(0)$$

$$= (w - v_1)(T).$$

Hence and by Theorem 2.19, it follows that

$$u_1(t) \leq w(t) \leq v_1(t), \quad t \in J.$$

Assume that

$$u_{n-1}(t) \leq w(t) \leq v_{n-1}(t), \quad t \in J,$$

for some $n \in \mathbb{N}$, $n \geq 2$. We will prove that

$$u_n(t) \leq w(t) \leq v_n(t), \quad t \in J. \tag{2.42}$$

We have

$$(w - u_n)^{\Delta}(t) = w^{\Delta}(t) - u_n^{\Delta}(t)$$

$$\geq f(t, w(t), Tw(t), Sw(t))$$

$$-f(t, u_{n-1}(t), Tu_{n-1}(t), Su_{n-1}(t))$$

$$\geq -a(t)(w - u_{n-1})(t) - b(t)T(w - u_{n-1})(t)$$

$$-c(t)S(w - u_{n-1})(t),$$

$$\begin{aligned}
(w-u_n)(t_j^+)-(w-u_n)(t_j) &= w(t_j^+)-w(t_j)-(u_n(t_j^+)-u_n(t_j))\\
&= I_j(w(t_j))+L_ju_n(t_j)-I_j(u_{n-1}(t_j))-L_ju_{n-1}(t_j)\\
&\geq -L_j(w(t_j)-u_{n-1}(t_j))+L_j(u_n(t_j)-u_{n-1}(t_j))\\
&= -L_j(w(t_j)-u_n(t_j)),\\
(w-u_n)(0) &= w(0)-u_n(0)\\
&= (w-u_n)(T),\\
(w-v_n)^{\Delta}(t) &= w^{\Delta}(t)-v_n^{\Delta}(t)\\
&\leq f(t,w(t),Tw(t),Sw(t))\\
&\quad -f(t,v_{n-1}(t),Tv_{n-1}(t),Sv_{n-1}(t))\\
&\leq -a(t)(w-v_{n-1})(t)-b(t)T(w-v_{n-1})(t)\\
&\quad -c(t)S(w-v_{n-1})(t),\\
(w-v_n)(t_j^+)-(w-v_n)(t_j) &= w(t_j^+)-w(t_j)-(v_n(t_j^+)-v_n(t_j))\\
&= I_j(w(t_j))+L_jv_n(t_j)-I_j(v_{n-1}(t_j))-L_jv_{n-1}(t_j)\\
&\geq -L_j(w(t_j)-v_{n-1}(t_j))+L_j(v_n(t_j)-v_{n-1}(t_j))\\
&= -L_j(w(t_j)-v_n(t_j)),\\
(w-v_n)(0) &= w(0)-v_n(0)\\
&= (w-v_n)(T).
\end{aligned}$$

Hence and by Theorem 2.19, it follows that (2.42) holds for any $n \in \mathbb{N}$. Letting $n \to \infty$ in (2.42), we get

$$u^*(t) \le w(t) \le v^*(t), \quad t \in J.$$

This completes the proof.

2.6 Multiple Positive Solutions for First Order Impulsive Integral Boundary Value Problems

Let $\mathbb{T}$ be a time scale with forward jump operator and delta differentiation operator σ and Δ, respectively. Consider the following first order nonlinear impulsive integral boundary value problem

$$\begin{aligned} x^{\Delta}(t)+p(t)x(\sigma(t)) &= f(t,x(\sigma(t))), \quad t \in J=[0,T]\backslash\{t_1,\ldots,t_m\}, \\ x(t_j^+) &= x(t_j^-)+I_j(x(t_j)), \quad j \in \{1,\ldots,m\}, \\ \alpha x(0)-\beta x(\sigma(T)) &= \int_0^{\sigma(T)} g(s)x(s)\Delta s, \end{aligned} \tag{2.43}$$

where

(B23) $p \in \mathscr{C}([0,\sigma(T)])$, $p \ge 0$ on $[0,\sigma(T)]$,
(B24) $I_j \in \mathscr{C}(\mathbb{R})$, $I_j \ge 0$ on $[0,\infty)$, $j \in \{1,\ldots,m\}$,
(B25) g is a non-negative integrable function on $[0,\sigma(T)]$,
(B26) α and β are non-negative constants,

$$\Gamma = \alpha - \beta e_p(0,\sigma(T)) - \int_0^{\sigma(T)} g(s)e_p(0,s)\Delta s > 0,$$

(B27) $f \in \mathscr{C}([0,\sigma(T)] \times \mathbb{R})$, $f \ge 0$ on $[0,\sigma(T)] \times [0,\infty)$,

$$0 < t_1 < \ldots < t_m < T,$$

t_j, $j \in \{1,\ldots,m\}$, are right-dense. Let $PC([0,\sigma(T)])$ and $PC^{\Delta}([0,\sigma(T)])$ be as in Section 2.2

Lemma 2.13. *Suppose (B23), (B25), (B26), $h \in \mathscr{C}([0,\sigma(T)])$, $v_j \in \mathbb{R}$, $j \in \{1,\ldots,m\}$. Then $x \in PC([0,\sigma(T)])$ is a solution of the integral equation*

$$x(t) = \int_0^{\sigma(T)} G(t,s)h(s)\Delta s + \sum_{j=1}^m G(t,t_j)v_j, \quad t \in [0,\sigma(T)], \tag{2.44}$$

where

$$G(t,s)=\begin{cases}\frac{1}{\Gamma}e_{\ominus p}(t,s)\left(\alpha-\int_0^{\sigma(s)}g(\tau)e_{\ominus p}(\tau,0)\Delta\tau\right) & 0\le s\le t\le\sigma(T)\\ \frac{1}{\Gamma}e_{\ominus p}(t,s)\left(\beta e_{\ominus p}(\sigma(T),0)+\int_{\sigma(s)}^{\sigma(T)}g(r)e_{\ominus p}(r,0)\Delta r\right) & 0\le t\le s\le\sigma(T)\end{cases}$$

if and only if $x\in PC^{\Delta}([0,\sigma(T)])$ *is a solution of the BVP*

$$\begin{aligned} x^{\Delta}(t)+p(t)x(\sigma(t)) &= h(t), \quad t\in J,\\ x(t_j^+) &= x(t_j^-)+v_j, \quad j\in\{1,\dots,m\},\\ \alpha x(0)-\beta x(\sigma(T)) &= \int_0^{\sigma(T)}g(s)x(s)\Delta s. \end{aligned}\tag{2.45}$$

Remark 2.3. Suppose that x satisfies (2.44). Then

$$\begin{aligned} x(t) &= \int_0^{\sigma(T)}G(t,s)h(s)\Delta s+\sum_{j=1}^{m}G(t,t_j)v_j\\ &= \int_0^{t}G(t,s)h(s)\Delta s+\int_t^{\sigma(T)}G(t,s)h(s)\Delta s\\ &\quad+\sum_{\{j:t_j<t\}}G(t,t_j)v_j+\sum_{\{j:t\le t_j\}}G(t,t_j)v_j\\ &= \frac{1}{\Gamma}\int_0^{t}e_{\ominus p}(t,s)\left(\alpha-\int_0^{\sigma(s)}g(\tau)e_{\ominus p}(\tau,0)\Delta\tau\right)h(s)\Delta s\\ &\quad+\frac{1}{\Gamma}\int_t^{\sigma(T)}e_{\ominus p}(t,s)\left(\beta e_{\ominus p}(\sigma(T),0)+\int_{\sigma(s)}^{\sigma(T)}g(r)e_{\ominus p}(r,0)\Delta r\right)h(s)\Delta s\\ &\quad+\sum_{\{j:t_j<t\}}\frac{1}{\Gamma}e_{\ominus p}(t,t_j)\left(\alpha-\int_0^{\sigma(t_j)}g(\tau)e_{\ominus p}(\tau,0)\Delta\tau\right)v_j\\ &\quad+\sum_{\{j:t\le t_j\}}\frac{1}{\Gamma}e_{\ominus p}(t,t_j)\left(\beta e_{\ominus p}(\sigma(T),0)+\int_{\sigma(t_j)}^{\sigma(T)}g(r)e_{\ominus p}(r,0)\Delta r\right)v_j\\ &= \frac{\alpha}{\Gamma}\int_0^{t}e_{\ominus p}(t,s)h(s)\Delta s-\frac{1}{\Gamma}\int_0^{t}e_{\ominus p}(t,s)\int_0^{\sigma(s)}g(\tau)e_{\ominus p}(\tau,0)h(s)\Delta\tau\Delta s\\ &\quad+\frac{\beta}{\Gamma}e_{\ominus p}(\sigma(T),0)\int_t^{\sigma(T)}e_{\ominus p}(t,s)h(s)\Delta s\\ &\quad+\frac{1}{\Gamma}\int_t^{\sigma(T)}\int_{\sigma(s)}^{\sigma(T)}e_{\ominus p}(t,s)g(r)e_{\ominus p}(r,0)h(s)\Delta r\Delta s \end{aligned}$$

$$+\frac{\alpha}{\Gamma}\sum_{\{j:t_j<t\}} e_{\ominus p}(t,t_j)-\frac{1}{\Gamma}\sum_{\{j:t_j<t\}} e_{\ominus p}(t,t_j)\left(\int_0^{\sigma(t_j)} g(\tau)e_{\ominus p}(\tau,0)\Delta\tau\right)v_j$$

$$+\frac{\beta e_{\ominus p}(\sigma(T),0)}{\Gamma}\sum_{\{j:t\le t_j\}} e_{\ominus p}(t,t_j)v_j+\frac{1}{\Gamma}\sum_{\{j:t\le t_j\}} e_{\ominus p}(t,t_j)$$

$$\times\left(\int_{\sigma(t_j)}^{\sigma(T)} g(r)e_{\ominus p}(r,0)\Delta r\right)v_j$$

$$=\frac{\alpha}{\Gamma}\int_0^t e_{\ominus p}(t,s)h(s)\Delta s-\frac{1}{\Gamma}\int_0^t e_{\ominus p}(t,s)\int_0^{\sigma(T)} g(\tau)e_{\ominus p}(\tau,0)h(s)\Delta\tau\Delta s$$

$$+\frac{1}{\Gamma}\int_0^t e_{\ominus p}(t,s)\int_{\sigma(s)}^{\sigma(T)} g(\tau)e_{\ominus p}(\tau,0)h(s)\Delta\tau\Delta s$$

$$+\frac{1}{\Gamma}\int_t^{\sigma(T)} e_{\ominus p}(t,s)\int_{\sigma(s)}^{\sigma(T)} g(\tau)e_{\ominus p}(\tau,0)h(s)\Delta\tau\Delta s$$

$$+\frac{\beta}{\Gamma}e_{\ominus p}(\sigma(T),0)\int_t^{\sigma(T)} e_{\ominus p}(t,s)h(s)\Delta s$$

$$+\frac{\alpha}{\Gamma}\sum_{\{j:t_j<t\}} e_{\ominus p}(t,t_j)-\frac{1}{\Gamma}\sum_{\{j:t_j<t\}} e_{\ominus p}(t,t_j)\left(\int_0^{\sigma(t_j)} g(\tau)e_{\ominus p}(\tau,0)\Delta\tau\right)v_j$$

$$+\frac{\beta e_{\ominus p}(\sigma(T),0)}{\Gamma}\sum_{\{j:t\le t_j\}} e_{\ominus p}(t,t_j)v_j+\frac{1}{\Gamma}\sum_{\{j:t\le t_j\}} e_{\ominus p}(t,t_j)$$

$$\times\left(\int_{\sigma(t_j)}^{\sigma(T)} g(r)e_{\ominus p}(r,0)\Delta r\right)v_j$$

$$=\frac{\alpha}{\Gamma}\int_0^t e_{\ominus p}(t,s)h(s)\Delta s-\frac{1}{\Gamma}\int_0^t e_{\ominus p}(t,s)\int_0^{\sigma(T)} g(\tau)e_{\ominus p}(\tau,0)h(s)\Delta\tau\Delta s$$

$$+\frac{1}{\Gamma}\int_0^{\sigma(T)} e_{\ominus p}(t,s)\int_{\sigma(s)}^{\sigma(T)} g(\tau)e_{\ominus p}(\tau,0)h(s)\Delta\tau\Delta s$$

$$+\frac{\beta}{\Gamma}e_{\ominus p}(\sigma(T),0)\int_t^{\sigma(T)} e_{\ominus p}(t,s)h(s)\Delta s$$

$$+\frac{\alpha}{\Gamma}\sum_{\{j:t_j<t\}} e_{\ominus p}(t,t_j)-\frac{1}{\Gamma}\sum_{\{j:t_j<t\}} e_{\ominus p}(t,t_j)\left(\int_0^{\sigma(t_j)} g(\tau)e_{\ominus p}(\tau,0)\Delta\tau\right)v_j$$

$$+\frac{\beta e_{\ominus p}(\sigma(T),0)}{\Gamma}\sum_{\{j:t\le t_j\}} e_{\ominus p}(t,t_j)v_j+\frac{1}{\Gamma}\sum_{\{j:t\le t_j\}} e_{\ominus p}(t,t_j)$$

$$\times\left(\int_{\sigma(t_j)}^{\sigma(T)} g(r)e_{\ominus p}(r,0)\Delta r\right)v_j$$

$$= \frac{\alpha}{\Gamma}\int_0^t e_{\ominus p}(t,s)h(s)\Delta s - \frac{1}{\Gamma}\int_0^t e_{\ominus p}(t,s)\int_0^{\sigma(T)} g(\tau)e_{\ominus p}(\tau,0)h(s)\Delta\tau\Delta s$$

$$+\frac{1}{\Gamma}\int_0^{\sigma(T)} e_{\ominus p}(t,s)\int_{\sigma(s)}^{\sigma(T)} g(\tau)e_{\ominus p}(\tau,0)h(s)\Delta\tau\Delta s$$

$$+\frac{\beta}{\Gamma}e_{\ominus p}(\sigma(T),0)\int_0^{\sigma(T)} e_{\ominus p}(t,s)h(s)\Delta s$$

$$+\frac{\alpha}{\Gamma}\sum_{\{j:t_j<t\}} e_{\ominus p}(t,t_j) - \frac{1}{\Gamma}\sum_{\{j:t_j<t\}} e_{\ominus p}(t,t_j)\left(\int_0^{\sigma(T)} g(\tau)e_{\ominus p}(\tau,0)\Delta\tau\right)v_j$$

$$+\frac{1}{\Gamma}\sum_{\{j:t_j<t\}} e_{\ominus p}(t,t_j)\left(\int_{\sigma(t_j)}^{\sigma(T)} g(\tau)e_{\ominus p}(\tau,0)\Delta\tau\right)v_j$$

$$+\frac{\beta e_{\ominus p}(\sigma(T),0)}{\Gamma}\sum_{\{j:t\le t_j\}} e_{\ominus p}(t,t_j)v_j + \frac{1}{\Gamma}\sum_{\{j:t\le t_j\}} e_{\ominus p}(t,t_j)$$

$$\times\left(\int_{\sigma(t_j)}^{\sigma(T)} g(r)e_{\ominus p}(r,0)\Delta r\right)v_j$$

$$= \frac{\alpha}{\Gamma}\int_0^t e_{\ominus p}(t,s)h(s)\Delta s - \frac{1}{\Gamma}\int_0^t e_{\ominus p}(t,s)\int_0^{\sigma(T)} g(\tau)e_{\ominus p}(\tau,0)h(s)\Delta\tau\Delta s$$

$$+\frac{1}{\Gamma}\int_0^{\sigma(T)} e_{\ominus p}(t,s)\int_{\sigma(s)}^{\sigma(T)} g(\tau)e_{\ominus p}(\tau,0)h(s)\Delta\tau\Delta s$$

$$+\frac{\beta}{\Gamma}e_{\ominus p}(\sigma(T),0)\int_t^{\sigma(T)} e_{\ominus p}(t,s)h(s)\Delta s + \frac{\alpha}{\Gamma}\sum_{\{j:t_j<t\}} e_{\ominus p}(t,t_j)v_j$$

$$-\frac{1}{\Gamma}\sum_{\{j:t_j<t\}} e_{\ominus p}(t,t_j)\left(\int_0^{\sigma(T)} g(\tau)e_{\ominus p}(\tau,0)\Delta\tau\right)v_j$$

$$+\frac{1}{\Gamma}\sum_{j=1}^{m} e_{\ominus p}(t,t_j)\left(\int_{\sigma(t_j)}^{\sigma(T)} g(\tau)e_{\ominus p}(\tau,0)\Delta\tau\right)v_j$$

$$+\frac{\beta e_{\ominus p}(\sigma(T),0)}{\Gamma}\sum_{\{j:t_j<t\}} e_{\ominus p}(t,t_j)v_j.$$

Proof. We have that $x \in PC([0,\sigma(T)])$ is a solution of the impulsive dynamic equation

$$\begin{aligned} x^{\Delta}(t)+p(t)x(\sigma(t)) &= h(t), \quad t\in J, \\ x(t_j^+) &= x(t_j^-)+v_j, \quad j\in\{1,\ldots,m\}, \end{aligned} \tag{2.46}$$

if and only if $x \in PC([0,\sigma(T)])$ and satisfies the integral equation

$$\begin{aligned} x(t) = &\int_0^t e_{\ominus p}(t,s)h(s)\Delta s + e_{\ominus p}(t,0)x(0) \\ &+ \sum_{\{j:t_j<t\}} e_{\ominus p}(t,t_j)v_j, \quad t\in[0,\sigma(T)]. \end{aligned} \tag{2.47}$$

1. Let $x \in PC^{\Delta}([0,\sigma(T)])$ be a solution of equation (2.45). Then x can be represented in the form (2.47). Hence,

$$\begin{aligned} x(\sigma(T)) = &\int_0^{\sigma(T)} e_{\ominus p}(\sigma(T),s)h(s)\Delta s + e_{\ominus p}(\sigma(T),0)x(0) \\ &+ \sum_{\{j:t_j<\sigma(T)\}} e_{\ominus p}(\sigma(T),t_j)v_j, \end{aligned}$$

$$\begin{aligned} \beta x(\sigma(T)) = &\beta\int_0^{\sigma(T)} e_{\ominus p}(\sigma(T),s)h(s)\Delta s + \beta e_{\ominus p}(\sigma(T),0)x(0) \\ &+\beta \sum_{\{j:t_j<\sigma(T)\}} e_{\ominus p}(\sigma(T),t_j)v_j, \end{aligned}$$

and

$$\begin{aligned} \alpha x(0)-\beta x(\sigma(T)) &= \alpha x(0) - \beta\int_0^{\sigma(T)} e_{\ominus p}(\sigma(T),s)h(s)\Delta s - \beta e_{\ominus p}(\sigma(T),0)x(0) \\ &\quad -\beta \sum_{\{j:t_j<\sigma(T)\}} e_{\ominus p}(\sigma(T),t_j)v_j \\ &= (\alpha - \beta e_{\ominus p}(\sigma(T),0))x(0) - \beta\int_0^{\sigma(T)} e_{\ominus p}(\sigma(T),s)h(s)\Delta s \\ &\quad -\beta \sum_{\{j:t_j<\sigma(T)\}} e_{\ominus p}(\sigma(T),t_j)v_j \end{aligned} \tag{2.48}$$

$$= \int_0^{\sigma(T)} g(s)x(s)\Delta s$$

$$= \int_0^{\sigma(T)} g(s)\Bigg(\int_0^s e_{\ominus p}(s,\tau)h(\tau)\Delta\tau + e_{\ominus p}(s,0)x(0)$$

$$+ \sum_{\{j:t_j<s\}} e_{\ominus p}(s,t_j)v_j \Bigg)\Delta s$$

$$= \int_0^{\sigma(T)} \int_0^s g(s)e_{\ominus p}(s,\tau)h(\tau)\Delta\tau\Delta s + \int_0^{\sigma(T)} g(s)e_{\ominus p}(s,0)x(0)\Delta s$$

$$+ \int_0^{\sigma(T)} g(s)\left(\sum_{\{j:t_j<s\}} e_{\ominus p}(s,t_j)v_j \right)\Delta s,$$

whereupon,

$$\left(\alpha - \beta e_{\ominus p}(\sigma(T),0) - \int_0^{\sigma(T)} g(s)e_{\ominus p}(s,0)\Delta s\right)x(0)$$

$$= \int_0^{\sigma(T)} \int_0^s g(s)e_{\ominus p}(s,\tau)h(\tau)\Delta\tau\Delta s + \int_0^{\sigma(T)} g(s)\left(\sum_{\{j:t_j<s\}} e_{\ominus p}(s,t_j)v_j \right)\Delta s$$

$$+\beta \int_0^{\sigma(T)} e_{\ominus p}(\sigma(T),s)h(s)\Delta s + \beta \sum_{\{j:t_j<\sigma(T)\}} e_{\ominus p}(\sigma(T),t_j)v_j$$

and

$$x(0) = \frac{1}{\Gamma}\Bigg(\int_0^{\sigma(T)} \int_0^s g(s)e_{\ominus p}(s,\tau)h(\tau)\Delta\tau\Delta s$$

$$+ \int_0^{\sigma(T)} g(s)\left(\sum_{\{j:t_j<s\}} e_{\ominus p}(s,t_j)v_j \right)\Delta s$$

$$+\beta \int_0^{\sigma(T)} e_{\ominus p}(\sigma(T),s)h(s)\Delta s + \beta \sum_{\{j:t_j<\sigma(T)\}} e_{\ominus p}(\sigma(T),t_j)v_j \Bigg).$$

Note that

$$\int_0^{\sigma(T)} g(s)\left(\sum_{\{j:t_j<s\}} e_{\ominus p}(s,t_j)v_j \right)\Delta s = \int_0^{t_1} g(s)\left(\sum_{\{j:t_j<s\}} e_{\ominus p}(s,t_j)v_j \right)\Delta s$$

$$
\begin{aligned}
&+\int_{t_1}^{t_2} g(s)\left(\sum_{\{j:t_j<s\}} e_{\ominus p}(s,t_j)v_j\right)\Delta s \\
&+\int_{t_2}^{t_3} g(s)\left(\sum_{\{j:t_j<s\}} e_{\ominus p}(s,t_j)v_j\right)\Delta s \\
&+\cdots \\
&+\int_{t_m}^{\sigma(T)} g(s)\left(\sum_{\{j:t_j<s\}} e_{\ominus p}(s,t_j)v_j\right)\Delta s \\
=&\int_{t_1}^{t_2} g(s)e_{\ominus p}(s,t_1)v_1\Delta s \\
&+\int_{t_2}^{t_3} g(s)\left(e_{\ominus p}(s,t_1)v_1+e_{\ominus p}(s,t_2)v_2\right)\Delta s \\
&+\cdots \\
&+\int_{t_m}^{\sigma(T)} g(s)\Big(e_{\ominus p}(s,t_1)v_1+e_{\ominus p}(s,t_2)v_2 \\
&+\cdots+e_{\ominus p}(s,t_m)v_m\Big)\Delta s \\
=&\, v_1\int_{t_1}^{\sigma(T)} g(s)e_{\ominus p}(s,t_1)\Delta s \\
&+v_2\int_{t_2}^{\sigma(T)} g(s)e_{\ominus p}(s,t_2)\Delta s \\
&+\cdots \\
&+v_m\int_{t_m}^{\sigma(T)} g(s)e_{\ominus p}(s,t_m)\Delta s \\
=&\sum_{j=1}^{m} v_j\int_{t_j}^{\sigma(T)} g(s)e_{\ominus p}(s,t_j)\Delta s.
\end{aligned}
$$

Therefore,

$$
x(0)=\frac{1}{\Gamma}\Big(\int_0^{\sigma(T)}\int_{\sigma(s)}^{\sigma(T)} g(s)e_{\ominus p}(s,\tau)h(\tau)\Delta\tau\Delta s+\sum_{j=1}^{m} v_j\int_{t_j}^{\sigma(T)} g(s)e_{\ominus p}(s,t_j)\Delta s
$$

$$
\begin{aligned}
&+ \beta \int_0^{\sigma(T)} e_{\ominus p}(\sigma(T),s)h(s)\Delta s + \beta \sum_{\{j:t_j<\sigma(T)\}} e_{\ominus p}(\sigma(T),t_j)v_j\Bigg) \\
&= \frac{1}{\Gamma}\Bigg(\beta \int_0^{\sigma(T)} e_{\ominus p}(\sigma(T),s)h(s)\Delta s \\
&+ \int_0^{\sigma(T)}\int_0^{\sigma(T)} g(s)e_{\ominus p}(s,\tau)h(\tau)\Delta\tau\Delta s - \int_0^{\sigma(T)}\int_0^{\sigma(s)} g(s)e_{\ominus p}(s,\tau)h(\tau)\Delta\tau\Delta s \\
&+ \sum_{j=1}^{m} v_j\left(\int_{t_j}^{\sigma(T)} g(s)e_{\ominus p}(s,t_j)\Delta s + \beta e_{\ominus p}(\sigma(T),t_j)\right)\Bigg).
\end{aligned}
$$

Then

$$
\begin{aligned}
x(t) &= \frac{e_{\ominus p}(t,0)}{\Gamma}\Bigg(\beta \int_0^{\sigma(T)} e_{\ominus p}(\sigma(T),s)h(s)\Delta s \\
&+ \int_0^{\sigma(T)}\int_0^{\sigma(T)} g(s)e_{\ominus p}(s,\tau)h(\tau)\Delta\tau\Delta s - \int_0^{\sigma(T)}\int_0^{\sigma(s)} g(s)e_{\ominus p}(s,\tau)h(\tau)\Delta\tau\Delta s \\
&+ \sum_{j=1}^{m} v_j\left(\int_{t_j}^{\sigma(T)} g(s)e_{\ominus p}(s,t_j)\Delta s + \beta e_{\ominus p}(\sigma(T),t_j)\right)\Bigg) \\
&+ \int_0^{t} e_{\ominus p}(t,s)h(s)\Delta s + \sum_{\{j:t_j<t\}} e_{\ominus p}(t,t_j)v_j \\
&= \frac{e_{\ominus p}(t,0)}{\Gamma}\Bigg(\beta \int_0^{\sigma(T)} e_{\ominus p}(\sigma(T),s)h(s)\Delta s \\
&+ \int_0^{\sigma(T)}\int_0^{\sigma(T)} g(s)e_{\ominus p}(s,\tau)h(\tau)\Delta\tau\Delta s - \int_0^{\sigma(T)}\int_0^{\sigma(s)} g(s)e_{\ominus p}(s,\tau)h(\tau)\Delta\tau\Delta s \\
&+ \sum_{j=1}^{m} v_j\left(\int_{t_j}^{\sigma(T)} g(s)e_{\ominus p}(s,t_j)\Delta s + \beta e_{\ominus p}(\sigma(T),t_j)\right)\Bigg) \\
&+ \frac{1}{\Gamma}\Bigg(\alpha \int_0^{t} e_{\ominus p}(t,s)h(s)\Delta s - \beta e_p(0,\sigma(T)) \int_0^{t} e_{\ominus p}(t,s)h(s)\Delta s \\
&- \left(\int_0^{\sigma(T)} g(s)e_p(0,s)\Delta s\right)\int_0^{t} e_{\ominus p}(t,s)h(s)\Delta s \\
&+ \alpha \sum_{\{j:t_j<t\}} e_{\ominus p}(t,t_j)v_j - \beta e_p(0,\sigma(T)) \sum_{\{j:t_j<t\}} e_{\ominus p}(t,t_j)v_j
\end{aligned}
$$

$$- \left(\int_0^{\sigma(T)} g(s) e_p(0,s) \Delta s \right) \sum_{\{j: t_j < t\}} e_{\ominus p}(t,t_j) v_j \Bigg)$$

$$= \frac{\alpha}{\Gamma} \int_0^t e_{\ominus p}(t,s) h(s) \Delta s - \frac{1}{\Gamma} \int_0^{\sigma(T)} \int_0^t e_{\ominus p}(t,s) e_{\ominus p}(\tau,0) g(\tau) h(s) \Delta s \Delta \tau$$

$$+ \frac{1}{\Gamma} \int_0^{\sigma(T)} \int_{\sigma(s)}^{\sigma(T)} g(s) e_{\ominus p}(s,0) e_{\ominus p}(t,\tau) h(\tau) \Delta \tau \Delta s$$

$$+ \frac{\beta}{\Gamma} e_{\ominus p}(\sigma(T),0) \int_0^{\sigma(T)} e_{\ominus p}(t,s) h(s) \Delta s + \frac{\alpha}{\Gamma} \sum_{\{j: t_j < t\}} e_{\ominus p}(t,t_j) v_j$$

$$+ \frac{1}{\Gamma} e_{\ominus p}(t,0) \sum_{j=1}^m v_j \left(\int_{t_j}^{\sigma(T)} g(s) e_{\ominus p}(s,t_j) \Delta s \right)$$

$$+ \frac{\beta}{\Gamma} e_{\ominus p}(\sigma(T),0) \sum_{\{j: t_j < t\}} e_{\ominus p}(t,t_j) v_j$$

$$- \frac{1}{\Gamma} \left(\int_0^{\sigma(T)} g(s) e_p(0,s) \Delta s \right) \sum_{\{j: t_j < t\}} e_{\ominus p}(t,t_j) v_j$$

$$= \int_0^{\sigma(T)} G(t,s) h(s) \Delta s + \sum_{j=1}^m G(t,t_j) v_j,$$

i.e., x satisfy (2.44).

2. Let $x \in PC([0,\sigma(T)])$ satisfy (2.44). Define

$$H(s) = e_p(t,0) G(t,s), \quad t,s \in [0,\sigma(T)].$$

Then

$$H(s) = \begin{cases} \frac{1}{\Gamma} e_p(s,0) \left(\alpha - \int_0^{\sigma(s)} g(r) e_p(0,r) \Delta r \right) & 0 \le s \le t \le \sigma(T), \\ \frac{1}{\Gamma} e_p(s,0) \left(\beta e_p(0,\sigma(T)) + \int_{\sigma(s)}^{\sigma(T)} g(r) e_p(0,r) \Delta r \right) & 0 \le t \le s \le \sigma(T). \end{cases}$$

Then

$$x(t) e_p(t,0) = \int_0^{\sigma(T)} H(s) h(s) \Delta s + \sum_{j=1}^m H(t_j) v_j, \quad t \in [0,\sigma(T)].$$

Hence,

$$(x e_p(\cdot,0))^{\Delta}(t) = \left(\int_0^{\sigma(T)} H(s) h(s) \Delta s \right)^{\Delta}$$

$$
\begin{aligned}
&= \left(\int_0^t H(s)h(s)\Delta s + \int_t^{\sigma(T)} H(s)h(s)\Delta s\right)^{\Delta} \\
&= \left(\frac{1}{\Gamma}\int_0^t e_p(s,0)\left(\alpha - \int_0^{\sigma(s)} g(r)e_p(0,r)\Delta r\right)h(s)\Delta s\right)^{\Delta} \\
&\quad + \left(\frac{1}{\Gamma}\int_t^{\sigma(T)} e_p(s,0)\biggl(\beta e_p(0,\sigma(T))\right. \\
&\quad \left. + \int_{\sigma(s)}^{\sigma(T)} g(r)e_p(0,r)\Delta r\biggr)h(s)\Delta s\right)^{\Delta} \\
&= \frac{1}{\Gamma}e_p(t,0)\left(\alpha - \int_0^{\sigma(t)} g(r)e_p(0,r)\Delta r\right)h(t) \\
&\quad - \frac{1}{\Gamma}e_p(t,0)\left(\beta e_p(0,\sigma(T)) + \int_{\sigma(t)}^{\sigma(T)} g(r)e_p(0,r)\Delta r\right)h(t) \\
&= \frac{1}{\Gamma}e_p(t,0)\left(\alpha - \beta e_p(0,\sigma(T)) - \int_0^{\sigma(T)} g(r)e_p(0,r)\Delta r\right)h(t) \\
&= e_p(t,0)h(t), \quad t \in [0,\sigma(T)].
\end{aligned}
$$

Hence and

$$
x^{\Delta}(t)e_p(t,0) + p(t)e_p(t,0)x(\sigma(t)) = e_p(t,0)h(t), \quad t \in [0,\sigma(T)],
$$

or

$$
x^{\Delta}(t) + p(t)x(\sigma(t)) = h(t), \quad t \in [0,\sigma(T)].
$$

Next,

$$
\begin{aligned}
x(t_k^+) &= \int_0^{\sigma(T)} G(t_k^+,s)h(s)\Delta s + \sum_{j=1}^m G(t_k^+,t_j)v_j \\
&= \int_0^{t_k} G(t_k^+,s)h(s)\Delta s + \int_{t_k}^{\sigma(T)} G(t_k^+,s)h(s)\Delta s \\
&\quad + \sum_{\{j:t_j<t_k^+\}} G(t_k^+,t_j)v_j + \sum_{\{j:t_j\geq t_k^+\}} G(t_k^+,t_j)v_j, \\
x(t_k^-) &= \int_0^{\sigma(T)} G(t_k^-,s)h(s)\Delta s + \sum_{j=1}^m G(t_k^-,t_j)v_j \\
&= \int_0^{t_k} G(t_k^-,s)h(s)\Delta s + \int_{t_k}^{\sigma(T)} G(t_k^-,s)h(s)\Delta s
\end{aligned}
$$

$$+\sum_{\{j:t_j<t_k^-\}} G(t_k^-,t_j)v_j + \sum_{\{j:t_j\geq t_k^-\}} G(t_k^-,t_j)v_j,$$

$$\begin{aligned}
x(t_k^+)-x(t_k^-) &= \left(G(t_k^+,t_k)-G(t_k^-,t_k)\right)v_k \\
&= \frac{1}{\Gamma}\Bigg(\alpha-\int_0^{\sigma(t_k)} g(r)e_p(0,r)\Delta r-\beta e_p(0,\sigma(T)) \\
&\quad -\int_{\sigma(t_k)}^{\sigma(T)} g(r)e_p(0,r)\Delta r\Bigg)v_k \\
&= \frac{1}{\Gamma}\left(\alpha-\int_0^{\sigma(T)} g(r)e_p(0,r)\Delta r-\beta e_p(0,\sigma(T))\right)v_k \\
&= v_k.
\end{aligned}$$

Also, by (2.48), we get

$$\begin{aligned}
\alpha x(0)-\beta x(\sigma(T)) &= \int_0^{\sigma(T)} g(s)\Bigg(\int_0^s e_{\ominus p}(s,\tau)h(\tau)\Delta\tau+e_{\ominus p}(s,0)x(0) \\
&\quad +\sum_{\{j:t_j<s\}} e_{\ominus p}(s,t_j)v_j\Bigg)\Delta s \\
&= \int_0^{\sigma(T)} g(s)x(s)\Delta s.
\end{aligned}$$

This completes the proof.

Lemma 2.14. *Suppose* $(B23)$, $(B25)$ *and* $(B26)$. *Then*

$$A\leq G(t,s)\leq B,\quad t,s\in[0,\sigma(T)],$$

where

$$\begin{aligned}
A &= \frac{\beta}{\Gamma}\left(e_p(0,\sigma(T))\right)^2, \\
B &= \frac{1}{\Gamma}e_p(\sigma(T),0)\left(\alpha+\beta e_p(0,\sigma(T))+\int_0^{\sigma(T)} g(s)e_p(0,s)\Delta s\right).
\end{aligned}$$

Proof. By $(B26)$, we get

$$\alpha-\int_0^{\sigma(T)} g(s)e_p(0,s)\Delta s\geq\beta e_p(0,\sigma(T)).$$

By $(B23)$, we obtain

$$e_p(s,0) \geq 1, \quad s \in [0,\sigma(T)],$$

$$e_p(0,t) \geq e_p(0,\sigma(T)), \quad t \in [0,\sigma(T)].$$

Then

$$G(t,s) = \begin{cases} \frac{1}{\Gamma} e_{\ominus p}(t,s)\left(\alpha - \int_0^{\sigma(s)} g(\tau) e_{\ominus p}(\tau,0)\Delta\tau\right) & 0 \leq s \leq t \leq \sigma(T) \\ \frac{1}{\Gamma} e_{\ominus p}(t,s)\left(\beta e_{\ominus p}(\sigma(T),0) + \int_{\sigma(s)}^{\sigma(T)} g(r) e_{\ominus p}(r,0)\Delta r\right) \\ 0 \leq t \leq s \leq \sigma(T) \end{cases}$$

$$\geq \begin{cases} \frac{1}{\Gamma} e_p(s,0) e_p(0,t)\left(\alpha - \int_0^{\sigma(T)} g(\tau) e_{\ominus p}(\tau,0)\Delta\tau\right) & 0 \leq s < t \leq \sigma(T) \\ \frac{1}{\Gamma} e_p(s,0) e_p(0,t)\left(\beta e_{\ominus p}(\sigma(T),0)\right) & 0 \leq t \leq s \leq \sigma(T) \end{cases}$$

$$\geq \begin{cases} \frac{1}{\Gamma} e_p(0,\sigma(T))(\beta e_p(0,\sigma(T))) & 0 \leq s < t \leq \sigma(T) \\ \frac{\beta}{\Gamma}\left(e_p(0,\sigma(T))\right)^2 & 0 \leq t \leq s \leq \sigma(T) \end{cases}$$

$$= A, \quad t,s \in [0,\sigma(T)].$$

Next, by $(B23)$, we have

$$e_p(s,0) \leq e_p(\sigma(T),0), \quad s \in [0,\sigma(T)],$$

$$e_p(0,t) \leq 1, \quad t \in [0,\sigma(T)],$$

and

$$G(t,s) = \begin{cases} \frac{1}{\Gamma} e_{\ominus p}(t,s)\left(\alpha - \int_0^{\sigma(s)} g(\tau) e_{\ominus p}(\tau,0)\Delta\tau\right) & 0 \leq s \leq t \leq \sigma(T) \\ \frac{1}{\Gamma} e_{\ominus p}(t,s)\left(\beta e_{\ominus p}(\sigma(T),0) + \int_{\sigma(s)}^{\sigma(T)} g(r) e_{\ominus p}(r,0)\Delta r\right) \\ 0 \leq t \leq s \leq \sigma(T) \end{cases}$$

$$= \begin{cases} \frac{1}{\Gamma} e_p(s,0) e_p(0,t) \left(\alpha - \int_0^{\sigma(s)} g(r) e_p(0,r) \Delta r \right) & 0 \le s < t \le \sigma(T) \\ \frac{1}{\Gamma} e_p(s,0) e_p(0,t) \left(\beta e_p(0,\sigma(T)) + \int_{\sigma(s)}^{\sigma(T)} g(r) e_p(0,r) \Delta r \right) \\ 0 \le t \le s \le \sigma(T) \end{cases}$$

$$\le \begin{cases} \frac{1}{\Gamma} e_p(\sigma(T),0) \left(\alpha + \int_0^{\sigma(T)} g(r) e_p(0,r) \Delta r + \beta e_p(0,\sigma(T)) \right) \\ 0 \le s < t \le \sigma(T) \\ \frac{1}{\Gamma} e_p(\sigma(T),0) \left(\alpha + \int_0^{\sigma(T)} g(r) e_p(0,r) \Delta r + \beta e_p(0,\sigma(T)) \right) \\ 0 \le t \le s \le \sigma(T) \end{cases}$$

$$= B, \quad t,s \in [0,\sigma(T)].$$

This completes the proof.

For $x \in PC([0,\sigma(T)])$, define the operator

$$\Phi x(t) = \int_0^{\sigma(T)} G(t,s) f(s, x(\sigma(s))) \Delta s + \sum_{j=1}^{m} G(t,t_j) I_j(x(t_j)), \quad t \in [0,\sigma(T)].$$

Lemma 2.15. *The operator $\Phi : PC([0,\sigma(T)]) \to PC([0,\sigma(T)])$ is completely continuous.*

Proof. Let $\{x_n\}_{n \in \mathbb{N}}$ be a sequence of elements of $PC([0,\sigma(T)])$ such that $x_n \to x$, as $n \to \infty$, in $PC([0,\sigma(T)])$, for some $x \in PC([0,\sigma(T)])$. Then

$$\begin{aligned} |\Phi x_n(t) - \Phi x(t)| &= \Bigg| \int_0^{\sigma(T)} G(t,s) f(s, x_n(\sigma(s))) \Delta s + \sum_{j=1}^{m} G(t,t_j) I_j(x(t_j)) \\ &\quad - \int_0^{\sigma(T)} G(t,s) f(s, x(\sigma(s))) \Delta s - \sum_{j=1}^{m} G(t,t_j) I_j(x(t_j)) \Bigg| \\ &= \Bigg| \int_0^{\sigma(T)} G(t,s) (f(s, x_n(\sigma(s))) - f(s, x(\sigma(s)))) \Delta s \\ &\quad + \sum_{j=1}^{m} G(t,t_j) (I_j(x_n(t_j)) - I_j(x(t_j))) \Bigg| \\ &\le \left| \int_0^{\sigma(T)} G(t,s) (f(s, x_n(\sigma(s))) - f(s, x(\sigma(s)))) \Delta s \right| \end{aligned}$$

$$
\begin{aligned}
&+\left|\sum_{j=1}^{m} G(t,t_j)(I_j(x_n(t_j))-I_j(x(t_j)))\right| \\
&\leq \int_0^{\sigma(T)} G(t,s)|f(s,x_n(\sigma(s)))-f(s,x(\sigma(s)))|\Delta s \\
&\quad+\sum_{j=1}^{m} G(t,t_j)|I_j(x_n(t_j))-I_j(x(t_j))| \\
&\leq B\bigg(\int_0^{\sigma(T)} |f(s,x_n(\sigma(s)))-f(s,x(\sigma(s)))|\Delta s \\
&\quad+\sum_{j=1}^{m} |I_j(x_n(t_j))-I_j(x(t_j))|\bigg) \\
&\to 0, \quad \text{as} \quad n\to\infty.
\end{aligned}
$$

Hence,

$$\|\Phi x_n-\Phi x\|\to 0, \quad \text{as} \quad n\to\infty.$$

Let

$$C=\{x\in PC([0,\sigma(T)]):\|x\|\leq\gamma\}$$

for some positive constant γ. Then there exists a positive constant γ_1 such that, for $x\in C$, we have

$$|f(t,x(\sigma(t)))|\leq\gamma_1, \quad t\in[0,\sigma(T)],$$

$$|I_j(x(t_j))|\leq\gamma_1, \quad j\in\{1,\ldots,m\}.$$

Hence, for $x\in C$, we have

$$
\begin{aligned}
|\Phi x(t)| &= \left|\int_0^{\sigma(T)} G(t,s)f(s,x(\sigma(s)))\Delta s+\sum_{j=1}^{m} G(t,t_j)I_j(x(t_j))\right| \\
&\leq \left|\int_0^{\sigma(T)} G(t,s)f(s,x(\sigma(s)))\Delta s\right|+\left|\sum_{j=1}^{m} G(t,t_j)I_j(x(t_j))\right| \\
&\leq \int_0^{\sigma(T)} G(t,s)|f(s,x(\sigma(s)))|\Delta s+\sum_{j=1}^{m} G(t,t_j)|I_j(x(t_j))| \\
&\leq B\gamma_1\sigma(T)+Bm\gamma_1, \quad t\in[0,\sigma(T)].
\end{aligned}
$$

Then, for $x\in C$, we have

$$\|\Phi x\| \le B\gamma_1(m+\sigma(T)).$$

Let $t^1, t^2 \in [0, \sigma(T)]$ be arbitrarily chosen. Then, for $x \in PC([0, \sigma(T)])$, we have

$$\begin{aligned}
|\Phi x(t^1) - \Phi x(t^2)| &= \Bigg| \int_0^{\sigma(T)} G(t^1, s) f(s, x(\sigma(s)))\Delta s + \sum_{j=1}^m G(t^1, t_j) I_j(x(t_j)) \\
&\quad - \int_0^{\sigma(T)} G(t^2, s) f(s, x(\sigma(s)))\Delta s - \sum_{j=1}^m G(t^2, t_j) I_j(x(t_j)) \Bigg| \\
&= \Bigg| \int_0^{\sigma(T)} (G(t^1, s) - G(t^2, s)) f(s, x(\sigma(s)))\Delta s \\
&\quad + \sum_{j=1}^m (G(t^1, t_j) - G(t^2, t_j)) I_j(x(t_j)) \Bigg| \\
&\le \int_0^{\sigma(T)} |G(t^1, s) - G(t^2, s)| |f(s, x(\sigma(s)))| \Delta s \\
&\quad + \sum_{j=1}^m |G(t^1, t_j) - G(t^2, t_j)| |I_j(x(t_j))| \\
&\to 0, \quad \text{as} \quad t^1 \to t^2.
\end{aligned}$$

Now, using the Arzela-Ascoli theorem, we conclude that $\Phi : PC([0, \sigma(T)]) \to PC([0, \sigma(T)])$ is completely continuous. This completes the proof.

Let $\delta = \dfrac{A}{B}$ and

$$K = \{x \in PC([0, \sigma(T)]) : x(t) \ge \delta \|x\|, \quad t \in [0, \sigma(T)]\}.$$

Lemma 2.16. *We have $\Phi : K \to K$.*

Proof. Let $x \in K$ be arbitrarily chosen. Then

$$\begin{aligned}
\Phi x(t) &= \int_0^{\sigma(T)} G(t, s) f(s, x(\sigma(s)))\Delta s + \sum_{j=1}^m G(t, t_j) I_j(x(t_j)) \\
&\le B\left(\int_0^{\sigma(T)} f(s, x(\sigma(s)))\Delta s + \sum_{j=1}^m I_j(x(t_j)) \right), \quad t \in [0, \sigma(T)],
\end{aligned}$$

whereupon,

$$\|\Phi x\| \le B\left(\int_0^{\sigma(T)} f(s, x(\sigma(s)))\Delta s + \sum_{j=1}^m I_j(x(t_j)) \right).$$

Next,

$$\begin{aligned}
\Phi x(t) &= \int_0^{\sigma(T)} G(t,s) f(s,x(\sigma(s)))\Delta s + \sum_{j=1}^{m} G(t,t_j) I_j(x(t_j)) \\
&\geq A\left(\int_0^{\sigma(T)} f(s,x(\sigma(s)))\Delta s + \sum_{j=1}^{m} I_j(x(t_j))\right) \\
&\geq \frac{A}{B}\|\Phi x\| \\
&= \delta\|\Phi x\|, \quad t \in [0,\sigma(T)],
\end{aligned}$$

i.e., $\Phi x \in K$. This completes the proof.

Let

$$\max f_0 = \lim_{x\to 0} \max_{t\in[0,T]} \frac{f(t,x)}{x}, \quad \min f_0 = \lim_{x\to 0} \min_{t\in[0,T]} \frac{f(t,x)}{x},$$

$$\max f_\infty = \lim_{x\to \infty} \max_{t\in[0,T]} \frac{f(t,x)}{x}, \quad \min f_\infty = \lim_{x\to \infty} \min_{t\in[0,T]} \frac{f(t,x)}{x},$$

$$I_{j0} = \lim_{x\to 0} \frac{I_j(x)}{x}, \quad I_{j\infty} = \lim_{x\to\infty} \frac{I_j(x)}{x}, \quad j \in \{1,\ldots,m\}.$$

Theorem 2.22. *Assume* $(B23)$-$(B26)$ *and*

$$\max f_0 = 0, \quad \min f_\infty = \infty, \quad I_{j0} = 0, \quad j \in \{1,\ldots,m\}.$$

Then problem (2.43) *has at least one positive solution.*

Proof. By Lemma 2.15, it follows that $\Phi : PC([0,\sigma(T)]) \to PC([0,\sigma(T)])$ is completely continuous. From Lemma 2.16, we have that $\Phi : K \to K$. Since

$$\max f_0 = 0 \quad \text{and} \quad I_{j0} = 0, \quad j \in \{1,\ldots,m\},$$

for $\varepsilon \in \left(0,(B\sigma(T)+Bm)^{-1}\right]$, there exists a positive constant r_1 such that

$$f(t,x) \leq \varepsilon x, \quad I_j(x) \leq \varepsilon x, \quad j \in \{1,\ldots,m\}, \quad x \in \Omega_{r_1},$$

$t \in [0,\sigma(T)]$, where

$$\Omega_{r_1} = \{x \in K : \|x\| < r_1\}.$$

Because $\min f_\infty = \infty$, for $M \geq (A\sigma(T)\delta)^{-1}$, there exists a positive constant $r_2 \geq \dfrac{r_1}{\delta}$ such that

$$f(t,x) \geq Mx, \quad x \in \Omega_{r_2}, \quad t \in [0,\sigma(T)],$$

where

$$\Omega_{r_2} = \{x \in K : \|x\| < r_2\}.$$

For $x \in K \bigcap \partial \Omega_1$, we have

$$\begin{aligned}
\Phi x(t) &= \int_0^{\sigma(T)} G(t,s) f(s, x(\sigma(s)))\Delta s + \sum_{j=1}^{m} G(t,t_j) I_j(x(t_j)) \\
&\leq B\left(\int_0^{\sigma(T)} f(s, x(\sigma(s)))\Delta s + \sum_{j=1}^{m} I_j(x(t_j))\right) \\
&\leq \varepsilon B\left(\int_0^{\sigma(T)} x(\sigma(s))\Delta s + \sum_{j=1}^{m} x(t_j)\right) \\
&\leq \varepsilon B(\sigma(T) + m)\|x\| \\
&\leq \|x\|, \quad t \in [0, \sigma(T)].
\end{aligned}$$

Then

$$\|\Phi x\| \leq \|x\|, \quad x \in K \bigcap \partial \Omega_{r_1}.$$

For $x \in K \bigcap \partial \Omega_{r_2}$, we have

$$x(\sigma(s)) \geq \delta \|x\|, \quad s \in [0, T],$$

and

$$\begin{aligned}
\Phi x(t) &= \int_0^{\sigma(T)} G(t,s) f(s, x(\sigma(s)))\Delta s + \sum_{j=1}^{m} G(t,t_j) I_j(x(t_j)) \\
&\geq \int_0^{\sigma(T)} G(t,s) f(s, x(\sigma(s)))\Delta s \\
&\geq A \int_0^{\sigma(T)} f(s, x(\sigma(s)))\Delta s \\
&\geq AM \int_0^{\sigma(T)} x(\sigma(s))\Delta s \\
&\geq AM\delta \|x\| \sigma(T) \\
&\geq \|x\|, \quad t \in [0, \sigma(T)].
\end{aligned}$$

Therefore,

$$\|\Phi x\| \geq \|x\|, \quad x \in K \bigcap \partial \Omega_{r_2}.$$

Hence and applying the Guo-Krasnosel'skii Theorem 1.11, we conclude that Φ has at least one fixed point in $K\bigcap(\overline{\Omega_{r_2}}\backslash\Omega_{r_1})$. This completes the proof.

Theorem 2.23. *Assume* $(B23)$-$(B27)$,

$$\max f_\infty = 0, \quad \min f_0 = \infty, \quad I_{j\infty} = 0, \quad j \in \{1,\ldots,m\}.$$

Then problem (2.43) *has at least one positive solution.*

Proof. By Lemma 2.15, we have that $\Phi : PC([0,\sigma(T)]) \to PC([0,\sigma(T)])$ is completely continuous. From Lemma 2.16, it follows that $\Phi : K \to K$. Since

$$\max f_\infty = 0 \quad \text{and} \quad I_{j\infty} = 0, \quad j \in \{1,\ldots,m\},$$

for $\varepsilon \in \left(0,(B\sigma(T)+Bm)^{-1}\right]$, it follows that there exists a positive constant r_1 such that

$$f(t,x) \le \varepsilon x, \quad I_j(x) \le \varepsilon x, \quad j \in \{1,\ldots,m\}, \quad x \in [\delta r_1,\infty), \quad t \in [0,\sigma(T)].$$

Since $\min f_0 = \infty$, for $M > (A\sigma(T)\delta)^{-1}$, there exists a positive constant r_2 such that $r_2 < \delta r_1$ and

$$f(t,x) \ge Mx, \quad x \in (0,r_2], \quad t \in [0,\sigma(T)].$$

Let

$$\Omega_1 = \{x \in K : \|x\| \le r_1\},$$

$$\Omega_2 = \{x \in K : \|x\| < r_2\}.$$

We have

$$\overline{\Omega_2} \subset \Omega_1.$$

For $x \in K\bigcap\partial\Omega_1$, we get

$$\begin{aligned}
\Phi x(t) &= \int_0^{\sigma(T)} G(t,s)f(s,x(\sigma(s)))\Delta s + \sum_{j=1}^m G(t,t_j)I_j(x(t_j)) \\
&\le B\left(\int_0^{\sigma(T)} f(s,x(\sigma(s)))\Delta s + \sum_{j=1}^m I_j(x(t_j))\right) \\
&\le B\left(\varepsilon\int_0^{\sigma(T)} x(\sigma(s))\Delta s + \varepsilon\sum_{j=1}^m x(t_j)\right) \\
&\le B\varepsilon(\sigma(T)+m)\|x\| \\
&\le \|x\|, \quad t \in [0,\sigma(T)],
\end{aligned}$$

whereupon,

$$\|\Phi x\| \leq \|x\|.$$

For $x \in K\bigcap \partial\Omega_2$, we obtain

$$x(\sigma(s)) \geq \delta\|x\|, \quad s \in [0,T],$$

and

$$\begin{aligned}
\Phi x(t) &= \int_0^{\sigma(T)} G(t,s)f(s,x(\sigma(s)))\Delta s + \sum_{j=1}^{m} G(t,t_j)I_j(x(t_j)) \\
&\geq \int_0^{\sigma(T)} G(t,s)f(s,x(\sigma(s)))\Delta s \\
&\geq A\int_0^{\sigma(T)} f(s,x(\sigma(s)))\Delta s \\
&\geq AM\int_0^{\sigma(T)} x(\sigma(s))\Delta s \\
&\geq MA\delta\sigma(T)\|x\| \\
&\geq \|x\|, \quad t \in [0,\sigma(T)],
\end{aligned}$$

whereupon,

$$\|\Phi x\| \geq \|x\|.$$

Hence and applying the Guo-Krasnosel'skii Theorem 1.11, it follows that the operator Φ has a fixed point in $K\bigcap(\overline{\Omega_2}\backslash\Omega_1)$. This completes the proof.

Theorem 2.24. *Assume* (B23)-(B27),

$$\min f_0 = \infty, \quad \min f_\infty = \infty,$$

there exists a positive constant R *such that*

$$f(t,x) < \frac{R}{2B\sigma(T)}, \quad t \in [0,\sigma(T)], \quad 0 < x \leq R,$$

$$I_j(x) \leq \frac{x}{2Bm}, \quad x \in (0,\infty), \quad j \in \{1,\dots,m\}.$$

Then problem (2.43) *has at least two positive solutions.*

Proof. Let

$$\Omega_R = \{x \in K : \|x\| < R\}.$$

Then, for $x \in K\bigcap \partial\Omega_R$, we get

$$\begin{aligned}
\Phi x(t) &= \int_0^{\sigma(T)} G(t,s)f(s,x(\sigma(s)))\Delta s + \sum_{j=1}^{m} G(t,t_j)I_j(x(t_j)) \\
&\le B\int_0^{\sigma(T)} f(s,x(\sigma(s)))\Delta s + B\sum_{j=1}^{m} I_j(x(t_j)) \\
&\le \frac{R}{2} + \frac{1}{2m}\sum_{j=1}^{m} x(t_j) \\
&\le \frac{R}{2} + \frac{1}{2}\|x\| \\
&\le \frac{R}{2} + \frac{R}{2} \\
&= R \\
&= \|x\|, \quad t \in [0,\sigma(T)],
\end{aligned}$$

whereupon,

$$\|\Phi x\| \le \|x\|.$$

Since $\min f_0 = \infty$, for $M > (A\sigma(T)\delta)^{-1}$, there exists a positive constant $R_1 < \delta R$ such that

$$f(t,x) \ge Mx, \quad x \in [0,R_1], \quad t \in [0,\sigma(T)].$$

Let

$$\Omega_{R_1} = \{x \in K : \|x\| < R_1\}.$$

Then, for $x \in K\bigcap \partial\Omega_{R_1}$, we have

$$x(t) \ge \delta\|x\|, \quad t \in [0,\sigma(T)],$$

and

$$\begin{aligned}
\Phi x(t) &= \int_0^{\sigma(T)} G(t,s)f(s,x(\sigma(s)))\Delta s + \sum_{j=1}^{m} G(t,t_j)I_j(x(t_j)) \\
&\ge \int_0^{\sigma(T)} G(t,s)f(s,x(\sigma(s)))\Delta s \\
&\ge A\int_0^{\sigma(T)} f(s,x(\sigma(s)))\Delta s \\
&\ge AM\int_0^{\sigma(T)} x(\sigma(s))\Delta s
\end{aligned}$$

$$\geq AM\delta\sigma(T)\|x\|$$

$$\geq \|x\|, \quad t\in[0,\sigma(T)],$$

whereupon,

$$\|\Phi x\| \geq \|x\|.$$

Because $\min f_\infty = \infty$, for $N \geq (A\sigma(T)\delta)^{-1}$, there exists a positive constant $R_2 > \frac{R}{\delta}$ such that

$$f(t,x) \geq Nx, \quad t\in[0,\sigma(T)], \quad x\in[\delta R_2,\infty).$$

Let

$$\Omega_{R_2} = \{x\in K : \|x\| < R_2\}.$$

Then, for $x \in K\bigcap\partial\Omega_{R_2}$, we obtain

$$x(t) \geq \delta\|x\|, \quad t\in[0,\sigma(T)],$$

and

$$\begin{aligned}\Phi x(t) &= \int_0^{\sigma(T)} G(t,s)f(s,x(\sigma(s)))\Delta s + \sum_{j=1}^m G(t,t_j)I_j(x(t_j))\\ &\geq \int_0^{\sigma(T)} G(t,s)f(s,x(\sigma(s)))\Delta s\\ &\geq A\int_0^{\sigma(T)} f(s,x(\sigma(s)))\Delta s\\ &\geq AN\int_0^{\sigma(T)} x(\sigma(s))\Delta s\\ &\geq AN\delta\sigma(T)\|x\|\\ &\geq \|x\|, \quad t\in[0,\sigma(T)],\end{aligned}$$

whereupon,

$$\|\Phi x\| \geq \|x\|.$$

Note that

$$\overline{\Omega_{R_1}} \subset \overline{\Omega_R} \subset \Omega_{R_2}.$$

Hence and applying the Guo-Krasnosel'skii Theorem 1.11, it follows that Φ has at least one fixed point in $K\bigcap(\overline{\Omega_R}\backslash\Omega_{R_1})$ and Φ has at least one fixed point in $K\bigcap(\overline{\Omega_{R_2}}\backslash\Omega_R)$. This completes the proof.

Theorem 2.25. *Assume* $(B23)$-$(B27)$ *and there exists a positive constant* r *such that*

$$f(t,x) \geq \frac{r}{A\sigma(T)}, \quad x \in (0,r], \quad t \in [0,\sigma(T)],$$

$$\max f_0 = 0, \quad \max f_\infty = 0, \quad I_{j0} = 0, \quad I_{j\infty} = 0, \quad j \in \{1,\ldots,m\}.$$

Then problem (2.43) *has at least two positive solutions.*

Proof. Let

$$\Omega_r = \{x \in K : \|x\| < r\}.$$

Then, for $x \in K \bigcap \partial\Omega_r$, we get

$$\begin{aligned}
\Phi x(t) &= \int_0^{\sigma(T)} G(t,s) f(s,x(\sigma(s)))\Delta s + \sum_{j=1}^m G(t,t_j) I_j(x(t_j)) \\
&\geq \int_0^{\sigma(T)} G(t,s) f(s,x(\sigma(s)))\Delta s \\
&\geq A \int_0^{\sigma(T)} f(s,x(\sigma(s)))\Delta s \\
&\geq r \\
&= \|x\|, \quad t \in [0,\sigma(T)],
\end{aligned}$$

whereupon,

$$\|\Phi x\| \geq \|x\|.$$

Since

$$\max f_0 = 0, \quad I_{j0} = 0, \quad j \in \{1,\ldots,m\},$$

for $\varepsilon \leq (B\sigma(T) + Bm)^{-1}$, there exists a positive constant $r_1 < \delta r$ such that

$$f(t,x) \leq \varepsilon x, \quad I_j(x) \leq \varepsilon x, \quad x \in (0,r_1], \quad t \in [0,\sigma(T)], \quad j \in \{1,\ldots,m\}.$$

Let

$$\Omega_{r_1} = \{x \in K : \|x\| < r_1\}.$$

Then, for $x \in K \bigcap \partial\Omega_{r_1}$, we have

$$\begin{aligned}
\Phi x(t) &= \int_0^{\sigma(T)} G(t,s) f(s,x(\sigma(s)))\Delta s + \sum_{j=1}^m G(t,t_j) I_j(x(t_j)) \\
&\leq B\left(\int_0^{\sigma(T)} f(s,x(\sigma(s)))\Delta s + \sum_{j=1}^m I_j(x(t_j))\right) \\
&\leq \varepsilon B\left(\int_0^{\sigma(T)} x(\sigma(s))\Delta s + \sum_{j=1}^m x(t_j)\right)
\end{aligned}$$

$$\le \varepsilon B(\sigma(T)+m)\|x\|$$

$$\le \|x\|, \quad t \in [0, \sigma(T)],$$

whereupon,

$$\|\Phi x\| \le \|x\|.$$

Because

$$\max f_\infty = 0, \quad I_{j\infty} = 0, \quad j \in \{1, \ldots, m\},$$

for $\varepsilon_1 \le (B\sigma(T)+Bm)^{-1}$, there exists a positive constant $r_2 > \frac{r}{\delta}$ such that

$$f(t,x) \le \varepsilon_1 x, \quad I_j(x) \le \varepsilon_1 x, \quad x \in [\delta r_2, \infty), \quad j \in \{1, \ldots, m\}, \quad t \in [0, \sigma(T)].$$

Let

$$\Omega_{r_2} = \{x \in K : \|x\| < r_2\}.$$

Then, for $x \in K \bigcap \partial\Omega_{r_2}$, we get

$$\begin{aligned}
\Phi x(t) &= \int_0^{\sigma(T)} G(t,s) f(s, x(\sigma(s)))\Delta s + \sum_{j=1}^m G(t,t_j) I_j(x(t_j)) \\
&\le B\left(\int_0^{\sigma(T)} f(s, x(\sigma(s)))\Delta s + \sum_{j=1}^m I_j(x(t_j))\right) \\
&\le \varepsilon_1 B\left(\int_0^{\sigma(T)} x(\sigma(s))\Delta s + \sum_{j=1}^m x(t_j)\right) \\
&\le \varepsilon_1 B(\sigma(T)+m)\|x\| \\
&\le \|x\|, \quad t \in [0, \sigma(T)],
\end{aligned}$$

whereupon,

$$\|\Phi x\| \le \|x\|.$$

Note that

$$\overline{\Omega_{r_1}} \subset \overline{\Omega_r} \subset \Omega_{r_2}.$$

Hence and applying the Guo-Krasnosel'skii fixed point Theorem 1.11, it follows that the operator Φ has at least one fixed point in $K \bigcap (\overline{\Omega_r} \backslash \Omega_{r_1})$ and Φ has at least one fixed point in $K \bigcap (\overline{\Omega_{r_2}} \backslash \Omega_r)$. This completes the proof.

Exercise 2.7. Assume $(B23)$-$(B27)$ and there exists a positive constant r such that $f(t,x) > \frac{r}{A\sigma(T)}$ for $0 < x \le r$, $t \in [0, \sigma(T)]$,

$$\max f_0 = 0, \quad \max f_\infty = 0, \quad I_{j0} = 0, \quad j \in \{1, \ldots, m\},$$

there exists a positive constant d such that $I_j(x) \le \frac{x}{2Bm}$, $x \ge d$, $j \in \{1, \ldots, m\}$. Prove that problem (2.43) has at least two positive solutions.

Theorem 2.26. *Assume (B23)-(B27) and there exist numbers d, a and c with $0 < d < a < \frac{a}{\delta} < c$ such that*

$$\max_{t\in[0,\sigma(T)]} f(t,x) \le \frac{d}{2B\sigma(T)}, \quad I_j(x) \le \frac{d}{2Bm}, \quad j \in \{1, \ldots, m\}, \quad x \in (0, d],$$

$$\max_{t\in[0,\sigma(T)]} f(t,x) \le \frac{c}{2B\sigma(T)}, \quad I_j(x) \le \frac{c}{2Bm}, \quad j \in \{1, \ldots, m\}, \quad x \in (0, c],$$

$$\min_{t\in[0,\sigma(T)]} f(t,x) > \frac{a}{2A\sigma(T)}, \quad I_j(x) > \frac{a}{2Am}, \quad j \in \{1, \ldots, m\}, \quad x \in \left[a, \frac{a}{\delta}\right].$$

Then problem (2.43) has at least three positive solutions.

Proof. For $x \in K$, define

$$\alpha(x) = \min_{t\in[0,\sigma(T)]} x(t).$$

We have

$$\alpha(x) \ge 0, \quad x \in K.$$

For $\lambda \in [0, 1]$ and for any $x, y \in K$, we obtain

$$\begin{aligned}
\alpha(\lambda x + (1-\lambda)y) &= \min_{t\in[0,\sigma(T)]} (\lambda x(t) + (1-\lambda)y(t)) \\
&\ge \lambda \min_{t\in[0,\sigma(T)]} x(t) + (1-\lambda) \min_{t\in[0,\sigma(T)]} y(t) \\
&= \lambda\alpha(x) + (1-\lambda)\alpha(y).
\end{aligned}$$

Also, if $\{x_n\}_{n\in\mathbb{N}} \subset K$ and $x_n \to x$, as $n \to \infty$, in K, then

$$\begin{aligned}
\alpha(x_n) &= \min_{t\in[0,\sigma(T)]} x_n(t) \\
&\to \min_{t\in[0,\sigma(T)]} x(t) \\
&= \alpha(x), \quad \text{as} \quad n \to \infty.
\end{aligned}$$

Therefore, α is a non-negative continuous concave functional on K and $\alpha(x) \le \|x\|$, $x \in K$. Assume that there exists a positive constant r such that

$$\max_{t\in[0,\sigma(T)]} f(t,x) \le \frac{r}{2B\sigma(T)}, \quad I_j(x) \le \frac{r}{2Bm}, \quad j \in \{1, \ldots, m\}, \quad x \in (0, r].$$

Then, for $x \in K_r$, we get

$$\begin{aligned}
\Phi x(t) &= \int_0^{\sigma(T)} G(t,s)f(s,x(\sigma(s)))\Delta s + \sum_{j=1}^{m} G(t,t_j)I_j(x(t_j)) \\
&\le B\frac{r}{2B\sigma(T)}\sigma(T) + Bm\frac{r}{2Bm} \\
&= \frac{r}{2} + \frac{r}{2} \\
&= r, \quad t \in [0,\sigma(T)],
\end{aligned}$$

whereupon,

$$\|\Phi x\| \le r \quad \text{or} \quad \Phi x \in K_r.$$

Therefore, $\Phi : \overline{K_c} \to K_c$ and $\Phi : \overline{K_d} \to K_d$. Let $b = \frac{a}{\delta}$. Note that

$$a < \frac{(1+\delta)a}{2\delta} < \frac{a}{\delta}$$

and

$$\begin{aligned}
\alpha\left(\frac{(1+\delta)a}{2\delta}\right) &= \frac{(1+\delta)a}{2\delta} \\
&> a.
\end{aligned}$$

Consequently,

$$\frac{(1+\delta)a}{2\delta} \in \{x \in K(\alpha,a,b) : \alpha(x) > a\},$$

i.e.,

$$\{x \in K(\alpha,a,b) : \alpha(x) > a\} \neq \emptyset.$$

Next, for $x \in K(\alpha,a,b)$, we get

$$\begin{aligned}
\Phi x(t) &= \int_0^{\sigma(T)} G(t,s)f(s,x(\sigma(s)))\Delta s + \sum_{j=1}^{m} G(t,t_j)I_j(x(t_j)) \\
&\ge A\int_0^{\sigma(T)} f(s,x(\sigma(s)))\Delta s + A\sum_{j=1}^{m} I_j(x(t_j)) \\
&> A\frac{a}{2A\sigma(T)}\sigma(T) + Am\frac{a}{2Am} \\
&= \frac{a}{2} + \frac{a}{2}
\end{aligned}$$

$$= a, \quad t \in [0, \sigma(T)],$$

whereupon,

$$\alpha(\Phi x) > a.$$

Now, suppose that $x \in K(\alpha, a, c)$ and $\|\Phi x\| > b = \frac{a}{\delta}$. Then

$$\begin{aligned}
\alpha(\Phi x) &= \min_{t\in[0,\sigma(T)]} \left(\int_0^{\sigma(T)} G(t,s) f(s, x(\sigma(s)))\Delta s + \sum_{j=1}^{m} G(t,t_j) I_j(x(t_j)) \right) \\
&\geq A \int_0^{\sigma(T)} f(s, x(\sigma(s)))\Delta s + A \sum_{j=1}^{m} I_j(x(t_j)) \\
&\geq \frac{A}{B} \left(\int_0^{\sigma(T)} G(t,s) f(s, x(\sigma(s)))\Delta s + \sum_{j=1}^{m} G(t,t_j) I_j(x(t_j)) \right) \\
&= \frac{A}{B} \Phi x(t), \quad t \in [0, \sigma(T)].
\end{aligned}$$

Consequently,

$$\begin{aligned}
\alpha(\Phi(x)) &\geq \frac{A}{B} \max_{t\in[0,\sigma(T)]} \Phi x(t) \\
&= \frac{A}{B} \|\Phi x\| \\
&> \delta b \\
&= a.
\end{aligned}$$

From here and from Theorem 1.14, it follows that problem (2.43) has at least three positive solutions x_1, x_2, x_3 such that

$$\|x_1\| < d, \quad \min_{t\in[0,\sigma(T)]} x_2(t) > a, \quad \|x_3\| > d, \quad \min_{t\in[0,\sigma(T)]} x_3(t) < a.$$

This completes the proof.

Exercise 2.8. Let $\mathbb{T} = [0,1] \bigcup [2,3]$, where $[0,1]$ and $[2,3]$ are the real-valued intervals. Prove that the PBVP

$$x^{\Delta}(t) + x(\sigma(t)) = f(t, x(\sigma(t))), \quad t \in [0,3] \setminus \left\{ \frac{1}{3}, \frac{1}{2}, \frac{5}{2} \right\},$$

$$x\left(\frac{1}{3}^{+}\right) - x\left(\frac{1}{3}^{-}\right) = I\left(x\left(\frac{1}{3}\right)\right)$$

$$x\left(\frac{1}{2}^{+}\right)-x\left(\frac{1}{2}^{-}\right)=I\left(x\left(\frac{1}{2}\right)\right)$$

$$x\left(\frac{5}{2}^{+}\right)-x\left(\frac{5}{2}^{-}\right)=I\left(x\left(\frac{5}{2}\right)\right)$$

$$x(0)-x(3)=0,$$

where

$$f(t,x(\sigma(t)))=\begin{cases}(x(\sigma(t)))^2+\dfrac{e^2-1}{5184e^4(e^2+2)^2}, & x\in\left[0,\dfrac{e^2(e^4-1)}{48}\right]\\[2ex] \dfrac{e^2-2}{18e^6(e^2+2)}x(\sigma(t))+\dfrac{e^4(e^4-4)^2}{48}, & x\in\left[\dfrac{e^2(e^4-1)}{48},\infty\right),\end{cases}$$

$$I(x)=\begin{cases}x^2+\dfrac{e^2-1}{5184e^4(e^2+2)^2}, & x\in\left[0,\dfrac{e^2(e^4-1)}{48}\right]\\[2ex] \dfrac{e^2-2}{18e^6(e^2+2)}x+\dfrac{e^4(e^4-4)^2}{48}, & x\in\left[\dfrac{e^2(e^4-1)}{48},\infty\right),\end{cases}$$

has at least three positive solutions.

Hint. Choose

$$d=\frac{e^2-2}{24e^2(e^2+2)},\quad a=\frac{e^2-2}{24}$$

and c sufficiently large.

2.7 Advanced Practical Problems

Problem 2.1. Let $\mathbb{T}=[0,1]\bigcup[3,4]$, where $[0,1]$ and $[3,4]$ are the real-valued intervals. Investigate for existence of solutions for the following impulsive dynamic equation

$$y^{\nabla}(t)=t^2+t+1+4(y(t)),\quad t\in[0,1]\bigcup(3,4],$$

$$y(3^+)=2y(3)+1,$$

$$B(y(0),y(4))=0.$$

Problem 2.2. Let $\mathbb{T}=[0,1]\bigcup[2,4]\bigcup[5,6]\bigcup[7,8]$, where $[0,1]$, $[2,4]$, $[5,6]$ and $[7,8]$ are the real-valued intervals. Consider the periodic boundary value problem

$$
\begin{aligned}
x^{\Delta}(t) &= -2tx^{\sigma}(t) + \frac{1}{t^2+1}, \quad t \in \mathbb{T}\backslash\{2,4\},\\
x(2+) &= \frac{1}{1+(x(2))^2},\\
x(5+) &= \frac{2}{4+2(x(5))^2},\\
x(7+) &= \frac{1}{1+3(x(7))^2},\\
x(0) &= x(8).
\end{aligned}
$$

Find constants l_1 and l_2 as in Theorem 2.2 so that the considered PBVP to has a unique solution.

Problem 2.3. Prove that problem

$$
\begin{aligned}
x^{\Delta}(t) + p(t)x(\sigma(t)) &= 12(x(\sigma(t)))^{\frac{1}{21}} + (x(\sigma(t)))^{\frac{1}{7}}, \quad t \in J_0,\\
x(t_k^+) &= x(t_k^-) + 4(x(t_k^-))^{\frac{1}{17}} + 7(x(t_k^-))^{\frac{1}{11}}, \quad k \in \{1,\dots,m\},\\
x(0) &= x(\sigma(T)),
\end{aligned}
$$

where p satisfies $(B24)$ and has at least one positive solution.

Problem 2.4. Find conditions for f, I_k, $k \in \{1,\dots,m\}$, so that problem

$$
\begin{aligned}
x^{\Delta}(t) &= f(t,x(\sigma(t))), \quad t \in J_0,\\
x(t_k^+) &= x(t_k) + I_k(x(t_k)), \quad k \in \{1,\dots,m\},\\
x(0) &= x(\sigma(T)),
\end{aligned}
$$

has at least one positive solution.

Hint. Take a function $a \in \mathscr{R}$ arbitrarily and let

$$
g(t,x(\sigma(t))) = a(t)x(\sigma(t)) + f(t,x(\sigma(t))), \quad t \in J_0.
$$

Thus we get problem

$$
\begin{aligned}
x^{\Delta}(t) + a(t)x(\sigma(t)) &= g(t,x(\sigma(t))), \quad t \in J_0,\\
x(t_k^+) &= x(t_k) + I_k(x(t_k)), \quad k \in \{1,\dots,m\},\\
x(0) &= x(\sigma(T)),
\end{aligned}
$$

i.e., the considered problem can be reduced to problem (2.14), (2.15).

Problem 2.5. Let $\mathbb{T} = \bigcup_{n=0}^{\infty} [2n, 2n+1]$. Prove that the following PBVP

$$x^{\Delta}(t) + p(t)x(\sigma(t)) = f(t, x(\sigma(t))), \quad t \in [0,3] \backslash \{\frac{1}{2}\},$$

$$x\left(\frac{1}{2}^{+}\right) - x\left(\frac{1}{2}^{-}\right) = I\left(x\left(\frac{1}{2}\right)\right),$$

$$\alpha x(0) - \beta x(4) = \int_0^4 g(s)x(s)\Delta s,$$

has at least one positive solution, where

1.

$$T = 3, \quad p(t) = t,$$

$$f(t, x(\sigma(t))) = (t+1)(x(\sigma(t)))^4,$$

$$I(x) = x^3, \quad \alpha = 1, \quad \beta = \frac{1}{2},$$

$$g(t) = \begin{cases} t & t \in [0,1] \\ 0 & t \notin [0,1], \end{cases}$$

2.

$$T = 3, \quad p(t) = t,$$

$$f(t, x(\sigma(t))) = (t+1)(x(\sigma(t)))^{\frac{1}{3}},$$

$$I(x) = x^{\frac{1}{2}}, \quad \alpha = 1, \quad \beta = \frac{1}{2},$$

$$g(t) = \begin{cases} 1 & t \in \{1,3\} \\ 0 & \text{otherwise.} \end{cases}$$

2.8 Notes and References

This chapter investigates some boundary value problems for first order impulsive dynamic equations. Criteria are given under which the investigated boundary value problems have at least one, at least two and least three solutions. The material in this chapter can be found in [6], [12], [26], [27], [33], [34], [40], [38], [59], [60], and references therein.

Chapter 3
The Green Function for Linear Second Order Dynamic Equations

Let $\mathbb{T}$ be a time scale with delta differentiation operator and forward jump operator Δ and σ, respectively.

3.1 Cauchy Function

Suppose that

(C1) $p, q \in \mathscr{C}_{rd}(\mathbb{T})$, $p(t) \neq 0$ for any $t \in \mathbb{T}$.

Let $\mathbb{D}$ be the set of all functions $x : \mathbb{T} \to \mathbb{R}$ such that $x^\Delta : \mathbb{T}^\kappa \to \mathbb{R}$ exists and it is continuous and $(px^\Delta)^\Delta : \mathbb{T}^{\kappa^2} \to \mathbb{R}$ is rd-continuous. For a function $x \in \mathbb{D}$, define the operator

$$Lx(t) = (px^\Delta)^\Delta(t) + q(t)x^\sigma(t), \quad t \in \mathbb{T}. \tag{3.1}$$

Theorem 3.1. *The operator $L : \mathbb{D} \to \mathscr{C}_{rd}$ is a linear operator.*

Proof. Let $x, y \in \mathbb{D}$ and $\alpha, \beta \in \mathbb{R}$ be arbitrarily chosen. Then

$$\begin{aligned}
L(\alpha x + \beta y)(t) &= \left(p(\alpha x + \beta y)^\Delta\right)^\Delta(t) + q(\alpha x + \beta y)^\sigma(t) \\
&= \left(p(\alpha x^\Delta + \beta y^\Delta)\right)^\Delta(t) + q(\alpha x^\sigma + \beta y^\sigma)(t) \\
&= \alpha\left(px^\Delta\right)^\Delta(t) + \alpha q x^\sigma(t) \\
&\quad + \alpha\left(py^\Delta\right)^\Delta(t) + \alpha q y^\sigma(t) \\
&= \alpha Lx(t) + \beta Ly(t), \quad t \in \mathbb{T}.
\end{aligned}$$

This completes the proof.

DOI: 10.1201/9781003173557-3

Now, consider equation

$$Lx = 0. \tag{3.2}$$

Definition 3.1. We say that $x : \mathbb{T} \times \mathbb{T}^{\kappa^2} \to \mathbb{R}$ is the Cauchy function for (3.2) if for each $s \in \mathbb{T}^{\kappa^2}$ the function $x(\cdot, s)$ is the solution of the IVP

$$Lx(\cdot, s) = 0,$$

$$x(\sigma(s), s) = 0,$$

$$x^{\Delta}(\sigma(s), s) = \frac{1}{p(\sigma(s))}.$$

Example 3.1. Let $q = 0$ and

$$x(t, s) = \int_{\sigma(s)}^{t} \frac{1}{p(\tau)} \Delta\tau, \quad t \in \mathbb{T}, \quad s \in \mathbb{T}^{\kappa^2}. \tag{3.3}$$

Then

$$x(\sigma(s), s) = 0,$$

$$x^{\Delta}(t, s) = \frac{1}{p(t)},$$

$$x^{\Delta}(\sigma(s), s) = \frac{1}{p(\sigma(s))},$$

$$(px^{\Delta})^{\Delta}(t) = 0, \quad t \in \mathbb{T}.$$

Thus, (3.3) is the Cauchy function for equation

$$(px^{\Delta})^{\Delta} = 0.$$

If $p = 1$, then the Cauchy function is

$$x(t, s) = t - \sigma(s), \quad t \in \mathbb{T}, \quad s \in \mathbb{T}^{\kappa^2}.$$

Example 3.2. Let $\mathbb{T} = 2^{\mathbb{N}_0}$,

$$p(t) = \frac{1}{1 + t + t^2},$$

$$q(t) = 0, \quad t \in \mathbb{T}.$$

Here

$$\sigma(t) = 2t,$$

$$\mu(t) = t, \quad t \in \mathbb{T}.$$

Let

$$g(t) = \frac{1}{7}t^3 + \frac{1}{3}t^2 + t, \quad t \in \mathbb{T}.$$

Then

$$\begin{aligned} g^{\Delta}(t) &= \frac{1}{7}\left((\sigma(t))^2 + t\sigma(t) + t^2\right) + \frac{1}{3}(\sigma(t) + t) + 1 \\ &= \frac{1}{7}(4t^2 + 2t^2 + t^2) + \frac{1}{3}(2t + t) + 1 \\ &= t^2 + t + 1, \quad t \in \mathbb{T}. \end{aligned}$$

Hence,

$$\begin{aligned} x(t,s) &= \int_{\sigma(s)}^{t} \frac{1}{p(\tau)} \Delta\tau \\ &= \int_{2s}^{t} (\tau^2 + \tau + 1)\Delta\tau \\ &= \int_{2s}^{t} g^{\Delta}(\tau)\Delta\tau \\ &= g(\tau)\Big|_{\tau=2s}^{\tau=t} \\ &= g(t) - g(2s) \\ &= \frac{1}{7}t^3 + \frac{1}{3}t^2 + t - \frac{1}{7}(2s)^3 - \frac{1}{3}(2s)^2 - 2s \\ &= \frac{1}{7}(t-2s)(t^2 + 2st + 4s^2) + \frac{1}{3}(t-2s)(t+2s) + t - 2s \\ &= (t-2s)\left(\frac{1}{7}t^2 + \frac{2}{7}st + \frac{4}{7}s^2 + \frac{1}{3}t + \frac{2}{3}s + 1\right), \quad t \in \mathbb{T}, \quad s \in \mathbb{T}^{\kappa^2}, \end{aligned}$$

is the Cauchy function for equation (3.2).

Exercise 3.1. Let $\mathbb{T} = \mathbb{N}_0^2$,

$$p(t) = \frac{1}{1+t},$$

$$q(t) = 0, \quad t \in \mathbb{T}.$$

Find the Cauchy function for equation (3.2).

Example 3.3. Let $\mathbb{T} = 2\mathbb{Z}$. Consider the following equation

$$\frac{24}{7}x^{\Delta^2} + 7x^{\sigma} = 0 \tag{3.4}$$

subject to the conditions

$$\begin{aligned} x(\sigma(s),s) &= 0, \\ x^{\Delta}(\sigma(s),s) &= \frac{7}{24}, \end{aligned} \tag{3.5}$$

where $s \in \mathbb{T}$ is fixed. Here

$$p(t) = \frac{24}{7},$$

$$q(t) = 7,$$

$$\sigma(t) = t+2,$$

$$\mu(t) = 2, \quad t \in \mathbb{T}.$$

Equation (3.4) can be rewritten in the form

$$\frac{24}{7}x^{\Delta^2} + 14x^{\Delta} + 7x = 0.$$

The characteristic equation is

$$\frac{24}{7}r^2 + 14r + 7 = 0.$$

Its roots are

$$\begin{aligned} r_{1,2} &= \frac{-7 \pm \sqrt{49-24}}{\frac{24}{7}} \\ &= \frac{-7 \pm 5}{\frac{24}{7}} \end{aligned}$$

and

$$\begin{aligned} r_1 &= -\frac{12}{\frac{24}{7}} \\ &= -\frac{7}{2}, \\ r_2 &= \frac{-2}{\frac{24}{7}} \\ &= -\frac{7}{12}. \end{aligned}$$

The general solution of equation (3.4) is given in the form

$$x(t,s) = c_1 e_{-\frac{7}{2}}(t,s) + c_2 e_{-\frac{7}{12}}(t,s),$$

where c_1 and c_2 are constants which will be determined using conditions (3.5). We have

$$\begin{aligned} x(\sigma(s),s) &= c_1 e_{-\frac{7}{2}}(\sigma(s),s) + c_2 e_{-\frac{7}{12}}(\sigma(s),s) \\ &= c_1 \left(e_{-\frac{7}{2}}(s,s) - 7 e_{-\frac{7}{2}}(s,s) \right) \\ &\quad + c_2 \left(e_{-\frac{7}{12}}(s,s) - \frac{7}{6} e_{-\frac{7}{12}}(s,s) \right) \\ &= -6c_1 - \frac{1}{6}c_2, \\ x^{\Delta}(t,s) &= -\frac{7}{2}c_1 e_{-\frac{7}{2}}(t,s) - \frac{7}{12}c_2 e_{-\frac{7}{12}}(t,s), \\ x^{\Delta}(\sigma(s),s) &= -\frac{7}{2}c_1 e_{-\frac{7}{2}}(\sigma(s),s) - \frac{7}{12}c_2 e_{-\frac{7}{12}}(\sigma(s),s) \\ &= -\frac{7}{2}c_1(-6) - \frac{7}{12}c_2\left(-\frac{1}{6}\right) \\ &= 21c_1 + \frac{7}{72}c_2. \end{aligned}$$

Now, using conditions (3.5), we arrive at the system

$$\begin{aligned} -6c_1 - \frac{1}{6}c_2 &= 0 \\ 21c_1 + \frac{7}{72}c_2 &= \frac{7}{24}. \end{aligned}$$

Then

$$c_1 = -\frac{1}{36}c_2$$

and

$$21\left(-\frac{1}{36}c_2\right) + \frac{7}{72}c_2 = \frac{7}{24}$$

or

$$\frac{(-42+7)}{72}c_2 = \frac{7}{24},$$

$$-35c_2 = 21,$$

$$c_2 = -\frac{3}{5},$$

$$c_1 = -\frac{1}{36}\left(-\frac{3}{5}\right)$$

$$= \frac{1}{60}.$$

Consequently,

$$x(t,s) = \frac{1}{60}e_{-\frac{7}{2}}(t,s) - \frac{3}{5}e_{-\frac{7}{12}}(t,s), \quad t \in \mathbb{T},$$

is the Cauchy function for equation (3.4).

Exercise 3.2. Let $\mathbb{T} = \mathbb{Z}$. Find the Cauchy function for equation

$$x^{\Delta^2} - 3x^{\sigma} = 0.$$

Theorem 3.2. *Assume* (C1). *Let* $f \in \mathscr{C}_{rd}$ *and* $x(t,s)$ *be the Cauchy function of* (3.2). *Then*

$$x(t) = \int_a^t x(t,s)f(s)\Delta s, \quad t \in \mathbb{T},$$

is the solution to the IVP

$$Lx = f(t), \quad t \in \mathbb{T},$$

$$x(a) = 0,$$

$$x^{\Delta}(a) = 0.$$

Proof. We have

$$x^{\Delta}(t) = x(\sigma(t),t)f(t) + \int_a^t x^{\Delta}(t,s)f(s)\Delta s$$

$$
\begin{aligned}
&= \int_a^t x^{\Delta}(t,s) f(s) \Delta s, \\
(px^{\Delta})(t) &= p(t) \int_a^t x^{\Delta}(t,s) f(s) \Delta s \\
&= \int_a^t (px^{\Delta}(\cdot,s))(t) f(s) \Delta s, \\
(px^{\Delta})^{\Delta}(t) &= p(\sigma(t)) x^{\Delta}(\sigma(t),t) f(t) \\
&\quad + \int_a^t (px^{\Delta}(\cdot,s))^{\Delta}(t) f(s) \Delta s \\
&= f(t) + \int_a^t (px^{\Delta}(\cdot,s))^{\Delta}(t) f(s) \Delta s, \\
Lx(t) &= (px^{\Delta})^{\Delta}(t) + q(t) x^{\sigma}(t) \\
&= f(t) + q(t) x^{\sigma}(t) \\
&\quad + \int_a^t (px^{\Delta}(\cdot,s))^{\Delta}(t) f(s) \Delta s \\
&= f(t) + q(t) \int_a^{\sigma(t)} x(\sigma(t),s) f(s) \Delta s \\
&\quad + \int_a^t (px^{\Delta}(\cdot,s))^{\Delta}(t) f(s) \Delta s \\
&= f(t) + \int_a^t \left(q(t) x(\sigma(t),s) + (px^{\Delta}(\cdot,s))^{\Delta}(t) \right) f(s) \Delta s \\
&\quad + q(t) \int_t^{\sigma(t)} x(\sigma(t),s) f(s) \Delta s \\
&= f(t) + q(t) \mu(t) x(\sigma(t),t) f(t) \\
&= f(t), \quad t \in \mathbb{T},
\end{aligned}
$$

and

$$
x(a) = 0,
$$

$$x^{\Delta}(a) = 0.$$

This completes the proof.

Theorem 3.3. *Assume* $(C1)$. *Let the Cauchy function* x *for equation* (3.2) *satisfy*

$$x(t,s) \geq 0, \quad t \geq \sigma(s).$$

If $u, v \in \mathbb{D}$ *are functions for which*

$$Lu(t) \geq Lv(t), \quad t \in [a,b],$$

$$u(a) = v(a),$$

$$u^{\Delta}(a) = v^{\Delta}(a),$$

then

$$u(t) \geq v(t), \quad t \in [a, \sigma^2(b)]. \tag{3.6}$$

Proof. Let

$$w(t) = u(t) - v(t), \quad t \in [a, \sigma^2(b)],$$

and

$$h(t) = Lw(t), \quad t \in [a,b].$$

Then

$$h(a) = h^{\Delta}(a) = 0$$

and by Theorem 3.2, it follows that

$$\begin{aligned} w(t) &= \int_a^t x(t,s)h(s)\Delta s \\ &= \int_a^{\rho(t)} x(t,s)h(s)\Delta s + \int_{\rho(t)}^t x(t,s)h(s)\Delta s \\ &= \int_a^{\rho(t)} x(t,s)h(s)\Delta s + \mu(\rho(t))x(t,\rho(t))h(\rho(t)) \\ &= \int_a^{\rho(t)} x(t,s)h(s)\Delta s \\ &\geq 0, \quad t \in [a, \sigma^2(b)]. \end{aligned}$$

Hence, we get (3.6). This completes the proof.

3.2 The Green Function for General Two Point Boundary Value Problems

Let $a, b \in \mathbb{T}$, $a < b$. Consider the following BVPs

$$\begin{aligned} Lx &= 0, \\ \alpha x(a) - \beta x^{\Delta}(a) &= 0, \\ \gamma x(\sigma^2(b)) + \delta x^{\Delta}(\sigma(b)) &= 0, \end{aligned} \tag{3.7}$$

and

$$\begin{aligned} Lx &= h(t), \\ \alpha x(a) - \beta x^{\Delta}(a) &= A, \\ \gamma x(\sigma^2(b)) + \delta x^{\Delta}(\sigma(b)) &= B, \end{aligned} \tag{3.8}$$

where

(C2) $h \in \mathscr{C}_{rd}(\mathbb{T})$,
(C3) α, β, γ, δ, A and B are given constants.

We will start our investigations in this section with the following existence and uniqueness result.

Theorem 3.4. *Assume* (C1)-(C3) *and assume that the homogeneous BVP* (3.7) *has only the trivial solution. Then the nonhomogeneous BVP* (3.8) *has a unique solution.*

Proof. Let u and v be two linearly independent solutions of equation (3.2). Then its general solution is given by the expression

$$x(t) = c_1 u(t) + c_2 v(t), \quad t \in [a, b], \tag{3.9}$$

where c_1 and c_2 are constants. Then (3.9) satisfies the boundary conditions in (3.7) if and only if

$$\begin{aligned} \alpha x(a) - \beta x^{\Delta}(a) &= 0 \\ \gamma x(\sigma^2(b)) + \delta x^{\Delta}(\sigma(b)) &= 0. \end{aligned} \tag{3.10}$$

The system (3.10) is equivalent to the system

$$\alpha c_1 u(a) + \alpha c_2 v(a) - \beta c_1 u^{\Delta}(a) - \beta c_2 v^{\Delta}(a) = 0$$

$$\gamma c_1 u(\sigma^2(b)) + \gamma c_2 v(\sigma^2(b)) + \delta c_1 u^{\Delta}(\sigma(b)) + \delta c_2 v^{\Delta}(\sigma(b)) = 0$$

or

$$\begin{aligned} c_1(\alpha u(a)-\beta u^{\Delta}(a))+c_2(\alpha v(a)-\beta v^{\Delta}(a)) &= 0 \\ c_1(\gamma u(\sigma^2(b))+\delta u^{\Delta}(\sigma(b)))+c_2(\gamma v(\sigma^2(b))+\delta v^{\Delta}(\sigma(b))) &= 0. \end{aligned} \tag{3.11}$$

Set

$$M=\begin{pmatrix} \alpha u(a)-\beta u^{\Delta}(a) & \alpha v(a)-\beta v^{\Delta}(a) \\ \gamma u(\sigma^2(b))+\delta u^{\Delta}(\sigma(b)) & \gamma v(\sigma^2(b))+\delta v^{\Delta}(\sigma(b)) \end{pmatrix}.$$

Then the system (3.11) can be written in the form

$$M\begin{pmatrix} c_1 \\ c_2 \end{pmatrix}=0. \tag{3.12}$$

Because we have assumed that (3.7) has the trivial solution only, using (3.9), we have that

$$(c_1,c_2)=(0,0).$$

Hence, (3.12) has the zero solution if and only if

$$\det M\neq 0. \tag{3.13}$$

Now, suppose that w is a solution of equation

$$Lx=h(t), \quad t\in[a,b].$$

Then, its general solution is given in the form

$$x(t)=a_1u(t)+a_2v(t)+w(t), \quad t\in[a,b], \tag{3.14}$$

where a_1 and a_2 are constants which will be determined below. Note that (3.14) satisfies the boundary conditions in (3.8) if and only if

$$\alpha a_1u(a)+\alpha a_2v(a)+w(a)-\beta a_1u^{\Delta}(a)-\beta a_2v^{\Delta}(a)-w^{\Delta}(a)=A$$

$$\gamma a_1u(\sigma^2(b))+\gamma a_2v(\sigma^2(b))+\gamma w(\sigma^2(b))+\delta a_1u^{\Delta}(\sigma(b))+\delta a_2v^{\Delta}(\sigma(b))$$

$$+\delta w(\sigma(b))=B$$

or

$$a_1(\alpha u(a)-\beta u^{\Delta}(a))+a_2(\alpha v(a)-\beta v^{\Delta}(a))=A-w(a)+w^{\Delta}(a)$$

$$a_1(\gamma u(\sigma^2(b))+\delta u^{\Delta}(\sigma(b)))+a_2(\gamma v(\sigma^2(b))+\delta v^{\Delta}(\sigma(b)))=B-\gamma w(\sigma^2(b))$$

$$-\delta w(\sigma(b)),$$

or

$$M\begin{pmatrix} a_1 \\ a_2 \end{pmatrix} = \begin{pmatrix} A - w(a) + w^{\Delta}(a) \\ B - \gamma w(\sigma^2(b)) - \delta w(\sigma(b)) \end{pmatrix}.$$

Now, using (3.13), the last system has a unique solution. From here, we conclude that (3.8) has a unique solution. This completes the proof.

Example 3.4. Let $q = 0$. Then

$$u(t) = 1 \quad \text{and} \quad v(t) = \int_a^t \frac{\Delta\tau}{p(\tau)}, \qquad t \in [a,b],$$

are linearly independent solutions of $Lx = 0$. In this case,

$$M = \begin{pmatrix} \alpha & -\dfrac{\beta}{p(a)} \\ \gamma\ \gamma \displaystyle\int_a^{\sigma^2(b)} \dfrac{\Delta\tau}{p(\tau)} + \dfrac{\delta}{p(\sigma(b))} \end{pmatrix}$$

and

$$\det M = \alpha\gamma \int_a^{\sigma^2(b)} \frac{\Delta\tau}{p(\tau)} + \frac{\alpha\delta}{p(\sigma(b))} + \frac{\gamma\beta}{p(a)}.$$

Hence, (3.7) has the trivial solution if and only if

$$\alpha\gamma \int_a^{\sigma^2(b)} \frac{\Delta\tau}{p(\tau)} + \frac{\alpha\delta}{p(\sigma(b))} + \frac{\gamma\beta}{p(a)} \neq 0.$$

Theorem 3.5. *Suppose (C1)-(C3). Assume that the BVP (3.7) has only the trivial solutions. For each fixed $s \in [a,b]$, let $u(\cdot,s)$ be the unique solution of the BVP*

$$\begin{aligned} Lu(\cdot,s) &= 0, \\ \alpha u(a,s) - \beta u^{\Delta}(a,s) &= 0, \\ \gamma u(\sigma^2(b),s) + \delta u^{\Delta}(\sigma(b),s) &= -\gamma x(\sigma^2(b),s) - \delta x^{\Delta}(\sigma(b),s), \end{aligned} \tag{3.15}$$

where $x(t,s)$ is the Cauchy function for (3.2). Then we define the Green function $G : [a,\sigma^2(b)] \times [a,b] \to \mathbb{R}$ for the BVP (3.7) by

$$G(t,s) = \begin{cases} u(t,s) & \text{if } t \leq s \\ v(t,s) & \text{if } t \geq \sigma(s), \end{cases}$$

where

$$v(t,s) = u(t,s) + x(t,s), \quad t \in [a,\sigma^2(b)], \quad s \in [a,b].$$

For each fixed $s \in [a,b]$, $v(\cdot,s)$ is a solution of equation (3.2) and satisfies the second boundary condition of (3.7). Moreover,

$$x(t) = \int_a^{\sigma(b)} G(t,s)h(s)\Delta s$$

is the solution of the BVP (3.8) *with* $A = B = 0$.

Proof. Let $s \in [a,b]$ be fixed. Since u and x are solutions of equation (3.2), we have that $v(\cdot, s)$ is a solution of equation (3.2). Next,

$$\begin{aligned}\gamma v(\sigma^2(b),s) + \delta v^{\Delta}(\sigma(b),s) &= \gamma u(\sigma^2(b),s) + \delta u^{\Delta}(\sigma(b),s) \\ &\quad + \gamma x(\sigma^2(b),s) + \delta x^{\Delta}(\sigma(b),s) \\ &= 0,\end{aligned}$$

i.e., v satisfies the second boundary condition of (3.7). Note that

$$\begin{aligned} x(t) &= \int_a^{\sigma(b)} G(t,s)h(s)\Delta s \\ &= \int_a^t G(t,s)h(s)\Delta s + \int_t^{\sigma(b)} G(t,s)h(s)\Delta s \\ &= \int_a^t v(t,s)h(s)\Delta s + \int_t^{\sigma(b)} u(t,s)h(s)\Delta s \qquad (3.16) \\ &= \int_a^t (u(t,s)+x(t,s))h(s)\Delta s + \int_t^{\sigma(b)} u(t,s)h(s)\Delta s \\ &= \int_a^{\sigma(b)} u(t,s)h(s)\Delta s + \int_a^t x(t,s)h(s)\Delta s, \quad t \in [a, \sigma^2(b)]. \end{aligned}$$

Let

$$z(t) = \int_a^t x(t,s)h(s)\Delta s, \quad t \in [a, \sigma^2(b)].$$

Then, by Theorem 3.2, it follows that

$$Lz = h(t), \quad t \in [a,b],$$

$$z(a) = z^{\Delta}(a) = 0.$$

Applying (3.16), we get

$$x(t) = \int_a^{\sigma(b)} u(t,s)h(s)\Delta s + z(t), \quad t \in [a,b],$$

and

$$\begin{aligned} Lx(t) &= \int_a^{\sigma(b)} Lu(t,s)h(s)\Delta s + Lz(t) \\ &= h(t), \quad t \in [a,b]. \end{aligned}$$

Now, we will show that x satisfies both boundary conditions of (3.8). We have

$$\begin{aligned} \alpha x(a) - \beta x^{\Delta}(a) &= \alpha \int_a^{\sigma(b)} G(a,s)h(s)\Delta s - \beta \int_a^{\sigma(b)} G^{\Delta}(a,s)h(s)\Delta s \\ &= \alpha \int_a^{\sigma(b)} u(a,s)h(s)\Delta s - \beta \int_a^{\sigma(b)} u^{\Delta}(a,s)h(s)\Delta s \\ &= \int_a^{\sigma(b)} (\alpha u(a,s) - \beta u^{\Delta}(a,s))h(s)\Delta s \\ &= 0 \end{aligned}$$

and by (3.16), we have

$$\begin{aligned} x(t) &= \int_a^{\sigma(b)} u(t,s)h(s)\Delta s + \int_a^t x(t,s)h(s)\Delta s \\ &= \int_a^{\sigma(b)} (v(t,s) - x(t,s))h(s)\Delta s + \int_a^t x(t,s)h(s)\Delta s \\ &= \int_a^{\sigma(b)} v(t,s)h(s)\Delta s - \int_a^{\sigma(b)} x(t,s)h(s)\Delta s - \int_t^a x(t,s)h(s)\Delta s \\ &= \int_a^{\sigma(b)} v(t,s)h(s)\Delta s - \int_t^{\sigma(b)} x(t,s)h(s)\Delta s \\ &= \int_a^{\sigma(b)} v(t,s)h(s)\Delta s + \int_{\sigma(b)}^t x(t,s)h(s)\Delta s, \quad t \in [a,\sigma^2(b)]. \end{aligned}$$

Let

$$w(t) = \int_{\sigma(b)}^t x(t,s)h(s)\Delta s, \quad t \in [a,\sigma^2(b)].$$

Then, applying Theorem 3.2, we get

$$Lw = h(t),$$

$$w(\sigma(b)) = w^{\Delta}(\sigma(b)) = 0.$$

Therefore,

$$w(\sigma^2(b)) = w(\sigma(b)) + \mu(\sigma(b))w^{\Delta}(\sigma(b))$$

$$= 0$$

and

$$\gamma x(\sigma^2(b)) + \delta x^\Delta(\sigma(b)) = \int_a^{\sigma(b)} \left(\gamma v(\sigma^2(b),s) + \delta v^\Delta(\sigma^2(b),s)\right) h(s)\Delta s$$

$$+\gamma w(\sigma^2(b)) + \delta w^\Delta(\sigma(b))$$

$$= 0.$$

This completes the proof.

Corollary 3.1. *Suppose* $(C1)$-$(C3)$ *and the BVP* (3.7) *has only the trivial solution. Let, also, there exists an* $s \in (a, \sigma^2(b))$ *such that* $\sigma(s) = s$. *Then*

$$G^\Delta(s^+, s) - G^\Delta(s^-, s) = \frac{1}{p(s)},$$

where G *is the function in Theorem 1.6.*

Proof. By the definition of the Green function, we get

$$G^\Delta(s^+, s) - G^\Delta(s^-, s) = v^\Delta(s^+, s) - u^\Delta(s^-, s)$$

$$= u^\Delta(s^+, s) + x^\Delta(s^+, s) - u^\Delta(s^-, s)$$

$$= x^\Delta(s^+, s)$$

$$= \frac{1}{p(s)}.$$

This completes the proof.

Theorem 3.6. *Assume* $(C1)$-$(C3)$ *and the BVP* (3.7) *has only the trivial solution. Let* ϕ *be the solution of the IVP*

$$L\phi = 0$$

$$\phi(a) = \beta$$

$$\phi^\Delta(a) = \alpha$$

and let ψ *be the solution to problem*

$$L\psi = 0$$

$$\psi(\sigma^2(b)) = \delta$$

$$\psi^{\Delta}(\sigma(b)) = -\gamma,$$

where α, β, γ and δ are constants. Then the Green function for the BVP (3.7) *is given by*

$$G(t,s) = \begin{cases} \frac{1}{c}\phi(t)\psi(\sigma(s)) & \text{if} \quad t \le s \\ \\ \frac{1}{c}\psi(t)\phi(\sigma(s)) & \text{if} \quad t \ge \sigma(s), \end{cases}$$

where $c = p(t)W(\phi,\psi)(t)$, $t \in [a,\sigma^2(b)]$, is a constant. Moreover, the Green function satisfies the symmetry condition

$$G(t,\sigma(s)) = G(\sigma(s),t), \quad s,t \in [a,b].$$

Proof. By the Abel theorem, we conclude that $c = p(t)W(\phi,\psi)(t)$, $t \in [a,\sigma^2(b)]$, is a constant. Note that

$$\alpha\phi(a) - \beta\phi^{\Delta}(a) = \alpha\beta - \beta\alpha$$

$$= 0,$$

$$\gamma\psi(\sigma^2(b)) + \delta\psi^{\Delta}(\sigma(b)) = \gamma\delta - \delta\gamma$$

$$= 0.$$

Therefore, ϕ and ψ satisfy the first and second boundary conditions of (3.7), respectively. Set

$$u(t,s) = \frac{1}{c}\phi(t)\psi(\sigma(s)),$$

$$v(t,s) = \frac{1}{c}\psi(t)\phi(\sigma(s)), \quad t \in [a,\sigma^2(b)], \quad s \in [a,b].$$

Note that $u(\cdot,s)$ and $v(\cdot,s)$ satisfy equation (3.2) on $[a,\sigma^2(b)]$ for each fixed $s \in [a,b]$. Also, for each fixed $s \in [a,b]$, we have

$$\alpha u(a,s) - \beta u^{\Delta}(a,s) = \frac{\alpha}{c}\phi(a)\psi(\sigma(s)) - \frac{\beta}{c}\phi^{\Delta}(a)\psi(\sigma(s))$$

$$= \frac{1}{c}(\alpha\phi(a) - \beta\phi^{\Delta}(a))\psi(\sigma(s))$$

$$= 0$$

and

$$\begin{aligned}
\gamma v(\sigma^2(b),s) + \delta v^{\Delta}(\sigma(b),s) &= \frac{\gamma}{c}\psi(\sigma^2(b))\phi(\sigma(s)) + \frac{\delta}{c}\psi^{\Delta}(\sigma(b))\phi(\sigma(s)) \\
&= \frac{1}{c}(\gamma\psi(\sigma^2(b)) + \delta\psi^{\Delta}(\sigma(b)))\phi(\sigma(s)) \\
&= 0.
\end{aligned}$$

Thus, for each fixed $s \in [a,b]$, u and v satisfy the first and second boundary conditions of (3.7). Let

$$\begin{aligned}
k(t,s) &= v(t,s) - u(t,s) \\
&= \frac{1}{c}(\phi(\sigma(s))\psi(t) - \phi(t)\psi(\sigma(s))).
\end{aligned}$$

We have that for each fixed $s \in [a,b]$, the function k satisfies equation (3.2). Next,

$$\begin{aligned}
k(\sigma(s),s) &= \frac{1}{c}(\phi(\sigma(s))\psi(\sigma(s)) - \phi(\sigma(s))\psi(\sigma(s))) \\
&= 0
\end{aligned}$$

and

$$\begin{aligned}
k^{\Delta}(t,s) &= \frac{1}{c}(\phi(\sigma(s))\psi^{\Delta}(t) - \phi^{\Delta}(t)\psi(\sigma(s))), \\
k^{\Delta}(\sigma(s),s) &= \frac{1}{c}(\phi(\sigma(s))\psi^{\Delta}(\sigma(s)) - \phi^{\Delta}(\sigma(s))\psi(\sigma(s))) \\
&= \frac{\phi(\sigma(s))\psi^{\Delta}(\sigma(s)) - \phi^{\Delta}(\sigma(s))\psi(\sigma(s))}{p(\sigma(s))W(\phi,\psi)(\sigma(s))} \\
&= \frac{1}{p(\sigma(s))}.
\end{aligned}$$

Therefore, $k(t,s) = x(t,s)$ is the Cauchy function for equation (3.2). Now, we will prove that $u(\cdot,s)$ satisfies the second boundary condition of (3.15) for each fixed $s \in [a,b]$. We have

$$\begin{aligned}
\gamma u(\sigma^2(b),s)+\delta u^{\Delta}(\sigma(b),s) &= \frac{\gamma}{c}\phi(\sigma^2(b))\psi(\sigma(s))+\frac{\delta}{c}\phi^{\Delta}(\sigma(b))\psi(\sigma(s))\\
&= \frac{1}{c}(\gamma\phi(\sigma^2(b))+\delta\phi^{\Delta}(\sigma(b)))\psi(\sigma(s))\\
&= 0\\
&= \gamma v(\sigma^2(b),s)+\delta v^{\Delta}(\sigma(b),s)\\
&\quad -\gamma x(\sigma^2(b),s)+\delta x^{\Delta}(\sigma(b),s).
\end{aligned}$$

Hence and by Theorem 3.5, we conclude that G is the Green function for the BVP (3.7). This completes the proof.

Example 3.5. Consider the BVP

$$\begin{aligned}
x^{\Delta^2} &= 0, \quad t\in[a,b],\\
x(a) &= 0,\\
\gamma x(\sigma^2(b))+\delta x^{\Delta}(\sigma(b)) &= 0.
\end{aligned}$$

We will construct the Green function. Firstly, we will construct the solution of the IVP

$$\begin{aligned}
\phi^{\Delta^2} &= 0, \quad t\in[a,b],\\
\phi(a) &= 0,\\
\phi^{\Delta}(a) &= 1,
\end{aligned}$$

in the form

$$\phi(t)=a_1t+a_2, \quad t\in[a,b],$$

where $a_1,a_2\in\mathbb{R}$. We have

$$\begin{aligned}
\phi(a) &= a_1a+a_2\\
&= 0,\\
\phi^{\Delta}(t) &= a_1,
\end{aligned}$$

$$\phi^{\Delta}(a) = 1$$

$$= a_1.$$

Thus,

$$a_1 = 1$$

$$a_2 = -a$$

and

$$\phi(t) = t - a, \quad t \in [a,b].$$

Now we will construct the solution of the IVP

$$\psi^{\Delta^2} = 0, \quad t \in [a,b],$$

$$\psi(\sigma^2(b)) = \delta,$$

$$\psi^{\Delta}(\sigma(b)) = -\gamma,$$

in the form

$$\psi(t) = b_1 t + b_2, \quad t \in [a,b],$$

where $b_1, b_2 \in \mathbb{R}$. We have

$$\psi(\sigma^2(b)) = b_1\sigma^2(b) + b_2$$

$$= \delta,$$

$$\psi^{\Delta}(t) = b,$$

$$\psi^{\Delta}(\sigma(b)) = b_1$$

$$= -\gamma.$$

So, we get the system

$$b_1 = -\gamma$$

$$b_1\sigma^2(b) + b_2 = \delta,$$

whereupon,

$$b_1 = -\gamma$$

$$b_2 = \delta + \gamma\sigma^2(b).$$

Therefore,

$$\psi(t) = -\gamma t + \delta + \gamma\sigma^2(b), \quad t \in [a,b].$$

Now, we will find $W(\phi,\psi)(t)$, $t \in [a,b]$. We have

$$W(\phi,\psi)(t) = \begin{vmatrix} \phi(t) & \psi(t) \\ \phi^\Delta(t) & \psi^\Delta(t) \end{vmatrix}$$

$$= \begin{vmatrix} t-a & -\gamma t + \delta + \gamma\sigma^2(b) \\ 1 & -\gamma \end{vmatrix}$$

$$= -\gamma t + \gamma a + \gamma t + \delta + \gamma\sigma^2(b)$$

$$= \gamma(a - \sigma^2(b)) - \delta.$$

Therefore,

$$G(t,s) = \begin{cases} \dfrac{(t-a)(-\gamma\sigma(s) + \delta + \gamma\sigma^2(b))}{\gamma(a-\sigma^2(b)) - \delta} & \text{if} \quad t \le s \\[2ex] \dfrac{(-\gamma t + \delta + \gamma\sigma^2(b))(\sigma(s) - a)}{\gamma(a-\sigma^2(b)) - \delta}, & \text{if} \quad \sigma(s) \le t \end{cases}$$

is the Green function for the considered BVP.

3.3 The Green Function for Dirichlet Problems

In this section, we will investigate the following Dirichlet problem

$$Lx = 0, \quad x(a) = x(\sigma^2(b)) = 0. \tag{3.17}$$

Theorem 3.7. *Assume (C1)-(C3) and the BVP (3.17) has only the trivial solution. Let $x(t,s)$ be the Cauchy function for (3.2). For each fixed $s \in \mathbb{T}$, let $u(\cdot,s)$ be the unique solution of the BVP*

$$Lu(\cdot,s) = 0$$

$$u(a,s) = 0$$

$$u(\sigma^2(b),s) = -x(\sigma^2(b),s).$$

Then

$$G(t,s) = \begin{cases} u(t,s) & \text{if} \quad t \le s \\ v(t,s) & \text{if} \quad t \ge \sigma(s), \end{cases} \tag{3.18}$$

where

$$v(t,s) = u(t,s) + x(t,s),$$

is the Green function for the BVP (3.17). *Moreover, for each* $s \in [a,b]$, $v(\cdot,s)$ *is a solution of equation* (3.2) *and*

$$v(\sigma^2(b),s) = 0.$$

Proof. We apply Theorem 3.5 for $\alpha = 1$, $\beta = 0$, $\gamma = 1$ and $\delta = 0$ and conclude that the function (3.18) is the Green function for (3.17). Next, since u and x satisfy (3.2), we obtain that v satisfies (3.2). Moreover,

$$\begin{aligned} v(\sigma^2(b),s) &= u(\sigma^2(b),s) + x(\sigma^2(b),s) \\ &= -x(\sigma^2(b),s) + x(\sigma^2(b),s) \\ &= 0. \end{aligned}$$

This completes the proof.

Example 3.6. Let

$$\mathbb{T} = \left\{0, \frac{1}{2}, 1, \frac{3}{2}, 2, \frac{5}{2}, \dots\right\}$$

and

$$Lx = x^{\Delta^2} - 2x^{\sigma} = 0. \tag{3.19}$$

Here

$$\begin{aligned} p(t) &= 1, \\ q(t) &= -2, \\ \sigma(t) &= t + \frac{1}{2}, \\ \mu(t) &= \frac{1}{2}, \quad t \in \mathbb{T}. \end{aligned}$$

The operator L can be rewritten in the following form

$$Lx(t) = x^{\Delta^2}(t) - 2(x(t) + \mu(t)x^{\Delta}(t))$$

$$= x^{\Delta^2}(t) - x^{\Delta}(t) - 2x(t), \quad t \in \mathbb{T}.$$

Fix $s \in \mathbb{T}$ arbitrarily. We will find the Cauchy function for L. The characteristic equation for equation (3.19) is as follows

$$r^2 - r - 2 = 0,$$

whereupon,

$$r_1 = 2, \quad r_2 = -1.$$

Then the general solution of equation (3.19) is given by

$$x(t,s) = c_1 e_2(t,s) + c_2 e_{-1}(t,s), \quad t \in \mathbb{T},$$

where c_1 and c_2 are constants which we will determine by the conditions

$$x(\sigma(s),s) = 0, \quad x^{\Delta}(\sigma(s),s) = \frac{1}{p(\sigma(s))}.$$

We have

$$x(\sigma(s),s) = c_1 e_2(\sigma(s),s) + c_2 e_{-1}(\sigma(s),s)$$

$$= c_1 e_2\left(s+\frac{1}{2},s\right) + c_2 e_{-1}\left(s+\frac{1}{2},s\right),$$

$$e_{-1}\left(s+\frac{1}{2},s\right) = e^{\int_s^{s+\frac{1}{2}} \frac{1}{\mu(\tau)} \log(1+\mu(\tau)(-1))\Delta\tau}$$

$$= e^{\int_s^{s+\frac{1}{2}} 2\log\left(1-\frac{1}{2}\right)\Delta\tau}$$

$$= e^{2\left(\log\frac{1}{2}\right)\frac{1}{2}}$$

$$= e^{\log\frac{1}{2}}$$

$$= \frac{1}{2}$$

and

$$e_2\left(s+\frac{1}{2},s\right) = e^{\int_s^{s+\frac{1}{2}} \frac{1}{\mu(\tau)} \log(1+\mu(\tau)2)\Delta\tau}$$

$$= e^{2\log\left(1+\frac{1}{2}\cdot 2\right)\frac{1}{2}}$$

$$= e^{\log 2}$$

$$= 2.$$

Hence,

$$\begin{aligned} x(\sigma(s),s) &= 2c_1 + \frac{1}{2}c_2 \\ &= 0 \end{aligned}$$

and

$$\begin{aligned} x^{\Delta}(t,s) &= 2c_1e_2(t,s) - c_2e_{-1}(t,s), \quad t \in \mathbb{T}, \\ x^{\Delta}(\sigma(s),s) &= 2c_1e_2(\sigma(s),s) - c_2e_{-1}(\sigma(s),s) \\ &= 4c_1 - \frac{1}{2}c_2 \\ &= 1. \end{aligned}$$

Thus, we get the system

$$\begin{aligned} 2c_1 + \frac{1}{2}c_2 &= 0 \\ 4c_1 - \frac{1}{2}c_2 &= 1. \end{aligned}$$

For its solution we get

$$\begin{aligned} c_1 &= \frac{1}{6}, \\ c_2 &= -\frac{2}{3}. \end{aligned}$$

Therefore,

$$x(t,s) = \frac{1}{6}e_2(t,s) - \frac{2}{3}e_{-1}(t,s), \quad t \in \mathbb{T},$$

is the Cauchy function for (3.19). Now, we consider the BVP

$$u^{\Delta^2} - 2u^{\sigma} = 0$$

$$u(0,s) = 0$$

$$u(\sigma^2(1),s) = -x(\sigma^2(1),s)$$

or

$$u^{\Delta^2} - 2u^\sigma = 0$$

$$u(0,s) = 0$$

$$u(2,s) = -\frac{1}{6}e_2(2,s) + \frac{2}{3}e_{-1}(2,s).$$

We will search its solution in the form

$$u(t,s) = a_1e_2(t,s) + a_2e_{-1}(t,s), \quad t \in \mathbb{T},$$

where a_1 and a_2 are constants. We have

$$u(0,s) = a_1e_2(0,s) + a_2e_{-1}(0,s)$$

$$= 0$$

and

$$u(2,s) = a_1e_2(2,s) + a_2e_{-1}(2,s)$$

$$= -\frac{1}{6}e_2(2,s) + \frac{2}{3}e_{-1}(2,s).$$

Thus, we get the system

$$a_1e_2(0,s) + a_2e_{-1}(0,s) = 0$$

$$a_1e_2(2,s) + a_2e_{-1}(2,s) = -\frac{1}{6}e_2(2,s) + \frac{2}{3}e_{-1}(2,s).$$

We have

$$a_1 = -a_2\frac{e_{-1}(0,s)}{e_2(0,s)}$$

and

$$-a_2e_2(2,s)\frac{e_{-1}(0,s)}{e_2(0,s)} + a_2e_{-1}(2,s) = -\frac{1}{6}e_2(2,s) + \frac{2}{3}e_{-1}(2,s)$$

or

$$-a_2 e_2(2,s)e_2(s,0)e_{-1}(0,s)+a_2 e_{-1}(2,s)=-\frac{1}{6}e_2(2,s)+\frac{2}{3}e_{-1}(2,s),$$

or

$$-a_2 e_2(2,0)e_{-1}(0,s)+a_2 e_{-1}(2,s)=-\frac{1}{6}e_2(2,s)+\frac{2}{3}e_{-1}(2,s),$$

or

$$a_2(e_{-1}(2,s)-e_2(2,0)e_{-1}(0,s))=-\frac{1}{6}e_2(2,s)+\frac{2}{3}e_{-1}(2,s),$$

or

$$a_2=\frac{-\frac{1}{6}e_2(2,s)+\frac{2}{3}e_{-1}(2,s)}{e_{-1}(2,s)-e_2(2,0)e_{-1}(0,s)}.$$

Hence,

$$\begin{aligned}a_1&=-\frac{e_1(0,s)}{e_2(0,s)}\left(\frac{-\frac{1}{6}e_2(2,s)+\frac{2}{3}e_{-1}(2,s)}{e_{-1}(2,s)-e_2(2,0)e_{-1}(0,s)}\right)\\&=\frac{\frac{1}{6}e_{-1}(0,s)e_2(2,s)-\frac{2}{3}e_{-1}(0,s)e_{-1}(2,s)}{e_2(0,s)e_{-1}(2,s)-e_{-1}(0,s)e_2(2,0)e_2(0,s)}\\&=\frac{\frac{1}{6}e_{-1}(0,s)e_2(2,s)-\frac{2}{3}e_{-1}(0,s)e_{-1}(2,s)}{e_2(0,s)e_{-1}(2,s)-e_{-1}(0,s)e_2(2,s)}.\end{aligned}$$

Therefore,

$$\begin{aligned}u(t,s)=&\frac{\frac{1}{6}e_{-1}(0,s)e_2(2,s)-\frac{2}{3}e_{-1}(0,s)e_{-1}(2,s)}{e_2(0,s)e_{-1}(2,s)-e_{-1}(0,s)e_2(2,s)}e_2(t,s)\\&+\frac{-\frac{1}{6}e_2(2,s)+\frac{2}{3}e_{-1}(2,s)}{e_{-1}(2,s)-e_2(2,0)e_{-1}(0,s)}e_{-1}(t,s),\quad t\in\mathbb{T},\end{aligned}$$

and

$$\begin{aligned}v(t,s)&=u(t,s)+x(t,s)\\&=\frac{\frac{1}{6}e_{-1}(0,s)e_2(2,s)-\frac{2}{3}e_{-1}(0,s)e_{-1}(2,s)}{e_2(0,s)e_{-1}(2,s)-e_{-1}(0,s)e_2(2,s)}e_2(t,s)\\&\quad+\frac{-\frac{1}{6}e_2(2,s)+\frac{2}{3}e_{-1}(2,s)}{e_{-1}(2,s)-e_2(2,0)e_{-1}(0,s)}e_{-1}(t,s)\\&\quad+\frac{1}{6}e_2(t,s)-\frac{2}{3}e_{-1}(t,s)\\&=\left(\frac{\frac{1}{6}e_{-1}(0,s)e_2(2,s)-\frac{2}{3}e_{-1}(0,s)e_{-1}(2,s)}{e_2(0,s)e_{-1}(2,s)-e_{-1}(0,s)e_2(2,s)}+\frac{1}{6}\right)e_2(t,s)\end{aligned}$$

$$+\left(\frac{-\frac{1}{6}e_2(2,s)+\frac{2}{3}e_{-1}(2,s)}{e_{-1}(2,s)-e_2(2,0)e_{-1}(0,s)}-\frac{2}{3}\right)e_{-1}(t,s), \quad t\in\mathbb{T},$$

and

$$G(t,s)=\begin{cases}\dfrac{\frac{1}{6}e_{-1}(0,s)e_2(2,s)-\frac{2}{3}e_{-1}(0,s)e_{-1}(2,s)}{e_2(0,s)e_{-1}(2,s)-e_{-1}(0,s)e_2(2,s)}e_2(t,s)\\[2ex] +\dfrac{-\frac{1}{6}e_2(2,s)+\frac{2}{3}e_{-1}(2,s)}{e_{-1}(2,s)-e_2(2,0)e_{-1}(0,s)}e_{-1}(t,s), \quad t\le s\\[2ex] \left(\dfrac{\frac{1}{6}e_{-1}(0,s)e_2(2,s)-\frac{2}{3}e_{-1}(0,s)e_{-1}(2,s)}{e_2(0,s)e_{-1}(2,s)-e_{-1}(0,s)e_2(2,s)}+\dfrac{1}{6}\right)e_2(t,s)\\[2ex] +\left(\dfrac{-\frac{1}{6}e_2(2,s)+\frac{2}{3}e_{-1}(2,s)}{e_{-1}(2,s)-e_2(2,0)e_{-1}(0,s)}-\dfrac{2}{3}\right)e_{-1}(t,s), \quad t\ge\sigma(s),\end{cases}$$

is the Green function of the BVP

$$x^{\Delta^2}-2x^{\sigma}=0$$

$$x(0)=x(2)=0.$$

Exercise 3.3. Let $\mathbb{T}=2^{\mathbb{N}_0}$. Find the Green functions for the following BVP

$$x^{\Delta^2}-12x^{\sigma}=0$$

$$x(1)=x(8)=0.$$

Theorem 3.8. *Assume* $(C1)$-$(C3)$ *and* $q=0$. *Then the Green function for the BVP* (3.17) *is given by*

$$G(t,s)=\begin{cases}-\dfrac{\int_a^t\frac{1}{p(\tau)}\Delta\tau\int_{\sigma(s)}^{\sigma^2(b)}\frac{1}{p(\tau)}\Delta\tau}{\int_a^{\sigma(b)}\frac{1}{p(\tau)}\Delta\tau} & \textit{if} \quad t\le s\\[3ex] -\dfrac{\int_a^{\sigma(s)}\frac{1}{p(\tau)}\Delta\tau\int_t^{\sigma^2(b)}\frac{1}{p(\tau)}\Delta\tau}{\int_a^{\sigma(b)}\frac{1}{p(\tau)}\Delta\tau} & \textit{if} \quad t\ge\sigma(s).\end{cases}$$

Proof. Fix $s\in\mathbb{T}$. In the case $q=0$ the Cauchy function is

$$x(t,s)=\int_{\sigma(s)}^{t}\frac{1}{p(\tau)}\Delta\tau.$$

We will search the function u in the following form

$$u(t,s) = c_1 \int_a^t \frac{1}{p(\tau)}\Delta\tau \int_{\sigma(s)}^t \frac{1}{p(\tau)}\Delta\tau, \quad t \in \mathbb{T},$$

where c_1 is a constant which will be determined by the condition

$$u(\sigma^2(b),s) = -x(\sigma^2(b),s).$$

We have

$$\begin{aligned} u(\sigma^2(b),s) &= c_1 \int_a^{\sigma(b)} \frac{1}{p(\tau)}\Delta\tau \int_{\sigma(s)}^{\sigma^2(b)} \frac{1}{p(\tau)}\Delta\tau \\ &= -\int_{\sigma(s)}^{\sigma^2(b)} \frac{1}{p(\tau)}\Delta\tau. \end{aligned}$$

Hence,

$$\begin{aligned} c_1 &= -\frac{\int_{\sigma(s)}^{\sigma^2(b)} \frac{1}{p(\tau)}\Delta\tau}{\int_a^{\sigma(b)} \frac{1}{p(\tau)}\Delta\tau \int_{\sigma(s)}^{\sigma^2(b)} \frac{1}{p(\tau)}\Delta\tau} \\ &= -\frac{1}{\int_a^{\sigma(b)} \frac{1}{p(\tau)}\Delta\tau}. \end{aligned}$$

Therefore,

$$u(t,s) = -\frac{\int_a^t \frac{1}{p(\tau)}\Delta\tau \int_{\sigma(s)}^t \frac{1}{p(\tau)}\Delta\tau}{\int_{\sigma(s)}^{\sigma^2(b)} \frac{1}{p(\tau)}\Delta\tau}, \quad t \leq s,$$

and

$$\begin{aligned} v(t,s) &= u(t,s) + x(t,s) \\ &= -\frac{\int_a^t \frac{1}{p(\tau)}\Delta\tau \int_{\sigma(s)}^t \frac{1}{p(\tau)}\Delta\tau}{\int_{\sigma(s)}^{\sigma^2(b)} \frac{1}{p(\tau)}\Delta\tau} + \int_{\sigma(s)}^t \frac{1}{p(\tau)}\Delta\tau \\ &= \frac{\left(-\int_a^t \frac{1}{p(\tau)}\Delta\tau + \int_{\sigma(s)}^t \frac{1}{p(\tau)}\Delta\tau\right) \int_{\sigma(s)}^t \frac{1}{p(\tau)}\Delta\tau}{\int_{\sigma(s)}^{\sigma^2(b)} \frac{1}{p(\tau)}\Delta\tau} \\ &= \frac{\left(-\int_a^{\sigma(s)} \frac{1}{p(\tau)}\Delta\tau - \int_{\sigma(s)}^t \frac{1}{p(\tau)}\Delta\tau + \int_{\sigma(s)}^t \frac{1}{p(\tau)}\Delta\tau\right) \int_{\sigma(s)}^t \frac{1}{p(\tau)}\Delta\tau}{\int_{\sigma(s)}^{\sigma^2(b)} \frac{1}{p(\tau)}\Delta\tau} \end{aligned}$$

$$= -\frac{\int_a^{\sigma(s)} \frac{1}{p(\tau)}\Delta\tau \int_{\sigma(s)}^t \frac{1}{p(\tau)}\Delta\tau}{\int_{\sigma(s)}^{\sigma^2(b)} \frac{1}{p(\tau)}\Delta\tau}, \quad t \geq \sigma(s).$$

This completes the proof.

Example 3.7. Let $\mathbb{T} = 2^{\mathbb{N}_0}$. Consider the BVP

$$\left(\frac{1}{7t^2+3t+1}u^\Delta\right)^\Delta = 0$$

$$u(1,2) = 0$$

$$u(16,2) = -4284.$$

Here

$$q(t) = 0,$$

$$p(t) = \frac{1}{7t^2+3t+1},$$

$$\sigma(t) = 2t, \quad t \in \mathbb{T},$$

$$a = 1,$$

$$b = 4,$$

$$s = 2.$$

Let

$$g(t) = t^3 + t^2 + t, \quad t \in \mathbb{T}.$$

Then

$$g^\Delta(t) = (\sigma(t))^2 + t\sigma(t) + t^2 + \sigma(t) + t + 1$$

$$= 4t^2 + 2t^2 + t^2 + 2t + t + 1$$

$$= 7t^2 + 3t + 1, \quad t \in \mathbb{T},$$

and

$$\int_{\sigma(s)}^t \frac{1}{p(\tau)}\Delta\tau = \int_4^t (7\tau^2 + 3\tau + 1)\Delta\tau$$

$$
\begin{aligned}
&= \int_4^t g^{\Delta}(\tau)\Delta\tau \\
&= g(\tau)\Big|_{\tau=4}^{\tau=t} \\
&= g(t) - 4 \\
&= t^3 + t^2 + t - 64 - 16 - 4 \\
&= t^3 + t^2 + t - 84, \quad t \in \mathbb{T}.
\end{aligned}
$$

Therefore, the Cauchy function is

$$
\begin{aligned}
x(t,2) &= t^3 + t^2 + t - 84, \\
x(\sigma^2(b),2) &= x(16,2) \\
&= 16^3 + 16^2 + 16 - 84 \\
&= 4096 + 256 + 16 - 84 \\
&= 4368 - 84 \\
&= 4284.
\end{aligned}
$$

Note that

$$
\begin{aligned}
\int_a^t \frac{1}{p(\tau)}\Delta\tau &= \int_1^t g^{\Delta}(\tau)\Delta\tau \\
&= g(t) - g(1) \\
&= t^3 + t^2 + t - 3, \\
\int_{\sigma(s)}^{\sigma^2(b)} \frac{1}{p(\tau)}\Delta\tau &= x(\sigma^2(b),2) \\
&= 4284,
\end{aligned}
$$

$$\begin{aligned}\int_a^{\sigma(b)} \frac{1}{p(\tau)}\Delta\tau &= \int_1^{16} g^{\Delta}(\tau)\Delta\tau\\ &= g(16)-g(1)\\ &= 4368-3\\ &= 4365,\end{aligned}$$

$$\begin{aligned}\int_a^{\sigma(s)} \frac{1}{p(\tau)}\Delta\tau &= \int_1^{4} g^{\Delta}(\tau)\Delta\tau\\ &= g(4)-g(1)\\ &= 84-3\\ &= 81,\end{aligned}$$

$$\begin{aligned}\int_t^{\sigma^2(b)} \frac{1}{p(\tau)}\Delta\tau &= \int_t^{16} g^{\Delta}(\tau)\Delta\tau\\ &= g(16)-g(t)\\ &= 4368-t^3-t^2-t, \quad t\in\mathbb{T}.\end{aligned}$$

Thus,

$$\begin{aligned}u(t,2) &= -\frac{(t^3+t^2+t-3)(t^3+t^2+t-84)}{4284},\\ v(t,2) &= -\frac{81(t^3+t^2+t-84)}{4284}\\ &= -\frac{9(t^3+t^2+t-84)}{476}, \quad t\geq 4.\end{aligned}$$

Consequently,

$$G(t,s)=\begin{cases} -\dfrac{(t^3+t^2+t-3)(t^3+t^2+t-84)}{4284} & \text{if} \quad t\leq 2\\[2ex] -\dfrac{9(t^3+t^2+t-84)}{476} & \text{if} \quad t\geq 4\end{cases}$$

is the Green function.

Exercise 3.4. Let $\mathbb{T} = 2\mathbb{Z}$ and

$$q(t) = 0,$$

$$p(t) = \frac{1}{t^2+t+1}, \quad t \in \mathbb{T}.$$

Find the Green function for the BVP (3.17).

3.4 The Green Function for Focal Boundary Value Problems

In this section we will investigate the following focal BVP

$$\begin{aligned} Lx &= 0 \\ x(a) &= 0 \\ x^{\Delta}(\sigma(b)) &= 0. \end{aligned} \tag{3.20}$$

Theorem 3.9. *Suppose* $(C1)$-$(C3)$. *Assume that the BVP* (3.20) *has only the trivial solution. For each fixed* $s \in [a,b]$, *let* $u(\cdot,s)$ *be the solution of the BVP*

$$\begin{aligned} Lu(\cdot,s) &= 0 \\ u(a,s) &= 0 \\ u^{\Delta}(\sigma(b),s) &= -x^{\Delta}(\sigma(b),s), \end{aligned} \tag{3.21}$$

where $x(t,s)$ *is the Cauchy function for* (3.2). *Then*

$$G(t,s) = \begin{cases} u(t,s) & \text{if} \quad t \le s \\ u(t,s) + x(t,s) & \text{if} \quad t \ge \sigma(s) \end{cases} \tag{3.22}$$

is the Green function for the BVP (3.20).

Proof. We apply Theorem 3.5 for $\alpha = 1$, $\beta = 0$, $\gamma = 0$ and $\delta = 1$ and we get that the function (3.22) is the Green function for the BVP (3.20). This completes the proof.

Theorem 3.10. *Suppose* $(C1)$-$(C3)$ *and* $q = 0$. *Then the Green function for the BVP* (3.20) *is given by*

$$G(t,s) = \begin{cases} -\int_a^t \frac{1}{p(\tau)}\Delta\tau & \text{if} \quad t \le s \\ \\ -\int_a^{\sigma(s)} \frac{1}{p(\tau)}\Delta\tau & \text{if} \quad t \ge \sigma(s). \end{cases}$$

Proof. Fix $s \in \mathbb{T}$. Since $q(t) = 0$, $t \in \mathbb{T}$, the Cauchy function for (3.2) is

$$x(t,s) = \int_{\sigma(s)}^t \frac{1}{p(\tau)}\Delta\tau, \quad t \in \mathbb{T}.$$

We will search a solution u of the BVP (3.21) in the form

$$u(t,s) = c_1 \int_a^t \frac{1}{p(\tau)}\Delta\tau, \quad t \in \mathbb{T},$$

where c_1 is a constant which will be determined by the third equation of (3.21). We have

$$\begin{aligned} u^{\Delta}(t,s) &= \frac{c_1}{p(t)}, \quad t \in \mathbb{T}, \\ u^{\Delta}(\sigma(b),s) &= \frac{c_1}{p(\sigma(b))} \\ &= -x^{\Delta}(\sigma(b),s) \\ &= -\frac{1}{p(\sigma(b))}. \end{aligned}$$

Therefore,

$$c_1 = -1$$

and

$$u(t,s) = -\int_a^t \frac{1}{p(\tau)}\Delta\tau, \quad t \le s,$$

and

$$\begin{aligned} v(t,s) &= u(t,s) + x(t,s) \\ &= -\int_a^t \frac{1}{p(\tau)}\Delta\tau + \int_{\sigma(s)}^t \frac{1}{p(\tau)}\Delta\tau \\ &= -\int_a^t \frac{1}{p(\tau)}\Delta\tau - \int_t^{\sigma(s)} \frac{1}{p(\tau)}\Delta\tau \\ &= -\int_a^{\sigma(s)} \frac{1}{p(\tau)}\Delta\tau, \quad t \ge \sigma(s). \end{aligned}$$

This completes the proof.

Example 3.8. Let $\mathbb{T} = 2\mathbb{Z}$. Consider the BVP

$$\left(\frac{1}{2t+1}u^{\Delta}\right)^{\Delta} = 0$$

$$u(0) = 0$$

$$u^{\Delta}(6) = 0.$$

Here

$$\sigma(t) = t+2,$$

$$q(t) = 0,$$

$$p(t) = \frac{1}{2t+1}, \quad t \in \mathbb{T},$$

$$a = 0,$$

$$b = 4.$$

Let

$$g(t) = t^2 - t + 1, \quad t \in \mathbb{T}.$$

Then

$$\begin{aligned} g^{\Delta}(t) &= \sigma(t) + t - 1 \\ &= t + 2 + t - 1 \\ &= 2t + 1, \quad t \in \mathbb{T}. \end{aligned}$$

Then

$$\begin{aligned} \frac{1}{p(t)} &= 2t+1 \\ &= g^{\Delta}(t), \end{aligned}$$

$$-\int_a^t \frac{1}{p(\tau)}\Delta\tau = -\int_0^t g^{\Delta}(\tau)\Delta\tau$$

$$= -g(t) + g(0)$$

$$= -t^2 + t - 1 + 1$$

$$= -t^2 + t,$$

$$-\int_a^{\sigma(s)} \frac{1}{p(\tau)} \Delta\tau = -\int_0^{s+2} \frac{1}{p(\tau)} \Delta\tau$$

$$= -g(s+2) + g(0)$$

$$= -(s+2)^2 + (s+2) - 1 + 1$$

$$= -s^2 - 4s - 4 + s + 2$$

$$= -s^2 - 3s - 2, \quad t, s \in \mathbb{T}.$$

Therefore, the Green function

$$G(t,s) = \begin{cases} -t^2 + t & \text{if} \quad t \le s \\ \\ -s^2 - 3s - 2 & \text{if} \quad t \ge s+2, \quad t, s \in \mathbb{T}. \end{cases}$$

Exercise 3.5. Let $\mathbb{T} = \mathbb{N}_0^3$ and

$$q(t) = 0,$$

$$p(t) = \frac{1}{t + t^2}, \quad t \in \mathbb{T}.$$

Find the Green function for the BVP (3.20).

Exercise 3.6. Let $p = 1$ and $q = 0$. Prove that the Green function of the BVP (3.20) is given by

$$G(t,s) = \begin{cases} a - t & \text{if} \quad t \le s \\ \\ a - \sigma(s) & \text{if} \quad t \ge \sigma(s). \end{cases}$$

3.5 The Green Function for Periodic Boundary Value Problems

Definition 3.2. Let $T \in \mathbb{T}$, $T \ge 0$. We say that the time scale is a periodic time scale with a period $\sigma^2(T)$ if $t \in \mathbb{T}$ implies $t + \sigma^2(T) \in \mathbb{T}$.

Below we will suppose that $\mathbb{T}$ is a periodic time scale with a period $\sigma^2(T)$. In this section, we will investigate the following periodic BVP

$$\begin{aligned} Lx &= 0 \\ x(a) &= x(a+\sigma^2(T)) \\ x^{\Delta}(a) &= x^{\Delta}(a+\sigma(T)). \end{aligned} \tag{3.23}$$

Theorem 3.11. *Suppose* $(C1)$-$(C3)$ *and* $A, B \in \mathbb{R}$, $h \in \mathscr{C}_{rd}(\mathbb{T})$. *Assume that the BVP* (3.23) *has only the trivial solution. Then the nonhomogeneous BVP*

$$\begin{aligned} Lx &= h(t) \\ x(a) &= x(a+\sigma^2(T)) \\ x^{\Delta}(a) &= x^{\Delta}(a+\sigma(T)) \end{aligned} \tag{3.24}$$

has a unique solution.

Proof. Let x_1 and x_2 be linearly independent solutions of equation (3.2). Then the general solution of the BVP (3.23) has the form

$$x(t) = c_1 x_1(t) + c_2 x_2(t), \quad t \in \mathbb{T},$$

where c_1 and c_2 are constants. We have

$$x(a) = c_1 x_1(a) + c_2 x_2(a),$$

$$x(a+\sigma^2(T)) = c_1 x_1(a+\sigma^2(T)) + c_2 x_2(a+\sigma^2(T)),$$

$$x^{\Delta}(t) = c_1 x_1^{\Delta}(t) + c_2 x_2^{\Delta}(t),$$

$$x^{\Delta}(a) = c_1 x_1^{\Delta}(a) + c_2 x_2^{\Delta}(a),$$

$$x^{\Delta}(a+\sigma(T)) = c_1 x_1^{\Delta}(a+\sigma(T)) + c_2 x_2^{\Delta}(a+\sigma(T)).$$

Hence,

$$c_1(x_1(a) - x_1(a+\sigma^2(T))) + c_2(x_2(a) - x_2(a+\sigma^2(T))) = 0$$

$$c_1(x_1^{\Delta}(a) - x_1^{\Delta}(a+\sigma(T))) + c_2(x_2^{\Delta}(a) - x_2^{\Delta}(a+\sigma(T))) = 0.$$

Set

$$M = \begin{pmatrix} x_1(a) - x_1(a+\sigma^2(T)) & x_2(a) - x_2(a+\sigma^2(T)) \\ x_1^\Delta(a) - x_1^\Delta(a+\sigma(T)) & x_2^\Delta(a) - x_2^\Delta(a+\sigma(T)) \end{pmatrix}.$$

Thus, we get the system

$$M \begin{pmatrix} c_1 \\ c_2 \end{pmatrix} = \begin{pmatrix} 0 \\ 0 \end{pmatrix}.$$

Because the BVP (3.23) has only the trivial solution, we conclude that

$$\det M \neq 0. \tag{3.25}$$

Let, now, u_0 be a fixed solution of the nonhomogeneous equation

$$Lx = h(t), \quad t \in \mathbb{T}.$$

The general solution of the BVP (3.24) has the form

$$u(t) = c_1 x_1(t) + c_2 x_2(t) + u_0(t), \quad t \in \mathbb{T}.$$

We have

$$u(a) = c_1 x_1(a) + c_2 x_2(a) + u_0(a),$$

$$u(a+\sigma^2(T)) = c_1 x_1(a+\sigma^2(T)) + c_2 x_2(a+\sigma^2(T)) + u_0(a+\sigma^2(T)),$$

$$u^\Delta(t) = c_1 x_1^\Delta(t) + c_2 x_2^\Delta(t) + u_0^\Delta(t),$$

$$u^\Delta(a) = c_1 x_1^\Delta(a) + c_2 x_2^\Delta(a) + u_0^\Delta(a),$$

$$u^\Delta(a+\sigma(T)) = c_1 x_1^\Delta(a+\sigma(T)) + c_2 x_2^\Delta(a+\sigma(T)) + u_0^\Delta(a+\sigma(T)).$$

By the boundary conditions of (3.24), we arrive at the system

$$a_1(x_1(a) - x_1(a+\sigma^2(T))) + a_2(x_2(a) - x_2(a+\sigma^2(T))) = A - u_0(a) - u_0(a+\sigma^2(T))$$

$$a_1(x_1^\Delta(a) - x_1^\Delta(a+\sigma(T))) + a_2(x_2^\Delta(a) - x_2^\Delta(a+\sigma(T))) = B - u_0^\Delta(a) - u_0^\Delta(a+\sigma(T)),$$

which can be written in the form

$$M \begin{pmatrix} c_1 \\ c_2 \end{pmatrix} = \begin{pmatrix} A - u_0(a) - u_0(a+\sigma^2(T)) \\ B - u_0^\Delta(a) - u_0^\Delta(a+\sigma(T)) \end{pmatrix}.$$

Because (3.25), we conclude that there exists a unique solution (c_1, c_2) and hence, the BVP (3.24) has a unique solution. This completes the proof.

Theorem 3.12. *Assume* $(C1)$-$(C3)$ *and the homogeneous BVP* (3.23) *has only the trivial solution. For each fixed* $s \in \mathbb{T}$, *suppose that* $u(\cdot,s)$ *is the solution of the BVP*

$$\begin{aligned} Lu(\cdot,s) &= 0 \\ u(a,s) - u(a+\sigma^2(T),s) &= x(a+\sigma^2(T),s) \\ u^{\Delta}(a,s) - u^{\Delta}(a+\sigma(T),s) &= x^{\Delta}(a+\sigma(T),s), \end{aligned} \tag{3.26}$$

where $x(t,s)$ *is the Cauchy function for equation* (3.2). *Then the function*

$$G(t,s) = \begin{cases} u(t,s) & \text{if} \quad t \le s \\ v(t,s) = u(t,s) + x(t,s) & \text{if} \quad t \ge \sigma(s) \end{cases} \tag{3.27}$$

is the Green function for (3.23). *If* $h \in \mathscr{C}_{rd}(\mathbb{T})$, *then*

$$x(t) = \int_a^{\sigma(T)} G(t,s)h(s)\Delta s \tag{3.28}$$

is the unique solution of the nonhomogeneous BVP (3.24) *with* $A = B = 0$. *Moreover,* $v(\cdot,s)$ *is the solution of* (3.23) *and*

$$u(a,s) = v(a+\sigma^2(T),s),$$

$$u^{\Delta}(a,s) = v^{\Delta}(a+\sigma(T),s).$$

Proof. Since $u(t,s)$ and $x(t,s)$ are solutions of equation (3.2), then $v(t,s)$ satisfies equation (3.2). Next,

$$\begin{aligned} u(a,s) &= u(a+\sigma^2(T),s) + x(a+\sigma^2(T),s) \\ &= v(a+\sigma^2(T),s), \\ u^{\Delta}(a,s) &= u^{\Delta}(a+\sigma(T),s) + x^{\Delta}(a+\sigma(T),s) \\ &= v^{\Delta}(a+\sigma(T),s). \end{aligned}$$

Now, we suppose that G is defined by (3.27). Consider the function x, defined by (3.28). We have

$$x(t) = \int_a^{\sigma(T)} G(t,s)h(s)\Delta s$$

$$= \int_a^t G(t,s)h(s)\Delta s + \int_t^{\sigma(T)} G(t,s)h(s)\Delta s$$

$$= \int_a^t (u(t,s)+x(t,s))h(s)\Delta s + \int_t^{\sigma(T)} u(t,s)h(s)\Delta s$$

$$= \int_a^t u(t,s)h(s)\Delta s + \int_t^{\sigma(T)} u(t,s)h(s)\Delta s + \int_a^t x(t,s)h(s)\Delta s$$

$$= \int_a^{\sigma(T)} u(t,s)h(s)\Delta s + \int_a^t x(t,s)h(s)\Delta s, \quad t \in \mathbb{T}.$$

Let

$$z(t) = \int_a^t x(t,s)h(s)\Delta s, \quad t \in \mathbb{T}.$$

By Theorem 3.2, it follows that

$$Lz = h(t)$$

$$z(a) = 0$$

$$z^{\Delta}(a) = 0.$$

Hence,

$$Lx(t) = \int_a^{\sigma(T)} Lu(t,s)h(s)\Delta s + Lz(t)$$

$$= h(t), \quad t \in \mathbb{T}.$$

Moreover,

$$x(a) = \int_a^{\sigma(T)} u(a,s)h(s)\Delta s + z(a)$$

$$= \int_a^{\sigma(T)} v(a+\sigma^2(T),s)h(s)\Delta s$$

$$= \int_a^{\sigma(T)} G(a+\sigma^2(T),s)h(s)\Delta s$$

$$= x(a+\sigma^2(T))$$

and

$$x^{\Delta}(a) = \int_a^{\sigma(T)} u^{\Delta}(a,s)h(s)\Delta s$$

$$= \int_a^{\sigma(T)} v^{\Delta}(a+\sigma(T),s)h(s)\Delta s$$

$$= \int_a^{\sigma(T)} G^{\Delta}(a+\sigma(T),s)h(s)\Delta s$$

$$= x^{\Delta}(a+\sigma(T)).$$

This completes the proof.

Exercise 3.7. Let $\mathbb{T}=\mathbb{Z}$. Find the Green function for the BVP (3.23) in the case when

$$p(t)=t, \quad q(t)=\frac{1}{1+t+t^2}, \quad t\in\mathbb{T}, \quad T+1.$$

3.6 The Green Function for Three Point Boundary Value Problems

In this section we will investigate some three point boundary value problems. We will start with the following three point BVP

$$\begin{aligned} Lx &= h(t) \\ \alpha x(a)-\beta x^{\Delta}(a) &= kx(\eta) \\ \gamma x(\sigma^2(b))+\delta x^{\Delta}(\sigma(b)) &= -\gamma x(\sigma^2(b),s)-\delta x^{\Delta}(\sigma(b),s), \end{aligned} \tag{3.29}$$

where $x(t,s)$ is the Cauchy function for (3.2), p and q satisfy $(C1)$, h satisfies $(C2)$, α, β, γ and δ satisfy $(C3)$ and

(C4) $(\gamma,\delta)\neq(0,0)$, $k\in(a,\sigma^2(b))$, $\gamma+\delta\neq 0$, $k\gamma(a-\eta)+a\delta(k-\alpha)-\beta\gamma\neq 0$.

Theorem 3.13. *Suppose $(C1)$-$(C4)$. Then the solution for the BVP (3.29) is represented in the form*

$$x(t)=\int_a^{\sigma(b)} K(t,s)h(s)\Delta s, \quad t\in[a,\sigma^2(b)],$$

where

$$K(t,s)=G(t,s)+\frac{-k(\gamma+\delta)a+k\gamma t}{k\gamma(a-\eta)+a\delta(k-\alpha)-\beta\gamma}G(\eta,s),$$

$t\in[a,\sigma^2(b)]$, and $G(t,s)$ is the Green function, defined in Theorem 3.5.

Proof. By Theorem 3.5, it follows that the function

$$w(t) = \int_a^{\sigma(b)} G(t,s)h(s)\Delta s, \quad t \in [a, \sigma^2(b)],$$

is the solution of the BVP (3.15). We will search a solution of the BVP (3.29) in the form

$$x(t) = w(t) + (a_1 + a_2 t)w(\eta),$$

where a_1 and a_2 are constants which will be determined by the second and third equations of (3.29). We have

$$x(a) = w(a) + (a_1 + aa_2)w(\eta),$$

$$x^{\Delta}(t) = w^{\Delta}(t) + a_2 w(\eta),$$

$$x^{\Delta}(a) = w^{\Delta}(a) + a_2 w(\eta).$$

From here,

$$\begin{aligned} \alpha x(a) - \beta x^{\Delta}(a) &= \alpha w(a) + \alpha(a_1 + aa_2)w(\eta) \\ &\quad - \beta w^{\Delta}(a) - \beta a_2 w(\eta) \\ &= (\alpha a_1 + a_2(\alpha a - \beta))w(\eta). \end{aligned}$$

Next,

$$\begin{aligned} x(\eta) &= w(\eta) + (a_1 + a_2\eta)w(\eta) \\ &= (1 + a_1 + a_2\eta)w(\eta). \end{aligned}$$

Hence, using the second equation of (3.29), we get

$$\alpha x(a) - \beta x^{\Delta}(a) = kx(\eta) \quad \Longleftrightarrow$$

$$(\alpha(a_1 + aa_2) - \beta a_2)w(\eta) = k(1 + a_1 + a_2\eta)w(\eta) \quad \Longleftrightarrow$$

$$\alpha(a_1 + aa_2) - \beta a_2 = k(1 + a_1 + a_2\eta) \quad \Longleftrightarrow$$

$$a_1(\alpha - k) + a_2(\alpha a - \beta - k\eta) = k.$$

Moreover,

$$x(\sigma^2(b)) = w(\sigma^2(b)) + (a_1 + aa_2)w(\eta),$$

$$x^{\Delta}(\sigma(b)) = w^{\Delta}(\sigma(b)) + a_2 w(\eta),$$

whereupon,

$$\begin{aligned}\gamma x(\sigma^2(b))+\delta x^{\Delta}(\sigma(b)) &= \gamma w(\sigma^2(b))+\gamma(a_1+aa_2)w(\eta)\\&\quad+\delta w^{\Delta}(\sigma(b))+\delta a_2 w(\eta)\\&=(\gamma a_1+(\gamma+\delta)aa_2)w(\eta)\\&\quad-\gamma x(\sigma^2(b),s)-\delta x^{\Delta}(\sigma(b),s).\end{aligned}$$

Now, using the last equation and the third equation of (3.29), we arrive at equation

$$(\gamma a_1+(\gamma+\delta)aa_2)w(\eta)=0,$$

or

$$\gamma a_1+(\gamma+\delta)aa_2=0.$$

Therefore, we get the system

$$\begin{aligned}a_1(\alpha-k)+a_2(\alpha a-\beta-k\eta) &= k\\ \gamma a_1+(\gamma+\delta)aa_2 &= 0.\end{aligned}$$

From it, we obtain

$$\begin{aligned}a_1 &= -\frac{\gamma+\delta}{\gamma}aa_2\\ -aa_2\frac{\gamma+\delta}{\gamma}(\alpha-k)+a_2(\alpha a-\beta-k\eta) &= k,\end{aligned}$$

whereupon,

$$a_2\left((\alpha-k)\frac{\gamma+\delta}{\gamma}a+\alpha a-\beta-k\eta\right)=k,$$

or

$$\begin{aligned}a_1 &= -\frac{k(\gamma+\delta)a}{k\gamma(a-\eta)+a\delta(k-\alpha)-\beta\gamma}\\ a_2 &= \frac{k\gamma}{k\gamma(a-\eta)+a\delta(k-\alpha)-\beta\gamma}.\end{aligned}$$

Consequently,

$$x(t)=w(t)+\left(\frac{-k(\gamma+\delta)a+k\gamma t}{k\gamma(a-\eta)+a\delta(k-\alpha)-\beta\gamma}\right)w(\eta)$$

$$= \int_a^{\sigma(b)} \left(G(t,s) + \frac{-k(\gamma+\delta)a + k\gamma t}{k\gamma(a-\eta) + a\delta(k-\alpha) - \beta\gamma} G(\eta,s) \right) h(s)\Delta s$$

$$= \int_a^{\sigma(b)} K(t,s)h(s)\Delta s, \quad t \in [a, \sigma^2(b)].$$

This completes the proof.

Now, we consider the BVP

$$\begin{aligned} Lx &= h(t) \\ x(a) &= kx(\eta) \\ x(\sigma^2(b)) &= 0, \end{aligned} \tag{3.30}$$

where

(C5) $k \in \mathbb{R}$, $\eta \in (a, \sigma^2(b))$, $a - k\eta - (1-k)\sigma^2(b) \neq 0$.

Theorem 3.14. *Assume* (C1)-(C3), (C5) *and* $q = 0$. *Then the solution x of the BVP* (3.30) *is given the form*

$$x(t) = \int_a^{\sigma(b)} K(t,s)h(s)\Delta s, \quad t \in [a, \sigma^2(b)],$$

where

$$K(t,s) = G(t,s) + \left(\frac{\sigma^2(b)k}{a - k\eta - (1-k)\sigma^2(b)} + \frac{k}{a - k\eta - (1-k)\sigma^2(b)} t \right) G(\eta,s),$$
$$t \in [a, \sigma^2(b)],$$

where $G(t,s)$ is the Green function, given in Theorem 3.8.

Proof. By Theorem 3.8, it follows that

$$w(t) = \int_a^{\sigma(b)} G(t,s)h(s)\Delta s, \quad t \in [a, \sigma^2(b)],$$

is the solution of the first equation of (3.30) that satisfies the boundary conditions

$$w(a) = w(\sigma^2(b)) = 0.$$

We will search a solution of the BVP (3.30) in the form

$$x(t) = w(t) + (a_1 + a_2 t)w(\eta), \quad t \in [a, \sigma^2(b)],$$

where a_1 and a_2 are constants which will be determined below. We have

$$
\begin{aligned}
x(a) &= w(a) + (a_1 + aa_2)w(\eta) \\
&= (a_1 + aa_2)w(\eta), \\
x(\eta) &= w(\eta) + (a_1 + \eta a_2)w(\eta) \\
&= (1 + a_1 + \eta a_2)w(\eta), \\
x(\sigma^2(b)) &= w(\sigma^2(b)) + (a_1 + \sigma^2(b)a_2)w(\eta) \\
&= (a_1 + \sigma^2(b)a_2)w(\eta).
\end{aligned}
$$

Thus,

$$
\begin{gathered}
x(a) = kx(\eta) \quad \Longleftrightarrow \\
(a_1 + aa_2)w(\eta) = k(1 + a_1 + \eta a_2)w(\eta) \quad \Longleftrightarrow \\
a_1 + aa_2 = k(1 + a_1 + \eta a_2) \quad \Longleftrightarrow \\
a_1(1-k) + a_2(a - k\eta) = k,
\end{gathered}
$$

and

$$
\begin{gathered}
x(\sigma^2(b)) = 0 \quad \Longleftrightarrow \\
(a_1 + \sigma^2(b)a_2)w(\eta) = 0 \quad \Longleftrightarrow \\
a_1 + \sigma^2(b)a_2 = 0,
\end{gathered}
$$

and we arrive at the system

$$
\begin{aligned}
a_1(1-k) + a_2(1 - k\eta) &= k \\
a_1 + \sigma^2(b)a_2 &= 0.
\end{aligned}
$$

Then

$$
\begin{aligned}
a_1 &= -\sigma^2(b)a_2 \\
-(1-k)\sigma^2(b)a_2 + a_2(a - k\eta) &= k
\end{aligned}
$$

or

$$
a_1 = -\frac{\sigma^2(b)k}{a - k\eta - (1-k)\sigma^2(b)}
$$

$$a_2 = \frac{k}{a-k\eta-(1-k)\sigma^2(b)}.$$

Therefore,

$$x(t) = w(t) + \left(-\frac{\sigma^2(b)k}{a-k\eta-(1-k)\sigma^2(b)} + \frac{k}{a-k\eta-(1-k)\sigma^2(b)}t\right) w(\eta)$$

$$= \int_a^{\sigma(b)} G(t,s)h(s)\Delta s$$

$$+ \int_a^{\sigma(b)} \left(-\frac{\sigma^2(b)k}{a-k\eta-(1-k)\sigma^2(b)} + \frac{k}{a-k\eta-(1-k)\sigma^2(b)}t\right) G(\eta,s)h(s)\Delta s$$

$$= \int_a^{\sigma(b)} \Bigg(G(t,s)$$

$$+ \left(-\frac{\sigma^2(b)k}{a-k\eta-(1-k)\sigma^2(b)} + \frac{k}{a-k\eta-(1-k)\sigma^2(b)}t\right) G(\eta,s)\Bigg) h(s)\Delta s,$$

$t \in [a, \sigma^2(b)]$, is the solution of the BVP (3.30). This completes the proof.

Example 3.9. Let $\mathbb{T} = 2^{\mathbb{N}_0}$. Consider the BVP

$$\left(\frac{1}{3t+1}x^{\Delta}\right)^{\Delta} = 0, \quad t \in [1,16],$$

$$x(1) = 3x(4),$$

$$x(16) = 0.$$

Here

$$\sigma(t) = 2t,$$

$$q(t) = 0,$$

$$p(t) = \frac{1}{3t+1}, \quad t \in \mathbb{T},$$

$$a = 1,$$

$$b = 4,$$

$$\eta = 4,$$

$$k = 3.$$

Let

$$g(t) = t^2 + t, \quad t \in \mathbb{T}.$$

We have

$$\begin{aligned}
\frac{1}{p(t)} &= 3t + 1, \\
g^{\Delta}(t) &= \sigma(t) + t + 1 \\
&= 2t + t + 1 \\
&= 3t + 1, \\
\int_a^t \frac{1}{p(\tau)} \Delta\tau &= \int_1^t g^{\Delta}(\tau) \Delta\tau \\
&= g(t) - g(1) \\
&= t^2 + t - 2, \\
\int_{\sigma(s)}^{\sigma^2(b)} \frac{1}{p(\tau)} \Delta\tau &= \int_{2s}^{16} g^{\Delta}(\tau) \Delta\tau \\
&= g(16) - g(2s) \\
&= 256 + 16 - 4s^2 - 2s \\
&= 272 - 4s^2 - 2s, \\
\int_a^{\sigma(b)} \frac{1}{p(\tau)} \Delta\tau &= \int_1^8 g^{\Delta}(\tau) \Delta\tau \\
&= g(8) - g(1) \\
&= 64 + 8 - 2 \\
&= 70,
\end{aligned}$$

$$\begin{aligned}
\int_t^{\sigma^2(b)} \frac{1}{p(\tau)}\Delta\tau &= \int_t^{16} g^{\Delta}(\tau)\Delta\tau \\
&= g(16) - g(t) \\
&= 256 + 16 - t^2 - t \\
&= 272 - t^2 - t, \\
\int_a^{\sigma(s)} \frac{1}{p(\tau)}\Delta\tau &= \int_1^{2s} g^{\Delta}(\tau)\Delta\tau \\
&= g(2s) - g(1) \\
&= 4s^2 + 2s - 2, \quad t, s \in \mathbb{T}.
\end{aligned}$$

Therefore,

$$G(t,s) = \begin{cases} -\dfrac{(t^2+t-2)(272-4s^2-2s)}{70} & \text{if} \quad t \le s \\[2ex] -\dfrac{(4s^2+2s-2)(272-t^2-t)}{70} & \text{if} \quad t \ge 2s \end{cases}$$

and

$$\begin{aligned}
K(t,s) &= G(t,s) + \left(-\frac{48}{1-12+2\cdot 16} + \frac{3}{1-12+2\cdot 16}t\right) G(\eta,s) \\
&= G(t,s) + \left(-\frac{48}{21} + \frac{3}{21}t\right) G(\eta,s) \\
&= G(t,s) + \left(-\frac{16}{7} + \frac{1}{7}t\right) G(\eta,s), \quad t, s \in \mathbb{T},
\end{aligned}$$

is the Green function of the considered BVP.

Exercise 3.8. Let $\mathbb{T} = 3\mathbb{Z}$. Find the Green function for the BVP

$$\left(\frac{1}{t^4+t^2+1}x^{\Delta}\right)^{\Delta} = 0$$

$$x(0) = \frac{1}{2}x(9)$$

$$x(15) = 0.$$

Now, we consider the BVP

$$\begin{aligned} Lx &= h(t) \\ x(a) &= kx(\eta) \\ x^{\Delta}(\sigma(b)) &= 0, \end{aligned} \tag{3.31}$$

where p and q satisfy $(C1)$, h satisfies $(C2)$, and

(C6) $k \neq 1$.

Theorem 3.15. *Suppose* $(C1)$, $(C2)$ *and* $(C6)$. *Then the solution of the BVP* (3.31) *can be represented in the following form*

$$x(t) = \int_a^{\sigma(b)} K(t,s)h(s)\Delta s,$$

where

$$K(t,s) = G(t,s) + \frac{k}{1-k} G(\eta,s), \quad t,s \in \mathbb{T},$$

where $G(t,s)$ *is the Green function, defined in Theorem 3.9.*

Proof. By Theorem 3.9, it follows that

$$w(t) = \int_a^{\sigma(b)} G(t,s)h(s)\Delta s, \quad t \in [a,\sigma(b)],$$

is the solution of equation (3.2) that satisfies the boundary conditions

$$w(a) = w^{\Delta}(\sigma(b)) = 0.$$

We will search the solution of the BVP (3.31) in the form

$$x(t) = w(t) + (a_1 + a_2 t)w(\eta), \quad t \in [a,\sigma(b)],$$

where a_1 and a_2 are real constants which will be determined below. We have

$$\begin{aligned} x(a) &= w(a) + (a_1 + aa_2)w(\eta) \\ &= (a_1 + aa_2)w(\eta), \\ x(\eta) &= w(\eta) + (a_1 + \eta a_2)w(\eta) \\ &= (1 + a_1 + \eta a_2)w(\eta). \end{aligned}$$

Hence and by the second equation of (3.31), we obtain

$$(a_1 + aa_2)w(\eta) = k(1 + a_1 + \eta a_2)w(\eta)$$

or

$$a_1 + aa_2 = k(1 + a_1 + \eta a_2),$$

or

$$(1-k)a_1 + (a - k\eta)a_2 = k.$$

Next,

$$x^{\Delta}(t) = w^{\Delta}(t) + a_2 w(\eta),$$

$$\begin{aligned} x^{\Delta}(\sigma(b)) &= w^{\Delta}(\sigma(b)) + a_2 w(\eta) \\ &= a_2 w(\eta) \\ &= 0. \end{aligned}$$

Therefore, $a_2 = 0$ and $a_1 = \dfrac{k}{1-k}$. Consequently,

$$\begin{aligned} x(t) &= w(t) + \frac{k}{1-k} w(\eta) \\ &= \int_a^{\sigma(b)} G(t,s)h(s)\Delta s + \int_a^{\sigma(b)} \frac{k}{1-k} G(\eta,s)h(s)\Delta s \\ &= \int_a^{\sigma(b)} \left(G(t,s) + \frac{k}{1-k} G(\eta,s) \right) h(s)\Delta s, \quad t \in [a, \sigma(b)]. \end{aligned}$$

This completes the proof.

Now, we consider the BVP

$$\begin{aligned} Lx &= h(t) \\ \alpha x(a) - \beta x^{\Delta}(a) &= k_1 x(\eta) + k_2 x^{\Delta}(\eta) \\ \gamma x(\sigma^2(b)) + \delta x^{\Delta}(\sigma(b)) &= 0, \end{aligned} \tag{3.32}$$

where p and q satisfy $(C1)$, h satisfies $(C2)$ and

(C7) $k_1, k_2 \in \mathbb{R}$ and either

$$\gamma(a\alpha - \beta - k_1\eta - k_2) \neq (\alpha - k_1)(\gamma\sigma^2(b) + \delta)$$

if $\gamma \neq 0$, or $\alpha \neq k_1$ if $\gamma = 0$.

Theorem 3.16. *Suppose* (C1), (C2) *and* (C7). *Then the solution of the BVP* (3.32) *can be represented in the form*

$$x(t) = \int_a^{\sigma(b)} K(t,s)h(s)\Delta s, \quad t \in [a, \sigma^2(b)],$$

where

$$K(t,s) = G(t,s) + \Bigg(- \frac{\gamma\sigma^2(b) + \delta}{\gamma(a\alpha - \beta - k_1\eta - k_2) - (\alpha - k_1)(\gamma\sigma^2(b) + \delta)}$$

$$+ \frac{\gamma k_1}{\gamma(a\alpha - \beta - k_1\eta - k_2) - (\alpha - k_1)(\gamma\sigma^2(b) + \delta)} t \Bigg) G(\eta, s)$$

if $\gamma \neq 0$, *and*

$$K(t,s) = G(t,s) + \frac{k_1}{\alpha - k_1} G(\eta, s)$$

if $\gamma = 0$. *Here* $G(t,s)$ *is the Green function, defined in Theorem 3.5.*

Proof. By Theorem 3.5, it follows that the function

$$w(t) = \int_a^{\sigma(b)} G(t,s)h(s)\Delta s, \quad t \in [a, \sigma^2(b)],$$

is the unique solution of the BVP (3.8) for $A = B = 0$. We will search a solution to the BVP (3.32) in the form

$$x(t) = w(t) + (a_1 + a_2 t)w(\eta) + (a_3 + a_4 t)w^\Delta(\eta), \quad t \in [a, \sigma^2(b)],$$

where a_1, a_2, a_3, a_4 are real constants which will be determined below. We have

$$x(a) = w(a) + (a_1 + a a_2)w(\eta) + (a_3 + a a_4)w^\Delta(\eta),$$

$$x(\eta) = w(\eta) + (a_1 + \eta a_2)w(\eta) + (a_3 + \eta a_4)w^\Delta(\eta)$$

$$= (1 + a_1 + \eta a_2)w(\eta) + (a_3 + \eta a_4)w^\Delta(\eta),$$

$$x(\sigma^2(b)) = w(\sigma^2(b)) + (a_1 + \sigma^2(b)a_2)w(\eta)$$

$$+ (a_3 + \sigma^2(b)a_4)w^\Delta(\eta),$$

$$x^\Delta(t) = w^\Delta(t) + a_2 w(\eta) + a_4 w^\Delta(\eta),$$

$$x^{\Delta}(a) = w^{\Delta}(a) + a_2 w(\eta) + a_4 w^{\Delta}(\eta),$$

$$x^{\Delta}(\eta) = w^{\Delta}(\eta) + a_2 w(\eta) + a_4 w^{\Delta}(\eta),$$

$$x^{\Delta}(\sigma(b)) = w^{\Delta}(\sigma(b)) + a_2 w(\eta) + a_4 w^{\Delta}(\eta).$$

Then

$$\begin{aligned} \alpha x(a) - \beta x^{\Delta}(a) &= \alpha w(a) + \alpha(a_1 + a a_2) w(\eta) + \alpha(a_3 + a a_4) w^{\Delta}(\eta) \\ &\quad -\beta w^{\Delta}(a) - \beta a_2 w(\eta) - \beta a_4 w^{\Delta}(\eta) \\ &= (\alpha a_1 + (a\alpha - \beta) a_2) w(\eta) + (\alpha a_3 + (a\alpha - \beta) a_4) w^{\Delta}(\eta) \end{aligned}$$

and

$$\begin{aligned} k_1 x(\eta) + k_2 x^{\Delta}(\eta) &= k_1(1 + a_1 + \eta a_2) w(\eta) + k_1(a_3 + \eta a_4) w^{\Delta}(\eta) \\ &\quad + k_2 w^{\Delta}(\eta) + k_2 a_2 w(\eta) + k_2 a_4 w^{\Delta}(\eta) \\ &= (a_1 k_1 + (k_1 \eta + k_2) a_2 + k_1)\, w(\eta) \\ &\quad + (k_1 a_3 + (k_1 \eta + k_2) a_4)\, w^{\Delta}(\eta). \end{aligned}$$

Hence and by the second equation of (3.32), we get

$$\begin{aligned} &(\alpha a_1 + (a\alpha - \beta) a_2) w(\eta) + (\alpha a_3 + (a\alpha - \beta) a_4) w^{\Delta}(\eta) \\ &= (a_1 k_1 + (k_1 \eta + k_2) a_2 + k_1)\, w(\eta) \\ &\quad + (k_1 a_3 + (k_1 \eta + k_2) a_4)\, w^{\Delta}(\eta), \end{aligned}$$

whereupon,

$$\alpha a_1 + (a\alpha - \beta) a_2 = a_1 k_1 + (k_1 \eta + k_2) a_2 + k_1$$

$$\alpha a_3 + (a\alpha - \beta) a_4 = a_3 k_1 + (k_1 \eta + k_2) a_4$$

or

$$(\alpha - k_1) a_1 + (a\alpha - \beta - k_1 \eta - k_2) a_2 = k_1$$

$$(\alpha - k_1)a_3 + (a\alpha - \beta - k_1\eta - k_2)a_4 = 0.$$

Next,

$$\begin{aligned} 0 &= \gamma x(\sigma^2(b)) + \delta x^{\Delta}(\sigma(b)) \\ &= \gamma w(\sigma^2(b)) + \gamma\left(a_1 + \sigma^2(b)a_2\right) w(\eta) \\ &\quad + \gamma\left(a_3 + \sigma^2(b)a_4\right) w^{\Delta}(\eta) \\ &\quad + \delta w^{\Delta}(\sigma(b)) + a_2\delta w(\eta) + a_4\delta w^{\Delta}(\eta) \\ &= \left(\gamma a_3 + (\gamma\sigma^2(b) + \delta)a_4\right) w^{\Delta}(\eta) \\ &\quad + \left(\gamma a_1 + (\gamma\sigma^2(b) + \delta)a_2\right) w(\eta) \end{aligned}$$

implies that

$$\gamma a_3 + (\gamma\sigma^2(b) + \delta)a_4 = 0$$

$$\gamma a_1 + (\gamma\sigma^2(b) + \delta)a_2 = 0.$$

Therefore, we get the system

$$\alpha a_1 + (a\alpha - \beta)a_2 = a_1k_1 + (k_1\eta + k_2)a_2 + k_1$$

$$\alpha a_3 + (a\alpha - \beta)a_4 = a_3k_1 + (k_1\eta + k_2)a_4$$

$$\gamma a_3 + (\gamma\sigma^2(b) + \delta)a_4 = 0$$

$$\gamma a_1 + (\gamma\sigma^2(b) + \delta)a_2 = 0.$$

1. Let $\gamma \neq 0$. Then

$$a_3 = -\frac{\gamma\sigma^2(b) + \delta}{\gamma}a_4$$

$$a_1 = -\frac{\gamma\sigma^2(b) + \delta}{\gamma}a_2.$$

Hence,

$$\left(-\frac{(\alpha-k_1)(\gamma\sigma^2(b)+\delta)}{\gamma}+a\alpha-\beta-k_1\eta-k_2\right)a_2=k_1$$

$$\left(-\frac{(\alpha-k_1)(\gamma\sigma^2(b)+\delta)}{\gamma}+a\alpha-\beta-k_1\eta-k_2\right)a_4=0$$

and

$$a_2=\frac{\gamma k_1}{\gamma(a\alpha-\beta-k_1\eta-k_2)-(\alpha-k_1)(\gamma\sigma^2(b)+\delta)}$$

$$a_4=0$$

$$a_3=0$$

$$a_1=-\frac{\gamma\sigma^2(b)+\delta}{\gamma(a\alpha-\beta-k_1\eta-k_2)-(\alpha-k_1)(\gamma\sigma^2(b)+\delta)}.$$

Therefore,

$$\begin{aligned}x(t)&=w(t)+\Bigg(-\frac{\gamma\sigma^2(b)+\delta}{\gamma(a\alpha-\beta-k_1\eta-k_2)-(\alpha-k_1)(\gamma\sigma^2(b)+\delta)}\\&\quad+\frac{\gamma k_1}{\gamma(a\alpha-\beta-k_1\eta-k_2)-(\alpha-k_1)(\gamma\sigma^2(b)+\delta)}t\Bigg)w(\eta)\\&=\int_a^{\sigma(b)}G(t,s)h(s)\Delta s\\&\quad+\int_a^{\sigma(b)}G(\eta,s)\Bigg(-\frac{\gamma\sigma^2(b)+\delta}{\gamma(a\alpha-\beta-k_1\eta-k_2)-(\alpha-k_1)(\gamma\sigma^2(b)+\delta)}\\&\quad+\frac{\gamma k_1}{\gamma(a\alpha-\beta-k_1\eta-k_2)-(\alpha-k_1)(\gamma\sigma^2(b)+\delta)}t\Bigg)h(s)\Delta s,\end{aligned}$$

$t\in[a,\sigma^2(b)]$.

2. Let $\gamma=0$. Then

$$a_2=a_3=a_4=0$$

and

$$a_1=\frac{k_1}{\alpha-k_1},$$

and

$$x(t)=w(t)+\frac{k_1}{\alpha-k_1}w(\eta)$$

$$= \int_a^{\sigma(b)} G(t,s)h(s)\Delta s + \frac{k_1}{\alpha - k_1}\int_a^{\sigma(b)} G(\eta,s)h(s)\Delta s$$

$$= \int_a^{\sigma(b)} \left(G(t,s) + \frac{k_1}{\alpha - k_1}G(\eta,s)\right) h(s)\Delta s, \quad t \in [a,\sigma^2(b)].$$

This completes the proof.

Exercise 3.9. Find the Green function for the following BVP

$$Lx = h(t)$$

$$\alpha x(a) - \beta x^{\Delta}(a) = 0$$

$$\gamma x(\sigma^2(b)) + \delta x^{\Delta}(\sigma(b)) = k_1 x(\eta) + k_2 x^{\Delta}(\eta).$$

3.7 The Green Function for $\Delta - \nabla$ Boundary Value Problems

Consider the second order homogeneous linear dynamic equation

$$-(py^{\Delta})^{\nabla}(t) + q(t)y(t) = 0, \tag{3.33}$$

where

(C8) $q \in \mathscr{C}(\mathbb{T})$, $p : \mathbb{T} \to \mathbb{R}$ is ∇-differentiable on $\mathbb{T}_\kappa$, $p(t) \neq 0$, $t \in \mathbb{T}$, and p^{∇} : $\mathbb{T}_\kappa \to \mathbb{R}$ is continuous.

Definition 3.3. A function $y : \mathbb{T} \to \mathbb{R}$ is said to be a solution of equation (3.33) provided

1. y is Δ-differentiable,
2. $y^{\Delta} : \mathbb{T}^\kappa \to \mathbb{R}$ is ∇-differentiable,
3. $(y^{\Delta})^{\nabla} : \mathbb{T}^\kappa \bigcap \mathbb{T}_\kappa \to \mathbb{R}$ is continuous,
4. (3.33) holds for any $t \in \mathbb{T}^\kappa \bigcap \mathbb{T}_\kappa$.

We define the quasi-Δ-derivative of y at $t \in \mathbb{T}^\kappa$ as follows.

$$y^{[\Delta]}(t) = p(t)y^{\Delta}(t).$$

Theorem 3.17. *Let $t_0 \in \mathbb{T}^\kappa$ be fixed and $c_0, c_1 \in \mathbb{R}$. Then equation (3.33) has a unique solution y such that*

$$y(t_0) = c_0, \quad y^{[\Delta]}(t_0) = c_1. \tag{3.34}$$

Proof. For $r \in \mathbb{T}$, $r \geq t_0$, let $A(r)$ be the statement: there is a unique function y_r defined on $[\rho(t_0), \sigma(r)]$ and satisfying (3.33) for $t \in [t_0, r]$ and the initial condition (3.34).

1. The statement $A(r)$ is true at $r = t_0$. If t_0 is right-dense or left-dense, then $A(t_0)$ has no sense. Let t_0 is right-scattered and left-scattered. In this case we have

$$y^{\Delta}(t_0) = \frac{y(\sigma(t_0)) - y(t_0)}{\sigma(t_0) - t_0} = \frac{y(\sigma(t_0)) - c_0}{\sigma(t_0) - t_0},$$

$$c_1 = p(t_0)y^{\Delta}(t_0) = p(t_0)\frac{y(\sigma(t_0)) - c_0}{\sigma(t_0) - t_0},$$

whereupon,

$$\frac{c_1(\sigma(0) - t_0)}{p(t_0)} + c_0 = y(\sigma(t_0))$$

and

$$y^{\Delta}(t_0) = \frac{c_1}{p(t_0)}.$$

Note that

$$y^{\Delta}(\rho(t_0)) = \frac{y(t_0) - y(\rho(t_0))}{t_0 - \rho(t_0)},$$

$$(py^{\Delta})^{\nabla}(t_0) = \frac{p(\rho(t_0))y^{\Delta}(\rho(t_0)) - p(t_0)y^{\Delta}(t_0)}{\rho(t_0) - t_0} = \frac{p(\rho(t_0))\frac{y(t_0) - y(\rho(t_0))}{t_0 - \rho(t_0)} - c_1}{\rho(t_0) - t_0} = \frac{-p(\rho(t_0))(c_0 - y(\rho(t_0))) + c_1(t_0 - \rho(t_0))}{(t_0 - \rho(t_0))^2}.$$

Hence,

$$0 = q(t_0)y(t_0) - (py^{\Delta})^{\nabla}(t_0) = c_0 q(t_0) - \frac{-p(\rho(t_0))(c_0 - y(\rho(t_0))) + c_1(t_0 - \rho(t_0))}{(t_0 - \rho(t_0))^2}$$

and

$$\frac{-p(\rho(t_0))(c_0 - y(\rho(t_0))) + c_1(t_0 - \rho(t_0))}{(t_0 - \rho(t_0))^2} = c_0 q(t_0),$$

or

$$-p(\rho(t_0))(c_0 - y(\rho(t_0))) + c_1(t_0 - \rho(t_0)) = c_0 q(t_0)(t_0 - \rho(t_0))^2,$$

or

$$-c_0 p(\rho(t_0)) + p(\rho(t_0))y(\rho(t_0)) = (t_0 - \rho(t_0))(c_0 q(t_0)(t_0 - \rho(t_0)) - c_1),$$

or

$$p(\rho(t_0))y(\rho(t_0)) = (t_0 - \rho(t_0))(c_0 q(t_0)(t_0 - \rho(t_0)) - c_1) + c_0 p(\rho(t_0)),$$

or

$$y(\rho(t_0)) = \frac{(t_0 - \rho(t_0))(c_0 q(t_0)(t_0 - \rho(t_0)) - c_1) + c_0 p(\rho(t_0))}{p(\rho(t_0))}.$$

Thus, we find the solution $y_{t_0} : [\rho(t_0), \sigma(t_0)] \to \mathbb{R}$ and the statement $A(r)$ is true.

2. Let r be right-scattered and $A(r)$ be true. Then the IVP (3.33), (3.34) has a unique solution $y_r : [\rho(t_0), \sigma(r)] \to \mathbb{R}$. Define the function $y_{\sigma(r)} : [\rho(t_0), \sigma^2(r)] \to \mathbb{R}$ as follows

$$y_{\sigma(r)}(t) = y_r(t), \quad t \in [\rho(t_0), \sigma(r)],$$

and

$$\begin{aligned} y_{\sigma(r)}(\sigma^2(r)) &= y_{\sigma(r)}(\sigma(r)) + (\sigma^2(r) - \sigma(r))y^{\Delta}_{\sigma(r)}(\sigma(r)) \\ &= y_r(\sigma(r)) + (\sigma^2(r) - \sigma(r))y^{\Delta}_r(\sigma(r)). \end{aligned}$$

By (3.33), we have

$$-(py^{\Delta})^{\nabla}(\sigma(r)) + q(\sigma(r))y_r(\sigma(r)) = 0,$$

whereupon,

$$\begin{aligned} 0 = &-\frac{p(\sigma(r))y^{\Delta}_r(\sigma(r)) - p(r)y^{\Delta}(r)}{\sigma(r) - r} \\ &+ q(\sigma(r))y_r(\sigma(r)) \end{aligned}$$

or

$$(\sigma(r) - r)q(\sigma(r))y_r(\sigma(r)) = p(\sigma(r))y^{\Delta}_r(\sigma(r)) - p(r)y^{\Delta}_r(r)$$

and

$$y^{\Delta}_r(\sigma(r)) = \frac{(\sigma(r) - r)q(\sigma(r))y_r(\sigma(r)) + p(r)y^{\Delta}_r(r)}{p(\sigma(r))}.$$

Therefore,

$$y_{\sigma(r)}(\sigma^2(r)) = y_r(\sigma(r)) + (\sigma^2(r) - \sigma(r))\left(\frac{p(r)}{p(\sigma(r))} y_r^{\Delta}(r) + \frac{(\sigma(r) - r)q(\sigma(r))y_r(\sigma(r))}{p(\sigma(r))}\right).$$

Thus, $y_{\sigma(r)}$ is the unique solution of the IVP on $[\rho(t_0), \sigma^2(r)]$.

3. Let r_0 be right-dense and $A(r_0)$ be true. Take $r_1 > r_0$. We will prove that $A(r)$ is true for any $r \in [r_0, r_1]$. Note that the IVP (3.33), (3.34) for $t \in [r_0, r_1]$ is equivalent to the integral equation

$$y(t) = y_{r_0}(r_0) + y_{r_0}^{[\Delta]}(r_0) \int_{r_0}^{t} \frac{\Delta\tau}{p(\tau)} + \int_{r_0}^{t} \frac{1}{p(\tau)} \left(\int_{r_0}^{\tau} q(s)y(s)\nabla s\right) \Delta\tau, t \in [r_0, r_1].$$

By the method of the successive approximations, the last integral equation has a unique solution for any $r \in [r_0, r_1]$.

4. Let r_0 be left-dense and $A(r)$ be true for any $r < r_0$. We will show that $A(r_0)$ is true. Note that for any $r < r_0$ the IVP (3.33), (3.34) has a unique solution $y_r(t)$ defined by

$$y_r(t) = c_0 + c_1 \int_{r_0}^{t} \frac{\Delta\tau}{p(\tau)} + \int_{r_0}^{t} \frac{1}{p(\tau)} \left(\int_{r_0}^{\tau} q(s)y_r(s)\nabla s\right) \Delta\tau, t \in [t_0, r].$$

Hence,

$$\begin{aligned} y_{r_0}(t) &= \lim_{r \to r_0} y_r(t) \\ &= c_0 + c_1 \int_{r_0}^{t} \frac{\Delta\tau}{p(\tau)} \\ &\quad + \int_{r_0}^{t} \frac{1}{p(\tau)} \left(\int_{r_0}^{\tau} q(s)y_{r_0}(s)\nabla s\right) \Delta\tau, t \in [t_0, r]. \end{aligned}$$

Therefore, y_{r_0} is the unique solution of the IVP (3.33), (3.34) for $t \in [t_0, r_0]$, i.e., $A(r_0)$ is true.

Now, applying the induction principle, we conclude that the IVP (3.33), (3.34) has a unique solution on $[t_0, \infty)$. By the backward induction principle, it can be shown, similarly as above, that the IVP (3.33), (3.34) has a unique solution on $(-\infty, t_0]$. This completes the proof.

Now we will reduce equation (3.33) to a system. For this aim, we set

$$u(t) = p(t)y^{\Delta}(t)$$

or

$$y^{\Delta}(t) = \frac{u(t)}{p(t)}.$$

Hence,

$$\begin{aligned} u^{\nabla}(t) &= (py^{\Delta})^{\nabla}(t) \\ &= q(t)y(t). \end{aligned}$$

Now, using that

$$y^{\nabla}(t) = y^{\Delta}(\rho(t)),$$

we get

$$\begin{aligned} y^{\nabla}(t) &= y^{\Delta}(\rho(t)) \\ &= \frac{u(\rho(t))}{p(\rho(t))} \\ &= \frac{1}{p(\rho(t))}\left(u(t) + (\rho(t) - t)u^{\nabla}(t)\right) \\ &= \frac{1}{p(\rho(t))}u(t) + \frac{(\rho(t) - t)q(t)}{p(\rho(t))}y(t). \end{aligned}$$

So, we get the system

$$\begin{aligned} y^{\nabla}(t) &= \frac{(\rho(t) - t)q(t)}{p(\rho(t))}y(t) + \frac{1}{p(\rho(t))}u(t) \\ u^{\nabla}(t) &= q(t)y(t). \end{aligned}$$

Now, using that

$$u^{\Delta}(t) = u^{\nabla}(\sigma(t)),$$

we obtain

$$\begin{aligned} u^{\Delta}(t) &= q(\sigma(t))y(\sigma(t)) \\ &= q(\sigma(t))(y(t) + (\sigma(t) - t)y^{\Delta}(t)) \\ &= q(\sigma(t))\left(y(t) + (\sigma(t) - t)\frac{u(t)}{p(t)}\right) \end{aligned}$$

$$= q(\sigma(t))y(t) + (\sigma(t)-t)\frac{q(\sigma(t))}{p(t)}u(t).$$

Thus, we get the system

$$y^{\Delta}(t) = \frac{1}{p(t)}u(t)$$

$$u^{\Delta}(t) = q(\sigma(t))y(t) + (\sigma(t)-t)\frac{q(\sigma(t))}{p(t)}u(t).$$

Definition 3.4. Let $y, z : \mathbb{T} \to \mathbb{R}$. We define the Wronskian of y and z as follows

$$W(y,z) = y(t)z^{[\Delta]}(t) - y^{[\Delta]}(t)z(t)$$

$$= p(t)(y(t)z^{\Delta}(t) - y^{\Delta}(t)z(t)).$$

Theorem 3.18. *The Wronskian of any two solutions of equation* (3.33) *is independent of t.*

Proof. Let y and z be any two solutions of equation (3.33). Then

$$\begin{aligned}
(W(y,z))^{\nabla}(t) &= \left(yz^{[\Delta]} - y^{[\Delta]}z\right)^{\nabla}(t) \\
&= y^{\nabla}(t)z^{[\Delta]}(\rho(t)) + y(t)(z^{[\Delta]})^{\nabla}(t) \\
&\quad - y^{[\Delta]}(\rho(t))z^{\nabla}(t) - (y^{[\Delta]})^{\nabla}(t)z(t) \\
&= y^{\nabla}(t)p(\rho(t))z^{\Delta}(\rho(t)) + y(t)(pz^{\Delta})^{\nabla}(t) \\
&\quad - p(\rho(t))y^{\Delta}(\rho(t))z^{\nabla}(t) - z(t)(py^{\Delta})^{\nabla}(t) \\
&= q(t)y(t)z(t) - q(t)z(t)y(t) \\
&= 0, \quad t \in \mathbb{T}_{\kappa} \bigcap \mathbb{T}^{\kappa}.
\end{aligned}$$

Therefore, $W(y,z)$ is a constant on $\mathbb{T}^{\kappa}$. This completes the proof.

Corollary 3.2. *If y and z are both solutions of equation* (3.33)*, then either $W(y_1,y_2)(t) = 0$ or $W(y_1,y_2)(t) \neq 0$ for any $t \in \mathbb{T}^{\kappa}$.*

Corollary 3.3. *Any two solutions y and z of equation* (3.33) *are linearly independent if and only if their Wronskian is non-zero.*

Definition 3.5. We say that two solutions y_1 and y_2 of equation (3.33) form a fundamental set of solutions of equation (3.33) if they are linearly independent.

Now, we consider the nonhomogeneous equation

$$-(py^{\Delta})^{\nabla}(t)+q(t)y(t)=h(t), \tag{3.35}$$

where $h:\mathbb{T}\to\mathbb{R}$ is a continuous function.

Theorem 3.19. *Let y_1 and y_2 form a fundamental set of solutions for equation* (3.33) *and $w=W(y_1,y_2)$. Then the general solution of equation* (3.35) *has the representation*

$$y(t)=a_1y_1(t)+a_2y_2(t)+\frac{1}{w}\int_{\rho(t_0)}^{t}(y_1(t)y_2(s)-y_1(s)y_2(t))h(s)\nabla s,$$

where a_1 and a_2 are constants.

Proof. Let

$$z(t)=\frac{1}{w}\int_{\rho(t_0)}^{t}(y_1(t)y_2(s)-y_1(s)y_2(t))h(s)\nabla s.$$

It is enough to prove that the function z satisfies equation (3.35). We have

$$\begin{aligned}
z^{\Delta}(t) &= \frac{1}{w}(y_1(\sigma(t))y_2(\sigma(t))-y_1(\sigma(t))y_2(\sigma(t)))h(\sigma(t))\\
&\quad+\frac{1}{w}\int_{\rho(t_0)}^{t}(y_1^{\Delta}(t)y_2(s)-y_1(s)y_2^{\Delta}(t))h(s)\nabla s,\\
p(t)z^{\Delta}(t) &= \frac{1}{w}\int_{\rho(t_0)}^{t}\left(p(t)y_1^{\Delta}(t)y_2(s)-p(t)y_1(s)y_2^{\Delta}(t)\right)h(s)\nabla s\\
&= \frac{1}{w}\int_{\rho(t_0)}^{t}\left(y_1^{[\Delta]}(t)y_2(s)-y_1(s)y_2^{[\Delta]}(t)\right)h(s)\nabla s
\end{aligned}$$

and

$$\begin{aligned}
(pz^{\Delta})^{\nabla}(t) &= \frac{1}{w}\left(y_1^{[\Delta]}(\rho(t))y_2(t)-y_1(t)y_2^{[\Delta]}(\rho(t))\right)h(t)\\
&\quad+\frac{1}{w}\int_{\rho(t_0)}^{t}\left(\left(y_1^{[\Delta]}\right)^{\nabla}(t)y_2(s)-y_1(s)\left(y_2^{[\Delta]}\right)^{\nabla}(t)\right)h(s)\nabla s.
\end{aligned} \tag{3.36}$$

Note that, if $\rho(t)=t$, then

$$\begin{aligned}
y_1^{[\Delta]}(\rho(t))y_2(t)-y_1(t)y_2^{[\Delta]}(\rho(t)) &= y_1^{[\Delta]}(t)y_2(t)-y_1(t)y_2^{[\Delta]}(t)\\
&= p(t)(y_1^{\Delta}(t)y_2(t)-y_1(t)y_2^{\Delta}(t))
\end{aligned}$$

$$= -W(y_1,y_2)(t)$$

$$= -w.$$

If $\rho(t) < t$, then

$$\begin{aligned}
y_1^{[\Delta]}(\rho(t))y_2(t) - y_1(t)y_2^{[\Delta]}(\rho(t)) &= p(\rho(t))\left(y_1^{\Delta}(\rho(t))y_2(t) - y_1(t)y_2^{\Delta}(\rho(t))\right) \\
&= p(\rho(t))\Big(y_1^{\Delta}(\rho(t))\Big(y_2(\rho(t)) \\
&\quad +(t-\rho(t))y_2^{\Delta}(\rho(t))\Big) \\
&\quad - \left(y_1(\rho(t)) + (t-\rho(t))y_1^{\Delta}(\rho(t))\right) y_2^{\Delta}(\rho(t))\Big) \\
&= p(\rho(t))\left(y_1^{\Delta}(\rho(t))y_2(\rho(t)) - y_1(\rho(t))y_2^{\Delta}(\rho(t))\right) \\
&= -W(y_1,y_2)(\rho(t)) \\
&= -w.
\end{aligned}$$

Hence and by (3.36), we arrive at

$$\begin{aligned}
(pz^{\Delta})^{\nabla}(t) &= -h(t) + \frac{1}{w}\int_{\rho(t_0)}^{t}\left(\left(y_1^{[\Delta]}\right)^{\nabla}(t)y_2(s) - y_1(s)\left(y_2^{[\Delta]}\right)^{\nabla}(t)\right)h(s)\nabla s \\
&= -h(t) + \frac{q(t)}{w}\int_{\rho(t_0)}^{t}\left(y_1(t)y_2(s) - y_1(s)y_2(t)\right)h(s)\nabla s \\
&= -h(t) + q(t)z(t),
\end{aligned}$$

i.e., z satisfies (3.35). This completes the proof.

Now, suppose that $a \in \mathbb{T}_{\kappa}$ and $b \in \mathbb{T}^{\kappa}$. Consider equation (3.35) subject to the boundary conditions

$$\begin{aligned}
\alpha y(\rho(a)) - \beta y^{[\Delta]}(\rho(a)) &= 0 \\
\gamma y(b) + \delta y^{[\Delta]}(b) &= 0,
\end{aligned} \tag{3.37}$$

where

(C8) $\alpha, \beta, \gamma, \delta \in \mathbb{R}$, $|\alpha| + |\beta| \neq 0$, $|\gamma| + |\delta| \neq 0$.

Let ϕ and ψ be solutions of the IVP

$$\begin{aligned} -\left(p\phi^{\Delta}\right)^{\nabla}(t)+q(t)\phi(t) &= 0, \quad t\in[a,b],\\ \phi(\rho(a))=\beta, \quad \phi^{[\Delta]}(\rho(a)) &= \alpha, \end{aligned} \tag{3.38}$$

and

$$\begin{aligned} -\left(p\psi^{\Delta}\right)^{\nabla}(t)+q(t)\psi(t) = 0, \quad t\in[a,b],\\ \psi(b)=\delta, \quad \psi^{[\Delta]}(b) = -\gamma. \end{aligned} \tag{3.39}$$

Set

$$\begin{aligned} D &= -W(\phi,\psi)\\ &= \phi^{[\Delta]}(t)\psi(t)-\phi(t)\psi^{[\Delta]}(t), \quad t\in[\rho(a),b]. \end{aligned}$$

Since the Wronskian of any two solutions of equation (3.33) is independent of t, we have

$$\phi^{[\Delta]}(\rho(a))\psi(\rho(a))-\phi(\rho(a))\psi^{[\Delta]}(\rho(a))=\phi^{[\Delta]}(b)\psi(b)-\phi(b)\psi^{[\Delta]}(b).$$

Theorem 3.20. *$D\neq 0$ if and only if equation* (3.33) *has only the trivial solution satisfying the boundary conditions* (3.37).

Proof. Let $D=0$. Then ψ is a nontrivial solution of the BVP (3.33), (3.37). Assume that $D\neq 0$. Then ϕ and ψ form a fundamental set of solutions of equation (3.33). Then any solution y of (3.33), (3.37) will have the form

$$y(t)=a_1\phi(t)+a_2\psi(t),$$

where a_1 and a_2 are constants. Note that

$$\begin{aligned} y(\rho(a)) &= a_1\phi(\rho(a))+a_2\psi(\rho(a)),\\ y^{[\Delta]}(t) &= a_1\phi^{[\Delta]}(t)+a_2\psi^{[\Delta]}(t),\\ y^{[\Delta]}(\rho(a)) &= a_1\phi^{[\Delta]}(\rho(a))+a_2\psi^{[\Delta]}(\rho(a)),\\ y(b) &= a_1\phi(b)+a_2\psi(b),\\ y^{[\Delta]}(b) &= a_1\phi^{[\Delta]}(b)+a_2\psi^{[\Delta]}(b) \end{aligned}$$

and using (3.38) and (3.39), we find

$$0=\alpha y(\rho(a))-\beta y^{[\Delta]}(\rho(a))$$

$$
\begin{aligned}
&= \alpha\left(a_1\phi(\rho(a)) + a_2\psi(\rho(a))\right)\\
&\quad -\beta\left(a_1\phi^{[\Delta]}(\rho(a)) + a_2\psi^{[\Delta]}(\rho(a))\right)\\
&= a_1\left(\alpha\phi(\rho(a)) - \beta\phi^{[\Delta]}(\rho(a))\right)\\
&\quad +a_2\left(\alpha\psi(\rho(a)) - \beta\psi^{[\Delta]}(\rho(a))\right)\\
&= a_2\left(\alpha\psi(\rho(a)) - \beta\psi^{[\Delta]}(\rho(a))\right)\\
&= a_2 D
\end{aligned}
$$

and

$$
\begin{aligned}
0 &= \gamma y(b) + \delta y^{[\Delta]}(b)\\
&= \gamma(a_1\phi(b) + a_2\psi(b)) + \delta\left(a_1\phi^{[\Delta]}(b) + a_2\psi^{[\Delta]}(b)\right)\\
&= a_1\left(\gamma\phi(b) + \delta\phi^{[\Delta]}(b)\right)\\
&\quad +a_2\left(\gamma\psi(b) + \delta\psi^{[\Delta]}(b)\right)\\
&= a_1\left(\gamma\phi(b) + \delta\phi^{[\Delta]}(b)\right)\\
&= a_1 D,
\end{aligned}
$$

whereupon,

$$a_1 = a_2 = 0.$$

Thus, y is a trivial solution of (3.33), (3.37). This completes the proof.

Theorem 3.21. *Let $D \neq 0$. Then the BVP* (3.35), (3.37) *has a unique solution y for which the representation*

$$y(t) = \int_{\rho(a)}^{b} G(t,s)h(s)\nabla s$$

holds, where

$$G(t,s) = \frac{1}{D}\begin{cases} \phi(t)\psi(s) & \rho(a) \le t \le s \le \sigma(b)\\ \phi(s)\psi(t) & \rho(a) \le s \le t \le \sigma(b)\end{cases}$$

is the Green function, ϕ and ψ are the solutions of the IVPs (3.33), (3.38) *and* (3.33), (3.39), *respectively.*

Proof. Since $D \neq 0$, we have that ϕ and ψ form a fundamental set of solutions. Hence, applying Theorem 3.19, we have that the general solution of (3.35) is given by the representation

$$\begin{aligned} y(t) &= a_1\phi(t) + a_2\psi(t) \\ &\quad + \frac{1}{D}\int_{\rho(a)}^{t} (\phi(s)\psi(t) - \phi(t)\psi(s))h(s)\nabla s, \end{aligned}$$

where a_1 and a_2 are constants. We have

$$\begin{aligned} y^{[\Delta]}(t) &= a_1\phi^{[\Delta]}(t) + a_2\psi^{[\Delta]}(t) \\ &\quad + \frac{p(t)}{D}(\phi(\rho(t))\psi(\rho(t)) - \phi(\rho(t))\psi(\rho(t)))h(\rho(t)) \\ &\quad + \frac{1}{D}\int_{\rho(a)}^{t} \left(\phi(s)\psi^{[\Delta]}(t) - \phi^{[\Delta]}(t)\psi(s)\right)h(s)\nabla s \\ &= a_1\phi^{[\Delta]}(t) + a_2\psi^{[\Delta]}(t) \\ &\quad + \frac{1}{D}\int_{\rho(a)}^{t} \left(\phi(s)\psi^{[\Delta]}(t) - \phi^{[\Delta]}(t)\psi(s)\right)h(s)\nabla s. \end{aligned}$$

Hence,

$$\begin{aligned} 0 &= \alpha y(\rho(a)) - \beta y^{[\Delta]}(\rho(a)) \\ &= \alpha a_1\phi(\rho(a)) + \alpha a_2\phi(\rho(a)) \\ &\quad - \beta a_1\phi^{[\Delta]}(\rho(a)) - \beta a_2\phi^{[\Delta]}(\rho(a)) \\ &= \alpha a_1\beta + \alpha a_2\psi(\rho(s)) - \alpha a_1\beta - \beta a_2\psi^{[\Delta]}(\rho(a)) \\ &= a_2\left(\alpha\psi(\rho(a)) - \beta\psi^{[\Delta]}(\rho(a))\right), \end{aligned}$$

whereupon,

$$a_2 = 0.$$

Next,

$$
\begin{aligned}
0 &= \gamma a_1 \phi(b) + \frac{\gamma}{D}\int_{\rho(a)}^{b} (\phi(s)\psi(b) - \phi(b)\psi(s))h(s)\nabla s \\
&\quad + \delta a_1 \phi^{[\Delta]}(b) + \frac{\delta}{D}\int_{\rho(a)}^{b} \left(\phi(s)\psi^{[\Delta]}(b) - \phi^{[\Delta]}(b)\psi(s)\right) h(s)\nabla s \\
&= a_1 \left(\gamma\phi(b) + \delta\phi^{[\Delta]}(b)\right) \\
&\quad + \frac{\gamma}{D}\int_{\rho(a)}^{b} (\delta\phi(s) - \phi(b)\psi(s))\, h(s)\nabla s \\
&\quad + \frac{\delta}{D}\int_{\rho(a)}^{b} \left(-\gamma\phi(s) - \phi^{[\Delta]}(b)\psi(s)\right) h(s)\nabla s \\
&= a_1 \left(\gamma\phi(b) + \delta\phi^{[\Delta]}(b)\right) \\
&\quad - \frac{\gamma\phi(b) + \delta\phi^{[\Delta]}(b)}{D}\int_{\rho(a)}^{b} \psi(s)h(s)\nabla s.
\end{aligned}
$$

Therefore,

$$
\begin{aligned}
a_1 &= \frac{\gamma\phi(b) + \delta\phi^{[\Delta]}(b)}{D\left(\gamma\phi(b) + \delta\phi^{[\Delta]}(b)\right)}\int_{\rho(a)}^{b} \psi(s)h(s)\nabla s \\
&= \frac{1}{D}\int_{\rho(a)}^{b} \psi(s)h(s)\nabla s.
\end{aligned}
$$

Consequently,

$$
\begin{aligned}
y(t) &= \frac{\phi(t)}{D}\int_{\rho(a)}^{b} \psi(s)h(s)\nabla s \\
&\quad + \frac{1}{D}\int_{\rho(a)}^{t} (\phi(s)\psi(t) - \phi(t)\psi(s))\, h(s)\nabla s \\
&= \frac{\phi(t)}{D}\int_{\rho(a)}^{t} \psi(s)h(s)\nabla s + \frac{\phi(t)}{D}\int_{t}^{b} \psi(s)h(s)\nabla s \\
&\quad + \frac{1}{D}\int_{\rho(a)}^{t} (\phi(s)\psi(t) - \phi(t)\psi(s))\, h(s)\nabla s \\
&= \frac{1}{D}\int_{\rho(a)}^{t} \phi(s)\psi(t)h(s)\nabla s + \frac{1}{D}\int_{t}^{b} \phi(t)\psi(s)h(s)\nabla s,
\end{aligned}
$$

$t \in [\rho(a), b]$, and

$$y^{[\Delta]}(t) = \frac{1}{D}\int_{\rho(a)}^{t} \phi(s)\psi^{[\Delta]}(t)h(s)\nabla s$$
$$-\frac{1}{D}\int_{t}^{b} \phi^{[\Delta]}(t)\psi(s)h(s)\nabla s, \quad t \in [\rho(a), b].$$

Then

$$\begin{aligned}\alpha y(\rho(a)) - \beta y^{[\Delta]}(\rho(a)) &= \frac{1}{D}\int_{\rho(a)}^{b} \alpha\phi(\rho(a))\psi(s)h(s)\nabla s \\ &\quad -\frac{1}{D}\int_{\rho(a)}^{b} \beta\phi^{[\Delta]}(\rho(a))\psi(s)h(s)\nabla s \\ &= 0\end{aligned}$$

and

$$\begin{aligned}\gamma y(b) + \delta y^{[\Delta]}(b) &= \frac{1}{D}\int_{\rho(a)}^{b} \gamma\phi(s)\psi(b)h(s)\nabla s \\ &\quad +\frac{1}{D}\int_{\rho(a)}^{b} \delta\phi^{[\Delta]}(b)\psi(s)h(s)\nabla s \\ &= 0.\end{aligned}$$

Therefore, y is a solution of the BVP (3.35), (3.37). This completes the proof.

Now, we consider the BVP (3.35), (3.37) under the assumptions

(C9) $p(t) \geq 0$, $q(t) \geq 0$, $t \in [\rho(a), b]$,
(C10) $\alpha, \beta, \gamma, \delta \geq 0$, $\alpha + \beta > 0$, $\gamma + \delta > 0$,

$$\int_{t_0}^{b} \frac{1}{p(\tau)} \left(\int_{t_0}^{\tau} q(s)\nabla s \right) \Delta\tau < 1.$$

Below, we suppose that ϕ and ψ are the solutions of the IVPs (3.33), (3.38) and (3.33), (3.39), respectively.

Theorem 3.22. *Suppose* (C9) *and* (C10)*. Then*

1. $\phi(t) \geq 0$, $t \in [\rho(a), \sigma(b)]$, $\psi(t) \geq 0$, $t \in [\rho(a), b]$,
2. $\phi(t) > 0$, $t \in [\rho(a), \sigma(b)]$, $\psi(t) > 0$, $t \in [\rho(a), b]$,
3. $\phi^{[\Delta]}(t) \geq 0$, $t \in [\rho(a), b]$, $\psi^{[\Delta]}(t) \geq 0$, $t \in [\rho(a), b]$.

Proof. For $t \in [\rho(a), b]$, define the statement

$$A(t) : \phi(t) \geq 0, \quad \phi^{[\Delta]}(t) \geq 0, \quad t \in [\rho(a), b].$$

1. $A(\rho(a))$ is true. Indeed,

$$\phi(\rho(a)) = \beta$$

$$> 0,$$

$$\phi^{[\Delta]}(\rho(a)) = \alpha$$

$$\geq 0.$$

2. Let t be right-scattered and $A(t)$ be true. We will prove that $A(\sigma(t))$ is true. We have

$$\phi(\sigma(t)) = \phi(t) + (\sigma(t) - t)\phi^{\Delta}(t)$$

$$\geq 0.$$

By the definition for the ∇ derivative, we have

$$\phi^{[\Delta]}(t) = \phi^{[\Delta]}(\rho(t)) + (t - \rho(t))\left(\phi^{[\Delta]}\right)^{\nabla}(t)$$

$$= \phi^{[\Delta]}(\rho(t)) + (t - \rho(t))q(t)\phi(t).$$

Hence,

$$\phi^{[\Delta]}(\sigma(t)) = \phi^{[\Delta]}(t) + (\sigma(t) - t)q(\sigma(t))\phi(\sigma(t))$$

$$\geq 0.$$

3. Let t_0 be right-dense, $A(t_0)$ be true and $t_1 \in [\rho(a), b]$ be such that $t_1 > t_0$ and sufficiently close to t_0. We will prove that $A(t)$ is true for $t \in [t_0, t_1]$. By (3.33), we get

$$\phi^{[\Delta]}(t) = \phi^{[\Delta]}(t_0) + \int_{t_0}^{t} q(s)\phi(s)\nabla s$$

and

$$\phi(t) = \phi(t_0) + \phi^{[\Delta]}(t_0)\int_{t_0}^{t} \frac{\Delta\tau}{p(\tau)} + \int_{t_0}^{t} \frac{1}{p(\tau)}\left(\int_{t_0}^{\tau} q(s)\phi(s)\nabla s\right)\Delta\tau. \tag{3.40}$$

We will prove that equation (3.40) has a unique solution. Define

$$y_0(t) = \phi(t_0) + \phi^{[\Delta]}(t_0)\int_{t_0}^{t}\frac{\Delta\tau}{p(\tau)},$$

$$y_j(t) = \int_{t_0}^{t}\frac{1}{p(\tau)}\left(\int_{t_0}^{\tau} q(s)y_{j-1}(s)\nabla s\right)\Delta\tau, \quad j\in\mathbb{N}.$$

Set

$$M_0 = \phi(t_0) + \phi^{[\Delta]}(t_0)\int_{t_0}^{t}\frac{\Delta\tau}{p(\tau)},$$

$$M_1 = \int_{t_0}^{t}\frac{1}{p(\tau)}\left(\int_{t_0}^{\tau} q(s)\nabla s\right)\Delta\tau.$$

We have

$$y_0(t) \le M_0,$$

$$\begin{aligned}
y_1(t) &= \int_{t_0}^{t}\frac{1}{p(\tau)}\left(\int_{t_0}^{\tau} q(s)y_0(s)\nabla s\right)\Delta\tau \\
&\le M_0\int_{t_0}^{t}\frac{1}{p(\tau)}\left(\int_{t_0}^{\tau} q(s)\nabla s\right)\Delta\tau \\
&= M_0M_1, \quad t\in[t_0,t_1].
\end{aligned}$$

Assume

$$y_n(t) \le M_oM_1^n, \quad t\in[t_0,t_1].$$

We will prove that

$$y_{n+1}(t) \le M_0M_1^{n+1}, \quad t\in[t_0,t_1].$$

Really, we have

$$\begin{aligned}
y_{n+1}(t) &= \int_{t_0}^{t}\frac{1}{p(\tau)}\left(\int_{t_0}^{\tau} q(s)y_n(s)\nabla s\right)\Delta\tau \\
&\le M_0M_1^n\int_{t_0}^{t}\frac{1}{p(\tau)}\left(\int_{t_0}^{\tau} q(s)\nabla s\right)\Delta\tau \\
&= M_0M_1^{n+1}, \quad t\in[t_0,t_1].
\end{aligned}$$

Hence,

$$\begin{aligned}
\sum_{n=0}^{\infty} y_n(t) &\le M_0\sum_{n=0}^{\infty} M_1^n \\
&= \frac{M_0}{1-M_1}, \quad t\in[t_0,t_1].
\end{aligned}$$

Therefore, $\sum\limits_{n=0}^{\infty} y_n(t)$ is uniformly convergent with respect to $t\in[t_0,t_1]$. Let

$$y(t)=\sum_{n=0}^{\infty} y_n(t),\quad t\in[t_0,t_1].$$

We have

$$y(t)\geq y_0(t),\quad t\in[t_0,t_1]. \tag{3.41}$$

Now, assume that (3.40) has two solutions y^1 and y^2 on $[t_0,t_1]$. Then

$$\begin{aligned}
|y^1(t)-y^2(t)| &= \left|\int_{t_0}^{t}\frac{1}{p(\tau)}\left(\int_{t_0}^{\tau} q(s)(y^1(s)-y^2(s))\nabla s\right)\Delta\tau\right|\\
&\leq \int_{t_0}^{t}\frac{1}{p(\tau)}\left(\int_{t_0}^{\tau} q(s)|y^1(s)-y^2(s)|\nabla s\right)\Delta\tau\\
&\leq \left(\int_{t_0}^{t}\frac{1}{p(\tau)}\left(\int_{t_0}^{\tau} q(s)\nabla s\right)\Delta\tau\right)\|y^1-y^2\|\\
&\leq M_1\|y^1-y^2\|,\quad t\in[t_0,t_1].
\end{aligned}$$

Therefore,

$$\|y^1-y^2\|\leq M_1\|y^1-y^2\|,$$

whereupon, $y^1\equiv y^2$ on $[t_0,t_1]$. Hence and by (3.41), we conclude that

$$\phi(t)\geq \phi^{[\Delta]}(t_0)+\int_{t_0}^{t} q(s)\phi(s)\nabla s,\quad t\in[t_0,t_1].$$

Therefore,

$$\phi(t)\geq 0,\quad t\in[t_0,t_1]$$

and using that $\phi^{[\Delta]}(t_0)\geq 0$, we arrive at

$$\begin{aligned}
\phi^{[\Delta]}(t) &= \phi^{[\Delta]}(t_0)+\int_{t_0}^{t} q(s)\phi(s)\nabla s\\
&\geq 0,\quad t\in[t_0,t_1].
\end{aligned}$$

4. Let $t\in[\rho(a),b]$ be right-dense and $A(s)$ be true for all $s<t$, i.e.,

$$\phi(s)\geq 0,$$

$$\phi^{[\Delta]}(s)\geq 0,\quad s\in[\rho(a),t].$$

Passing to the limit as $s\to t$ and using the continuity of ϕ and $\phi^{[\Delta]}$, we get

$$\phi(t) \geq 0,$$

$$\phi^{[\Delta]}(t) \geq 0.$$

Thus, $A(t)$ is true.

Now, applying the induction principle, we conclude that the statement $A(t)$ is true for any $t \in [\rho(a), b]$. Note that

$$\phi^{[\Delta]}(t) = \alpha + \int_{\rho(a)}^{t} q(s)\phi(s)\nabla s,$$

$$\phi(t) = \beta + \alpha \int_{\rho(a)}^{t} \frac{\Delta\tau}{p(\tau)}$$

$$+ \int_{\rho(a)}^{t} \frac{1}{p(\tau)} \left(\int_{\rho(a)}^{\tau} q(s)\phi(s)\nabla s \right) \Delta\tau.$$

By the last equation, we get

$$\phi(t) > 0, \quad t \in (\rho(a), b].$$

Using the backward induction principle, we get the statements for ψ. This completes the proof.

Below, we suppose $(C9)$ and $(C10)$. Since $\phi(t) \geq 0, t \in [\rho(a), \sigma(b)]$, we have

$$D \geq \alpha\gamma + \beta\delta + \delta\alpha \int_{\rho(a)}^{b} \frac{\Delta\tau}{p(\tau)}.$$

If $q \equiv 0$ on $[\rho(a), b]$, then

$$D = \alpha\gamma + \beta\delta + \delta\alpha \int_{\rho(a)}^{b} \frac{\Delta\tau}{p(\tau)}.$$

Thus, we conclude

1. if q is not identically zero on $[\rho(a), b]$, then $D > 0$,
2. if q is identically zero on $[\rho(a), b]$, then $D > 0$ if and only if $\alpha + \gamma > 0$,
3. if $q \equiv 0$ on $[\rho(a), b]$ and $\alpha + \gamma > 0$, then

   a. $G(t,s) \geq 0, t, s \in [\rho(a), b]$,
   b. $G(t,s) > 0, t, s \in [\rho(a), b]$,
   c. if $\beta > 0$ and $\delta > 0$, we have $G(t,s) > 0$ for $t, s \in [a, b]$.

3.8 Advanced Practical Problems

Problem 3.1. Let $\mathbb{T} = 3^{\mathbb{N}_0}$ and

$$q(t) = 0,$$

$$p(t) = t^3 + t + 1, \quad t \in \mathbb{T}.$$

Find the Cauchy function for equation (3.2).

Problem 3.2. Let $\mathbb{T} = 4\mathbb{Z}$. Find the Cauchy function for equation

$$x^{\Delta^2} - 3x^{\sigma} = 0.$$

Problem 3.3. Let $\mathbb{T} = 4\mathbb{Z}$. Find the Green function for the following BVP

$$2x^{\Delta^2} - 7x^{\sigma} = 0$$

$$x(0) = x(32) = 0.$$

Problem 3.4. Let $\mathbb{T} + 3^{\mathbb{N}_0}$,

$$q(t) = 0,$$

$$p(t) = t + 1, \quad t \in \mathbb{T}.$$

Find the Green function for the BVP (3.17).

Problem 3.5. Let $\mathbb{T} = 4^{\mathbb{N}_0}$ and

$$q(t) = 0,$$

$$p(t) = t^2 + t + 1, \quad t \in \mathbb{T}.$$

Find the Green function for the BVP (3.20).

Problem 3.6. Find the Green function for the BVP (3.23) in the following cases:

1. $\mathbb{T} = 2\mathbb{Z}$, $p(t) = t$, $q(t) = 0$, $T = 4$.
2. $\mathbb{T} = 3\mathbb{Z}$, $p(t) = 1 + t$, $q(t) = 0$, $T = 3$.

Problem 3.7. Let $\mathbb{T} = 3^{\mathbb{N}_0}$. Find the Green function for the BVP

$$\left(t^2 + t + 1)x^{\Delta}\right)^{\Delta} = 0$$

$$x(1) = x(15)$$

$$x(27) = 0.$$

3.9 Notes and References

This chapter gives a definition for the Cauchy function and investigates some of its properties. The Green function is constructed for general two point, Dirichlet, focal and periodic BVPs. They are also constructed for some three point BVPs. Some of the results in this chapter can be found in [4], [19] and [20].

Chapter 4
Linear Second Order Eigenvalue Problems

4.1 First Eigenvalue Problem

Let $\mathbb{T}$ be a time scale with forward jump operator, backward jump operator, delta differentiation operator and nabla differentiation operator σ, ρ, Δ and ∇, respectively. Let, also, $[a,b] \subseteq \mathbb{T}$. Consider the following eigenvalue problem

$$\begin{aligned} u^{\Delta^2} + qu^\sigma + \lambda u^\sigma &= 0, \\ \alpha u(\rho(a)) + \beta u^\Delta(\rho(a)) &= 0, \\ \gamma u(b) + \delta u^\Delta(b) &= 0, \end{aligned} \tag{4.1}$$

where $q : \mathbb{T} \to \mathbb{R}$ is rd-continuous, α, β, γ, $\delta \in \mathbb{R}$ satisfy the condition $(\alpha,\beta) \neq (0,0)$, $(\gamma,\delta) \neq (0,0)$, $\lambda \in \mathbb{R}$ is a parameter. We define the operators

$$L(u) = u^{\Delta^2} + qu^\sigma,$$

$$R_a(u) = \alpha u(\rho(a)) + \beta u^\Delta(\rho(a)),$$

$$R_b(u) = \gamma u(b) + \delta u^\Delta(b).$$

Then problem (4.1) can be rewritten in the form

$$\begin{aligned} L(u) + \lambda u^\sigma &= 0 \quad \text{on} \quad (a,b), \\ R_a(u) = R_b(u) &= 0. \end{aligned} \tag{4.2}$$

Definition 4.1. A number $\lambda \in \mathbb{R}$ is called an eigenvalue of (4.2) provided there exists a nontrivial solution u of the boundary value problem (4.2). Such an u is called eigenfunction corresponding to the eigenvalue λ.

DOI: 10.1201/9781003173557-4

Definition 4.2. We define a delta inner product, shortly inner product, of u and v by

$$\langle u, v\rangle = \int_{\rho(a)}^{b} u(t)v(t)\Delta t. \tag{4.3}$$

We say that u and v are orthogonal and we write $u \perp v$ provided $\langle u, v\rangle = 0$. The norm of u is defined by

$$\|u\| = \sqrt{\langle u, u\rangle}. \tag{4.4}$$

Exercise 4.1. Prove that (4.3) satisfies all axioms for an inner product.

Exercise 4.2. Prove that (4.4) satisfies all axioms for a norm.

Example 4.1. Let $\mathbb{T} = \mathbb{Z}$, $u(t) = t$, $v(t) = t+2$, $t \in \mathbb{T}$, $a = 1$, $b = 3$. We will find the inner product of u and v, $\langle u, v\rangle$. We have

$$\sigma(t) = t+1, \quad \rho(t) = t-1, \quad t \in \mathbb{T}.$$

Let

$$h(t) = \frac{1}{3}t^3 + \frac{1}{2}t^2 - \frac{5}{6}t, \quad t \in \mathbb{T}.$$

Then

$$\begin{aligned}
h^{\Delta}(t) &= \frac{1}{3}\left((\sigma(t))^2 + t\sigma(t) + t^2\right) + \frac{1}{2}(\sigma(t)+t) - \frac{5}{6} \\
&= \frac{1}{3}\left((t+1)^2 + t(t+1) + t^2\right) + \frac{1}{2}(t+1+t) - \frac{5}{6} \\
&= \frac{1}{3}(t^2 + 2t + 1 + t^2 + t + t^2) + \frac{1}{2}(2t+1) - \frac{5}{6} \\
&= \frac{1}{3}(3t^2 + 3t + 1) + t + \frac{1}{2} - \frac{5}{6} \\
&= t^2 + t + \frac{1}{3} + t - \frac{1}{3} \\
&= t^2 + 2t, \quad t \in \mathbb{T}.
\end{aligned}$$

Hence,

$$\begin{aligned}
\langle u, v\rangle &= \int_0^3 u(t)v(t)\Delta t \\
&= \int_0^3 t(t+2)\Delta t
\end{aligned}$$

$$= \int_0^3 \left(t^2 + 2t\right) \Delta t$$

$$= \int_0^3 h^{\Delta}(t) \Delta t$$

$$= h(t)\Big|_{t=0}^{t=3}$$

$$= 11.$$

Example 4.2. Let $\mathbb{T} = 2^{\mathbb{N}_0} \bigcup \{0\}$, $a = 1$, $b = 2$,

$$u(t) = t + 1, \quad v(t) = t^2 + 2t - 2, \quad t \in \mathbb{T}.$$

We will find the inner product of u and v, $\langle u, v \rangle$. We have

$$\sigma(t) = 2t, \quad t \in 2^{\mathbb{N}_0}, \quad \sigma(0) = 1,$$

$$\rho(t) = \frac{1}{2}t, \quad t \in 2^{\mathbb{N}}, \quad \rho(1) = \rho(0) = 0.$$

Then

$$\langle u, v \rangle = \int_0^2 (t+1)\left(t^2 + 2t - 2\right) \Delta t$$

$$= \mu(t)(t+1)\left(t^2 + 2t - 2\right)\Big|_{t=0} + \mu(t)(t+1)\left(t^2 + 2t - 2\right)\Big|_{t=1}$$

$$= -2 + 2$$

$$= 0.$$

Example 4.3. Let $\mathbb{T} = 3^{\mathbb{N}_0}$, $u(t) = 3t - 1$, $v(t) = t + 4$, $t \in \mathbb{T}$, $a = 3$, $b = 9$. We will find the inner product of u and v, $\langle u, v \rangle$. Here

$$\sigma(t) = 3t, \quad t \in \mathbb{T}, \quad \rho(t) = \frac{1}{3}t, \quad t \in 3^{\mathbb{N}}, \quad \rho(1) = 1.$$

Then

$$\langle u, v \rangle = \int_1^9 u(t)v(t)\Delta t$$

$$= \int_1^9 (3t - 1)(t + 4)\Delta t$$

$$= \mu(t)\,(3t-1)\,(t+4)\Big|_{t=1} + \mu(t)\,(3t-1)\,(t+4)\Big|_{t=3}$$

$$= 20+336$$

$$= 356.$$

Exercise 4.3. Let $\mathbb{T} = 2^{\mathbb{N}_0}$, $u(t) = t^2+2t$, $v(t) = t^3-1$, $t \in \mathbb{T}$, $a=2$, $b=6$. Find the inner product of u and v, $\langle u, v\rangle$.

Answer. 2016.

With $\mathbb{D}$ we will denote the set of all functions $u : \mathbb{T} \to \mathbb{R}$ such that $u^{\Delta} : \mathbb{T}^{\kappa} \to \mathbb{R}$ is continuous and $u^{\Delta^2} : \mathbb{T}^{\kappa^2} \to \mathbb{R}$ is rd-continuous.

Theorem 4.1. *If $u, v \in \mathbb{D}$, then*

$$\langle u^{\sigma}, L(v)\rangle - \langle v^{\sigma}, L(u)\rangle = W(u,v)(b) - W(u,v)\,(\rho(a)).$$

Here $W(u,v)(\cdot)$ is the Wronskian of u and v.

Proof. We have

$$\begin{aligned} \langle u^{\sigma}, L(v)\rangle - \langle v^{\sigma}, L(u)\rangle &= \int_{\rho(a)}^{b} u^{\sigma}(t)\left(v^{\Delta^2}(t) + q(t)v^{\sigma}(t)\right)\Delta t \\ &\quad - \int_{\rho(a)}^{b} v^{\sigma}(t)\left(u^{\Delta^2}(t) + q(t)u^{\sigma}(t)\right)\Delta t \qquad (4.5) \\ &= \int_{\rho(a)}^{b}\left(u^{\sigma}(t)v^{\Delta^2}(t) - v^{\sigma}(t)u^{\Delta^2}(t)\right)\Delta t. \end{aligned}$$

Note that

$$\begin{aligned} (W(u,v))^{\Delta}(t) &= \left(\det\begin{pmatrix} u & v \\ u^{\Delta} & v^{\Delta}\end{pmatrix}\right)^{\Delta}(t) \\ &= \left(uv^{\Delta} - vu^{\Delta}\right)^{\Delta}(t) \\ &= u^{\Delta}(t)v^{\Delta}(t) + u^{\sigma}(t)v^{\Delta^2}(t) \\ &\quad - v^{\Delta}(t)u^{\Delta}(t) - v^{\sigma}(t)u^{\Delta^2}(t) \\ &= u^{\sigma}(t)v^{\Delta^2}(t) - v^{\sigma}(t)u^{\Delta^2}(t), \quad t \in [\rho(a), b]. \end{aligned}$$

From here and from (4.5), we get

$$\langle u^\sigma, L(v)\rangle - \langle v^\sigma, L(u)\rangle = \int_{\rho(a)}^{b} (W(u,v)(t))^{\Delta}(t)\Delta t$$

$$= W(u,v)(b) - W(u,v)(\rho(a)).$$

This completes the proof.

Theorem 4.2. *Let $R_a(u) = R_a(v) = 0$ and $(\alpha,\beta) \neq (0,0)$. Then $W(u,v)(\rho(a)) = 0$.*

Proof. We have

$$\begin{aligned} W(u,v)(\rho(a)) &= \det\begin{pmatrix} u(\rho(a)) & v(\rho(a)) \\ u^{\Delta}(\rho(a)) & v^{\Delta}(\rho(a)) \end{pmatrix} \\ &= u(\rho(a))v^{\Delta}(\rho(a)) - u^{\Delta}(\rho(a))v(\rho(a)). \end{aligned} \tag{4.6}$$

Since $R_a(u) = R_a(v) = 0$, we find

$$\begin{cases} \alpha u(\rho(a)) + \beta u^{\Delta}(\rho(a)) = 0 \\ \alpha v(\rho(a)) + \beta v^{\Delta}(\rho(a)) = 0. \end{cases}$$

Let $\alpha \neq 0$. Then

$$u(\rho(a)) = -\frac{\beta}{\alpha}u^{\Delta}(\rho(a)) \quad \text{and} \quad v(\rho(a)) = -\frac{\beta}{\alpha}v^{\Delta}(\rho(a)).$$

Hence and by (4.6), we get

$$\begin{aligned} W(u,v)(\rho(a)) &= -\frac{\beta}{\alpha}u^{\Delta}(\rho(a))v^{\Delta}(\rho(a)) + \frac{\beta}{\alpha}u^{\Delta}(\rho(a))v^{\Delta}(\rho(a)) \\ &= 0. \end{aligned}$$

Let $\beta \neq 0$. Then

$$u^{\Delta}(\rho(a)) = -\frac{\alpha}{\beta}u(\rho(a)) \quad \text{and} \quad v^{\Delta}(\rho(a)) = -\frac{\alpha}{\beta}v(\rho(a)).$$

Hence and by (4.6), we get

$$\begin{aligned} W(u,v)(\rho(a)) &= -\frac{\alpha}{\beta}u(\rho(a))v(\rho(a)) + \frac{\alpha}{\beta}u(\rho(a))v(\rho(a)) \\ &= 0. \end{aligned}$$

This completes the proof.

Exercise 4.4. Let $R_b(u) = R_b(v) = 0$ and $(\gamma,\delta) \neq (0,0)$. Prove that $W(u,v)(b) = 0$.

By Theorem 4.1, Theorem 4.2 and Exercise 4.4, we conclude that problem (4.2) is self-adjoint. Therefore, all eigenvalues of problem (4.2) are real.
Below we assume

$$\begin{cases} \beta > \alpha\mu(\rho(a)) \quad \text{if} \quad \beta \neq \alpha\mu(\rho(a)); \quad \alpha = -1 \quad \text{if} \quad \beta = \alpha\mu(\rho(a)), \\ \delta > 0 \quad \text{if} \quad \delta \neq 0; \quad \gamma = 1 \quad \text{if} \quad \delta = 0, \end{cases}$$

and

$$\begin{cases} \text{if } a \text{ is right-scattered, then it is also left-scattered,} \\ \text{if } b \text{ is left-scattered, then it is also right-scattered.} \end{cases}$$

Theorem 4.3. *For every $\lambda \in \mathbb{R}$ problem*

$$Lu + \lambda u^{\sigma} = 0, \tag{4.7}$$

$$u(\rho(a)) = \beta, \quad u^{\Delta}(\rho(a)) = -\alpha, \tag{4.8}$$

has unique solution $u : \mathbb{T} \to \mathbb{R}$.

Proof. Observe that equation (4.7) is equivalent to the system

$$\begin{cases} u^{\Delta} = v \\ v^{\Delta} = -(q+\lambda)u^{\sigma}. \end{cases}$$

Let

$$z = \begin{pmatrix} u \\ v \end{pmatrix}.$$

Then

$$\begin{aligned} z^{\Delta} &= \begin{pmatrix} u^{\Delta} \\ v^{\Delta} \end{pmatrix} \\ &= \begin{pmatrix} v \\ -(q+\lambda)u^{\sigma} \end{pmatrix} \\ &= \begin{pmatrix} 0 & 1 \\ -(q+\lambda) & 0 \end{pmatrix} \begin{pmatrix} u^{\sigma} \\ v \end{pmatrix} \\ &= \begin{pmatrix} 0 & 1 \\ -(q+\lambda) & 0 \end{pmatrix} \left(\begin{pmatrix} 1 & 0 \\ 0 & 0 \end{pmatrix} z^{\sigma} + \begin{pmatrix} 0 & 0 \\ 0 & 1 \end{pmatrix} z \right). \end{aligned}$$

Let

$$\mathscr{M} = \begin{pmatrix} 0 & 0 \\ 1 & 0 \end{pmatrix}, \quad \mathscr{H} = \begin{pmatrix} 0 & 1 \\ -(q+\lambda) & 0 \end{pmatrix}, \quad \mathscr{I} = \begin{pmatrix} 0 & 1 \\ -1 & 0 \end{pmatrix}.$$

Then

$$
\begin{aligned}
\mathscr{M}\mathscr{M}^{\star} &= \begin{pmatrix} 0 & 0 \\ 1 & 0 \end{pmatrix} \begin{pmatrix} 0 & 1 \\ 0 & 0 \end{pmatrix} \\
&= \begin{pmatrix} 0 & 0 \\ 0 & 1 \end{pmatrix}, \\
\mathscr{M}^{\star}\mathscr{M} &= \begin{pmatrix} 0 & 1 \\ 0 & 0 \end{pmatrix} \begin{pmatrix} 0 & 0 \\ 1 & 0 \end{pmatrix} \\
&= \begin{pmatrix} 1 & 0 \\ 0 & 0 \end{pmatrix}, \\
\mathscr{M}\mathscr{M}^{\star} + \mathscr{M}^{\star}\mathscr{M} &= \begin{pmatrix} 0 & 0 \\ 0 & 1 \end{pmatrix} + \begin{pmatrix} 1 & 0 \\ 0 & 0 \end{pmatrix} \\
&= \begin{pmatrix} 1 & 0 \\ 0 & 1 \end{pmatrix},
\end{aligned}
$$

and equation (4.7) can be rewritten in the form

$$
z^{\Delta} = \mathscr{H}\left(\mathscr{M}^{\star}\mathscr{M}z^{\sigma} + \mathscr{M}\mathscr{M}^{\star}z\right). \tag{4.9}
$$

Observe that

$$
\begin{aligned}
\mathscr{H}^{\star}\mathscr{I} + \mathscr{I}\mathscr{H} &= \begin{pmatrix} 0 & -(q+\lambda) \\ 1 & 0 \end{pmatrix} \begin{pmatrix} 0 & 1 \\ -1 & 0 \end{pmatrix} + \begin{pmatrix} 0 & 1 \\ -1 & 0 \end{pmatrix} \begin{pmatrix} 0 & 1 \\ -(q+\lambda) & 0 \end{pmatrix} \\
&= \begin{pmatrix} q+\lambda & 0 \\ 0 & 1 \end{pmatrix} + \begin{pmatrix} -(q+\lambda) & 0 \\ 0 & -1 \end{pmatrix} \\
&= \begin{pmatrix} 0 & 0 \\ 0 & 0 \end{pmatrix}.
\end{aligned}
$$

Hence, $\mathscr{H}$ is Hamiltonian. Also,

$$
\begin{aligned}
I - \mu\mathscr{H}\mathscr{M}^{\star}\mathscr{M} &= \begin{pmatrix} 1 & 0 \\ 0 & 1 \end{pmatrix} - \mu \begin{pmatrix} 0 & 1 \\ -(q+\lambda) & 0 \end{pmatrix} \begin{pmatrix} 1 & 0 \\ 0 & 0 \end{pmatrix} \\
&= \begin{pmatrix} 1 & 0 \\ 0 & 1 \end{pmatrix} - \mu \begin{pmatrix} 0 & 0 \\ -(q+\lambda) & 0 \end{pmatrix} \\
&= \begin{pmatrix} 1 & 0 \\ \mu(q+\lambda) & 1 \end{pmatrix},
\end{aligned}
$$

which is invertible. Therefore, $\mathscr{H}$ is Hamiltonian with respect to $\mathbb{T}$. Consequently, the system (4.9) is a Hamiltonian system. By Theorem 7.17, [9], pp. 294, the system (4.9) is also a symplectic system with

$$S = (I - \mu \mathscr{H}\mathscr{M}^{\star}\mathscr{M})^{-1}\mathscr{H}$$

$$= \begin{pmatrix} 1 & 0 \\ -\mu(q+\lambda) & 1 \end{pmatrix} \begin{pmatrix} 0 & 1 \\ -(q+\lambda) & 0 \end{pmatrix}$$

$$= \begin{pmatrix} 0 & 1 \\ -(q+\lambda) & -\mu(q+\lambda) \end{pmatrix}.$$

The system (4.9) can be rewritten in the form

$$z^{\Delta} = \mathscr{H}\left(\mu \mathscr{M}^{\star}\mathscr{M}z^{\Delta} + \mathscr{M}^{\star}\mathscr{M}z + \mathscr{M}\mathscr{M}^{\star}z\right)$$

$$= \mu \mathscr{H}\mathscr{M}^{\star}\mathscr{M}z^{\Delta} + \mathscr{H}z,$$

or

$$(I - \mu \mathscr{H}\mathscr{M}^{\star}\mathscr{M})z^{\Delta} = \mathscr{H}z,$$

or

$$z^{\Delta} = (I - \mu \mathscr{H}\mathscr{M}^{\star}\mathscr{M})^{-1}\mathscr{H}z,$$

or

$$z^{\Delta} = Sz.$$

Therefore, problem (4.7), (4.8) is equivalent to problem

$$z^{\Delta} = Sz, \quad z(\rho(a)) = \begin{pmatrix} \beta \\ -\alpha \end{pmatrix}. \tag{4.10}$$

By Theorem 7.11, [9], pp. 292, it follows that S is regressive. Hence and by Theorem 5.8, [9], pp. 190, it follows that problem (4.10) has a unique solution $z : \mathbb{T} \to \mathbb{R}^2$. This completes the proof.

For any $\lambda \in \mathbb{R}$ with $v(\cdot,\lambda)$ we will denote the unique solution of problem (4.7), (4.8). We set

$$\Lambda(\lambda) = R_b(v(\cdot,\lambda)).$$

Theorem 4.4. λ *is an eigenvalue of problem* (4.2) *if and only if* $\Lambda(\lambda) = 0$.

Proof. 1. Let $\Lambda(\lambda) = 0$. Then $R_b(v(\cdot,\lambda)) = 0$, i.e.,

$$\gamma v(b,\lambda) + \delta v^{\Delta}(b,\lambda) = 0. \tag{4.11}$$

Note that $v(\cdot,\lambda)$ satisfies equation (4.7). Also, we have $R_b(v(\cdot,\lambda)) = 0$ and

$$R_a(v(\cdot,\lambda)) = \alpha v(\rho(a),\lambda) + \beta v^{\Delta}(\rho(a),\lambda)$$

$$= \alpha\beta - \alpha\beta = 0.$$

Consequently, λ is an eigenvalue of problem (4.2).

2. Let λ be an eigenvalue of problem (4.2) corresponding to the eigenfunction $u(\cdot,\lambda)$. If we assume that $u(\rho(a),\lambda) = u^\Delta(\rho(a),\lambda) = 0$, then $u(\cdot,\lambda)$ will be a solution of problem

$$L(u) + \lambda u^\sigma = 0, \quad u(\rho(a),\lambda) = u^\Delta(\rho(a),\lambda) = 0.$$

Hence, using Theorem 4.3, we get that $u(\cdot,\lambda) \equiv 0$. This is a contradiction because $u(\cdot,\lambda)$ is a nontrivial solution to problem (4.2). Therefore, $u(\rho(a),\lambda) \neq 0$ or $u^\Delta(\rho(a),\lambda) \neq 0$. Assume that

$$\beta u(\rho(a),\lambda) - \alpha u^\Delta(\rho(a),\lambda) = 0.$$

Then, using that $R_a(u(\cdot,\lambda)) = 0$, we get the system

$$\begin{cases} \beta u(\rho(a),\lambda) - \alpha u^\Delta(\rho(a),\lambda) = 0 \\ \alpha u(\rho(a),\lambda) + \beta u^\Delta(\rho(a),\lambda) = 0. \end{cases}$$

Since $\alpha^2 + \beta^2 \neq 0$, we obtain that $u(\rho(a),\lambda) = u^\Delta(\rho(a),\lambda) = 0$, which is a contradiction. Consequently,

$$\beta u(\rho(a),\lambda) - \alpha u^\Delta(\rho(a),\lambda) \neq 0.$$

Consider the function

$$v(\cdot,\lambda) = \frac{\alpha^2+\beta^2}{\beta u(\rho(a),\lambda) - \alpha u^\Delta(\rho(a),\lambda)} u(\cdot,\lambda).$$

Then

$$\begin{aligned} L(v) &= \frac{\alpha^2+\beta^2}{\beta u(\rho(a),\lambda) - \alpha u^\Delta(\rho(a),\lambda)} L(u) \\ &= 0. \end{aligned}$$

Also,

$$\begin{aligned} R_b(v(\cdot,\lambda)) &= \frac{\alpha^2+\beta^2}{\beta u(\rho(a),\lambda) - \alpha u^\Delta(\rho(a),\lambda)} R_b(u(\cdot,\lambda)) \\ &= 0. \end{aligned}$$

Assume that $\alpha \neq 0$. Using that $R_a(u(\cdot,\lambda)) = 0$, we obtain

$$u(\rho(a),\lambda) = -\frac{\beta}{\alpha} u^\Delta(\rho(a),\lambda).$$

Hence,

$$\begin{aligned}
v(\rho(a),\lambda) &= \frac{\alpha^2+\beta^2}{\beta u(\rho(a),\lambda)-\alpha u^{\Delta}(\rho(a),\lambda)}u(\rho(a),\lambda)\\
&= \frac{\alpha^2+\beta^2}{-\frac{\beta^2}{\alpha}u^{\Delta}(\rho(a),\lambda)-\alpha u^{\Delta}(\rho(a),\lambda)}\left(-\frac{\beta}{\alpha}u^{\Delta}(\rho(a),\lambda)\right)\\
&= \beta
\end{aligned}$$

and

$$\begin{aligned}
v^{\Delta}(\rho(a),\lambda) &= \frac{\alpha^2+\beta^2}{\beta u(\rho(a),\lambda)-\alpha u^{\Delta}(\rho(a),\lambda)}u^{\Delta}(\rho(a),\lambda)\\
&= \frac{\alpha^2+\beta^2}{-\frac{\beta^2}{\alpha}u^{\Delta}(\rho(a),\lambda)-\alpha u^{\Delta}(\rho(a),\lambda)}u^{\Delta}(\rho(a),\lambda)\\
&= -\alpha.
\end{aligned}$$

Assume that $\beta \neq 0$. Then, again using $R_a(u(\cdot,\lambda)) = 0$, we find

$$u^{\Delta}(\rho(a),\lambda) = -\frac{\alpha}{\beta}u(\rho(a),\lambda).$$

From here,

$$\begin{aligned}
v(\rho(a),\lambda) &= \frac{\alpha^2+\beta^2}{\beta u(\rho(a),\lambda)-\alpha u^{\Delta}(\rho(a),\lambda)}u(\rho(a),\lambda)\\
&= \frac{\alpha^2+\beta^2}{\beta u(\rho(a),\lambda)+\frac{\alpha^2}{\beta}u(\rho(a),\lambda)}u(\rho(a),\lambda)\\
&= \beta
\end{aligned}$$

and

$$\begin{aligned}
v^{\Delta}(\rho(a),\lambda) &= \frac{\alpha^2+\beta^2}{\beta u(\rho(a),\lambda)-\alpha u^{\Delta}(\rho(a),\lambda)}u^{\Delta}(\rho(a),\lambda)\\
&= \frac{\alpha^2+\beta^2}{\beta u(\rho(a),\lambda)+\frac{\alpha^2}{\beta}u(\rho(a),\lambda)}\left(-\frac{\alpha}{\beta}u(\rho(a),\lambda)\right)\\
&= -\alpha.
\end{aligned}$$

Consequently, $v(\cdot,\lambda)$ satisfies problem (4.7), (4.8). Therefore,

$$\Lambda(\lambda) = R_b(v(\cdot,\lambda))$$

$$= R_b\left(\frac{\alpha^2+\beta^2}{\beta u(\rho(a),\lambda)-\alpha u^\Delta(\rho(a),\lambda)}u(\cdot,\lambda)\right)$$

$$= \frac{\alpha^2+\beta^2}{\beta u(\rho(a),\lambda)-\alpha u^\Delta(\rho(a),\lambda)}R_b(u(\cdot,\lambda))$$

$$= 0.$$

This completes the proof.

Theorem 4.5. *All eigenvalues of* (4.2) *are simple.*

Proof. Let λ be an arbitrary eigenvalue of problem (4.2) corresponding to the eigenfunctions $u_1, \ldots, u_k$. We set

$$v_l(\cdot,\lambda) = \frac{\alpha^2+\beta^2}{\beta u_l(\rho(a),\lambda)-\alpha u_l^\Delta(\rho(a),\lambda)}u_l(\cdot,\lambda), \quad l\in\{1,\ldots,k\}.$$

As in the proof of Theorem 4.4, we have that $v_l(\cdot,\lambda)$, $l\in\{1,\ldots,k\}$, are solutions to problem (4.7), (4.8). Hence and by Theorem 4.3, it follows that there exist constants $c_2, \ldots, c_k$, such that

$$v_l(\cdot,\lambda) = c_l v_1(\cdot,\lambda), \quad l\in\{2,\ldots,.k\}.$$

Hence,

$$\frac{\alpha^2+\beta^2}{\beta u_l(\rho(a),\lambda)-\alpha u_l^\Delta(\rho(a),\lambda)}u_l(\cdot,\lambda) = c_l\frac{\alpha^2+\beta^2}{\beta u_1(\rho(a),\lambda)-\alpha u_1^\Delta(\rho(a),\lambda)}u_1(\cdot,\lambda),$$

or

$$u_l(\cdot,\lambda) = c_l\frac{\beta u_l(\rho(a),\lambda)-\alpha u_l^\Delta(\rho(a),\lambda)}{\beta u_1(\rho(a),\lambda)-\alpha u_1^\Delta(\rho(a),\lambda)}u_1(\cdot,\lambda), \quad l\in\{2,\ldots,k\}.$$

This completes the proof.

Theorem 4.6. *Let u_1 and u_2 be eigenfunctions of problem* (4.2) *corresponding to different eigenvalues λ_1 and λ_2, respectively. Then*

$$\langle u_1^\sigma, u_2^\sigma\rangle = 0.$$

Proof. Since u_1 and u_2 are eigenfunctions of problem (4.2) corresponding to the eigenvalues λ_1 and λ_2, respectively, we have

$$L(u_1) = -\lambda_1 u_1^\sigma \quad \text{and} \quad L(u_2) = -\lambda_2 u_2^\sigma.$$

Then

$$\langle u_1^\sigma, L(u_2)\rangle = -\lambda_2\langle u_1^\sigma, u_2^\sigma\rangle$$

and

$$\langle u_2^\sigma, L(u_1)\rangle = -\lambda_1 \langle u_2^\sigma, u_1^\sigma\rangle$$

$$= -\lambda_1 \langle u_1^\sigma, u_2^\sigma\rangle.$$

Hence,

$$\begin{aligned}\langle u_1^\sigma, L(u_2)\rangle - \langle u_2^\sigma, L(u_1)\rangle &= -\lambda_2 \langle u_1^\sigma, u_2^\sigma\rangle + \lambda_1 \langle u_1^\sigma, u_2^\sigma\rangle \\ &= (\lambda_1 - \lambda_2)\langle u_1^\sigma, u_2^\sigma\rangle.\end{aligned} \tag{4.12}$$

Because $R_a(u_1) = R_a(u_2) = 0$, using Theorem 4.2, we obtain

$$W(u_1, u_2)(\rho(a)) = 0. \tag{4.13}$$

Since $R_b(u_1) = R_b(u_2) = 0$, using Exercise 4.4, we have

$$W(u_1, u_2)(b) = 0.$$

Hence and by (4.12), (4.13), using Theorem 4.1, we get

$$(\lambda_1 - \lambda_2)\langle u_1^\sigma, u_2^\sigma\rangle = \langle u_1^\sigma, L(u_2)\rangle - \langle u_2^\sigma, L(u_1)\rangle$$

$$= W(u_1, u_2)(b) - W(u_1, u_2)(\rho(a))$$

$$= 0.$$

Because $\lambda_1 \neq \lambda_2$, we get

$$\langle u_1^\sigma, u_2^\sigma\rangle = 0.$$

This completes the proof.

Below we suppose

$$\lim_{\nu\to\lambda} \langle v^\sigma(\cdot,\nu), v^\sigma(\cdot,\lambda)\rangle = \|v^\sigma(\cdot,\lambda)\|^2. \tag{4.14}$$

Theorem 4.7. *For all $t \in [\rho(a), b]$ and $\lambda \in \mathbb{R}$, we have*

$$v(t,\lambda)\frac{\partial}{\partial\lambda}v^\Delta(t,\lambda) - v^\Delta(t,\lambda)\frac{\partial}{\partial\lambda}v(t,\lambda) = -\int_{\rho(a)}^t (v^\sigma(\tau,\lambda))^2 \Delta\tau. \tag{4.15}$$

Proof. Let λ, $\nu \in \mathbb{R}$, $\lambda \neq \nu$. Then

$$\Big(v(t,\lambda)\Big(v^\Delta(t,\lambda) - v^\Delta(t,\nu)\Big) - v^\Delta(t,\lambda)(v(t,\lambda) - v(t,\nu))\Big)^\Delta$$

$$= \Big(v^\Delta(t,\lambda)v(t,\nu) - v(t,\lambda)v^\Delta(t,\nu)\Big)^\Delta$$

$$= v^{\sigma}(t,\nu)\, v^{\Delta^2}(t,\lambda) + v^{\Delta}(t,\lambda)\, v^{\Delta}(t,\nu)$$

$$-v^{\sigma}(t,\lambda)\, v^{\Delta^2}(t,\nu) - v^{\Delta}(t,\lambda)\, v^{\Delta}(t,\nu)$$

$$= v^{\sigma}(t,\nu)\, v^{\Delta^2}(t,\lambda) - v^{\sigma}(t,\lambda)\, v^{\Delta^2}(t,\nu) \quad ((4.7))$$

$$= v^{\sigma}(t,\nu)\left(-\lambda v^{\sigma}(t,\lambda) - q v^{\sigma}(t,\lambda)\right)$$

$$-v^{\sigma}(t,\lambda)\left(-\nu v^{\sigma}(t,\nu) - q v^{\sigma}(t,\nu)\right)$$

$$= (\nu - \lambda)\, v^{\sigma}(t,\lambda)\, v^{\sigma}(t,\nu).$$

Hence,

$$\left(v(t,\lambda)\frac{v^{\Delta}(t,\lambda) - v^{\Delta}(t,\nu)}{\nu-\lambda} - v^{\Delta}(t,\lambda)\frac{v(t,\lambda) - v(t,\nu)}{\nu-\lambda}\right)^{\Delta} = v^{\sigma}(t,\lambda)\, v^{\sigma}(t,\nu).$$

Now, we integrate the last equality from $\rho(a)$ to t and we get

$$v(t,\lambda)\frac{v^{\Delta}(t,\lambda) - v^{\Delta}(t,\nu)}{\nu-\lambda} - v^{\Delta}(t,\lambda)\frac{v(t,\lambda) - v(t,\nu)}{\nu-\lambda}$$

$$-v(\rho(a),\lambda)\frac{v^{\Delta}(\rho(a),\lambda) - v^{\Delta}(\rho(a),\nu)}{\nu-\lambda} + v^{\Delta}(\rho(a),\lambda)\frac{v(\rho(a),\lambda) - v(\rho(a),\nu)}{\nu-\lambda}$$

$$= \int_{\rho(a)}^{t} v^{\sigma}(\tau,\lambda)\, v^{\sigma}(\tau,\nu)\,\Delta\tau$$

or, using (4.8),

$$v(t,\lambda)\frac{v^{\Delta}(t,\lambda) - v^{\Delta}(t,\nu)}{\nu-\lambda} - v^{\Delta}(t,\lambda)\frac{v(t,\lambda) - v(t,\nu)}{\nu-\lambda} - \alpha\frac{\beta-\beta}{\nu-\lambda} - \beta\frac{-\alpha+\alpha}{\nu-\lambda}$$

$$= \int_{\rho(a)}^{t} v^{\sigma}(\tau,\lambda)\, v^{\sigma}(\tau,\nu)\,\Delta\tau,$$

or

$$v(t,\lambda)\frac{v^{\Delta}(t,\lambda) - v^{\Delta}(t,\nu)}{\nu-\lambda} - v^{\Delta}(t,\lambda)\frac{v(t,\lambda) - v(t,\nu)}{\nu-\lambda} = \int_{\rho(a)}^{t} v^{\sigma}(\tau,\lambda)\, v^{\sigma}(\tau,\nu)\,\Delta\tau.$$

We take $\nu \to \lambda$ in the last equality and using (4.14), we get the desired result. This completes the proof.

Corollary 4.1. *We have that* $\frac{v^{\Delta}(t,\lambda)}{v(t,\lambda)}$ *is strictly decreasing in* λ *function for any* $t \in (a,b]$ *whenever* $v(t,\lambda) \neq 0$.

Proof. Since

$$\frac{\partial}{\partial\lambda}\left(\frac{v^{\Delta}(t,\lambda)}{v(t,\lambda)}\right) = \frac{v(t,\lambda)\frac{\partial}{\partial\lambda}v^{\Delta}(t,\lambda) - v^{\Delta}(t,\lambda)\frac{\partial}{\partial\lambda}v(t,\lambda)}{(v(t,\lambda))^2},$$

using Theorem 4.7, we get that

$$\frac{\partial}{\partial\lambda}\left(\frac{v^{\Delta}(t,\lambda)}{v(t,\lambda)}\right) < 0$$

for any $t \in (a,b]$ whenever $v(t,\lambda) \neq 0$. This completes the proof.

Corollary 4.2. *Let* $v(b,\lambda) = 0$. *Then* $\frac{\partial}{\partial\lambda}v(b,\lambda) \neq 0$ *and* $v^{\Delta}(b,\lambda) \neq 0$.

Proof. We put $t = b$ in (4.15) and using that $v(b,\lambda) = 0$, we find

$$v^{\Delta}(b,\lambda)\frac{\partial}{\partial\lambda}v(b,\lambda) = \int_{\rho(a)}^{b} (v^{\sigma}(\tau,\lambda))^2 \Delta\tau.$$

Hence,

$$v^{\Delta}(b,\lambda)\frac{\partial}{\partial\lambda}v(b,\lambda) \neq 0.$$

This completes the proof.

Corollary 4.3. *All eigenvalues of* (4.2) *are isolated.*

Proof. By Theorem 4.4, it follows that λ is an eigenvalue of (4.2) if and only if $\Lambda(\lambda) = 0$. Suppose that λ is an eigenvalue of (4.2). Then $\Lambda(\lambda) = 0$ or

$$\gamma v(b,\lambda) + \delta v^{\Delta}(b,\lambda) = 0. \tag{4.16}$$

Assume that $v(b,\lambda) = 0$. Then, by Corollary 4.2, we get $v^{\Delta}(b,\lambda) \neq 0$ and $\frac{\partial}{\partial\lambda}v(b,\lambda) \neq 0$. If $\delta = 0$, then $\gamma = 1$ and by (4.16), we get $v(b,\lambda) = 0$. Hence, $\frac{\partial}{\partial\lambda}v(b,\lambda) = 0$. This is a contradiction. Therefore, $\delta > 0$. From here and from (4.16), we find $v^{\Delta}(b,\lambda) = 0$. This contradicts. Therefore, $v(b,\lambda) \neq 0$. Hence, $\Lambda(\lambda) = 0$ if and only if

$$\gamma + \delta\frac{v^{\Delta}(b,\lambda)}{v(b,\lambda)} = 0.$$

By Corollary 4.1, we have that $\frac{\partial}{\partial\lambda}\left(\frac{v^{\Delta}(b,\lambda)}{v(b,\lambda)}\right) < 0$. Therefore, λ is an isolated zero of $\Lambda(\lambda)$. This completes the proof.

Definition 4.3. We say that a solution u of problem (4.7), (4.8) has a generalized zero at t if $u(t) = 0$.

We set

$$\lambda^\star = -\left(1 + \sup_{t\in[\rho(a),b]} q(t)\right).$$

Since $q : \mathbb{T} \to \mathbb{R}$ is a rd-continuous function, it is a regulated function. Because every regulated function on a compact interval is bounded, we conclude that q is a bounded function on $[\rho(a),b]$. Consequently, $\lambda^\star > -\infty$. Also, for all $t \in [\rho(a),b]$ and for all $\lambda \le \lambda^\star$, we have

$$\begin{aligned}\lambda + q(t) &\le \lambda + \sup_{t\in[\rho(a),b]} q(t)\\ &\le \lambda^\star + \sup_{t\in[\rho(a),b]} q(t)\\ &= -1.\end{aligned}$$

Theorem 4.8. *Let $\lambda \le \lambda^*$ and u be a nontrivial solution of equation* (4.7). *If u has for some $t_1 \in [\rho(a),b]$ a generalized zero, then u^Δ has no generalized zeros in $[t_1, \sigma(t_1)]$.*

Proof. Because u is a nontrivial solution of equation (4.7) and $u(t_1) = 0$, we conclude that $u^\Delta(t_1) \neq 0$. We have

$$u^{\Delta^2}(t_1) = -(\lambda + q(t_1))\,u(\sigma(t_1))$$

and using that $\lambda + q(t_1) \le -1$, we obtain

$$\begin{aligned}u^\Delta(t_1)\,u^\Delta(\sigma(t_1)) &= u^\Delta(t_1)\left(\mu(t_1)\,u^{\Delta^2}(t_1) + u^\Delta(t_1)\right)\\ &= \left(u^\Delta(t_1)\right)^2 + \mu(t_1)\,u^\Delta(t_1)\,u^{\Delta^2}(t_1)\\ &= \left(u^\Delta(t_1)\right)^2 - (\lambda + q(t_1))\,\mu(t_1)\,u^\Delta(t_1)\,u(\sigma(t_1))\\ &= \left(u^\Delta(t_1)\right)^2 - (\lambda + q(t_1))\,(u(\sigma(t_1)) - u(t_1))\,u(\sigma(t_1))\\ &= \left(u^\Delta(t_1)\right)^2 - (\lambda + q(t_1))\,(u(\sigma(t_1)))^2\\ &\ge \left(u^\Delta(t_1)\right)^2 + (u(\sigma(t_1)))^2\end{aligned}$$

$$> 0.$$

Therefore, $u^{\Delta}(t_1) \neq 0$ and $u^{\Delta}(\sigma(t_1)) \neq 0$. This completes the proof.

Theorem 4.9. *Let $\lambda \le \lambda^\star$ and u be a nontrivial solution of equation* (4.7). *Then the function uu^{Δ} is strictly increasing on $[\rho(a), b]$.*

Proof. We have

$$\begin{aligned}
\left(uu^{\Delta}\right)^{\Delta}(t) &= \left(u^{\Delta}(t)\right)^2 + u(\sigma(t))\, u^{\Delta^2}(t) \\
&= \left(u^{\Delta}(t)\right)^2 - (\lambda + q(t))\,(u(\sigma(t)))^2 \\
&\ge \left(u^{\Delta}(t)\right)^2 + (u(\sigma(t)))^2 \\
&\ge 0
\end{aligned} \tag{4.17}$$

for any $t \in [\rho(a), b]$. Assume that $u^{\Delta}(t_1) = 0$ and $u(\sigma(t_1)) = 0$ for some $t_1 \in [\rho(a), b]$. If $\sigma(t_1) = t_1$, then $u^{\Delta}(t_1) = 0$ and $u(t_1) = 0$, which is impossible because u is a nontrivial solution of equation (4.7). If $\sigma(t_1) > t_1$, then $\mu(t_1) > 0$ and

$$\begin{aligned}
0 &= u(\sigma(t_1)) \\
&= \mu(t_1)\, u^{\Delta}(t_1) + u(t_1) \\
&= u(t_1),
\end{aligned}$$

which is again impossible because u is a nontrivial solution of equation (4.7). Consequently,

$$\left(u^{\Delta}(t), u(\sigma(t))\right) \neq (0,0)$$

for any $t \in [\rho(a), b]$. Hence and by (4.17), we get that

$$\left(uu^{\Delta}\right)^{\Delta}(t) > 0$$

for any $t \in [\rho(a), b]$. This completes the proof.

Theorem 4.10. *Let $\lambda \le \lambda^\star$ and u be a nontrivial solution of equation* (4.7). *If $\left(uu^{\Delta}\right)(t_1) \ge 0$ for some $t_1 \in [\rho(a), b]$, then u has not generalized zeros in $(t_1, b]$.*

Proof. We have

$$\begin{aligned}
u(t)u(\sigma(t)) &= u(t)\left(\mu(t)u^{\Delta}(t) + u(t)\right) \\
&= \mu(t)u(t)u^{\Delta}(t) + (u(t))^2.
\end{aligned} \tag{4.18}$$

By Theorem 4.9, it follows that uu^{Δ} is a strictly increasing function on $[\rho(a), b]$. Therefore,

$$u(t)u^{\Delta}(t) > u(t_1)u^{\Delta}(t_1) \tag{4.19}$$
$$\geq 0$$

for any $t \in (t_1, b]$. Let $t_2 \in (t_1, b]$ be arbitrarily chosen.

1. If $\mu(t_2) > 0$, then, using (4.19), we get

$$\mu(t_2)u(t_2)u^{\Delta}(t_2) > 0,$$

   and from (4.18), we find

$$u(t_2)u(\sigma(t_2)) > 0.$$

   Hence, $u(t_2) \neq 0$.
2. Let $\mu(t_2) = 0$. Assume that $u(t_2) = 0$. Then $u(t_2)u^{\Delta}(t_2) = 0$. This is a contradiction, because $uu^{\Delta} > 0$ for any $t \in (t_1, b]$. Therefore, $u(t_2) \neq 0$.

Because $t_2 \in (t_1, b]$ was arbitrarily chosen and we obtain that $u(t_2) \neq 0$, we conclude that $u(t) \neq 0$ for any $t_2 \in (t_1, b]$. This completes the proof.

Theorem 4.11. *Let $\lambda \leq \lambda^{\star}$ and u be a nontrivial solution of equation* (4.7). *Then $[\rho(a), b]$ contains at most one generalized zero of u.*

Proof. Let $t_1, t_2 \in [\rho(a), b]$ be such that $t_1 < t_2$ and

$$u(t_1) = u(t_2) = 0.$$

Since $u(t_1)u^{\Delta}(t_1) = 0$, by Theorem 4.10, it follows that u has no generalized zeros in $(t_1, b]$. This is a contradiction, because $t_2 \in (t_1, b]$ and $u(t_2) = 0$. This completes the proof.

With $n(\lambda)$ we will denote the number of generalized zeros of $v(\cdot, \lambda)$ in the open interval (a, b).

Theorem 4.12. *Let $\lambda \leq \lambda^{\star}$. If $n(\lambda) = 0$, then $\lim_{\nu \to -\infty} n(\nu) = 0$.*

Proof. Let

$$\kappa(\lambda) = \min_{t \in [\rho(a), \sigma(b)]} (v(t, \lambda))^2.$$

Then $\kappa(\lambda) > 0$. Let $\nu \leq \lambda$. We have

$$\begin{aligned}\left(vv^{\Delta}\right)^{\Delta}(t, \nu) &= \left(v^{\Delta}(t, \nu)\right)^2 + v^{\sigma}(t, \nu)v^{\Delta^2}(t, \nu) \\ &= \left(v^{\Delta}(t, \nu)\right)^2 - (\nu + q(t))(v^{\sigma}(t, \nu))^2.\end{aligned}$$

We integrate the last equality with respect to t, $t \in (\rho(a), b]$, and we get

$$v(t,\nu)v^{\Delta}(t,\nu)-v(\rho(a),\nu)v^{\Delta}(\rho(a),\nu)$$

$$=\int_{\rho(a)}^{t}\left(\left(v^{\Delta}(\tau,\nu)\right)^{2}-(\nu+q(\tau))(v^{\sigma}(\tau,\nu))^{2}\right)\Delta\tau$$

or

$$\begin{aligned}
v(t,\nu)v^{\Delta}(t,\nu) &= -\alpha\beta+\int_{\rho(a)}^{t}\left(\left(v^{\Delta}(\tau,\nu)\right)^{2}-(\nu+q(\tau))(v^{\sigma}(\tau,\nu))^{2}\right)\Delta\tau \\
&\geq -\alpha\beta-\int_{\rho(a)}^{t}(\nu+q(\tau))(v^{\sigma}(\tau,\nu))^{2}\Delta\tau \\
&\geq -\alpha\beta+\int_{\rho(a)}^{t}(-\nu+\lambda^{\star}+1)(v^{\sigma}(\tau,\nu))^{2}\Delta\tau \\
&\geq -\alpha\beta+\int_{\rho(a)}^{t}(-\nu+\lambda^{\star})(v^{\sigma}(\tau,\nu))^{2}\Delta\tau \\
&\geq -\alpha\beta+\int_{\rho(a)}^{t}(\lambda^{\star}-\nu)\kappa(\nu)\Delta\tau \\
&= -\alpha\beta+(\lambda^{\star}-\nu)\kappa(\nu)(t-\rho(a)) \\
&> 0
\end{aligned} \tag{4.20}$$

for small enough ν. Therefore, $\lim_{\nu\to-\infty} n(\nu)=0$. This completes the proof.

Theorem 4.13. *We have*

$$n(\lambda)\to 0 \quad \textit{and} \quad \Lambda(\lambda)\to\infty,$$

as $\lambda\to-\infty$.

Proof. We take $\lambda<\lambda^{\star}$. By Theorem 4.11, it follows that $n(\lambda^{\star})\in\{0,1\}$.

1. Let $n(\lambda^{\star})=0$. Then, using Theorem 4.12, it follows that $n(\lambda)\to 0$, as $\lambda\to-\infty$.
2. Let $n(\lambda^{\star})=1$. Then there exists $t_1\in(a,b)$ such that $v(t_1,\lambda^{\star})=0$. By Theorem 4.9 we have that $\left(vv^{\Delta}\right)(\cdot,\lambda^{\star})$ is a strictly increasing function on $[\rho(a),b]$. Hence,

   $$v(\rho(a),\lambda^{\star})v^{\Delta}(\rho(a),\lambda^{\star})<0,$$

   or

   $$-\alpha\beta<0,$$

   or

   $$\alpha\beta>0.$$

By Theorem 4.11 we have that the interval $[\rho(a), b]$ contains at most one zero of $v(\cdot, \lambda^\star)$. Since $t_1 \in (a, b)$ and $v(t_1, \lambda^\star) = 0$, we conclude that there exists $t_2 \in [\rho(a), b]$, $t_2 < t_1$, such that the interval $[\rho(a), t_2]$ contains no generalized zeros of $v(\cdot, \lambda^\star)$. Let $I = [\rho(a), t_2]$. If we assume that for some $t_3 \in I$ we have that $\left(vv^\Delta\right)(t_3, \lambda^\star) \geq 0$, by Theorem 4.10, it follows that v has no generalized zeros in $(t_3, b]$. This is a contradiction, because $t_1 \in (t_3, b]$ and $v(t_1, \lambda^\star) = 0$. Consequently, $\left(vv^\Delta\right)(t, \lambda^\star) < 0$ for any $t \in I$. Hence, $\frac{v^\Delta(t, \lambda^\star)}{v(t, \lambda^\star)} < 0$ for any $t \in I$. Now, we assume that

$$\frac{v^\Delta(t, \lambda)}{v(t, \lambda)} \in \left(\frac{v^\Delta(t, \lambda^\star)}{v(t, \lambda^\star)}, 0\right)$$

for any $t \in I$. Using Corollary 4.1, we get

$$\begin{aligned} 1 + \mu(t)\frac{v^\Delta(t, \lambda)}{v(t, \lambda)} &> 1 + \mu(t)\frac{v^\Delta(t, \lambda^\star)}{v(t, \lambda^\star)} \\ &= \frac{v^\sigma(t, \lambda^\star)}{v(t, \lambda^\star)}. \end{aligned} \tag{4.21}$$

Since $v(t, \lambda^\star)$ is a continuous function in $t \in [\rho(a), b]$ and $v(t_1, \lambda^\star) = 0$, then $\mathrm{sign} v(t, \lambda^\star) = \mathrm{sign} v^\sigma(t, \lambda^\star)$ for any $t \in I$. Therefore, $\frac{v^\sigma(t, \lambda^\star)}{v(t, \lambda^\star)} > 0$ for any $t \in I$. Hence and by (4.21), we conclude that

$$1 + \mu(t)\frac{v^\Delta(t, \lambda^\star)}{v(t, \lambda^\star)} > 0$$

for any $t \in I$. Note that

$$\begin{aligned} \left(\frac{v^\Delta(t, \lambda)}{v(t, \lambda)}\right)^\Delta &= \frac{v^{\Delta^2}(t, \lambda)v(t, \lambda) - \left(v^\Delta(t, \lambda)\right)^2}{v(t, \lambda)v^\sigma(t, \lambda)} \\ &= \frac{-(\lambda + q(t))v(t, \lambda)v^\sigma(t, \lambda) - \left(v^\Delta(t, \lambda)\right)^2}{v(t, \lambda)v^\sigma(t, \lambda)} \\ &= -(\lambda + q(t)) - \left(\frac{v^\Delta(t, \lambda)}{v(t, \lambda)}\right)^2 \frac{1}{\frac{v^\sigma(t, \lambda)}{v(t, \lambda)}} \end{aligned} \tag{4.22}$$

$$= -(\lambda + q(t)) - \left(\frac{v^\Delta(t, \lambda)}{v(t, \lambda)}\right)^2 \frac{1}{1 + \mu(t)\frac{v^\Delta(t, \lambda)}{v(t, \lambda)}}$$

$$
\begin{aligned}
&= (\lambda^\star - \lambda) - (\lambda^\star + q(t)) - \left(\frac{v^\Delta(t,\lambda)}{v(t,\lambda)}\right)^2 \frac{1}{1+\mu(t)\frac{v^\Delta(t,\lambda)}{v(t,\lambda)}} \\
&\geq (\lambda^\star - \lambda) - (\lambda^\star + q(t)) - \left(\frac{v^\Delta(t,\lambda^\star)}{v(t,\lambda^\star)}\right)^2 \frac{1}{1+\mu(t)\frac{v^\Delta(t,\lambda^\star)}{v(t,\lambda^\star)}} \\
&= \lambda^\star - \lambda - (\lambda^\star + q(t)) - \left(\frac{v^\Delta(t,\lambda^\star)}{v(t,\lambda^\star)}\right)^2 \frac{1}{\frac{v^\sigma(t,\lambda^\star)}{v(t,\lambda^\star)}} \\
&= \lambda^\star - \lambda - (\lambda^\star + q(t)) - \frac{(v^\Delta(t,\lambda^\star))^2}{v(t,\lambda^\star)v^\sigma(t,\lambda^\star)} \\
&= \lambda^\star - \lambda + \frac{-(\lambda^\star + q(t))v(t,\lambda^\star)v^\sigma(t,\lambda^\star) - (v^\Delta(t,\lambda^\star))^2}{v(t,\lambda^\star)v^\sigma(t,\lambda^\star)} \\
&= \lambda^\star - \lambda + \frac{v^{\Delta^2}(t,\lambda^\star)v(t,\lambda^\star) - (v^\Delta(t,\lambda^\star))^2}{v(t,\lambda^\star)v^\sigma(t,\lambda^\star)} \\
&= (\lambda^\star - \lambda) + \left(\frac{v^\Delta(t,\lambda^\star)}{v(t,\lambda^\star)}\right)^\Delta .
\end{aligned}
$$

We integrate the last inequality from $\rho(a)$ to t_2 and we get

$$
\begin{aligned}
\frac{v^\Delta(t_2,\lambda)}{v(t_2,\lambda)} &= -\frac{\beta}{\alpha} + (\lambda^\star - \lambda)(t_2 - \rho(a)) + \frac{v^\Delta(t_2,\lambda^\star)}{v(t_2,\lambda^\star)} - \frac{v^\Delta(\rho(a),\lambda^\star)}{v(\rho(a),\lambda^\star)} \\
&= (\lambda^\star - \lambda)(t_2 - \rho(a)) + \frac{v^\Delta(t_2,\lambda^\star)}{v(t_2,\lambda^\star)} \\
&\to \infty, \quad \text{as} \quad \lambda \to -\infty.
\end{aligned}
$$

This is a contradiction. Consequently,

$$
\frac{v^\Delta(t,\lambda)}{v(t,\lambda)} \notin \left(\frac{v^\Delta(t,\lambda^\star)}{v(t,\lambda^\star)}, 0\right)
$$

for some $t \in I$ and for some $\lambda < \lambda^\star$. Since $\frac{v^\Delta(t,\lambda)}{v(t,\lambda)}$ is a strictly decreasing function in λ for any $t \in [\rho(a), b]$, we conclude that there is an $s \in I$ and $\lambda_1 < \lambda^\star$ such that $v^\Delta(s,\lambda_1) = 0$ and $v(s,\lambda_1) \neq 0$. Hence and by Theorem 4.10, it follows that $v(\cdot,\lambda_1)$ has no generalized zeros in (s,b). Assume that $v(\cdot,\lambda_1)$ has a generalized zero in $[\rho(a), s)$. Then there exists $s_1 \in [\rho(a), s)$ such that $v(s_1,\lambda_1)v^\Delta(s_1,\lambda_1) = 0$. By Theorem 4.9 we have that $\left(vv^\Delta\right)(\cdot,\lambda_1)$ is a strictly increasing function in $[\rho(a), b]$. Then, for $s > s_1$ we have $v(s, \lambda_1) v^{\Delta}(s, \lambda_1) > 0$, which is a contradiction. Therefore, $v(\cdot, \lambda_1)$ has no generalized zeros in $[\rho(a), s)$. Hence, $v(\cdot, \lambda_1)$ has no generalized zeros in $[\rho(a), b]$. Then $n(\lambda_1) = 0$. From here and from Theorem 4.12, we get that $n(\lambda) \to 0$, as $\lambda \to -\infty$.

Now, we take $\nu < \lambda_1$ small enough. Then, using (4.20), we get

$$v(b, \nu) v^{\Delta}(b, \nu) \geq -\alpha\beta + (\lambda^{\star} - \nu)\kappa(\nu)(b - \rho(a)).$$

Hence,

$$v(b, \nu) v^{\Delta}(b, \nu) \to \infty, \quad \text{as} \quad \nu \to -\infty.$$

Assume that $v(t, \nu) < 0$ for any $t \in [\rho(a), b]$. Then $v(\rho(a), \nu) = \beta < 0$. Since

$$v^{\Delta^2}(t, \nu) = -(\nu + q(t)) v^{\sigma}(t, \nu),$$

then $v^{\Delta^2}(t, \nu) \geq 0$ for any $t \in [\rho(a), b]$, i.e., v^{Δ} is an increasing function in $[\rho(a), b]$. Above we have proved that there is an $s \in [\rho(a), b]$ such that $v^{\Delta}(s, \nu) = 0$. Hence,

$$v^{\Delta}(\rho(a), \nu) = -\alpha < 0,$$

i.e., $\alpha > 0$. Because $\beta \neq \alpha\mu(\rho(a))$, we have that $\beta > \alpha\mu(\rho(a))$. This is a contradiction. Therefore, $v(t, \nu) > 0$ for any $t \in [\rho(a), b]$. Since

$$-v^{\Delta}(b, \nu) = \alpha + \int_{\rho(a)}^{b} (\nu + q(t)) v^{\sigma}(t, \nu) \Delta t,$$

we conclude that

$$v^{\Delta}(b, \nu) \to \infty, \quad \text{as} \quad \nu \to -\infty.$$

Because

$$v(b, \nu) = \beta + \int_{\rho(a)}^{b} v^{\Delta}(\tau, \nu) \Delta\tau,$$

we obtain

$$v(b, \nu) \to \infty, \quad \text{as} \quad \nu \to -\infty.$$

As we have proved (4.22), we have

$$\left(\frac{v^{\Delta}(t, \nu)}{v(t, \nu)}\right)^{\Delta} \geq (\lambda^{\star} - \nu) + \left(\frac{v^{\Delta}(t, \lambda^{\star})}{v(t, \lambda^{\star})}\right)^{\Delta}$$

for any $t \in [\rho(a), b]$. The last inequality we integrate from $\rho(a)$ to b and we get

$$\frac{v^{\Delta}(b, \nu)}{v(b, \nu)} \geq (\lambda^{\star} - \nu)(b - \rho(a)) + \frac{v^{\Delta}(b, \lambda^{\star})}{v(b, \lambda^{\star})}.$$

Hence,

$$\frac{v^{\Delta}(b,\nu)}{v(b,\nu)} \to \infty, \quad \text{as} \quad \nu \to -\infty.$$

1. Let $\delta > 0$. Then

$$\Lambda(\nu) = v(b,\nu)\left(\gamma + \delta \frac{v^{\Delta}(b,\nu)}{v(b,\nu)}\right)$$

$$\to \infty, \quad \text{as} \quad \nu \to -\infty.$$

2. Let $\delta = 0$. Then $\gamma = 1$ and

$$\Lambda(\nu) = v(b,\nu)$$

$$\to \infty, \quad \text{as} \quad \nu \to -\infty.$$

This completes the proof.

Note that the eigenvalues of problem (4.2) can be arranged as follows

$$-\infty < \lambda_1 < \lambda_2 < \ldots.$$

If (4.2) has k eigenvalues only, then we will write $\lambda_l = \infty$ for $l > k$.
For $x \in \mathbb{R}$ we will use the following notation

$$x^{\dagger} = \begin{cases} \frac{1}{x} & \text{if} \quad x \neq 0 \\ 0 & \text{if} \quad x = 0. \end{cases}$$

Theorem 4.14 (Picone's Identity). *Let $t \in [\rho(a), \rho(b)]$ and*

$$Ker u_1(t) \subset Ker u_2(t) \quad \text{and} \quad Ker u_1^{\sigma}(t) \subset Ker u_2^{\sigma}(t),$$

where u_1 solves (4.7) and $u_2 \in \mathbb{C}^2_{rd}$. Then at t we have

$$-\left(L(u_2) + \lambda u_2^{\sigma}\right) u_2^{\sigma} = \left(u_2 u_1^{\dagger} W(u_2,u_1)\right)^{\Delta} + \left(u_1 u_1^{\sigma}\right)^{\dagger} \left(W(u_2,u_1)\right)^2.$$

Proof. 1. Let $(u_1 u_1^{\sigma})(t) \neq 0$. We have

$$\begin{aligned} u_2^{\sigma}(t)L(u_1)(t) - u_1^{\sigma}(t)L(u_2)(t) &= u_2^{\sigma}(t)\left(u_1^{\Delta^2}(t) - (\lambda + q(t))\,u_1^{\sigma}(t)\right) \\ &\quad - u_1^{\sigma}(t)\left(u_2^{\Delta^2}(t) - (\lambda + q(t))\,u_2^{\sigma}(t)\right) \\ &= u_2^{\sigma}(t)u_1^{\Delta^2}(t) - u_1^{\sigma}(t)u_2^{\Delta^2}(t) \end{aligned}$$

$$= \left(u_2 u_1^\Delta - u_1 u_2^\Delta\right)^\Delta(t)$$

$$= (W(u_2,u_1))^\Delta(t).$$

Hence,

$$\begin{aligned}
\left(u_2 u_1^\dagger W(u_2,u_1)\right)^\Delta &= \left(\frac{u_2}{u_1} W(u_2,u_1)\right)^\Delta \\
&= \frac{u_2^\sigma}{u_1^\sigma} W^\Delta(u_2,u_1) + \left(\frac{u_2}{u_1}\right)^\Delta W(u_2,u_1) \\
&= \frac{u_2^\sigma}{u_1^\sigma}\left(u_2^\sigma L(u_1) - u_1^\sigma L(u_2)\right) \\
&\quad + \frac{u_2^\Delta u_1 - u_2 u_1^\Delta}{u_1 u_1^\sigma} W(u_2,u_1) \\
&= \frac{u_2^\sigma}{u_1^\sigma}\left(-\lambda u_1^\sigma u_2^\sigma - u_1^\sigma L(u_2)\right) \\
&\quad - \frac{(W(u_2,u_1))^2}{u_1 u_1^\sigma} \\
&= -u_2^\sigma\left(\lambda u_2^\sigma + L(u_2)\right) - \frac{(W(u_2,u_1))^2}{u_1 u_1^\sigma} \\
&= -u_2^\sigma\left(\lambda u_2^\sigma + L(u_2)\right) - \left(u_1 u_1^\sigma\right)^\dagger (W(u_2,u_1))^2,
\end{aligned}$$

whereupon, we get the desired result.

2. Let $(u_1 u_1^\sigma)(t) = 0$. Hence, $(u_2 u_2^\sigma)(t) = 0$.

   a. Let t be right-scattered. Then

$$\begin{aligned}
W(u_2,u_1) &= u_2 u_1^\Delta - u_2^\Delta u_1 \\
&= u_2 \frac{u_1^\sigma - u_1}{\mu} - u_1 \frac{u_2^\sigma - u_2}{\mu} \\
&= \frac{u_2 u_1^\sigma - u_1 u_2^\sigma}{\mu} \\
&= 0,
\end{aligned}$$

$$
\begin{aligned}
\left(u_2u_1^{\dagger}W(u_2,u_1)\right)^{\Delta}+(u_1u_1^{\sigma})^{\dagger}(W(u_2,u_1))^2 &= \left(u_2u_1^{\dagger}W(u_2,u_1)\right)^{\Delta}\\
&= \frac{\left(u_2u_1^{\dagger}W(u_2,u_1)\right)^{\sigma}-u_2u_1^{\dagger}W(u_2,u_1)}{\mu}\\
&= \frac{\left(u_2u_1^{\dagger}\right)^{\sigma}W^{\sigma}(u_2,u_1)}{\mu}\\
&= \left(u_2u_1^{\dagger}\right)^{\sigma}\frac{W^{\sigma}(u_2,u_1)-W(u_2,u_1)}{\mu}\\
&= \left(u_2u_1^{\dagger}\right)^{\sigma}W^{\Delta}(u_2,u_1)\\
&= \left(u_2u_1^{\dagger}\right)^{\sigma}\left(L(u_1)u_2^{\sigma}-L(u_2)u_1^{\sigma}\right)\\
&= \left(u_2u_1^{\dagger}\right)^{\sigma}\left(-\lambda u_1^{\sigma}u_2^{\sigma}-L(u_2)u_1^{\sigma}\right)\\
&= -u_2^{\sigma}\left(\lambda u_2^{\sigma}+L(u_2)\right)
\end{aligned}
$$

at t.

b. Let t be left-scattered or left-dense. If t is left-scattered, then $u_2u_1^{\dagger}W(u_2,u_1)$ is a continuous function at t and $u_2u_1^{\dagger}W(u_2,u_1)=0$ at t. If t is left-dense, by L'Hôpital's rule, we have

$$
\begin{aligned}
\lim_{s\to t^-}\left(u_2u_1^{\dagger}W(u_2,u_1)\right)(s) &= \frac{u_2^{\Delta}(t)}{u_1^{\Delta}(t)}W(u_2,u_1)(t)\\
&= 0\\
&= \left(u_2u_1^{\dagger}W(u_2,u_1)\right)(t),
\end{aligned}
$$

i.e., $u_2u_1^{\dagger}W(u_2,u_1)$ is continuous at t. Hence,

$$
\begin{aligned}
\mu\left(u_2u_1^{\dagger}W(u_2,u_1)\right)^{\Delta} &= \left(u_2u_1^{\dagger}W(u_2,u_1)\right)^{\sigma}-u_2u_1^{\dagger}W(u_2,u_1)\\
&= \left(u_2u_1^{\dagger}W(u_2,u_1)\right)^{\sigma}\\
&= u_2^{\sigma}\left(u_1^{\sigma}\right)^{\dagger}W^{\sigma}(u_2,u_1)
\end{aligned}
$$

$$= u_2^\sigma \left(u_1^\sigma\right)^\dagger \left(W^\sigma\left(u_2,u_1\right) - W\left(u_2,u_1\right)\right).$$

Therefore,

$$\begin{aligned}
\left(u_2u_1^\dagger W\left(u_2,u_1\right)\right)^\Delta &= u_2^\sigma \left(u_1^\sigma\right)^\dagger \left(W\left(u_2,u_1\right)\right)^\Delta \\
&= u_2^\sigma \left(u_1^\sigma\right)^\dagger \left(L\left(u_1\right)u_2^\sigma - L\left(u_2\right)u_1^\sigma\right) \\
&= u_2^\sigma \left(u_1^\sigma\right)^\dagger \left(-\lambda u_1^\sigma u_2^\sigma - L\left(u_2\right)u_1^\sigma\right) \\
&= -u_2^\sigma \left(u_1^\sigma\right)^\dagger u_1^\sigma \left(\lambda u_2^\sigma + L\left(u_2\right)\right) \\
&= -u_2^\sigma \left(\lambda u_2^\sigma + L\left(u_2\right)\right) \\
&= -u_2^\sigma \left(\lambda u_2^\sigma + L\left(u_2\right)\right) - \left(u_1u_1^\sigma\right)^\dagger \left(W\left(u_2,u_1\right)\right)^2
\end{aligned}$$

at t.

c. Let t be right-dense. Then

$$u_1(t) = u_1\left(\sigma(t)\right) = 0 \quad \text{and} \quad u_2(t) = u_2\left(\sigma(t)\right) = 0.$$

Hence, using L'Hôpital's rule, we get

$$\begin{aligned}
\left(u_2u_1^\dagger W\left(u_2,u_1\right)\right)^\Delta &= \lim_{s\to t} \frac{\left(u_2u_1^\dagger W\left(u_2,u_1\right)\right)(t) - \left(u_2u_1^\dagger W\left(u_2,u_1\right)\right)(s)}{t-s} \\
&= \lim_{s\to t} \frac{\left(u_2u_1^\dagger W\left(u_2,u_1\right)\right)(s)}{s-t} \\
&= \lim_{s\to t}\left(u_2u_1^\dagger\right)(s) \lim_{s\to t} \frac{W\left(u_2,u_1\right)(s)}{s-t} \\
&= \frac{u_2^\Delta(t)}{u_1^\Delta(t)} \lim_{s\to t} \frac{W\left(u_2,u_1\right)(s) - W\left(u_2,u_1\right)(t)}{s-t} \\
&= \frac{u_2^\Delta(t)}{u_1^\Delta(t)} W^\Delta\left(u_2,u_1\right)(t) \\
&= \frac{u_2^\Delta(t)}{u_1^\Delta(t)} \left(L\left(u_1\right)u_2^\sigma - L\left(u_2\right)u_1^\sigma\right)
\end{aligned}$$

$$= \frac{u_2^\Delta(t)}{u_1^\Delta(t)}\left(-\lambda u_1^\sigma u_2^\sigma - L(u_2)u_1^\sigma\right)$$

$$= 0$$

$$= -\left(L(u_2) + \lambda u_2^\sigma\right)u_2^\sigma - \left(u_1 u_1^\sigma\right)^\dagger \left(W(u_2,u_1)\right)^2.$$

This completes the proof.

Theorem 4.15 (Extended Piconi's Identity). *Let $k \in \mathbf{N}$. Suppose that u solves equation (4.7). Let u_ν be solutions of equation*

$$L(u_\nu) + \rho_\nu u_\nu^\sigma = 0 \quad \textit{with} \quad \rho_\nu \in \mathbb{R}, \quad 1 \le \nu \le k.$$

Next, let $y \in \mathbb{C}_{rd}^2$ satisfy $y^\sigma \perp u_\nu^\sigma$, $1 \le \nu \le k$, and suppose $u_\nu^\sigma \perp u_\mu^\sigma$, $1 \le \nu < \mu \le k$, $\|u_\nu^\sigma\| = 1$, $1 \le \nu \le k$. Put $\widetilde{u}_1 = \sum_{\nu=1}^{k} \beta_\nu u_\nu$ with $\beta_\nu \in \mathbb{R}$, $1 \le \nu \le k$, and $\widetilde{u} = y + \widetilde{u}_1$. Suppose that $Ker u(t) \subset Ker y(t)$ for all $t \in [\rho(a), b]$. Then

$$-\langle L(y) + \lambda y^\sigma, y^\sigma\rangle = \int_{\rho(a)}^{b} \left(\left(uu^\sigma\right)^\dagger \left(W(\widetilde{u},u)\right)^2\right)(t)\Delta t$$

$$+ \sum_{\nu=1}^{k} \beta_\nu^2 (\lambda - \rho_\nu) + T(b) - T(\rho(a)),$$

where

$$T = \widetilde{u}u^\dagger W(\widetilde{u},u) + W(\widetilde{u},y).$$

Proof. We apply Piconi's identity for $\widetilde{u}$ and u and we get

$$\left(\widetilde{u}u^\dagger W(\widetilde{u},u)\right)^\Delta = -\left(uu^\sigma\right)^\dagger \left(W(\widetilde{u},u)\right)^2 - \left(L(\widetilde{u}) + \lambda \widetilde{u}^\sigma\right)\widetilde{u}^\sigma.$$

The last equality we integrate from $\rho(a)$ to b and we get

$$\left(\widetilde{u}u^\dagger W(\widetilde{u},u)\right)(b) - \left(\widetilde{u}u^\dagger W(\widetilde{u},u)\right)(\rho(a))$$

$$= -\int_{\rho(a)}^{b} \left(\left(uu^\sigma\right)^\dagger \left(W(\widetilde{u},u)\right)^2\right)(t)\Delta t - \int_{\rho(a)}^{b} \left(\left(L(\widetilde{u}) + \lambda \widetilde{u}^\sigma\right)\widetilde{u}^\sigma\right)(t)\Delta t,$$

or

$$\left(\widetilde{u}u^\dagger W(\widetilde{u},u)\right)(\rho(a)) - \left(\widetilde{u}u^\dagger W(\widetilde{u},u)\right)(b) - \int_{\rho(a)}^{b} \left(\left(uu^\sigma\right)^\dagger \left(W(\widetilde{u},u)\right)^2\right)(t)\Delta t$$

$$= \langle L(\widetilde{u}) + \lambda \widetilde{u}^\sigma, \widetilde{u}^\sigma\rangle$$

$$= \left\langle L(y) + \sum_{\nu=1}^{k} \beta_\nu L(u_\nu) + \lambda y^\sigma + \lambda \widetilde{u_1}^\sigma, y^\sigma + \widetilde{u_1}^\sigma \right\rangle$$

$$= \langle L(y) + \lambda y^\sigma, y^\sigma \rangle + \langle L(y) + \lambda y^\sigma, \widetilde{u_1}^\sigma \rangle$$

$$+ \left\langle \sum_{\nu=1}^{k} \beta_\nu L(u_\nu) + \lambda \widetilde{u_1}^\sigma, y^\sigma + \widetilde{u_1}^\sigma \right\rangle$$

$$= \langle L(y) + \lambda y^\sigma, y^\sigma \rangle + \langle L(y), \widetilde{u_1}^\sigma \rangle + \lambda \langle y^\sigma, \widetilde{u_1}^\sigma \rangle$$

$$+ \left\langle \sum_{\nu=1}^{k} \beta_\nu L(u_\nu), y^\sigma \right\rangle + \left\langle \sum_{\nu=1}^{k} \beta_\nu L(u_\nu), \widetilde{u_1}^\sigma \right\rangle$$

$$+ \langle \lambda \widetilde{u_1}^\sigma, y^\sigma \rangle + \langle \lambda \widetilde{u_1}^\sigma, \widetilde{u_1}^\sigma \rangle$$

$$= \langle L(y) + \lambda y^\sigma, y^\sigma \rangle + \langle L(y), \widetilde{u_1}^\sigma \rangle + \lambda \sum_{\nu=1}^{k} \beta_\nu \langle y^\sigma, u_\nu^\sigma \rangle$$

$$- \sum_{\nu=1}^{k} \beta_\nu \rho_\nu \langle u_\nu^\sigma, y^\sigma \rangle - \sum_{\nu=1}^{k} \beta_\nu \rho_\nu \langle u_\nu^\sigma, \widetilde{u_1}^\sigma \rangle$$

$$+ \sum_{\nu=1}^{k} \lambda \beta_\nu \langle u_\nu^\sigma, y^\sigma \rangle + \lambda \langle \widetilde{u_1}^\sigma, \widetilde{u_1}^\sigma \rangle$$

$$= \langle L(y) + \lambda y^\sigma, y^\sigma \rangle + \langle L(y), \widetilde{u_1}^\sigma \rangle$$

$$- \sum_{\nu=1}^{k} \beta_\nu \rho_\nu \langle u_\nu^\sigma, \widetilde{u_1}^\sigma \rangle + \lambda \langle \widetilde{u_1}^\sigma, \widetilde{u_1}^\sigma \rangle$$

$$= \langle L(y) + \lambda y^\sigma, y^\sigma \rangle + \langle L(y), \widetilde{u_1}^\sigma \rangle$$

$$- \sum_{\nu=1}^{k} \beta_\nu \rho_\nu \left\langle u_\nu^\sigma, \sum_{l=1}^{k} \beta_l u_l^\sigma \right\rangle + \lambda \langle \widetilde{u_1}^\sigma, \widetilde{u_1}^\sigma \rangle$$

$$= \langle L(y) + \lambda y^\sigma, y^\sigma \rangle + \langle L(y), \widetilde{u_1}^\sigma \rangle - \sum_{\nu=1}^{k} \beta_\nu^2 \rho_\nu$$

$$+ \lambda \left\langle \sum_{\nu=1}^{k} \beta_\nu u_\nu^\sigma, \sum_{l=1}^{k} \beta_l u_l^\sigma \right\rangle$$

$$= \langle L(y) + \lambda y^\sigma, y^\sigma \rangle + \langle L(y), \widetilde{u_1}^\sigma \rangle - \sum_{\nu=1}^{k} \beta_\nu^2 \rho_\nu + \lambda \sum_{\nu=1}^{k} \beta_\nu^2$$

$$= \langle L(y) + \lambda y^{\sigma}, y^{\sigma} \rangle + \langle L(y), \widetilde{u_1}^{\sigma} \rangle + \sum_{\nu=1}^{k} \beta_{\nu}^{2} (\lambda - \rho_{\nu})$$

$$= \langle L(y) + \lambda y^{\sigma}, y^{\sigma} \rangle + \sum_{\nu=1}^{k} \beta_{\nu}^{2} (\lambda - \rho_{\nu})$$

$$+ \langle y^{\sigma}, L(\widetilde{u_1}) \rangle + W(\widetilde{u_1}, y)(b) - W(\widetilde{u_1}, y)(\rho(a))$$

$$= \langle L(y) + \lambda y^{\sigma}, y^{\sigma} \rangle + \sum_{\nu=1}^{k} \beta_{\nu}^{2} (\lambda - \rho_{\nu})$$

$$+ \left\langle y^{\sigma}, \sum_{\nu=1}^{k} \beta_{\nu} L(u_{\nu}) \right\rangle + W(\widetilde{u_1}, y)(b) - W(\widetilde{u_1}, y)(\rho(a))$$

$$= \langle L(y) + \lambda y^{\sigma}, y^{\sigma} \rangle + \sum_{\nu=1}^{k} \beta_{\nu}^{2} (\lambda - \rho_{\nu}) - \sum_{\nu=1}^{k} \beta_{\nu} \rho_{\nu} \langle y^{\sigma}, u_{\nu}^{\sigma} \rangle$$

$$+ W(\widetilde{u_1}, y)(b) - W(\widetilde{u_1}, y)(\rho(a))$$

$$= \langle L(y) + \lambda y^{\sigma}, y^{\sigma} \rangle + \sum_{\nu=1}^{k} \beta_{\nu}^{2} (\lambda - \rho_{\nu})$$

$$+ W(\widetilde{u_1}, y)(b) - W(\widetilde{u_1}, y)(\rho(a)),$$

whereupon, we get the desired result. This completes the proof.

Let $\lambda_1^{\star} < \lambda_2^{\star} < \ldots$ be the eigenvalues of problem

$$L(u) + \lambda u^{\sigma} = 0, \quad R_a(u) = u(b) = 0, \tag{4.23}$$

with corresponding normalized eigenfunctions $x_{\nu}^{\star}$ with $\|x_{\nu}^{\star}\| = 1$.

Theorem 4.16. *If $\lambda \in \left(\lambda_k^{\star}, \lambda_{k+1}^{\star}\right]$, then $n(\lambda) = k$. Moreover, define a $k \times k$-matrix $\mathscr{Z}^{\star} = \left(z_{\mu\nu}^{\star}\right)$ as follows*

$$z_{\mu\nu}^{\star} = \begin{cases} x_{\nu}^{\star}(t_{\mu}) & \text{if } t_{\mu} \text{ is a zero of } v(\cdot, \lambda) \\ W(v, x_{\nu}^{\star})(t_{\mu}) & \text{if } t_{\mu} \text{ is a node of } v(\cdot, \lambda), \end{cases}$$

where $t_1 < t_2 < \ldots < t_k$ are the generalized zeros of $v(\cdot, \lambda)$. Then $\mathscr{Z}^{\star}$ is invertible.

Proof. We take $y = 0$, $\rho_{\nu} = \lambda_{\nu}^{\star}$, $u_{\nu} = x_{\nu}^{\star}$, $\nu \in \{1, \ldots, k\}$, $u = v(\cdot, \lambda)$ in Theorem 4.15. Let $\widetilde{u} = \sum_{\nu=1}^{k} c_{\nu} x_{\nu}^{\star}$, $c_{\nu} \in \mathbb{R}$, $\nu \in \{1, \ldots, k\}$. Note that

$$T(\rho(a)) = \left(\widetilde{u}v^{\dagger}W(\widetilde{u},v)\right)(\rho(a)) + W(\widetilde{u},0)(\rho(a))$$

$$= \left(\widetilde{u}v^{\dagger}W(\widetilde{u},v)\right)(\rho(a)).$$

If $v(\rho(a),\lambda) = 0$, then $T(\rho(a)) = 0$. If $v(\rho(a),\lambda) \neq 0$, then

$$T(\rho(a)) = \frac{\widetilde{u}(\rho(a))}{v(\rho(a),\lambda)} \begin{vmatrix} \widetilde{u}(\rho(a)) & v(\rho(a),\lambda) \\ \widetilde{u}^{\Delta}(\rho(a)) & v^{\Delta}(\rho(a),\lambda) \end{vmatrix}$$

$$= \frac{\widetilde{u}(\rho(a))}{v(\rho(a),\lambda)} \begin{vmatrix} \sum_{\nu=1}^{k} c_{\nu} x_{\nu}^{\star}(\rho(a)) & \beta \\ \sum_{\nu=1}^{k} c_{\nu} x_{\nu}^{\star\Delta}(\rho(a)) & -\alpha \end{vmatrix}$$

$$= \begin{cases} \dfrac{\widetilde{u}(\rho(a))}{v(\rho(a),\lambda)} \begin{vmatrix} -\dfrac{\beta}{\alpha} \sum_{\nu=1}^{k} c_{\nu} x_{\nu}^{\star\Delta}(\rho(a)) & \beta \\ \sum_{\nu=1}^{k} c_{\nu} x_{\nu}^{\star\Delta}(\rho(a)) & -\alpha \end{vmatrix} & \text{if} \quad \alpha \neq 0 \\[2ex] \dfrac{\widetilde{u}(\rho(a))}{v(\rho(a),\lambda)} \begin{vmatrix} \sum_{\nu=1}^{k} c_{\nu} x_{\nu}^{\star}(\rho(a)) & \beta \\ -\dfrac{\alpha}{\beta} \sum_{\nu=1}^{k} c_{\nu} x_{\nu}^{\star}(\rho(a)) & -\alpha \end{vmatrix} & \text{if} \quad \beta \neq 0 \end{cases}$$

$$= 0.$$

Also,

$$T(b) = \left(\widetilde{u}v^{\dagger}W(\widetilde{u},v)\right)(b) + W(\widetilde{u},0)(b)$$

$$= \left(\widetilde{u}v^{\dagger}W(\widetilde{u},v)\right)(b).$$

If $v(b,\lambda) = 0$, then $T(b) = 0$. If $v(b,\lambda) \neq 0$, then, using that $x_{\nu}^{\star}(b) = 0$,

$$T(b) = \frac{\sum_{\nu=1}^{k} c_{\nu} u_{\nu}^{\star}(b)}{v(b,\lambda)} \begin{vmatrix} \sum_{\nu=1}^{k} c_{\nu} x_{\nu}^{\star}(b) & v(b,\lambda) \\ \sum_{\nu=1}^{k} c_{\nu} x_{\nu}^{\star\Delta}(b) & v^{\Delta}(b,\lambda) \end{vmatrix}$$

$$= 0.$$

Suppose that v has no generalized zeros in (a,b). We apply Theorem 4.15 for v and $\widetilde{u}$ with $c_{\nu} = 1$, $\nu \in \{1,\ldots,k\}$, and we get

$$0 = \int_{\rho(a)}^{b} \left((vv^{\sigma})^{\dagger} (W(\widetilde{u}, v))^2 \right)(t)\Delta t + \sum_{\nu=1}^{k} c_{\nu}^2 (\lambda - \lambda_{\nu}^{\star}),$$

which is impossible. Therefore, v has at least one generalized zero in (a,b), say that

$$t_1 < t_2 < \ldots < t_p$$

are all of them. Define a $p \times k$-matrix $\mathscr{Z}^{\star}$ with $z_{\mu\nu}^{\star}$as above. We take $\begin{pmatrix} \beta_1 \\ \vdots \\ \beta_k \end{pmatrix} \in$ $\operatorname{Ker} \mathscr{Z}^{\star}$. Hence, using that $\mathscr{Z}^{\star} \begin{pmatrix} \beta_1 \\ \vdots \\ \beta_k \end{pmatrix} = 0$ and the definition of $\mathscr{Z}^{\star}$, we get

$$\begin{aligned} \widetilde{\widetilde{u}}(t_{\mu}) &= \sum_{\nu=1}^{k} \beta_{\nu} x_{\nu}^{\star}(t_{\mu}) \\ &= 0 \end{aligned}$$

for all zeros t_{μ} of $v(\cdot, \lambda)$ and $W\left(\widetilde{\widetilde{u}}, v\right)(t_{\mu}) = 0$ for all nodes t_{μ} of v. Then, applying Theorem 4.15 for v and $\widetilde{\widetilde{u}}$ with $\rho_{\nu} = \beta_{\nu}$, $\nu \in \{1, \ldots, k\}$, we get

$$0 = \int_{\rho(a)}^{b} \left((vv^{\sigma})^{\dagger} \left(W\left(\widetilde{\widetilde{u}}, v\right)\right)^2 \right)(t)\Delta t + \sum_{\nu=1}^{k} \beta_{\nu}^2 (\lambda - \lambda_{\nu}^{\star}),$$

whereupon,

$$\begin{aligned} 0 &\leq \sum_{\nu=1}^{k} \beta_{\nu}^2 (\lambda - \lambda_{\nu}^{\star}) \\ &= -\int_{\rho(a)}^{b} \left((vv^{\sigma})^{\dagger} \left(W\left(\widetilde{\widetilde{u}}, v\right)\right)^2 \right)(t)\Delta t \\ &\leq 0, \end{aligned}$$

which is possible if $\beta_1, \ldots, \beta_k = 0$. Therefore, $\operatorname{Ker} \mathscr{Z}^{\star} = \{0\}$. Thus $p \geq k$. Assume that $p \geq k+1$. Because k was arbitrarily chosen, we obtain that $n(\lambda) \geq k+1$ for any k. Hence, $\lim\limits_{\lambda \to -\infty} n(\lambda) \geq 1$. This contradicts with Theorem 4.13. Consequently, $p < k+1$. From here, $p = k$ and $\mathscr{Z}^{\star}$ is invertible. This completes the proof.

Theorem 4.17. *Let $\lambda_k^{\star} < \infty$. Then*

$$\lambda_{k+1}^{\star} \leq \inf_{\substack{y(\rho(a)) = y(b) = 0 \\ y^{\sigma} \perp x_{\nu}^{\star\sigma}, 1 \leq \nu \leq k}} R(y) \quad \textit{and} \quad \lambda_{k+1}^{\star} \leq \inf_{\substack{R_a(y) = R_b(y) = 0 \\ y^{\sigma} \perp x_{\nu}^{\star\sigma}, 1 \leq \nu \leq k}} R(y),$$

where $R(\cdot)$ is the Rayleigh quotient

$$R(y) = -\frac{\langle L(y), y^\sigma \rangle}{\|y^\sigma\|^2}, \quad y \neq 0.$$

Proof. Let $y \in \mathcal{C}^2_{rd}$ be such that

$$y(\rho(a))R_a(y) = y(b)R_b(y) = 0, \quad y \neq 0,$$

$y_\nu = x^\star_\nu$, $\rho_\nu = \lambda^\star_\nu$, $\lambda \in \left(\lambda^\star_k, \lambda^\star_{k+1}\right]$. Then, as in the proof of Theorem 4.16, we have $T(\rho(a)) = T(b) = 0$. Also, again by the proof of Theorem 4.16, we have that there are $\beta_1, \ldots, \beta_k$ such that $\widetilde{\widetilde{u}}(t_\mu) = 0$ for all zeros t_μ of v and $W\left(\widetilde{\widetilde{u}}, v\right)(t_\mu) = 0$ for all nodes t_μ of v. Hence and by Theorem 4.15, we have

$$\begin{aligned} -\langle L(y) + \lambda y^\sigma, y^\sigma \rangle &= \int_{\rho(a)}^b \left((uu^\sigma)^\dagger \left(W\left(\widetilde{\widetilde{u}}, u\right)\right)^2\right)(t)\Delta t + \sum_{\nu=1}^k \beta_\nu^2 (\lambda - \lambda^\star_\nu) \\ &\geq 0. \end{aligned}$$

Therefore,

$$-\langle L(y), y^\sigma \rangle \geq \lambda \langle y^\sigma, y^\sigma \rangle.$$

Because $\lambda \in \left(\lambda^\star_k, \lambda^\star_{k+1}\right]$ was arbitrarily chosen, we get the desired inequalities. This completes the proof.

Theorem 4.18 (Oscillation Theorem). *The eigenvalues of* (4.2) *can be arranged as*

$$-\infty < \lambda_1 < \lambda_2 < \ldots,$$

and an eigenfunction corresponding to λ_{k+1} has exactly k generalized zeros in the open interval (a, b).

Proof. Let $y = 0$, $u_\nu = x^\star_\nu$, $\rho_\nu = \lambda_\nu$, $k \in \mathbb{N}_0$, $\lambda = \lambda_{k+1}$ if $\lambda_{k+1} < \infty$. Then, using the proof of Theorem 4.16, we have $T(\rho(a)) = T(b) = 0$, $n(\lambda) = k$ and $\mathscr{L}^\star$ is invertible. This completes the proof.

Theorem 4.19 (Comparison Theorem). *If $k \in \mathbf{N}_0$ with $\lambda^\star_k < \infty$, then*

$$\lambda_k \leq \lambda^\star_k < \lambda_{k+1}.$$

Proof. This result follows from Theorem 4.18 and Theorem 4.16. This completes the proof.

Theorem 4.20 (Rayleigh's Principle). *We have*

$$\lambda^\star_{k+1} = \min_{\substack{R_a(y) = R_b(y) = 0 \\ \langle y^\sigma, x^\sigma_\nu \rangle = 0, 1 \leq \nu \leq k}} R(y),$$

where x_ν are the normalized eigenfunctions corresponding to λ_ν, $1 \le \nu \le k$.

Proof. Let $y(\rho(a))R_a(y) = y(b)R_b(y) = 0$, $y \neq 0$, $\rho_\nu = \lambda_\nu$, $k \in \mathbb{N}_0$, $\lambda = \lambda_{k+1}$ if $\lambda_{k+1} < \infty$. Then, as in the proof of Theorem 4.16, we have that $T(\rho(a)) = T(b) = 0$ and

$$\lambda^\star_{k+1} \le \inf_{\substack{y(\rho(a))R_a(y) = y(b)R_b(y) = 0 \\ \langle y^\sigma, x^\sigma_\nu\rangle = 0,\, 1 \le \nu \le k}} R(y).$$

Note that

$$\begin{aligned} R(x_{k+1}) &= -\frac{\langle L(x_{k+1}), x^\sigma_{k+1}\rangle}{\|x^\sigma_{k+1}\|^2} \\ &= \lambda_{k+1} \\ &\le \lambda^*_{k+1}. \end{aligned}$$

This completes the proof.

4.2 Second Eigenvalue Problem

In this section we consider problem

$$-\left(pu^\Delta\right)^\nabla(t) + q(t)u(t) = \lambda u(t), \quad t \in (a,b], \tag{4.24}$$

$$u(a) - hp(a)u^\Delta(a) = 0, \quad u(b) + Hp(b)u^\Delta(b) = 0, \tag{4.25}$$

where p is continuous on $[a,b]$ and continuously ∇-differentiable on $(a,b]$, $p(t) > 0$ for $t \in [a,b]$, q is piecewise continuous on $[a,b]$, $q(t) \ge 0$ for $t \in [a,b]$, h and H are non-negative real numbers. By $\mathscr{H}$ we denote the Hilbert space of all real ∇-measurable functions $u : (a,b] \to \mathbb{R}$ such that $u(b) = 0$ in the case b is left-scattered and $H = 0$, and

$$\int_a^b (u(t))^2 \nabla t < \infty,$$

with the inner product

$$\langle u, v\rangle = \int_a^b u(t)v(t)\nabla t,$$

and the norm

$$\|u\| = \sqrt{\langle u, u\rangle}$$

$$= \left(\int_a^b (u(t))^2 \nabla t \right)^{\frac{1}{2}}.$$

By $\mathscr{D}$ we denote the set of all functions $u \in \mathscr{H}$ with the following three properties.

1. u is continuous on $[a, \sigma(b)]$.
2. u^Δ is defined on $[a,b]$ and the conditions (4.25) hold.
3. pu^Δ is ∇-differentiable on $(a,b]$ and $\left(pu^\Delta\right)^\nabla(t) \in \mathscr{H}$.

Note that $\mathscr{D}$ is dense in $\mathscr{H}$. We define the operator $A : \mathscr{D} \to \mathscr{H}$ as follows

$$(Au)(t) = -\left(pu^\Delta\right)^\nabla(t) + q(t)u(t), \quad t \in (a,b], \quad u \in \mathscr{D}.$$

Definition 4.4. A complex number λ is called an eigenvalue of problem (4.24), (4.25) if there exists $u \in \mathscr{D}$ that satisfies equation (4.24). The function u is called an eigenfunction of problem (4.24), (4.25), corresponding to the eigenvalue λ.

Note that the eigenvalue problem (4.24), (4.25) is equivalent to problem

$$Au = \lambda u, \quad u \in \mathscr{D}, \quad u \neq 0. \tag{4.26}$$

Theorem 4.21. *Let p be continuous on $[a,b]$ and continuously ∇-differentiable on $(a,b]$, $p(t) > 0$ for $t \in [a,b]$, q is piecewise continuous on $[a,b]$, $q(t) \geq 0$ for $t \in [a,b]$, h and H are non-negative real numbers. Then for any $u, v \in \mathscr{D}$ we have*

$$\langle Au, v \rangle = \langle u, Av \rangle \tag{4.27}$$

and

$$\begin{aligned} \langle Au, u \rangle &= h\left(p(a)u^\Delta(a)\right)^2 + H\left(p(b)u^\Delta(b)\right)^2 \\ &\quad + \int_a^b p(t)\left(u^\Delta(t)\right)^2 \Delta t + \int_a^b q(t)(u(t))^2 \nabla t. \end{aligned} \tag{4.28}$$

Proof. Let $u, v \in \mathscr{D}$. Then

$$\begin{aligned} \langle Au, v \rangle &= \int_a^b (Au)(t)v(t)\nabla t \\ &= -\int_a^b \left(pu^\Delta\right)^\nabla(t)v(t)\nabla t + \int_a^b q(t)u(t)v(t)\nabla t \\ &= -p(t)u^\Delta(t)v(t)\Big|_{t=a}^{t=b} + \int_a^b p(t)u^\Delta(t)v^\Delta(t)\Delta t \\ &\quad + \int_a^b q(t)u(t)v(t)\nabla t \end{aligned}$$

$$
\begin{aligned}
&= -p(t)u^{\Delta}(t)v(t)\Big|_{t=a}^{t=b} + p(t)u(t)v^{\Delta}(t)\Big|_{t=a}^{t=b} \\
&\quad - \int_a^b \left(pv^{\Delta}\right)^{\nabla}(t)u(t)\nabla t + \int_a^b q(t)u(t)v(t)\nabla t \\
&= -p(b)u^{\Delta}(b)v(b) + p(a)u^{\Delta}(a)v(a) + p(b)u(b)v^{\Delta}(b) \\
&\quad - p(a)u(a)v^{\Delta}(a) + \int_a^b \left(-\left(pv^{\Delta}\right)^{\nabla}(t) + q(t)v(t)\right)u(t)\nabla t \\
&= -p(b)u^{\Delta}(b)v(b) + p(a)u^{\Delta}(a)v(a) - H\left(p(b)\right)^2 u^{\Delta}(b)v^{\Delta}(b) \\
&\quad - hp(a)u^{\Delta}(a)v^{\Delta}(a) + \int_a^b (Av)(t)u(t)\nabla t \\
&= -p(b)u^{\Delta}(b)\left(v(b) + Hp(b)v^{\Delta}(b)\right) \\
&\quad + p(a)u^{\Delta}(a)\left(v(a) - hv^{\Delta}(a)\right) + \langle u, Av\rangle \\
&= \langle u, Av\rangle .
\end{aligned}
$$

In particular, we have

$$
\begin{aligned}
\langle Au, u\rangle &= -p(t)u^{\Delta}(t)u(t)\Big|_{t=a}^{t=b} + \int_a^b p(t)\left(u^{\Delta}(t)\right)^2 \Delta t \\
&\quad + \int_a^b q(t)\left(u(t)\right)^2 \nabla t \\
&= h\left(p(a)\right)^2\left(u^{\Delta}(a)\right)^2 + H\left(p(b)\right)^2 u^{\Delta}(b) \\
&\quad + \int_a^b p(t)\left(u^{\Delta}(t)\right)^2 \Delta t + \int_a^b q(t)\left(u(t)\right)^2 \nabla t.
\end{aligned}
$$

This completes the proof.

By (4.27) and (4.28), it follows that A is symmetric and

$$\langle Au, u\rangle > 0$$

for all $u \in \mathscr{D}$, $u \neq 0$. Therefore, all eigenvalues of the operator A are real and positive. If λ_1 and λ_2 are two eigenvalues of A such that $\lambda_1 \neq \lambda_2$, and u_1, u_2 are eigenfunctions corresponding to λ_1 and λ_2, respectively, then $u_1 \perp u_2$. Really, we have

$$\begin{aligned}\langle Au_1,u_2\rangle &= \lambda_1 \langle u_1,u_2\rangle\\ &= \langle u_1,Au_2\rangle\\ &= \lambda_2 \langle u_1,u_2\rangle,\end{aligned}$$

whereupon,

$$(\lambda_1-\lambda_2)\langle u_1,u_2\rangle=0 \quad \text{and} \quad \langle u_1,u_2\rangle=0.$$

Since equation (4.24) cannot have two linearly independent solutions satisfying the conditions (4.25), then all eigenvalues of (4.24), (4.25) are simple, i.e., to each eigenvalue there corresponds a single eigenfunction up to a constant factor. Let

$$\operatorname{Ker}A=\{u\in\mathscr{D}:Au=0\}.$$

Suppose that $u\in\operatorname{Ker}A$. Using (4.28), we get that $u^{\Delta}(t)=0$ for $t\in[a,b]$. Hence, u is a constant on $[a,b]$. By the first condition of (4.25), we obtain that $u(t)=0$ for $t\in[a,b]$. Therefore, $\operatorname{Ker}A=\{0\}$. Hence, there exists A^{-1}. To present A^{-1}, we introduce the Green function

$$G(t,s)=-\frac{1}{w}\begin{cases} y(t)z(s) & \text{if} \quad t\le s\\ y(s)z(t) & \text{if} \quad s\le t,\end{cases}$$

where y and z are solutions of equation

$$-\left(pu^{\Delta}\right)^{\nabla}(t)+q(t)u(t)=0, \quad t\in(a,b],$$

satisfying the conditions

$$y(a)=h, \quad p(a)y^{\Delta}(a)=1, \quad z(b)=H, \quad p(b)z^{\Delta}(b)=-1,$$

and w is the Wronskian of the solutions y and z, i.e.,

$$\begin{aligned} w &= p(a)\begin{vmatrix} y(a) & z(a)\\ y^{\Delta}(a) & z^{\Delta}(z)\end{vmatrix}\\ &= \begin{vmatrix} h & z(a)\\ 1 & p(a)z^{\Delta}(a)\end{vmatrix}\\ &= -z(a)+hp(a)z^{\Delta}(a)\\ &= p(b)\begin{vmatrix} y(b) & z(b)\\ y^{\Delta}(b) & z^{\Delta}(b)\end{vmatrix}\end{aligned}$$

$$= \begin{vmatrix} y(b) & H \\ p(b)y^{\Delta}(b) & -1 \end{vmatrix}$$

$$= -y(b) - Hp(b)y^{\Delta}(b)$$

$$= p(t) \begin{vmatrix} y(t) & z(t) \\ y^{\Delta}(t) & z^{\Delta}(t) \end{vmatrix}$$

$$= p(t)y(t)z^{\Delta}(t) - p(t)y^{\Delta}(t)z(t), \quad t \in [a,b].$$

If we assume that $w = 0$, then $y \in \mathscr{D}$ and $Ay = 0$, i.e., $y \in \mathrm{Ker}A$. This is a contradiction because $\mathrm{Ker}A = \{0\}$ and $p(a)y^{\Delta}(a) = 1$. Thus,

$$\left(A^{-1}f\right)(t) = \int_a^b G(t,s)f(s)\nabla s, \quad t \in [a,b],$$

for any $f \in \mathscr{H}$. From here and from the definition of the Green function G, we conclude that A^{-1} is a completely continuous self-adjoint operator in $\mathscr{H}$. Let $B = A^{-1}$ and $\mu = \dfrac{1}{\lambda}$. Then problem (4.26) is equivalent to problem

$$Bg = \mu g, \quad g \in \mathscr{H}, \quad g \neq 0.$$

In other words, if λ is an eigenvalue and $u \in \mathscr{D}$ is a corresponding eigenfunction for A, then $\mu = \dfrac{1}{\lambda}$ is an eigenvalue for B with the same corresponding eigenfunction u. Conversely, if $\mu \neq 0$ is an eigenvalue and $u \in \mathscr{H}$ is a corresponding eigenfunction for B, then $u \in \mathscr{D}$ and $\lambda = \dfrac{1}{\mu}$ is an eigenvalue for A with the same eigenfunction u. If $\mu = 0$, then $Bu = 0$ and hence, $u = ABu = 0$. Therefore, $\mu = 0$ cannot be an eigenvalue for B. Applying the Hilbert-Schmidt Theorem to the operator B, we conclude that there exists an orthonormal system $\{\phi_k\}_{k=1}^{N}$ of eigenfunctions corresponding to eigenvalues $\{\lambda_k\}_{k=1}^{N}$, where $N = \dim \mathscr{H}$, and any function $f \in \mathscr{H}$ can be expanded in eigenfunctions ϕ_k in the form

$$f(t) = \sum_{k=1}^{N} c_k \phi_k(t), \quad t \in [a,b],$$

where c_k are the Fourier coefficients of f defined by

$$c_k = \int_a^b f(t)\phi_k(t)\nabla t.$$

In the case $N = \infty$, we have

$$\lim_{n\to\infty} \int_a^b \left(f(t) - \sum_{k=1}^{n} c_k \phi_k(t)\right)^2 \nabla t = 0. \tag{4.29}$$

Note that

$$\int_a^b \left(f(t)-\sum_{k=1}^n c_k\phi_k(t)\right)^2 \nabla t = \int_a^b \left((f(t))^2 - 2\sum_{k=1}^n c_k\phi_k(t)f(t)\right.$$
$$\left. + \sum_{k,m=1}^n c_k c_m \phi_k(t)\phi_m(t)\right)\nabla t$$
$$= \int_a^b (f(t))^2\nabla t - 2\sum_{k=1}^n c_k \int_a^b f(t)\phi_k(t)\nabla t + \sum_{k,m=1}^n c_k c_m \int_a^b \phi_k(t)\phi_m(t)\nabla t$$
$$= \int_a^b (f(t))^2\nabla t - 2\sum_{k=1}^n c_k^2 + \sum_{k=1}^n c_k^2$$
$$= \int_a^b (f(t))^2\nabla t - \sum_{k=1}^n c_k^2.$$

Hence and by (4.29), we derive the Parseval equality

$$\int_a^b (f(t))^2\nabla t = \sum_{k=1}^N c_k^2.$$

Theorem 4.22. *Suppose* $dim\mathscr{H} = \infty$. *Let* $f : [a,b] \to \mathbb{R}$ *be a function such that it has* Δ*-derivative* f^Δ *everywhere on* $[a,b]$*, except at a finite number of points* t_1, t_2, $\ldots$, t_m *belonging to* (a,b)*, the* Δ*-derivative being continuous everywhere except at these points at which* f^Δ *has finite limits from the left and right. Besides assume that* f *satisfies the boundary conditions*

$$f(a) - hp(a)f^\Delta(a) = 0, \quad f(b) + Hp(b)f^\Delta(b) = 0. \tag{4.30}$$

Then the series

$$\sum_{k=1}^n c_k\phi_k(t), \quad t \in [a,b],$$

where

$$c_k = \int_a^b f(t)\phi_k(t)\nabla t, \tag{4.31}$$

converges uniformly on $[a,b]$ *to the function* f.

Proof. First for the simplicity, we assume that the function f is Δ-differentiable on $[a,b]$ and f^Δ is continuous on $[a,b]$. Let $n \in \mathbb{N}$ be arbitrarily chosen. Consider the functional

$$J(u) = h\left(p(a)u^\Delta(a)\right)^2 + H\left(p(b)u^\Delta(b)\right)^2$$

$$+\int_a^b p(t)\left(u^{\Delta}(t)\right)^2\Delta t+\int_a^b q(t)\left(u(t)\right)^2\nabla t,\quad u\in\mathscr{H}.$$

We have that $J(u)\geq 0$ for any $u\in\mathscr{H}$. We substitute in the functional $J(u)$ the function

$$u(t)=f(t)-\sum_{k=1}^{n}c_k\phi_k(t),\quad t\in[a,b],$$

where c_k are defined by (4.31). Then

$$\begin{aligned}
J\left(f(t)-\sum_{k=1}^{n}c_k\phi_k(t)\right) &= h\left(p(a)f^{\Delta}(a)-p(a)\sum_{k=1}^{n}c_k\phi_k^{\Delta}(a)\right)^2\\
&+H\left(p(b)f^{\Delta}(b)-p(b)\sum_{k=1}^{n}c_k\phi_k^{\Delta}(b)\right)^2\\
&+\int_a^b p(t)\left(f^{\Delta}(t)-\sum_{k=1}^{n}c_k\phi_k^{\Delta}(t)\right)^2\Delta t\\
&+\int_a^b q(t)\left(f(t)-\sum_{k=1}^{n}c_k\phi_k(t)\right)^2\nabla t\\
&= h\left(p(a)f^{\Delta}(a)\right)^2+H\left(p(b)f^{\Delta}(b)\right)^2\\
&-2\sum_{k=1}^{n}c_k\left(h\left(p(a)\right)^2f^{\Delta}(a)\phi_k^{\Delta}(a)+H\left(p(b)\right)^2f^{\Delta}(b)\phi_k^{\Delta}(b)\right)\\
&+\sum_{k,l=1}^{n}c_kc_l\left(h\left(p(a)\right)^2\phi_k^{\Delta}(a)\phi_l^{\Delta}(a)+H\left(p(b)\right)^2\phi_k^{\Delta}(b)\phi_l^{\Delta}(b)\right)\\
&+\int_a^b p(t)\left(f^{\Delta}(t)\right)^2\Delta t+\int_a^b q(t)\left(f(t)\right)^2\nabla t\\
&-2\sum_{k=1}^{n}c_k\left(\int_a^b p(t)f^{\Delta}(t)\phi_k^{\Delta}(t)\Delta t+\int_a^b q(t)f(t)\phi_k(t)\nabla t\right)\\
&+\sum_{k,l=1}^{n}c_kc_l\left(\int_a^b p(t)\phi_k^{\Delta}(t)\phi_l^{\Delta}(t)\Delta t+\int_a^b q(t)\phi_k(t)\phi_l(t)\nabla t\right).
\end{aligned}\tag{4.32}$$

Now, applying integration by parts and using the boundary conditions (4.30), we get

$$
\begin{aligned}
&\int_a^b p(t)f^{\Delta}(t)\phi_k^{\Delta}(t)\Delta t + \int_a^b q(t)f(t)\phi_k(t)\nabla t \\
&= p(t)f(t)\phi_k^{\Delta}(t)\Big|_{t=a}^{t=b} - \int_a^b f(t)\left(p\phi_k^{\Delta}\right)^{\nabla}(t)\nabla t \\
&\quad + \int_a^b q(t)f(t)\phi_k(t)\nabla t \\
&= p(b)f(b)\phi_k^{\Delta}(b) - p(a)f(a)\phi_k^{\Delta}(a) + \lambda_k \int_a^b f(t)\phi_k(t)\nabla t \\
&= -H(p(b))^2 f^{\Delta}(b)\phi_k^{\Delta}(b) - h(p(a))^2 f^{\Delta}(a)\phi_k^{\Delta}(a) + c_k\lambda_k,
\end{aligned}
$$

$$
\begin{aligned}
&\int_a^b p(t)\phi_k^{\Delta}(t)\phi_l^{\Delta}(t)\Delta t + \int_a^b q(t)\phi_k(t)\phi_l(t)\nabla t \\
&= p(t)\phi_k(t)\phi_l^{\Delta}(t)\Big|_{t=a}^{t=b} - \int_a^b \phi_k(t)\left(p\phi_l^{\Delta}\right)^{2}(t)\nabla t \\
&\quad + \int_a^b q(t)\phi_k(t)\phi_l(t)\nabla t \\
&= p(b)\phi_k(b)\phi_l^{\Delta}(b) - p(a)\phi_k(a)\phi_l^{\Delta}(a) + \lambda_l \int_a^b \phi_k(t)\phi_l(t)\nabla t \\
&= -h(p(a))^2 \phi_k^{\Delta}(a)\phi_l^{\Delta}(a) - H(p(b))^2 \phi_k^{\Delta}(b)\phi_l^{\Delta}(b) + \lambda_l\delta_{kl},
\end{aligned}
$$

where δ_{kl} is the Kronecker symbol. Hence and by (4.32), we get

$$
\begin{aligned}
&J\left(f(t) - \sum_{k=1}^{n} c_k\phi_k(t)\right) \\
&= h\left(p(a)f^{\Delta}(a)\right)^2 + H\left(p(b)f^{\Delta}(b)\right)^2 \\
&\quad -2\sum_{k=1}^{n} c_k\left(h(p(a))^2 f^{\Delta}(a)\phi_k^{\Delta}(a) + H(p(b))^2 f^{\Delta}(b)\phi_k^{\Delta}(b)\right) \\
&\quad + \sum_{k,l=1}^{n} c_k c_l\left(h(p(a))^2 \phi_k^{\Delta}(a)\phi_l^{\Delta}(a) + H(p(b))^2 \phi_k^{\Delta}(b)\phi_l^{\Delta}(b)\right) \\
&\quad + \int_a^b p(t)\left(f^{\Delta}(t)\right)^2 \Delta t + \int_a^b q(t)(f(t))^2 \nabla t \\
&\quad +2\sum_{k=1}^{n} c_k\left(H(p(b))^2 f^{\Delta}(b)\phi_k^{\Delta}(b) + h(p(a))^2 f^{\Delta}(a)\phi_k^{\Delta}(a)\right)
\end{aligned}
$$

$$
\begin{aligned}
&-2\sum_{k=1}^{n}\lambda_k c_k^2\\
&-\sum_{k,l=1}^{n} c_k c_l \left(h\,(p(a))^2\,\phi_k^{\Delta}(a)\phi_l^{\Delta}(a)+H\,(p(b))^2\,\phi_k^{\Delta}(b)\phi_l^{\Delta}(b)\right)\\
&+\sum_{k,l=1}^{n}\lambda_l c_k c_l \delta_{kl}\\
=\;& h\left(p(a)f^{\Delta}(a)\right)^2+H\left(p(b)f^{\Delta}(b)\right)^2\\
&+\int_a^b p(t)\left(f^{\Delta}(t)\right)^2\Delta t+\int_a^b q(t)\,(f(t))^2\,\nabla t-\sum_{k=1}^{n}\lambda_k c_k^2.
\end{aligned}
$$

From here, using that

$$
J\left(f(t)-\sum_{k=1}^{n} c_k\phi_k(t)\right)\ge 0,
$$

we obtain the inequality

$$
\begin{aligned}
\sum_{k=1}^{n}\lambda_k c_k^2 \le\;& h\left(p(a)f^{\Delta}(a)\right)^2+H\left(p(b)f^{\Delta}(b)\right)^2\\
&+\int_a^b p(t)\left(f^{\Delta}(t)\right)^2\Delta t+\int_a^b q(t)\,(f(t))^2\,\nabla t.
\end{aligned}
$$

Because $n\in\mathbb{N}$ was arbitrarily chosen, from the last inequality, it follows

$$
\begin{aligned}
\sum_{k=1}^{\infty}\lambda_k c_k^2 \le\;& h\left(p(a)f^{\Delta}(a)\right)^2+H\left(p(b)f^{\Delta}(b)\right)^2\\
&+\int_a^b p(t)\left(f^{\Delta}(t)\right)^2\Delta t+\int_a^b q(t)\,(f(t))^2\,\nabla t
\end{aligned}
\tag{4.33}
$$

and the series $\sum_{k=1}^{\infty}\lambda_k c_k^2$ is convergent. Note that the proof of (4.33) is entirely unchanged if we assume that the function f satisfies only the conditions stated in the theorem. Really, when we integrate by parts, it is sufficient to integrate over the intervals on which f^{Δ} is continuous and then add all these integrals. Now, we will show that the series

$$
\sum_{k=1}^{\infty}\left|c_k\phi_k(t)\right|,\quad t\in[a,b],
\tag{4.34}
$$

is uniformly convergent on $[a,b]$. Because $\phi_k=\lambda_k A^{-1}\phi_k$ for any $k\in\mathbb{N}$, we have

$$\phi_k(t) = \lambda_k \int_a^b G(t,s)\phi_k(s)\nabla s, \quad t \in [a,b],$$

for any $k \in \mathbb{N}$. Let

$$g_k(t) = \int_a^b G(t,s)\phi_k(s)\nabla s, \quad t \in [a,b],$$

for any $k \in \mathbb{N}$. Then

$$\phi_k(t) = \lambda_k g_k(t), \quad t \in [a,b],$$

for any $k \in \mathbb{N}$. Therefore, the series (4.34) can be rewritten in the form

$$\sum_{k=1}^{\infty} \lambda_k |c_k g_k(t)|, \quad t \in [a,b]. \tag{4.35}$$

Observe that g_k are the Fourier coefficients of $G(t,s)$ as a function of s. Hence and by (4.33), we get

$$\begin{aligned}\sum_{k=1}^{\infty} \lambda_k (g_k(t))^2 &\leq h\left(p(a)G^{\Delta_s}(t,a)\right)^2 + H\left(p(b)G^{\Delta_s}(t,b)\right)^2 \\ &+ \int_a^b p(s)\left(G^{\Delta_s}(t,s)\right)^2 \Delta s + \int_a^b q(s)\,(G(t,s))^2 \nabla s, \quad t \in [a,b],\end{aligned}$$

where G^{Δ_s} is the delta derivative of $G(t,s)$ with respect to s. By the definition of $G(t,s)$, it follows that there exists a positive constant M such that

$$\begin{aligned}&h\left(p(a)G^{\Delta_s}(t,a)\right)^2 + H\left(p(b)G^{\Delta_s}(t,b)\right)^2 \\ &+ \int_a^b p(s)\left(G^{\Delta_s}(t,s)\right)^2 \Delta s + \int_a^b q(s)\,(G(t,s))^2 \nabla s \leq M, \quad t \in [a,b].\end{aligned}$$

Therefore,

$$\sum_{k=1}^{\infty} \lambda_k (g_k(t))^2 \leq M$$

for any $t \in [a,b]$. Hence, using the Cauchy-Schwartz inequality, we have

$$\begin{aligned}\sum_{k=m}^{m+p} \lambda_k |c_k g_k(t)| &= \sum_{k=m}^{m+p} \sqrt{\lambda_k}\,|c_k| \left|\sqrt{\lambda_k} g_k(t)\right| \\ &\leq \left(\sum_{k=m}^{m+p} \lambda_k c_k^2\right)^{\frac{1}{2}} \left(\sum_{k=m}^{m+p} \lambda_k (g_k(t))^2\right)^{\frac{1}{2}}\end{aligned}$$

$$\leq \sqrt{M}\left(\sum_{k=m}^{m+p}\lambda_k c_k^2\right)^{\frac{1}{2}}, \quad t\in[a,b],$$

for any $m,p\in\mathbb{N}$. From here, using that the series $\sum_{k=1}^{\infty}\lambda_k c_k^2$ is convergent, we conclude that the series $\sum_{k=1}^{\infty}\lambda_k\left|c_k g_k(t)\right|$ is uniformly convergent on $[a,b]$. Consequently, the series $\sum_{k=1}^{\infty}c_k\phi_k(t)$ is uniformly convergent on $[a,b]$. Let

$$f_1(t)=\sum_{k=1}^{\infty}c_k\phi_k(t), \quad t\in[a,b].$$

Hence, multiplying both sides of the last equality by ϕ_l and then ∇-integrating term-by-term, we get

$$\int_a^b f_1(t)\phi_l(t)\nabla t=c_l, \quad l\in\mathbb{N}.$$

Therefore, the Fourier coefficients for f and f_1 are the same. Then the Fourier coefficients for $f-f_1$ are equal to zero. Hence and by the Parseval equality, we obtain

$$\int_a^b(f(t)-f_1(t))^2\nabla t=0.$$

Therefore, $f(t)=f_1(t)$ for any $t\in[a,b]$. This completes the proof.

4.3 Third Eigenvalue Problem

Let $a=\inf\mathbb{T}$ and $b=\sup\mathbb{T}$. In this section we assume that $-\infty<a<\rho(b)\leq b<\infty$.

Consider the weighted, linear, eigenvalue problem

$$-\left(pu^{\Delta}\right)^{\Delta}(t)+q(t)u^{\sigma}(t)=\lambda w(t)u^{\sigma}(t) \quad \text{on} \quad \mathbb{T}^{\kappa^2}, \tag{4.36}$$

$$c_{00}u(a)+c_{01}u^{\Delta}(a)=0, \quad c_{10}u(\rho(b))+c_{11}u^{\Delta}(\rho(b))=0, \tag{4.37}$$

where $p\in\mathbb{H}^1(\mathbb{T}^{\kappa})$, $q,w\in\mathbb{L}^2(\mathbb{T}^{\kappa})$ and $p\geq c>0$ on $\mathbb{T}^{\kappa}$, $C\geq w\geq c>0$ on $\mathbb{T}^{\kappa^2}$, for some constants $C>c>0$, $\lambda\in\mathbb{R}$ is a parameter, c_{ij} are given real constants, $i,j\in\{0,1\}$, $(c_{00},c_{01})\neq(0,0)$, $(c_{10},c_{11})\neq(0,0)$. Define

$$X=\left\{u\in\mathbb{H}^2(\mathbb{T}):u \quad \text{satisfies} \quad (4.37)\right\}$$

with a norm $\|\cdot\|_X=\|\cdot\|_{2,\mathbb{T}}$, and $Z=\mathbb{L}^2(\mathbb{T}^{\kappa})$ with a norm $\|\cdot\|_Z=\|\cdot\|_{\mathbb{T}^{\kappa}}$. We also define inner products on Z by

$$\langle u,v\rangle = \int_a^{\rho(b)} uv\Delta, \quad \langle u,v\rangle_w = \int_a^{\rho(b)} uvw^{-1}\Delta, \quad u,v\in Z, \tag{4.38}$$

Exercise 4.5. Prove that (4.38) satisfy the axioms for an inner product.

Note that with the inner product $\langle\cdot,\cdot\rangle$, the space Z is a Hilbert space and the norm $\|\cdot\|_Z$ is the norm induced by this inner product. We will also use the norm $\|\cdot\|_w = \langle u,u\rangle_w$, $u\in Z$. The boundedness assumptions on w ensure that the norms $\|\cdot\|_Z$ and $\|\cdot\|_w$ are equivalent.

Definition 4.5. We say that $y,z\in Z$ are w-orthogonal if $\langle u,v\rangle_w = 0$ and a set $\{z_n : n\geq 1\}$ in Z is w-orthonormal if $\langle z_m,z_n\rangle = \delta_{mn}$, for all $m,n\geq 1$, where δ_{mn} is the Kronecker delta.

Define the function

$$\mathbf{1}_t(s) = \begin{cases} 1 & \text{if} \quad s=t \\ 0 & \text{if} \quad s\in\mathbb{T}\backslash\{t\}. \end{cases}$$

If t is isolated, then $\mathbf{1}_t\in\mathbb{C}(\mathbb{T})$. Now, we define the operator $I^\sigma:\mathbb{C}(\mathbb{T})\to\mathbb{C}_{rd}(\mathbb{T})$ by

$$I^\sigma(u) = u\circ\sigma, \quad u\in\mathbb{C}(\mathbb{T}).$$

For any $u_1,u_2\in\mathbb{C}(\mathbb{T})$ and $\alpha_1,\alpha_2\in\mathbb{R}$, we have

$$\begin{aligned} I^\sigma(\alpha_1u_1+\alpha_2u_2) &= (\alpha_1u_1+\alpha_2u_2)\circ\sigma \\ &= (\alpha_1u_1)\circ\sigma + (\alpha_2u_2)\circ\sigma \\ &= \alpha_1(u_1\circ\sigma)+\alpha_2(u_2\circ\sigma) \\ &= \alpha_1I^\sigma(u_1)+\alpha_2I^\sigma(u_2). \end{aligned}$$

Next,

$$\begin{aligned} |I^\sigma(u)|_{0,\mathbb{T}} &= \sup_{t\in\mathbb{T}}|u(\sigma(t))| \\ &\leq \sup_{t\in\mathbb{T}}|u(t)| \\ &= |u|_{0,\mathbb{T}}. \end{aligned}$$

So, the operator $I^\sigma:\mathbb{C}(\mathbb{T})\to\mathbb{C}_{rd}(\mathbb{T})$ is a linear, bounded operator. We define the operator $I^\sigma_{XZ}:X\to Z$ by

$$I^\sigma_{XZ}(u) = (I^\sigma(u))\Big|_{\mathbb{T}^{\kappa^2}}, \quad u\in X,$$

where $\left.(I^{\sigma}(u))\right|_{\mathbb{T}^{\kappa^2}}$ denotes the restriction of the function $I^{\sigma}(u) \in \mathbb{C}_{rd}(\mathbb{T})$ to the set $\mathbb{T}^{\kappa^2}$. Therefore, $\left.(I^{\sigma}(u))\right|_{\mathbb{T}^{\kappa^2}} \in Z$.

Theorem 4.23. *We have*

$$Ker(I^{\sigma}) = \mathbb{C}(\mathbb{T}) \bigcap span\{\mathbf{1}_a\}.$$

Proof. Suppose that there exists $u \in \mathbb{C}(\mathbb{T})$ such that $u \not\equiv 0$ and $I^{\sigma}(u) = 0$. Then there exists $t_0 \in \mathbb{T}$ such that $u(t_0) \neq 0$. If $\rho(t_0) < t_0$, then $I^{\sigma}(u)(\rho(t_0)) = u(t_0) \neq 0$, which is a contradiction with the choice of u. Let $\rho(t_0) = t_0$. Then there exists $\delta > 0$ such that for $t \in \mathbb{T}$ and $t_0 - \delta < t < t_0$, we have $u(t) \neq 0$. Hence, for $t \in \mathbb{T}$ and $t_0 - \delta < t < t_0$, we have $I^{\sigma}(t) = u(t_0) \neq 0$, which is a contradiction. Therefore, $u \equiv 0$ on $\mathbb{T}\backslash\{a\}$. If $t_0 = a$ and a is right-dense, and $u(a) \neq 0$, there exists a $\delta > 0$ such that for $t \in \mathbb{T}$ and $t_0 < t < t_0 + \delta$, we have $u(t) \neq 0$. From here, $I^{\sigma}(u)(t) \neq 0$ for $t \in \mathbb{T}$ and $t_0 < t < t_0 + \delta$. This is a contradiction.Therefore,, if a is right-dense, we have $u(a) = 0$. If $t_0 = a$ is right-scattered, then $\mathbf{1}_a \in \mathbb{C}(\mathbb{T})$ and $\mathrm{Ker}(I^{\sigma}) = \mathrm{span}\{\mathbf{1}_a\}$. This completes the proof.

Exercise 4.6. Prove that

$$\mathrm{Ker}(I^{\sigma}_{XZ}) = X \bigcap \mathrm{span}\{\mathbf{1}_a, \mathbf{1}_b\}.$$

Theorem 4.24. *If a is a right-dense point in $\mathbb{T}$, then $I^{\sigma} : \mathbb{C}(\mathbb{T}) \to \mathbb{C}_{rd}(\mathbb{T})$ is injective.*

Proof. Since a is a right-dense point in $\mathbb{T}$, we have that $\mathbf{1}_a \notin \mathbb{C}(\mathbb{T})$. Therefore, $\mathrm{Ker}(I^{\sigma}) = \{0\}$. Hence, $I^{\sigma} : \mathbb{C}(\mathbb{T}) \to \mathbb{C}_{rd}(\mathbb{T})$ is injective. This completes the proof.

If a is right-scattered, then

$$u^{\Delta}(a) = \frac{u(\sigma(a)) - u(a)}{\sigma(a) - a}.$$

Hence, when a is right-scattered, we have

$$c_{00}u(a) + c_{01}u^{\Delta}(a) = c_{00}u(a) + \frac{c_{01}}{\mu(a)}u(\sigma(a)) - \frac{c_{01}}{\mu(a)}u(a).$$

Let

$$a_0 = c_{00} - \frac{c_{01}}{\mu(a)}, \quad a_1 = \frac{c_{01}}{\mu(a)},$$

when a is right-scattered. Then

$$c_{00}u(a) + c_{01}u^{\Delta}(a) = a_0u(a) + a_1u(\sigma(a)),$$

when a is right-scattered. If b is left-scattered, then

$$u^{\Delta}(\rho(b)) = \frac{u(b) - u(\rho(b))}{b - \rho(b)}.$$

Hence, when b is left-scattered, we have

$$c_{10}u(\rho(b))+c_{11}u^{\Delta}(\rho(b)) = c_{10}u(\rho(b))+\frac{c_{11}}{b-\rho(b)}u(b) - \frac{c_{11}}{b-\rho(b)}u(\rho(b)).$$

Let

$$b_0 = c_{10}-\frac{c_{11}}{b-\rho(b)}, \quad b_1 = \frac{c_{11}}{b-\rho(b)},$$

when b is left-scattered. Then, if b is left-scattered,

$$c_{10}u(\rho(b))+c_{11}u^{\Delta}(\rho(b)) = b_0u(\rho(b))+b_1u(b).$$

Therefore,, if a is right-scattered or b is left-scattered, the boundary conditions at a or b can be written in the following form

$$a_0u(a)+a_1u(\sigma(a)) = 0 \quad \text{or} \quad b_0u(\rho(b))+b_1u(b) = 0,$$

respectively. From now on, we make the following assumptions.

(H1) If a is right-scattered, then $a_0 \neq 0$.
(H2) If b is left-scattered, then $b_1 \neq 0$.

Theorem 4.25. *The operator I^{σ}_{XZ} is injective and compact.*

Proof. 1. If a is right-dense, then $\mathbf{1}_a \notin \mathbb{C}(\mathbb{T})$. Hence, $\mathbf{1}_a \notin X$.
2. If a is right-scattered, then, using (H1), we have that $\mathbf{1}_a \notin X$.
3. If b is left-dense, then $\mathbf{1}_b \notin \mathbb{C}(\mathbb{T})$ and hence, $\mathbf{1}_b \notin X$.
4. If b is left-scattered, then, using (H2), we have that $\mathbf{1}_b \notin X$.

Consequently, $\mathrm{Ker}\,(I^{\sigma}_{XZ}) = \{0\}$. Therefore, I^{σ}_{XZ} is injective. By Theorem A.16, it follows that I^{σ}_{XZ} is compact. This completes the proof.

Now, we define the operator $L : X \to Z$ by

$$L(u) = -\left(pu^{\Delta}\right)^{\Delta_g} + qu^{\sigma}, \quad u \in X.$$

For any $u_1, u_2 \in X$ and for any $\alpha_1, \alpha_2 \in \mathbb{R}$, we have

$$\begin{aligned} L(\alpha_1u_1+\alpha_2u_2) &= -\left(p(\alpha_1u_1+\alpha_2u_2)^{\Delta}\right)^{\Delta_g} \\ &\quad + q(\alpha_1u_1+\alpha_2u_2)^{\sigma} \\ &= -\left(\alpha_1pu_1^{\Delta}+\alpha_2pu_2^{\Delta}\right)^{\Delta_g} \\ &\quad + \alpha_1qu_1^{\sigma}+\alpha_2qu_2^{\sigma} \end{aligned}$$

$$
\begin{aligned}
&= -\alpha_1 \left(pu_1^{\Delta}\right)^{\Delta_g} + \alpha_1 q u_1^{\sigma} \\
&\quad -\alpha_2 \left(pu_2^{\Delta}\right)^{\Delta_g} + \alpha_2 q u_2^{\sigma} \\
&= \alpha_1 L(u_1) + \alpha_2 L(u_2),
\end{aligned}
$$

i.e., $L : X \to Z$ is a linear operator. Note that for $u, v \in \mathbb{L}^2(\mathbb{T})(u, v \in \mathbb{L}^2(\mathbb{T}^{\kappa}))$ we have

$$
\|uv\|_{\mathbb{T}} \leq \|u\|_{\mathbb{T}} \|v\|_{\mathbb{T}} \quad (\|uv\|_{\mathbb{T}^{\kappa}} \leq \|u\|_{\mathbb{T}^{\kappa}} \|v\|_{\mathbb{T}^{\kappa}}).
$$

Then, for $u \in X$, we have

$$
L(u) = -p^{\Delta_g} u^{\Delta} - p^{\sigma} u^{\Delta\Delta_g} + q u^{\sigma}
$$

and

$$
\begin{aligned}
\|L(u)\|_Z &= \| - p^{\Delta_g} u^{\Delta} - p^{\sigma} u^{\Delta\Delta_g} + q u^{\sigma} \|_{\mathbb{T}^{\kappa}} \\
&\leq \|p^{\Delta_g} u^{\Delta}\|_{\mathbb{T}^{\kappa}} + \|p^{\sigma} u^{\Delta\Delta_g}\|_{\mathbb{T}^{\kappa}} + \|q u^{\sigma}\|_{\mathbb{T}^{\kappa}} \\
&\leq \|p^{\Delta_g}\|_{\mathbb{T}^{\kappa}} \|u^{\Delta}\|_{\mathbb{T}^{\kappa}} + \|p^{\sigma}\|_{\mathbb{T}^{\kappa}} \|u^{\Delta\Delta_g}\|_{\mathbb{T}^{\kappa}} \\
&\quad + \|q\|_{\mathbb{T}^{\kappa}} \|u^{\sigma}\|_{\mathbb{T}^{\kappa}} \\
&\leq \|p\|_{1,\mathbb{T}^{\kappa}} \|u\|_{2,\mathbb{T}} + \|p\|_{1,\mathbb{T}^{\kappa}} \|u\|_{2,\mathbb{T}} \\
&\quad + \|q\|_{\mathbb{T}^{\kappa}} \|u\|_{2,\mathbb{T}} \\
&= \left(2\|p\|_{1,\mathbb{T}^{\kappa}} + \|q\|_{\mathbb{T}^{\kappa}}\right) \|u\|_{2,\mathbb{T}} \\
&= \left(2\|p\|_{1,\mathbb{T}^{\kappa}} + \|q\|_{\mathbb{T}^{\kappa}}\right) \|u\|_X,
\end{aligned}
$$

i.e., the operator $L : X \to Z$ is a bounded operator.

Theorem 4.26. *For any $u, v \in X$, we have*

$$
\langle L(u), v^{\sigma} \rangle = \langle u^{\sigma}, L(v) \rangle. \tag{4.39}
$$

Proof. For $u, v \in X$, we have

$$
\begin{aligned}
\langle L(u), v^{\sigma} \rangle &= \int_a^{\rho(b)} L(u) v^{\sigma} \Delta \\
&= \int_a^{\rho(b)} \left(-\left(pu^{\Delta}\right)^{\Delta_g} + qu^{\sigma} \right) v^{\sigma} \Delta \\
&= -\int_a^{\rho(b)} \left(pu^{\Delta}\right)^{\Delta_g} v^{\sigma} \Delta + \int_a^{\rho(b)} qu^{\sigma} v^{\sigma} \Delta \\
&= -pu^{\Delta} v \Big|_a^{\rho(b)} + \int_a^{\rho(b)} pu^{\Delta} v^{\Delta} \Delta \\
&\quad + \int_a^{\rho(b)} qu^{\sigma} v^{\sigma} \Delta \\
&= -p(\rho(b)) u^{\Delta}(\rho(b)) v(\rho(b)) + p(a) u^{\Delta}(a) v(a) \\
&\quad + \int_a^{\rho(b)} pu^{\Delta} v^{\Delta} \Delta + \int_a^{\rho(b)} qu^{\sigma} v^{\sigma} \Delta
\end{aligned}
\tag{4.40}
$$

and

$$
\begin{aligned}
\langle u^{\sigma}, L(v) \rangle &= \int_a^{\rho(b)} u^{\sigma} L(v) \Delta \\
&= \int_a^{\rho(b)} u^{\sigma} \left(-\left(pv^{\Delta}\right)^{\Delta_g} + qv^{\sigma} \right) \Delta \\
&= -\int_a^{\rho(b)} \left(pv^{\Delta}\right)^{\Delta_g} u^{\sigma} \Delta + \int_a^{\rho(b)} qu^{\sigma} v^{\sigma} \Delta \\
&= -pv^{\Delta} u \Big|_a^{\rho(b)} + \int_a^{\rho(b)} pv^{\Delta} u^{\Delta} \Delta \\
&\quad + \int_a^{\rho(b)} qu^{\sigma} v^{\sigma} \Delta \\
&= -p(\rho(b)) v^{\Delta}(\rho(b)) u(\rho(b)) + p(a) v^{\Delta}(a) u(a) \\
&\quad + \int_a^{\rho(b)} pv^{\Delta} u^{\Delta} \Delta + \int_a^{\rho(b)} qu^{\sigma} v^{\sigma} \Delta.
\end{aligned}
\tag{4.41}
$$

Let $c_{00} \neq 0$ and $c_{10} \neq 0$. Then

$$
u(a) = -\frac{c_{01}}{c_{00}} u^{\Delta}(a), \quad v(a) = -\frac{c_{01}}{c_{00}} v^{\Delta}(a),
$$

$$u(\rho(b)) = -\frac{c_{11}}{c_{10}} u^{\Delta}(\rho(b)), \quad v(\rho(b)) = -\frac{c_{11}}{c_{10}} v^{\Delta}(\rho(b)).$$

Hence,

$$-p(\rho(b))u^{\Delta}(\rho(b))v(\rho(b)) + p(a)u^{\Delta}(a)v(a)$$

$$= \frac{c_{11}}{c_{10}} p(\rho(b))u^{\Delta}(\rho(b))v^{\Delta}(\rho(b)) - \frac{c_{01}}{c_{00}} u^{\Delta}(a)v^{\Delta}(a),$$

$$-p(\rho(b))v^{\Delta}(\rho(b))u(\rho(b)) + p(a)v^{\Delta}(a)u(a)$$

$$= \frac{c_{11}}{c_{10}} p(\rho(b))v^{\Delta}(\rho(b))u^{\Delta}(\rho(b)) - \frac{c_{01}}{c_{00}} p(a)u^{\Delta}(a)v^{\Delta}(a)$$

and using (4.40), (4.41), we get (4.39).

Let $c_{00} \neq 0$ and $c_{11} \neq 0$. Then

$$u(a) = -\frac{c_{01}}{c_{00}} u^{\Delta}(a), \quad v(a) = -\frac{c_{01}}{c_{00}} v^{\Delta}(a),$$

$$u^{\Delta}(\rho(b)) = -\frac{c_{10}}{c_{11}} u(\rho(b)), \quad v^{\Delta}(\rho(b)) = -\frac{c_{10}}{c_{11}} v(\rho(b)).$$

Hence,

$$-p(\rho(b))u^{\Delta}(\rho(b))v(\rho(b)) + p(a)u^{\Delta}(a)v(a)$$

$$= \frac{c_{10}}{c_{11}} p(\rho(b))u(\rho(b))v(\rho(b)) - \frac{c_{01}}{c_{00}} p(a)u^{\Delta}(a)v^{\Delta}(a),$$

$$-p(\rho(b))v^{\Delta}(\rho(b))u(\rho(b)) + p(a)v^{\Delta}(a)u(a)$$

$$= \frac{c_{10}}{c_{11}} p(\rho(b))v(\rho(b))u(\rho(b)) - \frac{c_{01}}{c_{00}} u^{\Delta}(a)v^{\Delta}(a)$$

and using (4.40), (4.41), we get (4.39).

Let $c_{01} \neq 0$ and $c_{10} \neq 0$. Then

$$u^{\Delta}(a) = -\frac{c_{00}}{c_{01}} u(a), \quad v^{\Delta}(a) = -\frac{c_{00}}{c_{01}} v(a),$$

$$u(\rho(b)) = -\frac{c_{11}}{c_{10}} u^{\Delta}(\rho(b)), \quad v(\rho(b)) = -\frac{c_{11}}{c_{10}} v^{\Delta}(\rho(b)).$$

Hence,

$$-p(\rho(b))u^{\Delta}(\rho(b))v(\rho(b)) + p(a)u^{\Delta}(a)v(a)$$

$$= \frac{c_{11}}{c_{10}} p(\rho(b)) u^{\Delta}(\rho(b)) v^{\Delta}(\rho(b)) - \frac{c_{00}}{c_{01}} p(a)u(a)v(a),$$

$$-p(\rho(b)) v^{\Delta}(\rho(b)) u(\rho(b)) + p(a)v^{\Delta}(a)u(a)$$

$$= \frac{c_{11}}{c_{10}} p(\rho(b)) u^{\Delta}(\rho(b)) v^{\Delta}(\rho(b)) - \frac{c_{00}}{c_{01}} p(a)u(a)v(a)$$

and using (4.40), (4.41), we get (4.39).

Let $c_{01} \neq 0$ and $c_{11} \neq 0$. Then

$$u^{\Delta}(a) = -\frac{c_{00}}{c_{01}} u(a), \quad v^{\Delta}(a) = -\frac{c_{00}}{c_{01}} v(a),$$

$$u^{\Delta}(\rho(b)) = -\frac{c_{10}}{c_{11}} u(\rho(b)), \quad v^{\Delta}(\rho(b)) = -\frac{c_{10}}{c_{11}} v(\rho(b)).$$

Hence,

$$-p(\rho(b)) u^{\Delta}(\rho(b)) v(\rho(b)) + p(a)u^{\Delta}(a)v(a)$$

$$= \frac{c_{10}}{c_{11}} p(\rho(b)) u(\rho(b)) v(\rho(b)) - \frac{c_{00}}{c_{01}} p(a)u(a)v(a),$$

$$-p(\rho(b)) v^{\Delta}(\rho(b)) u(\rho(b)) + p(a)u(a)v^{\Delta}(a)$$

$$= \frac{c_{10}}{c_{11}} p(\rho(b)) u(\rho(b)) v(\rho(b)) - \frac{c_{00}}{c_{01}} p(a)u(a)v(a)$$

and using (4.40), (4.41), we get (4.39). This completes the proof.

Theorem 4.27. *For any $u \in X$ there exists a constant D such that*

$$\langle L(u), u^{\sigma} \rangle \geq \frac{c}{2} \|u^{\Delta}\|^2_{\mathbb{T}^{\kappa}} + D\|u^{\sigma}\|^2_{\mathbb{T}^{\kappa}}.$$

Remark 4.1. The constant D in Theorem 4.27 need not be positive.

Proof. Let $u \in X$. Then

$$\begin{aligned}\langle L(u),u^\sigma\rangle &= \int_a^{\rho(b)} L(u)u^\sigma\Delta \\ &= \int_a^{\rho(b)}\left(-\left(pu^\Delta\right)^{\Delta_g}+qu^\sigma\right)u^\sigma\Delta \\ &= -\int_a^{\rho(b)}\left(pu^\Delta\right)^{\Delta_g}u^\sigma\Delta+\int_a^{\rho(b)}q\left(u^\sigma\right)^2\Delta \\ &= -pu^\Delta u\Big|_a^{\rho(b)}+\int_a^{\rho(b)}p\left(u^\Delta\right)^2\Delta \\ &\quad+\int_a^{\rho(b)}q\left(u^\sigma\right)^2\Delta \\ &\geq -p(\rho(b))u^\Delta(\rho(b))u(\rho(b))+p(a)u^\Delta(a)u(a) \\ &\quad+c\|u^\Delta\|^2_{\mathbb{T}^\kappa}-c_1\|u^\sigma\|^2_{\mathbb{T}^\kappa}\end{aligned} \tag{4.42}$$

for some positive constant c_1. Such constant exists because $q\in\mathbb{L}^2(\mathbb{T}^\kappa)$. Let

$$B_a=u^\Delta(a)u(a),\quad B_b=-u^\Delta(\rho(b))u(\rho(b)).$$

If $c_{01}=0$, using (4.37), we get $u(a)=0$ and $B_a=0$. If $c_{00}=0$, using (4.37), we get $u^\Delta(a)=0$ and $B_a=0$. Therefore, we suppose $c_{00}\neq 0$, $c_{01}\neq 0$. If $c_{10}=0$, using (4.37), we obtain $u^\Delta(\rho(b))=0$ and $B_b=0$. If $c_{11}=0$, using (4.37), we get $u(\rho(b))=0$ and $B_b=0$. Assume that $c_{10}\neq 0$ and $c_{11}\neq 0$. Hence, using (4.37), we find

$$B_a=-\frac{c_{00}}{c_{01}}(u(a))^2,\quad B_b=\frac{c_{10}}{c_{11}}(u(\rho(b)))^2.$$

Let

$$c_2=\max\left\{\left|\frac{c_{00}}{c_{01}}\right|,\quad\left|\frac{c_{10}}{c_{11}}\right|\right\}.$$

We have $c_2>0$. Consider the function

$$g(t)=(u(t))^2,\quad t\in\mathbb{T}^\kappa.$$

We apply Theorem A.18 for $\varepsilon=\dfrac{1}{\sqrt{8c_2}}$. Then there exists a positive constant c_3 such that

$$|u|_{0,\mathbb{T}^\kappa}\leq\frac{1}{\sqrt{8c_2}}\|u^\Delta\|_{\mathbb{T}^\kappa}+c_3\|u^\sigma\|_{\mathbb{T}^\kappa}.$$

Hence,

$$|u|^2_{0,\mathbb{T}^\kappa}\leq\left(\frac{1}{\sqrt{8c_2}}\|u^\Delta\|_{\mathbb{T}^\kappa}+c_3\|u^\sigma\|_{\mathbb{T}^\kappa}\right)^2$$

$$\leq 2\left(\frac{1}{\sqrt{8c_2}}\|u^{\Delta}\|_{\mathbb{T}^{\kappa}}\right)^2 + 2c_3^2\|u^{\sigma}\|_{\mathbb{T}^{\kappa}}^2$$

$$= \frac{1}{4c_2}\|u^{\Delta}\|_{\mathbb{T}^{\kappa}}^2 + 2c_3^2\|u^{\sigma}\|_{\mathbb{T}^{\kappa}}^2.$$

Therefore,

$$g(t) \leq \frac{1}{4c_2}\|u^{\Delta}\|_{\mathbb{T}^{\kappa}}^2 + 2c_3^2\|u^{\sigma}\|_{\mathbb{T}^{\kappa}}^2, \quad t \in \mathbb{T}^{\kappa}.$$

In particular,

$$g(a) \leq \frac{1}{4c_2}\|u^{\Delta}\|_{\mathbb{T}^{\kappa}}^2 + 2c_3^2\|u^{\sigma}\|_{\mathbb{T}^{\kappa}}^2,$$

$$g(\rho(b)) \leq \frac{1}{4c_2}\|u^{\Delta}\|_{\mathbb{T}^{\kappa}}^2 + 2c_3^2\|u^{\sigma}\|_{\mathbb{T}^{\kappa}}^2.$$

Then

$$|B_a| \leq c_2 g(t)$$

$$\leq \frac{1}{4}\|u^{\Delta}\|_{\mathbb{T}^{\kappa}}^2 + 2c_3^2 c_2\|u^{\sigma}\|_{\mathbb{T}^{\kappa}}^2,$$

$$|B_b| \leq c_2 g(t)$$

$$\leq \frac{1}{4}\|u^{\Delta}\|_{\mathbb{T}^{\kappa}}^2 + 2c_3^2 c_2\|u^{\sigma}\|_{\mathbb{T}^{\kappa}}^2.$$

From here,

$$B_a \geq -\frac{1}{4}\|u^{\Delta}\|_{\mathbb{T}^{\kappa}}^2 - 2c_3^2 c_2\|u^{\sigma}\|_{\mathbb{T}^{\kappa}}^2,$$

$$B_b \geq -\frac{1}{4}\|u^{\Delta}\|_{\mathbb{T}^{\kappa}}^2 - 2c_3^2 c_2\|u^{\sigma}\|_{\mathbb{T}^{\kappa}}^2,$$

$$p(a)B_a \geq -\frac{c}{4}\|u^{\Delta}\|_{\mathbb{T}^{\kappa}}^2 - 2cc_3^2 c_2\|u^{\sigma}\|_{\mathbb{T}^{\kappa}}^2,$$

$$p(\rho(b))B_b \geq -\frac{c}{4}\|u^{\Delta}\|_{\mathbb{T}^{\kappa}}^2 - 2cc_3^2 c_2\|u^{\sigma}\|_{\mathbb{T}^{\kappa}}^2.$$

From the last two inequalities and from (4.42), we get

$$\langle L(u), u^{\sigma}\rangle \geq -\frac{c}{4}\|u^{\Delta}\|_{\mathbb{T}^{\kappa}}^2 - 2cc_3^2 c_2\|u^{\sigma}\|_{\mathbb{T}^{\kappa}}^2$$

$$-\frac{c}{4}\|u^\Delta\|^2_{\mathbb{T}^\kappa} - 2cc_3^2c_2\|u^\sigma\|^2_{\mathbb{T}^\kappa}$$

$$+c\|u^\Delta\|^2_{\mathbb{T}^\kappa} - c_1\|u^\sigma\|^2_{\mathbb{T}^\kappa}$$

$$= \frac{c}{2}\|u^\Delta\|^2_{\mathbb{T}^\kappa} - \left(4cc_3^2c_2 + c_1\right)\|u^\sigma\|^2_{\mathbb{T}^\kappa}.$$

This completes the proof.

Theorem 4.28. *There exists a positive constant M such that*

$$\langle (L + MwI_{XZ}^\sigma)(u), u^\sigma \rangle > 0, \quad u \in X, \quad u \neq 0.$$

Hence, the operator $L + MwI_{XZ}^\sigma : X \to Z$ is injective.

Proof. We take a positive constant M such that

$$Mc > 4cc_3^2c_2 + c_1,$$

where the constants c_1, c_2 and c_3 are as in the proof of Theorem **??**. Hence, using the proof of Theorem **??**, we get

$$\langle (L + MwI_{XZ}^\sigma)(u), u^\sigma \rangle \geq \frac{c}{2}\|u^\Delta\|^2_{\mathbb{T}^\kappa}$$

$$+ \left(Mc - 4cc_3^2c_2 + c_1\right)\|u^\sigma\|^2_{\mathbb{T}^\kappa}$$

$$> 0$$

for any $u \in X$, $u \neq 0$. This completes the proof.

Remark 4.2. Without loss of generality, below we assume that M in Theorem 4.28 is equal to zero. Otherwise, we redefine q and λ to be $q + Mw$ and $\lambda + M$, respectively.

Theorem 4.29. *For any $h \in Z$ and $\tau \in \mathbb{T}^\kappa$, $\eta_1, \eta_2 \in \mathbb{R}$, the initial value problem*

$$\begin{aligned} -\left(pu^\Delta\right)^{\Delta_g} + qu^\sigma &= h \\ u(\tau) = \eta_1, \quad u^\Delta(\tau) &= \eta_2, \end{aligned} \tag{4.43}$$

has a unique solution $u \in \mathbb{H}^2(\mathbb{T})$.

Proof. Let

$$u_1 = u, \quad u_2 = pu^\Delta.$$

Then

$$u_1^{\Delta_g} = u^\Delta$$

$$
\begin{aligned}
&= \frac{1}{p} u_2,\\
u_2^{\Delta_g} &= \left(p u^{\Delta}\right)^{\Delta_g}\\
&= q u^{\sigma} - h\\
&= q u_1^{\sigma} - h\\
&= q\left(u_1 + \mu u_1^{\Delta}\right) - h\\
&= q\mu u_1^{\Delta} + q u_1 - h\\
&= q u_1 + \frac{q\mu}{p} u_2 - h.
\end{aligned}
$$

We define

$$
U = \begin{pmatrix} u_1 \\ u_2 \end{pmatrix}, \quad U_0 = \begin{pmatrix} \eta_1 \\ \eta_2 \end{pmatrix}, \quad R = \begin{pmatrix} 0 & \frac{1}{p} \\ q & \frac{q\mu}{p} \end{pmatrix}, \quad H = \begin{pmatrix} 0 \\ h \end{pmatrix}.
$$

Therefore, problem (4.43) is equivalent to problem

$$
U^{\Delta_g} = RU - H, \quad U(\tau) = U_0. \tag{4.44}
$$

We define a norm $\|R\|_{\mathbb{T}}$ of the matrix function R by

$$
r(t) = \frac{1}{p(t)} + |q(t)|\left(1 + \frac{\mu(t)}{p(t)}\right), \quad t \in \mathbb{T}, \quad \|R\|_{\mathbb{T}} = \|r\|_{\mathbb{T}}.
$$

Define the numbers

$$
\delta = \frac{1}{4\|R\|_{\mathbb{T}}^2}, \quad t_1 = \sup\{t \in \mathbb{T} : t \le \tau + \delta\}, \quad t_2 = \sigma(t_1).
$$

Then, either $t_2 \ge \tau + \delta$ or $t_2 = b$. Let

$$
\mathbb{T}_i = [\tau, t_i] \bigcap \mathbb{T}, \quad i = 1, 2.
$$

For any $U \in \mathbb{C}(\mathbb{T}_1) \times \mathbb{C}(\mathbb{T}_1)$, let

$$
|U(t)| = \max\{|u_1(t)|, |u_2(t)|\}, \quad t \in \mathbb{T}_1,
$$

$$|U|_{0,\mathbb{T}_1} = \sup\{|U(t)| : t \in \mathbb{T}_1\}.$$

Note that with the norm $|\cdot|_{0,\mathbb{T}_1}$, the space $\mathbb{C}(\mathbb{T}_1)\times\mathbb{C}(\mathbb{T}_1)$ is a Banach space. Now, we define

$$JU(t) = U_0 + \int_\tau^t (RU - H)\,\Delta, \quad t \in \mathbb{T}_1.$$

We have $J : \mathbb{C}(\mathbb{T}_1)\times\mathbb{C}(\mathbb{T}_1) \to \mathbb{C}(\mathbb{T}_1)\times\mathbb{C}(\mathbb{T}_1)$ and the solutions of problem (4.44) on $\mathbb{T}_1$ correspond to the fixed points of J. Also, for any $t \in \mathbb{T}_1$ and $U_1, U_2 \in \mathbb{C}(\mathbb{T}_1)\times \mathbb{C}(\mathbb{T}_1)$, we have

$$\begin{aligned}
|JU_1(t) - JU_2(t)| &= \left|\int_\tau^t R(U_1 - U_2)\Delta\right| \\
&\le \left(\int_\tau^t r\Delta\right)|U_1 - U_2|_{0,\mathbb{T}_1} \\
&\le \left(\int_\tau^t r^2\Delta\right)^{\frac{1}{2}}(t-\tau)^{\frac{1}{2}}|U_1 - U_2|_{0,\mathbb{T}_1} \\
&\le \|R\|_{\mathbb{T}}\delta^{\frac{1}{2}}|U_1 - U_2|_{0,\mathbb{T}_1} \\
&\le \frac{1}{2}|U_1 - U_2|_{0,\mathbb{T}_1}.
\end{aligned}$$

Therefore, J is a contraction mapping on $\mathbb{C}(\mathbb{T}_1)\times\mathbb{C}(\mathbb{T}_1)$. Thus J has a unique fixed point in $\mathbb{C}(\mathbb{T}_1)\times\mathbb{C}(\mathbb{T}_1)$. Hence, problem (4.44) has a unique solution on $\mathbb{T}_1$. If $t_1 < t_2$, then t_1 is a right-scattered, the solution can be extended to t_2 using that

$$U^{\Delta_g}(t_1) = \frac{U(\sigma(t_1)) - U(t_1)}{\sigma(t_1) - t_1}.$$

Because $\tau \in \mathbb{T}^\kappa$, $\eta_1, \eta_2 \in \mathbb{R}$ are arbitrary and δ is independent of τ, we conclude that (4.44) has a unique solution on $[\tau, b]\bigcap\mathbb{T}$. As in the above, we can construct a solution on $[a,\tau]\bigcap\mathbb{T}$. This completes the proof.

Let $\phi, \psi \in \mathbb{H}^2(\mathbb{T})$ be the unique solutions of equation $L(u) = 0$ on $\mathbb{T}$ satisfying the initial conditions

$$\phi(a) = c_{01}, \quad \phi^\Delta(a) = -c_{00},$$

$$\psi(\rho(b)) = c_{11}, \quad \psi^\Delta(\rho(b)) = -c_{10}.$$

By Theorem 4.29, it follows that these solutions exist. We have

$$c_{00}\phi(a) + c_{01}\phi^\Delta(a) = c_{00}c_{01} - c_{00}c_{01}$$

$$= 0,$$

$$c_{10}\psi(\rho(b)) + c_{11}\psi^{\Delta}(\rho(b)) = c_{10}c_{11} - c_{10}c_{11}$$

$$= 0.$$

Let

$$W = p\left(\phi^{\Delta}\psi - \phi\psi^{\Delta}\right).$$

Since $\phi, \psi \in \mathbb{H}^2(\mathbb{T})$, we have that $W \in \mathbb{H}^1(\mathbb{T})$.

Theorem 4.30. *W is a constant and $W \neq 0$.*

Proof. We have

$$\begin{aligned}
W^{\Delta_g} &= \left(p\left(\phi^{\Delta}\psi - \phi\psi^{\Delta}\right)\right)^{\Delta_g} \\
&= \left(p\phi^{\Delta}\psi - p\psi^{\Delta}\phi\right)^{\Delta_g} \\
&= \left(p\phi^{\Delta}\psi\right)^{\Delta_g} - \left(p\phi\psi^{\Delta}\right)^{\Delta_g} \\
&= \left(p\phi^{\Delta}\right)^{\Delta_g}\psi^{\sigma} + p\phi^{\Delta}\psi^{\Delta} \\
&\quad - \left(p\psi^{\Delta}\right)^{\Delta_g}\phi^{\sigma} - p\psi^{\Delta}\phi^{\Delta} \\
&= q\phi^{\sigma}\psi^{\sigma} - q\phi^{\sigma}\psi^{\sigma} \\
&= 0.
\end{aligned}$$

Therefore, W is a constant. Note that

$$\begin{aligned}
W(t) &= W(a) \\
&= p(a)\left(\phi^{\Delta}(a)\psi(a) - \phi(a)\psi^{\Delta}(a)\right) \\
&= p(a)\left(-c_{00}\psi(a) - c_{01}\psi^{\Delta}(a)\right) \\
&= W(\rho(b))
\end{aligned}$$

$$= p(\rho(b))\left(\phi^{\Delta}(\rho(b))\,\psi(\rho(b)) - \phi(\rho(b))\,\psi^{\Delta}(\rho(b))\right)$$

$$= p(\rho(b))\left(c_{11}\phi^{\Delta}(\rho(b)) + c_{10}\phi(\rho(b))\right).$$

Then $W = 0$ if and only if

$$c_{00}\psi(a) + c_{01}\psi^{\Delta}(a) = 0 \tag{4.45}$$

and

$$c_{10}\phi(\rho(b)) + c_{11}\phi^{\Delta}(\rho(b)) = 0. \tag{4.46}$$

Note that

$$c_{00}\phi(a) + c_{01}\phi^{\Delta}(a) = 0$$

and

$$c_{10}\psi(\rho(b)) + c_{11}\psi^{\Delta}(\rho(b)) = 0.$$

Since L is injective, using the uniqueness of the solution of the initial value problem, we have that ψ does not satisfy the condition (4.45) and ϕ does not satisfy (4.46). Therefore, $W \neq 0$. This completes the proof.

For $(t,s) \in \mathbb{T} \times \mathbb{T}$, we define

$$g(t,s) = \begin{cases} \frac{1}{W}\psi(t)\phi^{\sigma}(s) & \text{if} \quad t \geq \sigma(s) \\ \frac{1}{W}\phi(t)\psi^{\sigma}(s) & \text{if} \quad t \leq s. \end{cases}$$

We have $g(t,\cdot) \in \mathbb{C}_{rd}(\mathbb{T})$ for any $t \in \mathbb{T}$. Let

$$\begin{aligned} u(t) &= Gh(t) \\ &= \int_a^{\rho(b)} g(t,\cdot)h\Delta, \quad t \in \mathbb{T}, \quad h \in \mathbb{L}^2(\mathbb{T}^{\kappa}). \end{aligned} \tag{4.47}$$

We have

$$u(t) = \frac{1}{W}\psi(t)\int_a^t \phi^{\sigma}h\Delta + \frac{1}{W}\phi(t)\int_t^{\rho(b)} \psi^{\sigma}h\Delta$$

or

$$Wu(t) = \psi(t)\int_a^t \phi^{\sigma}h\Delta + \phi(t)\int_t^{\rho(b)} \psi^{\sigma}h\Delta.$$

Hence,

$$\begin{aligned} Wu^{\Delta}(t) &= \psi^{\Delta}(t)\int_a^t \phi^{\sigma}h\Delta + \psi^{\sigma}(t)\phi^{\sigma}(t)h(t) \\ &\quad + \phi^{\Delta}(t)\int_t^{\rho(b)} \psi^{\sigma}h\Delta - \phi^{\sigma}(t)\psi^{\sigma}(t)h(t) \end{aligned}$$

$$= \psi^\Delta(t)\int_a^t \phi^\sigma h\Delta + \phi^\Delta(t)\int_t^{\rho(b)} \psi^\sigma h\Delta.$$

Hence,

$$\begin{aligned} c_{00}Wu(a)+c_{01}Wu^\Delta(a) &= c_{00}\phi(a)\int_a^{\rho(b)} \psi^\sigma h\Delta + c_{01}\phi^\Delta(a)\int_a^{\rho(b)} \psi^\sigma h\Delta \\ &= \left(c_{00}\phi(a)+c_{01}\phi^\Delta(a)\right)\int_a^{\rho(b)} \psi^\sigma h\Delta \\ &= 0, \end{aligned}$$

$$\begin{aligned} c_{10}Wu(\rho(b))+c_{11}Wu^\Delta(\rho(b)) &= c_{10}\psi(\rho(b))\int_a^{\rho(b)} \phi^\sigma h\Delta + c_{11}\psi^\Delta(\rho(b)) \\ \times\int_a^{\rho(b)} \phi^\sigma h\Delta &= \left(c_{10}\psi(\rho(b))+c_{11}\psi^\Delta(\rho(b))\right)\int_a^{\rho(b)} \phi^\sigma h\Delta \\ &= 0, \end{aligned}$$

i.e., u satisfies (4.37). Next,

$$\begin{aligned} W\left(pu^\Delta\right)^{\Delta_g}(t) &= \left(p\psi^\Delta\right)^{\Delta_g}(t)\int_a^t \phi^\sigma h\Delta + \left(p\psi^\Delta\right)^\sigma(t)\phi^\sigma(t)h(t) \\ &\quad + \left(p\phi^\Delta\right)^{\Delta_g}(t)\int_t^{\rho(b)} \psi^\sigma h\Delta - \left(p\phi^\Delta\right)^\sigma(t)\psi^\sigma(t)h(t) \\ &= -Wh(t)+q(t)\psi^\sigma(t)\int_a^t \phi^\sigma h\Delta \qquad (4.48) \\ &\quad + q(t)\phi^\sigma(t)\int_t^{\rho(b)} \psi^\sigma h\Delta \\ &= -Wh(t)+q(t)\left(\psi^\sigma(t)\int_a^t \phi^\sigma h\Delta + \phi^\sigma(t)\int_t^{\rho(b)} \psi^\sigma h\Delta\right). \end{aligned}$$

If $t=\sigma(t)$, using (4.48), we have

$$W\left(pu^\Delta\right)^{\Delta_g}(t) = -Wh(t)+Wq(t)u(t).$$

If $t<\sigma(t)$, using (4.48), we obtain

$$W\left(pu^\Delta\right)^{\Delta_g}(t) = -Wh(t)+q(t)\left(\psi^\sigma(t)\int_a^{\sigma(t)} \phi^\sigma h\Delta - \psi^\sigma(t)\int_t^{\sigma(t)} \phi^\sigma h\Delta\right.$$

$$+\phi^\sigma(t)\int_{\sigma(t)}^{\rho(b)}\psi^\sigma h\Delta+\phi^\sigma(t)\int_t^{\sigma(t)}\psi^\sigma h\Delta\Big)$$

$$=-Wh(t)+q(t)\Big(\psi^\sigma(t)\int_a^{\sigma(t)}\phi^\sigma h\Delta-\phi^\sigma(t)\psi^\sigma(t)h(t)$$

$$+\phi^\sigma(t)\int_{\sigma(t)}^{\rho(b)}\psi^\sigma h\Delta+\phi^\sigma(t)\psi^\sigma(t)h(t)\Big)$$

$$=-Wh(t)+Wq(t)u^\sigma(t).$$

Consequently,, the function u, defined by (4.47), is a solution of equation $L(u)=h$. Note that $L^{-1}=G$ and $G^{-1}=L$.

Definition 4.6. A solution of problem (4.36), (4.37) will be defined to be a function $u\in X$ for which equation (4.36) holds on $\mathbb{T}^{\kappa^2}$.

The eigenvalue problem (4.36), (4.37) can be rewritten in the form

$$L(u)=\lambda w I_{XZ}^\sigma(u),\quad u\in X. \tag{4.49}$$

Substituting $u=Gz$, $z\in Z$, $\lambda=\dfrac{1}{\mu}$, and defining the operator $T=wI_{XZ}^\sigma G:Z\to X$, we get

$$L(Gz)=\frac{1}{\mu}wI_{XZ}^\sigma(Gz)$$

or

$$Tz=\mu z,\quad z\in Z. \tag{4.50}$$

Theorem 4.31. *The operator T is compact, self-adjoint and positive definite, with respect to the inner product $\langle\cdot,\cdot\rangle_w$, in the following senses*

$$\langle Tz_1,z_2\rangle_w=\langle z_1,Tz_2\rangle_w,\quad z_1,z_2\in Z,$$

$$\langle Tz,z\rangle_w>0,\quad z\in Z,z\neq 0.$$

The eigenvalue problems (4.49) *and* (4.50) *are equivalent, in the sense that λ is an eigenvalue of* (4.49) *if and only if $\mu=\dfrac{1}{\lambda}$ is an eigenvalue of* (4.50), *and the corresponding eigenfunctions u and z are related to each other by $u=Gz$.*

Proof. By Theorem 4.25, the operator $I_{XZ}^\sigma:X\to Z$ is injective and compact. Therefore, $T:Z\to Z$ is compact. Let now $z_1,z_2\in Z$ and

$$x_1=Gz_1,\quad x_2=Gz_2.$$

Then, using Theorem 4.26, we have

$$\begin{aligned}
\langle Tz_1, z_2\rangle_w &= \int_a^{\rho(b)} Tz_1 z_2 \frac{1}{w}\Delta \\
&= \int_a^b I_{XZ}^{\sigma}(Gz_1) z_2 \Delta \\
&= \langle I_{XZ}^{\sigma}(Gz_1), z_2\rangle \\
&= \langle I_{XZ}^{\sigma} x_1, z_2\rangle \\
&= \langle x_1^{\sigma}, z_2\rangle \\
&= \langle x_1^{\sigma}, L(x_2)\rangle \\
&= \langle L(x_1), x_2^{\sigma}\rangle \\
&= \langle z_1, I_{XZ}^{\sigma}(Gz_2)\rangle \\
&= \langle z_1, Tz_2\rangle_w.
\end{aligned}$$

Now, using that $\langle L(x_1), x_1^{\sigma}\rangle > 0$, $x_1 \in X$, $x_1 \neq 0$, we get

$$\begin{aligned}
\langle Tz_1, z_1\rangle_w &= \langle I_{XZ}^{\sigma}(Gz_1), z_1\rangle \\
&= \langle I_{XZ}^{\sigma}(x_1), L(x_1)\rangle \\
&= \langle x_1^{\sigma}, L(x_1)\rangle \\
&= \langle L(x_1), x_1^{\sigma}\rangle \\
&> 0
\end{aligned}$$

for $z_1 \in Z$, $z_1 \neq 0$. Note that I_{XZ}^{σ} and G are invertible operators.

1. Let u be an eigenfunction of problem (4.49) corresponding to the eigenvalue λ. Then, if $L(u) = z$, we find $u = Gz$ and

$$\begin{aligned}
z &= \lambda w I_{XZ}^{\sigma}(Gz) \\
&= \lambda Tz,
\end{aligned}$$

whereupon,

$$Tz = \frac{1}{\lambda} z.$$

Therefore, z is an eigenfunction of problem (4.50) corresponding to the eigenvalue $\frac{1}{\lambda}$.

2. Let z be an eigenfunction of problem (4.50) corresponding to the eigenvalue μ. Then, if $Gz = u$, we have $z = L(u)$ and

$$TL(u) = \mu L(u)$$

or

$$wI_{XZ}^{\sigma}(GL(u)) = \mu L(u),$$

or

$$wI_{XZ}^{\sigma}(u) = \mu L(u),$$

or

$$L(u) = \frac{1}{\mu} wI_{XZ}^{\sigma}(u).$$

Hence, u is an eigenfunction of problem (4.49) corresponding to the eigenvalue $\frac{1}{\mu}$. This completes the proof.

Theorem 4.32. *For each $k = 1, \ldots, N$, the operator T has an eigenvalue μ_k and a corresponding eigenfunction z_k such that the set of eigenfunctions $\{z_k\}$ is w-orthonormal and the following properties hold:*

1. *$\mu_k > \mu_{k+1}$. If $N = \infty$, then $\lim_{k \to \infty} \mu_k = 0$.*
2. *dim Ker$(T - \mu_k I_Z) = 1$.*
3. *For any $z \in Z$ we have*

$$z = \sum_{k=1}^{N} \langle z, z_k \rangle_w z_k. \tag{4.51}$$

Here I_Z is the identity operator on Z.

Proof. By the uniqueness of the solutions to the initial value problem for equation (4.36), and the equivalence of the eigenvalue problems (4.49) and (4.50), it follows that if λ_k is an eigenvalue, then dim Ker$(T - \mu_k I_Z) = 1$. Note that dim$Z = N$ and the set of eigenvalues of T spans Z. Hence, using that dim Ker$(T - \mu_k I_Z) = 1$, there must be exactly N eigenvalues μ_k, $k = 1, \ldots, N$, and a corresponding w-orthonormal set of eigenvectors

$$\{z_k : \quad k = 1, \ldots, N\}. \tag{4.52}$$

Then we order the eigenvalues so that $\mu_k > \mu_{k+1} > 0$. By the orthonormality of the set (4.52), it follows (4.51). This completes the proof.

Let $\{z_k\}$ be as in Theorem 4.32 and

$$\lambda_k = \frac{1}{\mu_k}, \quad u_k = Gz_k, \quad v_k = \frac{u_k}{\|wu_k^\sigma\|_w}.$$

Theorem 4.33. *For each $k = 1, \ldots, N$, λ_k is an eigenvalue of* (4.36)*, with corresponding eigenfunction $v_k \in X$ such that the set of eigenfunctions $\{wv_k^\sigma\}$ is w-orthonormal and the following properties hold.*

1. $\lambda_{k+1} > \lambda_k > 0$*. If $N = \infty$, then* $\lim_{k\to\infty} \lambda_k = \infty$.
2. *dim Ker*$(L - \lambda_k I_{XZ}^\sigma) = 1$.
3. *For any $z \in Z$ we have*

$$z = \sum_{k=1}^{N} \langle z, wv_k^\sigma \rangle_w v_k^\sigma. \tag{4.53}$$

Proof. By the equivalence of the eigenvalue problems (4.49), (4.50), we get

$$\begin{aligned} z_k &= L(u_k) \\ &= \lambda_k w u_k^\sigma. \end{aligned}$$

Hence,

$$\begin{aligned} 1 &= \|z_k\|_w \\ &= \lambda_k \|wu_k^\sigma\|_w. \end{aligned}$$

Therefore,

$$\begin{aligned} z_k &= w\lambda_k \|wu_k^\sigma\|_w \frac{u_k^\sigma}{\|wu_k^\sigma\|_w} \\ &= wv_k^\sigma. \end{aligned}$$

Thus, the set $\{wv_k^\sigma\}$ is w-orthonormal and, using (4.51), we find

$$\begin{aligned} z &= \sum_{k=1}^{N} \langle z, z_k \rangle_w z_k \\ &= \sum_{k=1}^{N} \langle z, wv_k^\sigma \rangle wv_k^\sigma, \end{aligned}$$

i.e., we get (4.53). This completes the proof.

Theorem 4.34 (Bessel's Equality). *Let $\{v_k\}$ be as in Theorem 4.33. For any $z \in Z$, we have*

$$\|z\|_w^2 = \sum_{k=1}^{N} \langle z, wv_k^\sigma \rangle_w^2$$

$$= \sum_{k=1}^{N} \langle z, v_k^\sigma \rangle^2.$$

Proof. By (4.53), we find

$$\begin{aligned}
\|z\|_w^2 &= \langle z, z\rangle_w \\
&= \langle \sum_{k=1}^{N} \langle z, w v_k^\sigma \rangle_w w v_k^\sigma, \sum_{l=1}^{N} \langle z, w v_l^\sigma \rangle_w w v_l^\sigma \rangle_w \\
&= \sum_{k,l=1}^{N} \langle z, w v_k^\sigma \rangle_w \langle z, w v_l^\sigma \rangle_w \langle w v_k^\sigma, w v_l^\sigma \rangle_w \\
&= \sum_{k=1}^{N} \langle z, w v_k^\sigma \rangle_w^2 \langle w v_k^\sigma, w v_k^\sigma \rangle_w \\
&= \sum_{k=1}^{N} \langle z, w v_k^\sigma \rangle_w^2 \\
&= \sum_{k=1}^{N} \langle z, v_k^\sigma \rangle^2.
\end{aligned}$$

This completes the proof.

Theorem 4.35. *Let $\{v_k\}$ be as in Theorem 4.33. For any $x \in X$, we have*

$$x = \sum_{k=1}^{N} \langle w x^\sigma, w v_k^\sigma \rangle_w v_k, \tag{4.54}$$

where the convergence is in the sense of the norm $|\cdot|_{0,\mathbb{T}}$.

Proof. First, we suppose that $N = \infty$. From equation

$$L(v_k) = \lambda_k w v_k^\sigma,$$

using Theorem 4.26, we get

$$\begin{aligned}\langle wx^\sigma, wv_k^\sigma\rangle &= \frac{1}{\lambda_k}\langle wx^\sigma, \lambda_k wv_k^\sigma\rangle_w \\ &= \frac{1}{\lambda_k}\langle wx^\sigma, L(v_k)\rangle_w \\ &= \frac{1}{\lambda_k}\langle x^\sigma, L(v_k)\rangle \\ &= \frac{1}{\lambda_k}\langle L(x), v_k^\sigma\rangle \\ &= \frac{1}{\lambda_k}\langle L(x), wv_k^\sigma\rangle_w.\end{aligned} \tag{4.55}$$

Next,

$$\frac{1}{\lambda_k}v_k = G(wv_k^\sigma)$$

and

$$\begin{aligned}\frac{1}{\lambda_k}v_k(t) &= \int_a^{\rho(b)} g(t,\cdot)wv_k^\sigma\Delta \\ &= \langle g(t,\cdot)w, v_k^\sigma\rangle, \quad t\in\mathbb{T}.\end{aligned}$$

Suppose that $t\in\mathbb{T}$ is arbitrarily chosen. Using Bessel's equality, we have

$$\begin{aligned}\sum_{k=1}^N \frac{1}{\lambda_k^2}(v_k(t))^2 &= \sum_{k=1}^N \langle g(t,\cdot)w, v_k^\sigma\rangle^2 \\ &= \|g(t,\cdot)w\|_w^2 \\ &\le c_1\end{aligned}$$

for some constant $c_1 > 0$. From the last inequality and from (4.55), we get for $n > m \ge 1$

$$\begin{aligned}&\left|\sum_{k=m}^n \langle wx^\sigma, wv_k^\sigma\rangle_w v_k(t)\right|^2 = \left|\sum_{k=m}^n \langle \lambda_k wx^\sigma, wv_k^\sigma\rangle_w \frac{1}{\lambda_k}v_k(t)\right| \\ &\le \sum_{k=m}^n \langle \lambda_k wx^\sigma, wv_k^\sigma\rangle_w^2 \sum_{k=m}^n \frac{1}{\lambda_k^2}(v_k(t))^2 \\ &\le c_1 \sum_{k=m}^n \langle L(x), wv_k^\sigma\rangle_w^2.\end{aligned}$$

By Bessel's equality, the series

$$\sum_{k=1}^{\infty} \langle L(x), wv_k^\sigma \rangle_w^2$$

converges. Since $t \in \mathbb{T}$ was arbitrarily chosen, the sequence of partial sums of the series in (4.55) forms a Cauchy sequence with respect to the norm $|\cdot|_{0,\mathbb{T}}$, and hence, the series converges with respect to this norm. Because $v_k \in X \subset \mathbb{C}(\mathbb{T})$ for any $k \in \mathbb{N}$, the sum of this series belongs to $\mathbb{C}(\mathbb{T})$. Now, we will prove that the series in (4.54) converges to x for $N < \infty$ or $N = \infty$. We define

$$z = wI_{XZ}^\sigma x \in Z.$$

We substitute it in (4.53) and we get

$$\begin{aligned} wI_{XZ}^\sigma x &= \sum_{k=1}^{N} \langle wI_{XZ}^\sigma x, wv_k^\sigma \rangle_w \\ &= \sum_{k=1}^{N} \langle wx^\sigma, wv_k^\sigma \rangle_w wv_k^\sigma \\ &= \sum_{k=1}^{N} \langle wx^\sigma, wv_k^\sigma \rangle_w wI_{XZ}^\sigma v_k \\ &= wI_{XZ}^\sigma \left(\sum_{k=1}^{N} \langle wx^\sigma, wv_k^\sigma \rangle_w v_k \right), \end{aligned}$$

whereupon,

$$wI_{XZ}^\sigma \left(x - \sum_{k=1}^{N} \langle wx^\sigma, wv_k^\sigma \rangle_w v_k \right) = 0.$$

Hence, using Theorem 4.25(I_{XZ}^σ is injective and compact), we get (4.54). This completes the proof.

4.4 Advanced Practical Problems

Problem 4.1. Let $\mathbb{T} = 2\mathbb{Z}$. Find a solution to the following eigenvalue problem:

$$u^{\Delta^2} + tu^\sigma + \lambda u^\sigma = 0$$

$$u(-2) = 0$$

$$2u(10) + 3u^\Delta(10) = 0.$$

Problem 4.2. Let $\mathbb{T} = 2^{\mathbb{N}_0}$. Find a solution to the following eigenvalue problem:

$$u^{\Delta^2} + \lambda u^{\sigma} = 0$$

$$u(2) + u^{\Delta}(2) = 0$$

$$u(16) = 0.$$

Problem 4.3. Let $\mathbb{T} = \mathbb{Z}$. Find a solution to the following eigenvalue problem:

$$u^{\Delta^2} + (1+\lambda)u^{\sigma} = 0$$

$$u(0) = 0$$

$$u(12) + u^{\Delta}(12) = 0.$$

Problem 4.4. Let $\mathbb{T} = 3^{\mathbb{N}_0}$. Find a solution to the following eigenvalue problem:

$$-u^{\Delta\nabla} = \lambda u$$

$$u(1) = 0$$

$$u(27) + u^{\Delta}(27) = 0.$$

Problem 4.5. Let $\mathbb{T} = 2^{\mathbb{N}_0}$. Find a solution to the following eigenvalue problem:

$$-u^{\Delta\nabla} + tu = \lambda u$$

$$u(1) - u^{\Delta}(1) = 0$$

$$u(32) = 0.$$

Problem 4.6. Let $\mathbb{T} = 3^{\mathbb{N}_0}$. Find a solution to the following eigenvalue problem:

$$-u^{\Delta\nabla} + \frac{1}{1+t}u = (1+\lambda)u$$

$$u(1) + 3u^{\Delta}(1) = 0$$

$$u(27) = 0.$$

4.5 Notes and References

In this chapter we introduce the Sturm-Liouville theory on time scales and investigate the first, second and third eigenvalue problems. Some of the results for the considered eigenvalue problems can be found in [9], [15], [31], [32], [47], [65].

Chapter 5
Boundary Value Problems for Nonlinear Second Order Dynamic Equations

Let $\mathbb{T}$ be a time scale with forward jump operator and delta differentiation operator σ and Δ, respectively.

5.1 Existence of Solutions of some Classes Second Order Boundary Value Problems on Regular Time Scales

In this section we suppose that the time scale $\mathbb{T}$ is a regular time scale, i.e., $\mathbb{T} = \mathbb{R}$ or all points of $\mathbb{T}$ are isolated. We will investigate for existence of solutions the following BVP

$$y^{\Delta^2} = f(t, y^{\sigma}), \quad t \in [a, b], \tag{5.1}$$

$$\begin{aligned} y(a) &= A, \\ y(\sigma^2(b)) &= B, \end{aligned} \tag{5.2}$$

where

(D1) $f \in \mathscr{C}([a, \sigma^2(b)] \times \mathbb{R})$,
(D2) $A, B \in \mathbb{R}$.

By the investigations in Chapter 3, the function

$$G(t,s) = \begin{cases} -\dfrac{(\sigma(s) - a)(\sigma^2(b) - t)}{\sigma^2(b) - a} & \text{if} \quad \sigma(s) \leq t \\ \\ -\dfrac{(t - a)(\sigma^2(b) - \sigma(s))}{\sigma^2(b) - a} & \text{if} \quad t \leq s \end{cases} \tag{5.3}$$

is the Green function for the BVP

DOI: 10.1201/9781003173557-5

$$\begin{aligned} y^{\Delta^2} &= 0 \\ y(a) &= 0 \\ y(\sigma^2(b)) &= 0. \end{aligned} \tag{5.4}$$

We will search a presentation of the solutions of the BVP (5.1), (5.2) in the following form

$$y(t) = \int_a^{\sigma(b)} G(t,s) f(s, y^\sigma(s)) \Delta s + a_1 t + a_2, \quad t \in [a, \sigma^2(b)],$$

where a_1 and a_2 are constants. By the first and second equations of (5.2), we get

$$\begin{aligned} y(a) &= a_1 a + a_2 \\ &= A, \\ y(\sigma^2(b)) &= a_1 \sigma^2(b) + a_2 \\ &= B, \end{aligned}$$

whereupon,

$$a_1(\sigma^2(b) - a) = B - A$$

or

$$a_1 = \frac{B-A}{\sigma^2(b) - a},$$

and

$$\begin{aligned} a_2 &= A - a a_1 \\ &= A - \frac{a(B-A)}{\sigma^2(b) - a} \\ &= \frac{A\sigma^2(b) - aA - aB + aA}{\sigma^2(b) - a} \\ &= \frac{A\sigma^2(b) - aB}{\sigma^2(b) - a}. \end{aligned}$$

Let

$$\phi(t) = \frac{B-A}{\sigma^2(b) - a} t + \frac{A\sigma^2(b) - aB}{\sigma^2(b) - a}, \quad t \in [a, \sigma^2(b)].$$

Then

$$y(t) = \int_a^{\sigma(b)} G(t,s) f(s, y^\sigma(s)) \Delta s + \phi(t), \quad t \in [a, \sigma^2(b)].$$

To prove existence of solutions of the BVP (5.1), (5.2) we will investigate a family of BVPs

$$\begin{aligned} y^{\Delta^2} &= \lambda f(t, y^{\sigma}), \quad t \in [a,b], \\ y(a) &= \lambda A, \\ y(\sigma^2(b)) &= \lambda B, \end{aligned} \tag{5.5}$$

where $\lambda \in [0,1]$. Let

$$\phi_\lambda(t) = \lambda \phi(t), \quad t \in [a, \sigma^2(b)].$$

Note that the solutions of the BVPs (5.5) can be represented in the form

$$y(t) = \phi_\lambda(t) + \lambda \int_a^{\sigma(b)} G(t,s) f(s, y^{\sigma}(s)) \Delta s, \quad t \in [a, \sigma^2(b)]. \tag{5.6}$$

Let

$$\beta = \left\{ |A|, |B|, \quad \frac{|B-A|}{\sigma^2(b)-a} \max\{|a|, |\sigma^2(b)|\} + \frac{|A\sigma^2(b) - aB|}{\sigma^2(b)-a} \right\}.$$

Lemma 5.1. *Suppose* $(D1)$ *and* $(D2)$. *Let* f *satisfy*

(D3) $\quad |f(t,p)| \le 2\alpha p f(t,p) + K, \quad t \in [a, \sigma^2(b)], \quad p \in \mathbb{R},$

where $\alpha \ge 0$ *and* $K \ge 0$ *are constants. If* $R > 0$ *satisfies the inequality*

(D4) $\quad R > \alpha \beta^2 + \beta + K \max_{t \in [a, \sigma^2(b)]} \int_a^{\sigma(b)} |G(t,s)| \Delta s,$

then any solution y *of the BVP* (5.5) *satisfies the inequality*

$$|y(t)| < R, \quad t \in [a, \sigma^2(b)], \quad \lambda \in [0,1].$$

Proof. Let y be a solution to the BVP (5.5) on $[a, \sigma^2(b)]$ for some $\lambda \in [0,1]$. Firstly, note that

$$\begin{aligned} |\phi(t)| &= \left| \frac{B-A}{\sigma^2(b)-a} t + \frac{A\sigma^2(b) - aB}{\sigma^2(b)-a} \right| \\ &\le \frac{|B-A|}{\sigma^2(b)-a} |t| + \frac{|A\sigma^2(b) - aB|}{\sigma^2(b)-a} \\ &\le \frac{|B-A|}{\sigma^2(b)-a} \max\{|a|, |\sigma^2(b)|\} + \frac{|A\sigma^2(b) - aB|}{\sigma^2(b)-a} \\ &= \beta, \quad t \in [a, \sigma^2(b)]. \end{aligned}$$

Then, using the representation (5.6), we arrive at

$$
\begin{aligned}
|y(t)| &= \left|\phi_\lambda(t)+\lambda\int_a^{\sigma(b)} G(t,s)f(s,y^\sigma(s))\Delta s\right| \\
&\le |\phi_\lambda(t)|+\lambda\left|\int_a^{\sigma(b)} G(t,s)f(s,y^\sigma(s))\Delta s\right| \\
&\le \lambda\int_a^{\sigma(b)} |G(t,s)||f(s,y^\sigma(s))|\Delta s+\lambda|\phi(t)| \\
&\le \lambda\int_a^{\sigma(b)} |G(t,s)||f(s,y^\sigma(s))|\Delta s+\beta \\
&\le \int_a^{\sigma(b)} |G(t,s)|\left(2\alpha\lambda f(s,y^\sigma(s))+\lambda K\right)\Delta s+\beta \\
&\le \int_a^{\sigma(b)} |G(t,s)|\left(2\alpha\lambda f(s,y^\sigma(s))+K\right)\Delta s+\beta, \quad t\in[a,\sigma^2(b)].
\end{aligned}
$$

Let

$$
r(t)=(y(t))^2, \quad t\in[a,\sigma^2(b)],
$$

and

$$
c(t)=\begin{cases} 1 \quad \text{if} \quad \mathbb{T}=\mathbb{R} \\ \dfrac{\mu^\sigma(t)}{\mu(t)} \quad \text{if} \quad \mathbb{T} \quad \text{is} \quad \text{isolated,} \end{cases}
$$

$t\in[a,\sigma^2(t)]$. Then

$$
\begin{aligned}
r^\Delta(t) &= y^\Delta(t)y(t)+y^\sigma(t)y^\Delta(t), \\
r^{\Delta^2}(t) &= y^{\Delta^2}(t)y^\sigma(t)+(y^\Delta(t))^2+y^{\Delta\sigma}(t)y^{\sigma\Delta}(t) \\
&\quad +y^{\Delta^2}(t)y^\sigma(t) \\
&= 2y^{\Delta^2}(t)y^\sigma(t)+(y^\Delta(t))^2+y^{\Delta\sigma}(t)y^{\sigma\Delta}(t),
\end{aligned}
$$

$t\in[a,\sigma^2(b)]$. Note that, if $\mathbb{T}$ is isolated, we have

$$
\begin{aligned}
y^{\sigma\Delta}(t) &= \left(y+\mu y^\Delta\right)^\Delta(t) \\
&= y^\Delta(t)+\mu^\sigma(t)y^{\Delta^2}(t)+\mu^\Delta(t)y^\Delta(t) \\
&= y^\Delta(t)+\mu^\sigma(t)y^{\Delta^2}(t)+\frac{\mu^\sigma(t)-\mu(t)}{\mu(t)}y^\Delta(t)
\end{aligned}
$$

$$= y^{\Delta}(t) + \mu^{\sigma}(t)y^{\Delta^2}(t) + \frac{\mu^{\sigma}(t)}{\mu(t)}y^{\Delta}(t) - y^{\Delta}(t)$$

$$= \frac{\mu^{\sigma}(t)}{\mu(t)}\left(y^{\Delta}(t) + \mu(t)y^{\Delta^2}(t)\right)$$

$$= \frac{\mu^{\sigma}(t)}{\mu(t)}y^{\Delta\sigma}(t), \quad t \in [a, \sigma^2(b)].$$

Therefore,

$$y^{\sigma\Delta}(t) = c(t)y^{\Delta\sigma}(t), \quad t \in [a, \sigma^2(b)],$$

and

$$r^{\Delta^2}(t) = 2y^{\Delta^2}(t)y^{\sigma}(t) + (y^{\Delta}(t))^2 + c(t)(y^{\Delta\sigma}(t))^2$$

$$\geq 2y^{\Delta^2}(t)y^{\sigma}(t)$$

$$= 2\lambda y^{\sigma}(t)f(t, y^{\sigma}(t)), \quad t \in [a, \sigma^2(b)].$$

Therefore,

$$|y(t)| \leq \int_a^{\sigma(b)} |G(t,s)|\left(\alpha r^{\Delta^2}(s) + K\right)\Delta s + \beta, \quad t \in [a, \sigma^2(b)]. \tag{5.7}$$

Observe that

$$\int_a^{\sigma(b)} |G(t,s)|r^{\Delta^2}(s)\Delta s = \int_a^t |G(t,s)|r^{\Delta^2}(s)\Delta s + \int_t^{\sigma^2(b)} |G(t,s)|r^{\Delta^2}(s)\Delta s$$

$$= \frac{\sigma^2(b) - t}{\sigma^2(b) - a}\int_a^t (\sigma(s) - a)r^{\Delta^2}(s)\Delta s$$

$$+ \frac{t - a}{\sigma^2(b) - a}\int_t^{\sigma^2(b)} (\sigma^2(b) - \sigma(s))r^{\Delta^2}(s)\Delta s$$

$$= \frac{\sigma^2(b) - t}{\sigma^2(b) - a}\left((s - a)r^{\Delta}(s)\Big|_{s=a}^{s=t} - \int_a^t r^{\Delta}(s)\Delta s\right)$$

$$+ \frac{t - a}{\sigma^2(b) - a}\left((\sigma^2(b) - s)r^{\Delta}(s)\Big|_{s=t}^{s=\sigma^2(b)} + \int_t^{\sigma^2(b)} r^{\Delta}(s)\Delta s\right)$$

$$= \frac{\sigma^2(b) - t}{\sigma^2(b) - a}\left((t - a)r^{\Delta}(t) - r(t) + r(a)\right)$$

$$
\begin{aligned}
&+\frac{t-a}{\sigma^2(b)-a}\left(-(\sigma^2(b)-t)r^{\Delta}(t)+r(\sigma^2(b))-r(t)\right)\\
&\le \frac{\sigma^2(b)-t}{\sigma^2(b)-a}\left((t-a)r^{\Delta}(t)+A^2\right)\\
&+\frac{t-a}{\sigma^2(b)-a}\left(-(\sigma^2(b)-t)r^{\Delta}(t)+B^2\right)\\
&\le \frac{t\sigma^2(b)-t^2-a\sigma^2(b)+at-t\sigma^2(b)+t^2+a\sigma^2(b)-at}{\sigma^2(b)-a}r^{\Delta}(t)\\
&+\frac{\sigma^2(b)-t}{\sigma^2(b)-a}\beta^2+\frac{t-a}{\sigma^2(b)-a}\beta^2\\
&=\beta^2.
\end{aligned}
$$

Now, applying (5.7), we obtain

$$
\begin{aligned}
|y(t)| &\le \alpha\beta^2+\beta+K\max_{t\in[a,\sigma^2(b)]}\int_a^{\sigma(b)}|G(t,s)|\Delta s\\
&< R, \quad t\in[a,\sigma^2(b)].
\end{aligned}
$$

This completes the proof.

Suppose that $\mathscr{C}([a,\sigma^2(b)])$ is endowed with the maximum norm. For $y\in\mathscr{C}([a,\sigma^2(b)])$, define the operator

$$
Ty(t)=\phi(t)+\int_a^{\sigma(b)}G(t,s)f(s,y^{\sigma}(s))\Delta s, \quad t\in[a,\sigma^2(b)].
$$

We have $T:\mathscr{C}([a,\sigma^2(b)])\to\mathscr{C}([a,\sigma^2(b)])$.

Lemma 5.2. *Suppose* $(D1)$ *and* $(D2)$*. Then* $T:\mathscr{C}([a,\sigma^2(b)])\to\mathscr{C}([a,\sigma^2(b)])$ *is a continuous operator.*

Proof. Firstly, note that

$$
|G(t,s)|\le\sigma(b)-a, \quad t,s\in[a,\sigma^2(b)].
$$

Take $\varepsilon>0$ arbitrarily. Let $\{y_n\}_{n\in\mathbb{N}}\subset\mathscr{C}([a,\sigma^2(b)])$, $y\in\mathscr{C}([a,\sigma^2(b)])$ and

$$
y_n\to y \quad \text{in} \quad \mathscr{C}([a,\sigma^2(b)]), \quad \text{as} \quad n\to\infty.
$$

Since $f\in\mathscr{C}([a,\sigma^2(b)]\times\mathbb{R})$, there is an $N\in\mathbb{N}$ such that

$$
|f(t,y_n^{\sigma}(t))-f(t,y^{\sigma}(t))|<\frac{\varepsilon}{(\sigma(b)-a)(\sigma^2(b)-a)}, \quad t\in[a,\sigma^2(b)],
$$

for any $n > N$. Hence,

$$\begin{aligned}
|Ty_n(t) - Ty(t)| &= \left| \phi(t) + \int_a^{\sigma(b)} G(t,s) f(s, y_n^\sigma(s)) \Delta s \right. \\
&\quad \left. -\phi(t) - \int_a^{\sigma(b)} f(s, y^\sigma(s)) \Delta s \right| \\
&= \left| \int_a^{\sigma(b)} G(t,s)(f(s, y_n^\sigma(s)) - f(s, y^\sigma(s))) \Delta s \right| \\
&\leq \int_a^{\sigma(b)} |G(t,s)||f(s, y_n^\sigma(s)) - f(s, y^\sigma(s))| \Delta s \\
&< (\sigma(b) - a)(\sigma^2(b) - a) \frac{\varepsilon}{(\sigma(b) - a)(\sigma^2(b) - a)} \\
&= \varepsilon
\end{aligned}$$

for any $t \in [a, \sigma^2(b)]$ and for any $n > N$. Thus, $T : \mathscr{C}([a, \sigma^2(b)]) \to \mathscr{C}([a, \sigma^2(b)])$ is a continuous operator. This completes the proof.

Lemma 5.3. *Suppose* $(D1)$ *and* $(D2)$. *Then* $T : \mathscr{C}([a, \sigma^2(b)]) \to \mathscr{C}([a, \sigma^2(b)])$ *is a completely continuous operator.*

Proof. Let $X \subset \mathscr{C}([a, \sigma^2(b)])$ is a bounded set. Then there exists a constant $L_1 > 0$ so that

$$|y(t)| \leq L_1, \quad t \in [a, \sigma^2(b)],$$

for any $y \in X$. Since $f \in \mathscr{C}([a, \sigma^2(b)] \times \mathbb{R})$, there exists a positive constant L_2 so that

$$|f(t,u)| \leq L_2, \quad t \in [a, \sigma^2(b)], \quad u \in [-L_1, L_1].$$

Then, for $y \in X$, we get

$$\begin{aligned}
|Ty(t)| &= \left| \phi(t) + \int_a^{\sigma(b)} G(t,s) f(s, y^\sigma(s)) \Delta s \right| \\
&\leq |\phi(t)| + \left| \int_a^{\sigma(b)} G(t,s) f(s, y^\sigma(s)) \Delta s \right| \\
&\leq \beta + \int_a^{\sigma(b)} |G(t,s)||f(s, y^\sigma(s))| \Delta s \\
&\leq \beta + (\sigma(b) - a)(\sigma^2(b) - a) L_2, \quad t \in [a, \sigma^2(b)].
\end{aligned}$$

Therefore, $T(X)$ is uniformly bounded. Now, take $t_1, t_2 \in [a, \sigma^2(b)]$ arbitrarily so that $t_1 < t_2$. Then, for $y \in \mathscr{C}([a, \sigma^2(b)])$, we have

$$
\begin{aligned}
|Ty(t_2) - Ty(t_1)| &= \left|\phi(t_2) + \int_a^{\sigma(b)} G(t_2,s) f(s, y^\sigma(s))\Delta s \right. \\
&\quad \left. -\phi(t_1) - \int_a^{\sigma(b)} G(t_1,s) f(s, y^\sigma(s))\Delta s\right| \\
&= \left|(\phi(t_2) - \phi(t_1)) + \int_a^{\sigma(b)} (G(t_2,s) - G(t_1,s)) f(s, y^\sigma(s))\Delta s\right| \\
&\leq |\phi(t_2) - \phi(t_1)| + \left|\int_a^{\sigma(b)} (G(t_2,s) - G(t_1,s)) f(s, y^\sigma(s))\Delta s\right| \\
&\leq |\phi(t_2) - \phi(t_1)| + \int_a^{\sigma(b)} |G(t_2,s) - G(t_1,s)||f(s, y^\sigma(s))|\Delta s.
\end{aligned}
$$

Consequently, $T(X)$ is equi-continuous. Now, applying the Arzela-Ascoli theorem, we conclude that $T : \mathscr{C}([a, \sigma^2(b)]) \to \mathscr{C}([a, \sigma^2(b)])$ is a completely continuous operator. This completes the proof.

Theorem 5.1. *Suppose* $(D1)$-$(D4)$. *Then the BVP* (5.1), (5.2) *has at least one solution.*

Proof. Let

$$
S = \{x \in \mathscr{C}([a, \sigma^2(b)]) : x = \lambda T x, \quad \lambda \in [0,1)\}.
$$

By Lemma 5.3, we have that $T : \mathscr{C}([a, \sigma^2(b)]) \to \mathscr{C}([a, \sigma^2(b)])$ is a completely continuous operator. By Lemma 5.1, it follows that

$$
|x(t)| < R, \quad t \in [a, \sigma^2(b)],
$$

for any $x \in \mathscr{C}([a, \sigma^2(b)])$. Hence and by Schaefer's fixed point Theorem 1.3, we conclude that the operator T has at least one fixed point $y \in \mathscr{C}([a, \sigma^2(b)])$ for which

$$
|y(t)| < R, \quad t \in [a, \sigma^2(b)].
$$

This completes the proof.

Example 5.1. Let $\mathbb{T} = \mathbb{Z}$. Consider the BVP

$$
y^{\Delta^2} = t(y^\sigma)^3
$$

$$
y(0) = 0
$$

$$
y(5) = 1.
$$

Here

$$
\sigma(t) = t + 1,
$$

$$f(t,p) = tp^3, \quad t \in [0,5], \quad p \in \mathbb{R},$$

$$a = 0,$$

$$b = 3,$$

$$A = 0,$$

$$B = 1$$

and

$$G(t,s) = \begin{cases} -\dfrac{(s+1)(5-t)}{5} & \text{if} \quad s+1 \leq t \\ -\dfrac{t(5-s)}{5} & \text{if} \quad t \leq s, \end{cases}$$

$t,s \in [0,5]$. Let

$$g(s) = \frac{s^2 - s}{2}, \quad s \in \mathbb{T}.$$

We have

$$\begin{aligned} g^{\Delta}(s) &= \frac{1}{2}(\sigma(s) + s - 1) \\ &= \frac{1}{2}(s+1+s-1) \\ &= s, \quad s \in \mathbb{T}, \end{aligned}$$

and

$$\begin{aligned} \int_a^{\sigma(b)} |G(t,s)|\Delta s &= \int_0^5 |G(t,s)|\Delta s \\ &= \int_0^t |G(t,s)|\Delta s + \int_t^5 |G(t,s)|\Delta s \\ &= \frac{5-t}{5}\int_0^t (s+1)\Delta s + \frac{t}{5}\int_t^5 (5-s)\Delta s \\ &= \frac{5-t}{5}\left(\int_0^t s\Delta s + t\right) + \frac{t}{5}\left(5(5-t) - \int_t^5 s\Delta s\right) \end{aligned}$$

$$
\begin{aligned}
&= \frac{5-t}{5}\left(\int_0^t g^{\Delta}(s)\Delta s + t\right) + \frac{t}{5}\left(25-5t-\int_t^5 g^{\Delta}(s)\Delta s\right)\\
&= \frac{5-t}{5}(g(t)-g(0)+t) + \frac{t}{5}(25-5t-g(5)+g(t))\\
&= \frac{5-t}{5}\left(\frac{t^2-t}{2}+t\right) + \frac{t}{5}\left(25-5t-10+\frac{t^2-t}{2}\right)\\
&= \frac{5-t}{10}(t^2+t) + \frac{t}{10}(30-10t+t^2-t)\\
&= \frac{5-t}{10}(t^2+t) + \frac{t}{10}(30-11t+t^2)\\
&= \frac{1}{10}(5t^2+5t-t^3-t^2+30t-11t^2+t^3)\\
&= \frac{1}{10}(-7t^2+35t)\\
&= \frac{7t}{10}(5-t), \quad t \in [0,5],
\end{aligned}
$$

and

$$
\begin{aligned}
\max_{t\in[0,5]} \int_a^{\sigma(b)} |G(t,s)|\Delta s &= \max_{t\in[0,5]}\left(\frac{7t}{10}(5-t)\right)\\
&= \frac{7}{10}\left(\frac{5}{2}\right)\left(5-\frac{5}{2}\right)\\
&= \frac{7}{4}\cdot\frac{5}{2}\\
&= \frac{35}{8}.
\end{aligned}
$$

Next,

$$
\begin{aligned}
|f(t,p)| &\le t(p^4+1)\\
&= pf(t,p)+t\\
&\le pf(t,p)+5, \quad t\in[0,5].
\end{aligned}
$$

Then

$$
\alpha = \frac{1}{2}, \quad K = 5.
$$

Moreover,

$$\beta = \max\{0,1,\frac{1}{5}\max\{0,5\}+0\}$$

$$= 1$$

and

$$R > \frac{1}{2}+1+5\cdot\frac{35}{8}$$

$$= \frac{3}{2}+\frac{175}{8}$$

$$= \frac{12+175}{8}$$

$$= \frac{187}{8}.$$

By Theorem 5.1, it follows that the considered BVP has at least one solution in $\mathscr{C}([0,5])$.

Exercise 5.1. Let $\mathbb{T}=2^{\mathbb{N}_0}$. Prove that the BVP

$$y^{\Delta^2} = t(y^{\sigma})^5,$$

$$y(1) = 1,$$

$$y(32) = -1$$

has at least one solution y for which

$$|y(t)| \leq 10000, \quad t \in [1,32].$$

Theorem 5.2. *Suppose* (*D*1). *Let* $R>0$ *be a constant such that*

$$uf(t,u) > 0, \quad t \in [a,b], \quad |u| > R.$$

If y *is a solution of equation* (5.1) *and* $|y(t)|$ *does not achieve its maximum value at* $t=a$ *or at* $t=\sigma^2(b)$, *then*

$$|y(t)| < R, \quad t \in [a,\sigma^2(b)].$$

Proof. Assume the contrary. Let

$$r(t) = (y(t))^2 - T, \quad t \in [a,\sigma^2(b)].$$

Then r must have a non-negative maximum. Since $|y(t)|$ does not achieve its maximum value at $t = a$ or at $t = \sigma^2(b)$, we have that there exists a $c \in (a, \sigma^2(b))$ such that the function r achieves its maximum value at $t = c$. Note that

$$r(t) \geq 0, \quad r(t) \leq r(c), \quad t \in (c, \sigma^2(b)). \tag{5.8}$$

Assume that

$$\rho(c) = c < \sigma(c).$$

If $r^{\Delta}(c) \geq 0$, then $r(\sigma(c)) \geq r(c)$, which contradicts with (5.8). If $r^{\Delta}(c) < 0$, then

$$\lim_{t \to c-} r^{\Delta}(t) = r^{\Delta}(c) < 0.$$

Therefore, there exists a $\delta > 0$ such that $r^{\Delta}(t) < 0$ for any $t \in (c - \delta, c]$. Hence, r is strictly decreasing on $(c - \delta, c]$. This is a contradiction because r achieves its maximum value at c. Therefore, c can not be simultaneously left-dense and right-scattered. Since r achieves its maximum value at c, we have

$$r^{\Delta}(\rho(c)) \leq 0.$$

Moreover,

$$\begin{aligned}
r^{\Delta}(t) &= y^{\Delta}(t)y^{\sigma}(t) + y(t)y^{\Delta}(t), \\
r^{\Delta^2}(t) &= y^{\Delta^2}(t)y^{\sigma}(t) + (y^{\Delta}(t))^2 \\
&\quad + y^{\Delta\sigma}(t)y^{\sigma\Delta}(t) + y^{\Delta^2}(t)y^{\sigma}(t) \\
&= 2y^{\Delta^2}(t)y^{\sigma}(t) + (y^{\Delta}(t))^2 + y^{\Delta\sigma}(t)y^{\sigma\Delta}(t) \\
&\geq 2f(t, y^{\sigma}(t))y^{\sigma}(t) + y^{\Delta\sigma}(t)y^{\sigma\Delta}(t), \quad t \in [a, \sigma^2(b)].
\end{aligned}$$

Hence,

$$r^{\Delta}(\rho(c)) \geq 2f(\rho(c), y^{\sigma}(\rho(c)))y^{\sigma}(\rho(c)) + y^{\Delta\sigma}(\rho(c))y^{\sigma\Delta}(\rho(c)).$$

Since c cannot be simultaneously left-dense and right-scattered, we have the following cases:

Case 1. Let

$$\rho(c) < c = \sigma(c).$$

Then

$$y^{\Delta\sigma}(\rho(c)) = y^{\Delta}(\sigma(\rho(c)))$$

$$= y^{\Delta}(c),$$

$$\begin{aligned} y^{\sigma\Delta}(c) &= \frac{y^{\sigma^2}(\rho(c)) - y^{\sigma}(\rho(c))}{\sigma(\rho(c)) - \rho(c)} \\ &= \frac{y^{\sigma}(\sigma(\rho(c))) - y(\sigma(\rho(c)))}{c - \rho(c)} \\ &= \frac{y^{\sigma}(c) - y(c)}{c - \rho(c)} \\ &= 0. \end{aligned}$$

Thus,

$$y^{\Delta\sigma}(\rho(c))y^{\sigma\Delta}(c) = 0.$$

Case 2. Let

$$\rho(c) < c < \sigma(c).$$

Then

$$\begin{aligned} y^{\Delta\sigma}(\rho(c)) &= y^{\Delta}(\sigma(\rho(c))) \\ &= y^{\Delta}(c) \end{aligned}$$

and using the computation in the previous case, we obtain

$$\begin{aligned} y^{\sigma\Delta}(\rho(c)) &= \frac{y^{\sigma}(c) - y(c)}{c - \rho(c)} \\ &= \frac{y^{\sigma}(c) - y(c)}{\sigma(c) - c} \cdot \frac{\sigma(c) - c}{c - \rho(c)} \\ &= y^{\Delta}(c)\frac{\sigma(c) - c}{c - \rho(c)}. \end{aligned}$$

Therefore,

$$\begin{aligned} y^{\Delta\sigma}(\rho(c))y^{\sigma\Delta}(\rho(c)) &= (y^{\Delta}(c))^2\frac{\sigma(c) - c}{c - \rho(c)} \\ &\geq 0. \end{aligned}$$

Case 3. Let

$$\rho(c) = c = \sigma(c).$$

We have

$$y^{\Delta\sigma}(\rho(c)) = y^{\Delta}(c),$$

$$\begin{aligned} y^{\sigma\Delta}(c) &= \lim_{t\to c}\frac{y^{\sigma}(t)-y^{\sigma}(c)}{t-c} \\ &= \lim_{t\to c}\frac{y(t)-y(c)}{t-c} \\ &= y^{\Delta}(c). \end{aligned}$$

Therefore,

$$\begin{aligned} y^{\sigma\Delta}(\rho(c))y^{\Delta\sigma}(\rho(c)) &= (y^{\Delta}(c))^2 \\ &\geq 0. \end{aligned}$$

Consequently,

$$\begin{aligned} r^{\Delta^2}(\rho(c)) &\geq 2f(\rho(c), y^{\sigma}(\rho(c)))y^{\sigma}(\rho(c)) \\ &> 0. \end{aligned}$$

This is a contradiction. This completes the proof.

Corollary 5.1. *Suppose* $(D5)$ *and* f *and* R *satisfy all conditions of Theorem 5.2. Let, also,* $|A|, |B| < R$. *Then any solution* y *of the BVP* (5.1), (5.2) *satisfies*

$$|y(t)| < R, \quad t \in [a, \sigma^2(b)].$$

Theorem 5.3. *Suppose* $(D5)$ *and* f *and* R *satisfy the conditions of Theorem 5.2. Consider equation* (5.1) *subject to the boundary conditions*

$$\begin{aligned} \alpha y(a) - \beta y^{\Delta}(a) &= C \\ \gamma y(\sigma^2(b)) + \delta y^{\Delta}(\sigma(b)) &= D, \end{aligned} \tag{5.9}$$

where

(D5) $\alpha, \beta, \gamma, \delta > 0.$

Then any solution of the BVP (5.1), (5.9) *satisfies the estimate*

$$|y(t)| < \max\left\{\frac{|C|}{\alpha}, \frac{|D|}{\gamma}, R\right\} + 1, \quad t \in [a, \sigma^2(b)].$$

Proof. Let

$$M = \max\left\{\frac{|C|}{\alpha}, \frac{|D|}{\gamma}, R\right\}.$$

Set

$$r(t) = (y(t))^2 - (M+1)^2, \quad t \in [a, \sigma^2(b)].$$

Assume that y has a non-negative maximum at $t = a$. Then

$$\begin{aligned} r^\Delta(t) &= y^\Delta(t) y^\sigma(t) + y^\Delta(t) y(t) \\ &= y^\Delta(t)\left(y^\sigma(t) + y(t)\right), \quad t \in [a, \sigma^2(b)], \end{aligned}$$

and

$$\begin{aligned} 0 &\geq r^\Delta(a) \\ &= y^\Delta(a)(y^\sigma(a) + y(a)) \\ &= y^\Delta(a)\left(y(a) + \mu(a) y^\Delta(a) + y(a)\right) \\ &= 2y(a) y^\Delta(a) + \mu(a)(y^\Delta(a))^2. \end{aligned}$$

From here, we conclude that

$$\begin{aligned} 2y(a) y^\Delta(a) &\leq -\mu(a)(y^\Delta(a))^2 \\ &\leq 0 \end{aligned}$$

and

$$\begin{aligned} 0 &\geq \beta y(a) y^\Delta(a) \\ &= y(a)(\alpha y(a) - C) \\ &= \alpha(y(a))^2 - C y(a) \\ &= \alpha(y(a))^2\left(1 - \frac{C}{\alpha y(a)}\right). \end{aligned}$$

Therefore,

$$\frac{C}{\alpha y(a)} \geq 1$$

or

$$y(a) \leq \frac{C}{\alpha}$$

$$\leq M.$$

Now, we assume that y has a non-negative maximum at $t = \sigma^2(b)$. Then

$$\begin{aligned} r^{\Delta}(\sigma(b)) &= 2y(\sigma^2(b))y^{\Delta}(\sigma(b)) - \mu(\sigma(b))(y^{\Delta}(\sigma(b)))^2 \\ &\geq 0 \end{aligned}$$

and

$$\begin{aligned} 2y(\sigma^2(b))y^{\Delta}(\sigma(b)) &\geq \mu(\sigma(b))(y^{\Delta}(\sigma(b)))^2 \\ &\geq 0. \end{aligned}$$

Therefore,

$$\begin{aligned} 0 &\leq \delta y(\sigma^2(b))y^{\Delta}(\sigma(b)) \\ &= y(\sigma^2(b))\left(D - \gamma y(\sigma^2(b))\right) \\ &= \gamma(y(\sigma^2(b)))^2\left(\frac{D}{\gamma y(\sigma^2(b))} - 1\right) \end{aligned}$$

and then

$$\frac{D}{\gamma y(\sigma^2(b))} - 1 \geq 0,$$

or

$$\begin{aligned} y(\sigma^2(b)) &\leq \frac{D}{\gamma} \\ &\leq M. \end{aligned}$$

If y achieves a non-negative maximum value at a point in $(a, \sigma^2(b))$, then, using Theorem 5.2, we get

$$|y(t)| \leq R, \quad t \in [a, \sigma^2(b)].$$

This completes the proof.

Corollary 5.2. *Assume* $(D1)$, $(D5)$ *and* R, f *satisfy all conditions of Theorem 5.2. If* $|B| < R$, *then any solution* y *of* (5.1) *subject to the boundary condition*

$$\begin{aligned} \alpha y(a) - \beta y^{\Delta}(a) &= C \\ y(\sigma^2(b)) &= B \end{aligned} \tag{5.10}$$

satisfies the estimate

$$|y(t)| \leq \max\left\{\frac{|C|}{\alpha}, R\right\} + 1, \quad t \in [a, \sigma^2(b)].$$

Corollary 5.3. *Assume* $(D1)$, $(D5)$ *and* R, f *satisfy all conditions of Theorem 5.2. If* $|A| < R$, *then any solution* y *of* (5.1) *subject to the boundary condition*

$$\begin{aligned} y(a) &= A \\ \gamma y(\sigma^2(b)) + \delta y^{\Delta}(\sigma(b)) &= D \end{aligned} \tag{5.11}$$

satisfies the estimate

$$|y(t)| \leq \max\left\{\frac{|D|}{\gamma}, R\right\} + 1, \quad t \in [a, \sigma^2(b)].$$

For our next result we need the following fixed point theorem.

Theorem 5.4 (Leray-Shauder Fixed Point Theorem). *Let* Ω *be an open, convex and bounded subset of a Banach space* X *and let* $T : \overline{\Omega} \to X$ *be a compact operator. If* $y \neq \lambda T y$ *for all* $y \in \partial\Omega$ *and all* $\lambda \in [0,1]$, *then* $y = Ty$ *for some* $y \in \Omega$.

Theorem 5.5. *Suppose* $(D1)$, $(D2)$ *and let* R *and* f *satisfy all conditions of Theorem 5.2. If* $|A|, |B| < R$, *then the BVP* (5.1), (5.2) *has at least one solution* y *such that*

$$|y(t)| < R, \quad t \in [a, \sigma^2(b)].$$

Proof. Let

$$S = \{y : y \in \mathscr{C}([a, \sigma^2(b)]), y^{\Delta^2} \in \mathscr{C}_{rd}([a,b])\},$$

$$\Omega = \{y \in S : \|y\| < R\}.$$

Note that

$$\lambda y f(t,y) > 0, \quad \text{for} \quad |y| > R,$$

$$|\lambda A|, \quad |\lambda B| < R, \quad \lambda \in (0,1].$$

Hence and by Theorem 5.2, it follows that for any solution y of the BVP

$$y^{\Delta^2}(t) = \lambda f(t, y(\sigma(t))), \quad t \in [a,b],$$

$$y(a) = \lambda A,$$

$$y(\sigma^2(b)) = \lambda B,$$

we have

$$|y(t)| < R, \quad t \in [a, \sigma^2(b)].$$

Therefore,

$$y \neq \lambda T y, \quad y \in \partial\Omega, \quad \lambda \in [0,1].$$

Hence and by Leray-Schauder's fixed point theorem, it follows that the BVP (5.1), (5.2) has at least one solution y for which

$$|y(t)| < R, \quad t \in [a, \sigma^2(b)].$$

This completes the proof.

Note that the function

$$G_1(t,s) = \begin{cases} -\dfrac{(\beta+(t-a)\alpha)(\delta+(\sigma^2(b)-\sigma(s))\gamma)}{\alpha\gamma(\sigma^2(b)-a)+\alpha\delta+\beta\gamma} & \text{if} \quad t \leq s \\ -\dfrac{(\beta+(\sigma(s)-a)\alpha)(\delta+(\sigma^2(b)-t)\gamma)}{\alpha\gamma(\sigma^2(b)-a)+\alpha\delta+\beta\gamma} & \text{if} \quad \sigma(s) \leq t \end{cases}$$

is the Green function for the BVP

$$y^{\Delta^2} = 0, \quad t \in [a,b],$$

$$\alpha y(a) - \beta y^{\Delta}(a) = 0,$$

$$\gamma y(\sigma^2(b)) + \delta y^{\Delta}(\sigma(b)) = 0.$$

We will search a presentation of the solution y of the BVP (5.1), (5.9) in the following form

$$y(t) = \int_a^{\sigma(b)} G_1(t,s) f(s, y^{\sigma}(s)) \Delta s + a_1 t + a_2, \quad t \in [a, \sigma^2(b)],$$

where $a_1, a_2 \in \mathbb{R}$ will be determined by using the boundary conditions (5.9). We have

$$C = \alpha(a_1 a + a_2) - \beta a_1$$

$$D = \gamma(a_1 \sigma^2(b) + a_2) + \delta a_1$$

or

$$C = (a\alpha - \beta)a_1 + \alpha a_2$$

$$D = (\gamma\sigma^2(b) + \delta)a_1 + \gamma a_2,$$

whereupon,

$$C\gamma - D\alpha = ((a\alpha-\beta)\gamma - \alpha(\gamma\sigma^2(b)+\delta))a_1$$
$$= (-\alpha\gamma(\sigma^2(b)-a)-\alpha\delta-\beta\gamma)a_1$$

or

$$a_1 = \frac{D\alpha - C\gamma}{\alpha\gamma(\sigma^2(b)-a)+\alpha\delta+\beta\gamma},$$

and

$$\begin{aligned} a_2 &= \frac{C-(a\alpha-\beta)a_1}{\alpha} \\ &= \frac{C - \frac{(a\alpha-\beta)(D\alpha-C\gamma)}{\alpha\gamma(\sigma^2(b)-a)+\alpha\delta+\beta\gamma}}{\alpha} \\ &= \frac{C\gamma\alpha\sigma^2(b) - C\gamma\alpha a + C\gamma\beta + C\alpha\delta - a\alpha^2 D + a\alpha C\gamma + \beta D\alpha - \beta C\gamma}{\alpha(\alpha\gamma(\sigma^2(b)-a)+\alpha\delta+\beta\gamma)} \\ &= \frac{C(\gamma\sigma^2(b)+\delta)+D(\beta-a\alpha)}{\alpha\gamma(\sigma^2(b)-a)+\alpha\delta+\beta\gamma}. \end{aligned}$$

Consequently,

$$\begin{aligned} y(t) &= \int_a^{\sigma(b)} G_1(t,s)f(s,y^\sigma(s))\Delta s + \frac{D\alpha - C\gamma}{\alpha\gamma(\sigma^2(b)-a)+\alpha\delta+\beta\gamma}t \\ &\quad + \frac{C(\gamma\sigma^2(b)+\delta)+D(\beta-a\alpha)}{\alpha\gamma(\sigma^2(b)-a)+\alpha\delta+\beta\gamma}, \end{aligned}$$

$t \in [a,\sigma^2(b)]$. For $y \in \mathscr{C}([a,\sigma^2(b)])$, define the operator

$$\begin{aligned} T_1y(t) &= \int_a^{\sigma(b)} G_1(t,s)f(s,y^\sigma(s))\Delta s + \frac{D\alpha - C\gamma}{\alpha\gamma(\sigma^2(b)-a)+\alpha\delta+\beta\gamma}t \\ &\quad + \frac{C(\gamma\sigma^2(b)+\delta)+D(\beta-a\alpha)}{\alpha\gamma(\sigma^2(b)-a)+\alpha\delta+\beta\gamma}, \end{aligned}$$

$t \in [a,\sigma^2(b)]$.

Exercise 5.2. Prove that $T_1 : \mathscr{C}([a,\sigma^2(b)]) \to \mathscr{C}([a,\sigma^2(b)])$ is a completely continuous operator.

Theorem 5.6. *Suppose* (*D*1), (*D*5) *and let f and R satisfy all conditions of Theorem* 5.2. *Then the BVP* (5.1), (5.9) *has at least one solution y satisfying*

$$|y(t)| < \max\left\{\frac{|C|}{\alpha}, \frac{|D|}{\gamma}, R\right\} + 1, \quad t \in [a, \sigma^2(b)].$$

Proof. Let S be the set as in the proof of Theorem 5.5. Define

$$\Omega_1 = \left\{ y \in S : |y(t)| \le \max\left\{\frac{|C|}{\alpha}, \frac{|D|}{\gamma}, R\right\} + 1, \quad t \in [a, \sigma^2(b)] \right\}.$$

To prove that T_1 has a fixed point in Ω_1, consider equation

$$y = \lambda T_1 y, \quad \lambda \in [0,1],$$

which is equivalent to the BVP

$$\begin{aligned} y^{\Delta^2} &= \lambda f(t, y^\sigma), \quad t \in [a,b], \\ \alpha y(a) - \beta y^\Delta(a) &= \lambda C, \\ \gamma y(\sigma^2(b)) + \delta y^\Delta(\sigma(b)) &= \lambda D. \end{aligned} \tag{5.12}$$

By Theorem 5.3, it follows that any solution of the BVP (5.12) satisfies

$$|y(t)| < \max\left\{\frac{|C|}{\alpha}, \frac{|D|}{\gamma}, R\right\} + 1, \quad t \in [a, \sigma^2(b)].$$

Therefore, for any solution y of the BVP (5.12), we have that $y \notin \partial\Omega_1$. Hence and by the Leray-Schauder fixed point theorem, it follows that the operator T_1 has a fixed point $y \in \Omega_1$ and

$$|y(t)| < \max\left\{\frac{|C|}{\alpha}, \frac{|D|}{\gamma}, R\right\} + 1, \quad t \in [a, \sigma^2(b)].$$

This completes the proof.

Now, by the investigations of Chapter 3, we have that the function

$$G_2(t,s) = \begin{cases} -\dfrac{(\beta - (t-a)\alpha)(\sigma^2(b) - \sigma(s))}{\alpha(\sigma^2(b) - a) + \beta} & \text{if} \quad t \le s \\[2ex] -\dfrac{(\beta + (\sigma(s) - a)\alpha)(\sigma^2(b) - t)}{\alpha(\sigma^2(b) - a) + \beta} & \text{if} \quad \sigma(s) \le t \end{cases}$$

is the Green function for the BVP

$$y^{\Delta^2} = 0, \quad t \in [a,b],$$

$$\alpha y(a) - \beta y^\Delta(a) = 0,$$

$$y(\sigma^2(b)) = 0.$$

We will search an integral representation of the solutions of the BVP (5.1), (5.10), in the form

$$y(t) = \int_a^{\sigma(b)} G_2(t,s) f(s, y^\sigma(s)) \Delta s + a_1 t + a_2, \quad t \in [a, \sigma^2(b)],$$

where $a_1, a_2 \in \mathbb{R}$ will be determined by the boundary conditions (5.10). We have

$$\begin{aligned} C &= \alpha y(a) - y^\Delta(a) \\ &= \alpha(a a_1 + a_2) - \beta a_1 \\ &= (\alpha a - \beta) a_1 + \alpha a_2 \end{aligned}$$

and

$$B = a_1 \sigma^2(b) + a_2.$$

Thus, we get the system

$$\begin{aligned} (\alpha a - \beta) a_1 + \alpha a_2 &= C \\ a_1 \sigma^2(b) + a_2 &= B. \end{aligned}$$

Hence,

$$(\alpha(\sigma^2(b) - a) + \beta) a_1 = \alpha B - C$$

or

$$a_1 = \frac{\alpha B - C}{\alpha(\sigma^2(b) - a) + \beta},$$

and

$$\begin{aligned} a_2 &= B - \sigma^2(b) a_1 \\ &= B - \frac{\alpha \sigma^2(b) B - C}{\alpha(\sigma^2(b) - a) + \beta} \\ &= \frac{\alpha \sigma^2(b) B - \alpha a B + \beta B - \alpha \sigma^2(b) B + \sigma^2(b) C}{\alpha(\sigma^2(b) - a) + \beta} \\ &= \frac{\sigma^2(b) C + (\beta - \alpha a) B}{\alpha(\sigma^2(b) - a) + \beta}. \end{aligned}$$

Therefore,

$$y(t) = \int_a^{\sigma(b)} G_2(t,s) f(s, y^\sigma(s)) \Delta s + \frac{\alpha B - C}{\alpha(\sigma^2(b) - a) + \beta} t$$

$$+ \frac{\sigma^2(b) C + (\beta - \alpha a) B}{\alpha(\sigma^2(b) - a) + \beta}, \quad t \in [a, \sigma^2(b)].$$

For $y \in \mathscr{C}([a, \sigma^2(b)])$, define the operator

$$T_2 y(t) = \int_a^{\sigma(b)} G_2(t,s) f(s, y^\sigma(s)) \Delta s + \frac{\alpha B - C}{\alpha(\sigma^2(b) - a) + \beta} t$$

$$+ \frac{\sigma^2(b) C + (\beta - \alpha a) B}{\alpha(\sigma^2(b) - a) + \beta}, \quad t \in [a, \sigma^2(b)].$$

Exercise 5.3. Prove that $T_2 : \mathscr{C}([a, \sigma^2(b)]) \to \mathscr{C}([a, \sigma^2(b)])$ is a completely continuous operator.

Theorem 5.7. *Suppose* $(D1)$, $(D5)$ *and let* f *and* R *satisfy the conditions of Theorem 5.2. Then the BVP* (5.1), (5.10) *has at least one solution satisfying*

$$|y(t)| < \max\left\{\frac{|C|}{\alpha}, R\right\} + 1, \quad t \in [a, \sigma^2(b)].$$

Proof. Define

$$\Omega_2 = \left\{ y \in S : \|y\| < \max\left\{\frac{|C|}{\alpha}, R\right\} + 1 \right\}.$$

Consider equation

$$y = \lambda T_2 y, \quad \lambda \in [0,1], \quad y \in \mathscr{C}([a, \sigma^2(b)]),$$

which is equivalent to the BVP

$$\begin{aligned} y^{\Delta^2} &= \lambda f(t, y^\sigma), \quad t \in [a,b], \\ \alpha y(a) - \beta y^\Delta(a) &= \lambda C, \\ y(\sigma^2(b)) &= \lambda B. \end{aligned} \tag{5.13}$$

By Corollary 5.2, it follows that for any solution $y \in \mathscr{C}([a, \sigma^2(b)])$ of the BVP (5.13), we have

$$\|y\| < \max\left\{\frac{|C|}{\alpha}, R\right\} + 1.$$

Thus, for any solution y of the BVP (5.13), we have that $y \notin \partial\Omega_2$. Hence and by the Leray-Schauder fixed point Theorem 5.4, it follows that the BVP (5.1), (5.10) has at least one solution in Ω_2. This completes the proof.

By the investigations in Chapter 3, we have that the function

$$G_3(t,s) = \begin{cases} \dfrac{(t-a)(-\gamma\sigma(s)+\delta+\gamma\sigma^2(b))}{\gamma(\sigma^2(b)-a)+\delta} & \text{if} \quad t \le s \\ \\ \dfrac{(-\gamma t+\delta+\gamma\sigma^2(b))(\sigma(s)-a)}{\gamma(\sigma^2(b)-a)+\delta}, & \text{if,} \quad \sigma(s) \le t, \end{cases}$$

is the Green function for the BVP

$$y^{\Delta^2} = 0, \quad t \in [a,b],$$

$$y(a) = 0,$$

$$\gamma y(\sigma^2(b)) + \delta y^{\Delta}(\sigma(b)) = 0.$$

We will find an integral representation of the solutions of the BVP (5.1), (5.11) in the form

$$y(t) = \int_a^{\sigma(b)} G_3(t,s) f(s, y^{\sigma}(s))\Delta s + a_1 t + a_2, \quad t \in [a, \sigma^2(b)],$$

where $a_1, a_2 \in \mathbb{R}$ will be determined by the boundary conditions (5.11). We have

$$A = y(a)$$

$$= a_1 a + a_2,$$

$$D = \gamma y(\sigma^2(b)) + \delta y^{\Delta}(\sigma(b))$$

$$= \gamma(a_1\sigma^2(b) + a_2) + \delta a_1$$

$$= a_1(\gamma\sigma^2(b) + \delta) + \gamma a_2.$$

Thus, we get the system

$$a_1 a + a_2 = A$$

$$a_1(\gamma\sigma^2(b) + \delta) + \gamma a_2 = D.$$

Hence,

$$a_1(\gamma(\sigma^2(b)-a)+\delta)=D-\gamma A$$

or

$$a_1=\frac{D-\gamma A}{\gamma(\sigma^2(b)-a)+\delta},$$

and

$$\begin{aligned}a_2&=A-aa_1\\&=A-\frac{aD-a\gamma A}{\gamma(\sigma^2(b)-a)+\delta}\\&=\frac{A\gamma\sigma^2(b)-Aa\gamma+A\delta-aD+a\gamma A}{\gamma(\sigma^2(b)-a)+\delta}\\&=\frac{A(\gamma\sigma^2(b)+\delta)-aD}{\gamma(\sigma^2(b)-a)+\delta}.\end{aligned}$$

Consequently,

$$\begin{aligned}y(t)=&\int_a^{\sigma(b)}G_3(t,s)f(s,y^\sigma(s))\Delta s+\frac{D-\gamma A}{\gamma(\sigma^2(b)-a)+\delta}t\\&+\frac{A(\gamma\sigma^2(b)+\delta)-aD}{\gamma(\sigma^2(b)-a)+\delta},\quad t\in[a,\sigma^2(b)].\end{aligned}$$

For $y\in\mathscr{C}([a,\sigma^2(b)])$, define the operator

$$\begin{aligned}T_3y(t)=&\int_a^{\sigma(b)}G_3(t,s)f(s,y^\sigma(s))\Delta s+\frac{D-\gamma A}{\gamma(\sigma^2(b)-a)+\delta}t\\&+\frac{A(\gamma\sigma^2(b)+\delta)-aD}{\gamma(\sigma^2(b)-a)+\delta},\quad t\in[a,\sigma^2(b)].\end{aligned}$$

Exercise 5.4. Prove that the operator $T_3:\mathscr{C}([a,\sigma^2(b)])\to\mathscr{C}([a,\sigma^2(b)])$ is a completely continuous operator.

Theorem 5.8. *Suppose* (*D*1), (*D*5) *and let f and R satisfy the conditions of Theorem 5.2. Then the BVP* (5.1), (5.11) *has at least one solution satisfying*

$$|y(t)|<\max\left\{\frac{|D|}{\gamma},R\right\}+1,\quad t\in[a,\sigma^2(b)].$$

Proof. Define

$$\Omega_3=\left\{y\in S:\|y\|<\max\left\{\frac{|D|}{\gamma},R\right\}+1\right\}.$$

Consider equation

$$y = \lambda T_3 y, \quad \lambda \in [0,1], \quad y \in \mathscr{C}([a,\sigma^2(b)]),$$

which is equivalent to the BVP

$$\begin{aligned} y^{\Delta^2} &= \lambda f(t,y^\sigma), \quad t \in [a,b], \\ y(a) &= \lambda A, \\ y(\sigma^2(b)) + \delta y^\Delta(\sigma(b)) &= \lambda D. \end{aligned} \tag{5.14}$$

By Corollary 5.3, it follows that for any solution $y \in \mathscr{C}([a,\sigma^2(b)])$ of the BVP (5.14), we have

$$\|y\| < \max\left\{\frac{|D|}{\gamma}, R\right\} + 1.$$

Thus, for any solution y of the BVP (5.14), we have that $y \notin \partial\Omega_3$. Hence and by the Leray-Schauder fixed point Theorem 5.4, it follows that the BVP (5.1), (5.11) has at least one solution in Ω_3. This completes the proof.

Example 5.2. Let $\mathbb{T} = \mathbb{Z}$. Consider the BVP

$$y^{\Delta^2} = y^\sigma, \quad t \in [0,8],$$

$$y(0) = 1,$$

$$y(10) = 2.$$

Here

$$f(t,y) = y,$$

$$\sigma(t) = t+1, \quad t \in \mathbb{T}, \quad y \in \mathbb{R},$$

$$a = 0,$$

$$b = 8,$$

$$A = 1,$$

$$B = 2.$$

Take $R > 2$. Then

$$y f(t,y) = y^2$$

$$> 0, \quad |y| \geq R.$$

Hence and by Corollary 5.3, it follows that the considered BVP has at least one solution y for which

$$|y(t)| < R, \quad t \in [0,10].$$

Exercise 5.5. Let $\mathbb{T} = 2\mathbb{Z}$. Prove that the BVP

$$y^{\Delta^2} = 2ty^2, \quad t \in [2,14],$$

$$y(2) = 2,$$

$$y(18) = 3,$$

has at least one solution y satisfying

$$|y(t)| < 5, \quad t \in [2,18].$$

Example 5.3. Let $\mathbb{T} = 2^{\mathbb{N}_0}$. Consider the BVP

$$y^{\Delta^2} = y^{\sigma 3} + y^{\sigma 5}, \quad t \in [1,16],$$

$$y(1) - y^{\Delta}(1) = 3,$$

$$2y(64) + 3y^{\Delta}(32) = -1.$$

Here

$$f(t,y) = y^3 + y^5,$$

$$\sigma(t) = 2t,$$

$$a = 1,$$

$$b = 16,$$

$$\alpha = 1,$$

$$\beta = 1,$$

$$\gamma = 2,$$

$$\delta = 3,$$

$$C = 3,$$

$$D = -1.$$

Take $R = 1$. Then

$$\begin{aligned} yf(t,y) &= y(y^3 + y^5) \\ &= y^4 + y^6 \\ &> 0, \quad |y| > 1, \quad t \in [1,16]. \end{aligned}$$

Next,

$$\begin{aligned} \max\left\{\frac{|C|}{\alpha}, \frac{|D|}{\gamma}, R\right\} + 1 &= \max\left\{3, \frac{1}{2}, 1\right\} + 1 \\ &= 3 + 1 \\ &= 4. \end{aligned}$$

Hence and by Theorem 5.3, it follows that any solution of the considered BVP satisfies

$$|y(t)| < 4, \quad t \in [1,64].$$

Exercise 5.6. Let $\mathbb{T} = 2^{\mathbb{N}_0}$. Prove that any solution y of the BVP

$$y^{\Delta^2} = \frac{y^\sigma}{1 + y^{\sigma 4}}, \quad t \in [1,8],$$

$$y(1) - 2y^\Delta(1) = 1,$$

$$2y(32) + 4y^\Delta(16) = 0,$$

satisfies

$$|y(t)| < 4, \quad t \in [1,32].$$

Example 5.4. Let $\mathbb{T} = \mathbb{Z}$. Consider the BVP

$$y^{\Delta^2} = (1+t)y^{\sigma 3}, \quad t \in [0,10],$$

$$y(0) - y^{\Delta}(0) = 1,$$

$$y(12) = 1.$$

Here

$$f(t,y) = (1+t)y^3, \quad t \in [0,10], \quad y \in \mathbb{R},$$

$$\alpha = 1,$$

$$\beta = 1,$$

$$a = 0,$$

$$b = 10,$$

$$C = 1,$$

$$B = 1.$$

Take $R > 1$. Then

$$yf(t,y) = (1+t)y^4$$

$$> 0 \quad \text{for} \quad |y| \geq 1.$$

Note that

$$\max\left\{\frac{|C|}{\alpha}, R\right\} + 1 = \max\{1,1\} + 1$$

$$= 1 + 1$$

$$= 2.$$

Hence and applying Corollary 5.3, it follows that any solution y of the considered BVP satisfies the estimate

$$|y(t)| \leq 2, \quad t \in [0,12].$$

Exercise 5.7. Let $\mathbb{T} = 4^{\mathbb{N}_0}$. Prove that any solution y of the BVP

$$y^{\Delta^2} = (1+t^2)\frac{y^{\sigma 7}}{1+y^{\sigma 8}}, \quad t \in [1,16],$$

$$y(1) - y^{\Delta}(1) = 2,$$

$$y(256) = 3,$$

satisfies the estimate

$$|y(t)| \le 7, \quad t \in [1,256].$$

Example 5.5. Let $\mathbb{T} = 2\mathbb{Z}$. Consider the BVP

$$y^{\Delta^2} = \left(\frac{t+1}{1+t^2}\right)\frac{y^{\sigma 5}}{1+y^{\sigma 10}}, \quad t \in [0,16],$$

$$y(0) = 1,$$

$$2y(20) + 4y^{\Delta}(18) = 5.$$

Here

$$f(t,y) = \left(\frac{t+1}{1+t^2}\right)\frac{y^5}{1+y^{10}}, \quad t \in [0,20], \quad y \in \mathbb{R},$$

$$\sigma(t) = t+2, \quad t \in \mathbb{T},$$

$$A = 1,$$

$$D = 5,$$

$$a = 0,$$

$$b = 16,$$

$$\gamma = 2,$$

$$\delta = 4.$$

Take $R = 1$. Then

$$\begin{aligned} yf(t,y) &= \left(\frac{1+t}{1+t^2}\right)\frac{y^6}{1+y^{10}} \\ &> 0, \quad t\in[0,20], \quad |y|>1. \end{aligned}$$

Note that

$$\begin{aligned} \max\left\{\frac{|D|}{\gamma}, R\right\}+1 &= \max\left\{\frac{5}{2}, 1\right\}+1 \\ &= \frac{5}{2}+1 \\ &= \frac{7}{2}. \end{aligned}$$

Hence and applying Corollary 5.3, it follows that any solution y of the considered BVP satisfies the estimate

$$|y(t)| < \frac{7}{2}, \quad t\in[0,20].$$

Exercise 5.8. Let $\mathbb{T}=2^{\mathbb{N}_0}$. Prove that any solution y of the BVP

$$\begin{aligned} y^{\Delta^2} &= \left(\frac{1+t+t^2}{1+t^2+t^4}\right)\frac{y^{\sigma 7}}{1+y^{\sigma 14}}, \quad t\in[1,16], \\ y(1) &= 3, \\ 5y(64)+7y^{\Delta}(32) &= 18, \end{aligned}$$

satisfies the estimate

$$|y(t)| \le 5, \quad t\in[1,64].$$

Theorem 5.9. *Suppose* $(D1)$, $(D2)$ *and* $R>0$ *is such that*

$$yf(t,y)>0, \quad t\in[a,b], \quad \|y\|=R.$$

If $|A|, |B| < R$, *then any solution* y *of the BVP* (5.1), (5.2) *satisfies the estimate*

$$|y(t)| < R, \quad t\in[a,\sigma^2(b)].$$

Proof. Define

$$m(t,y^{\sigma}) = \begin{cases} \dfrac{R}{\|y^{\sigma}\|} f\left(t, \dfrac{Ry^{\sigma}}{\|y^{\sigma}\|}\right) & \text{for } \|y^{\sigma}\| \ge R \\ f(t,y^{\sigma}) \quad \text{for } \|y^{\sigma}\| \le R \end{cases}$$

and, for $y \in \mathscr{C}([a,\sigma^2(b)])$, define the operator

$$Ly(t) = \int_a^{\sigma(b)} G(t,s)m(s,y^\sigma(s))\Delta s$$

$$+\frac{B-A}{\sigma^2(b)-a}t+\frac{A\sigma^2(b)-aB}{\sigma^2(b)-a},$$

$t \in [a,\sigma^2(b)]$, where G is the Green function defined by (5.3). We have that $L : \mathscr{C}([a,\sigma^2(b)]) \to \mathscr{C}([a,\sigma^2(b)])$ is a completely continuous operator. Let

$$M = \max_{t\in[a,\sigma^2(b)]} \int_a^{\sigma(b)} |G(t,s)|\Delta s,$$

$$N = \max_{t\in[a,\sigma^2(b)]} \left(\frac{B-A}{\sigma^2(b)-a}t+\frac{A\sigma^2(b)-aB}{\sigma^2(b)-a}\right),$$

$$E = \max_{(t,y)\in[a,\sigma^2(b)]\times[-R,R]} m(t,y).$$

Also, define the set

$$\Omega = \{y \in S : \|y\| < ME+N+1\},$$

where S is the set as in the proof of Theorem 5.5. Note that, for any $\lambda \in [0,1]$, we have

$$\|\lambda Ly\| \le \lambda(ME+N)$$

$$\le ME+N$$

$$< ME+N+2.$$

Therefore, equation

$$y^{\Delta^2} = m(t,y^\sigma), \quad t \in [a,b], \tag{5.15}$$

subject to the boundary conditions (5.2), has a solution in Ω. Next, for $\|y\| \ge R$, we have

$$y^\sigma m(t,y^\sigma) = pf(t,p)$$

$$> 0,$$

where $p = \dfrac{Ry^\sigma}{\|y^\sigma\|}$. Consequently, all solutions of (5.15), (5.2) satisfy

$$|y(t)| < R, \quad t \in [a,\sigma^2(b)],$$

and they are solutions of the BVP (5.1), (5.2). This completes the proof.

Exercise 5.9. Suppose $(D1)$, $(D5)$ and let

$$yf(t,y) > 0, \quad t \in [a,b], \quad \|y\| = R.$$

Prove that the BVP (5.1), (5.9) has at least one solution y for which

$$|y(t)| < R, \quad t \in [a, \sigma^2(b)].$$

5.2 Lower and Upper Solutions Method for Nonlinear Second Order Boundary Value Problems

In this section, we will investigate equation

$$u^{\Delta^2}(t) + f(t,u(t)) = 0, \quad t \in \mathbb{T}\backslash\{a,b\}, \tag{5.16}$$

subject to the following boundary conditions

$$\begin{aligned} u^{\Delta}(a) &= 0, \\ u(b) &= 0, \end{aligned} \tag{5.17}$$

where

(D6) $\quad f \in \mathscr{C}(\mathbb{T} \times \mathbb{R})$.

Definition 5.1. Let $\alpha : \mathbb{T} \to \mathbb{R}$. We call α a lower solution of the BVP (5.16), (5.17) if

$$\alpha^{\Delta^2}(t) + f(t,\alpha(t)) \geq 0, \quad t \in \mathbb{T}\backslash\{a,b\},$$

satisfying

$$\alpha^{\Delta}(a) \geq 0, \quad \alpha(b) \leq 0.$$

Definition 5.2. Let $\beta : \mathbb{T} \to \mathbb{R}$. We call β a upper solution of the BVP (5.16), (5.17) if

$$\beta^{\Delta^2}(t) + f(t,\beta(t)) \leq 0, \quad t \in \mathbb{T}\backslash\{a,b\},$$

satisfying

$$\beta^{\Delta}(a) \leq 0, \quad \beta(b) \geq 0.$$

For our next result we need the following fixed point theorem.

Theorem 5.10 (Brouwer Fixed Point Theorem). *Any continuous map* $f : \overline{\mathbf{B}}(m,a) \longmapsto \overline{\mathbf{B}}(m,a)$ *admits at least one fixed point in* $\overline{\mathbf{B}}(m,a)$. *Here* $B(m,a)$ *is a ball with a center* a *and a radius* m

Theorem 5.11 (Lower and Upper Solution Method). *Let α and β be lower and upper solutions of the BVP (5.16), (5.17), respectively, where $\alpha \le \beta$ on $\mathbb{T}$. Then the BVP (5.16), (5.17) has a solution u such that*

$$\alpha(t) \le u(t) \le \beta(t), \quad t \in \mathbb{T}.$$

Proof. Define

$$\widetilde{f}(t,x) = \begin{cases} f(t,\beta(t)) - \dfrac{x-\beta(t)}{x-\beta(t)+1}, & x > \beta(t) \\ f(t,x), \quad \alpha(t) \le x \le \beta(t) \\ f(t,\alpha(t)) - \dfrac{\alpha(t)-x}{\alpha(t)-x+1}, & x < \alpha(t), \end{cases}$$

for $t \in \mathbb{T}$ and $x \in \mathbb{R}$. We have that $\widetilde{f} \in \mathscr{C}(\mathbb{T} \times \mathbb{R})$ and there exists a positive constant M such that

$$|\widetilde{f}(t,x)| \le M$$

for any $t \in \mathbb{T}$ and any $x \in \mathbb{R}$. Consider equation

$$u^{\Delta^2}(t) + \widetilde{f}(t,u(t)) = 0, \quad t \in \mathbb{T}\backslash\{a,b\}, \tag{5.18}$$

subject to the boundary conditions (5.17). Define the set

$$E = \{u : \mathbb{T} \to \mathbb{R} : u^{\Delta}(a) = u(b) = 0\},$$

endowed with the norm

$$\|u\| = \sup_{t\in\mathbb{T}} |u(t)|.$$

For $u \in E$, define the operator

$$Tu(t) = \int_t^b \int_a^\tau \widetilde{f}(s,u(s))\Delta s\Delta\tau, \quad t \in \mathbb{T}.$$

For $u \in E$, we have

$$Tu(b) = 0,$$

$$(Tu)^{\Delta}(t) = -\int_a^t \widetilde{f}(s,u(s))\Delta s, \quad t \in \mathbb{T},$$

$$(Tu)^{\Delta}(a) = 0.$$

Therefore, $T : E \to E$ and it is a continuous operator. Take $r > M(b-a)^2$. Define

$$B(r) = \{u \in E : \|u\| < r\}.$$

For $u \in \overline{B}(r)$, we have

$$\begin{aligned}|Tu(t)| &= \left|\int_t^b \int_a^\tau \widetilde{f}(s,u(s))\Delta s\Delta\tau\right| \\ &\leq \int_t^b \int_a^\tau |\widetilde{f}(s,u(s))|\Delta s\Delta\tau \\ &\leq M(b-a)^2 \\ &< r.\end{aligned}$$

Thus, $T(\overline{B}(r)) \subseteq \overline{B}(r)$. Hence and by the Brouwer fixed point theorem, we conclude that the operator T has a fixed point $u \in \overline{B}(r)$, and this fixed point is a solution to the BVP (5.18), (5.17). Let now,

$$v(t) = u(t) - \beta(t), \quad t \in \mathbb{T}.$$

Assume that

$$\max\{v(t) : t \in \mathbb{T}\} = v(l) > 0.$$

Using (5.17) and the definition of β, we conclude that $l \in \mathbb{T}\backslash\{a,b\}$. Thus,

$$v^\Delta(\rho(l)) \geq 0, \quad v^\Delta(l) \leq 0, \quad v^{\Delta^2}(\rho(l)) \leq 0.$$

Hence,

$$\begin{aligned}0 &\geq u^{\Delta^2}(\rho(l)) - \beta^{\Delta^2}(\rho(l)) \\ &= -\widetilde{f}(l,u(l)) - \beta^{\Delta^2}(\rho(l)) \\ &= -f(l,\beta(l)) + \frac{u(l)-\beta(l)}{u(l)-\beta(l)+1} - \beta^{\Delta^2}(\rho(l)) \\ &\geq \frac{v(l)}{v(l)+1} \\ &> 0.\end{aligned}$$

This is a contradiction. Consequently,

$$u(t) \leq \beta(t), \quad t \in \mathbb{T}.$$

Let now,

$$w(t) = \alpha(t) - u(t), \quad t \in \mathbb{T}.$$

Assume that

$$\max\{w(t) : t \in \mathbb{T}\} = w(m) > 0.$$

Then, by the boundary conditions (5.17) and the definition of β, we conclude that $m \in \mathbb{T}\backslash\{a, b\}$. Also,

$$w^{\Delta}(\rho(m)) \geq 0, \quad w^{\Delta}(m) \leq 0, \quad w^{\Delta^2}(\rho(m)) \leq 0.$$

Hence,

$$\begin{aligned}
0 &\geq \alpha^{\Delta^2}(\rho(m)) - u^{\Delta^2}(\rho(m)) \\
&= \alpha^{\Delta^2}(\rho(m)) + \widetilde{f}(m, u(m)) \\
&= \alpha^{\Delta^2}(\rho(m)) + f(m, \alpha(m)) + \frac{\alpha(m) - u(m)}{\alpha(m) - u(m) + 1} \\
&\geq \frac{w(m)}{w(m) + 1} \\
&> 0.
\end{aligned}$$

This is a contradiction and

$$u(t) \geq \alpha(t), \quad t \in \mathbb{T}.$$

This completes the proof.

Example 5.6. Let $\mathbb{T} = 2^{\mathbb{N}_0}$. Then, using Theorem 5.11, it follows that the BVP

$$\begin{aligned}
u^{\Delta^2}(t) + \frac{u(t)}{1 + (u(t))^2} &= 0, \quad t \in \mathbb{T}\backslash\{1, 8\}, \\
u^{\Delta}(1) &= 0, \\
u(8) &= 0,
\end{aligned}$$

has at least one solution.

5.3 Existence of Symmetric Positive Solutions of Nonlinear Second Order m-Point Boundary Value Problems-I

In this section we will investigate the m-point BVP

$$u^{\Delta\nabla}(t)+q(t)f(u(t))=0, \quad t\in[a,b], \tag{5.19}$$

satisfying the boundary conditions

$$\begin{aligned} u^{\Delta}(a) &= \sum_{j=1}^{m-2} b_j u^{\Delta}(\xi_j),\\ u(b) &= \sum_{j=1}^{m-2} a_j u(\xi_j), \end{aligned} \tag{5.20}$$

where

(D7) $m\geq 2$, $\xi_j\in(a,\rho(b))$,

$$a<\xi_1<\ldots<\xi_{m-2}<\rho(b).$$

(D8) $a_j\geq 0$, $b_j\geq 0$, $j\in\{1,\ldots,m-2\}$, satisfying

$$0<\sum_{j=1}^{m-2} a_j<1,$$

$$0\leq\sum_{j=1}^{m-2} b_j<1,$$

(D9) $f:\mathbb{R}\to[0,\infty)$, $f\in\mathscr{C}(\mathbb{R})$.
(D10) $q:\mathbb{T}\to[0,\infty)$, $q\in\mathscr{C}_{ld}(\mathbb{T})$.

Definition 5.3. A function $u:\mathbb{T}\to\mathbb{R}$ is said to be a solution of the BVP (5.19), (5.20) if u is Δ-differentiable on $\mathbb{T}$, u^{Δ} is ∇-differentiable on $\mathbb{T}^{\kappa}\bigcap\mathbb{T}_{\kappa}$ and $u^{\Delta\nabla}:\mathbb{T}^{\kappa}\bigcap\mathbb{T}_{\kappa}\to\mathbb{R}$ is continuous and u satisfies (5.19), (5.20).

Remark 5.1. Suppose $(D9)$ and $(D10)$. Then

$$\begin{aligned}\left(\int_a^t (t-s)q(s)f(u(s))\nabla s\right)^{\Delta} &= \int_a^t q(s)f(u(s))\nabla s\\ &\quad +(\sigma(t)-\sigma(t))q(t)f(u(t))\end{aligned}$$

$$= \int_a^t q(s)f(u(s))\nabla s, \quad t \in \mathbb{T}.$$

Thus,

$$\int_a^t (t-s)q(s)f(u(s))\nabla s = \int_a^t \int_a^\tau q(s)f(u(s))\nabla s\Delta\tau, \quad t \in \mathbb{T}.$$

Lemma 5.4. *Suppose* (*D*7)-(*D*10). *Then u is a solution of the BVP* (5.19), (5.20) *if and only if u is a solution to the following integral equation.*

$$\begin{aligned} u(t) &= -\int_a^t (t-s)q(s)f(u(s))\nabla s - (t-a)\frac{\sum\limits_{j=1}^{m-2} b_j \int_a^{\xi_j} q(s)f(u(s))\nabla s}{1-\sum\limits_{j=1}^{m-2} b_j} \\ &\quad + \frac{1}{1-\sum\limits_{j=1}^{m-2} a_j}\Bigg(\int_a^b (b-s)q(s)f(u(s))\nabla s \\ &\quad - \sum_{j=1}^{m-2} a_j \int_a^{\xi_j} (\xi_j - s)q(s)f(u(s))\nabla s \\ &\quad + \frac{\sum\limits_{j=1}^{m-2} b_j \int_a^{\xi_j} q(s)f(u(s))\nabla s}{1-\sum\limits_{j=1}^{m-2} b_j}\left(b-a-\sum_{j=1}^{m-2} a_j\xi_j\right)\Bigg), \end{aligned} \tag{5.21}$$

$t \in [a,b]$.

Proof. 1. Let u be a solution to the BVP (5.19), (5.20). Then, we integrate equation (5.19) from a to t and we get

$$u^\Delta(t) = u^\Delta(a) - \int_a^t q(s)f(u(s))\nabla s, \quad t \in [a,b]. \tag{5.22}$$

Now, we integrate the last equation from a to t and we find

$$\begin{aligned} u(t) &= u(a) + (t-a)u^\Delta(a) - \int_a^t \int_a^\tau q(s)f(u(s))\nabla s\Delta\tau \\ &= u(a) + (t-a)u^\Delta(a) - \int_a^t (t-s)q(s)f(u(s))\nabla s, \quad t \in [a,b]. \end{aligned} \tag{5.23}$$

We put $t = \xi_j$, $j \in \{1,\dots,m-2\}$, and we obtain

$$u^\Delta(\xi_j) = u^\Delta(a) - \int_a^{\xi_j} q(s)f(u(s))\nabla s,$$

$$b_j u^{\Delta}(\xi_j) = u^{\Delta}(a) b_j - b_j \int_a^{\xi_j} q(s) f(u(s)) \nabla s, \quad j \in \{1, \ldots, m-2\},$$

$$\sum_{j=1}^{m-2} b_j u^{\Delta}(\xi_j) = u^{\Delta}(a) \sum_{j=1}^{m-2} b_j - \sum_{j=1}^{m-2} b_j \int_a^{\xi_j} q(s) f(u(s)) \nabla s$$
$$= u^{\Delta}(a),$$

whereupon,

$$u^{\Delta}(a) \left(1 - \sum_{j=1}^{m-2} b_j \right) = - \sum_{j=1}^{m-2} b_j \int_a^{\xi_j} q(s) f(u(s)) \nabla s,$$

or

$$u^{\Delta}(a) = - \frac{\sum_{j=1}^{m-2} b_j \int_a^{\xi_j} q(s) f(u(s)) \nabla s}{1 - \sum_{j=1}^{m-2} b_j}.$$

Now, we set $t = b$ and $t = \xi_j$, $j \in \{1, \ldots, m-2\}$, into (5.23) and we find

$$u(b) = u(a) + (b-a) u^{\Delta}(a) - \int_a^b (b-s) q(s) f(u(s)) \nabla s,$$

and

$$u(\xi_j) = u(a) + (\xi_j - a) u^{\Delta}(a) - \int_a^{\xi_j} (\xi_j - s) q(s) f(u(s)) \nabla s, \quad j \in \{1, \ldots, m-2\},$$

and

$$a_j u(\xi_j) = a_j u(a) + a_j (\xi_j - a) u^{\Delta}(a)$$
$$- a_j \int_a^{\xi_j} (\xi_j - s) q(s) f(u(s)) \nabla s, \quad j \in \{1, \ldots, m-2\},$$

whereupon,

$$u(b) = u(a) + (b-a) u^{\Delta}(a) - \int_a^b (b-s) q(s) f(u(s)) \nabla s$$
$$= u(a) \sum_{j=1}^{m-2} a_j + u^{\Delta}(a) \sum_{j=1}^{m-2} a_j (\xi_j - a)$$

$$-\sum_{j=1}^{m-2} a_j \int_a^{\xi_j} (\xi_j - s) q(s) f(u(s)) \nabla s,$$

and

$$u(a)\left(1-\sum_{j=1}^{m-2} a_j\right) = u^{\Delta}(a)\left(\sum_{j=1}^{m-2} a_j(\xi_j - a) - (b-a)\right)$$
$$-\sum_{j=1}^{m-2} a_j \int_a^{\xi_j} (\xi_j - s) q(s) f(u(s)) \nabla s$$
$$+\int_a^b (b-s) q(s) f(u(s)) \nabla s,$$

and

$$u(a) = -\frac{\left(\sum\limits_{j=1}^{m-2} a_j(\xi_j - a) - (b-a)\right)\left(\sum\limits_{j=1}^{m-2} b_j \int_a^{\xi_j} q(s) f(u(s)) \nabla s\right)}{\left(1-\sum\limits_{j=1}^{m-2} a_j\right)\left(1-\sum\limits_{j=1}^{m-2} b_j\right)}$$
$$-\frac{\sum\limits_{j=1}^{m-2} a_j \int_a^{\xi_j} (\xi_j - s) q(s) f(u(s)) \nabla s}{1-\sum\limits_{j=1}^{m-2} a_j}$$
$$+\frac{\int_a^b (b-s) q(s) f(u(s)) \nabla s}{1-\sum\limits_{j=1}^{m-2} a_j}.$$

Consequently,

$$u(t) = -\int_a^t (t-s) q(s) f(u(s)) \nabla s$$
$$+\frac{1}{1-\sum\limits_{j=1}^{m-2} a_j}\left(\frac{\left((b-a)-\sum\limits_{j=1}^{m-2} a_j(\xi_j - a)\right)\left(\sum\limits_{j=1}^{m-2} b_j \int_a^{\xi_j} q(s) f(u(s)) \nabla s\right)}{1-\sum\limits_{j=1}^{m-2} b_j}\right.$$
$$-\sum_{j=1}^{m-2} a_j \int_a^{\xi_j} (\xi_j - s) q(s) f(u(s)) \nabla s$$

$$+\int_a^b (b-s)q(s)f(u(s))\nabla s\Bigg)$$

$$-(t-a)\frac{\sum\limits_{j=1}^{m-2} b_j \int_a^{\xi_j} q(s)f(u(s))\nabla s}{1-\sum\limits_{j=1}^{m-2} b_j}, \quad t\in[a,b].$$

Therefore, u satisfies (5.21).

2. Let u satisfy (5.21). We Δ-differentiate (5.21) with respect to t and we find

$$u^{\Delta}(t) = -\int_a^t q(s)f(u(s))\nabla s - \frac{\sum\limits_{j=1}^{m-2} b_j \int_a^{\xi_j} q(s)f(u(s))\nabla s}{1-\sum\limits_{j=1}^{m-2} b_j}, \tag{5.24}$$

$t\in[a,b]$. Now, we ∇-differentiate the last equation with respect to t and we obtain

$$u^{\Delta\nabla}(t) = -q(t)f(u(t)), \quad t\in[a,b].$$

We set $t=a$ and $t=\xi_j$, $j\in\{1,\ldots,m-2\}$, into (5.24) and we find

$$u^{\Delta}(a) = -\frac{\sum\limits_{j=1}^{m-2} b_j \int_a^{\xi_j} q(s)f(u(s))\nabla s}{1-\sum\limits_{j=1}^{m-2} b_j},$$

$$u^{\Delta}(\xi_j) = -\int_a^{\xi_j} q(s)f(u(s))\nabla s - \frac{\sum\limits_{j=1}^{m-2} b_j \int_a^{\xi_j} q(s)f(u(s))\nabla s}{1-\sum\limits_{j=1}^{m-2} b_j},$$

$$b_j u^{\Delta}(\xi_j) = -b_j\int_a^{\xi_j} q(s)f(u(s))\nabla s - b_j\frac{\sum\limits_{j=1}^{m-2} b_j \int_a^{\xi_j} q(s)f(u(s))\nabla s}{1-\sum\limits_{j=1}^{m-2} b_j},$$

$$\sum_{j=1}^{m-2} b_j u^{\Delta}(\xi_j) = -\sum_{l=1}^{m-2} b_l \int_a^{\xi_l} q(s)f(u(s))\nabla s$$

$$-\sum_{j=1}^{m-2} b_j \frac{\sum\limits_{l=1}^{m-2} b_l \int_a^{\xi_l} q(s)f(u(s))\nabla s}{1-\sum_{j=1}^{m-2} b_j}$$

$$= -\sum_{l=1}^{m-2} b_l \int_a^{\xi_l} q(s)f(u(s))\nabla s \left(1 + \frac{\sum\limits_{j=1}^{m-2} b_j}{1 - \sum\limits_{l=1}^{m-2} b_l} \right)$$

$$= -\frac{\sum\limits_{l=1}^{m-2} b_l \int_a^{\xi_l} q(s)f(u(s))\nabla s}{1 - \sum\limits_{l=1}^{m-2} b_l}$$

$$= u^{\Delta}(a).$$

Now, we put $t = b$ and $t = \xi_j$, $j \in \{1, \ldots, m-2\}$, into (5.21) and we find

$$u(b) = -\int_a^b (b-s)q(s)f(u(s))\nabla s$$

$$+\frac{1}{1-\sum\limits_{l=1}^{m-2} a_l} \left(\frac{\left(b - a - \sum\limits_{j=1}^{m-2} a_j(\xi_j - a)\right)\left(\sum\limits_{j=1}^{m-2} b_j \int_a^{\xi_j} q(s)f(u(s))\nabla s\right)}{1 - \sum\limits_{j=1}^{m-2} b_j} \right.$$

$$-\sum_{j=1}^{m-2} a_j \int_a^{\xi_j} (\xi_j - s)q(s)f(u(s))\nabla s$$

$$\left. + \int_a^b (b-s)q(s)f(u(s))\nabla s \right)$$

$$-(b-a)\frac{\sum\limits_{j=1}^{m-2} b_j \int_a^{\xi_j} q(s)f(u(s))\nabla s}{1 - \sum\limits_{j=1}^{m-2} b_j}$$

$$= \frac{\sum\limits_{j=1}^{m-2} a_j}{1 - \sum\limits_{j=1}^{m-2} a_j} \int_a^b (b-s)q(s)f(u(s))\nabla s$$

$$+\frac{1}{1-\sum\limits_{j=1}^{m-2} a_j} \left(-\frac{\left(\sum\limits_{j=1}^{m-2} a_j(\xi_j - a)\right)\left(\sum\limits_{j=1}^{m-2} b_j \int_a^{\xi_j} q(s)f(u(s))\nabla s\right)}{1 - \sum\limits_{j=1}^{m-2} b_j} \right.$$

$$-\sum_{j=1}^{m-2} a_j \int_a^{\xi_j} (\xi_j - s) q(s) f(u(s)) \nabla s$$

$$+\frac{(b-a)\left(\sum\limits_{j=1}^{m-2} b_j \int_a^{\xi_j} q(s) f(u(s)) \nabla s\right)\sum\limits_{j=1}^{m-2} a_j}{1-\sum\limits_{j=1}^{m-2} b_j}\Bigg)$$

and

$$a_j u(\xi_j) = -a_j \int_a^{\xi_j} (\xi_j - s) q(s) f(u(s)) \nabla s + \frac{a_j}{1-\sum\limits_{j=1}^{m-2} a_j}$$

$$\times\left(\frac{\left(b-a-\sum\limits_{j=1}^{m-2} a_j(\xi_j - a)\right)\left(\sum\limits_{j=1}^{m-2} b_j \int_a^{\xi_j} q(s) f(u(s)) \nabla s\right)}{1-\sum\limits_{j=1}^{m-2} b_j}\right.$$

$$-\sum_{j=1}^{m-2} a_j \int_a^{\xi_j} (\xi_j - s) q(s) f(u(s)) \nabla s$$

$$\left.+\int_a^b (b-s) q(s) f(u(s)) \nabla s\right)$$

$$-a_j(\xi_j - a)\frac{\sum\limits_{j=1}^{m-2} b_j \int_a^{\xi_j} q(s) f(u(s)) \nabla s}{1-\sum\limits_{j=1}^{m-2} b_j},$$

$$\sum_{j=1}^{m-2} a_j u(\xi_j) = -\sum_{j=1}^{m-2} a_j \int_a^{\xi_j} (\xi_j - s) q(s) f(u(s)) \nabla s + \frac{\sum\limits_{j=1}^{m-2} a_j}{1-\sum\limits_{j=1}^{m-2} a_j}$$

$$\times\left(\frac{\left(b-a-\sum\limits_{j=1}^{m-2} a_j(\xi_j - a)\right)\left(\sum\limits_{j=1}^{m-2} b_j \int_a^{\xi_j} q(s) f(u(s)) \nabla s\right)}{1-\sum\limits_{j=1}^{m-2} b_j}\right.$$

$$-\sum_{j=1}^{m-2} a_j \int_a^{\xi_j} (\xi_j - s) q(s) f(u(s)) \nabla s$$

$$+\int_a^b (b-s)q(s)f(u(s))\nabla s\Bigg)$$

$$-\sum_{j=1}^{m-2} a_j(\xi_j-a)\frac{\sum\limits_{j=1}^{m-2} b_j\int_a^{\xi_j} q(s)f(u(s))\nabla s}{1-\sum\limits_{j=1}^{m-2} b_j}$$

$$=-\left(1+\frac{\sum\limits_{j=1}^{m-2} a_j}{1-\sum\limits_{j=1}^{m-2} a_j}\right)\sum_{j=1}^{m-2} a_j\int_a^{\xi_j}(\xi_j-s)q(s)f(u(s))\nabla s$$

$$+\frac{\sum\limits_{j=1}^{m-2} a_j}{1-\sum\limits_{j=1}^{m-2} a_j}\int_a^b (b-s)q(s)f(u(s))\nabla s$$

$$+\frac{\sum\limits_{j=1}^{m-2} b_j\int_a^{\xi_j} q(s)f(u(s))\nabla s}{1-\sum\limits_{j=1}^{m-2} b_j}\left(\left(b-a-\sum_{j=1}^{m-2} a_j(\xi_j-a)\right)\frac{\sum\limits_{j=1}^{m-2} a_j}{1-\sum\limits_{j=1}^{m-2} a_j}\right.$$

$$\left.-\sum_{j=1}^{m-2} a_j(\xi_j-a)\right)$$

$$=-\frac{\sum\limits_{j=1}^{m-2} a_j\int_a^{\xi_j}(\xi_j-s)q(s)f(u(s))\nabla s}{1-\sum\limits_{j=1}^{m-2} a_j}$$

$$+\frac{\sum\limits_{j=1}^{m-2} a_j}{1-\sum\limits_{j=1}^{m-2} a_j}\int_a^b (b-s)q(s)f(u(s))\nabla s$$

$$-\frac{\left(\sum\limits_{j=1}^{m-2} b_j\int_a^{\xi_j} q(s)f(u(s))\nabla s\right)\left(\sum\limits_{j=1}^{m-2} a_j(\xi_j-a)\right)}{\left(1-\sum\limits_{j=1}^{m-2} b_j\right)\left(1-\sum\limits_{j=1}^{m-2} a_j\right)}$$

$$+\frac{(b-a)\left(\sum\limits_{j=1}^{m-2} b_j \int_a^{\xi_j} q(s)f(u(s))\nabla s\right)\sum\limits_{j=1}^{m-2} a_j}{\left(1-\sum\limits_{j=1}^{m-2} a_j\right)\left(1-\sum\limits_{j=1}^{m-2} b_j\right)}$$

$$= u(b).$$

Therefore, the function u satisfies (5.19), (5.20). This completes the proof.

Suppose that $\mathscr{C}([a,b])$ is endowed with the maximum norm. For $u \in \mathscr{C}([a,b])$, define the operator

$$Tu(t) = -\int_a^t (t-s)q(s)f(u(s))\nabla s - (t-a)\frac{\sum\limits_{j=1}^{m-2} b_j \int_a^{\xi_j} q(s)f(u(s))\nabla s}{1-\sum\limits_{j=1}^{m-2} b_j}$$

$$+\frac{1}{1-\sum\limits_{j=1}^{m-2} a_j}\left(\int_a^b (b-s)q(s)f(u(s))\nabla s\right.$$

$$-\sum_{j=1}^{m-2} a_j \int_a^{\xi_j} (\xi_j - s)q(s)f(u(s))\nabla s$$

$$\left.+\frac{\sum\limits_{j=1}^{m-2} b_j \int_a^{\xi_j} q(s)f(u(s))\nabla s}{1-\sum\limits_{j=1}^{m-2} b_j}\left(b-a-\sum_{j=1}^{m-2} a_j\xi_j\right)\right), \quad t \in [a,b].$$

Lemma 5.5. *Suppose (D7)-(D10). Then the operator $T : \mathscr{C}([a,b]) \to \mathscr{C}([a,b])$ is a continuous operator.*

Proof. Let $\{u_n\}_{n\in\mathbb{N}} \subseteq \mathscr{C}([a,b])$, $u \in \mathscr{C}([a,b])$, and

$$u_n \to u \quad \text{in} \quad \mathscr{C}([a,b]), \quad \text{as} \quad n \to \infty.$$

Since $q \in \mathscr{C}_{rd}([a,b])$, there exists a positive constant M such that

$$0 \le q(t) \le M, \quad t \in [a,b].$$

Take $\varepsilon > 0$ arbitrarily and set

$$P = \max\left\{(b-a)^2 M, M\left((b-a)\frac{\sum\limits_{j=1}^{m-2} b_j(\xi_j - a)}{1-\sum\limits_{j=1}^{m-2} b_j}\right.\right.$$

$$+\frac{1}{1-\sum\limits_{j=1}^{m-2} a_j}\left((b-a)^2 M+\sum_{j=1}^{m-2} a_j(\xi_j-a)^2\right.$$

$$\left.\left.\left.\left.+\frac{\sum\limits_{j=1}^{m-2} b_j(\xi_j-a)}{1-\sum\limits_{j=1}^{m-2} b_j}\left(b-a+\sum_{j=1}^{m-2} a_j\xi_j\right)\right)\right)\right)\right\}.$$

There exists a constant $N=N(\varepsilon)>0$ so that

$$|f(u_n(t))-f(u(t))|<\frac{\varepsilon}{2P},\quad t\in[a,b],\quad n>N.$$

Hence,

$$|Tu_n(t)-Tu(t)|=\left|-\int_a^t (t-s)q(s)f(u_n(s))\nabla s-(t-a)\frac{\sum\limits_{j=1}^{m-2} b_j\int_a^{\xi_j} q(s)f(u_n(s))\nabla s}{1-\sum\limits_{j=1}^{m-2} b_j}\right.$$

$$+\frac{1}{1-\sum\limits_{j=1}^{m-2} a_j}\left(\int_a^b (b-s)q(s)f(u_n(s))\nabla s\right.$$

$$-\sum_{j=1}^{m-2} a_j\int_a^{\xi_j} (\xi_j-s)q(s)f(u_n(s))\nabla s$$

$$\left.+\frac{\sum\limits_{j=1}^{m-2} b_j\int_a^{\xi_j} q(s)f(u_n(s))\nabla s}{1-\sum\limits_{j=1}^{m-2} b_j}\left(b-a-\sum_{j=1}^{m-2} a_j\xi_j\right)\right)$$

$$+\int_a^t (t-s)q(s)f(u(s))\nabla s+(t-a)\frac{\sum\limits_{j=1}^{m-2} b_j\int_a^{\xi_j} q(s)f(u(s))\nabla s}{1-\sum\limits_{j=1}^{m-2} b_j}$$

$$-\frac{1}{1-\sum\limits_{j=1}^{m-2} a_j}\left(\int_a^b (b-s)q(s)f(u(s))\nabla s\right.$$

$$-\sum_{j=1}^{m-2} a_j\int_a^{\xi_j} (\xi_j-s)q(s)f(u(s))\nabla s$$

$$
+\frac{\sum_{j=1}^{m-2} b_j \int_a^{\xi_j} q(s) f(u(s)) \nabla s}{1-\sum_{j=1}^{m-2} b_j}\left(b-a-\sum_{j=1}^{m-2} a_j \xi_j\right)\Bigg)\Bigg|
$$

$$
\begin{aligned}
=\Bigg| & -\int_a^t (t-s) q(s)(f(u_n(s))-f(u(s)))\nabla s \\
& -(t-a)\frac{\sum_{j=1}^{m-2} b_j \int_a^{\xi_j} q(s)(f(u_n(s))-f(u(s)))\nabla s}{1-\sum_{j=1}^{m-2} b_j} \\
& +\frac{1}{1-\sum_{j=1}^{m-2} a_j}\Bigg(\int_a^b (b-s) q(s)(f(u_n(s))-f(u(s)))\nabla s \\
& -\sum_{j=1}^{m-2} a_j \int_a^{\xi_j} (\xi_j-s) q(s)(f(u_n(s))-f(u(s)))\nabla s \\
& +\frac{\sum_{j=1}^{m-2} b_j \int_a^{\xi_j} q(s)(f(u_n(s))-f(u(s)))\nabla s}{1-\sum_{j=1}^{m-2} b_j}\left(b-a-\sum_{j=1}^{m-2} a_j \xi_j\right)\Bigg)\Bigg| \\
\le & \left|\int_a^t (t-s) q(s)(f(u_n(s))-f(u(s)))\nabla s\right| \\
& +(t-a)\left|\frac{\sum_{j=1}^{m-2} b_j \int_a^{\xi_j} q(s)(f(u_n(s))-f(u(s)))\nabla s}{1-\sum_{j=1}^{m-2} b_j}\right| \\
& +\frac{1}{1-\sum_{j=1}^{m-2} a_j}\Bigg(\left|\int_a^b (b-s) q(s)(f(u_n(s))-f(u(s)))\nabla s\right| \\
& +\sum_{j=1}^{m-2} a_j \left|\int_a^{\xi_j} (\xi_j-s) q(s)(f(u_n(s))-f(u(s)))\nabla s\right| \\
& +\frac{\sum_{j=1}^{m-2} b_j \left|\int_a^{\xi_j} q(s)(f(u_n(s))-f(u(s)))\nabla s\right|}{1-\sum_{j=1}^{m-2} b_j}\left(\left|b-a-\sum_{j=1}^{m-2} a_j \xi_j\right|\right)\Bigg)
\end{aligned}
$$

$$
\begin{aligned}
&\leq \int_a^t (t-s)q(s)|f(u_n(s))-f(u(s))|\nabla s \\
&+(t-a)\frac{\sum\limits_{j=1}^{m-2} b_j \int_a^{\xi_j} q(s)|f(u_n(s))-f(u(s))|\nabla s}{1-\sum\limits_{j=1}^{m-2} b_j} \\
&+\frac{1}{1-\sum\limits_{j=1}^{m-2} a_j}\Bigg(\int_a^b (b-s)q(s)|f(u_n(s))-f(u(s))|\nabla s \\
&+\sum_{j=1}^{m-2} a_j \int_a^{\xi_j} (\xi_j-s)q(s)|f(u_n(s))-f(u(s))|\nabla s \\
&+\frac{\sum\limits_{j=1}^{m-2} b_j \int_a^{\xi_j} q(s)|f(u_n(s))-f(u(s))|\nabla s}{1-\sum\limits_{j=1}^{m-2} b_j}\left(b-a+\sum_{j=1}^{m-2} a_j\xi_j\right)\Bigg) \\
&\leq \frac{\varepsilon}{2P}M(b-a)^2+\frac{\varepsilon}{2P}\Bigg(M\Bigg((b-a)\frac{\sum\limits_{j=1}^{m-2} b_j(\xi_j-a)}{1-\sum\limits_{j=1}^{m-2} b_j} \\
&+\frac{1}{1-\sum\limits_{j=1}^{m-2} a_j}\Bigg((b-a)^2M+\sum_{j=1}^{m-2} a_j(\xi_j-a)^2 \\
&+\frac{\sum\limits_{j=1}^{m-2} b_j(\xi_j-a)}{1-\sum\limits_{j=1}^{m-2} b_j}\left(b-a+\sum_{j=1}^{m-2} a_j\xi_j\right)\Bigg)\Bigg)\Bigg) \\
&\leq \frac{\varepsilon}{2}+\frac{\varepsilon}{2} \\
&= \varepsilon, \quad t\in[a,b], \quad n>N.
\end{aligned}
$$

Consequently, $T:\mathscr{C}([a,b])\to\mathscr{C}([a,b])$ is a continuous operator. This completes the proof.

Theorem 5.12. *Suppose* (*D*7)-(*D*10) *and*

$$
0\leq f(x)\leq Q, \quad x\in\mathbb{R},
$$

where Q is a positive constant. Then the BVP (5.19), (5.20) *has at least one solution.*

Proof. Let

$$R = MQ\left((b-a)^2 + \left((b-a)\frac{\sum\limits_{j=1}^{m-2} b_j(\xi_j - a)}{1-\sum\limits_{j=1}^{m-2} b_j} \right.\right.$$

$$+\frac{1}{1-\sum\limits_{j=1}^{m-2} a_j}\left((b-a)^2 M + \sum_{j=1}^{m-2} a_j(\xi_j-a)^2 \right.$$

$$\left.\left.\left.\left. +\frac{\sum\limits_{j=1}^{m-2} b_j(\xi_j - a)}{1-\sum\limits_{j=1}^{m-2} b_j}\left(b-a+\sum_{j=1}^{m-2} a_j\xi_j \right)\right)\right)\right)\right),$$

where

$$0 \le q(t) \le M, \quad t\in[a,b],$$

M is a positive constant. Define

$$E = \{u \in \mathscr{C}([a,b]) : |u(t)| < R, \quad t\in[a,b]\}.$$

For $u \in \overline{E}$, we have

$$|Tu(t)| = \left| -\int_a^t (t-s)q(s)f(u(s))\nabla s - (t-a)\frac{\sum\limits_{j=1}^{m-2} b_j \int_a^{\xi_j} q(s)f(u(s))\nabla s}{1-\sum\limits_{j=1}^{m-2} b_j} \right.$$

$$+\frac{1}{1-\sum\limits_{j=1}^{m-2} a_j}\left(\int_a^b (b-s)q(s)f(u(s))\nabla s \right.$$

$$-\sum_{j=1}^{m-2} a_j \int_a^{\xi_j} (\xi_j - s)q(s)f(u(s))\nabla s$$

$$\left.\left. +\frac{\sum\limits_{j=1}^{m-2} b_j \int_a^{\xi_j} q(s)f(u(s))\nabla s}{1-\sum\limits_{j=1}^{m-2} b_j}\left(b-a-\sum_{j=1}^{m-2} a_j\xi_j \right)\right)\right|$$

$$
\begin{aligned}
&\leq \left|\int_a^t (t-s)q(s)f(u(s))\nabla s\right| + (t-a)\left|\frac{\sum_{j=1}^{m-2} b_j \int_a^{\xi_j} q(s)f(u(s))\nabla s}{1-\sum_{j=1}^{m-2} b_j}\right| \\
&\quad + \frac{1}{1-\sum_{j=1}^{m-2} a_j}\Bigg(\left|\int_a^b (b-s)q(s)f(u(s))\nabla s\right| \\
&\quad + \sum_{j=1}^{m-2} a_j\left|\int_a^{\xi_j} (\xi_j-s)q(s)f(u(s))\nabla s\right| \\
&\quad + \frac{\sum_{j=1}^{m-2} b_j\left|\int_a^{\xi_j} q(s)f(u(s))\nabla s\right|}{1-\sum_{j=1}^{m-2} b_j}\left(\left|b-a-\sum_{j=1}^{m-2} a_j\xi_j\right|\right)\Bigg) \\
&\leq \int_a^t (t-s)q(s)|f(u(s))|\nabla s + (t-a)\frac{\sum_{j=1}^{m-2} b_j \int_a^{\xi_j} q(s)|f(u(s))|\nabla s}{1-\sum_{j=1}^{m-2} b_j} \\
&\quad + \frac{1}{1-\sum_{j=1}^{m-2} a_j}\Bigg(\int_a^b (b-s)q(s)|f(u(s))|\nabla s \\
&\quad + \sum_{j=1}^{m-2} a_j\int_a^{\xi_j} (\xi_j-s)q(s)|f(u(s))|\nabla s \\
&\quad + \frac{\sum_{j=1}^{m-2} b_j \int_a^{\xi_j} q(s)|f(u(s))|\nabla s}{1-\sum_{j=1}^{m-2} b_j}\left(b-a+\sum_{j=1}^{m-2} a_j\xi_j\right)\Bigg) \\
&\leq MQ\Bigg((b-a)^2 + \Bigg((b-a)\frac{\sum_{j=1}^{m-2} b_j(\xi_j-a)}{1-\sum_{j=1}^{m-2} b_j} \\
&\quad + \frac{1}{1-\sum_{j=1}^{m-2} a_j}\Bigg((b-a)^2 M + \sum_{j=1}^{m-2} a_j(\xi_j-a)^2
\end{aligned}
$$

$$
+\frac{\sum_{j=1}^{m-2} b_j(\xi_j-a)}{1-\sum_{j=1}^{m-2} b_j}\left(b-a+\sum_{j=1}^{m-2} a_j\xi_j\right)\Bigg)\Bigg)\Bigg)
$$

$$
= R.
$$

Therefore, $T:\overline{E}\to\overline{E}$. By Lemma 5.5, it follows that $T:\overline{E}\to\overline{E}$ is a continuous operator. Hence and by the Brouwer fixed point Theorem 5.13, it follows that the operator T has a fixed point in E. By Lemma 5.4, we conclude that u is a solution to the BVP (5.19), (5.20). This completes the proof.

For $u\in\mathscr{C}([a,b])$, define the operators

$$
Au(t) = -\int_0^t (t-s)q(s)f(u(s))\nabla s,
$$

$$
\begin{aligned}
Bu(t) &= -(t-a)\frac{\sum_{j=1}^{m-2} b_j\int_a^{\xi_j} q(s)f(u(s))\nabla s}{1-\sum_{j=1}^{m-2} b_j}\\
&\quad+\frac{1}{1-\sum_{j=1}^{m-2} a_j}\Bigg(\int_a^b (b-s)q(s)f(u(s))\nabla s\\
&\quad-\sum_{j=1}^{m-2} a_j\int_a^{\xi_j} (\xi_j-s)q(s)f(u(s))\nabla s\\
&\quad+\frac{\sum_{j=1}^{m-2} b_j\int_a^{\xi_j} q(s)f(u(s))\nabla s}{1-\sum_{j=1}^{m-2} b_j}\left(b-a-\sum_{j=1}^{m-2} a_j\xi_j\right)\Bigg),\quad t\in[a,b].
\end{aligned}
$$

Lemma 5.6. *Suppose* (D9) *and* (D10). *Then the operator* $A:\mathscr{C}([a,b])\to\mathscr{C}([a,b])$ *is a continuous operator.*

Proof. Let M be the constant by the proof of Lemma 5.5. Let, also, $\{u_n\}_{n\in\mathbb{N}}\subseteq\mathscr{C}([a,b])$, $u\in\mathscr{C}([a,b])$, and

$$
u_n\to u \quad \text{in} \quad \mathscr{C}([a,b]), \quad \text{as} \quad n\to\infty.
$$

Take $\varepsilon>0$ arbitrarily. Then there exists an $N=N(\varepsilon)>0$ such that

$$
|f(u_n(t))-f(u(t))|<\frac{\varepsilon}{M(b-a)^2}, \quad t\in[a,b], \quad n>N.
$$

Hence,

$$\begin{aligned}
|Au_n(t) - Au(t)| &= \left| -\int_a^t (t-s)q(s)f(u_n(s))\nabla s \right. \\
&\quad \left. + \int_a^t (t-s)q(s)f(u(s))\nabla s \right| \\
&= \left| \int_a^t (t-s)q(s)(f(u_n(s)) - f(u(s)))\nabla s \right| \\
&\le \int_a^t (t-s)|q(s)||f(u_n(s)) - f(u(s))|\nabla s \\
&< \frac{\varepsilon}{M(b-a)^2} M(b-a)^2 \\
&= \varepsilon, \quad t \in [a,b], \quad n > N.
\end{aligned}$$

Thus, $A : \mathscr{C}([a,b]) \to \mathscr{C}([a,b])$ is a continuous operator. This completes the proof.

Lemma 5.7. *Suppose (D9)-(D10). Then $A : \mathscr{C}([a,b]) \to \mathscr{C}([a,b])$ is a completely continuous operator.*

Proof. Let M be the constant as in the proof of Lemma 5.5. Let, also, $X \subseteq \mathscr{C}([a,b])$ be a bounded set. Then there exists a constant $L_1 > 0$ so that

$$|u(t)| \le L_1, \quad t \in [a,b],$$

for any $u \in X$. Since $f \in \mathscr{C}([a,b])$, there exists a constant $L_2 > 0$ so that

$$|f(u(t))| \le L_2, \quad t \in [a,b],$$

for any $u \in X$. Take $u \in X$ arbitrarily. Then

$$\begin{aligned}
|Au(t)| &= \left| -\int_a^t (t-s)q(s)f(u(s))\nabla s \right| \\
&\le \int_a^t (t-s)|q(s)||f(u(s))|\nabla s \\
&\le L_2 M(b-a)^2, \quad t \in [a,b].
\end{aligned}$$

Therefore, $A(X)$ is uniformly bounded. Now, we take $t_1, t_2 \in [a,b]$, $t_1 < t_2$, and $u \in X$. Then

$$|Au(t_2) - Au(t_1)| = \left| -\int_a^{t_2} (t_2 - s)q(s)f(u(s))\nabla s \right.$$

$$
\begin{aligned}
&+\int_a^{t_1}(t_1-s)q(s)f(u(s))\nabla s\Big| \\
&=\Big|-\int_a^{t_1}(t_2-s)q(s)f(u(s))\nabla s \\
&-\int_{t_1}^{t_2}(t_2-s)q(s)f(u(s))\nabla s \\
&+\int_a^{t_1}(t_1-s)q(s)f(u(s))\nabla s\Big| \\
&=\Big|-\int_a^{t_1}(t_2-t_1)q(s)f(u(s))\nabla s \\
&-\int_{t_1}^{t_2}(t_2-s)q(s)f(u(s))\nabla s\Big| \\
&\leq\Big|\int_a^{t_1}(t_2-t_1)q(s)f(u(s))\nabla s\Big| \\
&+\Big|\int_{t_1}^{t_2}(t_2-s)q(s)f(u(s))\nabla s\Big| \\
&\leq\int_a^{t_1}(t_2-t_1)|q(s)||f(u(s))|\nabla s \\
&+\int_{t_1}^{t_2}(t_2-s)|q(s)||f(u(s))|\nabla s \\
&\leq ML_2(b-a)(t_2-t_1)+ML_2(t_2-t_1)(b-a) \\
&=2ML_2(b-a)(t_2-t_1).
\end{aligned}
$$

Therefore, $A(X)$ is equi-continuous. Now, applying the Arzela-Ascoli theorem, we conclude that the operator $A:\mathscr{C}([a,b])\to\mathscr{C}([a,b])$ is a completely continuous operator. This completes the proof.

Let

$$
P_1=M\left((b-a)\frac{\sum\limits_{j=1}^{m-2}b_j(\xi_j-a)}{1-\sum\limits_{j=1}^{m-2}b_j}\right.
$$

$$+\frac{1}{1-\sum\limits_{j=1}^{m-2}a_j}\Bigg((b-a)^2M+\sum_{j=1}^{m-2}a_j(\xi_j-a)^2$$

$$+\frac{\sum\limits_{j=1}^{m-2}b_j(\xi_j-a)}{1-\sum\limits_{j=1}^{m-2}b_j}\left(b-a+\sum_{j=1}^{m-2}a_j\xi_j\right)\Bigg)\Bigg),$$

where M is the constant defined in the proof of Lemma 5.5.

Lemma 5.8. *Suppose (D7)-(D10). Let, also,*

(D11) $|f(u)-f(v)|\leq L|u-v|$, $u,v\in\mathbb{R}$, $f(0)=0$, *where L is a positive constant so that*

$$LP_1<1.$$

Then $B:\mathscr{C}([a,b])\to\mathscr{C}([a,b])$ is a contraction.

Proof. Let $u,v\in\mathscr{C}([a,b])$ be arbitrarily chosen. Then

$$|Bu_1(t)-Bu_2(t)|=\Bigg|-(t-a)\frac{\sum\limits_{j=1}^{m-2}b_j\int_a^{\xi_j}q(s)f(u(s))\nabla s}{1-\sum\limits_{j=1}^{m-2}b_j}$$

$$+\frac{1}{1-\sum\limits_{j=1}^{m-2}a_j}\Bigg(\int_a^b(b-s)q(s)f(u(s))\nabla s$$

$$-\sum_{j=1}^{m-2}a_j\int_a^{\xi_j}(\xi_j-s)q(s)f(u(s))\nabla s$$

$$+\frac{\sum\limits_{j=1}^{m-2}b_j\int_a^{\xi_j}q(s)f(u(s))\nabla s}{1-\sum\limits_{j=1}^{m-2}b_j}\left(b-a-\sum_{j=1}^{m-2}a_j\xi_j\right)\Bigg)$$

$$+(t-a)\frac{\sum\limits_{j=1}^{m-2}b_j\int_a^{\xi_j}q(s)f(v(s))\nabla s}{1-\sum\limits_{j=1}^{m-2}b_j}$$

$$+\frac{1}{1-\sum\limits_{j=1}^{m-2}a_j}\Bigg(\int_a^b(b-s)q(s)f(v(s))\nabla s$$

$$
\begin{aligned}
&-\sum_{j=1}^{m-2} a_j \int_a^{\xi_j} (\xi_j - s) q(s) f(v(s)) \nabla s \\
&+\frac{\sum\limits_{j=1}^{m-2} b_j \int_a^{\xi_j} q(s) f(v(s)) \nabla s}{1-\sum\limits_{j=1}^{m-2} b_j} \left(b - a - \sum_{j=1}^{m-2} a_j \xi_j \right) \Bigg) \Bigg| \\
=& \Bigg| -(t-a) \frac{\sum\limits_{j=1}^{m-2} b_j \int_a^{\xi_j} q(s)(f(u(s)) - f(v(s)) \nabla s}{1-\sum\limits_{j=1}^{m-2} b_j} \\
&+\frac{1}{1-\sum\limits_{j=1}^{m-2} a_j} \Bigg(\int_a^b (b-s) q(s)(f(u(s)) - f(v(s))) \nabla s \\
&-\sum_{j=1}^{m-2} a_j \int_a^{\xi_j} (\xi_j - s) q(s)(f(u(s)) - f(v(s))) \nabla s \\
&+\frac{\sum\limits_{j=1}^{m-2} b_j \int_a^{\xi_j} q(s)(f(u(s)) - f(v(s))) \nabla s}{1-\sum\limits_{j=1}^{m-2} b_j} \left(b - a - \sum_{j=1}^{m-2} a_j \xi_j \right) \Bigg) \Bigg| \\
\leq& \left| (t-a) \frac{\sum\limits_{j=1}^{m-2} b_j \int_a^{\xi_j} q(s)(f(u(s)) - f(v(s))) \nabla s}{1-\sum\limits_{j=1}^{m-2} b_j} \right| \\
&+\left| \frac{1}{1-\sum\limits_{j=1}^{m-2} a_j} \Bigg(\int_a^b (b-s) q(s)(f(u(s)) - f(v(s))) \nabla s \right| \\
&+\left| \sum_{j=1}^{m-2} a_j \int_a^{\xi_j} (\xi_j - s) q(s)(f(u(s)) - f(v(s))) \nabla s \right| \\
&+\left| \frac{\sum\limits_{j=1}^{m-2} b_j \int_a^{\xi_j} q(s)(f(u(s)) - f(v(s))) \nabla s}{1-\sum\limits_{j=1}^{m-2} b_j} \left(b - a - \sum_{j=1}^{m-2} a_j \xi_j \right) \Bigg) \right|
\end{aligned}
$$

$$\le (t-a)\frac{\sum_{j=1}^{m-2} b_j \int_a^{\xi_j} q(s)|f(u(s))-f(v(s))|\nabla s}{1-\sum_{j=1}^{m-2} b_j}$$

$$+\frac{1}{1-\sum_{j=1}^{m-2} a_j}\Bigg(\int_a^b (b-s)q(s)|f(u(s))-f(v(s))|\nabla s$$

$$+\sum_{j=1}^{m-2} a_j \int_a^{\xi_j} (\xi_j-s)q(s)|f(u(s))-f(v(s))|\nabla s$$

$$+\frac{\sum_{j=1}^{m-2} b_j \int_a^{\xi_j} q(s)|f(u(s))-f(v(s))|\nabla s}{1-\sum_{j=1}^{m-2} b_j}\left(b-a+\sum_{j=1}^{m-2} a_j\xi_j\right)\Bigg)$$

$$\le L(t-a)\frac{\sum_{j=1}^{m-2} b_j \int_a^{\xi_j} q(s)|u(s)-v(s)|\nabla s}{1-\sum_{j=1}^{m-2} b_j}\Bigg|$$

$$+L\frac{1}{1-\sum_{j=1}^{m-2} a_j}\Bigg(\int_a^b (b-s)q(s)|u(s)-v(s)|\nabla s$$

$$+L\sum_{j=1}^{m-2} a_j \int_a^{\xi_j} (\xi_j-s)q(s)|u(s)-v(s)|\nabla s$$

$$+L\frac{\sum_{j=1}^{m-2} b_j \int_a^{\xi_j} q(s)|u(s)-v(s)|\nabla s}{1-\sum_{j=1}^{m-2} b_j}\left(b-a+\sum_{j=1}^{m-2} a_j\xi_j\right)\Bigg)$$

$$\le LM\Bigg((b-a)\frac{\sum_{j=1}^{m-2} b_j(\xi_j-a)}{1-\sum_{j=1}^{m-2} b_j}$$

$$+\frac{1}{1-\sum_{j=1}^{m-2} a_j}\Bigg((b-a)^2M+\sum_{j=1}^{m-2} a_j(\xi_j-a)^2$$

$$+\frac{\sum_{j=1}^{m-2} b_j(\xi_j - a)}{1-\sum_{j=1}^{m-2} b_j}\left(b-a+\sum_{j=1}^{m-2} a_j\xi_j\right)\Bigg)\Bigg)\|u-v\|$$

$$= LP_1\|u-v\|, \quad t \in [a,b].$$

From here,

$$\|Bu - Bv\| \leq LP_1\|u-v\|.$$

Therefore,, $B : \mathscr{C}([a,b]) \to \mathscr{C}([a,b])$ is a contraction.

Take $r > 0$ arbitrarily and define

$$X = \{u \in \mathscr{C}([a,b]) : |u(t)| \leq r, \quad t \in [a,b]\}.$$

Note that X is a closed, convex, nonempty subset of the Banach space $\mathscr{C}([a,b])$.

Lemma 5.9. *Suppose* $(D7)$-$(D11)$ *and*

(D12) $\quad L(P_1 + (b-a)^2 M) < 1.$

Then $Au + Bv \in X$ *for any* $u, v \in X$.

Proof. Take $u, v \in X$ arbitrarily. Then

$$|Au(t) + Bv(t)| = \Bigg| - \int_a^t (t-s)q(s)f(u(s))\nabla s - (t-a)\frac{\sum_{j=1}^{m-2} b_j \int_a^{\xi_j} q(s)f(v(s))\nabla s}{1-\sum_{j=1}^{m-2} b_j}$$

$$+\frac{1}{1-\sum_{j=1}^{m-2} a_j}\Bigg(\int_a^b (b-s)q(s)f(v(s))\nabla s$$

$$-\sum_{j=1}^{m-2} a_j \int_a^{\xi_j} (\xi_j - s)q(s)f(v(s))\nabla s$$

$$+\frac{\sum_{j=1}^{m-2} b_j \int_a^{\xi_j} q(s)f(v(s))\nabla s}{1-\sum_{j=1}^{m-2} b_j}\left(b-a-\sum_{j=1}^{m-2} a_j\xi_j\right)\Bigg)\Bigg|$$

$$\leq \left|\int_a^t (t-s)q(s)f(u(s))\nabla s\right| + \left|(t-a)\frac{\sum_{j=1}^{m-2} b_j \int_a^{\xi_j} q(s)f(v(s))\nabla s}{1-\sum_{j=1}^{m-2} b_j}\right|$$

$$+\frac{1}{1-\sum_{j=1}^{m-2}a_j}\left(\left|\int_a^b (b-s)q(s)f(v(s))\nabla s\right|\right.$$

$$+\left|\sum_{j=1}^{m-2}a_j\int_a^{\xi_j}(\xi_j-s)q(s)f(v(s))\nabla s\right|$$

$$\left.+\left|\frac{\sum_{j=1}^{m-2}b_j\int_a^{\xi_j}q(s)f(v(s))\nabla s}{1-\sum_{j=1}^{m-2}b_j}\left(b-a-\sum_{j=1}^{m-2}a_j\xi_j\right)\right|\right)$$

$$\leq \int_a^t (t-s)q(s)|f(u(s))|\nabla s+(t-a)\frac{\sum_{j=1}^{m-2}b_j\int_a^{\xi_j}q(s)|f(v(s))|\nabla s}{1-\sum_{j=1}^{m-2}b_j}$$

$$+\frac{1}{1-\sum_{j=1}^{m-2}a_j}\left(\int_a^b (b-s)q(s)|f(v(s))|\nabla s\right.$$

$$+\sum_{j=1}^{m-2}a_j\int_a^{\xi_j}(\xi_j-s)q(s)|f(v(s))|\nabla s$$

$$\left.+\frac{\sum_{j=1}^{m-2}b_j\int_a^{\xi_j}q(s)|f(v(s))|\nabla s}{1-\sum_{j=1}^{m-2}b_j}\left(b-a+\sum_{j=1}^{m-2}a_j\xi_j\right)\right)$$

$$\leq L\left(\int_a^t (t-s)q(s)|u(s)|\nabla s+(t-a)\frac{\sum_{j=1}^{m-2}b_j\int_a^{\xi_j}q(s)|v(s)|\nabla s}{1-\sum_{j=1}^{m-2}b_j}\right.$$

$$+\frac{1}{1-\sum_{j=1}^{m-2}a_j}\left(\int_a^b (b-s)q(s)|v(s)|\nabla s\right.$$

$$+\sum_{j=1}^{m-2}a_j\int_a^{\xi_j}(\xi_j-s)q(s)|v(s)|\nabla s$$

$$
+\frac{\sum_{j=1}^{m-2} b_j \int_a^{\xi_j} q(s)|v(s)|\nabla s}{1-\sum_{j=1}^{m-2} b_j}\left(b-a+\sum_{j=1}^{m-2} a_j\xi_j\right)\Bigg)\Bigg)
$$

$$
\begin{aligned}
&\leq L\Bigg((b-a)^2M+\int_a^t (t-s)q(s)|f(u(s))|\nabla s \\
&\quad +(t-a)\frac{\sum_{j=1}^{m-2} b_j \int_a^{\xi_j} q(s)|f(v(s))|\nabla s}{1-\sum_{j=1}^{m-2} b_j} \\
&\quad +\frac{1}{1-\sum_{j=1}^{m-2} a_j}\Bigg(\int_a^b (b-s)q(s)|f(v(s))|\nabla s \\
&\quad +\sum_{j=1}^{m-2} a_j \int_a^{\xi_j} (\xi_j-s)q(s)|f(v(s))|\nabla s \\
&\quad +\frac{\sum_{j=1}^{m-2} b_j \int_a^{\xi_j} q(s)|f(v(s))|\nabla s}{1-\sum_{j=1}^{m-2} b_j}\left(b-a+\sum_{j=1}^{m-2} a_j\xi_j\right)\Bigg)\Bigg)r \\
&= L\left((b-a)^2M+P_1\right)r \\
&< r, \quad t\in[a,b].
\end{aligned}
$$

Therefore, $Au+Bv\in X$. This completes the proof.

For our next result we need the following fixed point theorem.

Theorem 5.13 (Krasnosel'skii Fixed Point Theorem). *Let M be a closed convex nonempty subset of a Banach space $(\mathscr{B},\|\cdot\|)$. Suppose that*

1. *$A: M\to\mathscr{B}$ is completely continuous,*
2. *$B: M\to\mathscr{B}$ is a contraction,*
3. *$x,y\in M$ implies $Ax+By\in M$.*

Then the map $A+B$ has a fixed point in M.

Theorem 5.14. *Suppose $(D7)-(D12)$. Then the BVP* (5.19), (5.20) *has at least one solution.*

Proof. By Lemma 5.7, it follows that the operator $A : X \to \mathscr{C}([a,b])$ is a completely continuous operator. By Lemma 5.8, we have that $B : X \to \mathscr{C}([a,b])$ is a contraction. Now, applying Lemma 5.9 and the Krasnosel'skii fixed point theorem,

we conclude that the operator $A+B$ has a fixed point in X. Hence and by Lemma 5.4, we conclude that the BVP (5.19), (5.20) has at least one positive solution. This completes the proof.

Example 5.7. Let $\mathbb{T} = \mathbb{Z}$. Consider the BVP

$$u^{\Delta\nabla} + \left(\frac{1}{600(1+t^2)}\right)\frac{(u(t))^2}{10000+(u(t))^2} = 0, \quad t \in [0,10],$$

$$u^{\Delta}(0) = \frac{1}{4}u^{\Delta}(1) + \frac{1}{4}u^{\Delta}(2),$$

$$u(10) = \frac{1}{3}u(1) + \frac{1}{3}u(2).$$

Here

$$\sigma(t) = t+1,$$

$$q(t) = \frac{1}{600(1+t^2)},$$

$$f(x) = \frac{x^2}{10000+x^2}, \quad x \in \mathbb{R}, \quad t \in \mathbb{T},$$

$$m = 4,$$

$$a = 0,$$

$$b = 10,$$

$$\xi_1 = 1,$$

$$\xi_2 = 2,$$

$$b_1 = \frac{1}{4},$$

$$b_2 = \frac{1}{4},$$

$$a_1 = \frac{1}{3},$$

$$a_2 = \frac{1}{3}.$$

Then

$$\begin{aligned}
b_1 + b_2 &= \frac{1}{4} + \frac{1}{4} \\
&= \frac{1}{2}, \\
a_1 + a_2 &= \frac{1}{3} + \frac{1}{3} \\
&= \frac{2}{3}, \\
1 - (b_1 + b_2) &= 1 - \frac{1}{2} \\
&= \frac{1}{2}, \\
1 - (a_1 + a_2) &= 1 - \frac{2}{3} \\
&= \frac{1}{3}, \\
b - a &= 10, \\
\xi_1 - a &= 1, \\
\xi_2 - a &= 2,
\end{aligned}$$

and

$$\begin{aligned}
a_1(\xi_1 - a)^2 + a_2(\xi_2 - a)^2 &= \frac{1}{2} + \frac{1}{2} \cdot 4 \\
&= \frac{1}{2} + 2 \\
&= \frac{5}{2},
\end{aligned}$$

$$\begin{aligned} a_1\xi_1 + a_2\xi_2 &= \frac{1}{3} + \frac{2}{3} \\ &= 1, \\ b_1(\xi_1 - a) + b_2(\xi_2 - a) &= \frac{1}{4} + \frac{1}{2} \\ &= \frac{3}{4}, \end{aligned}$$

and

$$0 \le q(t) \le \frac{1}{600}, \quad t \in \mathbb{T},$$

i.e., $M = \dfrac{1}{600}$, and

$$\begin{aligned} M_1 &= M\left((b-a)^2 + (b-a)\frac{\sum_{j=1}^{m-2} b_j(\xi_j - a)}{1 - \sum_{j=1}^{m-2} b_j} \right. \\ &\quad + \frac{1}{1 - \sum_{j=1}^{m-2} a_j}\left((b-a)^2 + \sum_{j=1}^{m-2} a_j(\xi_j - a)^2 \right. \\ &\quad \left.\left. + \frac{\sum_{j=1}^{m-2} b_j(\xi_j - a)}{1 - \sum_{j=1}^{m-2} b_j}\left(b - a + \sum_{j=1}^{m-2} a_j\xi_j \right)\right)\right) \\ &= \frac{1}{600}\left(10^2 + 10 \cdot \frac{\frac{3}{4}}{\frac{1}{2}} \right. \\ &\quad \left. + \frac{1}{\frac{1}{3}}\left(10^2 + \frac{5}{2} + \frac{\frac{3}{4}}{\frac{1}{2}}(10-1) \right)\right) \\ &= \frac{1}{600}\left(100 + 15 + 3\left(100 + \frac{5}{2} + \frac{27}{2} \right)\right) \\ &= \frac{1}{600}(425 + 48) \\ &= \frac{463}{600}. \end{aligned}$$

Note that

$$\begin{aligned}\frac{d}{dx}f(x) &= \frac{d}{dx}\left(\frac{x^2}{10000+x^2}\right)\\ &= \frac{2x(10000+x^2)-2x(x^2)}{(10000+x^2)^2}\\ &= \frac{20000x}{(10000+x^2)^2}, \quad x\in\mathbb{R}.\end{aligned}$$

Thus,

$$\begin{aligned}0 &\leq \left|\frac{d}{dx}f(x)\right|\\ &= \left|\frac{20000x}{(10000+x^2)^2}\right|\\ &= \frac{200|100x|}{(10000+x^2)(10000+x^2)}\\ &\leq \frac{200}{10000+x^2}\\ &\leq \frac{200}{10000}\\ &= \frac{1}{50}\end{aligned}$$

and $L = \frac{1}{50}$. Hence,

$$\begin{aligned}LM_1 &= \frac{1}{5}\left(\frac{463}{600}\right)\\ &= \frac{463}{3000}\\ &< 1.\end{aligned}$$

From here, it follows that the considered BVP has at least one solution.

Exercise 5.10. Let $\mathbb{T} = 2^{\mathbb{N}_0}$. Prove that the BVP

$$u^{\Delta\nabla} + \left(\frac{1}{1000+t^4}\right)\frac{(u(t))^3}{10^4+(u(t))^6} = 0, \quad t\in[1,16],$$

$$u^{\Delta}(1) = \frac{1}{8}u^{\Delta}(2) + \frac{1}{8}u^{\Delta}(4) + \frac{1}{16}u^{\Delta}(8),$$

$$u(16) = \frac{1}{24}u(2) + \frac{1}{48}u(4) + \frac{1}{24}u(8),$$

has at least one solution.

Lemma 5.10. *Suppose (D7)-(D10). Then the operator $B : \mathscr{C}([a,b]) \to \mathscr{C}([a,b])$ is a completely continuous operator.*

Proof. Let M be the constant as in the proof of Lemma 5.5. Let, also, $X \subseteq \mathscr{C}([a,b])$ be a bounded set. Then there exists a constant $L_1 > 0$ so that

$$|u(t)| \le L_1, \quad t \in [a,b],$$

for any $u \in X$. Since $f \in \mathscr{C}([a,b])$, there exists a constant $L_2 > 0$ so that

$$|f(u(t))| \le L_2, \quad t \in [a,b],$$

for any $u \in X$. Take $u \in X$ arbitrarily. Then

$$\begin{aligned}
|Bu(t)| = \Bigg| &-(t-a)\frac{\sum_{j=1}^{m-2} b_j \int_a^{\xi_j} q(s)f(u(s))\nabla s}{1-\sum_{j=1}^{m-2} b_j} \\
&+\frac{1}{1-\sum_{j=1}^{m-2} a_j}\Bigg(\int_a^b (b-s)q(s)f(u(s))\nabla s \\
&-\sum_{j=1}^{m-2} a_j \int_a^{\xi_j} (\xi_j - s)q(s)f(u(s))\nabla s \\
&+\frac{\sum_{j=1}^{m-2} b_j \int_a^{\xi_j} q(s)f(u(s))\nabla s}{1-\sum_{j=1}^{m-2} b_j}\left(b-a-\sum_{j=1}^{m-2} a_j\xi_j\right)\Bigg)\Bigg| \\
\le\ & (t-a)\frac{\sum_{j=1}^{m-2} b_j \left|\int_a^{\xi_j} q(s)f(u(s))\nabla s\right|}{1-\sum_{j=1}^{m-2} b_j} \\
&+\frac{1}{1-\sum_{j=1}^{m-2} a_j}\Bigg(\left|\int_a^b (b-s)q(s)f(u(s))\nabla s\right|
\end{aligned}$$

$$+\sum_{j=1}^{m-2} a_j \left| \int_a^{\xi_j} (\xi_j - s) q(s) f(u(s)) \nabla s \right|$$

$$+\frac{\sum\limits_{j=1}^{m-2} b_j \left| \int_a^{\xi_j} q(s) f(u(s)) \nabla s \right|}{1-\sum\limits_{j=1}^{m-2} b_j} \left(b - a + \sum_{j=1}^{m-2} a_j \xi_j \right) \Bigg)$$

$$\le (t-a) \frac{\sum\limits_{j=1}^{m-2} b_j \int_a^{\xi_j} |q(s)||f(u(s))| \nabla s}{1-\sum\limits_{j=1}^{m-2} b_j}$$

$$+\frac{1}{1-\sum\limits_{j=1}^{m-2} a_j} \Bigg(\int_a^b (b-s)|q(s)||f(u(s))| \nabla s$$

$$+\sum_{j=1}^{m-2} a_j \int_a^{\xi_j} (\xi_j - s)|q(s)||f(u(s))| \nabla s$$

$$+\frac{\sum\limits_{j=1}^{m-2} b_j \int_a^{\xi_j} |q(s)||f(u(s))| \nabla s}{1-\sum\limits_{j=1}^{m-2} b_j} \left(b - a + \sum_{j=1}^{m-2} a_j \xi_j \right) \Bigg)$$

$$\le L_2 M \Bigg((b-a) \frac{\sum\limits_{j=1}^{m-2} b_j (\xi_j - a)}{1-\sum\limits_{j=1}^{m-2} b_j}$$

$$+\frac{1}{1-\sum\limits_{j=1}^{m-2} a_j} \Bigg((b-a)^2 + \sum_{j=1}^{m-2} a_j (\xi_j - a)^2$$

$$+\frac{\sum\limits_{j=1}^{m-2} b_j (\xi_j - a)}{1-\sum\limits_{j=1}^{m-2} b_j} \left(b - a + \sum_{j=1}^{m-2} a_j \xi_j \right) \Bigg) \Bigg), \quad t \in [a,b].$$

Thus, $B(X)$ is uniformly bounded. Let now, $t_1, t_2 \in [a,b]$, $t_1 < t_2$. Take $u \in X$ arbitrarily. Then

$$
\begin{aligned}
|Bu(t_1)-Bu(t_2)| &= \Bigg| -(t_1-a)\frac{\sum_{j=1}^{m-2} b_j \int_a^{\xi_j} q(s)f(u(s))\nabla s}{1-\sum_{j=1}^{m-2} b_j} \\
&\quad + \frac{1}{1-\sum_{j=1}^{m-2} a_j}\Bigg(\int_a^b (b-s)q(s)f(u(s))\nabla s \\
&\quad - \sum_{j=1}^{m-2} a_j \int_a^{\xi_j} (\xi_j-s)q(s)f(u(s))\nabla s \\
&\quad + \frac{\sum_{j=1}^{m-2} b_j \int_a^{\xi_j} q(s)f(u(s))\nabla s}{1-\sum_{j=1}^{m-2} b_j}\left(b-a-\sum_{j=1}^{m-2} a_j\xi_j\right)\Bigg) \\
&\quad -(t_2-a)\frac{\sum_{j=1}^{m-2} b_j \int_a^{\xi_j} q(s)f(u(s))\nabla s}{1-\sum_{j=1}^{m-2} b_j} \\
&\quad - \frac{1}{1-\sum_{j=1}^{m-2} a_j}\Bigg(\int_a^b (b-s)q(s)f(u(s))\nabla s \\
&\quad - \sum_{j=1}^{m-2} a_j \int_a^{\xi_j} (\xi_j-s)q(s)f(u(s))\nabla s \\
&\quad + \frac{\sum_{j=1}^{m-2} b_j \int_a^{\xi_j} q(s)f(u(s))\nabla s}{1-\sum_{j=1}^{m-2} b_j}\left(b-a-\sum_{j=1}^{m-2} a_j\xi_j\right)\Bigg)\Bigg| \\
&= (t_2-t_1)\frac{\sum_{j=1}^{m-2} b_j \int_a^{\xi_j} q(s)f(u(s))\nabla s}{1-\sum_{j=1}^{m-2} b_j} \\
&\le ML_2\frac{\sum_{j=1}^{m-2} b_j(\xi_j-a)}{1-\sum_{j=1}^{m-2} b_j}(t_2-t_1).
\end{aligned}
$$

Consequently, $B(X)$ is equi-continuous. Now, applying the Arzela-Ascoli theorem, we conclude that $B:\mathscr{C}([a,b])\to\mathscr{C}([a,b])$ is a completely continuous operator. This completes the proof.

Lemma 5.11. *The operator* $T:\mathscr{C}([a,b])\to\mathscr{C}([a,b])$ *is a completely continuous operator.*

Proof. By Lemma 5.7, it follows that the operator $A:\mathscr{C}([a,b])\to\mathscr{C}([a,b])$ is a completely continuous operator. From Lemma 5.10, it follows that $B:\mathscr{C}([a,b])\to\mathscr{C}([a,b])$ is a completely continuous operator. Hence, $T=A+B:\mathscr{C}([a,b])\to\mathscr{C}([a,b])$ is a completely continuous operator. This completes the proof.

Let

$$M_1=M\Bigg((b-a)^2+(b-a)\frac{\sum\limits_{j=1}^{m-2}b_j(\xi_j-a)}{1-\sum\limits_{j=1}^{m-2}b_j}+\frac{1}{1-\sum\limits_{j=1}^{m-2}a_j}\Bigg((b-a)^2+\sum_{j=1}^{m-2}a_j(\xi_j-a)^2+\frac{\sum\limits_{j=1}^{m-2}b_j(\xi_j-a)}{1-\sum\limits_{j=1}^{m-2}b_j}\Big(b-a+\sum_{j=1}^{m-2}a_j\xi_j\Big)\Bigg)\Bigg).$$

Theorem 5.15. *Suppose* (*D*7)-(*D*12). *Then the BVP has at least one solution.*

Proof. Let $R>0$ be arbitrarily chosen. Define the set

$$\Omega=\{u\in\mathscr{C}([a,b]):\|u\|<R\}.$$

We have $0\in\Omega$ and, using Lemma 5.11, we conclude that $T:\overline{\Omega}\to\mathscr{C}([a,b])$ is a completely continuous operator. Next, for $u\in\partial\overline{\Omega}$ and $\lambda\in[0,1]$, we have $\|u\|=R$ and

$$|Tu(t)|=\lambda\Bigg(\Bigg|-\int_a^t(t-s)q(s)f(u(s))\nabla s-(t-a)\frac{\sum\limits_{j=1}^{m-2}b_j\int_a^{\xi_j}q(s)f(v(s))\nabla s}{1-\sum\limits_{j=1}^{m-2}b_j}$$

$$+\frac{1}{1-\sum\limits_{j=1}^{m-2}a_j}\Bigg(\int_a^b(b-s)q(s)f(u(s))\nabla s$$

$$-\sum_{j=1}^{m-2} a_j \int_a^{\xi_j} (\xi_j - s)q(s)f(u(s))\nabla s$$

$$+\frac{\sum\limits_{j=1}^{m-2} b_j \int_a^{\xi_j} q(s)f(u(s))\nabla s}{1-\sum\limits_{j=1}^{m-2} b_j}\left(b-a-\sum_{j=1}^{m-2} a_j\xi_j\right)\Bigg)\Bigg|\Bigg)$$

$$\leq \lambda\left(\left|\int_a^t (t-s)q(s)f(u(s))\nabla s\right| + \left|(t-a)\frac{\sum\limits_{j=1}^{m-2} b_j \int_a^{\xi_j} q(s)f(v(s))\nabla s}{1-\sum\limits_{j=1}^{m-2} b_j}\right|\right.$$

$$+\frac{1}{1-\sum\limits_{j=1}^{m-2} a_j}\left(\left|\int_a^b (b-s)q(s)f(u(s))\nabla s\right|\right.$$

$$+\left|\sum_{j=1}^{m-2} a_j \int_a^{\xi_j} (\xi_j - s)q(s)f(u(s))\nabla s\right|$$

$$\left.\left.+\left|\frac{\sum\limits_{j=1}^{m-2} b_j \int_a^{\xi_j} q(s)f(u(s))\nabla s}{1-\sum\limits_{j=1}^{m-2} b_j}\left(b-a-\sum_{j=1}^{m-2} a_j\xi_j\right)\right|\right)\right)$$

$$\leq \lambda\left(\int_a^t (t-s)q(s)|f(u(s))|\nabla s + (t-a)\frac{\sum\limits_{j=1}^{m-2} b_j \int_a^{\xi_j} q(s)|f(v(s))|\nabla s}{1-\sum\limits_{j=1}^{m-2} b_j}\right.$$

$$+\frac{1}{1-\sum\limits_{j=1}^{m-2} a_j}\left(\int_a^b (b-s)q(s)|f(u(s))|\nabla s\right.$$

$$+\sum_{j=1}^{m-2} a_j \int_a^{\xi_j} (\xi_j - s)q(s)|f(u(s))|\nabla s$$

$$\left.\left.+\frac{\sum\limits_{j=1}^{m-2} b_j \int_a^{\xi_j} q(s)|f(u(s))|\nabla s}{1-\sum\limits_{j=1}^{m-2} b_j}\left(b-a+\sum_{j=1}^{m-2} a_j\xi_j\right)\right)\right)$$

$$\leq \lambda L\Bigg(\int_a^t (t-s)q(s)|u(s)|\nabla s + (t-a)\frac{\sum_{j=1}^{m-2} b_j \int_a^{\xi_j} q(s)|u(s)|\nabla s}{1-\sum_{j=1}^{m-2} b_j}$$

$$+\frac{1}{1-\sum_{j=1}^{m-2} a_j}\Bigg(\int_a^b (b-s)q(s)|u(s)|\nabla s$$

$$+\sum_{j=1}^{m-2} a_j \int_a^{\xi_j} (\xi_j - s)q(s)|u(s)|\nabla s$$

$$+\frac{\sum_{j=1}^{m-2} b_j \int_a^{\xi_j} q(s)|u(s)|\nabla s}{1-\sum_{j=1}^{m-2} b_j}\Bigg(b-a-\sum_{j=1}^{m-2} a_j\xi_j\Bigg)\Bigg)\Bigg)$$

$$\leq LM_1R$$

$$< R, \quad t \in [a,b].$$

Therefore,

$$u \neq \lambda Tu, \quad \lambda \in [0,1], \quad u \in \partial\Omega.$$

Hence and by the Leray-Schauder fixed point 5.4, it follows that the operator T has a fixed point $u \in \Omega$. Now, applying Lemma 5.4, we conclude that u is a solution to the BVP (5.19), (5.20). This completes the proof.

Below, we will investigate the BVP (5.19), (5.20) for existence of multiple solutions. Define

$$P = \Big\{u \in \mathscr{C}([a,b]) : u \quad \text{is} \quad \text{concave, non-increasing} \quad \text{and}$$

$$\text{non-negative} \quad \text{on} \quad [a,b]\Big\}.$$

Lemma 5.12. *Let $u \in P$. Then*

$$u(t) \geq \frac{b-t}{b-a}\|u\|, \quad t \in [a,b].$$

Proof. Since u is non-increasing and non-negative on $[a,b]$, we have

$$u(a) \geq u(t)$$

$$\geq u(b), \quad t \in [a,b].$$

Thus,

$$\|u\| = \max_{t\in[a,b]} u(t)$$

$$= u(a).$$

Let

$$y(t) = u(t) - \frac{b-t}{b-a}\|u\|, \quad t \in [a,b]. \tag{5.25}$$

Then

$$y^{\Delta\nabla}(t) = u^{\Delta\nabla}(t)$$

$$\leq 0, \quad t \in [a,b],$$

and

$$y(a) = u(a) - \frac{b-a}{b-a}\|u\|$$

$$= u(a) - u(a)$$

$$= 0,$$

$$y(b) = u(b) - \frac{b-b}{b-a}\|u\|$$

$$= u(b)$$

$$\geq 0.$$

Let

$$G(t,s) = \frac{1}{b-a}\begin{cases} (s-a)(b-t) & a \leq s \leq t \leq b \\ (t-a)(b-s) & a \leq t \leq s \leq b. \end{cases}$$

Then

$$\int_a^b G(t,s)y^{\Delta\nabla}(s)\nabla s = \int_a^t G(t,s)y^{\Delta\nabla}(s)\nabla s + \int_t^b G(t,s)y^{\Delta\nabla}(s)\nabla s$$

$$= \frac{b-t}{b-a}\int_a^t (s-a)y^{\Delta\nabla}(s)\nabla s + \frac{t-a}{b-a}\int_t^b (b-s)y^{\Delta\nabla}(s)\nabla s$$

$$
\begin{aligned}
&= \frac{b-t}{b-a}\left((t-a)y^{\Delta}(t) - \int_a^t y^{\Delta}(s)\nabla s\right)\\
&\quad + \frac{t-a}{b-a}\left(-(b-t)y^{\Delta}(t) + \int_t^b y^{\Delta}(s)\nabla s\right)\\
&= -\frac{b-t}{b-a}\int_a^t y^{\Delta}(s)\nabla s + \frac{t-a}{b-a}\int_t^b y^{\Delta}(s)\nabla s\\
&= -\frac{b-t}{b-a}(y(t)-y(a)) + \frac{t-a}{b-a}(y(b)-y(t))\\
&= -\left(\frac{b-t}{b-a}+\frac{t-a}{b-a}\right)y(t) + \frac{t-a}{b-a}y(b)\\
&= -y(t) + \frac{t-a}{b-a}y(b), \quad t \in [a,b],
\end{aligned}
$$

whereupon,

$$
\begin{aligned}
y(t) &= \frac{t-a}{b-a}y(b) - \int_a^b G(t,s)y^{\Delta\nabla}(s)\nabla s\\
&\geq 0, \quad t \in [a,b].
\end{aligned}
$$

By the last inequality and (5.25), we get

$$y(t) \geq 0, \quad t \in [a,b],$$

or

$$u(t) \geq \frac{b-t}{b-a}\|u\|, \quad t \in [a,b].$$

This completes the proof.

Let

$$Y = \{t \in \mathbb{T} : a \leq t \leq \frac{b-a}{2}\}$$

and

$$h = \max Y.$$

Fix $r \in \mathbb{T}$ such that $0 < r < h$. For $u \in P$, define

$$
\begin{aligned}
\gamma(u) &= \min_{t\in[r,h]} u(t)\\
&= u(h),
\end{aligned}
$$

$$\theta(u) = \max_{t\in[h,b]} u(t)$$

$$= u(h)$$

and

$$\alpha(u) = \max_{t\in[r,b]} u(t)$$

$$= u(r).$$

For $u \in P$, we have

$$\gamma(u) = \theta(u)$$

$$\leq \alpha(u).$$

By Lemma 5.12, it follows that

$$\gamma(u) = u(h)$$

$$\geq \frac{b-h}{b-a}\|u\|, \quad u \in P.$$

Therefore,

$$\|u\| \leq \frac{b-a}{b-h}\gamma(u), \quad u \in P.$$

For $d > 0$, define

$$P(\gamma,d) = \{x \in P : \gamma(x) < d\}.$$

Let, also,

$$\lambda = \frac{b-h}{b-a}\sum_{j=1}^{m-2} a_j(b-\xi_j)\int_a^h q(s)\nabla s,$$

$$\eta = \frac{1}{1-\sum\limits_{j=1}^{m-2} a_j}\left(\int_a^b (b-s)q(s)\nabla s\right.$$

$$\left.+\frac{\sum\limits_{j=1}^{m-2} b_j \int_a^{\xi_j} q(s)\nabla s}{1-\sum\limits_{j=1}^{m-2} b_j}\left(b-a-\sum_{j=1}^{m-2} a_j\xi_j\right)\right),$$

$$\lambda_r = \frac{b-r}{b-a}\sum_{j=1}^{m-2} a_j(b-\xi_j)\int_r^b q(s)\nabla s.$$

Theorem 5.16. *Assume* $(D7)$-$(D10)$ *and*

$$0 < a^1 < \frac{\lambda_r}{\eta} b^1 < \frac{\lambda_r(b-a)}{\eta(b-h)} c_1, \quad b - a - \sum_{j=1}^{m-2} a_j \xi_j > 0,$$

and

(D13) $f(w) > \dfrac{c_1}{\lambda}$ *for* $c_1 \leq w \leq \dfrac{b-a}{b-h} c_1$,

(D14) $f(w) < \dfrac{b^1}{\eta}$ *for* $0 \leq w \leq \dfrac{b-a}{b-h} b^1$,

(D15) $f(w) > \dfrac{a^1}{\lambda_r}$ *for* $0 \leq w \leq a^1$.

Then the BVP (5.19), (5.20) *has at least two solutions* u_1 *and* u_2 *so that*

$$a^1 < \alpha(u_1) \quad with \quad \theta(u_1) < b^1 \tag{5.26}$$

and

$$b^1 < \theta(u_2) \quad with \quad \gamma(u_2) < c_1. \tag{5.27}$$

Proof. By Lemma 5.11, it follows that $T : P \to \mathscr{C}([a,b])$ is a completely continuous operator. Let $w \in \overline{P(\gamma, c_1)}$. Then

$$(Tu)^{\Delta\nabla}(t) = -q(t) f(u(t))$$

$$\leq 0, \quad t \in [a,b].$$

Thus, $Tu(\cdot)$ is a concave function on $[a,b]$ and $(Tu)^{\Delta}(\cdot)$ is non-increasing on $[a,b]$. Therefore,

$$(Tu)^{\Delta}(\xi_j) \leq (Tu)^{\Delta}(a), \quad j \in \{1, \ldots, m-2\},$$

and

$$(Tu)^{\Delta}(a) = \sum_{j=1}^{m-2} b_j (Tu)^{\Delta}(\xi_j)$$

$$\leq (Tu)^{\Delta}(a) \sum_{j=1}^{m-2} b_j,$$

whereupon,

$$(Tu)^{\Delta}(a) \left(1 - \sum_{j=1}^{m-2} b_j\right) \leq 0$$

and

$$(Tu)^{\Delta}(a) \leq 0.$$

Then

$$(Tu)^{\Delta}(t) \leq 0, \quad t \in [a,b].$$

Hence, $Tu(\cdot)$ is a non-increasing function on $[a,b]$ and

$$Tu(a) \geq Tu(t)$$

$$\geq Tu(b), \quad t \in [a,b],$$

and

$$Tu(\xi_j) \geq Tu(b), \quad j \in \{1,\ldots,m-2\},$$

and

$$Tu(b) = \sum_{j=1}^{m-2} a_j Tu(\xi_j)$$

$$\geq Tu(b) \sum_{j=1}^{m-2} a_j$$

and

$$\left(1 - \sum_{j=1}^{m-2} a_j\right) Tu(b) \geq 0.$$

Thus,

$$Tu(b) \geq 0$$

and

$$Tu(t) \geq Tu(b)$$

$$\geq 0, \quad t \in [a,b].$$

Therefore, $T : \overline{P(\gamma,c_1)} \to P$. Now, we take $u \in \partial P(\gamma,c_1)$ arbitrarily. Then

$$\gamma(u) = \min_{t\in[r,h]} u(t)$$

$$= u(h)$$

$$= c_1.$$

Thus,

$$u(t) \geq c_1, \quad t \in [a,h].$$

Since

$$\|u\| \leq \frac{b-a}{b-h}\gamma(u)$$

$$\leq \frac{b-a}{b-h} c_1,$$

we get

$$c_1 \leq u(t)$$

$$\leq \frac{b-a}{b-h} c_1, \quad t \in [r,h].$$

By $(D13)$, we obtain

$$f(w(t)) \geq \frac{c_1}{\lambda}, \quad t \in [r,h].$$

Next,

$$\begin{aligned}
\gamma(Tu) &= Tu(h) \\
&\geq \frac{b-h}{b-a} \|Tu\| \\
&= \frac{b-h}{b-a} Tu(a) \\
&= \frac{b-h}{b-a} \cdot \frac{1}{1-\sum\limits_{j=1}^{m-2} a_j} \Bigg(\int_a^b (b-s) q(s) f(u(s)) \nabla s \\
&\quad - \sum_{j=1}^{m-2} a_j \int_a^{\xi_j} (\xi_j - s) q(s) f(u(s)) \nabla s \\
&\quad + \frac{\sum\limits_{j=1}^{m-2} b_j \int_a^{\xi_j} q(s) f(u(s)) \nabla s}{1-\sum\limits_{j=1}^{m-2} b_j} \left(b-a-\sum_{j=1}^{m-2} a_j \xi_j \right) \Bigg) \\
&\geq \frac{b-h}{b-a} \Bigg(\sum_{j=1}^{m-2} a_j \int_a^b (b-s) q(s) f(u(s)) \nabla s \\
&\quad - \sum_{j=1}^{m-2} a_j \int_a^b (\xi_j - s) q(s) f(u(s)) \nabla s \Bigg) \\
&= \frac{b-h}{b-a} \sum_{j=1}^{m-2} a_j \int_a^b (b-\xi_j) q(s) f(u(s)) \nabla s
\end{aligned}$$

$$\geq \frac{b-h}{b-a}\sum_{j=1}^{m-2} a_j \int_a^h (b-\xi_j)q(s)f(u(s))\nabla s$$

$$\geq \frac{c_1}{\lambda}\cdot\frac{b-h}{b-a}\sum_{j=1}^{m-2} a_j(b-\xi_j)\int_a^b q(s)\nabla s$$

$$= c_1.$$

Let now, $u \in \partial P(\theta, b^1)$. Then

$$\theta(u) = \max_{t\in[h,b]} u(t)$$

$$= u(h)$$

$$= b^1.$$

Therefore,

$$0 \leq u(t) \leq b^1, \quad t \in [h,b],$$

and

$$b \leq u(t)$$

$$\leq \|u\|$$

$$= u(a), \quad t \in [a,b].$$

Observe that

$$\|u\| \leq \frac{b-a}{b-h}\gamma(u)$$

$$= \frac{b-a}{b-h}\theta(u)$$

$$= \frac{b-a}{b-h}b^1$$

$$\geq b^1, \quad u \in P.$$

Thus,

$$u(t) \leq \frac{b-a}{b-h}b^1, \quad t \in [a,b].$$

By $(D14)$, we get

$$f(u(t)) \le \frac{b^1}{\eta}, \quad t \in [a,b],$$

and

$$\begin{aligned}
\theta(Tu) &= Tu(h) \\
&\le Tu(a) \\
&= \frac{1}{1-\sum\limits_{j=1}^{m-2} a_j}\Bigg(\int_a^b (b-s)q(s)f(u(s))\nabla s \\
&\quad - \sum_{j=1}^{m-2} a_j \int_a^{\xi_j} (\xi_j - s)q(s)f(u(s))\nabla s \\
&\quad + \frac{\sum\limits_{j=1}^{m-2} b_j \int_a^{\xi_j} q(s)f(u(s))\nabla s}{1-\sum\limits_{j=1}^{m-2} b_j}\left(b-a-\sum_{j=1}^{m-2} a_j\xi_j\right)\Bigg) \\
&\le \frac{1}{1-\sum\limits_{j=1}^{m-2} a_j}\Bigg(\int_a^b (b-s)q(s)f(u(s))\nabla s \\
&\quad + \frac{\sum\limits_{j=1}^{m-2} b_j \int_a^{\xi_j} q(s)f(u(s))\nabla s}{1-\sum\limits_{j=1}^{m-2} b_j}\left(b-a-\sum_{j=1}^{m-2} a_j\xi_j\right)\Bigg) \\
&< \frac{b^1}{\eta}\cdot\frac{1}{1-\sum\limits_{j=1}^{m-2} a_j}\Bigg(\int_a^b (b-s)q(s)\nabla s \\
&\quad + \frac{\sum\limits_{j=1}^{m-2} b_j \int_a^{\xi_j} q(s)\nabla s}{1-\sum\limits_{j=1}^{m-2} b_j}\left(b-a-\sum_{j=1}^{m-2} a_j\xi_j\right)\Bigg) \\
&= b^1
\end{aligned}$$

Note that $\frac{a}{2} \in P(\alpha, a^1)$ and

$$\alpha(u) = \frac{a^1}{2}$$

$$< a^1.$$

Thus, $P(\alpha, a^1) \neq \emptyset$. Let $u \in \partial P(\alpha, a^1)$. Then

$$\alpha(u) = \max_{t\in[r,b]} u(t)$$

$$= u(r)$$

$$= a^1.$$

Hence,

$$0 \leq u(t) \leq a^1, \quad t \in [r,b].$$

Therefore,

$$f(u(t)) \geq \frac{a^1}{\lambda_r}, \quad t \in [r,b],$$

and

$$\alpha(Tu) = Tu(r)$$

$$\geq \frac{b-r}{b-a} Tu(a)$$

$$= \frac{b-r}{b-a} \cdot \frac{1}{1 - \sum\limits_{j=1}^{m-2} a_j} \Bigg(\int_a^b (b-s)q(s)f(u(s))\nabla s$$

$$- \sum_{j=1}^{m-2} a_j \int_a^{\xi_j} (\xi_j - s)q(s)f(u(s))\nabla s$$

$$+ \frac{\sum\limits_{j=1}^{m-2} b_j \int_a^{\xi_j} q(s)f(u(s))\nabla s}{1 - \sum\limits_{j=1}^{m-2} b_j} \left(b - a - \sum_{j=1}^{m-2} a_j \xi_j \right) \Bigg)$$

$$\geq \frac{b-r}{b-a} \Bigg(\sum_{j=1}^{m-2} a_j \int_a^b (b-s)q(s)f(u(s))\nabla s$$

$$- \sum_{j=1}^{m-2} a_j \int_a^b (\xi_j - s)q(s)f(u(s))\nabla s \Bigg)$$

$$
\begin{aligned}
&= \frac{b-r}{b-a}\sum_{j=1}^{m-2} a_j(b-\xi_j)\int_a^b q(s)f(u(s))\nabla s \\
&\geq \frac{b-r}{b-a}\sum_{j=1}^{m-2} a_j(b-\xi_j)\int_r^b q(s)f(u(s))\nabla s \\
&\geq \frac{a^1}{\lambda_r}\cdot\frac{b-r}{b-a}\sum_{j=1}^{m-2} a_j(b-\xi_j)\int_r^b q(s)\nabla s \\
&= a^1.
\end{aligned}
$$

Therefore, all conditions of the Avery-Henderson fixed point theorem are satisfied. Hence, the operator T has at least two fixed points u_1 and u_2 for which (5.26) and (5.27) hold. This completes the proof.

5.4 Existence of Symmetric Positive Solutions of Nonlinear Second Order m-Point Boundary Value Problems-II

In this section we will investigate the following BVP

$$
u^{\Delta\nabla}(t)+g(t)u^{\Delta}(t)+h(t)u(t)+\lambda q(t)f(t,u(t))=0, \quad t\in(a,b), \tag{5.28}
$$

$$
u(\rho(a))=0, \quad u(\sigma(b))=\sum_{j=1}^{m-2}\alpha_j u(\eta_j), \tag{5.29}
$$

where

(D16) $\alpha_j\geq 0$, $a<\eta_j<\eta_{j+1}<b$, $j\in\{1,\ldots,m-2\}$, $\lambda>0$,
(D17) $q\in\mathscr{C}([\rho(a),\sigma(b)])$, $q\geq 0$ on $[\rho(a),\sigma(b)]$, there exists a $t_0\in(\rho(a),\sigma(b))$, $q(t_0)>0$, q may be singular at $t=\rho(a)$, $t=\sigma(b)$,
(D18) $g\in\mathscr{C}([a,b])$, $g\geq 0$, $h\in\mathscr{C}([a,b])$, $h\leq 0$ on $[a,b]$.

Note that the BVPs

$$
\begin{aligned}
&\phi_1^{\Delta\nabla}(t)+g(t)\phi_1^{\Delta}(t)+h(t)\phi_1(t) = 0, \quad t\in(a,b), \\
&\phi_1(\rho(a))=0, \quad \phi_1(\sigma(b)) = 1,
\end{aligned} \tag{5.30}
$$

and

$$
\begin{aligned}
&\phi_2^{\Delta\nabla}(t)+g(t)\phi_2^{\Delta}(t)+h(t)\phi_2(t) = 0, \quad t\in(a,b), \\
&\phi_2(\rho(a))=1, \quad \phi_2(\sigma(b)) = 0
\end{aligned} \tag{5.31}
$$

have unique solutions ϕ_1 and ϕ_2, respectively, such that

1. ϕ_1 is strictly increasing on $[\rho(a),\sigma(b)]$,

2. ϕ_2 is strictly decreasing on $[\rho(a), \sigma(b)]$.

In addition, we suppose

(D19) $\quad 0 < \sum_{j=1}^{m-2} \alpha_j \phi_1(\eta_j) < 1.$

Lemma 5.13. *Assume (D18) and (D19). Let $y \in \mathscr{C}([\rho(a), \sigma(b)])$. Then the BVP*

$$\begin{aligned} x^{\Delta\nabla}(t) + g(t)x^{\Delta}(t) + h(t)x(t) + y(t) &= 0, \quad t \in (a,b), \\ x(\rho(a)) = 0, \quad x(\sigma(b)) &= \sum_{j=1}^{m-2} \alpha_j x(\eta_j) \end{aligned} \tag{5.32}$$

is equivalent to the integral equation

$$x(t) = \int_{\rho(a)}^{\sigma(b)} H(t,s)p(s)y(s)\nabla s + A\phi_1(t), \quad t \in [\rho(a), \sigma(b)], \tag{5.33}$$

where

$$p(t) = e_g(\rho(t), \rho(a)),$$

$$A = \frac{1}{1 - \sum_{j=1}^{m-2} \alpha_j \phi_1(\eta_j)} \sum_{j=1}^{m-2} \alpha_j \int_{\rho(a)}^{\sigma(b)} H(\eta_j, s)p(s)y(s)\nabla s,$$

$$H(t,s) = \frac{1}{\phi_1^{\Delta}(\rho(a))} \begin{cases} \phi_1(s)\phi_2(t) & s \le t \\ \phi_1(t)\phi_2(s) & t \le s. \end{cases}$$

Proof. 1. Let x be the solution of the BVP (5.32). We have

$$\begin{aligned} \begin{vmatrix} \phi_1(\rho(a)) & \phi_1^{\Delta}(\rho(a)) \\ \phi_2(\rho(a)) & \phi_2^{\Delta}(\rho(a)) \end{vmatrix} &= \phi_1(\rho(a))\phi_2^{\Delta}(\rho(a)) - \phi_2(\rho(a))\phi_1^{\Delta}(\rho(a)) \\ &= -\phi_1^{\Delta}(\rho(a)) \\ &\neq 0. \end{aligned}$$

Then by the variations of constants, we get that the solution of the BVP (5.32) can be represented in the form (5.33).

2. Let x satisfies (5.33). Then

$$x(t) = \int_{\rho(a)}^{\sigma(b)} H(t,s)p(s)y(s)\nabla s + A\phi_1(t)$$

$$= \int_{\rho(a)}^{t} H(t,s)p(s)y(s)\nabla s + \int_{t}^{\sigma(b)} H(t,s)p(s)y(s)\nabla s + A\phi_1(t)$$

$$= \frac{1}{\phi_1^{\Delta}(\rho(a))}\phi_2(t)\int_{\rho(a)}^{t} \phi_1(s)p(s)y(s)\nabla s$$

$$+\frac{1}{\phi_1^{\Delta}(\rho(a))}\phi_1(t)\int_{t}^{\sigma(b)} \phi_2(s)p(s)y(s)\nabla s + A_1\phi(t), \quad t \in [\rho(a),\sigma(b)].$$

Hence,

$$x^{\Delta}(t) = \frac{1}{\phi_1^{\Delta}(\rho(a))}\Bigg(\phi_2^{\Delta}(t)\int_{\rho(a)}^{t} \phi_1(s)p(s)y(s)\nabla s$$

$$+\phi_1(\sigma(t))\phi_2(\sigma(t))p(\sigma(t))y(\sigma(t))$$

$$+\phi_1^{\Delta}(t)\int_{t}^{\sigma(b)} \phi_2(s)p(s)y(s)\nabla s$$

$$-\phi_1(\sigma(t))\phi_2(\sigma(t))p(\sigma(t))y(\sigma(t))\Bigg) + A\phi_1^{\Delta}(t)$$

$$= \frac{1}{\phi_1^{\Delta}(\rho(a))}\Bigg(\phi_2^{\Delta}(t)\int_{\rho(a)}^{t} \phi_1(s)p(s)y(s)\nabla s$$

$$+\phi_1^{\Delta}(t)\int_{t}^{\sigma(b)} \phi_2(s)p(s)y(s)\nabla s\Bigg),$$

$$x^{\Delta\nabla}(t) = \frac{1}{\phi_1^{\Delta}(\rho(a))}\Bigg(\phi_2^{\Delta\nabla}(t)\int_{\rho(a)}^{\rho(t)} \phi_1(s)p(s)y(s)\nabla s + \phi_1(t)\phi_2^{\Delta}(t)p(t)y(t)$$

$$+\phi_1^{\Delta\nabla}(t)\int_{\rho(t)}^{\sigma(b)} \phi_2(s)p(s)y(s)\nabla s - \phi_1^{\Delta}(t)p(t)\phi_2(t)y(t)\Bigg)$$

$$+A\phi_1^{\Delta\nabla}(t), \quad t \in [a,b],$$

and

$$x^{\Delta\nabla}(t) + g(t)x^{\Delta}(t) + h(t)x(t)$$

$$= \frac{1}{\phi_1^{\Delta}(\rho(a))}\Bigg(\phi_2^{\Delta\nabla}(t)\int_{\rho(a)}^{\rho(t)} \phi_1(s)p(s)y(s)\nabla s + \phi_1(t)\phi_2^{\Delta}(t)p(t)y(t)$$

$$+\phi_1^{\Delta\nabla}(t)\int_{\rho(t)}^{\sigma(b)} \phi_2(s)p(s)y(s)\nabla s - \phi_1^{\Delta}(t)p(t)\phi_2(t)y(t)\Bigg)$$

$$+A\phi_1^{\Delta\nabla}(t)$$

$$
\begin{aligned}
&+\frac{1}{\phi_1^{\Delta}(\rho(a))}\left(g(t)\phi_2^{\Delta}(t)\int_{\rho(a)}^{t}\phi_1(s)p(s)y(s)\nabla s\right.\\
&\left.+g(t)\phi_1^{\Delta}(t)\int_{t}^{\sigma(b)}\phi_2(s)p(s)y(s)\nabla s\right)\\
&+Ag(t)\phi_1^{\Delta}(t)\\
&+\frac{1}{\phi_1^{\Delta}(\rho(a))}\left(h(t)\phi_1(t)\int_{t}^{\sigma(b)}\phi_2(s)p(s)y(s)\nabla s\right.\\
&\left.+h(t)\phi_2(t)\int_{\rho(a)}^{t}\phi_1(s)p(s)y(s)\nabla s\right)+Ah(t)\phi_1(t)\\
&=A\left(\phi_1^{\Delta\nabla}(t)+g(t)\phi_1^{\Delta}(t)+h(t)\phi_1(t)\right)\\
&+\left(\frac{1}{\phi_1^{\Delta}(\rho(a))}\int_{\rho(a)}^{t}\phi_1(s)p(s)y(s)\nabla s\right)\left(\phi_2^{\Delta\nabla}(t)+g(t)\phi_2^{\Delta}(t)+h(t)\phi_2(t)\right)\\
&+\frac{1}{\phi_1^{\Delta}(\rho(a))}\left(\phi_2^{\Delta\nabla}(t)\int_{t}^{\rho(t)}\phi_1(s)p(s)y(s)\nabla s+\phi_1^{\Delta\nabla}(t)\int_{\rho(t)}^{t}\phi_2(s)p(s)y(s)\nabla s\right)\\
&+\left(\frac{1}{\phi_1^{\Delta}(\rho(a))}\int_{t}^{\sigma(b)}\phi_2(s)p(s)y(s)\nabla s\right)\left(\phi_1^{\Delta\nabla}(t)+g(t)\phi_1^{\Delta}(t)+h(t)\phi_1(t)\right)\\
&+\frac{1}{\phi_1^{\Delta}(\rho(a))}\left(\phi_2^{\Delta}(t)\phi_1(t)-\phi_1^{\Delta}(t)\phi_2(t)\right)p(t)y(t)\\
&=\frac{1}{\phi_1^{\Delta}(\rho(a))}\left(\phi_2^{\Delta\nabla}(t)(\rho(t)-t)\phi_1(t)p(t)y(t)\right.\\
&\left.+\phi_1^{\Delta\nabla}(t)(t-\rho(t))\phi_2(t)p(t)y(t)\right)\\
&+\frac{1}{\phi_1^{\Delta}(\rho(a))}\left(\phi_2^{\Delta}(t)\phi_1(t)-\phi_1^{\Delta}(t)\phi_2(t)\right)p(t)y(t)\\
&=\frac{1}{\phi_1^{\Delta}(\rho(a))}p(t)y(t)\Big(\left(\phi_2^{\Delta}(t)\phi_1(t)-\phi_1^{\Delta}(t)\phi_2(t)\right)\\
&+(\rho(t)-t)\left(\phi_2^{\Delta\nabla}(t)\phi_1(t)-\phi_1^{\Delta\nabla}(t)\phi_2(t)\right)\Big)\\
&=\frac{1}{\phi_1^{\Delta}(\rho(a))}p(t)y(t)\Big(\left(\phi_2^{\Delta}(t)\phi_1(t)-\phi_1^{\Delta}(t)\phi_2(t)\right)
\end{aligned}
$$

$$
\begin{aligned}
&+(\rho(t)-t)\left(\phi_2^{\Delta}(t)\phi_1(t)-\phi_1^{\Delta}(t)\phi_2(t)\right)^{\nabla}\\
&-(\rho(t)-t)\left(\phi_2^{\Delta\rho}(t)\phi_1^{\nabla}(t)-\phi_1^{\Delta\rho}(t)\phi_2^{\nabla}(t)\right)\Big)\\
&=\frac{1}{\phi_1^{\Delta}(\rho(a))}p(t)y(t)\Big(\left(\phi_2^{\Delta}(t)\phi_1(t)-\phi_1^{\Delta}(t)\phi_2(t)\right)\\
&+(\rho(t)-t)\left(\phi_2^{\Delta}(t)\phi_1(t)-\phi_1^{\Delta}(t)\phi_2(t)\right)^{\nabla}\Big)\\
&=\frac{1}{\phi_1^{\Delta}(\rho(a))}p(t)y(t)\left(\phi_2^{\Delta}(\rho(t))\phi_1(\rho(t))-\phi_1^{\Delta}(\rho(t))\phi_2(\rho(t))\right)\\
&=\frac{1}{\phi_1^{\Delta}(\rho(a))}p(t)y(t)e_{\ominus g}(\rho(t),\rho(a))\left(-\phi_1^{\Delta}(\rho(a))\right)\\
&=-p(t)y(t)e_{\ominus g}(\rho(t),\rho(a))\\
&=-y(t),\quad t\in[\rho(a),\sigma(b)].
\end{aligned}
$$

Next,

$$
\begin{aligned}
H(\rho(a),s)&=\frac{1}{\phi_1^{\Delta}(\rho(a))}\phi_2(s)\phi_1(\rho(a))\\
&=0
\end{aligned}
$$

and

$$
\begin{aligned}
H(\sigma(b),s)&=\frac{1}{\phi_1^{\Delta}(\rho(a))}\phi_1(s)\phi_2(\sigma(b))\\
&=0.
\end{aligned}
$$

Then

$$
\begin{aligned}
x(\sigma(b))&=\int_{\rho(a)}^{\sigma(b)}H(\sigma(b),s)p(s)y(s)\nabla s+A\phi_1(\sigma(b))\\
&=A
\end{aligned}
$$

and

$$
\sum_{j=1}^{m-2}\alpha_j x(\eta_j)=\sum_{j=1}^{m-2}\alpha_j\left(\int_{\rho(a)}^{\sigma(b)}H(\eta_j,s)p(s)y(s)\nabla s+A\phi_1(\eta_j)\right)
$$

$$
\begin{aligned}
&= \sum_{j=1}^{m-2} \alpha_j \Bigg(\int_{\rho(a)}^{\sigma(b)} H(\eta_j,s)p(s)y(s)\nabla s \\
&\quad + \frac{1}{1-\sum_{j=1}^{m-2}\alpha_j\phi(\eta_j)} \sum_{j=1}^{m-2} \alpha_j\phi(\eta_j) \int_{\rho(a)}^{\sigma(b)} H(\eta_j,s)p(s)y(s)\nabla s \Bigg) \\
&= \frac{1}{1-\sum_{j=1}^{m-2}\alpha_j\phi(\eta_j)} \sum_{j=1}^{m-2} \alpha_j \int_{\rho(a)}^{\sigma(b)} H(\eta_j,s)p(s)y(s)\nabla s \\
&= A.
\end{aligned}
$$

This completes the proof.

Lemma 5.14. *The Green function $H(\cdot,\cdot)$ satisfies the following properties.*

1. $H(t,s) \le H(t,t)$.
2. $\dfrac{\phi_1^\Delta(\rho(a))}{\|\phi_1\|\|\phi_2\|} H(t,t)H(s,s) \le H(t,s) \le H(s,s)$
3. $H(t,t) \le \phi_1(t)\dfrac{\|\phi_2\|}{\phi_1^\Delta(\rho(a))}$.

Proof. 1. Since ϕ_1 is strictly increasing on $[\rho(a),\sigma(b)]$ and ϕ_2 is strictly decreasing on $[\rho(a),\sigma(b)]$, we get

$$
\begin{aligned}
H(t,s) &= \frac{1}{\phi_1^\Delta(\rho(a))} \begin{cases} \phi_1(s)\phi_2(t) & s \le t \\ \phi_1(t)\phi_2(s) & t \le s. \end{cases} \\
&\le \frac{1}{\phi_1^\Delta(\rho(a))} \phi_1(t)\phi_2(t) \\
&= H(t,t), \quad t,s \in [\rho(a),\sigma(b)].
\end{aligned}
$$

2. We have

$$
\begin{aligned}
\frac{\phi_1^\Delta(\rho(a))}{\|\phi_1\|\|\phi_2\|} H(t,t)H(s,s) &= \frac{\phi_1^\Delta(\rho(a))}{\|\phi_1\|\|\phi_2\|} \left(\frac{1}{\phi_1^\Delta(\rho(a))}\phi_1(t)\phi_2(t) \right) \\
&\quad \times \left(\frac{1}{\phi_1^\Delta(\rho(a))}\phi_1(s)\phi_2(s) \right) \\
&= \frac{1}{\phi_1^\Delta(\rho(a))} \begin{cases} \dfrac{\phi_1(s)\phi_2(s)\phi_1(t)\phi_2(t)}{\|\phi_1\|\|\phi_2\|} & s \le t \\[2ex] \dfrac{\phi_1(t)\phi_1(s)\phi_2(s)\phi_2(t)}{\|\phi_1\|\|\phi_2\|} & t \le s. \end{cases}
\end{aligned}
$$

$$\leq \frac{1}{\phi_1^{\Delta}(\rho(a))} \begin{cases} \dfrac{\phi_1(s)\phi_1(t)\phi_2(t)\phi_2(s)}{\phi_1(t)\phi_2(s)} & s \leq t \\ \dfrac{\phi_1(t)\phi_1(s)\phi_2(s)\phi_2(t)}{\phi_1(s)\phi_2(t)} & t \leq s. \end{cases}$$

$$= \frac{1}{\phi_1^{\Delta}(\rho(a))} \begin{cases} \phi_1(s)\phi_2(t) & s \leq t \\ \phi_1(t)\phi_2(s) & t \leq s. \end{cases}$$

$$= H(t,s)$$

$$\leq \frac{1}{\phi_1^{\Delta}(\rho(a))} \begin{cases} \phi_1(s)\phi_2(s) & s \leq t \\ \phi_1(s)\phi_2(s) & t \leq s. \end{cases}$$

$$= H(s,s), \quad t,s \in [\rho(a), \sigma(b)].$$

3. We have

$$H(t,s) \leq \frac{1}{\phi_1^{\Delta}(\rho(a))}\phi_1(t)\phi_2(t)$$
$$\leq \frac{1}{\phi_1^{\Delta}(\rho(a))}\phi_1(t)\|\phi_2\|, \quad t \in [\rho(a), \sigma(b)].$$

This completes the proof.

Set

$$G(t,s) = H(t,s) + \frac{1}{1 - \sum\limits_{j=1}^{m-2} \alpha_j \phi(\eta_j)} \sum_{j=1}^{m-2} \alpha_j H(\eta_j, s)\phi_1(t),$$

$t,s \in [\rho(a), \sigma(b)]$. Then the solution of the BVP (5.32) can be written in the form

$$x(t) = \lambda \int_{\rho(a)}^{\sigma(b)} G(t,s)p(s)y(s)\nabla s, \quad t \in [\rho(a), \sigma(b)],$$

and the solution of the BVP (5.28), (5.29) can be written in the form

$$u(t) = \lambda \int_{\rho(a)}^{\sigma(b)} G(t,s)q(s)f(s,u(s))\nabla s, \quad t \in [\rho(a), \sigma(b)].$$

Set

$$C_1 = 1 + \frac{\|\phi_1\|}{1 - \sum\limits_{j=1}^{m-2} \alpha_j \phi_1(\eta_j)} \sum_{j=1}^{m-2} \alpha_j,$$

$$C_2 = \frac{\phi_1^{\Delta}(\rho(a))}{\|\phi_1\| \|\phi_2\|} \frac{1}{1 - \sum\limits_{j=1}^{m-2} \alpha_j \phi_1(\eta_j)} \sum_{j=1}^{m-2} \alpha_j H(\eta_j, \eta_j),$$

$$C_3 = \frac{\|\phi_2\|}{\phi_1^{\Delta}(\rho(a))} + \frac{1}{1 - \sum\limits_{j=1}^{m-2} \alpha_j \phi_1(\eta_j)} \sum_{j=1}^{m-2} \alpha_j H(\eta_j, \eta_j).$$

Lemma 5.15. *The Green function $G(\cdot,\cdot)$ has the following properties.*

1. $C_2\phi_1(t)H(s,s) \le G(t,s) \le C_1 H(s,s)$.
2. $G(t,s) \le C_3\phi_1(t)$, $s,t \in [\rho(a), \sigma(b)]$.

Proof. 1. By Lemma 5.14, we find

$$\begin{aligned} G(t,s) &= H(t,s) + \frac{1}{1 - \sum\limits_{j=1}^{m-2} \alpha_j \phi(\eta_j)} \sum_{j=1}^{m-2} \alpha_j H(\eta_j, s)\phi_1(t) \\ &\le H(s,s) + \frac{1}{1 - \sum\limits_{j=1}^{m-2} \alpha_j \phi(\eta_j)} \sum_{j=1}^{m-2} \alpha_j H(s,s) \|\phi_1\| \\ &\le \left(1 + \frac{1}{1 - \sum\limits_{j=1}^{m-2} \alpha_j \phi(\eta_j)} \sum_{j=1}^{m-2} \alpha_j \|\phi_1\| \right) H(s,s) \\ &= C_1 H(s,s), \quad t,s \in [\rho(a), \sigma(b)]. \end{aligned}$$

Moreover,

$$\begin{aligned} G(t,s) &= H(t,s) + \frac{1}{1 - \sum\limits_{j=1}^{m-2} \alpha_j \phi(\eta_j)} \sum_{j=1}^{m-2} \alpha_j H(\eta_j, s)\phi_1(t) \\ &\ge \frac{1}{1 - \sum\limits_{j=1}^{m-2} \alpha_j \phi(\eta_j)} \sum_{j=1}^{m-2} \alpha_j H(\eta_j, s)\phi_1(t) \end{aligned}$$

$$\geq \frac{1}{1-\sum\limits_{j=1}^{m-2}\alpha_j\phi(\eta_j)}\sum_{j=1}^{m-2}\alpha_j\frac{\phi_1^\Delta(\rho(a))}{\|\phi_1\|\|\phi_2\|}H(\eta_j,\eta_j)H(s,s)\phi_1(t)$$

$$= C_2\phi_1(t)H(s,s), \quad t,s\in[\rho(a),\sigma(b)].$$

2. We have

$$\begin{aligned}
G(t,s) &= H(t,s)+\frac{1}{1-\sum\limits_{j=1}^{m-2}\alpha_j\phi(\eta_j)}\sum_{j=1}^{m-2}\alpha_jH(\eta_j,s)\phi_1(t)\\
&\leq H(t,t)+\frac{1}{1-\sum\limits_{j=1}^{m-2}\alpha_j\phi(\eta_j)}\sum_{j=1}^{m-2}\alpha_jH(\eta_j,\eta_j)\phi_1(t)\\
&\leq \frac{\|\phi_2\|}{\phi_1^\Delta(\rho(a))}\phi_1(t)+\frac{1}{1-\sum\limits_{j=1}^{m-2}\alpha_j\phi(\eta_j)}\sum_{j=1}^{m-2}\alpha_jH(\eta_j,\eta_j)\phi_1(t)\\
&= \left(\frac{\|\phi_2\|}{\phi_1^\Delta(\rho(a))}+\frac{1}{1-\sum\limits_{j=1}^{m-2}\alpha_j\phi(\eta_j)}\sum_{j=1}^{m-2}\alpha_jH(\eta_j,\eta_j)\right)\phi_1(t)\\
&= C_3\phi_1(t), \quad s,t\in[\rho(a),\sigma(b)].
\end{aligned}$$

This completes the proof.

Below we will suppose

(D19) $f\in\mathscr{C}([\rho(a),\sigma(b)]\times[0,\infty))$ and there exists a function $g_1\in\mathscr{C}([\rho(a),\sigma(b)])$, $g_1\geq 0$ on $[\rho(a),\sigma(b)]$, such that

$$f(t,u)\geq -g_1(t), \quad t\in[\rho(a),\sigma(b)], \quad u>0,$$

$f(\cdot,u)$ may be singular at $t=\rho(a)$, $t=\sigma(b)$ for $u\in\mathbb{R}$.

(D20) $f(t,0)>0$, $t\in[\rho(a),\sigma(b)]$,

(D21) there exists $[\theta_1,\theta_2]\subset(\rho(a),\sigma(b))$ such that

$$\lim_{u\to+\infty}\min_{t\in[\theta_1,\theta_2]}\frac{f(t,u)}{u}=\infty.$$

(D22)

$$\int_{\rho(a)}^{\sigma(b)}H(s,s)p(s)q(s)f(s,z)\nabla s<\infty, \quad z\in[0,m],$$

$m > 0$ is a constant.

For a function $y : \mathbb{R} \to \mathbb{R}$ define

$$y(t)^* = \begin{cases} y(t) & y(t) \geq 0 \\ 0 & y(t) < 0, \end{cases}$$

$t \in \mathbb{R}$. Suppose that v is a solution to the BVP

$$\begin{aligned} v^{\Delta\nabla}(t) + g(t)v^{\Delta}(t) + h(t)v(t) + \lambda q(t)g_1(t) &= 0, \quad t \in (a,b), \\ v(\rho(a)) = 0, \quad v(\sigma(b)) &= \sum_{j=1}^{m-2} \alpha_j v(\eta_j). \end{aligned} \tag{5.34}$$

Then v satisfies the integral equation

$$v(t) = \lambda \int_{\rho(a)}^{\sigma(b)} G(t,s)p(s)q(s)g_1(s)\nabla s, \quad t \in [\rho(a), \sigma(b)].$$

Lemma 5.16. *Suppose* (D16)- (D18). *Then*

$$v(t) \leq \lambda C_0 \phi_1(t), \quad t \in [\rho(a), \sigma(b)],$$

where

$$C_0 = C_3 \int_{\rho(a)}^{\sigma(b)} p(s)q(s)g_1(s)\nabla s.$$

Proof. By Lemma 5.15, we get

$$G(t,s) \leq C_3 \phi_1(t), \quad t,s \in [\rho(a), \sigma(b)].$$

Then

$$\begin{aligned} v(t) &= \lambda \int_{\rho(a)}^{\sigma(b)} G(t,s)p(s)q(s)g_1(s)\nabla s \\ &\leq C_3 \lambda \phi_1(t) \int_{\rho(a)}^{\sigma(b)} p(s)q(s)g_1(s)\nabla s \\ &= C_0 \lambda \phi_1(t), \quad t \in [\rho(a), \sigma(b)]. \end{aligned}$$

This completes the proof.

To prove existence of solutions of the BVP (5.28), (5.29), we will prove existence of solutions of the following BVP

$$x^{\Delta\nabla}(t)+g(t)x^{\Delta}(t)+h(t)x(t)+\lambda q(t)\left(f(t,(x(t)-v(t))^{*})+g_1(t)\right)=0,\quad t\in(a,b)$$

$$x(\rho(a))=0,\quad x(\sigma(b)) = \sum_{j=1}^{m-2}\alpha_j x(\eta_j). \tag{5.35}$$

Let $X=\mathscr{C}([\rho(a),\sigma(b)])$ be endowed with the maximum norm. Define

$$P=\left\{x\in X: x(t)\geq \frac{C_2}{C_1}\phi_1(t)\|x\|,\quad t\in[\rho(a),\sigma(b)]\right\}.$$

For $x\in X$, define the operator

$$Tx(t)=\lambda\int_{\rho(a)}^{\sigma(b)}G(t,s)p(s)q(s)\left(f(s,(x(s)-v(s))^{*})+g_1(s)\right)\nabla s,$$

$t\in[\rho(a),\sigma(b)]$.

Lemma 5.17. *Suppose* (*D*16)-(*D*22). *Then* $T:P\to P$.

Proof. Take $x\in P$ arbitrarily. Using Lemma 5.15, we get

$$\begin{aligned}Tx(t)&=\lambda\int_{\rho(a)}^{\sigma(b)}G(t,s)p(s)q(s)\left(f(s,(x(s)-v(s))^{*})+g_1(s)\right)\nabla s\\&\leq C_1\lambda\int_{\rho(a)}^{\sigma(b)}H(s,s)p(s)q(s)\left(f(s,(x(s)-v(s))^{*})+g_1(s)\right)\nabla s,\end{aligned}$$

$t\in[\rho(a),\sigma(b)]$. Then

$$\|Tx\|\leq\lambda C_1\int_{\rho(a)}^{\sigma(b)}H(s,s)p(s)q(s)\left(f(s,(x(s)-v(s))^{*})+g_1(s)\right)\nabla s.$$

Next,

$$\begin{aligned}Tx(t)&=\lambda\int_{\rho(a)}^{\sigma(b)}G(t,s)p(s)q(s)\left(f(s,(x(s)-v(s))^{*})+g_1(s)\right)\nabla s\\&\geq C_2\phi_1(t)\lambda\int_{\rho(a)}^{\sigma(b)}H(s,s)p(s)q(s)\left(f(s,(x(s)-v(s))^{*})+g_1(s)\right)\nabla s\\&\geq\frac{C_2}{C_1}\phi_1(t)\|Tx\|,\quad t\in[\rho(a),\sigma(b)].\end{aligned}$$

Consequently, $Tx\in P$ and $T:P\to P$. This completes the proof.

Lemma 5.18. *Suppose* (*D*16)-(*D*22). *Then* $T:P\to P$ *is a continuous operator.*

Proof. Let $\{x_n\}_{n\in\mathbb{N}}\subset P$, $x\in P$, be such that

$$x_n \to x \quad \text{in} \quad X, \quad \text{as} \quad n \to \infty.$$

Let, also,

$$M_1 = \max_{(T,s)\in[\rho(a),\sigma(b)]\times[\rho(a),\sigma(b)]} |G(t,s)|,$$

$$M_2 = \lambda \int_{\rho(a)}^{\sigma(b)} p(s)q(s)\nabla s.$$

Take $\varepsilon > 0$ arbitrarily. Since $f \in \mathscr{C}([\rho(a),\sigma(b)] \times [0,\infty))$, there exists an $N = N(\varepsilon) > 0$ so that

$$|f(s,(x_n(s)-v(s))^*) - f(s,(x(s)-v(s))^*)| < \frac{\varepsilon}{M_1M_2}, \quad s \in [\rho(a),\sigma(b)].$$

Thus,

$$\begin{aligned}
|Tx_n(t) - Tx(t)| &= \Big|\lambda \int_{\rho(a)}^{\sigma(b)} G(t,s)p(s)q(s)\left(f(s,(x_n(s)-v(s))^*) + g_1(s)\right)\nabla s \\
&\quad -\lambda \int_{\rho(a)}^{\sigma(b)} G(t,s)p(s)q(s)\left(f(s,(x(s)-v(s))^*) + g_1(s)\right)\nabla s\Big| \\
&= \Big|\lambda \int_{\rho(a)}^{\sigma(b)} G(t,s)p(s)q(s)\Big(f(s,(x_n(s)-v(s))^*) \\
&\quad -f(s,(x(s)-v(s))^*)\Big)\nabla s\Big| \\
&\le \lambda \int_{\rho(a)}^{\sigma(b)} |G(t,s)|p(s)q(s)\Big|f(s,(x_n(s)-v(s))^*) \\
&\quad -f(s,(x(s)-v(s))^*)\Big|\nabla s \\
&< \frac{\varepsilon}{M_1M_2}M_1M_2 \\
&= \varepsilon, \quad t \in [\rho(a),\sigma(b)], \quad n > N.
\end{aligned}$$

This completes the proof.

Lemma 5.19. *Suppose* $(D16)$-$(D17)$. *Then* $T : P \to P$ *is a completely continuous operator.*

Proof. Let $Y \subset P$ be a bounded set. Then there exists a constant $L_1 > 0$ so that

$$|x(t)| \le L_1, \quad t \in [\rho(a),\sigma(b)].$$

By the definition of the function v, it follows that there exists a positive constant L_2 so that

$$|v(t)| \le L_2, \quad t \in [\rho(a), \sigma(b)].$$

Since $f \in \mathscr{C}([\rho(a), \sigma(b)] \times [0, \infty))$, there exists a constant $L_3 > 0$ so that

$$|f(t, (u - v)^*)| \le L_3, \quad t \in [\rho(a), \sigma(b)], \quad u \in Y.$$

Now, using that $g_1 \in \mathscr{C}([\rho(a), \sigma(b)])$, we have that there exists a constant $L_4 > 0$ so that

$$g_1(t) \le L_4, \quad t \in [\rho(a), \sigma(b)].$$

Let M_1 and M_2 be the constants as in the proof of Lemma 5.18. Take $x \in Y$ arbitrarily. Then

$$\begin{aligned}
|Tx(t)| &= \left|\lambda \int_{\rho(a)}^{\sigma(b)} G(t,s)p(s)q(s)\left(f(s,(x(s)-v(s))^*) + g_1(s)\right)\nabla s\right| \\
&\le |\lambda| \int_{\rho(a)}^{\sigma(b)} |G(t,s)|p(s)|q(s)|\left(|f(s,(x(s)-v(s))^*)| + g_1(s)\right)\nabla s \\
&\le |\lambda| M_1(L_3 + L_4) \int_{\rho(a)}^{\sigma(b)} p(s)q(s)\nabla s \\
&= M_1 M_2 (L_3 + L_4), \quad t \in [\rho(a), \sigma(b)].
\end{aligned}$$

Thus,

$$\|Tx\| \le M_1 M_2 (L_3 + L_4)$$

and $T(Y)$ is uniformly bounded. Let now, $t_1, t_2 \in [\rho(a), \sigma(b)]$, $t_1 < t_2$, and $x \in Y$. Then

$$\begin{aligned}
|Tx(t_2) - Tx(t_1)| = \Bigg|&\lambda \int_{\rho(a)}^{t_2} G(t_2,s)p(s)q(s)\left(f(s,(x(s)-v(s))^*) + g_1(s)\right)\nabla s \\
&+\lambda \int_{t_2}^{\sigma(b)} G(t_2,s)p(s)q(s)\left(f(s,(x(s)-v(s))^*) + g_1(s)\right)\nabla s \\
&-\lambda \int_{\rho(a)}^{t_1} G(t_1,s)p(s)q(s)\left(f(s,(x(s)-v(s))^*) + g_1(s)\right)\nabla s \\
&-\lambda \int_{t_1}^{\sigma(b)} G(t_1,s)p(s)q(s)\left(f(s,(x(s)-v(s))^*) + g_1(s)\right)\nabla s\Bigg| \\
= \Bigg|&\lambda \int_{\rho(a)}^{t_1} (G(t_2,s) - G(t_1,s))p(s)q(s)\left(f(s,(x(s)-v(s))^*) + g_1(s)\right)\nabla s \\
&+\lambda \int_{t_2}^{\sigma(b)} (G(t_2,s) - G(t_1,s))p(s)q(s)\left(f(s,(x(s)-v(s))^*) + g_1(s)\right)\nabla s
\end{aligned}$$

$$
\begin{aligned}
&+\lambda \int_{t_1}^{t_2} G(t_2,s)p(s)q(s)\left(f(s,(x(s)-v(s))^*)+g_1(s)\right)\nabla s \\
&-\lambda \int_{t_1}^{t_2} G(t_1,s)p(s)q(s)\left(f(s,(x(s)-v(s))^*)+g_1(s)\right)\nabla s\Bigg| \\
\leq &\left|\lambda \int_{\rho(a)}^{t_1} (G(t_2,s)-G(t_1,s))p(s)q(s)\left(f(s,(x(s)-v(s))^*)+g_1(s)\right)\nabla s\right| \\
&+\left|\lambda \int_{t_2}^{\sigma(b)} (G(t_2,s)-G(t_1,s))p(s)q(s)\left(f(s,(x(s)-v(s))^*)+g_1(s)\right)\nabla s\right| \\
&+\left|\lambda \int_{t_1}^{t_2} G(t_2,s)p(s)q(s)\left(f(s,(x(s)-v(s))^*)+g_1(s)\right)\nabla s\right| \\
&+\left|\lambda \int_{t_1}^{t_2} G(t_1,s)p(s)q(s)\left(f(s,(x(s)-v(s))^*)+g_1(s)\right)\nabla s\right| \\
\leq &|\lambda| \int_{\rho(a)}^{t_1} |G(t_2,s)-G(t_1,s)|p(s)q(s)\left(f(s,(x(s)-v(s))^*)+g_1(s)\right)\nabla s \\
&+|\lambda| \int_{t_2}^{\sigma(b)} |G(t_2,s)-G(t_1,s)|p(s)q(s)\left(f(s,(x(s)-v(s))^*)+g_1(s)\right)\nabla s \\
&+|\lambda| \int_{t_1}^{t_2} |G(t_2,s)|p(s)q(s)\left(f(s,(x(s)-v(s))^*)+g_1(s)\right)\nabla s \\
&+\lambda \int_{t_1}^{t_2} |G(t_1,s)|p(s)q(s)\left(f(s,(x(s)-v(s))^*)+g_1(s)\right)\nabla s.
\end{aligned}
$$

From here, we conclude that $T(Y)$ is equi-continuous. Now, applying the Arzela-Ascoli theorem, we get that $T : P \to P$ is a completely continuous operator. This completes the proof.

Theorem 5.17. *Suppose* $(D16)$-$(D22)$. *Then there exists a constant* $\overline{\lambda} > 0$ *such that for any* $0 < \lambda \leq \overline{\lambda}$ *the BVP* (5.28), (5.29) *has at least one positive solution.*

Proof. Fix $\delta \in (0,1)$. By $(D20)$, it follows that there exists an $\varepsilon \in (0,1)$ such that

$$
f(t,z) \geq \delta f(t,0), \quad t \in [\rho(a),\sigma(b)], \quad z \in [0,\varepsilon].
$$

Set

$$
\overline{f}(\varepsilon) = \max_{(t,z)\in[\rho(a),\sigma(b)]\times[0,\varepsilon]} (f(t,z)+g(t)),
$$

$$
c = C_1 \int_{\rho(a)}^{\sigma(b)} H(s,s)p(s)q(s)\nabla s,
$$

$$\overline{\lambda} = \frac{\varepsilon}{2c\overline{f}(\varepsilon)}.$$

Take $0 < \lambda < \overline{\lambda}$. Note that

$$\frac{\varepsilon}{2c\overline{f}(\varepsilon)} > \lambda$$

or

$$\frac{\overline{f}(\varepsilon)}{\varepsilon} < \frac{1}{2c\lambda}$$

and

$$\lim_{z\to 0} \frac{\overline{f}(z)}{z} = \infty.$$

Hence, there exists an $R_0 \in (0,\varepsilon)$ such that

$$\frac{\overline{f}(R_0)}{R_0} = \frac{1}{2c\lambda}.$$

Set

$$U = \{x \in P : \|x\| < R_0\}.$$

Let $x \in \partial U$ and $\nu \in [0,1]$ be such that

$$\nu T x = x.$$

Then

$$\begin{aligned}
|\nu Tx(t)| &= \left| \nu\lambda \int_{\rho(a)}^{\sigma(b)} G(t,s)p(s)q(s)\left(f(s,(x(s)-v(s))^*) + g_1(s)\right)\nabla s \right| \\
&= \nu\lambda \int_{\rho(a)}^{\sigma(b)} G(t,s)p(s)q(s)\left(f(s,(x(s)-v(s))^*) + g_1(s)\right)\nabla s \\
&\leq C_1 \nu\lambda \int_{\rho(a)}^{\sigma(b)} H(s,s)p(s)q(s) \max_{(s,z)\in[\rho(a),\sigma(b)]\times[0,R_0]} (f(s,z)+g_1(s))\nabla s \\
&\leq C_1 \nu\lambda \int_{\rho(a)}^{\sigma(b)} H(s,s)p(s)q(s)\overline{f}(R_0)\nabla s \\
&\leq \lambda c\overline{f}(R_0), \quad t \in [\rho(a),\sigma(b)].
\end{aligned}$$

Hence,

$$\begin{aligned}
\frac{\overline{f}(R_0)}{R_0} &\geq \frac{1}{\lambda c} \\
&> \frac{1}{2\lambda c}
\end{aligned}$$

$$= \frac{\overline{f}(R_0)}{R_0}.$$

This is a contradiction. Thus,

$$x \neq \nu T x, \quad \nu \in [0,1], \quad x \in \partial U.$$

Hence and by the Leray-Schauder fixed point Theorem 5.4, it follows that the operator T has a fixed point $x \in \overline{U}$. Note that, if $x(s) - v(s) < 0$ for some $s \in [\rho(a), \sigma(b)]$, then

$$f(s,(x(s)-v(s))^*) + g_1(s) = g_1(s)$$

and if $x(s) - v(s) \geq 0$ for some $s \in [\rho(a), \sigma(b)]$, then

$$0 \leq x(s) - v(s) \leq R_0$$

and since $R_0 \in (0, \varepsilon)$, we get

$$f(s,(x(s)-v(s))^*) + g_1(s) \geq \delta f(s,0) + g_1(s)$$

$$\geq g_1(s).$$

Therefore,

$$x(t) = \lambda \int_{\rho(a)}^{\sigma(b)} G(t,s)p(s)q(s)\left(f(s,(x(s)-v(s))^*) + g_1(s)\right)\nabla s$$

$$\geq \lambda \int_{\rho(a)}^{\sigma(b)} G(t,s)p(s)q(s)g_1(s)\nabla s$$

$$= v(t), \quad t \in [\rho(a), \sigma(b)].$$

Let

$$u(t) = x(t) - v(t), \quad t \in [\rho(a), \sigma(b)].$$

Then

$$u^{\Delta\nabla}(t) + v^{\Delta\nabla}(t) + g(t)(u^{\Delta}(t) + v^{\Delta}(t)) + h(t)(u(t) + v(t))$$

$$+\lambda q(t)(f(t,u(t)) + g_1(t)) = 0, \quad t \in [\rho(a), \sigma(b)],$$

or

$$u^{\Delta\nabla}(t) + g(t)u^{\Delta}(t) + h(t)u(t) + \lambda q(t) f(t,u(t)) = 0, \quad t \in (a,b),$$

and

$$0 = x(\rho(a))$$

$$= u(\rho(a)) + v(\rho(a))$$

$$= u(\rho(a)),$$

$$x(\sigma(b)) = u(\sigma(b)) + v(\sigma(b))$$

$$= u(\sigma(b)) + \sum_{j=1}^{m-2} \alpha_j v(\eta_j)$$

$$= \sum_{j=1}^{m-2} \alpha_j x(\eta_j),$$

whereupon,

$$u(\sigma(b)) = \sum_{j=1}^{m-2} \alpha_j (x(\eta_j) - v(\eta_j))$$

$$= \sum_{j=1}^{m-2} \alpha_j u(\eta_j).$$

Therefore, u is a solution to the BVP (5.28), (5.29). For it we have

$$\|u\| \leq \|x\|$$

$$\leq R_0$$

$$< 1.$$

This completes the proof.

Theorem 5.18. *Suppose* $(D16)$-$(D22)$. *Then there exists an* $\lambda^* > 0$ *such that for any* $0 < \lambda \leq \lambda^*$ *the BVP* (5.28), (5.29) *has at least one positive solution.*

Proof. Let

$$r = \frac{C_1 C_3}{C_2} \int_{\rho(a)}^{\sigma(b)} q(s) g_1(s) \nabla s,$$

$$R_1 = \max\{1, r\},$$

$$\Omega_1 = \{x \in P : \|x\| < R_1\}.$$

Let

$$\lambda^* = \min\left\{1, R_1\left(C_1\int_{\rho(a)}^{\sigma(b)} H(s,s)p(s)q(s)\left(\max_{0\le z\le R_1} f(s,z)+g_1(s)\right)\nabla s\right)^{-1}\right\} \tag{5.36}$$

and $\lambda \in (0,\lambda^*)$. Take $x \in P\bigcap \partial\Omega_1$ arbitrarily. Then $\|x\| = R_1$ and

$$\begin{aligned} x(s)-v(s) &\le x(s) \\ &\le \|x\| \end{aligned}$$

and

$$\begin{aligned} |Tx(t)| &= \lambda \int_{\rho(a)}^{\sigma(b)} G(t,s)p(s)q(s)\left(f(s,(x(s)-v(s))^*)+g_1(s)\right)\nabla s \\ &\le \lambda C_1 \int_{\rho(a)}^{\sigma(b)} H(s,s)p(s)q(s)\left(f(s,(x(s)-v(s))^*)+g_1(s)\right)\nabla s \\ &\le \lambda C_1 \int_{\rho(a)}^{\sigma(b)} H(s,s)p(s)q(s)\left(\max_{0\le z\le R_1} f(s,z)+g_1(s)\right)\nabla s \\ &\le \lambda^* C_1 \int_{\rho(a)}^{\sigma(b)} H(s,s)p(s)q(s)\left(\max_{0\le z\le R_1} f(s,z)+g_1(s)\right)\nabla s \\ &\le R_1 \\ &= \|x\|, \quad t\in[\rho(a),\sigma(b)]. \end{aligned}$$

Hence,

$$\|Tx\| \le \|x\|, \quad x\in P\bigcap\partial\Omega_1.$$

Now, we choose the constant $N > 0$ so that

$$\frac{\lambda C_2^2 N}{2C_1}\int_{\theta_1}^{\theta_2} H(s,s)\phi_1(s)p(s)q(s)\nabla s \min_{\theta_1\le z\le\theta_2} \phi_1(t) \ge 1.$$

By $(D21)$, it follows that for any $t\in[\theta_1,\theta_2]$ there exists a constant $B>0$ such that

$$\frac{f(t,z)}{z} > N, \quad z > B.$$

Take

$$\begin{aligned} R_2 &= \max\left\{R_1+1, 2\lambda r, \frac{2C_1(B+1)}{C_2}\min_{\theta_1\le t\le\theta_2}\phi_1(t)\right\}, \\ \Omega_2 &= \{x\in P : \|x\| < R_2\} \end{aligned}$$

and $x \in P \bigcap \partial\Omega_2$. Then

$$\begin{aligned}
x(t)-v(t) &= x(t)-\lambda\int_{\rho(a)}^{\sigma(b)} G(t,s)p(s)q(s)g_1(s)\nabla s \\
&\geq x(t)-\lambda C_3\phi_1(t)\int_{\rho(a)}^{\sigma(b)} p(s)q(s)g_1(s)\nabla s \\
&\geq x(t)-\lambda\frac{C_1x(t)}{C_2\|x\|}C_3\int_{\rho(a)}^{\sigma(b)} p(s)q(s)g_1(s)\nabla s \\
&\geq x(t)-\lambda\frac{C_1x(t)}{C_2\|x\|}C_3\int_{\rho(a)}^{\sigma(b)} q(s)g_1(s)\nabla s \\
&= x(t)-\lambda\frac{x(t)}{\|x\|}r \\
&\geq x(t)-\frac{x(t)}{R_2}\lambda r \\
&= \left(1-\frac{\lambda r}{R_2}\right)x(t) \\
&\geq \frac{1}{2}x(t) \\
&\geq 0, \quad t\in[a,b].
\end{aligned}$$

Then

$$\begin{aligned}
\min_{t\in[\theta_1,\theta_2]}(x(t)-v(t)) &\geq \min_{t\in[\theta_1,\theta_2]}\left(\frac{1}{2}x(t)\right) \\
&\geq \min_{t\in[\theta_1,\theta_2]}\left(\frac{C_2}{2C_1}\phi_1(t)\|x\|\right) \\
&= \frac{R_2C_2\min\limits_{t\in[\theta_1,\theta_2]}\phi_1(t)}{2C_1} \\
&\geq B+1 \\
&> B.
\end{aligned}$$

Now,

$$|Tx(t)| \geq \max_{t\in[a,b]}\lambda\int_{\rho(a)}^{\sigma(b)} C_2\phi_1(t)H(s,s)p(s)q(s)\left(f(s,(x(s)-v(s))^*)+g_1(s)\right)\nabla s$$

$$
\begin{aligned}
&\geq \max_{t\in[a,b]} \lambda \int_{\rho(a)}^{\sigma(b)} C_2\phi_1(t)H(s,s)p(s)q(s)\left(f(s,(x(s)-v(s))^*)\right)\nabla s\\
&\geq \min_{t\in[\theta_1,\theta_2]} \lambda \int_{\theta_1}^{\theta_2} C_2\phi_1(t)H(s,s)p(s)q(s)\left(f(s,(x(s)-v(s)))\right)\nabla s\\
&\geq \min_{t\in[\theta_1,\theta_2]} \lambda \int_{\theta_1}^{\theta_2} C_2\phi_1(t)H(s,s)p(s)q(s)N(x(s)-v(s))\nabla s\\
&\geq \lambda C_2 \min_{t\in[\theta_1,\theta_2]} \phi_1(t) \int_{\theta_1}^{\theta_2} H(s,s)p(s)q(s)\frac{N}{2}x(s)\nabla s\\
&\geq \frac{\lambda C_2^2 N}{C_1} \min_{t\in[\theta_1,\theta_2]} \phi_1(t) \int_{\theta_1}^{\theta_2} H(s,s)p(s)q(s)\phi_1(s)\|x\|\nabla s\\
&\geq \|x\|, \quad t\in[\rho(a),\sigma(b)].
\end{aligned}
$$

Therefore,

$$
\|Tx\| \geq \|x\|, \quad x\in P\bigcap\partial\Omega_2.
$$

Now, applying the Guo-Krasnosel'skii fixed point Theorem 1.11, we conclude that the operator T has a fixed point x so that

$$
R_1 \leq \|x\| \leq R_2.
$$

Since $\|x\| > r$, we get

$$
\begin{aligned}
x(t)-v(t) &\geq \frac{C_2}{C_1}\phi_1(t)\|x\| - \lambda\int_{\rho(a)}^{\sigma(b)} G(t,s)p(s)q(s)g_1(s)\nabla s\\
&\geq \frac{C_2}{C_1}\phi_1(t)\|x\| - \lambda\phi_1(t)C_3\int_{\rho(a)}^{\sigma(b)} p(s)q(s)g_1(s)\nabla s\\
&\geq \frac{C_2}{C_1}\phi_1(t)\|x\| - \frac{C_2}{C_1}\phi_1(t)\lambda r\\
&\geq \frac{C_2}{C_1}\phi_1(t)r - \frac{C_2}{C_1}\phi_1(t)\lambda r\\
&= (1-\lambda)\frac{C_2}{C_1}r\phi_1(t)\\
&> 0, \quad t\in[\rho(a),\sigma(b)].
\end{aligned}
$$

Now, as in the proof of Theorem 5.17, we conclude that the function $u = x - v$ is a solution to the BVP (5.28), (5.29). This completes the proof.

Theorem 5.19. *Suppose* $(D16)$-$(D22)$. *Then the BVP* (5.28), (5.29) *has at least two positive solutions.*

Proof. Take $\delta \in (0,1)$. Let $\varepsilon \in \left(0, \frac{r}{2}\right)$ be such that

$$f(t,z) \geq \delta f(t,0), \quad t \in [\rho(a), \sigma(b)], \quad z \in [0, \varepsilon].$$

Let, also,

$$\overline{f}(\varepsilon) = \max_{(t,z)\in[\rho(a),\sigma(b)]\times[0,\varepsilon]} (f(t,z) + g_1(t)),$$

$$c = C_1 \int_{\rho(a)}^{\sigma(b)} H(s,s)p(s)q(s)\nabla s,$$

$$r = \frac{C_1 C_3}{C_2} \int_{\rho(a)}^{\sigma(b)} q(s)g_1(s)\nabla s,$$

$$\overline{\lambda} = \frac{\varepsilon}{2c\overline{f}(\varepsilon)}.$$

Set $R_1 = 1 + r$ and

$$\Omega = \{x \in P : \|x\| < R_1\},$$

$$\lambda^* = \min\left\{1, R_1\left(C_1 \int_{\rho(a)}^{\sigma(b)} H(s,s)p(s)q(s)\left(\max_{z\in[0,R_1]} f(s,z) + g_1(s)\right)\nabla s\right)^{-1}\right\}.$$

Let

$$0 < \lambda < \min\{\overline{\lambda}, \lambda^*\}.$$

Since $0 < \lambda < \overline{\lambda}$. Then, by Theorem 5.17, it follows that the BVP (5.28), (5.29) has at least one positive solution u_1 so that

$$\|u_1\| \leq R_0 < \frac{r}{2}.$$

By $0 < \lambda \leq \lambda^*$ and Theorem 5.18, it follows that the BVP (5.28), (5.29) has at least one positive solution u_2 such that

$$\begin{aligned} \|u_2\| &\geq \frac{R}{2} \\ &> \frac{r}{2}. \end{aligned}$$

This completes the proof.

Example 5.8. Let $\mathbb{T} = 2^{\mathbb{N}_0}$. Consider the BVP

$$u^{\Delta\nabla}(t)+(t+1)u^{\Delta}(t)+(t^2+1)u(t)+\lambda\left((u(t))^2-\frac{9}{2}u(t)+2\right)=0,$$

$t\in(2,16)$,

$$u(1)=0,$$

$$u(32)=u(4)+3u(8)+4u(16).$$

Here

$$a=2,$$

$$b=16,$$

$$\eta_1=4,$$

$$\eta_2=8,$$

$$\eta_3=16,$$

$$\alpha_1=1,$$

$$\alpha_2=3,$$

$$\alpha_3=4,$$

$$g(t)=t+1,$$

$$h(t)=t^2+1,$$

$$q(t)=1,$$

$$f(t,u)=u^2-\frac{9}{2}u+2,\quad t\in\mathbb{T},\quad u\in\mathbb{R}.$$

Take

$$g_1(t)=4,\quad t\in\mathbb{T}.$$

We have

$$
\begin{aligned}
f(t,u)+g_1(t) &= u^2 - \frac{9}{2}u + 2 + 4 \\
&= u^2 - \frac{9}{2}u + 6 \\
&> 0, \quad t \in \mathbb{T}, \\
f(t,0) &= 2 \\
&> 0, \\
\lim_{u\to\infty} \frac{f(t,u)}{u} &= \lim_{u\to\infty} \frac{u^2 - \frac{9}{2}u + 2}{u} \\
&= \lim_{u\to\infty} \left(u - \frac{9}{2} + \frac{2}{u} \right) \\
&= \infty.
\end{aligned}
$$

Let

$$
\begin{aligned}
\delta &= \frac{1}{4}, \\
\varepsilon &= \frac{1}{4}, \\
C &= \int_1^{32} C_1 H(s,s) p(s) \nabla s.
\end{aligned}
$$

Then

$$
\begin{aligned}
\max_{z\in[0,\varepsilon]} f(t,z) &= \max_{z\in[0,\varepsilon]} \left(z^2 - \frac{9}{2}z + 2 \right) \\
&= 2
\end{aligned}
$$

and

$$
\begin{aligned}
\overline{\lambda} &= \frac{1}{8c\left(\max_{z\in[0,\varepsilon]} f(t,z) + 4 \right)} \\
&= \frac{1}{8c(2+4)}
\end{aligned}
$$

$$= \frac{1}{48c}.$$

Hence and by Theorem 5.17, if

$$\lambda \in \left(0, \frac{1}{48c}\right),$$

the considered BVP has at least one positive solution u_1 so that

$$\|u_1\| \leq \frac{1}{4}.$$

Let

$$p = 4\frac{cC_3}{C_2} + 1,$$

$$R_1 = p.$$

Then

$$\int_{\rho(2)}^{\sigma(16)} C_1 H(s,s) p(s) \left(\max_{z \in [0,R_1]} f(s,z) + g_1(s) \right) \nabla s = \int_1^{32} C_1 H(s,s) \left(\frac{9}{2} + 4 \right) \nabla s$$

$$= \frac{17}{2} c.$$

Now, applying Theorem 5.18, we conclude that if

$$0 < \lambda < \min\left\{1, \frac{20}{17c}\right\},$$

then the considered BVP has at least one positive solution u_2 so that

$$\|u_2\| \geq p.$$

Exercise 5.11. Let $\mathbb{T} = 2\mathbb{Z}$. Prove that the BVP

$$u^{\Delta\nabla}(t) + (1+t^2)u^{\Delta}(t) + (1+t^4)u(t) + \lambda\left((u(t))^4 + \sin(2\pi u(t))\right) = 0,$$

$t \in [2, 18]$,

$$u(0) = 0,$$

$$u(20) = u(4) + \frac{1}{2}u(6) + \frac{1}{3}u(12),$$

has at least one positive solution.

5.5 Existence of Symmetric Positive Solutions for Second Order Boundary Value Problems with Integral Boundary Conditions

In this section we will investigate the BVP

$$u^{\Delta\nabla}(t)+f(t,u(t),u^{\Delta}(t))=0, \quad t\in(a,b), \tag{5.37}$$

$$\alpha u(a)-\beta\lim_{t\to a+}u^{\Delta}(t_{)}=\int_a^b h_1(s)u(s)\nabla s, \tag{5.38}$$

$$\alpha u(b)+\beta\lim_{t\to b-}u^{\Delta}(t)=\int_a^b h_2(s)u(s)\nabla s, \tag{5.39}$$

where

(D23) $\alpha,\beta>0$,
(D24) $h_1,h_2\in\mathscr{C}([a,b])$, $h_1\geq 0$, $h_2\geq 0$ on $[a,b]$, h_1 and h_2 are symmetric on $[a,b]$, $1-v_1-v_2>0$, where

$$\kappa=2\alpha\beta+\alpha^2(b-a),$$

$$k_1(t)=\frac{\beta+\alpha(b-t)}{\kappa},$$

$$k_2(t)=\frac{\beta+\alpha(t-a)}{\kappa},$$

$$v_1=\int_a^b h_1(t)k_1(t)\nabla t,$$

$$v_2=\int_a^b h_2(t)k_2(t)\nabla t,$$

(D25) $f\in\mathscr{C}([a,b]\times[0,\infty)\times\mathbb{R})$, $f\geq 0$ on $[a,b]\times[0,\infty)\times\mathbb{R}$, $f(\cdot,u,v)$ is symmetric on $[a,b]$ and

$$f(t,u,v)=f(t,u,-v), \quad (t,u,v)\in[a,b]\times[0,\infty)\times\mathbb{R},$$

(D26) $f(t,\cdot,u)$, $f(t,u,\cdot)$ are non-decreasing for each $(t,u,v)\in[a,b]\times[0,\infty)\times\mathbb{R}$.

Before deducting and proving the existence result for BVP (5.37), (5.38), we will investigate equation

$$-u^{\Delta\nabla}(t)=p(t), \quad t\in[a,b], \tag{5.40}$$

subject to the boundary conditions (5.38), (5.39).

Lemma 5.20. *Let* (D23) *and* (D24) *hold,* $\kappa\neq 0$ *and* $p\in\mathscr{C}([a,b])$. *Then any solution* $u\in\mathscr{C}([a,b])$ *of the BVP* (5.40), (5.38), (5.39) *can be represented in the form*

$$u(t) = \int_a^b H(t,s)p(s)\nabla s, \quad t \in [a,b],$$

where

$$H(t,s) = G(t,s) + B_1(t)\int_a^b G(s,\tau)h_1(\tau)\Delta\tau$$
$$+B_2(t)\int_a^b G(s,\tau)h_2(\tau)\nabla\tau,$$

$$G(t,s) = \frac{1}{\kappa}\begin{cases} (\beta+\alpha(s-a))(\beta+\alpha(b-t)) & a \le s \le t \le b \\ (\beta+\alpha(t-a))(\beta+\alpha(b-s)) & a \le t \le s \le b, \end{cases}$$

$$B_1(t) = \frac{(1-\nu_2)k_1(t)+\nu_2 k_2(t)}{1-\nu_1-\nu_2},$$

$$B_2(t) = \frac{(1-\nu_1)k_2(t)+\nu_1 k_1(t)}{1-\nu_1-\nu_2}, \quad t \in [a,b].$$

Proof. Suppose that u is a solution to the BVP (5.40), (5.38), (5.39). Set

$$A = \lim_{t\to a+} u^\Delta(t).$$

We ∇-integrate equation (5.40) from a to t and we get

$$-u^\Delta(t) + A = \int_a^t p(s)\nabla s, \quad t \in [a,b],$$

or

$$u^\Delta(t) = A - \int_a^t p(s)\nabla s, \quad t \in [a,b].$$

Now, we Δ-integrate the last equation from a to t and we find

$$u(t) = u(a) + A(t-a) - \int_a^t \int_a^s p(\tau)\nabla\tau\Delta s$$
$$= u(a) + A(t-a) - \int_a^t (t-s)p(s)\nabla s, \quad t \in [a,b].$$

By the boundary conditions, we find

$$\alpha u(a) - \beta A = \int_a^b h_1(s)u(s)\nabla s$$

and

$$\alpha u(b) + \beta \lim_{t\to b-} u^\Delta(t) = \alpha u(a) + \alpha A(b-a) - \alpha\int_a^b (b-s)p(s)\nabla s$$

$$+\beta A-\beta\int_a^b p(s)\nabla s$$

$$=\alpha u(a)+(\alpha(b-a)+\beta)A$$

$$-\alpha\int_a^b (b-s)p(s)\nabla s-\beta\int_a^b p(s)\nabla s,$$

whereupon,

$$\alpha u(a)+(\alpha(b-a)+\beta)A-\alpha\int_a^b (b-s)p(s)\nabla s-\beta\int_a^b p(s)\nabla s=\int_a^b h_2(s)u(s)\nabla s$$

or

$$\alpha u(a)+(\alpha(b-a)+\beta)A=\int_a^b h_2(s)u(s)\nabla s+\alpha\int_a^b (b-s)p(s)\nabla s$$

$$+\beta\int_a^b p(s)\nabla s.$$

Thus, we get the system

$$\alpha u(a)-\beta A=\int_a^b h_1(s)u(s)\nabla s$$

$$\alpha u(a)+(\alpha(b-a)+\beta)A=\int_a^b h_2(s)u(s)\nabla s+\alpha\int_a^b (b-s)p(s)\nabla s$$

$$+\beta\int_a^b p(s)\nabla s.$$

Then

$$(\alpha(b-a)+2\beta)A=\int_a^b h_2(s)u(s)\nabla s-\int_a^b h_1(s)u(s)\nabla s$$

$$+\alpha\int_a^b (b-s)p(s)\nabla s+\beta\int_a^b p(s)\nabla s$$

or

$$A=\frac{1}{\alpha(b-a)+2\beta}\left(\int_a^b h_2(s)u(s)\nabla s-\int_a^b h_1(s)u(s)\nabla s\right.$$

$$\left.+\alpha\int_a^b (b-s)p(s)\nabla s-\beta\int_a^b p(s)\nabla s\right)$$

and

$$
\begin{aligned}
u(a) = & \frac{\beta}{\alpha(\alpha(b-a)+2\beta)}\left(\int_a^b h_2(s)u(s)\nabla s - \int_a^b h_1(s)u(s)\nabla s\right. \\
& \left. +\alpha\int_a^b (b-s)p(s)\nabla s + \beta\int_a^b p(s)\nabla s\right) \\
& +\frac{1}{\alpha}\int_a^b h_1(s)u(s)\nabla s.
\end{aligned}
$$

Therefore,

$$
\begin{aligned}
u(t) = & \frac{\beta}{\alpha(\alpha(b-a)+2\beta)}\left(\int_a^b h_2(s)u(s)\nabla s - \int_a^b h_1(s)u(s)\nabla s\right. \\
& \left. +\alpha\int_a^b (b-s)p(s)\nabla s + \beta\int_a^b p(s)\nabla s\right) \\
& +\frac{1}{\alpha}\int_a^b h_1(s)u(s)\nabla s \\
& +\frac{1}{\alpha(b-a)+2\beta}\left(\int_a^b h_2(s)u(s)\nabla s - \int_a^b h_1(s)u(s)\nabla s\right. \\
& \left. +\alpha\int_a^b (b-s)p(s)\nabla s + \beta\int_a^b p(s)\nabla s\right)(t-a) \\
& -\int_a^t (t-s)p(s)\nabla s \\
= & \frac{\alpha(t-a)+\beta}{\alpha(\alpha(b-a)+2\beta)}\int_a^b (h_2(s)-h_1(s))u(s)\nabla s \\
& +\frac{\alpha(t-a)+\beta}{\alpha(b-a)+2\beta}\int_a^b (b-s)p(s)\nabla s \\
& +\frac{\beta(\alpha(t-a)+\beta)}{\alpha(\alpha(b-a)+2\beta)}\int_a^b p(s)\nabla s \\
& -\int_a^t (t-s)p(s)\nabla s \\
& +\frac{1}{\alpha}\int_a^b h_1(s)u(s)\nabla s \\
= & \frac{\alpha(t-a)+\beta}{\alpha(\alpha(b-a)+2\beta)}\int_a^b h_2(s)u(s)\nabla s
\end{aligned}
$$

$$
\begin{aligned}
&+\frac{\alpha(b-t)+\beta}{\alpha(\alpha(b-a)+2\beta)}\int_a^b h_1(s)u(s)\nabla s\\
&+\frac{\alpha(t-a)+\beta}{\alpha(b-a)+2\beta}\int_a^b (b-s)p(s)\nabla s\\
&+\frac{\beta(\alpha(t-a)+\beta)}{\alpha(\alpha(b-a)+2\beta)}\int_a^b p(s)\nabla s\\
&-\int_a^t (t-s)p(s)\nabla s\\
=&\frac{\beta+\alpha(t-a)}{\kappa}\int_a^b h_2(s)u(s)\nabla s+\frac{\alpha(b-t)+\beta}{\kappa}\int_a^b h_2(s)u(s)\nabla s\\
&+\frac{\alpha(\alpha(t-a)+\beta)}{\kappa}\int_a^t (b-s)p(s)\nabla s\\
&+\frac{\alpha(t-a)+\beta}{\alpha(b-a)+2\beta}\int_t^b (b-s)p(s)\nabla s\\
&+\frac{\beta(\alpha(t-a)+\beta)}{\kappa}\int_a^t p(s)\nabla s\\
&+\frac{\beta(\alpha(t-a)+\beta)}{\kappa}\int_t^b p(s)\nabla s-\int_a^t (t-s)p(s)\nabla s\\
=&\frac{\beta+\alpha(t-a)}{\kappa}\int_a^b h_2(s)u(s)\nabla s\\
&+\frac{\alpha(b-t)+\beta}{\kappa}\int_a^b h_1(s)u(s)\nabla s\\
&+\frac{\alpha(t-a)+\beta}{\kappa}\int_a^t (\beta+\alpha(b-s))p(s)\nabla s\\
&-\int_a^t (t-s)p(s)\nabla s\\
&+\frac{\alpha(t-a)+\beta}{\kappa}\int_t^b (\alpha(b-s)+\beta)p(s)\nabla s\\
=&\frac{\beta+\alpha(t-a)}{\kappa}\int_a^b h_2(s)u(s)_{\nabla}s+\frac{\alpha(b-t)+\beta}{\kappa}\int_a^b h_1(s)u(s)\nabla s\\
&+\frac{\beta+\alpha(b-t)}{\kappa}\int_a^t (\beta+\alpha(s-a))p(s)\nabla s
\end{aligned}
$$

$$
\begin{aligned}
&+\frac{\alpha(t-a)+\beta}{\kappa}\int_t^b(\alpha(b-s)+\beta)p(s)\nabla s\\
&=\frac{\beta+\alpha(t-a)}{\kappa}\int_a^b h_2(s)u(s)\nabla s+\frac{\beta+\alpha(b-t)}{\kappa}\int_a^b h_1(s)u(s)\nabla s\\
&+\int_a^b G(t,s)p(s)\nabla s\\
&=k_2(t)\int_a^b h_2(s)u(s)\nabla s+k_1(t)\int_a^b h_1(s)u(s)\nabla s\\
&+\int_a^b G(t,s)p(s)\nabla s,\quad t\in[a,b],
\end{aligned}
$$

i.e.,

$$
\begin{aligned}
u(t)&=k_2(t)\int_a^b h_2(s)u(s)\nabla s+k_1(t)\int_a^b h_1(s)u(s)\nabla s\\
&+\int_a^b G(t,s)p(s)\nabla s,\quad t\in[a,b].
\end{aligned}
\tag{5.41}
$$

Now, we multiply (5.41) with $h_1(t)$ and then we ∇-integrate from a to b and we find

$$
\begin{aligned}
\int_a^b u(t)h_1(t)\nabla t&=\left(\int_a^b h_2(s)u(s)\nabla s\right)\left(\int_a^b h_1(t)k_2(t)\nabla t\right)\\
&+\left(\int_a^b h_1(s)u(s)\nabla s\right)\left(\int_a^b h_1(t)k_1(t)\nabla t\right)\\
&+\int_a^b h_1(t)\int_a^b G(t,s)p(s)\nabla s\nabla t.
\end{aligned}
$$

Let

$$
A_1=\int_a^b h_1(t)\int_a^b G(t,s)p(s)\nabla s\nabla t.
$$

Since h_1 is symmetric on $[a,b]$, we have

$$
\begin{aligned}
\int_a^b h_1(t)k_1(t)\nabla t&=\int_a^b h_1(t)\frac{\beta+\alpha(b-t)}{\kappa}\nabla t\\
&=\frac{\beta}{\kappa}\int_a^b h_1(t)\nabla t+\frac{\alpha}{\kappa}\int_a^b(b-t)h_1(t)\nabla t\\
&=\frac{\beta}{\kappa}\int_a^b h_1(t)\nabla t+\frac{\alpha}{\kappa}\int_a^b(s-a)h_1(b+a-s)\nabla s\\
&=\frac{\beta}{\kappa}\int_a^b h_1(t)\nabla t+\frac{\alpha}{\kappa}\int_a^b(s-a)h_1(s)\nabla s
\end{aligned}
$$

$$= \int_a^b h_1(t)\frac{\beta+\alpha(t-a)}{\kappa}\nabla t$$

$$= \int_a^b h_1(t)k_2(t)\nabla t.$$

Consequently,

$$\int_a^b u(t)h_1(t)\nabla t = v_1\int_a^b h_2(s)u(s)\nabla s$$

$$+v_1\int_a^b h_1(s)u(s)\nabla s + A_1,$$

whereupon,

$$(1-v_1)\int_a^b u(s)h_1(s)\nabla s = v_1\int_a^b h_2(s)u(s)\nabla s + A_1$$

or

$$(1-v_1)\int_a^b u(s)h_1(s)\nabla s - v_1\int_a^b h_2(s)u(s)\nabla s = A_1.$$

Set

$$A_2 = \int_a^b h_2(t)\int_a^b G(t,s)p(s)\nabla s\nabla t.$$

Since h_2 is symmetric on $[a,b]$, As in the above, we have

$$\int_a^b h_2(t)k_1(t)\nabla t = \int_a^b h_2(t)k_2(t)\nabla t.$$

We multiply (5.41) with $h_2(t)$ and then we integrate from a to t and we find

$$\int_a^b u(t)h_2(t)\nabla t = \left(\int_a^b h_2(t)k_2(t)\nabla t\right)\left(\int_a^b h_2(s)u(s)\nabla s\right)$$

$$+\left(\int_a^b h_2(t)k_1(t)\nabla t\right)\left(\int_a^b h_1(s)u(s)\nabla s\right)$$

$$+\int_a^b h_2(t)\int_a^b G(t,s)p(s)\nabla s$$

$$= v_2\int_a^b h_2(s)u(s)\nabla s + v_2\int_a^b h_1(s)u(s)\nabla s + A_2,$$

whereupon,

$$(1-v_2)\int_a^b u(s)h_2(s)\nabla s - v_2\int_a^b h_1(s)u(s)\nabla s = A_2.$$

Therefore, we get the system

$$(1-v_1)\int_a^b u(s)h_1(s)\nabla s - v_1\int_a^b h_2(s)u(s)\nabla s = A_1$$

$$(1-v_2)\int_a^b u(s)h_2(s)\nabla s - v_2\int_a^b h_1(s)u(s)\nabla s = A_2.$$

We have

$$(1-v_1-v_2)\int_a^b h_2(s)u(s)\nabla s = A_2(1-v_1)+A_1v_2$$

or

$$\int_a^b h_2(s)u(s)\nabla s = \frac{h_2(1-v_1)+A_1v_2}{1-v_1-v_2},$$

and

$$(1-v_1-v_2)\int_a^b u(s)h_1(s)\nabla s = A_1(1-v_2)+A_2v_1,$$

or

$$\int_a^b u(s)h_1(s)\nabla s = \frac{A_1(1-v_2)+A_2v_1}{1-v_1-v_2}.$$

Now, we apply (5.41) and we obtain

$$\begin{aligned}
u(t) &= k_2(t)\frac{A_2(1-v_1)+A_1v_2}{1-v_1-v_2} + k_1(t)\frac{A_1(1-v_2)+A_2v_1}{1-v_1-v_2} \\
&\quad + \int_a^b G(t,s)p(s)\nabla s \\
&= A_2B_2(t)+A_1B_1(t)+\int_a^b G(t,s)p(s)\nabla s \\
&= \int_a^b H(t,s)p(s)\nabla s, \quad t\in[a,b].
\end{aligned}$$

This completes the proof.

Below, to ensure existence of symmetric solutions of the BVP (5.37), (5.38), (5.39), we suppose that

$$h_1(t) = h_2(t) = h(t), \quad t\in[a,b],$$

where h is a non-negative and symmetric function on $[a,b]$. Also, we suppose that

$$k_1(t) = k_2(t) = \frac{1}{2\alpha}, \quad t\in[a,b].$$

Then

$$\begin{aligned} v_1 &= v_2 \\ &= \frac{1}{2\alpha}\int_a^b h(s)\nabla s, \quad t \in [a,b], \end{aligned}$$

and the assumption $(D24)$ becomes the following expression.

(D24) $h \in \mathscr{C}([a,b])$, $h \geq 0$ and h is symmetric on $[a,b]$, $\alpha - v > 0$,

$$v = \int_a^b h(s)\nabla s.$$

We have that

$$\begin{aligned} v_1 &= \frac{1}{2\alpha}v, \\ B_1(t) &= B_2(t) \\ &= \frac{(1-v_1)k_2(t)+v_1k_2(t)}{1-v_1-v_2} \\ &= \frac{\frac{1}{2\alpha}}{1-\frac{1}{\alpha}v} \\ &= \frac{1}{2(\alpha-v)}. \end{aligned}$$

Set

$$B = \frac{1}{\alpha - v}.$$

Hence,

$$\begin{aligned} H(t,s) &= G(t,s) + B_1(t)\int_a^b G(s,\tau)h(\tau)\nabla\tau \\ &\quad + B_2(t)\int_a^b G(s,\tau)h(\tau)\nabla\tau \\ &= G(t,s) + \frac{1}{2(\alpha-v)}\int_a^b G(s,\tau)h(\tau)\nabla\tau \\ &\quad + \frac{1}{2(\alpha-v)}\int_a^b G(s,\tau)h(\tau)\nabla\tau \\ &= G(t,s) + \frac{1}{\alpha-v}\int_a^b G(s,\tau)h(\tau)\nabla\tau \end{aligned}$$

$$= G(t,s) + B\int_a^b G(s,\tau)h(\tau)\nabla\tau, \quad t,s \in [a,b].$$

Lemma 5.21. *Assume* (*D*23) *and* (*D*24). *Then*

1.

$$G(b+a-t, b+a-s) = G(t,s),$$

$$H(b+a-t, b+a-s) = H(t,s), \quad t,s \in [a,b].$$

2.

$$\frac{1}{\kappa}\beta^2 \le G(t,s) \le G(s,s) \le \frac{1}{\kappa}D,$$

$$\frac{1}{\kappa}\beta^2\gamma \le H(t,s) \le H(s,s) \le \frac{1}{\kappa}\gamma D, \quad t,s \in [a,b],$$

where

$$D = (\beta + \alpha(b-a))^2,$$

$$\gamma = 1 + B\nu.$$

Proof. 1. We will consider the following cases.

a. Let $a \le s \le t \le b$. Then

$$G(t,s) = \frac{1}{\kappa}(\beta + \alpha(t-a))(\beta + \alpha(b-s))$$

and

$$b+a-t \le b+a-s.$$

Hence,

$$\begin{aligned} G(b+a-t, b+a-s) &= \frac{1}{\kappa}(\beta + \alpha(b+a-t-a))(\beta + \alpha(b-b-a+s)) \\ &= \frac{1}{\kappa}(\beta + \alpha(b-t))(\beta + \alpha(s-a)) \\ &= G(t,s). \end{aligned}$$

b. $a \le t \le s \le b$. Then

$$G(t,s) = \frac{1}{\kappa}(\beta + \alpha(t-a))(\beta + \alpha(b-s))$$

and

$$b+a-t \geq b+a-s,$$

and

$$\begin{aligned} G(b+a-t,b+a-s) &= \frac{1}{\kappa}(\beta+\alpha(b+a-s-a))(\beta+\alpha(b-b-a+t)) \\ &= \frac{1}{\kappa}(\beta+\alpha(b-s))(\beta+\alpha(t-a)) \\ &= G(t,s). \end{aligned}$$

Therefore,

$$G(b+a-t,b+a-s) = G(t,s)$$

for any $t,s \in [a,b]$. Hence, using that h is symmetric on $[a,b]$, we get

$$\begin{aligned} H(b+a-t,b+a-s) &= G(b+a-t,b+a-s) \\ &\quad + B\int_a^b G(b+a-s,\tau)h(\tau)\nabla\tau \\ &= G(t,s) \\ &\quad + B\int_a^b G(b+a-s,b+a-\tau)h(b+a-\tau)\nabla\tau \\ &= G(t,s) + B\int_a^b G(s,\tau)h(\tau)\nabla\tau \\ &= H(t,s), \quad t,s \in [a,b]. \end{aligned}$$

2. We will consider the following cases.

a. Let $a \leq s \leq t \leq b$. Then

$$b-t \leq b-s, \quad t-a \geq s-a$$

and

$$\begin{aligned} G(t,s) &= \frac{1}{\kappa}(\beta+\alpha(s-a))(\beta+\alpha(b-t)) \\ &\leq \frac{1}{\kappa}(\beta+\alpha(s-a))(\beta+\alpha(b-s)) \\ &= G(s,s) \end{aligned}$$

$$\leq \frac{1}{\kappa}(\beta+\alpha(b-a))^2$$

$$= D$$

and

$$\begin{aligned} G(t,s) &= \frac{1}{\kappa}(\beta+\alpha(s-a))(\beta+\alpha(b-t)) \\ &\geq \frac{1}{\kappa}(\beta+\alpha(b-t))(\beta+\alpha(t-a)) \\ &= G(t,t) \\ &\geq \frac{\beta^2}{\kappa}. \end{aligned}$$

b. Let $a \leq t \leq s \leq b$. Then

$$b-t \geq b-s, \quad s-a \geq t-a,$$

and

$$\begin{aligned} G(t,s) &= \frac{1}{\kappa}(\beta+\alpha(t-a))(\beta+\alpha(b-s)) \\ &\leq \frac{1}{\kappa}(\beta+\alpha(s-a))(\beta+\alpha(b-s)) \\ &= G(s,s) \\ &\leq \frac{1}{\kappa}(\beta+\alpha(b-a))^2 \\ &= \frac{D}{\kappa} \end{aligned}$$

and

$$\begin{aligned} G(t,s) &= \frac{1}{\kappa}(\beta+\alpha(t-a))(\beta+\alpha(b-s)) \\ &\geq \frac{1}{\kappa}\beta^2. \end{aligned}$$

Consequently,

$$\frac{1}{\kappa}\beta^2 \leq G(t,s) \leq G(s,s) \leq \frac{1}{\kappa}D, \quad t,s \in [a,b].$$

Hence,

$$\begin{aligned}
H(t,s) &= G(t,s) + B\int_a^b G(s,\tau)h(\tau)\nabla\tau \\
&\leq G(s,s) + B\int_a^b G(s,\tau)h(\tau)\nabla\tau \\
&= H(s,s) \\
&\leq \frac{1}{\kappa}D + \frac{BD}{\kappa}\int_a^b h(\tau)\nabla\tau \\
&= \frac{1}{\kappa}D + \frac{BD}{\kappa}\nu \\
&= \frac{1}{\kappa}(1+B\nu)D \\
&= \frac{1}{\kappa}\gamma D, \quad t,s \in [a,b],
\end{aligned}$$

and

$$\begin{aligned}
H(t,s) &= G(t,s) + B\int_a^b G(s,\tau)h(\tau)\nabla\tau \\
&\geq \frac{\beta^2}{\gamma} + \frac{B\beta^2}{\gamma}\int_a^b h(\tau)\nabla\tau \\
&\geq \frac{\beta^2}{\gamma}, \quad t,s \in [a,b].
\end{aligned}$$

This completes the proof.

Let $E = \mathscr{C}([a,b])$ be endowed with the norm

$$\|u\| = \max\{\max_{t\in[a,b]} |u(t)|, \quad \max_{t\in[a,b]} |u^{\Delta}(t)|\}$$

and

$$P = \Big\{u \in E : u(t) \geq 0, \quad u \quad \text{is} \quad \text{symmetric} \quad \text{and} \quad \text{concave} \quad \text{on} \quad [a,b]\Big\}.$$

For $u \in P$, define the operator

$$Su(t) = \int_a^b H(t,s)f(s,u(s),u^{\Delta}(s))\nabla s, \quad t \in [a,b].$$

Lemma 5.22. *Suppose* (*D*23)-(*D*26). *Then* $S : P \to P$.

Proof. Fix $u \in P$ arbitrarily. Since $H(\cdot,\cdot) \geq 0$ on $[a,b] \times [a,b]$ and $f \geq 0$ on $[a,b] \times [0,\infty) \times \mathbb{R}$, we conclude that $Su \geq 0$ on $[a,b]$. Because $G(\cdot,s)$ is a concave function on $[a,b]$, we have that $H(\cdot,s)$ is a concave function on $[a,b]$ and hence, Su is a concave function on $[a,b]$. Now, we will prove that Su is a symmetric function on $[a,b]$. Since u is symmetric on $[a,b]$, we get

$$u(b+a-s) = u(s), \quad s \in [a,b].$$

Now, using the chain rule, we obtain

$$-u^{\Delta}(b+a-s) = u^{\Delta}(s), \quad s \in [a,b].$$

We have

$$\begin{aligned}
Su(b+a-t) &= \int_a^b H(b+a-t,s)f(s,u(s),u^{\Delta}(s))\nabla s \\
&= \int_a^b H(b+a-t,b+a-s)f(b+a-s,u(b+a-s),u^{\Delta}(b+a-s))\nabla s \\
&= \int_a^b H(t,s)f(b+a-s,u(s),u^{\Delta}(b+a-s))\nabla s \\
&= \int_a^b H(t,s)f(s,u(s),u^{\Delta}(b+a-s))\nabla s \\
&= \int_a^b H(t,s)f(s,u(s),-u^{\Delta}(s))\nabla s \\
&= \int_a^b H(t,s)f(s,u(s),u^{\Delta}(s))\nabla s \\
&= Su(t), \quad t \in [a,b].
\end{aligned}$$

This completes the proof.

Lemma 5.23. *Suppose* (*D*23)-(*D*26). *Then* $S : P \to P$ *is decreasing.*

Proof. Let $u_1, u_2 \in P$ be such that

$$u_1(t) \leq u_2(t), \quad t \in [a,b].$$

Hence, $u_2 - u_1 \in P$ and then $u_2 - u_1$ is a concave function on $[a,b]$. Then there exists an $\eta \in [a,b]$ so that $u_2 - u_1$ is a non-decreasing on $[a,\eta]$ and non-increasing on $[\eta,b]$. Therefore,

$$u_1^{\Delta}(t) \leq u_2^{\Delta}(t), \quad t \in [a,\eta],$$

$$u_2^{\Delta}(t) \leq u_1^{\Delta}(t), \quad t \in [\eta,b].$$

Here

$$\eta = \inf\{\delta \in [a,b] : \sup_{t\in[a,b]} (u_2 - u_1)(t) = (u_2 - u_1)(\delta)\}.$$

Therefore,

$$\begin{aligned}
Su_1(t) &= \int_a^b H(t,s)f(s,u_1(s),u_1^{\Delta}(s))\nabla s \\
&= \int_a^{\eta} H(t,s)f(s,u_1(s),u_1^{\Delta}(s))\nabla s \\
&\quad + \int_{\eta}^b H(t,s)f(s,u_1(s),u_1^{\Delta}(s))\nabla s \\
&\leq \int_a^{\eta} H(t,s)f(s,u_2(s),u_2^{\Delta}(s))\nabla s \\
&\quad + \int_{\eta}^b H(t,s)f(s,u_2(s),u_1^{\Delta}(s))\nabla s \\
&= \int_a^{\eta} H(t,s)f(s,u_2(s),u_2^{\Delta}(s))\nabla s \\
&\quad + \int_{\eta}^b H(t,s)f(s,u_2(s),-u_1^{\Delta}(s))\nabla s \\
&\leq \int_a^{\eta} H(t,s)f(s,u_2(s),u_2^{\Delta}(s))\nabla s \\
&\quad + \int_{\eta}^b H(t,s)f(s,u_2(s),-u_2^{\Delta}(s))\nabla s \\
&= \int_a^{\eta} H(t,s)f(s,u_2(s),u_2^{\Delta}(s))\nabla s \\
&\quad + \int_{\eta}^b H(t,s)f(s,u_2(s),u_2^{\Delta}(s))\nabla s \\
&= \int_a^b H(t,s)f(s,u_2(s),u_2^{\Delta}(s))\nabla s
\end{aligned}$$

$$= Su_2(t), \quad t \in [a,b].$$

This completes the proof.

Lemma 5.24. *Suppose* $(D23)$-$(D26)$. *Then* $S : P \to P$ *is a completely continuous operator.*

Proof. Let $X \subset P$ is a bounded set. Then there exists a positive constant L_1 so that

$$u(t) \leq L_1,$$

$$|u^{\Delta}(t)| \leq L_1, \quad t \in [a,b].$$

Since $f \in \mathscr{C}([a,b] \times [0,\infty) \times \mathbb{R})$, there exists a positive constant L_2 such that

$$|f(t,u,v)| \leq L_2, \quad t \in [a,b], \quad u \in [0,L_1], \quad v \in [-L_1,L_1].$$

Now, note that there exists a positive constant L_3 so that

$$|H^{\Delta}(t,s)| \leq L_3, \quad t,s \in [a,b].$$

Fix $u \in X$. Then

$$\begin{aligned}|Su(t)| &= \left| \int_a^b H(t,s) f(s,u(s),u^{\Delta}(s)) \nabla s \right| \\ &\leq \int_a^b H(t,s) |f(s,u(s),u^{\Delta}(s))| \nabla s \\ &\leq \frac{1}{\kappa} \gamma D L_2 (b-a), \quad t \in [a,b],\end{aligned}$$

and

$$\begin{aligned}|(Su)^{\Delta}(t)| &= \left| \int_a^b H^{\Delta}(t,s) f(s,u(s),u^{\Delta}(s)) \nabla s \right| \\ &\leq \int_a^b |H^{\Delta}(t,s)| |f(s,u(s),u^{\Delta}(s))| \nabla s \\ &\leq L_2 L_3 (b-a), \quad t \in [a,b].\end{aligned}$$

Thus, $S(X)$ is uniformly bounded. Let $t_1, t_2 \in [a,b]$, $t_1 < t_2$, Then

$$|Su(t_2) - Su(t_1)| = \left| \int_a^b H(t_2,s) f(s,u(s),u^{\Delta}(s)) \nabla s \right.$$

$$
\begin{aligned}
&\quad - \int_a^b H(t_1,s) f(s,u(s),u^\Delta(s))\nabla s \Big| \\
&= \Big| \int_a^{t_2} H(t_2,s) f(s,u(s),u^\Delta(s))\nabla s \\
&\quad + \int_{t_2}^b H(t_2,s) f(s,u(s),u^\Delta(s))\nabla s \\
&\quad - \int_a^{t_1} H(t_1,s) f(s,u(s),u^\Delta(s))\nabla s \\
&\quad - \int_{t_1}^b H(t_1,s) f(s,u(s),u^\Delta(s))\nabla s \Big| \\
&= \Big| \int_a^{t_1} (H(t_2,s)-H(t_1,s)) f(s,u(s),u^\Delta(s))\nabla s \\
&\quad + \int_{t_1}^{t_2} H(t_2,s) f(s,u(s),u^\Delta(s))\nabla s \\
&\quad + \int_{t_2}^b (H(t_2,s)-H(t_1,s)) f(s,u(s),u^\Delta(s))\nabla s \\
&\quad - \int_{t_1}^{t_2} H(t_1,s) f(s,u(s),u^\Delta(s))\nabla s \Big| \\
&\le \Big| \int_a^{t_1} (H(t_2,s)-H(t_1,s)) f(s,u(s),u^\Delta(s))\nabla s \Big| \\
&\quad + \Big| \int_{t_1}^{t_2} H(t_2,s) f(s,u(s),u^\Delta(s))\nabla s \Big| \\
&\quad + \Big| \int_{t_2}^b (H(t_2,s)-H(t_1,s)) f(s,u(s),u^\Delta(s))\nabla s \Big| \\
&\quad + \Big| \int_{t_1}^{t_2} H(t_1,s) f(s,u(s),u^\Delta(s))\nabla s \Big| \\
&\le \int_a^{t_1} |H(t_2,s)-H(t_1,s)| |f(s,u(s),u^\Delta(s))| \nabla s \\
&\quad + \int_{t_1}^{t_2} H(t_2,s) |f(s,u(s),u^\Delta(s))| \nabla s \\
&\quad + \int_{t_2}^b |H(t_2,s)-H(t_1,s)| |f(s,u(s),u^\Delta(s))| \nabla s
\end{aligned}
$$

$$+\int_{t_1}^{t_2} H(t_1,s)|f(s,u(s),u^\Delta(s))|\nabla s,$$

whereupon, we conclude that $S(X)$ is equi-continuous. Now, applying the Arzela-Ascoli theorem, we obtain that $S : P \to P$ is a completely continuous operator. This completes the proof.

Theorem 5.20. *Suppose* (*D*23)-(*D*26). *Let there exist positive constants* a_1 *and* c *such that* $a_1 < c$ *and*

$$\sup_{t\in[a,b]} f(t,c,c) \le a_1,$$

$$c \ge \max\left\{\frac{2\gamma(b-a)D}{\kappa}, \frac{D(b-a)}{\gamma}, \frac{\kappa-\alpha\beta}{\kappa}, \frac{(b-a)(\kappa-\alpha\beta)}{\kappa}\right\} a_1.$$

Then the BVP (5.37), (5.38), (5.39) *has at least one solution* $w^* \in P$ *such that*

$$\max_{t\in[a,b]} w^*(t) \le c, \quad \lim_{n\to\infty} S^n w_0 = w^*,$$

where

$$w_0(t) = (b-a)a_1 G(t,t) + \frac{\kappa(b-a)a_1 D}{\kappa}.$$

Proof. Let

$$P_c = \{w \in P : \|w\| \le c\}.$$

Take $w \in P_c$ arbitrarily. Then

$$\begin{aligned} 0 &\le w(t) \\ &\le \max_{t\in[a,b]} w(t) \\ &\le c, \end{aligned}$$

$$\begin{aligned} w^\Delta(t) &\le \max_{t\in[a,b]} |w^\Delta(t)| \\ &\le c, \quad t \in [a,b]. \end{aligned}$$

Then

$$\begin{aligned} Sw(t) &= \int_a^b H(t,s)f(s,w(s),w^\Delta(s))\nabla s \\ &\le \frac{\gamma D}{\kappa}\int_a^b f(s,c,c)\nabla s \\ &\le \frac{\gamma D}{\kappa} a_1(b-a) \end{aligned}$$

$$\leq c, \quad t \in [a,b],$$

and

$$\begin{aligned}
\left|(Sw)^{\Delta}(t)\right| &= \left|\int_a^b H^{\Delta}(t,s)f(s,w(s),w^{\Delta}(s))\nabla s\right| \\
&= \left|\int_a^b G^{\Delta}(t,s)f(s,w(s),w^{\Delta}(s))\nabla s\right| \\
&\leq \int_a^b |G^{\Delta}(t,s)|f(s,w(s),w^{\Delta}(s))\nabla s \\
&\leq \frac{(\alpha\beta+\alpha^2(b-a)}{\kappa}\int_a^b f(s,w(s),w^{\Delta}(s))\nabla s \\
&\leq \frac{(\alpha\beta+\alpha^2(b-a)}{\kappa}\int_a^b f(s,c,c)\nabla s \\
&\leq a_1\frac{(\alpha\beta+\alpha^2(b-a)}{\kappa}(b-a) \\
&= a_1\frac{(\kappa-\alpha\beta)(b-a)}{\kappa} \\
&\leq c, \quad t \in [a,b].
\end{aligned}$$

Consequently,

$$\|Sw\| \leq c$$

and $S(P_c) \subseteq P_c$. Next,

$$\begin{aligned}
w_0^{\Delta}(t) &= (b-a)a_1G^{\Delta}(t,t) \\
&= \frac{(b-a)a_1\left(\alpha^2(b+a)-\alpha^2(t+\sigma(t))\right)}{\kappa} \\
&\leq \frac{(b-a)^2a_1\alpha^2}{\kappa}.
\end{aligned}$$

If $b-a \leq 1$, then

$$\begin{aligned}
\frac{(b-a)^2a_1\alpha^2}{\kappa} &\leq \frac{(b-a)a_1\alpha^2}{\kappa} \\
&= \frac{(\kappa-2\alpha\beta)a_1}{\kappa}
\end{aligned}$$

$$\leq \frac{(\kappa - \alpha\beta)a_1}{\kappa}$$

$$\leq c.$$

If $b-a \geq 1$, then

$$\frac{(b-a)^2 a_1 \alpha^2}{\kappa} \leq \frac{(b-a)(b-a)^2 a_1 \alpha^2}{\kappa}$$

$$\leq \frac{D(b-a)a_1}{\kappa}$$

$$\leq c.$$

Consequently,

$$\max_{t \in [a,b]} \left| w_0^{\Delta}(t) \right| \leq c$$

and $w_0 \in P_c$. Set

$$w_{n+a}(t) = S^{n+a} w_0(t), \quad t \in [a,b], \quad n \in \mathbb{N}_0.$$

Since $S(P_c) \subseteq P_c$, then $S^n(P_c) \subseteq P_c$ and $w_n \in P_c$, $n \in \mathbb{N}_0$. Because $S : P \to P$ is a completely continuous operator, there exists a convergent subsequence $\{w_{n_k}\}_{k \in \mathbb{N}}$ of the sequence $\{w_n\}_{n \in \mathbb{N}}$ and a $w^* \in P_c$ so that

$$w_{n_k} \to w^* \quad \text{in} \quad P_c, \quad \text{as} \quad k \to \infty.$$

Note that

$$\begin{aligned}
w_1(t) &= S w_0(t) \\
&= \int_a^b H(t,s) f(s, w_0(s), w_0^{\Delta}(s)) \nabla s \\
&\quad + B \int_a^b \int_a^b G(s,\tau) h(\tau) f(s, w_0(s), w_0^{\Delta}(s)) \nabla \tau \nabla s \\
&\leq a_1 \int_a^b G(t,t) \nabla s + B a_1 \int_a^b \int_a^b G(s,s) h(\tau) \nabla \tau \nabla s \\
&= (b-a) a_1 G(t,t) + B v a_1 \int_a^b G(s,s) \nabla s \\
&\leq (b-a) a_1 G(t,t) + \frac{\gamma (b-a) a_1 D}{\kappa}
\end{aligned}$$

$$= w_0(t), \quad t \in [a,b].$$

Hence, using that $S : P \to P$ is non-decreasing, we find

$$Sw_1(t) \leq w_0(t), \quad t \in [a,b],$$

or

$$w_2(t) \leq w_1(t), \quad t \in [a,b].$$

By induction,

$$w_{n+a}(t) \leq w_n(t), \quad t \in [a,b].$$

Therefore,

$$w_n \to w^* \quad \text{in} \quad P_c, \quad \text{as} \quad n \to \infty.$$

Since $S : P \to P$ is a continuous operator, we conclude that

$$Sw^* = w^*$$

and w^* is a solution to the BVP (5.37), (5.38), (5.39). This completes the proof.

Exercise 5.12. Let $\mathbb{T} = \mathbb{Z}$. Prove that the BVP

$$u^{\Delta\nabla}(t) + t^2\left((u^{\Delta}(t))^2 + (u(t))^4 + 10^{-4}\right) = 0, \quad t \in (-3,3),$$

$$10u(-3) - 10 \lim_{t\to -3+} u^{\Delta}(t) = \int_{-3}^{3} \frac{1}{s^4+1} u(s)\nabla s,$$

$$10u(3) + 10 \lim_{t\to 3-} u^{\Delta}(t) = \int_{-3}^{3} \frac{1}{s^4+1} u(s)\nabla s$$

has at least one positive solution.

5.6 Existence of Solutions of Second Order p-Laplacian Boundary Value Problems

In this section we will investigate the following BVP

$$
\begin{aligned}
\left(\phi_p(u^{\Delta}(t))\right)^{\Delta} + g(t)f(t,u(t),u^{\Delta}(t)) = 0, \quad t \in (a,b), \\
u(a) &= \sum_{j=1}^{m-2} \alpha_j u(\xi_j), \\
\phi_p(u^{\Delta}(b)) &= \sum_{j=1}^{m-2} \beta_j \phi_p(u^{\Delta}(\xi_j)),
\end{aligned}
\tag{5.42}
$$

where

(D27) $\phi_p(u) = |u|^{p-2}u$, $p > 1$, $\frac{1}{p} + \frac{1}{q} = 1$, $m > 2$.

(D28) $a < \xi_1 < \xi_2 < \ldots < \xi_{m-2} < \rho(b)$, $\xi_j \in \mathbb{T}$, $\alpha_j, \beta_j \in [0,\infty)$, $j \in \{1,\ldots,m-2\}$,

$$0 \le \sum_{j=1}^{m-2} \alpha_j < 1,$$

$$0 \le \sum_{j=1}^{m-2} \beta_j < 1.$$

(D29) $g \in \mathscr{C}_{ld}([a,b])$, $g \ge 0$ on $[a,b]$,

$$0 < \int_{\eta}^{b} g(t)\Delta t < \infty,$$

where

$$\eta = \min\left\{t \in \mathbb{T} : \frac{b-a}{2} \le t \le b\right\}.$$

(D30) $f \in \mathscr{C}([a,b] \times [0,\infty) \times \mathbb{R})$, $f \ge 0$ on $(a,b) \times [0,\infty) \times \mathbb{R}$,

$$g(t)f(t,0,0) \not\equiv 0, \quad f(t,0,0) \ge 0, \quad t \in [a,b].$$

We will start our investigations with the following useful lemma.

Lemma 5.25. *Let*

$$\sum_{j=1}^{m-2} \alpha_j \ne 1, \quad \sum_{j=1}^{m-2} \beta_j \ne 1, \quad h \in \mathscr{C}_{ld}([a,b]).$$

Then the BVP

$$\begin{aligned}
\left(\phi_p(u^{\Delta}(t))\right)^{\Delta} + h(t) &= 0, \quad t \in (a,b),\\
u(a) &= \sum_{j=1}^{m-2} \alpha_j u(\xi_j),\\
\phi_p(u^{\Delta}(b)) &= \sum_{j=1}^{m-2} \beta_j \phi_p(u^{\Delta}(\xi_j))
\end{aligned} \tag{5.43}$$

has the unique solution

$$u(t) = \int_a^t \phi_q\left(A + \int_s^b h(\tau)\Delta\tau\right)\Delta s + B, \quad t \in [a,b], \tag{5.44}$$

where

$$A = \frac{\sum\limits_{j=1}^{m-2} \beta_j \int_{\xi_j}^b h(\tau)\Delta\tau}{1 - \sum\limits_{j=1}^{m-2} \beta_j},$$

$$B = \frac{\sum\limits_{j=1}^{m-2} \alpha_j \int_a^{\xi_j} \phi_q\left(A + \int_s^b h(\tau)\Delta\tau\right)\Delta s}{1 - \sum\limits_{j=1}^{m-2} \alpha_j}.$$

Proof. 1. Let u be a solution to the BVP (5.43). We integrate the first equation of (5.43) from t to b and we find

$$\phi_p(u^{\Delta}(b)) - \phi_p(u^{\Delta}(t)) + \int_t^b h(s)\Delta s = 0, \quad t \in [a,b],$$

or

$$\phi_p(u^{\Delta}(t)) = \phi_p(u^{\Delta}(b)) + \int_t^b h(s)\Delta s, \quad t \in [a,b].$$

Then

$$\phi_p(u^{\Delta}(\xi_j)) = \phi_p(u^{\Delta}(b)) + \int_{\xi_j}^b h(s)\Delta s,$$

$$\beta_j\phi_p(u^{\Delta}(\xi_j)) = \beta_j\phi_p(u^{\Delta}(b)) + \beta_j\int_{\xi_j}^b h(s)\Delta s,$$

$$\sum_{j=1}^{m-2} \beta_j\phi_p(u^{\Delta}(\xi_j)) = \sum_{j=1}^{m-2} \beta_j\phi_p(u^{\Delta}(b)) + \sum_{j=1}^{m-2} \beta_j\int_{\xi_j}^b h(s)\Delta s$$

$$= \phi_p(u^\Delta(b)),$$

whereupon,

$$\left(1-\sum_{j=1}^{m-2}\beta_j\right)\phi_p(u^\Delta(b)) = \sum_{j=1}^{m-2}\beta_j\int_{\xi_j}^b h(s)\Delta s$$

or

$$\phi_p(u^\Delta(b)) = \frac{\sum\limits_{j=1}^{m-2}\beta_j\int_{\xi_j}^b h(\tau)\Delta\tau}{1-\sum\limits_{j=1}^{m-2}\beta_j}$$

$$= A$$

and

$$\phi_p(u^\Delta(t)) = A + \int_t^b h(s)\Delta s, \quad t\in[a,b].$$

Hence,

$$u^\Delta(t) = \phi_q\left(A+\int_t^b h(s)\Delta s\right), \quad t\in[a,b].$$

Now, we integrate the last equation from a to t and we find

$$u(t)-u(a) = \int_a^t \phi_q\left(A+\int_s^b h(\tau)\Delta\tau\right)\Delta s, \quad t\in[a,b],$$

or

$$u(t) = u(a) + \int_a^t \phi_q\left(A+\int_s^b h(\tau)\Delta\tau\right)\Delta s, \quad t\in[a,b].$$

Hence,

$$u(\xi_j) = u(a) + \int_a^{\xi_j}\phi_q\left(A+\int_s^b h(\tau)\Delta\tau\right)\Delta s,$$

$$\alpha_j u(\xi_j) = u(a)\alpha_j + \alpha_j\int_a^{\xi_j}\phi_q\left(A+\int_s^b h(\tau)\Delta\tau\right)\Delta s,$$

$$\sum_{j=1}^{m-2}\alpha_j u(\xi_j) = u(a)\sum_{j=1}^{m-2}\alpha_j + \sum_{j=1}^{m-2}\alpha_j\int_a^{\xi_j}\phi_q\left(A+\int_s^b h(\tau)\Delta\tau\right)\Delta s$$

$$= u(a),$$

whereupon,

$$\left(1-\sum_{j=1}^{m-2}\alpha_j\right)u(a) = \sum_{j=1}^{m-2}\alpha_j\int_a^{\xi_j}\phi_q\left(A+\int_s^b h(\tau)\Delta\tau\right)\Delta s,$$

or

$$u(a) = \frac{\sum_{j=1}^{m-2} \alpha_j \int_a^{\xi_j} \phi_q\left(A + \int_s^b h(\tau)\Delta\tau\right)\Delta s}{1 - \sum_{j=1}^{m-2} \alpha_j}$$

$$= B.$$

Thus,

$$u(t) = \int_a^t \phi_q\left(A + \int_s^b h(\tau)\Delta\tau\right)\Delta s + B, \quad t \in [a,b],$$

i.e., u satisfies the integral equation (5.44).

2. Let u satisfy (5.44). Then

$$u(\xi_j) = \int_a^{\xi_j} \phi_q\left(A + \int_s^b h(\tau)\Delta\tau\right)\Delta s + B,$$

$$\alpha_j u(\xi_j) = \alpha_j \int_a^{\xi_j} \phi_q\left(A + \int_s^b h(\tau)\Delta\tau\right)\Delta s + B\alpha_j,$$

$$\begin{aligned}\sum_{j=1}^{m-2} \alpha_j u(\xi_j) &= \sum_{j=1}^{m-2} \alpha_j \int_a^{\xi_j} \phi_q\left(A + \int_s^b h(\tau)\Delta\tau\right)\Delta s + B\sum_{j=1}^{m-2} \alpha_j \\ &= \left(\sum_{k=1}^{m-2} \alpha_k\right) \frac{\sum_{j=1}^{m-2} \alpha_j \int_a^{\xi_j} \phi_q\left(A + \int_s^b h(\tau)\Delta\tau\right)\Delta s}{1 - \sum_{j=1}^{m-2} \alpha_j} \\ &\quad + \sum_{j=1}^{m-2} \alpha_j \int_a^{\xi_j} \phi_q\left(A + \int_s^b h(\tau)\Delta\tau\right)\Delta s \\ &= \frac{\sum_{j=1}^{m-2} \alpha_j \int_a^{\xi_j} \phi_q\left(A + \int_s^b h(\tau)\Delta\tau\right)\Delta s}{1 - \sum_{j=1}^{m-2} \alpha_j} \\ &= B \\ &= u(a).\end{aligned}$$

We differentiate (5.44) and we find

$$u^{\Delta}(t) = \phi_q\left(A + \int_t^b h(s)\Delta s\right), \quad t \in [a,b],$$

$$\phi_p(u^{\Delta}(t)) = A + \int_t^b h(s)\Delta s, \quad t \in [a,b]. \tag{5.45}$$

From here,

$$\phi_p(u^{\Delta}(b)) = A,$$

$$\phi_p(u^{\Delta}(\xi_j)) = A + \int_{\xi_j}^b h(s)\Delta s,$$

$$\beta_j\phi_p(u^{\Delta}(\xi_j)) = A\beta_j + \beta_j\int_{\xi_j}^b h(s)\Delta s,$$

$$\begin{aligned}
\sum_{j=1}^{m-2}\beta_j\phi_p(u^{\Delta}(\xi_j)) &= A\sum_{j=1}^{m-2}\beta_j + \sum_{j=1}^{m-2}\beta_j\int_{\xi_j}^b h(s)\Delta s \\
&= \left(\sum_{j=1}^{m-2}\beta_j\right)\frac{\sum_{j=1}^{m-2}\beta_j\int_{\xi_j}^b h(\tau)\Delta\tau}{1-\sum_{j=1}^{m-2}\beta_j} \\
&\quad + \sum_{j=1}^{m-2}\beta_j\int_{\xi_j}^b h(s)\Delta s \\
&= \frac{\sum_{j=1}^{m-2}\beta_j\int_{\xi_j}^b h(\tau)\Delta\tau}{1-\sum_{j=1}^{m-2}\beta_j} \\
&= A \\
&= \phi_p(u^{\Delta}(b)).
\end{aligned}$$

Now, we differentiate (5.45) with respect to t and we find

$$\left(\phi_p(u^{\Delta}(t))\right)^{\Delta} = -h(t), \quad t \in (a,b),$$

i.e., u satisfies (5.42). This completes the proof.

Lemma 5.26. *Suppose (D27), (D28) and $h \in \mathscr{C}_{ld}([a,b])$, $h > 0$ on $[a,b]$. Then any solution to the BVP (5.43) is non-negative on $[a,b]$.*

Proof. The assertion follows by Lemma **??**. This completes the proof.

Lemma 5.27. *Suppose (D27), (D28), $h \in \mathscr{C}_{ld}([a,b])$ and $h \geq 0$ on $[a,b]$. Then the unique solution u of the BVP (5.43) satisfies*

$$u(t) \geq \gamma \max_{t\in[a,b]} u(t),$$

where

$$\gamma = \frac{\sum_{j=1}^{m-2} \alpha_j(\xi_j - a)}{b - a - \sum_{j=1}^{m-2} \alpha_j(b - \xi_j)}.$$

Proof. Let

$$\phi(t) = \phi_q\left(A + \int_t^b h(s)\Delta s\right), \quad t \in [a,b].$$

We have

$$u^{\Delta}(t) = \phi(t)$$

$$\geq 0, \quad t \in [a,b],$$

and

$$u^{\Delta}(t_1) \leq u^{\Delta}(t_2), \quad t_1, t_2 \in [a,b], \quad t_1 \geq t_2.$$

Therefore, u^{Δ} is a decreasing function. Since $u^{\Delta} \geq 0$ on $[a,b]$, we have that u is concave down on $[a,b]$. Hence,

$$\frac{u(b) - u(a)}{b - a} \geq \frac{u(b) - u(\xi_j)}{b - \xi_j}, \quad j \in \{1, \ldots, m-2\},$$

and

$$(b - \xi_j)u(b) - (b - \xi_j)u(a) \geq (b - a)u(b) - (b - a)u(\xi_j)$$

or

$$(b - a)u(\xi_j) - (\xi_j - a)u(b) \geq (b - \xi_j)u(a), \quad j \in \{1, \ldots, m-2\},$$

and

$$(b - a)\alpha_j u(\xi_j) - u(b)\alpha_j(\xi_j - a) \geq \alpha_j(b - \xi_j)u(a), \quad j \in \{1, \ldots, m-2\}.$$

Then

$$(b-a)\sum_{j=1}^{m-2} \alpha_j u(\xi_j) - u(b)\sum_{j=1}^{m-2} \alpha_j(\xi_j - a) \geq u(a)\sum_{j=1}^{m-2} \alpha_j(b - \xi_j), \quad j \in \{1, \ldots, m-2\}.$$

Now, using the first boundary condition, we get

$$(b-a)u(a)-u(a)\sum_{j=1}^{m-2}\alpha_j(b-\xi_j)\geq u(b)\sum_{j=1}^{m-2}\alpha_j(\xi_j-a),$$

or

$$u(a)\left(b-a-\sum_{j=1}^{m-2}\alpha_j(b-\xi_j)\right)\geq u(b)\sum_{j=1}^{m-2}\alpha_j(\xi_j-a),$$

or

$$\begin{aligned}u(t)&\geq u(a)\\ &\geq u(b)\frac{\sum\limits_{j=1}^{m-2}\alpha_j(\xi_j-a)}{b-a-\sum\limits_{j=1}^{m-2}\alpha_j(b-\xi_j)}\\ &=\gamma u(b)\\ &=\gamma\max_{t\in[a,b]}u(t),\quad t\in[a,b].\end{aligned}$$

This completes the proof.

Let $E=\mathscr{C}_{ld}^1([a,b])$ be endowed with the norm

$$\|u\|=\max\{\max_{t\in[a,b]}|u(t)|,\quad \max_{t\in[a,b]}|u^{\Delta}(t)|\}.$$

Define

$$\begin{aligned}P=\{u\in E:u(t)\geq 0,\quad u^{\Delta}(t)\geq 0,\quad t\in[a,b],\\ u\quad\text{is}\quad\text{concave}\quad\text{on}\quad[a,b]\}.\end{aligned}$$

For $u\in E$, define the operator

$$\begin{aligned}Fu(t)=\int_a^t\phi_q\Bigg(&\frac{\sum\limits_{j=1}^{m-2}\beta_j\int_{\xi_j}^b g(s)f(s,u(s),u^{\Delta}(s))\Delta s}{1-\sum\limits_{j=1}^{m-2}\beta_j}\\ &+\int_s^b g(\tau)f(\tau,u(\tau),u^{\Delta}(\tau))\Delta\tau\Bigg)\Delta s\end{aligned}$$

$$+\frac{\sum_{j=1}^{m-2}\alpha_j}{1-\sum_{j=1}^{m-2}\alpha_j}\int_a^{\xi_j}\phi_q\Bigg(\frac{\sum_{j=1}^{m-2}\beta_j\int_{\xi_j}^b g(\tau)f(\tau,u(\tau),u^{\Delta}(\tau))\Delta\tau}{1-\sum_{j=1}^{m-2}\beta_j}$$

$$+\int_s^b g(\tau)f(\tau,u(\tau),u^{\Delta}(\tau))\Delta\tau\Bigg)\Delta s,\quad t\in[a,b].$$

Lemma 5.28. *Suppose* $(D27)$-$(D30)$. *Then* $F:P\to P$.

Proof. Let $u\in P$. By $(D29)$ and $(D30)$, we get

$$Fu(t)\geq 0,\quad t\in[a,b],$$

and

$$(Fu)^{\Delta}(t)=\phi_q\Bigg(\frac{\sum_{j=1}^{m-2}\beta_j\int_{\xi_j}^b g(s)f(s,u(s),u^{\Delta}(s))\Delta s}{1-\sum_{j=1}^{m-2}\beta_j}$$

$$+\int_t^b g(\tau)f(\tau,u(\tau),u^{\Delta}(\tau))\Bigg),$$

$$(Fu)^{\Delta}(t_1)\leq (Fu)^{\Delta}(t_2),\quad t,t_1,t_2\in[a,b],\quad t_1\geq t_2.$$

Therefore, $Fu\in P$. This completes the proof.

For $L\geq 0$, define

$$M(L)=\phi_q\left(\frac{L^2\sum_{j=1}^{m-2}\beta_j(b-\xi_j)}{1-\sum_{j=1}^{m-2}\beta_j}+L^2(b-a)\right)$$

$$\times\left(b-a+\frac{\sum_{j=1}^{m-2}\alpha_j(\xi_j-a)}{1-\sum_{j=1}^{m-2}\alpha_j}\right),$$

$$N(L)=\phi_q\left(\frac{L^2\sum_{j=1}^{m-2}\beta_j(b-\xi_j)}{1-\sum_{j=1}^{m-2}\beta_j}+L^2(b-a)\right).$$

Lemma 5.29. *Suppose* $(D27)$-$(D30)$. *If*

$$0 \le g(t) \le L, \quad t \in [a,b],$$

$$0 \le f(t,u,v) \le L, \quad t \in [a,b], \quad u \in [0,\infty), \quad v \in \mathbb{R},$$

then for any $u \in P$, we have

$$\|Fu\| \le \max\{M(L), N(L)\}.$$

Proof. Let $u \in P$ be arbitrarily chosen. Then

$$\begin{aligned}
Fu(t) &= \int_a^t \phi_q\Bigg(\frac{\sum\limits_{j=1}^{m-2}\beta_j\int_{\xi_j}^b g(s)f(s,u(s),u^\Delta(s))\Delta s}{1-\sum\limits_{j=1}^{m-2}\beta_j}\\
&\quad + \int_s^b g(\tau)f(\tau,u(\tau),u^\Delta(\tau)))\Delta\tau\Bigg)\Delta s\\
&\quad + \frac{\sum\limits_{j=1}^{m-2}\alpha_j}{1-\sum\limits_{j=1}^{m-2}\alpha_j}\int_a^{\xi_j}\phi_q\Bigg(\frac{\sum\limits_{j=1}^{m-2}\beta_j\int_{\xi_j}^b g(\tau)f(\tau,u(\tau),u^\Delta(\tau))\Delta\tau}{1-\sum\limits_{j=1}^{m-2}\beta_j}\\
&\quad + \int_s^b g(\tau)f(\tau,u(\tau),u^\Delta(\tau))\Delta\tau\Bigg)\Delta s\\
&\le \int_a^b \phi_q\left(\frac{L^2\sum\limits_{j=1}^{m-2}\beta_j(b-\xi_j)}{1-\sum\limits_{j=1}^{m-2}\beta_j} + L^2(b-a)\right)\Delta s\\
&\quad + \frac{\sum\limits_{j=1}^{m-2}\alpha_j}{1-\sum\limits_{j=1}^{m-2}\alpha_j}\int_a^{\xi_j}\phi_q\left(\frac{L^2\sum\limits_{j=1}^{m-2}\beta_j(b-\xi_j)}{1-\sum\limits_{j=1}^{m-2}\beta_j} + L^2(b-a)\right)\Delta s\\
&= \phi_q\left(\frac{L^2\sum\limits_{j=1}^{m-2}\beta_j(b-\xi_j)}{1-\sum\limits_{j=1}^{m-2}\beta_j} + L^2(b-a)\right)
\end{aligned}$$

$$\times \left(b-a+\frac{\sum\limits_{j=1}^{m-2}\alpha_j(\xi_j-a)}{1-\sum\limits_{j=1}^{m-2}\alpha_j} \right)$$

$$= M(L), \quad t \in [a,b],$$

and

$$(Fu)^{\Delta}(t) = \phi_q\Bigg(\frac{\sum\limits_{j=1}^{m-2}\beta_j \int_{\xi_j}^{b} g(s) f(s,u(s),u^{\Delta}(s))\Delta s}{1-\sum\limits_{j=1}^{m-2}\beta_j}$$

$$+ \int_t^b g(\tau) f(\tau,u(\tau),u^{\Delta}(\tau))\Delta\tau \Bigg)$$

$$\leq \phi_q\left(\frac{L^2\sum\limits_{j=1}^{m-2}\beta_j(b-\xi_j)}{1-\sum\limits_{j=1}^{m-2}\beta_j} + L^2(b-a) \right)$$

$$= N(L), \quad t \in [a,b].$$

Thus,

$$||Fu|| = \max\{M(L), N(L)\}.$$

This completes the proof.

Lemma 5.30. *Suppose* (*D*27)-(*D*30)*. Then* $F: P \to P$ *is a completely continuous operator.*

Proof. Let $X \subseteq P$ be a bounded set. Then there exists a constant $L_1 > 0$ so that

$$0 \leq u(t) \leq L_1,$$

$$|u^{\Delta}(t)| \leq L_1, \quad t \in [a,b].$$

Since $g \in \mathscr{C}_{ld}([a,b])$ and $f \in \mathscr{C}([a,b]\times[0,\infty)\times\mathbb{R})$, there exists a positive constant L so that

$$0 \leq g(t) \leq L$$

$$0 \leq f(t,u,v) \leq L, \quad t \in [a,b], \quad u \in [0,L_1], \quad v \in [-L_1,L_1].$$

Hence and by Lemma 5.29, we have that

$$\|u\| \leq \max\{M(L), N(L)\}.$$

Therefore, $F(X)$ is uniformly bounded. Let now, $t_1, t_2 \in [a,b]$, $t_1 < t_2$. Then, for $u \in X$, we find

$$\begin{aligned}
|Fu(t_2) - Fu(t_1)| = \Bigg| \int_a^{t_2} \phi_q\Bigg(&\frac{\sum_{j=1}^{m-2} \beta_j \int_{\xi_j}^b g(s) f(s, u(s), u^\Delta(s))\Delta s}{1 - \sum_{j=1}^{m-2} \beta_j} \\
&+ \int_s^b g(\tau) f(\tau, u(\tau), u^\Delta(\tau)))\Delta\tau \Bigg)\Delta s \\
&+ \frac{\sum_{j=1}^{m-2} \alpha_j}{1 - \sum_{j=1}^{m-2} \alpha_j} \int_a^{\xi_j} \phi_q\Bigg(\frac{\sum_{j=1}^{m-2} \beta_j \int_{\xi_j}^b g(\tau) f(\tau, u(\tau), u^\Delta(\tau))\Delta\tau}{1 - \sum_{j=1}^{m-2} \beta_j} \\
&+ \int_s^b g(\tau) f(\tau, u(\tau), u^\Delta(\tau))\Delta\tau \Bigg)\Delta s \\
&- \int_a^{t_1} \phi_q\Bigg(\frac{\sum_{j=1}^{m-2} \beta_j \int_{\xi_j}^b g(s) f(s, u(s), u^\Delta(s))\Delta s}{1 - \sum_{j=1}^{m-2} \beta_j} \\
&+ \int_s^b g(\tau) f(\tau, u(\tau), u^\Delta(\tau)))\Delta\tau \Bigg)\Delta s \\
&+ \frac{\sum_{j=1}^{m-2} \alpha_j}{1 - \sum_{j=1}^{m-2} \alpha_j} \int_a^{\xi_j} \phi_q\Bigg(\frac{\sum_{j=1}^{m-2} \beta_j \int_{\xi_j}^b g(\tau) f(\tau, u(\tau), u^\Delta(\tau))\Delta\tau}{1 - \sum_{j=1}^{m-2} \beta_j} \\
&+ \int_s^b g(\tau) f(\tau, u(\tau), u^\Delta(\tau))\Delta\tau \Bigg)\Delta s \Bigg| \\
= \int_{t_1}^{t_2} \phi_q\Bigg(&\frac{\sum_{j=1}^{m-2} \beta_j \int_{\xi_j}^b g(s) f(s, u(s), u^\Delta(s))\Delta s}{1 - \sum_{j=1}^{m-2} \beta_j} \\
&+ \int_s^b g(\tau) f(\tau, u(\tau), u^\Delta(\tau)))\Delta\tau \Bigg)\Delta s
\end{aligned}$$

$$\leq N(L)(t_2 - t_1).$$

Next, using that

$$(Fu)^{\Delta}(t) = \phi_q\left(\frac{\sum_{j=1}^{m-2} \beta_j \int_{\xi_j}^{b} g(s) f(s, u(s), u^{\Delta}(s)) \Delta s}{1 - \sum_{j=1}^{m-2} \beta_j} + \int_t^b g(\tau) f(\tau, u(\tau), u^{\Delta}(\tau))) \Delta \tau\right), \quad t \in [a,b],$$

is uniformly continuous on $[a,b]$, we find that $F(X)$ is equi-continuous. Hence and by the Arzela-Ascoli theorem, we conclude that $F : P \to P$ is a completely continuous operator. This completes the proof.

Theorem 5.21. *Suppose* (*D*27)-(*D*30) *and there exist positive constants* L_1 *and* k *so that*

$$\int_{\xi_{m-2}}^{b} g(\tau) f(\tau, u(\tau), u^{\Delta}(\tau)) \Delta \tau \geq L_1$$

for any $u \in P$ *and*

$$\max\{M(k), N(k)\} \leq k,$$

and

$$0 \leq g(t) \leq k, \quad t \in [a,b],$$

$$0 \leq f(t,u,v) \leq k, \quad t \in [a,b], \quad u \in [0,k], \quad v \in [-k,k].$$

Then the BVP (5.42) *has at least one positive solution.*

Proof. Take $k_1 > 0$ so that

$$k_1 \leq \max\left\{ \frac{\sum_{j=1}^{m-2} \alpha_j}{1 - \sum_{j=1}^{m-2} \alpha_j} (\xi_j - a) \phi_q \left(\frac{\sum_{j=1}^{m-2} \beta_j L_1}{1 - \sum_{j=1}^{m-2} \beta_j} \right), k \right\}.$$

Let

$$\Omega_1 = \{u \in P : \|u\| < k_1\}.$$

Take $u \in P \bigcap \partial \Omega_1$ arbitrarily. Then

$$\|u\| = k_1$$

and

$$\begin{aligned}
\|Fu\| &\geq \frac{\sum_{j=1}^{m-2}\alpha_j}{1-\sum_{j=1}^{m-2}\alpha_j}\int_a^{\xi_j}\phi_q\left(\frac{\sum_{j=1}^{m-2}\beta_j\int_{\xi_j}^b g(\tau)f(\tau,u(\tau),u^{\Delta}(\tau))\Delta\tau}{1-\sum_{j=1}^{m-2}\beta_j}\right.\\
&\quad\left.+\int_s^b g(\tau)f(\tau,u(\tau),u^{\Delta}(\tau))\Delta\tau\right)\Delta s\\
&\geq \frac{\sum_{j=1}^{m-2}\alpha_j}{1-\sum_{j=1}^{m-2}\alpha_j}\int_a^{\xi_j}\phi_q\left(\frac{\sum_{j=1}^{m-2}\beta_j\int_{\xi_j}^b g(\tau)f(\tau,u(\tau),u^{\Delta}(\tau))\Delta\tau}{1-\sum_{j=1}^{m-2}\beta_j}\right)\Delta s\\
&\geq \frac{\sum_{j=1}^{m-2}\alpha_j}{1-\sum_{j=1}^{m-2}\alpha_j}\int_a^{\xi_j}\phi_q\left(\frac{\sum_{j=1}^{m-2}\beta_j\int_{\xi_{m-2}}^b g(\tau)f(\tau,u(\tau),u^{\Delta}(\tau))\Delta\tau}{1-\sum_{j=1}^{m-2}\beta_j}\right)\Delta s\\
&\geq \frac{\sum_{j=1}^{m-2}\alpha_j}{1-\sum_{j=1}^{m-2}\alpha_j}(\xi_j-a)\phi_q\left(\frac{\sum_{j=1}^{m-2}\beta_jL_1}{1-\sum_{j=1}^{m-2}\beta_j}\right)\\
&\geq k_1\\
&= \|u\|.
\end{aligned}$$

Let

$$\Omega_2 = \{u \in P : \|u\| < k\}.$$

Take $u \in P\bigcap\partial\Omega_2$ arbitrarily. Then

$$\|u\| = k$$

and by Lemma 5.29, we get

$$\begin{aligned}
\|Fu\| &\leq \max\{M(k),N(k)\}\\
&\leq k\\
&= \|u\|.
\end{aligned}$$

Now, applying the Guo-Krasnosels'kii Theorem 1.11, we conclude that the operator F has a fixed point in $P\bigcap(\overline{\Omega}_2\backslash\Omega_1)$. This completes the proof.

Theorem 5.22. *Suppose* $(D27)$-$(D30)$ *and there exist positive constants* L_1, k *and* k_1 *so that*

$$\int_{\xi_{m-2}}^{b} g(\tau)f(\tau,u(\tau),u^{\Delta}(\tau))\Delta\tau \geq L_1$$

for any $u \in P$ *and*

$$\max\{M(k),N(k)\} \leq k_1 < k,$$

$$0 \leq g(t) \leq k, \quad t \in [a,b],$$

$$0 \leq f(t,u,v) \leq k, \quad t \in [a,b], \quad u \in [0,k], \quad v \in [-k,k],$$

$$k \leq \frac{\sum\limits_{j=1}^{m-2} \alpha_j}{1-\sum\limits_{j=1}^{m-2} \alpha_j}(\xi_j - a)\phi_q\left(\frac{\sum\limits_{j=1}^{m-2} \beta_j L_1}{1-\sum\limits_{j=1}^{m-2} \beta_j}\right).$$

Then the BVP (5.42) *has at least one positive solution.*

Proof. Let

$$\Omega_1 = \{u \in P : \|u\| < k_1\}.$$

Take $u \in P\bigcap \partial\Omega_1$ arbitrarily. Then

$$\|u\| = k_1$$

and by Lemma 5.29, we get

$$\begin{aligned}\|Fu\| &\leq \max\{M(k),N(k)\}\\ &\leq k_1\\ &= \|u\|.\end{aligned}$$

Let

$$\Omega_2 = \{u \in P : \|u\| < k\}.$$

Take $u \in P\bigcap \partial\Omega_2$ arbitrarily. Then

$$\|u\| = k$$

and

$$
\begin{aligned}
\|Fu\| &\geq \frac{\sum_{j=1}^{m-2} \alpha_j}{1-\sum_{j=1}^{m-2} \alpha_j} \int_a^{\xi_j} \phi_q\Bigg(\frac{\sum_{j=1}^{m-2} \beta_j \int_{\xi_j}^b g(\tau) f(\tau, u(\tau), u^{\Delta}(\tau)) \Delta \tau}{1-\sum_{j=1}^{m-2} \beta_j} \\
&\quad + \int_s^b g(\tau) f(\tau, u(\tau), u^{\Delta}(\tau)) \Delta \tau \Bigg) \Delta s \\
&\geq \frac{\sum_{j=1}^{m-2} \alpha_j}{1-\sum_{j=1}^{m-2} \alpha_j} \int_a^{\xi_j} \phi_q\left(\frac{\sum_{j=1}^{m-2} \beta_j \int_{\xi_j}^b g(\tau) f(\tau, u(\tau), u^{\Delta}(\tau)) \Delta \tau}{1-\sum_{j=1}^{m-2} \beta_j} \right) \Delta s \\
&\geq \frac{\sum_{j=1}^{m-2} \alpha_j}{1-\sum_{j=1}^{m-2} \alpha_j} \int_a^{\xi_j} \phi_q\left(\frac{\sum_{j=1}^{m-2} \beta_j \int_{\xi_{m-2}}^b g(\tau) f(\tau, u(\tau), u^{\Delta}(\tau)) \Delta \tau}{1-\sum_{j=1}^{m-2} \beta_j} \right) \Delta s \\
&\geq \frac{\sum_{j=1}^{m-2} \alpha_j}{1-\sum_{j=1}^{m-2} \alpha_j} (\xi_j - a) \phi_q\left(\frac{\sum_{j=1}^{m-2} \beta_j L_1}{1-\sum_{j=1}^{m-2} \beta_j} \right) \\
&\geq k \\
&= \|u\|.
\end{aligned}
$$

Now, applying the Guo-Krasnosels'kii Theorem 1.11, we conclude that the operator F has a fixed point in $P \bigcap (\overline{\Omega}_2 \backslash \Omega_1)$. This completes the proof.

Theorem 5.23. *Suppose (D27)-(D30) and there exist positive constants L_1, k, k_1 and k_2 so that*

$$\int_{\xi_{m-2}}^b g(\tau) f(\tau, u(\tau), u^{\Delta}(\tau)) \Delta \tau \geq L_1$$

for any $u \in P$ and

$$\max\{M(k), N(k)\} \leq k_1 < k,$$

$$k_3 > \max\{\max\{M(k), N(k)\}, k\},$$

$$0 \leq g(t) \leq k, \quad t \in [a,b],$$

$$0 \leq f(t,u,v) \leq k, \quad t \in [a,b], \quad u \in [0,k], \quad v \in [-k,k],$$

$$k \leq \frac{\sum\limits_{j=1}^{m-2} \alpha_j}{1 - \sum\limits_{j=1}^{m-2} \alpha_j} (\xi_j - a) \phi_q \left(\frac{\sum\limits_{j=1}^{m-2} \beta_j L_1}{1 - \sum\limits_{j=1}^{m-2} \beta_j} \right).$$

Then the BVP (5.42) *has at least two positive solutions.*

Proof. Let Ω_1 and Ω_2 be as in the proof of Theorem 5.22. By Theorem 5.22, it follows that the BVP (5.42) has at least one positive solution u_1 in $P \bigcap (\overline{\Omega}_2 \backslash \Omega_1)$. Define

$$\Omega_3 = \{u \in P : \|u\| < k_3\}.$$

For $u \in P \bigcap \partial \Omega_3$, we have

$$\|u\| = k_3$$

and using Lemma 5.29, we get

$$\begin{aligned} \|Fu\| &\leq \max\{M(k), N(k)\} \\ &\leq k_3 \\ &= \|u\|. \end{aligned}$$

Now, applying the Guo-Krasnosel'skii theorem, we conclude that the BVP (5.42) has at least one solution $u_2 \in P \bigcap (\overline{\Omega}_3 \backslash \Omega_2)$. This completes the proof.

For $u \in P$, let

$$\begin{aligned} \alpha(u) &= \max_{t \in [a,b]} |u(t)| \\ &= u(b), \\ \beta(u) &= \max_{t \in [a,b]} |u^{\Delta}(t)| \\ &= u^{\Delta}(a), \\ \psi(u) &= \min_{t \in [\eta,b]} u(t) \\ &= u(\eta). \end{aligned}$$

We have that $\alpha, \beta : P \to [0, \infty)$ are continuous convex functionals and

$$\|u\| = \max\{\alpha(u), \beta(u)\}.$$

We introduce the following quantities:

$$\lambda = \int_a^{\eta} \phi_q\left(\frac{\sum\limits_{j=1}^{m-2} \beta_j \int_{\xi_j}^b g(\tau)\Delta\tau}{1-\sum\limits_{j=1}^{m-2}\beta_j} + \int_{|eta}^b g(\tau)\Delta\tau \right)\Delta s,$$

$$L = \int_a^b \phi_q\left(\frac{\sum\limits_{j=1}^{m-2} \beta_j \int_{\xi_j}^b g(s)f(s,u(s),u^{\Delta}(s))\Delta s}{1-\sum\limits_{j=1}^{m-2}\beta_j} \right.$$

$$\left. + \int_s^b g(\tau)f(\tau,u(\tau),u^{\Delta}(\tau)))\Delta\tau \right)\Delta s$$

$$+ \frac{\sum\limits_{j=1}^{m-2}\alpha_j}{1-\sum\limits_{j=1}^{m-2}\alpha_j} \int_a^{\xi_j} \phi_q\left(\frac{\sum\limits_{j=1}^{m-2} \beta_j \int_{\xi_j}^b g(\tau)f(\tau,u(\tau),u^{\Delta}(\tau))\Delta\tau}{1-\sum\limits_{j=1}^{m-2}\beta_j} \right.$$

$$\left. + \int_s^b g(\tau)f(\tau,u(\tau),u^{\Delta}(\tau))\Delta\tau \right)\Delta s,$$

$$Q = \phi_q\left(\frac{\sum\limits_{j=1}^{m-2} \beta_j \int_{\xi_j}^b g(\tau)f(\tau,u(\tau),u^{\Delta}(\tau))\Delta\tau}{1-\sum\limits_{j=1}^{m-2}\beta_j} + \int_a^b g(\tau)f(\tau,u(\tau),u^{\Delta}(\tau))\Delta\tau \right).$$

For non-negative real numbers r, a and l, define the following sets.

$$P(\alpha,r;\beta,l) = \{u\in P : \alpha(u) < r, \quad \beta(u) < l\},$$

$$\overline{P}(\alpha,r;\beta,l) = \{u\in P : \alpha(u) \le r, \quad \beta(u) \le l\},$$

$$P(\alpha,r;\beta,l;\psi,a) = \{u\in P : \alpha(u) < r, \quad \beta(u) < l, \quad \psi(u) > a\},$$

$$\overline{P}(\alpha,r;\beta,l;\psi,a) = \{u\in P : \alpha(u) < r, \quad \beta(u) < l, \quad \psi(u) > a\}.$$

For our next result we need of the following fixed point theorem.

Theorem 5.24 (Bai-Ge Fixed Point Theorem). *Let P be a cone in a real Banach space E. Assume that the constants r_1, b, d, r_2, l_1 and l_2 satisfy*

$$0 < r_1 < b < d \le r_2$$

and

$$0 < l_1 \le l_2.$$

If there exists two non-negative continuous convex functionals α *and* β *on* P *and a non-negative continuous concave functional* ψ *on* P *so that*

1. *there exists* $M > 0$ *such that*

$$\|u\| \le M \max\{\alpha(u), \beta(u)\}, \quad u \in P,$$

2. $P(\alpha, r; \beta, l) \ne \emptyset$ *for any* $r > 0$ *and* $l > 0$,
3. $\psi(u) \le \alpha(u)$ *for any* $u \in \overline{P}(\alpha, r_2; \beta, l_2)$ *and if* $F : \overline{P}(\alpha, r_2; \beta, l_2) \to P(\alpha, r_2; \beta, l_2)$ *is a completely continuous operator such that*

   a. $\{u \in \overline{P}(\alpha, d; \beta, l_2; \psi, b) : \psi(u) > b\} \ne \emptyset$, $\psi(Fu) > b$ *for* $u \in \overline{P}(\alpha, d; \beta, l_2; \psi, b)$,
   b. $\alpha(Fu) < r_1$, $\beta(Fu) > l_1$ *for* $u \in \overline{P}(\alpha, r_1; \beta, l_1)$,
   c. $\psi(Fu) > b$ *for* $u \in \overline{P}(\alpha, r_2; \beta, l_2; \psi, b)$ *with* $\alpha(Fu) > d$,

then F *has at least three different fixed points* $u_1, u_2, u_3 \in \overline{P}(\alpha, r_2; \beta, l_2)$ *with*

$$u_1 \in P(\alpha, r_1; \beta, l_1),$$

$$u_2 \in \left\{\overline{P}(\alpha, r_2; \beta, l_2; \psi, b) : \psi(u) > b\right\},$$

$$u_3 \in \overline{P}(\alpha, r_2; \beta, l_2) \backslash \left(\overline{P}(\alpha, r_2; \beta, l_2; \psi, b) \bigcup \overline{P}(\alpha, r_1; \beta, l_1)\right).$$

Theorem 5.25. *Assume (D27)-(D30) and there exists*

$$a < r_1 < c < 2c \le r_2, \quad 0 < l_1 \le l_2$$

such that

$$\frac{c}{\lambda} \le \min\left\{\frac{r_2}{L}, \ \frac{l_2}{Q}\right\}$$

and let f *satisfy the following conditions.*

(D31)

$$f(t, w, u) \le \min\left\{\phi_p\left(\frac{r_2}{L}\right), \ \phi_p\left(\frac{l_2}{Q}\right)\right\}$$

for $(t, w, v) \in [a, b] \times [0, r_2] \times [-l_2, l_2]$,

(D32)

$$f(t, w, v) > \phi_p\left(\frac{c}{\lambda}\right)$$

for $(t, w, v) \in [\eta, b] \times [c, 2c] \times [-l_2, l_2]$.

(D33)

$$f(t, w, v) < \min\left\{\phi_p\left(\frac{r_1}{L}\right), \ \phi_p\left(\frac{l_1}{Q}\right)\right\}$$

for $(t,w,v) \in [a,b] \times [0,r_1] \times [-l_1,l_1]$.

Then the BVP (5.42) *has at least three positive solutions* u_1, u_2, u_3 *so that*

$$\max_{t\in[a,b]} u_1(t) < r_1, \quad \max_{t\in[a,b]} |u_1^{\Delta}(t)| < l_1,$$

$$c < \max_{t\in[\eta,b]} u_2(t) \leq \max_{t\in[a,b]} u_2(t) \leq r_2,$$

$$\max_{t\in[a,b]} |u_2^{\Delta}(t)| \leq l_2,$$

$$\max_{t\in[\eta,b]} u_3(t) < c, \quad r_1 < \max_{t\in[a,b]} u_3(t) < 2c,$$

$$l_1 < \max_{t\in[a,b]} |u_3^{\Delta}(t)| \leq l_2.$$

Proof. Let $u \in \overline{P}(\alpha, r_2; b, l_2)$. Then

$$\begin{aligned}\alpha(u) &= \max_{t\in[a,b]} u(t)\\ &\leq r_2,\\ \beta(u) &= \max_{t\in[a,b]} |u^{\Delta}(t)|\\ &\leq l_2.\end{aligned}$$

By the assumption $(D31)$, we have

$$f(t,u(t),u^{\Delta}(t)) \leq \min\left\{\phi_p\left(\frac{r_2}{L}\right), \quad \phi_p\left(\frac{l_2}{Q}\right)\right\}, \quad t\in[a,b].$$

Since $F : P \to P$, we get

$$\begin{aligned}\alpha(Fu) &= \max_{t\in[a,b]} Fu(t)\\ &= Fu(b)\\ &= \int_a^b \phi_q\Bigg(\frac{\sum\limits_{j=1}^{m-2} \beta_j \int_{\xi_j}^b g(s)f(s,u(s),u^{\Delta}(s))\Delta s}{1-\sum\limits_{j=1}^{m-2}\beta_j}\\ &\quad + \int_s^b g(\tau)f(\tau,u(\tau),u^{\Delta}(\tau))\Delta\tau\Bigg)\Delta s\end{aligned}$$

$$+\frac{\sum\limits_{j=1}^{m-2}\alpha_j}{1-\sum\limits_{j=1}^{m-2}\alpha_j}\int_a^{\xi_j}\phi_q\left(\frac{\sum\limits_{j=1}^{m-2}\beta_j\int_{\xi_j}^b g(\tau)f(\tau,u(\tau),u^\Delta(\tau))\Delta\tau}{1-\sum\limits_{j=1}^{m-2}\beta_j}\right.$$

$$\left.+\int_s^b g(\tau)f(\tau,u(\tau),u^\Delta(\tau))\Delta\tau\right)\Delta s$$

$$\leq\frac{r_2}{L}\left(\int_a^t\phi_q\left(\frac{\sum\limits_{j=1}^{m-2}\beta_j\int_{\xi_j}^b g(s)\Delta s}{1-\sum\limits_{j=1}^{m-2}\beta_j}\right.\right.$$

$$\left.+\int_s^b g(\tau)\Delta\tau\right)\Delta s$$

$$+\frac{\sum\limits_{j=1}^{m-2}\alpha_j}{1-\sum\limits_{j=1}^{m-2}\alpha_j}\int_a^{\xi_j}\phi_q\left(\frac{\sum\limits_{j=1}^{m-2}\beta_j\int_{\xi_j}^b g(\tau)\Delta\tau}{1-\sum\limits_{j=1}^{m-2}\beta_j}\right.$$

$$\left.\left.+\int_s^b g(\tau)\Delta\tau\right)\Delta s\right)$$

$$=r_2$$

and

$$\beta(Fu)=\max_{t\in[a,b]}|(Fu)^\Delta(t)|$$

$$=\phi_q\left(\frac{\sum\limits_{j=1}^{m-2}\beta_j\int_{\xi_j}^b g(s)f(s,u(s),u^\Delta(s))\Delta s}{1-\sum\limits_{j=1}^{m-2}\beta_j}\right.$$

$$\left.+\int_a^b g(\tau)f(\tau,u(\tau),u^\Delta(\tau))\Delta\tau\right)$$

$$\leq\frac{l_2}{Q}\phi_q\left(\frac{\sum\limits_{j=1}^{m-2}\beta_j\int_{\xi_j}^b g(s)\Delta s}{1-\sum\limits_{j=1}^{m-2}\beta_j}\right.$$

$$+\int_a^b g(\tau)\Delta\tau\Bigg)$$

$$= l_2.$$

Therefore, $F:\overline{P}(\alpha,r_2;\beta,l_2)\to\overline{P}(\alpha,r_2;\beta,l_2)$. Let now, $u\in\overline{P}(\alpha,r_1;\beta,l_1)$. Then, by $(D33)$, we get

$$f(t,u(t),u^{\Delta}(t))<\min\left\{\phi_p\left(\frac{r_1}{L}\right),\quad \phi_p\left(\frac{l_1}{Q}\right)\right\},\quad t\in[a,b].$$

Hence,

$$\begin{aligned}
\alpha(Fu) &= \max_{t\in[a,b]} Fu(t)\\
&= Fu(b)\\
&= \int_a^b \phi_q\Bigg(\frac{\sum_{j=1}^{m-2}\beta_j\int_{\xi_j}^b g(s)f(s,u(s),u^{\Delta}(s))\Delta s}{1-\sum_{j=1}^{m-2}\beta_j}\\
&\quad+\int_s^b g(\tau)f(\tau,u(\tau),u^{\Delta}(\tau))\Delta\tau\Bigg)\Delta s\\
&\quad+\frac{\sum_{j=1}^{m-2}\alpha_j}{1-\sum_{j=1}^{m-2}\alpha_j}\int_a^{\xi_j}\phi_q\Bigg(\frac{\sum_{j=1}^{m-2}\beta_j\int_{\xi_j}^b g(\tau)f(\tau,u(\tau),u^{\Delta}(\tau))\Delta\tau}{1-\sum_{j=1}^{m-2}\beta_j}\\
&\quad+\int_s^b g(\tau)f(\tau,u(\tau),u^{\Delta}(\tau))\Delta\tau\Bigg)\Delta s\\
&\le \frac{r_1}{L}\Bigg(\int_a^t \phi_q\Bigg(\frac{\sum_{j=1}^{m-2}\beta_j\int_{\xi_j}^b g(s)\Delta s}{1-\sum_{j=1}^{m-2}\beta_j}\\
&\quad+\int_s^b g(\tau)\Delta\tau\Bigg)\Delta s\\
&\quad+\frac{\sum_{j=1}^{m-2}\alpha_j}{1-\sum_{j=1}^{m-2}\alpha_j}\int_a^{\xi_j}\phi_q\Bigg(\frac{\sum_{j=1}^{m-2}\beta_j\int_{\xi_j}^b g(\tau)\Delta\tau}{1-\sum_{j=1}^{m-2}\beta_j}
\end{aligned}$$

$$
+\int_s^b g(\tau)\Delta\tau\Big)\Delta s\Big)
$$

$$
= r_1
$$

and

$$
\begin{aligned}
\beta(Fu) &= \max_{t\in[a,b]} |(Fu)^{\Delta}(t)| \\
&= \phi_q\left(\frac{\sum\limits_{j=1}^{m-2}\beta_j \int_{\xi_j}^b g(s)f(s,u(s),u^{\Delta}(s))\Delta s}{1-\sum\limits_{j=1}^{m-2}\beta_j}\right. \\
&\quad \left. +\int_a^b g(\tau)f(\tau,u(\tau),u^{\Delta}(\tau))\Delta\tau\right) \\
&\le \frac{l_1}{Q}\phi_q\left(\frac{\sum\limits_{j=1}^{m-2}\beta_j \int_{\xi_j}^b g(s)\Delta s}{1-\sum\limits_{j=1}^{m-2}\beta_j}\right. \\
&\quad \left. +\int_a^b g(\tau)\Delta\tau\right) \\
&= l_1.
\end{aligned}
$$

Therefore, $F:\overline{P}(\alpha,r_1;\beta,l_1)\to\overline{P}(\alpha,r_1;\beta,l_1)$. Now, we take $u(t)=2c$, $t\in[a,b]$. We have

$$
u(t)=2c\in\overline{P}(\alpha,2c;\beta,l_2;\psi,c)
$$

and

$$
\begin{aligned}
\psi(u) &= 2c \\
&> c.
\end{aligned}
$$

Therefore,

$$
\{u\in\overline{P}(\alpha,2c;\beta,l_2;\psi,c):\psi(u)>c\}\neq\emptyset.
$$

Moreover, for $u\in\overline{P}(\alpha,2e;\beta,l_2;\psi,c)$, we have

$$
c\le\psi(t)\le 2c,
$$

$$|u^{\Delta}(t)| \le l_2, \quad t \in [a,b],$$

and then

$$\begin{aligned}
\psi(Fu) &= \min_{t\in[\eta,b]} Fu(t) \\
&= Fu(\eta) \\
&= \int_a^{\eta} \phi_q\Bigg(\frac{\sum_{j=1}^{m-2} \beta_j \int_{\xi_j}^{b} g(s) f(s,u(s),u^{\Delta}(s))\Delta s}{1-\sum_{j=1}^{m-2}\beta_j} \\
&\quad + \int_s^b g(\tau) f(\tau,u(\tau),u^{\Delta}(\tau))\Delta\tau \Bigg)\Delta s \\
&\quad + \frac{\sum_{j=1}^{m-2}\alpha_j}{1-\sum_{j=1}^{m-2}\alpha_j} \int_a^{\xi_j} \phi_q\Bigg(\frac{\sum_{j=1}^{m-2} \beta_j \int_{\xi_j}^{b} g(\tau) f(\tau,u(\tau),u^{\Delta}(\tau))\Delta\tau}{1-\sum_{j=1}^{m-2}\beta_j} \\
&\quad + \int_s^b g(\tau) f(\tau,u(\tau),u^{\Delta}(\tau))\Delta\tau \Bigg)\Delta s \\
&\ge \int_a^{\eta} \phi_q\Bigg(\frac{\sum_{j=1}^{m-2} \beta_j \int_{\xi_j}^{b} g(s) f(s,u(s),u^{\Delta}(s))\Delta s}{1-\sum_{j=1}^{m-2}\beta_j} \\
&\quad + \int_s^b g(\tau) f(\tau,u(\tau),u^{\Delta}(\tau))\Delta\tau \Bigg)\Delta s \\
&\ge \frac{c}{\lambda} \int_a^{\eta} \phi_q\Bigg(\frac{\sum_{j=1}^{m-2} \beta_j \int_{\xi_j}^{b} g(s)\Delta s}{1-\sum_{j=1}^{m-2}\beta_j} \\
&\quad + \int_s^b g(\tau)\Delta\tau \Bigg)\Delta s \\
&= c.
\end{aligned}$$

So,

$$\psi(Fu) > c$$

for $u \in \overline{P}(\alpha, 2c; \beta, l_2; \psi, c)$. Now, we take $u \in \overline{P}(\alpha, r_2; \beta, l_2; \psi, c)$ and $\alpha(Fu) > 2c$. Then

$$\begin{aligned}
\psi(Fu) &= \min_{t\in[\eta,b]} Fu(t) \\
&= Fu(\eta) \\
&\geq \frac{\eta}{c} \max_{t\in[\eta,b]} Fu(t) \\
&> \frac{1}{2}\alpha(Fu) \\
&> c.
\end{aligned}$$

Now, applying the Bai-Ge fixed point theorem, we conclude that the operator F has three fixed points u_1, u_2, u_3 so that

$$\begin{aligned}
&u_1 \in P(\alpha, r_1; \beta, l_1), \\
&u_2 \in \left\{\overline{P}(\alpha, r_2; \beta, l_2; \psi, c) : \psi(u) > c\right\}, \\
&u_3 \in \overline{P}(\alpha, r_2; \beta, l_2) \backslash \left(\overline{P}(\alpha, r_2; \beta, l_2; \psi, c) \bigcup \overline{P}(\alpha, r_1; \beta, l_1)\right).
\end{aligned}$$

Moreover,

$$\alpha(u_3) \leq 2\psi(u_3)$$

and

$$\max_{t\in[a,b]} u_3(t) < 2c.$$

This completes the proof.

Example 5.9. Let $\mathbb{T} = \left\{1 - \left(\frac{1}{2}\right)^{\mathbb{N}_0}\right\} \bigcup [1,2]$, where $[1,2]$ is the real-valued interval. Consider the BVP

$$\begin{aligned}
&u^{\Delta^2}(t) + f(t, u(t), u^{\Delta}(t)) = 0, \quad t \in [0,2], \\
&u(0) = \frac{1}{2}u\left(\frac{1}{4}\right) + \frac{1}{6}u\left(\frac{3}{4}\right), \\
&u^{\Delta}(2) = \frac{1}{3}u^{\Delta}\left(\frac{1}{4}\right) + \frac{1}{6}u^{\Delta}\left(\frac{3}{4}\right),
\end{aligned}$$

where

$$f(t,w,v) = \begin{cases} \dfrac{t}{1000} + \dfrac{w^3}{3} + \dfrac{v^3}{1000000} & w \leq 3 \\ \dfrac{t}{1000} + 9 + \dfrac{v}{1000000} & w > 3. \end{cases}$$

Here

$$\begin{aligned} p &= q \\ &= 2, \\ \alpha_1 &= \frac{1}{2}, \\ \alpha_2 &= \frac{1}{6}, \\ \beta_1 &= \frac{1}{3}, \\ \beta_2 &= \frac{1}{6}, \\ \xi_1 &= \frac{1}{4}, \\ \xi_2 &= \frac{3}{4}, \\ a &= 0, \\ b &= 2, \\ g(t) &= 1, \quad t \in \mathbb{T}. \end{aligned}$$

We have

$$\begin{aligned} \alpha_1 + \alpha_2 &= \frac{1}{2} + \frac{1}{6} \\ &= \frac{2}{3} \\ &< 1, \end{aligned}$$

$$\begin{aligned}\beta_1+\beta_2 &= \frac{1}{3}+\frac{1}{6}\\ &= \frac{1}{2}\\ &< 1,\end{aligned}$$

$$f \geq 0 \quad \text{on} \quad [0,2]\times[0,\infty)\times\mathbb{R},$$

$$f(t,0,0) \not\equiv 0, \quad t\in[0,2].$$

Take

$$\begin{aligned}r_1 &= \frac{1}{2},\\ r_2 &= 140,\\ c &= 2,\\ l_1 &= \frac{1}{4},\\ l_2 &= 84.\end{aligned}$$

So,

$$0<r_1<c<2c<r_2$$

and

$$0<l_1<l_2.$$

Then

$$\begin{aligned}\eta &= \min\{t\in\mathbb{T}:\frac{b-a}{2}\leq t\leq b\}\\ &= \min\{t\in\mathbb{T}:1\leq t\leq 2\}\\ &= 1,\end{aligned}$$

$$\begin{aligned}\int_{\xi_1}^{2} g(\tau)\Delta\tau &= \int_{\frac{1}{4}}^{2}\Delta\tau\\ &= \frac{7}{4},\end{aligned}$$

$$\int_{\xi_2}^{2} g(\tau)\Delta\tau = \int_{\frac{3}{4}}^{2} \Delta\tau = \frac{5}{4},$$

$$\int_{a}^{b} g(\tau)\Delta\tau = \int_{0}^{2} \Delta\tau = 2,$$

$$\frac{\sum_{j=1}^{m-2} \beta_j \int_{\xi_j}^{b} g(\tau)\Delta\tau}{1-\sum_{j=1}^{m-2}\beta_j} = \frac{\frac{1}{3}\left(\frac{7}{4}\right)+\frac{1}{6}\left(\frac{5}{4}\right)}{1-\frac{1}{2}} = \frac{\frac{7}{12}+\frac{5}{24}}{\frac{1}{2}} = \frac{19}{12},$$

$$\frac{\sum_{j=1}^{m-2} \beta_j \int_{\xi_j}^{b} g(\tau)\Delta\tau}{1-\sum_{j=1}^{m-2}\beta_j} + \int_{a}^{b} g(\tau)\Delta\tau = \frac{19}{12}+2 = \frac{43}{12},$$

$$\frac{\sum_{j=1}^{m-2} \beta_j \int_{\xi_j}^{b} g(\tau)\Delta\tau}{1-\sum_{j=1}^{m-2}\beta_j} + \int_{\eta}^{b} g(\tau)\Delta\tau = \frac{19}{12}+1 = \frac{31}{12}$$

and

$$\lambda = \int_{a}^{\eta} \phi_q \left(\frac{\sum_{j=1}^{m-2} \beta_j \int_{\xi_j}^{b} g(\tau)\Delta\tau}{1-\sum_{j=1}^{m-2}\beta_j} + \int_{\eta}^{b} g(\tau)\Delta\tau \right) \Delta s$$

$$= \int_0^1 \frac{31}{12}\Delta\tau$$

$$= \frac{31}{12},$$

$$Q = \phi_q\left(\frac{\sum_{j=1}^{m-2}\beta_j\int_{\xi_j}^{b} g(\tau)\Delta\tau}{1-\sum_{j=1}^{m-2}\beta_j} + \int_a^b g(\tau)\Delta\tau\right)$$

$$= \frac{43}{12},$$

$$L = \int_a^b \phi_q\left(\frac{\sum_{j=1}^{m-2}\beta_j\int_{\xi_j}^{b} g(s)\Delta s}{1-\sum_{j=1}^{m-2}\beta_j}\right.$$

$$\left. + \int_s^b g(\tau)\Delta\tau\right)\Delta s$$

$$+ \frac{\sum_{j=1}^{m-2}\alpha_j}{1-\sum_{j=1}^{m-2}\alpha_j}\int_a^{\xi_j}\phi_q\left(\frac{\sum_{j=1}^{m-2}\beta_j\int_{\xi_j}^{b} g(\tau)\Delta\tau}{1-\sum_{j=1}^{m-2}\beta_j}\right.$$

$$\left. + \int_s^b g(\tau)\Delta\tau\right)\Delta s$$

$$\leq \int_a^b \phi_q\left(\frac{\sum_{j=1}^{m-2}\beta_j\int_{\xi_j}^{b} g(s)\Delta s}{1-\sum_{j=1}^{m-2}\beta_j}\right.$$

$$\left. + \int_a^b g(\tau)\Delta\tau\right)\Delta s$$

$$+ \frac{\sum_{j=1}^{m-2}\alpha_j}{1-\sum_{j=1}^{m-2}\alpha_j}\int_a^{\xi_j}\phi_q\left(\frac{\sum_{j=1}^{m-2}\beta_j\int_{\xi_j}^{b} g(\tau)\Delta\tau}{1-\sum_{j=1}^{m-2}\beta_j}\right.$$

$$\left. + \int_a^b g(\tau)\Delta\tau\right)\Delta s$$

$$
\begin{aligned}
&= \int_0^2 \frac{43}{12}\Delta s + \frac{\frac{1}{2}\int_0^{\frac{1}{4}} \frac{43}{12}\Delta s + \frac{1}{6}\int_0^{\frac{3}{4}} \frac{43}{12}\Delta s}{1-\frac{1}{2}-\frac{1}{6}} \\
&= \frac{43}{6} + \frac{\frac{1}{8}\left(\frac{43}{12}\right) + \frac{1}{6}\left(\frac{3}{4}\right)\left(\frac{43}{12}\right)}{1-\frac{2}{3}} \\
&= \frac{43}{6} + \frac{\frac{43}{96} + \frac{43}{96}}{\frac{1}{3}} \\
&= \frac{43}{6} + \frac{\frac{43}{48}}{\frac{1}{3}} \\
&= \frac{43}{6} + \frac{43}{16} \\
&= \frac{43(8+3)}{16} \\
&= \frac{473}{48} \\
&= \widetilde{L}.
\end{aligned}
$$

Hence, for $t \in [0,2]$, $w \in [0,140]$, $v \in [-80,80]$,

$$
\begin{aligned}
f(t,w,v) &\le \min\left\{\phi_p\left(\frac{r_2}{\widetilde{L}}\right), \phi_p\left(\frac{l_2}{Q}\right)\right\} \\
&\sim 14.2071 \\
&\le \min\left\{\phi_p\left(\frac{r_2}{L}\right), \phi_p\left(\frac{l_2}{Q}\right)\right\}
\end{aligned}
$$

and for $t \in [1,2]$, $w \in [2,4]$, $v \in [-80,80]$,

$$
\begin{aligned}
f(t,w,v) &> \phi_p\left(\frac{c}{\lambda}\right) \\
&\sim 0.7742,
\end{aligned}
$$

and for $t \in [0,2]$, $w \in \left[0, \frac{1}{2}\right]$, $v \in \left[-\frac{1}{4}, \frac{1}{4}\right]$,

$$
f(t,w,v) \le \min\left\{\phi_p\left(\frac{r_1}{\widetilde{L}}\right), \phi_p\left(\frac{l_1}{Q}\right)\right\}
$$

$$\sim 0.0507$$

$$\leq \min\left\{\phi_p\left(\frac{r_1}{L}\right), \phi_p\left(\frac{l_1}{Q}\right)\right\}.$$

Thus, by Theorem 5.26, the considered BVP has at least three solutions u_1, u_2 and u_3 for which

$$\max_{t\in[0,2]} u_1(t) < \frac{1}{2},$$

$$\max_{t\in[0,2]} |u_1^{\Delta}(t)| < \frac{1}{4},$$

$$2 < \min_{t\in[1,2]} u_2(t) \leq \max_{t\in[0,2]} u_2(t)$$

$$\leq 140,$$

$$\max_{t\in[0,80]} |u_2^{\Delta}(t)| \leq 80,$$

$$\min_{t\in[1,2]} u_3(t) < 2,$$

$$\frac{1}{2} < \max_{t\in[0,2]} u_3(t) < 4,$$

$$\frac{1}{4} < \max_{t\in[0,2]} |u_3^{\Delta}(t)| \leq 80.$$

Exercise 5.13. Let $\mathbb{T} = 2^{\mathbb{N}_0}$. Prove that the BVP

$$u^{\Delta^2}(t) + f(t, u(t), u^{\Delta}(t)) = 0, \quad t \in [1, 32],$$

$$u(1) = \frac{1}{3}u(4) + \frac{1}{6}u(8) + \frac{1}{20}u(16),$$

$$u^{\Delta}(32) = \frac{1}{9}u^{\Delta}(4) + \frac{1}{16}u^{\Delta}(16),$$

where

$$f(t, w, v) = \begin{cases} \frac{t^2}{10} + \frac{w^4}{36} + \frac{v^3}{10} & w \leq 5 \\ \frac{t^4}{10} + 11 + \frac{v}{10} & w > 5 \end{cases}$$

has at least three positive solutions.

Now, we consider the BVP

$$\begin{aligned}
\left(\phi_p(u^{\Delta}(t))\right)^{\nabla} + g(t)f(y,u(t),u^{\Delta}(t)) &= 0, \quad t \in (a,b), \\
\phi_p(u^{\Delta}(a)) &= \sum_{j=1}^{m-2} \alpha_j \phi_p(u^{\Delta}(\xi_j)), \\
u(b) &= \sum_{j=1}^{m-2} \beta_j u(\xi_j),
\end{aligned} \tag{5.46}$$

where ξ_j, α_j, β_j, $j \in \{1,\ldots,m-2\}$, p, g and f satisfy $(D27)$-$(D30)$. Define

$$P = \{u \in E : u(t) \geq 0, \quad u^{\Delta}(t) \leq 0, \quad t \in [a,b],$$

$$u \quad \text{is} \quad \text{concave} \quad \text{on} \quad [a,b]\}.$$

Exercise 5.14. Prove that u is a solution of the BVP (5.46) if and only if

$$\begin{aligned}
u(t) &= \int_a^t \phi_q\Bigg(\frac{\sum_{j=1}^{m-2} \alpha_j \int_a^{\xi_j} g(r)f(r,u(r),u^{\Delta}(r))\nabla r}{1-\sum_{j=1}^{m-2}\alpha_j} \\
&\quad + \int_a^s g(r)f(r,u(r),u^{\Delta}(r))\nabla r \Bigg)\Delta s \\
&\quad + \frac{\sum_{j=1}^{m-2}\beta_j}{1-\sum_{j=1}^{m-2}\beta_j} \int_{\xi_j}^b \phi_q\Bigg(\frac{\sum_{j=1}^{m-2} \alpha_j \int_a^{\xi_j} g(r)f(r,u(r),u^{\Delta}(r))\nabla r}{1-\sum_{j=1}^{m-2}\alpha_j} \\
&\quad + \int_a^s g(r)f(r,u(r),u^{\Delta}(r))\nabla r \Bigg)\Delta s, \quad t \in [a,b].
\end{aligned}$$

For $u \in E$, define the operator

$$\begin{aligned}
Au(t) &= \int_t^b \phi_q\Bigg(\frac{\sum_{j=1}^{m-2} \alpha_j \int_a^{\xi_j} g(r)f(r,u(r),u^{\Delta}(r))\nabla r}{1-\sum_{j=1}^{m-2}\alpha_j} \\
&\quad + \int_a^s g(r)f(r,u(r),u^{\Delta}(r))\nabla r \Bigg)\Delta s
\end{aligned}$$

$$+\frac{\sum\limits_{j=1}^{m-2}\beta_j}{1-\sum\limits_{j=1}^{m-2}\beta_j}\int_{\xi_j}^{b}\phi_q\Bigg(\frac{\sum\limits_{j=1}^{m-2}\alpha_j\int_a^{\xi_j}g(r)f(r,u(r),u^{\Delta}(r))\nabla r}{1-\sum\limits_{j=1}^{m-2}\alpha_j}$$

$$+\int_a^s g(r)f(r,u(r),u^{\Delta}(r))\nabla r\Bigg)\Delta s,\quad t\in[a,b].$$

Exercise 5.15. Prove that $A:P\to P$.

Exercise 5.16. Prove that $A:P\to P$ is a completely continuous operator.

For $u\in P$, define

$$\begin{aligned}\alpha(u)&=\max_{t\in[a,b]}u(t)\\&=u(a),\\ \beta(u)&=\max_{t\in[a,b]}|u^{\Delta}(t)|\\&=-u^{\Delta}(b),\\ \psi(u)&=\min_{t\in[a,b]}u(t)\\&=u(b).\end{aligned}$$

We have that $\alpha,\beta,\psi:P\to P$ are non-negative continuous convex functionals. Introduce the quantities

$$M=\frac{\sum\limits_{j=1}^{m-2}\beta_j}{1-\sum\limits_{j=1}^{m-2}\beta_j}\int_{\xi_j}^{b}\phi_q\Bigg(\frac{\sum\limits_{j=1}^{m-2}\alpha_j\int_a^{\xi_j}g(r)\nabla r}{1-\sum\limits_{j=1}^{m-2}\alpha_j}+\int_a^{\eta}g(r)\nabla r\Bigg)\Delta s,$$

$$N=\int_a^{b}\phi_q\Bigg(\frac{\sum\limits_{j=1}^{m-2}\alpha_j\int_a^{\xi_j}g(r)\nabla r}{1-\sum\limits_{j=1}^{m-2}\alpha_j}+\int_a^{s}g(r)\nabla r\Bigg)\Delta s$$

$$+\frac{\sum\limits_{j=1}^{m-2}\beta_j}{1-\sum\limits_{j=1}^{m-2}\beta_j}\int_{\xi_j}^{b}\phi_q\Bigg(\frac{\sum\limits_{j=1}^{m-2}\alpha_j\int_a^{\xi_j}g(r)\nabla r}{1-\sum\limits_{j=1}^{m-2}\alpha_j}+\int_a^{s}g(r)\nabla r\Bigg)\Delta s$$

$$Q = \phi_q\left(\frac{\sum\limits_{j=1}^{m-2} \alpha_j \int_a^{\xi_j} g(r)\nabla r}{1 - \sum\limits_{j=1}^{m-2} \alpha_j} + \int_a^b g(r)\nabla r \right).$$

Theorem 5.26. *Assume* (*D*27)-(*D*30) *and there exists*

$$a < r_1 < c < 2c \leq r_2, \quad 0 < l_1 \leq l_2$$

such that

$$\frac{c}{\lambda} \leq \min\left\{ \frac{r_2}{N}, \quad \frac{l_2}{Q} \right\}$$

and let f satisfy the following conditions.

(D34)

$$f(t,w,u) \leq \min\left\{ \phi_p\left(\frac{r_2}{N}\right), \quad \phi_p\left(\frac{l_2}{Q}\right) \right\}$$

for $(t,w,v) \in [a,b] \times [0,r_2] \times [-l_2,l_2]$,

(D35)

$$f(t,w,v) > \phi_p\left(\frac{c}{M}\right)$$

for $(t,w,v) \in [\eta,b] \times [c,2c] \times [-l_2,l_2]$.

(D36)

$$f(t,w,v) < \min\left\{ \phi_p\left(\frac{r_1}{N}\right), \quad \phi_p\left(\frac{l_1}{Q}\right) \right\}$$

for $(t,w,v) \in [a,b] \times [0,r_1] \times [-l_1,l_1]$.

Then the BVP (5.46) *has at least three positive solutions* u_1, u_2, u_3 *so that*

$$\max_{t\in[a,b]} u_1(t) < r_1, \quad \max_{t\in[a,b]} |u_1^{\Delta}(t)| < l_1,$$

$$c < \max_{t\in[\eta,b]} u_2(t) \leq \max_{t\in[a,b]} u_2(t) \leq r_2,$$

$$\max_{t\in[a,b]} |u_2^{\Delta}(t)| \leq l_2,$$

$$\max_{t\in[\eta,b]} u_3(t) < c, \quad r_1 < \max_{t\in[a,b]} u_3(t) < 2c,$$

$$l_1 < \max_{t\in[a,b]} |u_3^{\Delta}(t)| \leq l_2.$$

Proof. Let $u \in \overline{P}(\alpha, r_2; c, l_2)$. Then

$$\alpha(u) = \max_{t\in[a,b]} u(t)$$

$$\leq r_2,$$

$$\beta(u) = \max_{t\in[a,b]} |u^{\Delta}(t)|$$

$$\leq l_2.$$

By the assumption $(D31)$, we have

$$f(t,u(t),u^{\Delta}(t)) \leq \min\left\{\phi_p\left(\frac{r_2}{N}\right), \quad \phi_p\left(\frac{l_2}{Q}\right)\right\}, \quad t\in[a,b].$$

Since $F: P\to P$, we get

$$\begin{aligned}
\alpha(Au) &= \max_{t\in[a,b]} Au(t)\\
&= Au(b)\\
&= \int_a^b \phi_q\left(\frac{\sum\limits_{j=1}^{m-2}\alpha_j\int_a^{\xi_j} g(r)f(r,u(r),u^{\Delta}(r))\nabla r}{1-\sum\limits_{j=1}^{m-2}\alpha_j}\right.\\
&\quad \left.+\int_a^s g(r)f(r,u(r),u^{\Delta}(r))\nabla r\right)\Delta s\\
&\quad +\frac{\sum\limits_{j=1}^{m-2}\beta_j}{1-\sum\limits_{j=1}^{m-2}\beta_j}\int_{\xi_j}^b \phi_q\left(\frac{\sum\limits_{j=1}^{m-2}\alpha_j\int_a^{\xi_j} g(r)f(r,u(r),u^{\Delta}(r))\nabla r}{1-\sum\limits_{j=1}^{m-2}\alpha_j}\right.\\
&\quad \left.+\int_a^s g(r)f(r,u(r),u^{\Delta}(r))\nabla r\right)\Delta s\\
&\leq \frac{r_2}{N}\left(\int_a^b \phi_q\left(\frac{\sum\limits_{j=1}^{m-2}\alpha_j\int_a^{\xi_j} g(r)\nabla r}{1-\sum\limits_{j=1}^{m-2}\alpha_j}+\int_a^s g(r)\nabla r\right)\Delta s\right.\\
&\quad \left.+\frac{\sum\limits_{j=1}^{m-2}\beta_j}{1-\sum\limits_{j=1}^{m-2}\beta_j}\int_{\xi_j}^b \phi_q\left(\frac{\sum\limits_{j=1}^{m-2}\alpha_j\int_a^{\xi_j} g(r)\nabla r}{1-\sum\limits_{j=1}^{m-2}\alpha_j}+\int_a^s g(r)\nabla r\right)\Delta s\right)
\end{aligned}$$

$$= r_2$$

and

$$\begin{aligned}
\beta(Au) &= \max_{t\in[a,b]} |(Au)^{\Delta}(t)| \\
&= \phi_q\left(\frac{\sum\limits_{j=1}^{m-2} \alpha_j \int_a^{\xi_j} g(r) f(r,u(r),u^{\Delta}(r))\nabla r}{1-\sum\limits_{j=1}^{m-2} \alpha_j}\right. \\
&\quad \left. + \int_a^b g(r) f(r,u(r),u^{\Delta}(r))\nabla r\right) \\
&\le \frac{l_2}{Q}\phi_q\left(\frac{\sum\limits_{j=1}^{m-2} \alpha_j \int_a^{\xi_j} g(r)\nabla r}{1-\sum\limits_{j=1}^{m-2} \alpha_j} + \int_a^b g(r)\nabla r\right) \\
&= l_2.
\end{aligned}$$

Therefore, $A : \overline{P}(\alpha, r_2; \beta, l_2) \to \overline{P}(\alpha, r_2; \beta, l_2)$. Let now, $u \in \overline{P}(\alpha, r_1; \beta, l_1)$. Then, by $(D36)$, we get

$$f(t,u(t),u^{\Delta}(t)) < \min\left\{\phi_p\left(\frac{r_1}{N}\right), \quad \phi_p\left(\frac{l_1}{Q}\right)\right\}, \quad t \in [a,b].$$

Hence,

$$\begin{aligned}
\alpha(Au) &= \max_{t\in[a,b]} Au(t) \\
&= Au(a) \\
&= \int_a^b \phi_q\left(\frac{\sum\limits_{j=1}^{m-2} \alpha_j \int_a^{\xi_j} g(r) f(r,u(r),u^{\Delta}(r))\nabla r}{1-\sum\limits_{j=1}^{m-2} \alpha_j}\right. \\
&\quad \left. + \int_a^s g(r) f(r,u(r),u^{\Delta}(r))\nabla r\right)\Delta s
\end{aligned}$$

$$+\frac{\sum_{j=1}^{m-2}\beta_j}{1-\sum_{j=1}^{m-2}\beta_j}\int_{\xi_j}^{b}\phi_q\Bigg(\frac{\sum_{j=1}^{m-2}\alpha_j\int_a^{\xi_j}g(r)f(r,u(r),u^{\Delta}(r))\nabla r}{1-\sum_{j=1}^{m-2}\alpha_j}$$

$$+\int_a^s g(r)f(r,u(r),u^{\Delta}(r))\nabla r\Bigg)\Delta s$$

$$\leq \frac{r_1}{N}\Bigg(\int_a^b\phi_q\Bigg(\frac{\sum_{j=1}^{m-2}\alpha_j\int_a^{\xi_j}g(r)\nabla r}{1-\sum_{j=1}^{m-2}\alpha_j}+\int_a^s g(r)\nabla r\Bigg)\Delta s$$

$$+\frac{\sum_{j=1}^{m-2}\beta_j}{1-\sum_{j=1}^{m-2}\beta_j}\int_{\xi_j}^{b}\phi_q\Bigg(\frac{\sum_{j=1}^{m-2}\alpha_j\int_a^{\xi_j}g(r)\nabla r}{1-\sum_{j=1}^{m-2}\alpha_j}+\int_a^s g(r)\nabla r\Bigg)\Delta s\Bigg)$$

$$= r_1$$

and

$$\beta(Au) = \max_{t\in[a,b]}|(Au)^{\Delta}(t)|$$

$$= \phi_q\Bigg(\frac{\sum_{j=1}^{m-2}\alpha_j\int_a^{\xi_j}g(r)f(r,u(r),u^{\Delta}(r))\nabla r}{1-\sum_{j=1}^{m-2}\alpha_j}$$

$$+\int_a^b g(r)f(r,u(r),u^{\Delta}(r))\nabla r\Bigg)$$

$$\leq \frac{l_1}{Q}\phi_q\Bigg(\frac{\sum_{j=1}^{m-2}\alpha_j\int_a^{\xi_j}g(r)\nabla r}{1-\sum_{j=1}^{m-2}\alpha_j}+\int_a^b g(r)\nabla r\Bigg)$$

$$= l_1.$$

Therefore, $A:\overline{P}(\alpha,r_1;\beta,l_1)\to\overline{P}(\alpha,r_1;\beta,l_1)$. Now, we take $u(t)=2c$, $t\in[a,b]$. We have

$$u(t)=2c\in\overline{P}(\alpha,2c;\beta,l_2;\psi,c)$$

and

$$\psi(u) = 2c$$

$$> c.$$

Therefore,

$$\{u \in \overline{P}(\alpha, 2c; \beta, l_2; \psi, c) : \psi(u) > c\} \neq \emptyset.$$

Moreover, for $u \in \overline{P}(\alpha, 2c; \beta, l_2; \psi, c)$, we have

$$c \leq \psi(t) \leq 2c,$$

$$|u^{\Delta}(t)| \leq l_2, \quad t \in [a,b],$$

and then

$$\begin{aligned}
\psi(Au) &= \min_{t\in[\eta,b]} Au(t) \\
&= Au(b) \\
&= \frac{\sum\limits_{j=1}^{m-2} \beta_j}{1 - \sum\limits_{j=1}^{m-2} \beta_j} \int_{\xi_j}^{b} \phi_q\Bigg(\frac{\sum\limits_{j=1}^{m-2} \alpha_j \int_a^{\xi_j} g(r) f(r, u(r), u^{\Delta}(r)) \nabla r}{1 - \sum\limits_{j=1}^{m-2} \alpha_j} \\
&\quad + \int_a^s g(r) f(r, u(r), u^{\Delta}(r)) \nabla r \Bigg) \Delta s \\
&\geq \frac{\sum\limits_{j=1}^{m-2} \beta_j}{1 - \sum\limits_{j=1}^{m-2} \beta_j} \int_{\xi_j}^{b} \phi_q\Bigg(\frac{\sum\limits_{j=1}^{m-2} \alpha_j \int_a^{\xi_j} g(r) f(r, u(r), u^{\Delta}(r)) \nabla r}{1 - \sum\limits_{j=1}^{m-2} \alpha_j} \\
&\quad + \int_a^{\eta} g(r) f(r, u(r), u^{\Delta}(r)) \nabla r \Bigg) \Delta s \\
&\geq \frac{c}{M} \Bigg(\frac{\sum\limits_{j=1}^{m-2} \beta_j}{1 - \sum\limits_{j=1}^{m-2} \beta_j} \int_{\xi_j}^{b} \phi_q\Bigg(\frac{\sum\limits_{j=1}^{m-2} \alpha_j \int_a^{\xi_j} g(r) f(r, u(r), u^{\Delta}(r)) \nabla r}{1 - \sum\limits_{j=1}^{m-2} \alpha_j} \\
&\quad + \int_a^{\eta} g(r) f(r, u(r), u^{\Delta}(r)) \nabla r \Bigg) \Delta s \Bigg) \\
&= c.
\end{aligned}$$

So,

$$\psi(Au) > c$$

for $u \in \overline{P}(\alpha, 2c; \beta, l_2; \psi, c)$. Now, we take $u \in \overline{P}(\alpha, r_2; \beta, l_2; \psi, c)$ and $\alpha(Au) > 2c$. Then

$$\begin{aligned}
\psi(Au) &= \min_{t \in [\eta, b]} Au(t) \\
&= Au(b) \\
&\geq \frac{\eta}{c} \max_{t \in [\eta, b]} Au(t) \\
&> \frac{1}{2}\alpha(Au) \\
&> c.
\end{aligned}$$

Now, applying the Bai-Ge fixed point theorem, Theorem 5.24, we conclude that the operator A has three fixed points u_1, u_2, u_3 so that

$$\begin{aligned}
u_1 &\in P(\alpha, r_1; \beta, l_1), \\
u_2 &\in \left\{\overline{P}(\alpha, r_2; \beta, l_2; \psi, c) : \psi(u) > c\right\}, \\
u_3 &\in \overline{P}(\alpha, r_2; \beta, l_2) \backslash \left(\overline{P}(\alpha, r_2; \beta, l_2; \psi, c) \bigcup \overline{P}(\alpha, r_1; \beta, l_1)\right).
\end{aligned}$$

Moreover,

$$\alpha(u_3) \leq 2\psi(u_3)$$

and

$$\max_{t \in [a,b]} u_3(t) < 2c.$$

This completes the proof.

Exercise 5.17. Let $\mathbb{T} = 4^{\mathbb{N}_0}$. Prove that the BVP

$$\begin{aligned}
u^{\Delta\nabla}(t) + (t^2+1) f(t, u(t), u^{\Delta}(t)) &= 0, \quad t \in [1, 1024], \\
u(1024) &= \frac{1}{3}u(4) + \frac{1}{64}u(16) + \frac{1}{200}u(256), \\
u^{\Delta}(1) &= \frac{1}{96}u^{\Delta}(4) + \frac{1}{160}u^{\Delta}(256),
\end{aligned}$$

where

$$f(t,w,v) = \begin{cases} t + \frac{w^2}{6} + \frac{v^6}{10} & w \leq 1 \\ \frac{t^2}{10} + 10 + \frac{v^2}{10} & w > 1 \end{cases}$$

has at least three positive solutions.

5.7 Advanced Practical Problems

Problem 5.1. Let $\mathbb{T} = 3^{\mathbb{N}_0}$. Prove that the BVP

$$y^{\Delta^2} = (t^2 + t + 2)(y^\sigma)^7,$$

$$y(1) = 0,$$

$$y(243) = 1$$

has at least one solution y for which

$$|y(t)| \leq 1000000, \quad t \in [1, 243].$$

Problem 5.2. Let $\mathbb{T} = 4\mathbb{Z}$. Prove that the BVP

$$y^{\Delta^2} = \frac{2y + 3y^3}{1 + y^2}, \quad t \in [0, 16],$$

$$y(0) = -1,$$

$$y(24) = 3,$$

has at least one solution y satisfying

$$|y(t)| < 4, \quad t \in [0, 24].$$

Problem 5.3. Let $\mathbb{T} = 2\mathbb{Z}$. Prove that any solution y of the BVP

$$y^{\Delta^2} = \frac{y^5}{1 + 3y^2 + 9y^4}, \quad t \in [0, 16],$$

$$y(0) - 2y^\Delta(0) = 0,$$

$$y(20) + 2y^\Delta(20) = 0,$$

satisfies

$$|y(t)| < 2, \quad t \in [0,20].$$

Problem 5.4. Let $\mathbb{T} = 4\mathbb{Z}$. Prove that any solution y of the BVP

$$y^{\Delta^2} = (1+t+t^2)y^{\sigma 5}, \quad t \in [0,16],$$

$$y(0) - 3y^{\Delta}(0) = 5,$$

$$y(24) = 10,$$

satisfies the estimate

$$|y(t)| \leq 8, \quad t \in [0,24].$$

Problem 5.5. Let $\mathbb{T} = 3^{\mathbb{N}_0}$. Prove that any solution y of the BVP

$$y^{\Delta^2} = (1+t^2+t^4)\frac{y^{\sigma 9}}{1+y^{\sigma 9}+y^{\sigma 18}}, \quad t \in [1,81],$$

$$y(1) = 5,$$

$$2y(729) + 8y^{\Delta}(243) = -1,$$

satisfies the estimate

$$|y(t)| \leq 11, \quad t \in [1,729].$$

Problem 5.6. Let $\mathbb{T} = 3^{\mathbb{N}_0}$. Prove that the BVP

$$u^{\Delta\nabla} + \left(\frac{1}{(100+t^8)^3}\right)\frac{(u(t))^4}{20^8+(u(t))^4} = 0, \quad t \in [1,243],$$

$$u^{\Delta}(1) = \frac{1}{48}u^{\Delta}(3) + \frac{1}{48}u^{\Delta}(9) + \frac{1}{16}u^{\Delta}(27) + \frac{1}{100}u^{\Delta}(81),$$

$$u(243) = \frac{1}{240}u(3) + \frac{1}{48}u(9) + \frac{1}{24}u(27) + \frac{1}{96}u(81),$$

has at least one solution.

Problem 5.7. Let $\mathbb{T} = 3^{\mathbb{N}_0}$. Prove that the BVP

$$u^{\Delta\nabla}(t) + \frac{1}{1+t^2+t^4}u^{\Delta}(t) + (1+t^2+t^4+t^6)u(t) + \lambda\left((u(t))^2 + \cos(\pi u(t))\right) = 0,$$

$t \in [3,81]$,

$$u(1) = 0,$$

$$u(243) = \frac{1}{8}u(9) + \frac{1}{4}u(27) + \frac{1}{32}u(81),$$

has at least one positive solution.

Problem 5.8. Let $\mathbb{T} = 2\mathbb{Z}$. Prove that the BVP

$$u^{\Delta\nabla}(t) + t^4\left((u^\Delta(t))^4 + (u(t))^6 + 10^{-8}\right) = 0, \quad t \in (-16, 16),$$

$$7u(-16) - 7\lim_{t\to -16+} u^\Delta(t) = \int_{-16}^{16} \frac{1}{s^8+s^4+1}u(s)\nabla s,$$

$$7u(16) + 7\lim_{t\to 16-} u^\Delta(t) = \int_{-16}^{16} \frac{1}{s^8+s^4+1}u(s)\nabla s$$

has at least one positive solution.

Problem 5.9. Let $\mathbb{T} = 3^{\mathbb{N}_0}$. Prove that the BVP

$$u^{\Delta^2}(t) + f(t, u(t), u^\Delta(t)) = 0, \quad t \in [1, 243],$$

$$u(1) = \frac{1}{3}u(9) + \frac{1}{64}u(27) + \frac{1}{200}u(81),$$

$$u^\Delta(243) = \frac{1}{96}u^\Delta(27) + \frac{1}{160}u^\Delta(81),$$

where

$$f(t, w, v) = \begin{cases} \dfrac{t^7}{100} + \dfrac{w^8}{6} + \dfrac{v^7}{100} & w \le 1 \\[2ex] \dfrac{t^3}{10} + 100 + \dfrac{v^2}{1000} & w > 1 \end{cases}$$

has at least three positive solutions.

Problem 5.10. Let $\mathbb{T} = 3^{\mathbb{N}_0}$. Prove that the BVP

$$u^{\Delta\nabla}(t) + \frac{1}{1+t^2}f(t, u(t), u^\Delta(t)) = 0, \quad t \in [1, 243],$$

$$u(243) = \frac{2}{31}u(9) + \frac{1}{64}u(27) + \frac{1}{2}u(81),$$

$$u^\Delta(1) = \frac{1}{96}u^\Delta(27) + \frac{1}{8}u^\Delta(81),$$

where

$$f(t,w,v) = \begin{cases} \frac{t^4}{8} + \frac{w^2}{6} + \frac{v^4}{10} & w \leq 1 \\ \\ \frac{t^2}{10} + 10 + \frac{v^9}{100} & w > 1 \end{cases}$$

has at least three positive solutions.

5.8 Notes and References

This chapter investigates some classes of second order boundary value problems on time scales. There are deducted results for existence of one, two and three positive solutions for the considered boundary value problems. Some of the results in this chapter can be found in [3], [17], [18], [25], [29], [30], [41], [42], [44], [45], [48], [49], [55], [56], [58], [61] and [62].

Chapter 6
Nonlinear Second Order Eigenvalue Problems

Let $\mathbb{T}$ be a time scale with delta differentiation operator, nabla differentiation operator, forward jump operator and backward jump operator Δ, ∇, σ and ρ, respectively. Let $a, b \in \mathbb{T}$, $a < b$.

6.1 Solvability of Nonlinear Second Order Conjugate Eigenvalue Problems

In this section we will investigate the following nonlinear second order eigenvalue problem

$$\begin{aligned} y^{\Delta^2} + \lambda l(t) f(y(\sigma(t))) &= 0, \quad t \in [a,b], \\ y(a) = y(\sigma(b)) &= 0, \end{aligned} \tag{6.1}$$

where

(E1) $\lambda \in \mathbb{R}$, $l \in \mathscr{C}([a, \sigma(b)])$ is a non-negative function such that

$$0 < \int_a^{\sigma(b)} h(s) l(s) \Delta s < \infty,$$

where

$$h(t) = \min\{\sigma(t) - a, \quad \sigma(b) - \sigma(t)\}, \quad t \in [a,b].$$

(E2) $f \in \mathscr{C}([0,\infty))$ is a non-negative function such that

$$f_0 = \lim_{x \to 0+} \frac{f(x)}{x},$$

$$f_\infty = \lim_{x \to \infty} \frac{f(x)}{x}$$

both exist.

DOI: 10.1201/9781003173557-6

(E3) Let $\delta, w \in (a, \sigma(b))$ and $[\delta, w] \subset [a, \sigma(b)]$ be chosen so that

$$\int_{\delta}^{\sigma(w)} l(s)h(s)\Delta s > 0$$

and $\tau \in [\delta, w]$ is such that

$$g(\tau) = \max_{t \in [\delta, \sigma(w)]} g(t),$$

where

$$g(t) = \min\{t - a, \sigma(b) - t\}, \quad t \in [a, b].$$

By Chapter 3, it follows that the function

$$G(t,s) = \begin{cases} \dfrac{(t-a)(\sigma(b)-\sigma(s))}{\sigma(b)-a} & \text{if} \quad a \leq t \leq s \\[2ex] \dfrac{(\sigma(s)-a)(\sigma(b)-t)}{\sigma(b)-a} & \text{if} \quad \sigma(s) \leq t \leq \sigma(b) \end{cases}$$

is the Green function for the BVP

$$-y^{\Delta^2} = 0, \quad t \in [a, b],$$

$$y(a) = y(\sigma(b)) = 0.$$

Lemma 6.1. *We have*

$$\frac{1}{\sigma(b)-a} g(t)h(s) \leq G(t,s)$$

$$\leq \frac{(\sigma(s)-a)(\sigma(b)-\sigma(s))}{\sigma(b)-a}, \quad (t,s) \in [a, \sigma(b)] \times [a, b].$$

Proof. Fix $s \in [a, b]$.

1. Let $t \in [a, s]$. Then

$$G(t,s) = \frac{(t-a)(\sigma(b)-\sigma(s))}{\sigma(b)-\sigma(a)}$$

$$\leq \frac{(\sigma(s)-a)(\sigma(b)-\sigma(s))}{\sigma(b)-\sigma(a)}$$

and

$$G(t,s) = \frac{(t-a)(\sigma(b)-\sigma(s))}{\sigma(b)-\sigma(a)}$$

$$\geq \frac{1}{\sigma(b)-\sigma(a)} g(t)h(s).$$

2. Let $t \in (s, \sigma(b)]$. Then

$$G(t,s) = \frac{(\sigma(s)-a)(\sigma(b)-t)}{\sigma(b)-a}$$
$$\leq \frac{(\sigma(s)-a)(\sigma(b)-\sigma(s))}{\sigma(b)-a}$$

and

$$G(t,s) = \frac{(\sigma(s)-a)(\sigma(b)-t)}{\sigma(b)-a}$$
$$\geq \frac{1}{\sigma(b)-a} h(s)g(t).$$

This completes the proof.

Let $E = \mathscr{C}([a, \sigma^2(b)])$ be endowed with the maximum norm

$$\|x\| = \max_{t \in [a, \sigma^2(b)]} |x(t)|.$$

Define

$$P = \{x \in E : x(t) \geq \frac{1}{\sigma(b)-a} g(t)\|x\|, \quad t \in [a,b]\}.$$

For $y \in E$, define the operator

$$T_\lambda y(t) = \lambda \int_a^{\sigma(b)} G(t,s)l(s)f(y(\sigma(s)))\Delta s, \quad t \in [a, \sigma(b)].$$

Note that any fixed point of the operator T_λ is a solution to problem (6.1).

Lemma 6.2. *Let $l \in \mathscr{C}([a, \sigma(b)]$, $f \in \mathscr{C}([0, \infty))$ be non-negative functions. Then $T_\lambda : P \to P$.*

Proof. Take $y \in P$ arbitrarily. Then, using Lemma 6.1, we get

$$T_\lambda y(t) \leq \frac{\lambda}{\sigma(b)-a} \int_a^{\sigma(b)} (\sigma(s)-a)(\sigma(b)-\sigma(s))l(s)f(y(\sigma(s)))\Delta s, \quad t \in [a, \sigma(b)],$$

and

$$\|T_\lambda y\| \leq \frac{\lambda}{\sigma(b)-a} \int_a^{\sigma(b)} (\sigma(s)-a)(\sigma(b)-\sigma(s))l(s)f(y(\sigma(s)))\Delta s.$$

If

$$h(s) = \sigma(s) - a, \quad s \in [a,b],$$

using that

$$\sigma(b) - \sigma(s) \le \sigma(b) - s$$

$$\le \sigma(b) - a, \quad s \in [a,b],$$

we have

$$h(s) \ge \frac{(\sigma(s) - a)(\sigma(b) - \sigma(s))}{\sigma(b) - a}, \quad s \in [a,b].$$

If

$$h(s) = \sigma(b) - \sigma(s), \quad s \in [a,b],$$

using that

$$\sigma(s) - a \le \sigma(b) - a, \quad s \in [a,b],$$

we get

$$h(s) \ge \frac{(\sigma(s) - a)(\sigma(b) - \sigma(s))}{\sigma(b) - a}, \quad s \in [a,b].$$

Hence,

$$\begin{aligned} T_\lambda y(t) &\ge \frac{\lambda}{\sigma(b) - a} g(t) \int_a^{\sigma(b)} h(s) l(s) f(y(\sigma(s))) \Delta s \\ &\ge \frac{\lambda}{\sigma(b) - a} g(t) \int_a^{\sigma(b)} \frac{(\sigma(s) - a)(\sigma(b) - \sigma(s))}{\sigma(b) - a} l(s) f(y(\sigma(s))) \Delta s \\ &\ge \frac{1}{\sigma(b) - a} g(t) \|T_\lambda y\|, \quad t \in [a, \sigma(b)]. \end{aligned}$$

Therefore, $T_\lambda y \in P$. This completes the proof.

Lemma 6.3. *Let $l \in \mathscr{C}([a,\sigma(b)])$ and $f \in \mathscr{C}([0,\infty))$ be non-negative functions. Then $T_\lambda : P \to P$ is a completely continuous operator.*

Proof. Let $X \subset P$ be a bounded set. Then there exists a constant $L > 0$ so that

$$0 \le y(t) \le L, \quad t \in [a, \sigma(b)].$$

Since $l \in \mathscr{C}([a,\sigma(b)])$ and $f \in \mathscr{C}([0,\infty))$, there exists a constant $M > 0$ such that

$$l(t) \le M, \quad t \in [a, \sigma(b)],$$

$$f(u) \le M, \quad u \in [0, L].$$

Take $y \in X$ arbitrarily. Then

$$\begin{aligned} T_\lambda y(t) &= \lambda \int_a^{\sigma(b)} G(t,s)l(s)f(y(\sigma(s)))\Delta s \\ &\leq \lambda(\sigma(b)-a)^2 M^2, \quad t \in [a,\sigma(b)]. \end{aligned}$$

Therefore, $T_\lambda(X)$ is uniformly bounded. Let $t_1, t_2 \in [a,\sigma(b)]$, $t_1 < t_2$. Hence,

$$\begin{aligned} |T_\lambda y(t_2) - T_\lambda y(t_1)| &= \left| \lambda \int_a^{\sigma(b)} G(t_2,s)l(s)f(y(\sigma(s)))\Delta s \right. \\ &\quad \left. -\lambda \int_a^{\sigma(b)} G(t_1,s)l(s)f(y(\sigma(s)))\Delta s \right| \\ &= \left| \lambda \int_a^{t_2} \frac{(\sigma(s)-a)(\sigma(b)-t_2)}{\sigma(b)-a} l(s)f(y(\sigma(s)))\Delta s \right. \\ &\quad +\lambda \int_{t_2}^{\sigma(b)} \frac{(t_2-a)(\sigma(b)-\sigma(s))}{\sigma(b)-a} l(s)f(y(\sigma(s)))\Delta s \\ &\quad -\lambda \int_a^{t_1} \frac{(\sigma(s)-a)(\sigma(b)-t_1)}{\sigma(b)-a} l(s)f(y(\sigma(s)))\Delta s \\ &\quad \left. -\lambda \int_{t_1}^{\sigma(b)} \frac{(t_1-a)(\sigma(b)-\sigma(s))}{\sigma(b)-a} l(s)f(y(\sigma(s)))\Delta s \right| \\ &= \left| \lambda \int_a^{t_1} \frac{(\sigma(s)-a)(t_1-t_2)}{\sigma(b)-a} l(s)f(y(\sigma(s)))\Delta s \right. \\ &\quad +\lambda \int_{t_1}^{t_2} \frac{(\sigma(s)-a)(\sigma(b)-t_2)}{\sigma(b)-a} l(s)f(y(\sigma(s)))\Delta s \\ &\quad +\lambda \int_{t_2}^{\sigma(b)} \frac{(t_2-t_1)(\sigma(b)-\sigma(s))}{\sigma(b)-a} l(s)f(y(\sigma(s)))\Delta s \\ &\quad \left. -\lambda \int_{t_1}^{t_2} \frac{(t_1-a)(\sigma(b)-\sigma(s))}{\sigma(b)-a} l(s)f(y(\sigma(s)))\Delta s \right| \\ &\leq \left| \lambda \int_a^{t_1} \frac{(\sigma(s)-a)(t_1-t_2)}{\sigma(b)-a} l(s)f(y(\sigma(s)))\Delta s \right| \\ &\quad + \left| \lambda \int_{t_1}^{t_2} \frac{(\sigma(s)-a)(\sigma(b)-t_2)}{\sigma(b)-a} l(s)f(y(\sigma(s)))\Delta s \right| \\ &\quad + \left| \lambda \int_{t_2}^{\sigma(b)} \frac{(t_2-t_1)(\sigma(b)-\sigma(s))}{\sigma(b)-a} l(s)f(y(\sigma(s)))\Delta s \right| \\ &\quad + \left| \lambda \int_{t_1}^{t_2} \frac{(t_1-a)(\sigma(b)-\sigma(s))}{\sigma(b)-a} l(s)f(y(\sigma(s)))\Delta s \right| \end{aligned}$$

$$\leq \lambda \int_a^{t_1} \frac{(\sigma(s)-a)(t_2-t_1)}{\sigma(b)-a} l(s) f(y(\sigma(s)))\Delta s$$

$$+\lambda \int_{t_1}^{t_2} \frac{(\sigma(s)-a)(\sigma(b)-t_2)}{\sigma(b)-a} l(s) f(y(\sigma(s)))\Delta s$$

$$+\lambda \int_{t_2}^{\sigma(b)} \frac{(t_2-t_1)(\sigma(b)-\sigma(s))}{\sigma(b)-a} l(s) f(y(\sigma(s)))\Delta s$$

$$+\lambda \int_{t_1}^{t_2} \frac{(t_1-a)(\sigma(b)-\sigma(s))}{\sigma(b)-a} l(s) f(y(\sigma(s)))\Delta s$$

$$\leq \lambda(\sigma(b)-a)(t_2-t_1)M^2 + \lambda(\sigma(b)-a)(t_2-t_1)M^2$$

$$+\lambda(\sigma(b)-a)(t_2-t_1)M^2 + \lambda(\sigma(b)-a)(t_2-t_1)M^2$$

$$= 4\lambda(\sigma(b)-a)(t_2-t_1)M^2.$$

Consequently, $T_\lambda(X)$ is equi-continuous. Hence and by the Arzela-Ascoli theorem, it follows that $T_\lambda : P \to P$ is a completely continuous operator. This completes the proof.

Theorem 6.1. *Suppose* $(E1)$-$(E3)$ *and* $\sigma(b)$ *is right-dense. Then, for each* $\lambda > 0$ *satisfying*

$$\frac{(\sigma(b)-a)^2}{(\delta-a)g(\tau)f_\infty \int_\delta^w h(s)l(s)\Delta s} < \lambda$$

$$< \frac{\sigma(b)-a}{f_0 \int_a^{\sigma(b)} (\sigma(s)-a)(\sigma(b)-\sigma(s))l(s)\Delta s},$$

there exists at least one solution of problem (6.1).

Proof. Take $\varepsilon > 0$ so that

$$\frac{(\sigma(b)-a)^2}{(\delta-a)g(\tau)(f_\infty-\varepsilon) \int_\delta^w h(s)l(s)\Delta s} \leq \lambda$$

$$\leq \frac{\sigma(b)-a}{(f_0+\varepsilon) \int_a^{\sigma(b)} (\sigma(s)-a)(\sigma(b)-\sigma(s))l(s)\Delta s}.$$

By the definition of f_0, it follows that there exists a $H_1 > 0$ so that

$$f(x) \leq (f_0+\varepsilon)x, \quad x \in (0, H_1].$$

Define

$$\Omega_1 = \{x \in P : \|x\| < H_1\}.$$

Then, for $y \in P\bigcap \partial\Omega_1$, we have

$$\begin{aligned}
T_\lambda y(t) &= \lambda \int_a^{\sigma(b)} G(t,s)l(s)f(y(\sigma(s)))\Delta s \\
&\leq \frac{\lambda}{\sigma(b)-a}\int_a^{\sigma(b)} (\sigma(s)-a)(\sigma(b)-\sigma(s))l(s)f(y(\sigma(s)))\Delta s \\
&\leq \frac{\lambda_0(f_0+\varepsilon)}{\sigma(b)-a}\int_a^{\sigma(b)} (\sigma(s)-a)(\sigma(b)-\sigma(s))l(s)y(\sigma(s))\Delta s \\
&\leq \frac{\lambda(f_0+\varepsilon)}{\sigma(b)-a}\left(\int_a^{\sigma(b)} (\sigma(s)-a)(\sigma(b)-\sigma(s))l(s)\Delta s\right)\|y\| \\
&\leq \|y\|, \quad t \in [a,\sigma(b)].
\end{aligned}$$

Hence,

$$\|T_\lambda y\| \leq \|y\|, \quad y \in P\bigcap \partial\Omega_1.$$

Now, by the definition of f_∞, it follows that there exists an $H_2 > H_1$ such that

$$f(x) \geq (f_\infty - \varepsilon)x, \quad x \geq H_2.$$

Define

$$\Omega_2 = \{x \in P : \|x\| < H_2\}.$$

Take $y \in P\bigcap \partial\Omega_2$ arbitrarily. Then

$$\begin{aligned}
y(t) &\geq \frac{1}{\sigma(b)-a}g(t)\|y\| \\
&= \frac{1}{\sigma(b)-a}g(t)H_2, \quad t \in [a,\sigma(b)].
\end{aligned}$$

If $t \in [\delta, w]$ and

$$g(t) = t - a,$$

then

$$g(\tau) \geq \delta - a.$$

If $t \in [\delta, w]$ and

$$g(t) = \sigma(b) - t,$$

then

$$g(\tau) \geq \sigma(b) - a$$

$$\geq \delta - a.$$

Consequently,

$$y(t) \geq \frac{\delta - a}{\sigma(b) - a} H_2, \quad t \in [\delta, w].$$

From here,

$$\begin{aligned}
\|T_\lambda y\| &\geq T_\lambda y(\tau) \\
&= \lambda \int_a^{\sigma(b)} G(\tau, s) l(s) f(y(\sigma(s))) \Delta s \\
&\geq \lambda \int_\delta^w G(\tau, s) l(s) f(y(\sigma(s))) \Delta s \\
&\geq \frac{\lambda}{\sigma(b) - a} \int_\delta^w g(\tau) h(s) l(s) f(y(\sigma(s))) \Delta s \\
&\geq \frac{\lambda (f_\infty - \varepsilon)}{\sigma(b) - a} \int_\delta^w g(\tau) h(s) l(s) y(\sigma(s)) \Delta s \\
&\geq \frac{\lambda (f_\infty - \varepsilon)(\delta - a) H_2}{(\sigma(b) - a)^2} \int_\delta^w g(\tau) h(s) l(s) \Delta s \\
&\geq H_2 \\
&= \|y\|,
\end{aligned}$$

whereupon,

$$\|T_\lambda y\| \geq \|y\|, \quad y \in P \bigcap \partial \Omega_2.$$

Then, applying the Guo-Krasnosel'skii Theorem 1.11, it follows that the operator T_λ has a fixed point in P. This completes the proof.

Example 6.1. Let $\mathbb{T} = 2^{\mathbb{N}_0}$. Consider the eigenvalue problem

$$y^{\Delta^2} + \lambda (t+1) \frac{y(2t)\left(y(2t) + \frac{1}{336}\right)}{y(2t) + 2} = 0, \quad t \in [1, 4],$$

$$y(1) = y(8) = 0.$$

Here

$$\sigma(t) = 2t,$$

$$l(t) = t + 1, \quad t \in \mathbb{T},$$

$$f(u) = \frac{u\left(u+\frac{1}{336}\right)}{u+2}, \quad u \in \mathbb{R}.$$

Then

$$\begin{aligned}
f_0 &= \lim_{u\to 0+} \frac{f(u)}{u} \\
&= \lim_{u\to 0+} \frac{u+\frac{1}{336}}{u+2} \\
&= \frac{1}{672}, \\
f_\infty &= \lim_{u\to\infty} \frac{f(u)}{u} \\
&= \lim_{u\to\infty} \frac{u+\frac{1}{336}}{u+2} \\
&= 1, \\
h(1) &= \min\{\sigma(1)-1, \sigma(4)-\sigma(1)\} \\
&= \min\{2-1, 8-2\} \\
&= \min\{1, 6\} \\
&= 1, \\
h(2) &= \min\{\sigma(2)-1, \sigma(4)-\sigma(2)\} \\
&= \min\{4-1, 8-4\} \\
&= \min\{3, 4\} \\
&= 3, \\
h(4) &= \min\{\sigma(4)-1, \sigma(4)-\sigma(4)\}
\end{aligned}$$

$$= 0.$$

Then

$$\begin{aligned}\int_a^{\sigma(b)} h(s)l(s)\Delta s &= \int_1^8 h(s)l(s)\Delta s\\&= \mu(1)h(1)l(1)+\mu(2)h(2)l(2)+\mu(4)h(4)l(4)\\&= l(1)+6l(2)\\&= 2+18\\&= 20\\&> 0.\end{aligned}$$

Therefore, $(E1)$ and $(E2)$ hold. Take $\delta = 2$ and $w = 4$. Then $[\delta, w] \subset [a, \sigma(b)]$ and

$$\begin{aligned}\int_\delta^{\sigma(w)} l(s)h(s)\Delta s &= \int_2^8 l(s)h(s)\Delta s\\&= \mu(2)l(2)h(2)+\mu(4)l(4)h(4)\\&= 27\\&> 0,\end{aligned}$$

$$\begin{aligned}\int_\delta^{w} l(s)h(s)\Delta s &= \int_2^4 l(s)h(s)\Delta s\\&= \mu(2)l(2)h(2)\\&= 27\\&> 0,\end{aligned}$$

i.e., $(E3)$ holds. Moreover,

$$g(1) = \min\{1-1, 8-1\}$$

$$= 0,$$

$$\begin{aligned} g(2) &= \min\{2-1, 8-2\} \\ &= \min\{1, 6\} \\ &= 1, \\ g(4) &= \min\{4-1, 8-4\} \\ &= \min\{3, 4\} \\ &= 3. \end{aligned}$$

Then

$$\begin{aligned} g(\tau) &= \max_{t\in[1,4]} g(t) \\ &= \max\{g(1), g(2), g(4)\} \\ &= 3 \end{aligned}$$

and

$$\begin{aligned} \int_\delta^w h(s)l(s)\Delta s &= \int_1^2 h(s)l(s)\Delta s \\ &= \mu(1)h(1)l(1) \\ &= 2, \end{aligned}$$

and

$$\begin{aligned} \int_a^{\sigma(b)} (\sigma(s)-a)(\sigma(b)-\sigma(s))l(s)\Delta s &= \int_1^8 (2s-1)(8-2s)(s+1)\Delta s \\ &= \mu(1)(2-1)(8-2)(1+1) \\ &\quad +\mu(2)(4-1)(8-4)(2+1) \\ &\quad +\mu(4)(8-1)(8-8)(4+1) \end{aligned}$$

$$= 12 + 108$$

$$= 120,$$

and

$$\frac{(\sigma(b)-a)^2}{(\delta-a)g(\tau)f_\infty\int_\delta^w h(s)l(s)\Delta s} = \frac{49}{1\cdot 3\cdot 2} = \frac{49}{6},$$

$$\frac{\sigma(b)-a}{f_0\int_a^{\sigma(b)}(\sigma(s)-a)(\sigma(b)-\sigma(s))l(s)\Delta s} = \frac{7}{\frac{1}{672}\cdot 120} = \frac{196}{5}.$$

Now, by Theorem 6.1, it follows that if

$$\frac{49}{6} < \lambda < \frac{196}{5}$$

the considered eigenvalue problem has at least one positive solution.

Exercise 6.1. Let $\mathbb{T} = 3^{\mathbb{N}_0}$. Prove that the eigenvalue problem

$$y^{\Delta^2} + \lambda(t^2+3t+1)\frac{y(3t)\left((y(3t))^2+y(3t)+\frac{1}{10000}\right)}{(y(3t))^2+2} = 0, \quad t \in [1,27],$$

$$y(1) = y(81) = 0$$

has at least one positive solution.

Theorem 6.2. *Suppose* $(E1)$*-*$(E3)$ *and* $\sigma(b)$ *is right-dense. Then, for each* $\lambda > 0$ *satisfying*

$$\frac{(\sigma(b)-a)^2}{(\delta-a)g(\tau)f_0\int_\delta^w h(s)l(s)\Delta s} < \lambda < \frac{\sigma(b)-a}{f_\infty\int_a^{\sigma(b)}(\sigma(s)-a)(\sigma(b)-\sigma(s))l(s)\Delta s},$$

there exists at least one solution of problem (6.1).

Proof. Take $\eta > 0$ so that

$$\frac{(\sigma(b)-a)^2}{(\delta-a)g(\tau)(f_0-\eta)\int_\delta^w h(s)l(s)\Delta s}<\lambda$$

$$<\frac{\sigma(b)-a}{(f_\infty+\eta)\int_a^{\sigma(b)}(\sigma(s)-a)(\sigma(b)-\sigma(s))l(s)\Delta s}.$$

By the definition of f_0, it follows that there exists an $H_1>0$ so that

$$f(x)\geq(f_0-\eta)x,\quad 0\leq x\leq H_1.$$

Let

$$\Omega_1=\{x\in P:\|x\|<H_1\}.$$

Take $y\in P\bigcap\partial\Omega_1$. Then, for $t\in[\delta,w]$, we have

$$\begin{aligned}y(t)&\geq\frac{1}{\sigma(b)-a}g(t)\|y\|\\&\geq\frac{\delta-a}{\sigma(b)-a}\|y\|\\&=\frac{\delta-a}{\sigma(b)-a}H_1\end{aligned}$$

and

$$\begin{aligned}\|Ty\|&\geq Ty(\tau)\\&=\lambda\int_a^{\sigma(b)}G(\tau,s)l(s)f(y(\sigma(s)))\Delta s\\&\geq\lambda\int_\delta^w G(\tau,s)l(s)f(y(\sigma(s)))\Delta s\\&\geq\frac{\lambda}{\sigma(b)-a}\int_\delta^w h(s)l(s)f(y(\sigma(s)))\Delta s\\&\geq\frac{\lambda(f_0-\eta)}{\sigma(b)-a}\int_\delta^w h(s)l(s)y(\sigma(s))\Delta s\\&\geq\frac{\lambda(f_0-\eta)(\delta-a)H_1}{(\sigma(b)-a)^2}\int_\delta^w g(\tau)h(s)l(s)\Delta s\\&\geq H_1\\&=\|y\|,\end{aligned}$$

whereupon,

$$\|Ty\| \geq \|y\|, \quad y \in P \bigcap \partial \Omega_1.$$

By the definition of f_∞, it follows that there exists an $\widetilde{H}_2 > H_1$ so that

$$f(x) \leq (f_\infty + \eta)x, \quad x \geq \widetilde{H}_2.$$

1. Let f be bounded. Then there exists a constant $N > 0$ so that

$$f(x) \leq N, \quad x \in [0, \infty).$$

Take

$$H_2 = \max\left\{2\widetilde{H}_2, \frac{N}{f_\infty + \eta}\right\}.$$

Let

$$\Omega_2 = \{x \in P : \|x\| < H_2\}.$$

If $y \in P \bigcap \partial \Omega_2$, then

$$\begin{aligned}
T_\lambda y(t) &= \lambda \int_a^{\sigma(b)} G(t,s) l(s) f(y(\sigma(s))) \Delta s \\
&\leq N\lambda \int_a^{\sigma(b)} \frac{(\sigma(s) - a)(\sigma(b) - \sigma(s))}{\sigma(b) - a} l(s) \Delta s \\
&\leq \frac{N}{f_\infty + \eta} \\
&\leq H_2 \\
&= \|y\|, \quad t \in [a, \sigma(b)].
\end{aligned}$$

Therefore,

$$\|T_\lambda y\| \leq \|y\|, \quad y \in P \bigcap \partial \Omega_2.$$

2. Let f be unbounded. Then there exists a $H_2 > \max\{2H_1, \widetilde{H}_2\}$, and

$$f(x) \leq f(H_2), \quad x \in [0, H_2].$$

Hence,

$$\begin{aligned}
f(x) &\leq f(H_2) \\
&\leq (f_\infty + \eta) H_2, \quad x \in [0, H_2],
\end{aligned}$$

and

$$
\begin{aligned}
T_\lambda y(t) &= \lambda \int_a^{\sigma(b)} G(t,s)l(s)f(y(\sigma(s)))\Delta s \\
&\le \lambda(f_\infty+\eta)H_2 \int_a^{\sigma(b)} \frac{(\sigma(s)-a)(\sigma(b)-\sigma(s))}{\sigma(b)-a} l(s)\Delta s \\
&\le H_2 \\
&= \|y\|, \quad t\in[a,\sigma(b)].
\end{aligned}
$$

Consequently,

$$
\|T_\lambda y\| \le \|y\|, \quad y\in P\bigcap \partial\Omega_2.
$$

Then, applying the Guo-Krasnosel'skii Theorem 1.11, it follows that the operator T_λ has a fixed point in P. This completes the proof.

Exercise 6.2. Let $\mathbb{T}=\mathbb{Z}$. Prove that the eigenvalue problem

$$
y^{\Delta^2}+\lambda \frac{3t+1}{4t^2+t+2}\frac{y(t+1)\left(200(y(t+1))^4+(y(t+1))^2+1\right)}{(y(t+1))^4+2}=0, \quad t\in[0,10],
$$

$$
y(0)=y(11)=0
$$

has at least one positive solution.

Before our investigations of the eigenvalue problem (6.1), we will need the following estimate of the Green function G.

Lemma 6.4. *We have*

$$
G(t,s)\le G(\sigma(s),s), \quad t\in[a,\sigma(b)], \quad s\in[a,b],
$$

and

$$
G(t,s)\ge \frac{1}{4}G(\sigma(s),s), \quad t\in\left[\frac{\sigma(b)+3a}{4},\frac{3\sigma(b)+a}{4}\right], \quad s\in[a,b].
$$

Proof. We have

$$
G(\sigma(s),s)=\frac{(\sigma(s)-a)(\sigma(b)-\sigma(s))}{\sigma(b)-a}, \quad s\in[a,b],
$$

and

$$
\frac{G(t,s)}{G(\sigma(s),s)} = \begin{cases} \dfrac{t-a}{\sigma(s)-a} & a\le t\le s \\[2ex] \dfrac{\sigma(b)-t}{\sigma(b)-\sigma(s)} & \sigma(s)\le t\le \sigma(b) \end{cases}
$$

$$\leq 1, \quad t \in [a, \sigma(b)], \quad s \in [a, b].$$

For the second estimate of the Green function we will consider the following cases.

1. Let $t \in [a, s]$, $s \in [a, b]$. Then

$$\frac{G(t,s)}{G(\sigma(s),s)} = \frac{t-a}{\sigma(s)-a}$$
$$\geq \frac{t-a}{\sigma(b)-a}$$

and

$$\frac{t-a}{\sigma(b)-a} \geq \frac{1}{4} \iff$$
$$4t - 4a \geq \sigma(b) - a \iff$$
$$4t \geq 3a + \sigma(b) \iff$$
$$t \geq \frac{3a+\sigma(b)}{4}.$$

2. Let $t \in [\sigma(s), \sigma(b)]$. Then

$$\frac{G(t,s)}{G(\sigma(s),s)} = \frac{\sigma(b)-t}{\sigma(b)-\sigma(s)}$$
$$\geq \frac{\sigma(b)-t}{\sigma(b)-a}$$

and

$$\frac{\sigma(b)-t}{\sigma(b)-a} \geq \frac{1}{4} \iff$$
$$4\sigma(b) - 4t \geq \sigma(b) - a \iff$$
$$3\sigma(b) + a \geq 4t \iff$$
$$t \leq \frac{3\sigma(b)+a}{4}.$$

This completes the proof.

Now, assume

(E4) $\lambda \geq 0$, $f \in \mathscr{C}([0,\infty))$ is a non-decreasing function.
(E5) $l \in \mathscr{C}([a, \sigma(b)])$ and $l \not\equiv 0$ on $[a, \sigma(b)]$.
(E6) $f(0) \geq c > 0$.
(E7) $\lim_{u \to \infty} \frac{f(u)}{u} = \infty$.

(E8)

$$\delta_1 = \min\left\{t \in [a,\sigma(b)] : t \geq \frac{\sigma(b)+3a}{4}\right\},$$

$$w_1 = \max\left\{t \in [a,\sigma(b)] : t \leq \frac{3\sigma(b)+a}{4}\right\}$$

both exist and satisfy

$$\delta_1 < w_1$$

and if $\sigma(w_1) = b$, then $\sigma(w_1) < \sigma(b)$.

Let

$$\begin{aligned} h_1 &= \min_{s\in[a,\sigma(b)]} \frac{G(\sigma(w_1),s)}{G(\sigma(s),s)} \\ &= \min_{s\in[a,\sigma(b)]} \begin{cases} \dfrac{\sigma(w_1)-a}{\sigma(s)-a} & a \leq t \leq s \\ \\ \dfrac{\sigma(b)-\sigma(w_1)}{\sigma(b)-\sigma(s)} & \sigma(s) \leq t \leq \sigma(b) \end{cases} \end{aligned}$$

and

$$k = \min\left\{\frac{1}{4}, h_1\right\}.$$

Define

$$P_1 = \{u \in E : u(t) \geq 0, \quad t \in [a,\sigma^2(b)],$$

$$u(t) \geq k\|u\|, \quad t \in [\delta_1, w_1]\}.$$

Lemma 6.5. *We have $T_\lambda : P_1 \to P_1$.*

Proof. Let $y \in P_1$. Then

$$T_\lambda y(t) \geq 0, \quad t \in [a,\sigma(b)],$$

and

$$\begin{aligned} T_\lambda y(t) &= \lambda \int_a^{\sigma(b)} G(t,s)l(s)f(y(\sigma(s)))\Delta s \\ &\leq \frac{\lambda}{\sigma(b)-a}\int_a^{\sigma(b)} (\sigma(s)-a)(\sigma(b)-\sigma(s))l(s)f(y(\sigma(s)))\Delta s, \quad t \in [a,\sigma(b)]. \end{aligned}$$

Hence,

$$\|T_\lambda y\| \leq \frac{\lambda}{\sigma(b)-a}\int_a^{\sigma(b)} (\sigma(s)-a)(\sigma(b)-\sigma(s))l(s)f(y(\sigma(s)))\Delta s.$$

Moreover,

$$\begin{aligned}
T_\lambda y(t) &= \lambda \int_a^{\sigma(b)} G(t,s)l(s)f(y(\sigma(s)))\Delta s \\
&\geq \frac{1}{4}\lambda \int_a^{\sigma(b)} G(\sigma(s),s)l(s)f(y(\sigma(s)))\Delta s \\
&= \frac{1}{4}\frac{\lambda}{\sigma(b)-a}\int_a^{\sigma(b)} (\sigma(s)-a)(\sigma(b)-\sigma(s))l(s)f(y(\sigma(s)))\Delta s \\
&\geq \frac{1}{4}\|T_\lambda y\|, \quad t \in [\delta_1, \sigma(w_1)].
\end{aligned}$$

Consequently, $T_\lambda y \in P_1$ and $T_\lambda : P_1 \to P_1$. This completes the proof.

Theorem 6.3. *Suppose* $(E4)$-$(E6)$ *and* $(E8)$. *Then there exists a* $\lambda^* > 0$ *such that the operator* T_{λ^*} *has a fixed point in* $P_1 \backslash \{0\}$.

Proof. Let

$$e(t) = \int_a^{\sigma(b)} G(t,s)l(s)\Delta s, \quad t \in [a, \sigma(b)].$$

Using Lemma 6.4, we have

$$e(t) \leq \int_a^{\sigma(b)} G(\sigma(s),s)l(s)\Delta s, \quad t \in [a,\sigma(b)].$$

Then

$$\|e\| \leq \int_a^{\sigma(b)} G(\sigma(s),s)l(s)\Delta s.$$

Also,

$$\begin{aligned}
l(t) &\geq \frac{1}{4}\int_a^{\sigma(b)} G(\sigma(s),s)l(s)\Delta s \\
&\geq \frac{1}{4}\|e\|, \quad t \in [\delta_1, \sigma(w_1)].
\end{aligned}$$

Consequently, $e \in P_1$. Let

$$M = \max_{s\in[a,\sigma(b)]} f(e(\sigma(s))).$$

By $(E4)$ and $(E6)$, it follows that

$$M \geq f(0)$$

$$\geq c$$

$$> 0.$$

Let

$$\lambda^* = \frac{1}{M}.$$

Set

$$u_0(t) = e(t), \quad t \in [a, \sigma(b)].$$

Hence,

$$\begin{aligned}
u_0(t) &= e(t) \\
&= \int_a^{\sigma(b)} G(t,s)l(s)\Delta s \\
&= \lambda^* M \int_a^{\sigma(b)} G(t,s)l(s)\Delta s \\
&= \lambda^* \int_a^{\sigma(b)} G(t,s)l(s)M\Delta s \\
&\geq \lambda^* \int_a^{\sigma(b)} G(t,s)l(s)f(e(\sigma(s)))\Delta s \\
&= T_{\lambda^*}e(t) \\
&= T_{\lambda^*}u_0(t), \quad t \in [a, \sigma(b)].
\end{aligned}$$

Define

$$u_n(t) = T_{\lambda^*}u_{n-1}(t), \quad t \in [a, \sigma(b)], \quad n \in \mathbb{N}.$$

Now, using that f is a non-decreasing function, we get

$$\begin{aligned}
u_1(t) &= T_{\lambda^*}u_0(t) \\
&\leq u_0(t), \\
u_2(t) &= T_{\lambda^*}u_1(t) \\
&= \lambda^* \int_a^{\sigma(b)} G(t,s)l(s)f(u_1(\sigma(s)))\Delta s
\end{aligned}$$

$$\leq \lambda^* \int_a^{\sigma(b)} G(t,s)l(s)f(u_0(\sigma(s)))\Delta s$$

$$= u_1(t)$$

$$\leq u_0(t), \quad t \in [a, \sigma(b)].$$

Assume that

$$u_n(t) \leq u_{n-1}(t), \quad t \in [a, \sigma(b)], \tag{6.2}$$

for some $n \in \mathbb{N}$. We will prove that

$$u_{n+1}(t) \leq u_n(t), \quad t \in [a, \sigma(b)].$$

Indeed, we have

$$u_{n+1}(t) = T_{\lambda^*} u_n(t)$$

$$= \lambda^* \int_a^{\sigma(b)} G(t,s)l(s)f(u_n(\sigma(s)))\Delta s$$

$$= \lambda^* \int_a^{\sigma(b)} G(t,s)l(s)f(u_{n-1}(\sigma(s)))\Delta s$$

$$= u_n(t), \quad t \in [a, \sigma(b)].$$

Consequently, (6.2) holds for any $n \in \mathbb{N}$. Moreover,

$$u_n(t) = T_{\lambda^*} u_{n-1}(t)$$

$$= \lambda^* \int_a^{\sigma(b)} G(t,s)l(s)f(u_{n-1}(\sigma(s)))\Delta s$$

$$\geq \lambda^* \int_a^{\sigma(b)} G(t,s)l(s)f(0)\Delta s$$

$$\geq c\lambda^* \int_a^{\sigma(b)} G(t,s)l(s)\Delta s$$

$$= c\lambda^* u_0(t), \quad t \in [a, \sigma(b)].$$

Now, using the Lebesgue dominated theorem, we conclude that the sequence $\{u_n\}_{n\in\mathbb{N}_0}$ is a convergent sequence to an element $u \in P_1 \backslash \{0\}$. We have

$$u(t) = \lim_{n\to\infty} u_n(t)$$

$$
\begin{aligned}
&= \lim_{n\to\infty} \lambda^* \int_a^{\sigma(b)} G(t,s)l(s)f(u_{n-1}(\sigma(s)))\Delta s \\
&= \lambda^* \int_a^{\sigma(b)} G(t,s)l(s) \lim_{n\to\infty} f(u_{n-1}(\sigma(s)))\Delta s \\
&= \lambda^* \int_a^{\sigma(b)} G(t,s)l(s)f(u(\sigma(s)))\Delta s \\
&= T_{\lambda^*}u(t), \quad t \in [a,\sigma(b)],
\end{aligned}
$$

i.e., u is a fixed point of T_{λ^*}. This completes the proof.

Lemma 6.6. *Suppose* $(E4)$-$(E8)$ *and* $I \subset [d,\infty)$ *for some* $d > 0$. *Then there exists a constant* $c_1 > 0$ *so that for all* $\lambda \in I$ *and all possible fixed points* u *of* T_λ *of* λ, *one has* $\|u\| < c_1$.

Proof. Set

$$S_\lambda = \{u \in P_1 : T_\lambda u = u, \quad \lambda \in I\}.$$

If S_λ has a finite number of elements, then the assertion is evident. Assume that there is a sequence $\{u_n\}_{n\in\mathbb{N}_0}$ and $\{\lambda_n\}_{n\in\mathbb{N}_0} \subset I$ such that

$$T_{\lambda_n}u_n = u_n, \quad n \in \mathbb{N}_0,$$

and

$$\lim_{n\to\infty} \|u_n\| = \infty.$$

We have

$$u_n(t) \geq k\|u_n\|, \quad t \in [\delta_1, \sigma(w_1)].$$

We choose the constant $J_1 > 0$ so that

$$J_1 dk\|e\| > 1,$$

where e is the function defined as in the proof of Theorem 6.3. Let, also, $L_1 > 0$ be so that

$$f(u) > J_1 u$$

if $u > L_1$ and $t \in [\delta_1, \sigma(w_1)]$. Take $N \in \mathbb{N}$ such that

$$\|u_n\| > \frac{L_1}{k}, \quad n \geq N, \quad n \in \mathbb{N},$$

and

$$\lambda_n > d, \quad n \geq N, \quad n \in \mathbb{N}.$$

Then, for $t \in [\delta_1, \sigma(w_1)]$, we have

$$\begin{aligned}
\|u_n\| &= \|T_{\lambda_n} u_n\| \\
&= \|\lambda_n \int_a^{\sigma(b)} G(t,s) l(s) f(u_n(\sigma(s))) \Delta s\| \\
&\geq J_1 d \| \int_a^{\sigma(b)} G(t,s) l(s) u_n(\sigma(s)) \Delta s\| \\
&\geq J_1 d k \|u_n\| \| \int_a^{\sigma(b)} G(t,s) l(s) \Delta s\| \\
&= J_1 d k \|u_n\| \|e\| \\
&> \|u_n\| \quad n \geq N, \quad n \in \mathbb{N}.
\end{aligned}$$

This is a contradiction. This completes the proof.

Theorem 6.4. *Suppose* $(E4)$-$(E8)$ *and the operator* T_λ *has a fixed point at* $\hat{\lambda} > 0$. *Then for any* $\lambda_* \in (0, \hat{\lambda})$ *the operator* T_λ *has a fixed point at* λ^*.

Proof. Let $\hat{u}$ be the fixed point of the operator T_λ at $\hat{\lambda}$. Then

$$\begin{aligned}
\hat{u}(t) &= \hat{\lambda} \int_a^{\sigma(b)} G(t,s) l(s) f(\hat{u}(\sigma(s))) \Delta s \\
&\geq \lambda_* \int_a^{\sigma(b)} G(t,s) l(s) f(\hat{u}(\sigma(s))) \Delta s
\end{aligned}$$

for any $\lambda_* \in (0, \hat{\lambda})$. Set

$$\begin{aligned}
u_0(t) &= \hat{u}(t), \\
u_n(t) &= T_{\lambda_*} u_{n-1}(t), \quad t \in [a, \sigma(b)], \quad n \in \mathbb{N}.
\end{aligned}$$

Then, as in the proof of Theorem 6.3, we have

$$\begin{aligned}
c \lambda_* e(t) &\leq u_n(t) \\
&\leq u_{n-1}(t) \\
&\leq \dots \\
&\leq u_0(t).
\end{aligned}$$

Therefore, the sequence $\{u_n\}_{n\in\mathbb{N}}$ is a convergent sequence to the fixed point of T_λ at λ^*. This completes the proof.

Theorem 6.5. *Suppose* $(E4)$-$(E8)$. *Let*

$$\Lambda = \{\lambda > 0 : T_\lambda \quad \textit{has} \quad \textit{at} \quad \textit{least} \quad \textit{one} \quad \textit{fixed}$$

$$\textit{point} \quad \textit{at} \quad \lambda \quad \textit{in} \quad P_1\}.$$

Then Λ *is bounded.*

Proof. Assume that $\{u_n\}_{n\in\mathbb{N}_0} \subseteq P_1$ and $\{\lambda_n\}_{n\in\mathbb{N}_0} \subseteq \Lambda$ be such that

$$T_{\lambda_n} u_n = u_n$$

and

$$\lim_{n\to\infty} \lambda_n = \infty.$$

If we assume that there is a subsequence $\{u_{n_k}\}_{k\in\mathbb{N}_0}$ of the sequence $\{u_n\}_{n\in\mathbb{N}_0}$ such that

$$\lim_{k\to\infty} \|u_{n_k}\| = \infty,$$

then, using Lemma 6.6, we get a contradiction. Therefore, there exists a constant $H > 0$ such that

$$\|u_n\| \le H, \quad n \in \mathbb{N}_0.$$

We take $L_0 > 0$ so that

$$f(0) > L_0 H.$$

Hence, using that f is non-decreasing, we obtain

$$f(u_n) \ge f(0)$$

$$> L_0 H.$$

Then

$$H \ge u_n(t)$$

$$= \lambda_n \int_a^{\sigma(b)} G(t,s) l(s) f(u_n(\sigma(s)))\Delta s$$

$$> \lambda_n L_0 H \int_a^{\sigma(b)} G(t,s) l(s)\Delta s$$

$$\ge \lambda_n L_0 H k \|e\|.$$

Therefore,

$$\lambda_n L_0 k\|e\| < 1$$

and

$$L_0 k\|e\| \lim_{n\to\infty} \lambda_n \le 1,$$

which is a contradiction. Consequently, Λ is a bounded set. This completes the proof.

Theorem 6.6. *Suppose* $(E4)$-$(E8)$ *and* $\hat{\lambda} = \sup\Lambda$. *Then* $\Lambda = (0, \hat{\lambda}]$.

Proof. By Theorem 6.4, it follows that $(0, \hat{\lambda}) \subset \Lambda$. We will prove that $\hat{\lambda} \in \Lambda$. By the definition of $\hat{\lambda}$, it follows that there exists a non-decreasing sequence $\{\lambda_n\}_{n\in\mathbb{N}_0}$ such that

$$\lim_{n\to\infty} \lambda_n = \hat{\lambda}.$$

Let $\{u_n\}_{n\in\mathbb{N}_0}$ be a sequence of elements of P_1 so that

$$u_n = T_{\lambda_n} u_n, \quad n \in \mathbb{N}_0.$$

Since $\lambda_n \in (0, \hat{\lambda})$, $n \in \mathbb{N}_0$, by Lemma 6.6, we have that the sequence $\{u_n\}_{n\in\mathbb{N}_0}$ is uniformly bounded. Therefore, there exists a subsequence $\{u_{n_k}\}_{k\in\mathbb{N}_0}$ of the sequence $\{u_n\}_{n\in\mathbb{N}_0}$ that converges to an element $u \in P_1$. Then

$$\begin{aligned}
u(t) &= \lim_{k\to\infty} u_{n_k}(t) \\
&= \lim_{k\to\infty} \left(\lambda_{n_k} \int_a^{\sigma(b)} G(t,s) l(s) f(u_{n_k}(\sigma(s))) \Delta s \right) \\
&= \left(\lim_{k\to\infty} \lambda_{n_k} \right) \left(\lim_{k\to\infty} \int_a^{\sigma(b)} G(t,s) l(s) f(u_{n_k}(\sigma(s))) \Delta s \right) \\
&= \hat{\lambda} \int_a^{\sigma(b)} G(t,s) l(s) \lim_{k\to\infty} f(u_{n_k}(\sigma(s))) \Delta s \\
&= \hat{\lambda} \int_a^{\sigma(b)} G(t,s) l(s) f(u(\sigma(s))) \Delta s, \quad t \in [a, \sigma(b)].
\end{aligned}$$

Therefore, $\hat{\lambda} \in \Lambda$. This completes the proof.

For our next result we need the following fixed point index theorem.

Theorem 6.7 (Fixed Point Index Theorem). *Let* E *is a Banach space and* K *be a cone in* E. *For* $r > 0$, *define*

$$K_r = \{u \in K : \|u\| < r\}.$$

Assume that $A : \overline{K}_r \to K$ *is a completely continuous operator such that*

$$Ax \neq x, \quad x \in \partial K_r.$$

1. *If* $\|Ax\| \geq \|x\|$, $x \in \partial K_r$, *then*

$$i(A, K_r, K) = 0.$$

2. *If* $\|Ax\| \leq \|x\|$, $x \in \partial K_r$, *then*

$$i(A, K_r, K) = 1.$$

Theorem 6.8. *Assume* $(E4)$ - $(E8)$. *Then there exists a* $\lambda_* > 0$ *such that the eigenvalue problem* (6.1) *has at least two solutions for* $0 < \lambda < \lambda_*$, *has at least one solution for* $\lambda = \lambda_*$ *and has no positive solution for* $\lambda > \lambda_*$.

Proof. By Theorem 6.3, it follows that there exists a $\lambda_* > 0$ so that T_λ has a positive fixed point at $\lambda = \lambda_*$. Hence and by Theorem 6.4, it follows that T_λ has at least one fixed point at $\lambda \in (0, \lambda_*)$. For $\underline{\lambda} \in (0, \lambda_*)$, take $\delta_0 > 0$ such that

$$f\left((u_{\lambda_*} + \delta)(\sigma(s))\right) - f\left(u_{\lambda_*}(\sigma(s))\right) \leq f(0)\left(\frac{\lambda_*}{\underline{\lambda}} - 1\right)$$

for $s \in [a, \sigma(b)]$, $0 < \delta \leq \delta_0$. Then

$$\begin{aligned}
&\underline{\lambda} \int_a^{\sigma(b)} G(t,s)l(s)f\left((u_{\lambda_*} + \delta)(\sigma(s))\right)\Delta s \\
&\quad - \lambda_* \int_a^{\sigma(b)} G(t,s)l(s)f\left(u_{\lambda_*}(\sigma(s))\right)\Delta s \\
&= \underline{\lambda} \int_a^{\sigma(b)} G(t,s)l(s)\left(f\left((u_{\lambda_*} + \delta)(\sigma(s))\right) - f\left(u_{\lambda_*}(\sigma(s))\right)\right)\Delta s \\
&\quad - (\lambda_* - \underline{\lambda}) \int_a^{\sigma(b)} G(t,s)l(s)f\left(u_{\lambda_*}(\sigma(s))\right)\Delta s \\
&\leq (\lambda_* - \underline{\lambda}) f(0) \int_a^{\sigma(b)} G(t,s)l(s)\Delta s \\
&\quad - (\lambda_* - \underline{\lambda}) \int_a^{\sigma(b)} G(t,ls)l(s)f\left(u_{\lambda_*}(\sigma(s))\right)\Delta s \\
&= (\lambda_* - \underline{\lambda}) \int_a^{\sigma(b)} G(t,s)l(s)\left(f(0) - f\left(u_{\lambda_*}(\sigma(s))\right)\right)\Delta s \\
&\leq 0.
\end{aligned}$$

Therefore,

$$T_{\underline{\lambda}}(u_{\lambda_*} + \delta) \leq T_{\lambda_*}(u_{\lambda_*})$$

$$= u_{\lambda_*}$$

$$< u_{\lambda_*} + \delta.$$

Set

$$r = \|u_{\lambda_*} + \delta\|$$

and

$$P_r = \{u \in P : \|u\| < r\}.$$

We have $T_{\underline{\lambda}} : \overline{P}_r \to P$ is a completely continuous operator and

$$T_{\underline{\lambda}} u \neq u, \quad u \in \partial P_r.$$

Then, by the fixed point index theorem, we get

$$i(T_{\underline{\lambda}}, P_r, P) = 1.$$

Now, take $L_3 > r > 0$ so that

$$f(\hat{u}) \geq J_3 \hat{u}, \quad \hat{u} > L_3,$$

where J_3 is chosen so that

$$J_3 \underline{\lambda} k^2 \|e\| > 1.$$

Set

$$R = \frac{L_3}{k}$$

and

$$P_R = \{u \in P : \|u\| < R\}.$$

Then $T_{\underline{\lambda}} : P_R \to P$ is a completely continuous operator and

$$\begin{aligned}
T_{\underline{\lambda}} u(t) &= \underline{\lambda} \int_a^{\sigma(b)} G(t,s) l(s) f(u(\sigma(s))) \Delta s \\
&\geq \underline{\lambda} J_3 k \|u\| \int_a^{\sigma(b)} G(t,s) l(s) \Delta s \\
&= \underline{\lambda} J_3 k \|e\| \|u\| \\
&> \|u\|, \quad t \in [a, \sigma(b)].
\end{aligned}$$

Thus,

$$\|T_{\underline{\lambda}} u\| > \|u\|, \quad u \in \partial P_R.$$

Now, applying Theorem 6.7, we get

$$i(T_{\underline{\lambda}}, P_R, P) = 0.$$

Hence and by the additivity of the fixed point index (see the Appendix of this book)

$$i(T_{\underline{\lambda}}, P_R \backslash \overline{P \bigcap P_r}, P) = i(T_{\underline{\lambda}}, P_R, P) - i(T_{\underline{\lambda}}, P_r, P)$$

$$= -1.$$

Therefore, $T_{\underline{\lambda}}$ has a fixed point in $P_r \backslash \{0\}$ and another fixed point in $P_R \backslash \overline{P}_r$. Now, we choose $\lambda_* = \sup \Lambda$. Then, for $\lambda > \lambda_*$, the operator T_λ has no fixed point. This completes the proof.

Below we suppose

(E9) $f(\rho u) \geq \rho^\alpha f(u)$ for any $\rho \in (0,1)$, where $\alpha \in (0,1)$ is independent of ρ and u.

Let

$$P_+ = \{u \in \mathscr{C}([a, \sigma^2(b)]) : u(t) \geq 0, \quad t \in [a, \sigma^2(b)]\}.$$

For $u \in P_+$, define the operator

$$Tu(t) = \int_a^{\sigma(b)} G(t,s) l(s) f(u(\sigma(s))) \Delta s, \quad t \in [a, \sigma(b)].$$

Lemma 6.7. *Assume (E4)-(E6) and (E9). Then for any $u \in P_+ \backslash \{0\}$, there exist real numbers $W_u \geq w_u > 0$ such that*

$$w_u e(t) \leq Tu(t) \leq W_u e(t), \quad t \in [a, \sigma(b)],$$

where

$$e(t) = \int_a^{\sigma(b)} G(t,s) l(s) \Delta s.$$

Proof. Let $u \in P_+ \backslash \{0\}$ be arbitrarily chosen. By $(E4)$ we have that f is a non-decreasing function. Then

$$f(u(\sigma(s))) \leq f(\|u\|), \quad s \in [a, \sigma(b)],$$

and

$$Tu(t) = \int_a^{\sigma(b)} G(t,s) l(s) f(u(\sigma(s))) \Delta s$$

$$\leq f(\|u\|) \int_a^{\sigma(b)} G(t,s) l(s) \Delta s$$

$$= f(\|u\|) e(t), \quad t \in [a, \sigma(b)].$$

Set $W_u = f(\|u\|)$. Since $u \in P_+ \backslash \{0\}$, there exists an interval $(a_1, b_1) \subset [a, \sigma(b)]$ and a number $p > c$ such that

$$u(\sigma(t)) \geq p, \quad t \in (a_1, b_1).$$

Now, using $(E9)$, there exists an $s_0 > c$ and an $u^0 \in (0, \infty)$ such that $f(u^0) \geq s_0$. If $p \geq u^0$, then

$$\begin{aligned} f(u(\sigma(t)) &\geq f(p) \\ &\geq f(u^0) \\ &\geq s_0, \quad t \in (a_1, b_1). \end{aligned}$$

If $p < u^0$, then

$$\begin{aligned} f(u(\sigma(t))) &\geq f(p) \\ &= f\left(\frac{p}{u^0} u^0\right) \\ &\geq \left(\frac{p}{u^0}\right)^{\alpha} f(u^0) \\ &\geq \left(\frac{p}{u^0}\right)^{\alpha} s_0, \quad t \in (a_1, b_1). \end{aligned}$$

Set

$$w_u = \min\left\{c, \left(\frac{p}{u^0}\right)^{\alpha} s_0, s_0\right\}.$$

Then

$$\begin{aligned} Tu(t) &= \int_a^{\sigma(b)} G(t,s) l(s) f(u(\sigma(s))) \Delta s \\ &= \int_a^{a_1} G(t,s) l(s) f(u(\sigma(s))) \Delta s \\ &\quad + \int_{a_1}^{b_1} G(t,s) l(s) f(u(\sigma(s))) \Delta s \\ &\quad + \int_{b_1}^{\sigma(b)} G(t,s) l(s) f(u(\sigma(s))) \Delta s \\ &\geq c \int_a^{a_1} G(t,s) l(s) \Delta s \end{aligned}$$

$$
\begin{aligned}
&+\min\left\{\left(\frac{p}{u^0}\right)^{\alpha} s_0, s_0\right\}\int_{a_1}^{b_1} G(t,s)l(s)\Delta s \\
&+c\int_{b_1}^{\sigma(b)} G(t,s)l(s)\Delta s \\
&\geq w_u \int_a^b G(t,s)l(s)\Delta s \\
&= w_u e(t), \quad t\in[a,\sigma(b)].
\end{aligned}
$$

This completes the proof.

Theorem 6.9. *Suppose* $(E4)$-$(E6)$ *and* $(E9)$.

1. *Then the eigenvalue problem* (6.1) *has a unique solution* $u^* \in P_+\backslash\{0\}$ *such that*

$$
m_q e(t) \leq u^*(t) \leq M_Q e(t), \quad t\in[a,\sigma(b)].
$$

2. *For any* $u_0 \in P_+\backslash\{0\}$, *the sequence*

$$
u_n(t) = \int_a^{\sigma(b)} G(t,s)l(s)f(u_{n-1}(\sigma(s)))\Delta s, \quad t\in[a,\sigma(b)],
$$

converges to the unique solution u^* *of the eigenvalue problem* (6.1) *and*

$$
\|u_n - u^*\| = O\left(1-d^{\alpha^n}\right),
$$

where $0<d<1$ *is a positive constant.*

Proof. Let $u\in P_+\backslash\{0\}$, By $(E9)$, it follows that

$$
\begin{aligned}
T(\rho u(t)) &= \int_a^{\sigma(b)} G(t,s)l(s)f(\rho u(\sigma(s)))\Delta s \\
&\geq \rho^{\alpha}\int_a^{\sigma(b)} G(t,s)l(s)f(u(\sigma(s)))\Delta s \\
&= \rho^{\alpha} Tu(t), \quad t\in[a,\sigma(b)].
\end{aligned}
$$

Next, if $u_*, u_{**} \in P_+\backslash\{0\}$, $u_*(t) \leq u_{**}(t)$, $t\in[a,\sigma(b)]$, then, using that f is non-decreasing, we get

$$
\begin{aligned}
Tu_*(t) &= \int_a^{\sigma(b)} G(t,s)l(s)f(u_*(\sigma(s)))\Delta s \\
&\leq \int_a^{\sigma(b)} G(t,s)l(s)f(u_{**}(\sigma(s)))\Delta s
\end{aligned}
$$

$$= Tu_{**}(t), \quad t \in [a, \sigma(b)].$$

Moreover, by Lemma 6.7, it follows that there exist constants $0 < w_e \leq W_e$ such that

$$w_e e(t) \leq Te(t) \leq W_e e(t), \quad t \in [a, \sigma(b)].$$

Let

$$w = \max_{t \in [a,\sigma(b)]} W_e(t),$$

$$W = \min_{t \in [a,\sigma(b)]} W_e(t).$$

Take w_m and W_M so that

$$0 < w_m < \min\left\{1, w^{\frac{1}{a-\alpha}}\right\},$$

$$\infty > W_M > \max\left\{1, W^{\frac{1}{1-\alpha}}\right\}.$$

Let

$$u_0 = w_m e,$$

$$v_0 = W_M e,$$

$$u_n = Tu_{n-1},$$

$$v_n = Tv_{n-1}, \quad n \in \mathbb{N}.$$

Then

$$w_m e = u_0 \leq u_1 \leq \ldots \leq u_n \leq \ldots \leq v_n \leq \ldots v_1 \leq v_0 = W_M e.$$

Let

$$d = \frac{w_m}{W_M}.$$

Then

$$u_0 = dv_0.$$

Assume that

$$u_n \geq d^{\alpha^n} v_n \tag{6.3}$$

for some $n \in \mathbb{N}$. We will prove that

$$u_{n+1} \geq d^{\alpha^{n+1}} v_{n+1}.$$

Indeed,

$$\begin{aligned} u_{n+1} &= Tu_n \\ &\geq T\left(d^{\alpha^n} v_n\right) \\ &\geq d^{\alpha^{n+1}} Tv_n \\ &= d^{\alpha^{n+1}} v_{n+1}. \end{aligned}$$

Thus, (6.3) holds for any $n \in \mathbb{N}$. Next,

$$\begin{aligned} 0 &\leq u_{n+z} - u_n \\ &\leq v_n - u_n \\ &\leq \left(1 - d^{\alpha^n}\right) v_n \\ &\leq \left(1 - d^{\alpha^n}\right) v_0 \\ &\leq \left(1 - d^{\alpha^n}\right) W_M e \end{aligned}$$

and

$$\begin{aligned} \|v_{n+1} - u_n\| &\leq \|v_n - u_n\| \\ &\leq \left(1 - d^{\alpha^n}\right) W_M \|e\|, \end{aligned}$$

where $z \in \mathbb{N}$. Therefore, there exists an $u^* \in P_+ \backslash \{0\}$ such that

$$\begin{aligned} \lim_{n\to\infty} u_n(t) &= \lim_{n\to\infty} v_n(t) \\ &= u^*(t), \quad t \in [a, \sigma(b)], \end{aligned}$$

and u^* is a fixed point of T in $P_+ \backslash \{0\}$. This completes the proof.

Theorem 6.10. *Suppose* (*E*4)-(*E*6) *and* (*E*9). *Then the eigenvalue problem* (6.1) *has a unique solution for any* $\lambda > 0$. *Furthermore,*

1. u_λ *is non-decreasing in* $\lambda > 0$.
2. $u_\lambda \to 0$, *as* $\lambda \to 0$.
3. $u_\lambda \to \infty$, *as* $\lambda \to \infty$.
4. $u_\lambda \to u_{\lambda_0}$, *as* $\lambda \to \lambda_0$.

Proof. The proof follows by taking

$$T_\lambda u = \lambda T u, \quad u \in P_+, \quad \lambda \geq 0.$$

Exercise 6.3. Let $\mathbb{T} = 2\mathbb{Z}$. Prove that the eigenvalue problem

$$y^{\Delta^2} + \lambda\left(y^{\frac{1}{8}} + 2y^{\frac{1}{4}} + \frac{1}{2}\right) = 0, \quad t \in (0,8),$$

$$y(0) = y(10) = 0,$$

has at least one positive solution for any $\lambda > 0$.

6.2 Existence of Solutions of Iterative Second Order Eigenvalue Problems

In this section we will investigate the following second order iterative eigenvalue problem

$$\begin{aligned} u_i^{\Delta^2}(t) + \lambda_i a_i(t) f_i(u_{i+1}(\sigma(t))) &= 0, \quad i \in \{1,\dots,n\}, \quad t \in [a,b], \\ u_{n+1}(t) &= u_1(t), \quad t \in [a,b], \\ u_i(a) = u_i(\sigma^2(b)) &= 0, \quad i \in \{1,\dots,n\}, \end{aligned} \tag{6.4}$$

where

(E10) $f_i \in \mathscr{C}([0,\infty))$, $i \in \{1,\dots,n\}$, are non-negative functions.
(E11) $a_i \in \mathscr{C}([a,\sigma(b)])$, $i \in \{1,\dots,n\}$, are non-negative functions, $a_i \not\equiv 0$, $i \in \{1,\dots,n\}$, on $[a,\sigma(b)]$.
(E12) each of

$$f_{i0} = \lim_{x\to\infty} \frac{f_i(x)}{x}, \quad f_{i\infty} = \lim_{x\to\infty} \frac{f_i(x)}{x}, \quad i \in \{1,\dots,n\},$$

exists as positive numbers.

Note that the function

$$G(t,s) = \begin{cases} \dfrac{(t-a)(\sigma^2(b)-\sigma(s))}{\sigma^2(b)-a} & a \leq t \leq s \leq \sigma^2(b) \\[2ex] \dfrac{(\sigma(s)-a)(\sigma^2(b)-t)}{\sigma^2(b)-a} & a \leq \sigma(s) \leq t \leq \sigma^2(b). \end{cases}$$

We have

$$G(t,s) \le G(\sigma(s),s) = \frac{(\sigma^2(b)-\sigma(s))(\sigma(s)-a)}{\sigma^2(b)-a}, \quad t \in [a,\sigma^2(b)], \quad s \in [a,\sigma(b)]. \tag{6.5}$$

Let

$$I = \left[\frac{3a+\sigma^2(b)}{4}, \frac{a+3\sigma^2(b)}{4}\right],$$

$$k = \min\left\{\frac{1}{4}, \frac{\sigma^2(b)-a}{4(\sigma^2(b)-\sigma(a))}\right\}.$$

Next,

$$\frac{G(t,s)}{G(\sigma(s),s)} = \begin{cases} \dfrac{t-a}{\sigma(s)-a} & a \le t \le s \le \sigma^2(b) \\[2ex] \dfrac{\sigma^2(b)-t}{\sigma^2(b)-\sigma(s)} & a \le \sigma(s) \le t \le \sigma^2(b). \end{cases}$$

Let $t \in [a,s]$, $s \in [a,\sigma(b)]$. We have

$$\frac{t-a}{\sigma(s)-a} \ge \frac{t-a}{\sigma^2(b)-a}$$

and

$$\frac{t-a}{\sigma^2(b)-a} \ge \frac{1}{4} \iff$$

$$4t-4a \ge \sigma^2(b)-a \iff$$

$$4t \ge 3a+\sigma^2(b) \iff$$

$$t \ge \frac{3a+\sigma^2(b)}{4}$$

and

$$\frac{t-a}{\sigma^2(b)-a} \ge \frac{\sigma^2(b)-a}{4(\sigma^2(b)-a)} \iff$$

$$t-a \ge \frac{\sigma^2(b)-a}{4} \iff$$

$$4t-4a \ge \sigma^2(b)-a \iff$$

$$4t \ge \sigma^2(b)+3a \iff$$

$$t \ge \frac{\sigma^2(b)+3a}{4}.$$

Let $t \in [\sigma(s),\sigma(b)]$. Then

$$\frac{\sigma^2(b)-t}{\sigma^2(b)-\sigma(s)} \geq 1.$$

Therefore,

$$G(t,s) \geq kG(\sigma(s),s), \quad t \in [a,\sigma^2(b)], \quad s \in [a,\sigma(b)]. \tag{6.6}$$

Set

$$\xi = \min\left\{t \in \mathbb{T} : t \geq \frac{3a+\sigma^2(b)}{4}\right\},$$

$$w = \max\left\{t \in \mathbb{T} : t \leq \frac{a+3\sigma^2(b)}{4}\right\}.$$

We have

$$\frac{3a+\sigma^2(b)}{4} \leq \xi < w \leq \frac{a+3\sigma^2(b)}{4}.$$

Let $\tau_i \in [\xi, w]$ be defined so that

$$\int_\xi^w G(\tau_i,s)a_i(s)\Delta s = \max_{t\in[\xi,w]} \int_\xi^w G(t,s)a_i(s)\Delta s.$$

Define

$$l = \min_{s\in[a,\sigma(b)]} \frac{G(\sigma(w),s)}{G(\sigma(s),s)}$$

and

$$\gamma = \min\{k,l\}.$$

Let $\mathscr{C}([a,\sigma^2(b)])$ be endowed with the maximum norm and

$$P = \Big\{x \in \mathscr{C}([a,\sigma^2(b)]) : x(t) \geq 0, \quad t \in [a,\sigma^2(b)],$$

$$\min_{t\in[\xi,\sigma(w)]} x(t) \geq \gamma\|x\|\Big\}.$$

For $u \in P$, define the operator

$$Tu(t) = \lambda_1 \int_a^{\sigma(b)} G(t,s_1)a_1(s_1)f_1\Bigg(\lambda_2 \int_a^{\sigma(b)} G(\sigma(s_1),s_2)a_2(s_2)$$

$$\times f_2\Bigg(\lambda_3 \int_a^{\sigma(b)} G(\sigma(s_2),s_3)a_3(s_3)$$

$$\times f_{n-1}\Bigg(\lambda_n \int_a^{\sigma(b)} G(\sigma(s_{n-1},s_n)a_n(s_n)f_n(u(\sigma(s_n)))\Delta s_n\Bigg)$$

$$\cdots \Delta s_3\Big)\Delta s_2\Big)\Delta s_1, \quad t\in[a,\sigma^2(b)].$$

We have $T: P\to \mathscr{C}([a,\sigma^2(b)])$.

Lemma 6.8. *Suppose* $(E10)$-$(E12)$. *Then* $T: P\to P$.

Proof. Let $u\in P$ be arbitrarily chosen. Using (6.5), we get

$$\begin{aligned}
Tu(t) &= \lambda_1\int_a^{\sigma(b)} G(t,s_1)a_1(s_1)f_1\Bigg(\lambda_2\int_a^{\sigma(b)} G(\sigma(s_1),s_2)a_2(s_2)\\
&\quad\times f_2\Bigg(\lambda_3\int_a^{\sigma(b)} G(\sigma(s_2),s_3)a_3(s_3)\\
&\quad\times f_{n-1}\Bigg(\lambda_n\int_a^{\sigma(b)} G(\sigma(s_{n-1},s_n)a_n(s_n)f_n(u(\sigma(s_n)))\Delta s_n\Bigg)\\
&\quad\cdots\Delta s_3\Bigg)\Delta s_2\Bigg)\Delta s_1\\
&\le \lambda_1\int_a^{\sigma(b)} G(s,s_1)a_1(s_1)f_1\Bigg(\lambda_2\int_a^{\sigma(b)} G(\sigma(s_1),s_2)a_2(s_2)\\
&\quad\times f_2\Bigg(\lambda_3\int_a^{\sigma(b)} G(\sigma(s_2),s_3)a_3(s_3)\\
&\quad\times f_{n-1}\Bigg(\lambda_n\int_a^{\sigma(b)} G(\sigma(s_{n-1},s_n)a_n(s_n)f_n(u(\sigma(s_n)))\Delta s_n\Bigg)\\
&\quad\cdots\Delta s_3\Bigg)\Delta s_2\Bigg)\Delta s_1, \quad t\in[a,\sigma^2(b)],
\end{aligned}$$

whereupon,

$$\begin{aligned}
\|Tu\| &\le \lambda_1\int_a^{\sigma(b)} G(s,s_1)a_1(s_1)f_1\Bigg(\lambda_2\int_a^{\sigma(b)} G(\sigma(s_1),s_2)a_2(s_2)\\
&\quad\times f_2\Bigg(\lambda_3\int_a^{\sigma(b)} G(\sigma(s_2),s_3)a_3(s_3)\\
&\quad\times f_{n-1}\Bigg(\lambda_n\int_a^{\sigma(b)} G(\sigma(s_{n-1},s_n)a_n(s_n)f_n(u(\sigma(s_n)))\Delta s_n\Bigg)\\
&\quad\cdots\Delta s_3\Bigg)\Delta s_2\Bigg)\Delta s_1.
\end{aligned}$$

Now, using (6.6), we find

$$\begin{aligned}
Tu(t) &= \lambda_1 \int_a^{\sigma(b)} G(t,s_1)a_1(s_1)f_1\Bigg(\lambda_2 \int_a^{\sigma(b)} G(\sigma(s_1),s_2)a_2(s_2)\\
&\quad \times f_2\Bigg(\lambda_3 \int_a^{\sigma(b)} G(\sigma(s_2),s_3)a_3(s_3)\\
&\quad \times f_{n-1}\Bigg(\lambda_n \int_a^{\sigma(b)} G(\sigma(s_{n-1},s_n)a_n(s_n)f_n(u(\sigma(s_n)))\Delta s_n\Bigg)\\
&\quad \cdots \Delta s_3\Bigg)\Delta s_2\Bigg)\Delta s_1\\
&\geq \gamma\lambda_1 \int_a^{\sigma(b)} G(s,s_1)a_1(s_1)f_1\Bigg(\lambda_2 \int_a^{\sigma(b)} G(\sigma(s_1),s_2)a_2(s_2)\\
&\quad \times f_2\Bigg(\lambda_3 \int_a^{\sigma(b)} G(\sigma(s_2),s_3)a_3(s_3)\\
&\quad \times f_{n-1}\Bigg(\lambda_n \int_a^{\sigma(b)} G(\sigma(s_{n-1},s_n)a_n(s_n)f_n(u(\sigma(s_n)))\Delta s_n\Bigg)\\
&\quad \cdots \Delta s_3\Bigg)\Delta s_2\Bigg)\Delta s_1\\
&\geq \gamma\|Tu\|, \quad t\in[a,\sigma^2(b)].
\end{aligned}$$

Hence,

$$\min_{t\in[\xi,\sigma(w)]} Tu(t) \geq \gamma\|Tu\|,$$

i.e., $Tu \in P$ and $T: P \to P$. This completes the proof.

Exercise 6.4. Suppose $(E10)$-$(E12)$. Prove that $T: P \to P$ is a completely continuous operator.

Set

$$\begin{aligned}
L_1 &= \max_{i\in\{1,\dots,n\}} \left(\left(\gamma f_{i\infty} \int_\xi^w G(\tau_i,s)a_i(s)\Delta s\right)^{-1}\right),\\
L_2 &= \min_{i\in\{1,\dots,n\}} \left(\left(f_{i0} \int_a^{\sigma(b)} G(\sigma(s),s)a_i(s)\Delta s\right)^{-1}\right).
\end{aligned}$$

Theorem 6.11. *Suppose* $(E10)$-$(E12)$ *and*

$$L_1 < \lambda_i < L_2, \quad i\in\{1,\dots,n\}. \tag{6.7}$$

Then there exists an n-tuple $(u_1, u_2, \ldots, u_n)$ *satisfying* (6.4) *such that* $u_i(t) > 0$, $t \in (a, \sigma^2(b))$, $i \in \{1, \ldots, n\}$.

Proof. Take $\varepsilon > 0$ so that

$$\max_{i\in\{1,\ldots,n\}} \left(\left(\left(\gamma \int_\xi^w G(\tau_i, s) a_i(s) \Delta s \right) (f_{i\infty} - \varepsilon) \right)^{-1} \right) \le \min_{i\in\{1,\ldots,n\}} \lambda_i$$

and

$$\min_{i\in\{1,\ldots,n\}} \left(\left(\left(\gamma \int_\xi^w G(\tau_i, s) a_i(s) \Delta s \right) (f_{i0} + \varepsilon) \right)^{-1} \right) \le \max_{i\in\{1,\ldots,n\}} \lambda_i.$$

By the definition of f_{i0}, $i \in \{1, \ldots, n\}$, there exists an $H_1 > 0$ so that

$$f_i(x) \le (f_{i0} + \varepsilon)x, \quad x \in (0, H_1], \quad i \in \{1, \ldots, n\}.$$

Define

$$\Omega_1 = \{u \in P : \|u\| = H_1\}.$$

Let $u \in P \bigcap \partial\Omega_1$. Then $\|u\| = H_1$ and

$$\begin{aligned}
&\lambda_n \int_a^{\sigma(b)} G(\sigma(s_{n-1}), s_n) a_n(s_n) f_n(u(\sigma(s_n))) \Delta s_n \\
&\le \lambda_n \int_a^{\sigma(b)} G(\sigma(s-n), s_n) a_n(s_n) f_n(u(\sigma(s_n))) \Delta s_n \\
&\le \lambda_n \int_a^{\sigma(b)} G(\sigma(s_n), s_n) a_n(s_n) (f_{n0} + \varepsilon) \|u\| \Delta s_n \\
&\le \|u\| \\
&= H_1
\end{aligned}$$

and

$$\begin{aligned}
&\lambda_n \int_a^{\sigma(b)} G(\sigma(s_{n-2}), s_{n-1}) a_{n-1}(s_{n-1}) \\
&\quad \times f_{n-1} \left(\lambda_n \int_a^{\sigma(b)} G(\sigma(s_{n-1}), s_n) a_n(s_n) f_n(u(\sigma(s_n))) \Delta s_n \right) \Delta s_{n-1} \\
&\le \lambda_{n-1} \int_a^{\sigma(b)} G(\sigma(s_{n-2}), s_{n-1}) a_{n-1}(s_{n-1}) \\
&\quad \times \left(\lambda_n \int_a^{\sigma(b)} G(\sigma(s_{n-1}), s_n) a_n(s_{n-1}) f_n(u(\sigma(s_n))) \Delta s_n (f_{n-10} + \varepsilon) \right) \Delta s_n
\end{aligned}$$

$$
\begin{aligned}
&\le \lambda_{n-1}\int_a^{\sigma(b)} G(s_{n-1},s_{n-1})a_{n-1}(s_{n-1})\Delta s_{n-1}(f_{n-10}+\varepsilon)H_1\\
&\le H_1\\
&= \|u\|,
\end{aligned}
$$

and so on,

$$
\begin{aligned}
&\lambda_1\int_a^{\sigma(b)} G_1(t,s_1)a_1(s_1)f_1\left(\dots f_n(u(\sigma(s_n)))\Delta s_n\dots\right)\Delta s_1\\
&\le H_1\\
&= \|u\|, \quad t\in[a,\sigma^2(b)].
\end{aligned}
$$

Thus,

$$
Tu(t)\le H_1, \quad t\in[a,\sigma^2(b)],
$$

and

$$
\begin{aligned}
\|Tu\| &\le H_1\\
&= \|u\|, \quad u\in P\bigcap\partial\Omega_1.
\end{aligned}
$$

By the definition of $f_{i\infty}$, $i\in\{1,\dots,n\}$, there exists an $\overline{H}_2>0$ so that

$$
f_i(x)\ge(f_{i\infty}-\varepsilon)x, \quad x\ge\overline{H}_2, \quad \in\{1,\dots,n\}.
$$

Define

$$
H_2=\max\{2H_1,\overline{H}_2\},
$$

$$
\Omega_2=\{u\in P:\|u\|\le H_2\}.
$$

Take $u\in P\bigcap\partial\Omega_2$. Then $\|u\|=H_2$ and

$$
\begin{aligned}
&\lambda_n\int_a^{\sigma(b)} G(\sigma(s_{n-1}),s_n)a_n(s_n)f_n(u(\sigma(s_n)))\Delta s_n\\
&\ge \gamma\lambda_n\int_a^{\sigma(b)} G(\sigma(s_n),s_n)a_n(s_n)f_n(u(\sigma(s_n)))\Delta s_n
\end{aligned}
$$

$$\geq \gamma\lambda_n \int_a^{\sigma(b)} G(\sigma(s_n), s_n) a_n(s_n)(f_{n\infty} - \varepsilon)\|u\| \Delta s_n$$

$$\geq \|u\|$$

$$= H_2$$

and

$$\lambda_n \int_a^{\sigma(b)} G(\sigma(s_{n-2}), s_{n-1}) a_{n-1}(s_{n-1})$$

$$\times f_{n-1}\left(\lambda_n \int_a^{\sigma(b)} G(\sigma(s_{n-1}), s_n) a_n(s_n) f_n(u(\sigma(s_n)))\Delta s_n\right)\Delta s_{n-1}$$

$$\geq \lambda_{n-1} \int_a^{\sigma(b)} G(\sigma(s_{n-2}), s_{n-1}) a_{n-1}(s_{n-1})$$

$$\times\left(\lambda_n \int_a^{\sigma(b)} G(\sigma(s_{n-1}), s_n) a_n(s_{n-1}) f_n(u(\sigma(s_n)))\Delta s_n (f_{n-1\infty} - \varepsilon)\right)\Delta s_n$$

$$\leq \gamma\lambda_{n-1} \int_a^{\sigma(b)} G(s_{n-1}, s_{n-1}) a_{n-1}(s_{n-1})\Delta s_{n-1}(f_{n-1\infty} - \varepsilon) H_2$$

$$\geq H_2$$

$$= \|u\|,$$

and so on,

$$\lambda_1 \int_a^{\sigma(b)} G_1(t, s_1) a_1(s_1) f_1(\ldots f_n(u(\sigma(s_n))))\Delta s_n \ldots \Delta s_1$$

$$\geq H_2$$

$$= \|u\|, \quad t \in [a, \sigma^2(b)].$$

Thus,

$$Tu(t) \geq H_2, \quad t \in [a, \sigma^2(b)],$$

and

$$\|Tu\| \geq H_2$$

$$= \|u\|, \quad u \in P \bigcap \partial\Omega_2.$$

Now, applying the Guo-Krasnosel'skii fixed point Theorem 1.11, it follows that the operator T has at least one fixed point in $P\bigcap(\overline{\Omega}_2\backslash\Omega_1)$. This completes the proof.

Let ξ_i, $i\in\{1,\dots,n\}$, be defined by

$$\int_a^{\sigma(b)} G(\xi_i,s)a_i(s)\Delta s = \max_{t\in[a,\sigma^2(b)]}\int_a^{\sigma(b)} G(t,s)a_i(s)\Delta s,\quad i\in\{1,\dots,n\},$$

and

$$L_3 = \max_{i\in\{1,\dots,n\}}\left(\left(\gamma\int_\xi^w G(\tau_i,s)a_i(s)\Delta s f_{i0}\right)^{-1}\right),$$

$$L_4 = \max_{i\in\{1,\dots,n\}}\left(\left(\gamma\int_a^{\sigma(b)} G(\xi_i,s)a_i(s)\Delta s f_{i\infty}\right)^{-1}\right).$$

Theorem 6.12. *Suppose* $(E10)$-$(E12)$ *and each* λ_i, $i\in\{1,\dots,n\}$, *satisfy*

$$L_3<\lambda_i<L_4,\quad i\in\{1,\dots,n\}.$$

Then there exists an n-tuple $(u_1,u_2,\dots,u_n)$ *satisfying* (6.4) *such that* $u_i(t)>0$, $t\in(a,\sigma(b))$, $i\in\{1,\dots,n\}$.

Proof. Take $\varepsilon>0$ so that

$$\max_{i\in\{1,\dots,n\}}\left(\left(\gamma\int_\xi^w GG(\tau_i,s)a_i(s)\Delta s(f_{i0}-\varepsilon)\right)^{-1}\right)\le\min_{i\in\{1,\dots,n\}}\lambda_i,$$

$$\min_{i\in\{1,\dots,n\}}\left(\left(\gamma\int_a^{\sigma(b)} G(\xi_i,s)a_i(s)\Delta s(f_{i\infty}+\varepsilon)\right)^{-1}\right)\ge\max_{i\in\{1,\dots,n\}}\lambda_i.$$

By the definition of f_{i0}, $i\in\{1,\dots,n\}$, there exists an $\overline{H}_3>0$ such that

$$f_i(x)\ge(f_{i0}-\varepsilon)x,\quad x\in(0,\overline{H}_3],\quad i\in\{1,\dots,n\},$$

and

$$f_i(0)=0,\quad i\in\{1,\dots,n\}.$$

Thus, there exist $K_2,\dots,K_n$ so that

$$0<K_n<K_{n-1}<\dots<K_2<\overline{H}_3$$

and

$$\lambda_i f_i(t)n\le\frac{K_{i-1}}{\int_a^{\sigma(b)}G(\xi_i,s)a_i(s)\Delta s},\quad t\in[0,K_i],\quad i\in\{3,\dots,n\},$$

and

$$\lambda_2 f_2(t) \le \frac{\overline{H}_3}{\int_a^{\sigma(b)} G(\xi_2,s)a_2(s)\Delta s}, \quad t\in[0,K_2].$$

Let

$$\Omega_1 = \{u\in P: \|u\| < K_n\}.$$

Then, for $u\in P\bigcap \partial\Omega_1$, we have

$$\begin{aligned}
&\lambda_n \int_a^{\sigma(b)} G(\sigma(s_{n-1}),s_n)a_n(s_n)f_n(u(\sigma(s_n)))\Delta s_n\\
&\le \lambda_n \int_a^{\sigma(b)} G(\xi_n,s_n)a_n(s_n)f_n(u(\sigma(s_n)))\Delta s_n\\
&\le \frac{\int_a^{\sigma(b)} G(\xi_n,s_n)a_n(s_n)K_{n-1}\Delta s_n}{\int_a^{\sigma(b)} G(\xi_n,s_n)a_n(s_n)\Delta s_n}\\
&\le K_{n-1}
\end{aligned}$$

and so on,

$$\lambda_2 \int_a^{\sigma(b)} G(\sigma(s_1),s_2)a_2(s_2)f_2(\ldots)\Delta s_2 \le \overline{H}_3.$$

Then

$$\begin{aligned}
Tu(\tau_1) &= \lambda_1 \int_a^{\sigma(b)} G(\tau_1,s_1)a_1(s_1)f_1(\lambda_2\ldots)\Delta s_1\\
&\ge \lambda_1\gamma \int_\xi^{w} G(\tau_1,s_1)a_1(s_1)(f_{10}-\varepsilon)\|u\|\Delta s_1\\
&\ge \|u\|.
\end{aligned}$$

So,

$$\|Tu\| \ge \|u\|, \quad u\in P\bigcap\partial\Omega_1.$$

By the definition of $f_{i\infty}$, $i\in\{1,\ldots,n\}$, we have that f_i, $i\in\{1,\ldots,n\}$, are unbounded at ∞. Define

$$f_i^*(x) = \sup_{s\in[a,x]} f_i(s), \quad i\in\{1,\ldots,n\}.$$

We have that f_i^*, $i\in\{1,\ldots,n\}$, are non-decreasing and there exists an $\overline{H}_4$ such that

$$f_i^*(x) \ge (f_{i\infty}+\varepsilon)x. \quad x\ge \overline{H}_4, \quad i\in\{1,\ldots,n\}.$$

Then there exists an $H_4 > \max\{2\overline{H}_3, \overline{H}_4\}$ so that

$$f_i^*(x) \le f_i^*(H_4), \quad x\in(0,H_4].$$

Define

$$\Omega_2 = \{u \in P : \|u\| < H_4\}.$$

Take $u \in P\bigcap \partial\Omega_2$ arbitrarily. Then

$$\begin{aligned}
&\lambda_n \int_a^{\sigma(b)} G(\sigma(s_{n-1}), s_n) a_n(s_n) f_n(u(\sigma(s_n)))\Delta s_n \\
&\le \lambda_n \int_a^{\sigma(b)} G(\xi_n, s_n) a_n(s_n) f_n^*(u(\sigma(s_n)))\Delta s_n \\
&\le \lambda_n \int_a^{\sigma(b)} G(\xi_n, s_n) a_n(s_n) f_n^*(H_4)\Delta s_n \\
&\le \lambda_n (f_{n\infty} + \varepsilon) H_4 \int_a^{\sigma(b)} G(\sigma(s_{n-1}), s_n) a_n(s_n)\Delta s_n \\
&\le H_4 \\
&= \|u\|
\end{aligned}$$

and so on,

$$\lambda_2 \int_a^{\sigma(b)} G(\sigma(s_1), s_2) a_2(s_2) f_2(\dots)\Delta s_2 \le H_4.$$

Then

$$\begin{aligned}
Tu(\tau_1) &= \lambda_1 \int_a^{\sigma(b)} G(\tau_1, s_1) a_1(s_1) f_1(\lambda_2 \dots)\Delta s_1 \\
&\le \lambda_1 \int_a^{\sigma(b)} G(\tau_1, s_1) a_1(s_1) f_1^*(\lambda_2 \dots)\Delta s_1 \\
&\le \lambda_1 \int_a^{\sigma(b)} G(\tau_1, s_1) a_1(s_1) f_1^*(H_4)\Delta s_1 \\
&\le \lambda_1 \int_a^{\sigma(b)} G(\xi_1, s_1) a_1(s_1)\Delta s_1 (f_{1\infty} + \varepsilon) H_4 \\
&\le H_4 \\
&= \|u\|, \quad t \in [a, \sigma^2(b)].
\end{aligned}$$

Therefore,

$$\|Tu\| \le \|u\|, \quad u \in P\bigcap \partial\Omega_2.$$

Now, applying the Guo-Krasnosel'skii fixed point Theorem 1.11, it follows that the operator T has at least one fixed point in $P\bigcap\left(\overline{\Omega}_2\backslash\Omega_1\right)$. This completes the proof.

Example 6.2. Let $\mathbb{T}=\mathbb{Z}$. Consider the iterative eigenvalue problem

$$\begin{aligned} u_1^{\Delta^2}(t)+\lambda_1\frac{u_2(t+1)(u_2(t+1)+1)}{u_2(t+1)+2} &= 0, \\ u_2^{\Delta^2}(t)+\lambda_2\frac{u_3(t+1)\left((u_3(t+1))^2+1\right)}{(u_3(t+1))^2+2} &= 0, \\ u_3^{\Delta^2}(t)+\lambda_3\frac{u_1(t+1)(u_1(t+1)+2)}{u_1(t+1)+1} &= 0, \quad t\in[0,9], \\ u_i(0)=u_i(1) &= 0, \quad i\in\{1,2,3\}. \end{aligned}$$

Here

$$\begin{aligned} \sigma(t) &= t+1, \\ a_1(t) &= a_2(t) \\ &= a_3(t) \\ &= 1, \quad t\in\mathbb{T}, \\ f_1(x) &= \frac{x(x+1)}{x+2}, \\ f_2(x) &= \frac{x(x^2+1)}{x^2+2}, \\ f_3(x) &= \frac{x(x+2)}{x+1}, \quad x\in\mathbb{R}. \end{aligned}$$

We have

$$\begin{aligned} f_{10} &= \lim_{x\to 0}\frac{f_1(x)}{x} \\ &= \lim_{x\to 0}\frac{x+1}{x+2} \\ &= \frac{1}{2}, \end{aligned}$$

$$
\begin{aligned}
f_{20} &= \lim_{x\to 0} \frac{f_2(x)}{x} \\
&= \lim_{x\to 0} \frac{x^2+1}{x^2+2} \\
&= \frac{1}{2}, \\
f_{30} &= \lim_{x\to 0} \frac{f_3(x)}{x} \\
&= \lim_{x\to 0} \frac{x+2}{x+1} \\
&= 2, \\
f_{1\infty} &= \lim_{x\to\infty} \frac{f_1(x)}{x} \\
&= \lim_{x\to\infty} \frac{x+1}{x+2} \\
&= 1, \\
f_{2\infty} &= \lim_{x\to\infty} \frac{f_2(x)}{x} \\
&= \lim_{x\to\infty} \frac{x^2+1}{x^2+2} \\
&= 1, \\
f_{3\infty} &= \lim_{x\to\infty} \frac{f_3(x)}{x} \\
&= \lim_{x\to\infty} \frac{x^2+1}{x^2+2} \\
&= 1.
\end{aligned}
$$

The Green function is

$$G(t,s) = \begin{cases} \dfrac{t(10-s)}{11} & 0 \le t \le s \le 11 \\ \\ \dfrac{(s+1)(11-t)}{11} & 0 \le s+1 \le t \le 11. \end{cases}$$

Next,

$$\begin{aligned} I &= \left[\frac{11}{4}, \frac{33}{4}\right], \\ k &= \min\left\{\frac{1}{4}, \frac{11}{44}\right\} \\ &= \frac{1}{4}, \\ \xi &= \min\{t \in \mathbb{T} : t \ge \frac{11}{4}\} \\ &= 3, \\ w &= \max\{t \in \mathbb{T} : t \le \frac{33}{4}\} \\ &= 8. \end{aligned}$$

Let

$$\begin{aligned} g_1(t) &= \frac{t^2+t}{2}, \\ g_2(t) &= \frac{21t-t^2}{2}, \quad t \in \mathbb{T}. \end{aligned}$$

Then

$$\begin{aligned} g_1^{\Delta}(t) &= \frac{1}{2}(\sigma(t)+t+1) \\ &= \frac{1}{2}(t+1+t+1) \\ &= t+1, \\ g_2^{\Delta}(t) &= \frac{1}{2}(21-\sigma(t)-t) \end{aligned}$$

$$= \frac{1}{2}(21-t-1-t)$$

$$= \frac{1}{2}(20-2t)$$

$$= 10-t, \quad t \in \mathbb{T}.$$

Consequently,

$$\begin{aligned}
\int_{\xi}^{w} G(t,s)a_i(s)\Delta s &= \int_3^8 G(t,s)\Delta s \\
&= \int_3^t G(t,s)\Delta s + \int_t^8 G(t,s)\Delta s \\
&= \int_3^t \frac{(s+1)(11-t)}{11}\Delta s + \int_t^8 \frac{t(10-s)}{11}\Delta s \\
&= \frac{11-t}{11}\int_3^t (s+1)\Delta s + \frac{t}{11}\int_t^8 (10-s)\Delta s \\
&= \frac{11-t}{11}\int_3^t g_1^{\Delta}(s)\Delta s + \frac{t}{11}\int_t^8 g_2^{\Delta}(s)\Delta s \\
&= \frac{11-t}{11}(g_1(t)-g_1(3)) + \frac{t}{11}(g_2(8)--g_2(t)) \\
&= \frac{11-t}{11}\left(\frac{t^2+t}{2}-6\right) + \frac{t}{11}\left(\frac{168-64}{2}-\frac{21t-t^2}{2}\right) \\
&= \frac{11-t}{11}\left(\frac{t^2+t-12}{2}\right) + \frac{t}{11}\left(\frac{104-21t+t^2}{2}\right) \\
&= \frac{11t^2+11t-132-t^3-t^2+12t+104t-21t^2+t^3}{22} \\
&= \frac{-11t^2+127t-132}{22}, \quad i \in \{1,2,3\}, \quad t \in [0,10],
\end{aligned}$$

and

$$\begin{aligned}
\max_{t\in[3,8]} \int_{\xi}^{w} G(t,s)a_i(s)\Delta s &= \frac{-275+635-132}{22} \\
&= \frac{360-132}{22}
\end{aligned}$$

$$= \frac{228}{22}$$

$$= \frac{114}{11},$$

and

$$l = \min_{s\in[0,10]} \frac{G(9,s)}{G(s+1,s)}$$

$$= \frac{9}{s+1}$$

$$= 9,$$

$$\gamma = \min\{k,l\}$$

$$= \min\left\{\frac{1}{4},4\right\}$$

$$= \frac{1}{4}.$$

Hence,

$$L_1 = \left(\frac{1}{4}\cdot\frac{114}{11}\right)^{-1}$$

$$= \frac{22}{57}.$$

Also,

$$\int_a^{\sigma(b)} G(\sigma(s),s)a_i(s)\Delta s = \int_0^9 \frac{(s+1)(10-s)}{11}\Delta s$$

$$= \frac{1}{11}(1\cdot 10+2\cdot 9+3\cdot 8+4\cdot 7+5\cdot 6+6\cdot 5+7\cdot 6+8\cdot 7)$$

$$= \frac{1}{11}(10+18+24+28+30+30+42+56)$$

$$= \frac{238}{11},$$

$$L_2 = \min\left\{\frac{11}{124},\frac{11}{476}\right\}$$

$$= \frac{11}{476}.$$

Then, by Theorem 6.11, it follows that if

$$\frac{11}{476} < \lambda_i < \frac{22}{57}, \quad i \in \{1,2,3\},$$

the considered iterative eigenvalue problem has at least one non-negative solution.

Exercise 6.5. Let $\mathbb{T} = 2^{\mathbb{N}_0}$. Consider the iterative eigenvalue problem

$$u_1^{\Delta^2}(t) + \lambda_1 \frac{1}{1+t^2} \frac{u_2(2t)((u_2(2t))^4+7)}{(u_2(2t))^4+2} = 0,$$

$$u_2^{\Delta^2}(t) + \lambda_2 \frac{t^2+1}{t^4+10} \frac{u_3(2t)\left((u_3(2t))^2+1\right)}{(u_3(2t))^2+2} = 0,$$

$$u_3^{\Delta^2}(t) + \lambda_3 \frac{1}{1+t^4} \frac{u_4(2t)(u_4(2t)+21)}{u_4(2t)+10} = 0,$$

$$u_4^{\Delta^2}(t) + \lambda_4 \frac{1}{1+t} \frac{u_1(2t)((u_1(2t))^6+1)}{(u_1(2t))^6+12} = 0, \quad t \in [1,16],$$

$$u_i(1) = u_i(32) = 0, \quad i \in \{1,2,3,4\}.$$

Find $L_1, L_2 > 0$, so that the considered iterative eigenvalue problem to have at least one non-negative solution for any

$$L_1 < \lambda_i < L_2, \quad i \in \{1,2,3,4\}.$$

6.3 Existence of Solutions of Nonlinear Second Order n-Point Eigenvalue Problems

In this section we investigate the following n-point eigenvalue problem

$$\begin{aligned} (py^{\nabla})^{\Delta}(t) - q(t)y(t) + \lambda h(t) f(y(t)) &= 0, \quad t_1 < t < t_n, \\ \alpha y(t_1) - \beta p(t_1) y^{\nabla}(t_1) &= \sum_{j=2}^{n-1} a_j y(t_j), \\ \gamma y(t_n) + \delta p(t_n) y^{\nabla}(t_n) &= \sum_{j=2}^{n-1} b_j y(t_j), \end{aligned} \tag{6.8}$$

where

(E13) $n \geq 3$, $p, q : [t_1, t_n] \to (0, \infty)$, $p \in \mathscr{C}([t_1, t_n])$, there exists p^Δ on $[t_1, t_n]$ and $p^\Delta \in \mathscr{C}([t_1, t_n])$, $q \in \mathscr{C}([t_1, t_n])$,
(E14) $t_1 < t_2 < \ldots < t_n$,
(E15) $\alpha, \beta, \gamma, \delta \in [0, \infty)$, $\alpha\gamma + \alpha\delta + \beta\gamma > 0$, $a_j, b_j \in [0, \infty)$, $j \in \{2, \ldots, n-1\}$,
(E16) $f \in \mathscr{C}([0, \infty))$ is a non-negative function such that

$$f_0 = \lim_{y \to 0} \frac{f(y)}{y},$$

$$f_\infty = \lim_{y \to \infty} \frac{f(y)}{y},$$

exist,
(E17) $h \in \mathscr{C}([t_1, t_n])$ is a non-negative function such that there is a $t_* \in (\sigma(t_1), \rho(t_n)]$ so that $h(t_*) > 0$.

Let ϕ and ψ be the solutions of the IVPs

$$(p\phi^\nabla)^\Delta(t) - q(t)\phi(t) = 0, \quad t \in [t_1, t_n],$$

$$\phi(t_n) = \delta, \quad p(t_n)\phi^\nabla(t_n) = -\gamma,$$

and

$$(p\psi^\nabla)^\Delta(t) - q(t)\psi(t) = 0, \quad t \in [t_1, t_n],$$

$$\psi(t_1) = \beta, \quad p(t_1)\psi^\nabla(t_1) = \alpha.$$

Let, also,

$$\begin{aligned} d &= \alpha\phi(t_1) - \beta p(t_1)\phi^\nabla(t_1) \\ &= \gamma\psi(t_n) + \delta p(t_n)\psi^\nabla(t_n). \end{aligned}$$

By the investigations in Chapter 3, it follows

1. $d \neq 0$ if and only if the BVP

$$\begin{aligned} (py^\nabla)^\Delta(t) - q(t)y(t) + u(t) &= 0, \quad t_1 < t < t_n, \\ \alpha y(t_1) - \beta p(t_1)y^\nabla(t_1) &= 0, \\ \gamma y(t_n) + \delta p(t_n)y^\nabla(t_n) &= 0, \end{aligned} \tag{6.9}$$

where u is a continuous function on $[\rho(t_1), t_n]$, has only the trivial solution.

2. If $d \neq 0$, then the solution (6.9) satisfies equation

$$y(t) = \int_{t_1}^{t_n} G(t,s)u(s)\Delta s, \quad t \in [\rho(t_1), t_n],$$

where

$$G(t,s) = \begin{cases} \psi(t)\phi(s) & \rho(t_1) \le t \le s \le t_n \\ \psi(s)\phi(t) & \rho(t_1) \le s \le t \le t_n. \end{cases}$$

3. If $(E13)$-$(E15)$ hold, then

$$\psi^{\Delta}(t) \ge 0, \quad t \in [\rho(t_1), t_n], \quad \psi(t) > 0, \quad t \in [\rho(t_1), t_n],$$
$$p(t)\psi^{\nabla}(t) \ge 0, \quad t \in [\rho(t_1), t_n], \quad \phi(t) \ge 0, \quad t \in [\rho(t_1), t_n],$$
$$\phi(t) > 0, \quad t \in [\rho(t_1), t_n], \quad p(t)\phi^{\nabla}(t) \le 0, \quad t \in [\rho(t_1), t_n].$$

Set

$$D = \begin{vmatrix} -\sum_{j=2}^{n-1} a_j \psi(t_j) & d - \sum_{j=2}^{n-1} a_j \phi(t_j) \\ d - \sum_{j=2}^{n-1} b_j \psi(t_j) & -\sum_{j=2}^{n-1} b_j \phi(t_j) \end{vmatrix},$$

$$A(u) = \frac{1}{D} \begin{vmatrix} \sum_{j=2}^{n-1} a_j \int_{t_1}^{t_n} G(t_j,s)u(s)\Delta s & d - \sum_{j=2}^{n-1} a_j \phi(t_j) \\ \sum_{j=2}^{n-1} b_j \int_{t_1}^{t_n} G(t_j,s)u(s)\Delta s & -\sum_{j=2}^{n-1} b_j \phi(t_j) \end{vmatrix},$$

$$B(u) = \frac{1}{D} \begin{vmatrix} -\sum_{j=2}^{n-1} a_j \psi(t_j) & \sum_{j=2}^{n-1} a_j \int_{t_1}^{t_n} G(t_j,s)u(s)\Delta s \\ d - \sum_{j=2}^{n-1} b_j \psi(t_j) & \sum_{j=2}^{n-1} b_j \int_{t_1}^{t_n} G(t_j,s)u(s)\Delta s \end{vmatrix}.$$

By the definitions of $A(u)$ and $B(u)$, it follows that they satisfy the system

$$\begin{aligned} &-A(u)\sum_{j=2}^{n-1} a_j \psi(t_j) + B(u)\left(\alpha\phi(t_1) - \beta p(t_1)\phi^{\nabla}(t_1) - \sum_{j=2}^{n-1} a_j \phi(t_j)\right) \\ &= \sum_{j=2}^{n-1} a_j \int_{t_1}^{t_n} G(t_j,s)u(s)\Delta s \\ &A(u)\left(\gamma\psi(t_n) + \delta p(t_n)\psi^{\nabla}(t_n) - \sum_{j=2}^{n-1} b_j \psi(t_j)\right) - B(u)\sum_{j=2}^{n-1} b_j \phi(t_j) \\ &= \sum_{j=2}^{n-1} b_j \int_{t_1}^{t_n} G(t_j,s)u(s)\Delta s. \end{aligned} \tag{6.10}$$

Lemma 6.9. *Assume (E13)-(E15), $D \neq 0$ and $u \in \mathscr{C}_{rd}([t_1,t_n])$. If y satisfies the integral equation*

$$y(t) = \int_{t_1}^{t_n} G(t,s)u(s)\Delta s + A(u)\psi(t) + B(u)\phi(t), \quad t \in [\rho(t_1),t_n], \tag{6.11}$$

then y is a solution of the BVP

$$\begin{aligned}
(py^{\nabla})^{\Delta}(t) - q(t)y(t) + u(t) &= 0, \quad t_1 < t < t_n \\
\alpha y(t_1) - \beta p(t_1)y^{\nabla}(t_1) &= \sum_{j=2}^{n-1} a_j y(t_j), \\
\gamma y(t_n) + \delta p(t_n)y^{\nabla}(t_n) &= \sum_{j=2}^{n-1} b_j y(t_j).
\end{aligned} \tag{6.12}$$

Proof. Let y satisfy (6.11). Then

$$\begin{aligned}
y(t) &= \frac{1}{d}\int_{t_1}^{t} \phi(t)\psi(s)u(s)\Delta s + \frac{1}{d}\int_{t}^{t_n} \psi(t)\phi(s)u(s)\Delta s \\
&\quad + A(u)\psi(t) + B(u)\phi(t), \quad t \in [\rho(t_1),t_n].
\end{aligned}$$

Then

$$\begin{aligned}
y^{\nabla}(t) &= \frac{1}{d}\phi(\rho(t))\psi(\rho(t))u(\rho(t)) + \frac{1}{d}\int_{t_1}^{t} \phi^{\nabla}(t)\psi(s)u(s)\Delta s \\
&\quad - \frac{1}{d}\phi(\rho(t))\psi(\rho(t))u(\rho(t)) + \frac{1}{d}\int_{t}^{t_n} \psi^{\nabla}(t)\phi(s)u(s)\Delta s \\
&\quad + A(u)\psi^{\nabla}(t) + B(u)\phi^{\nabla}(t) \\
&= \frac{1}{d}\int_{t_1}^{t} \phi^{\nabla}(t)\psi(s)u(s)\Delta s + \frac{1}{d}\int_{t}^{t_n} \psi^{\nabla}(t)\phi(s)u(s)\Delta s \\
&\quad + A(u)\psi^{\nabla}(t) + B(u)\phi^{\nabla}(t), \\
p(t)y^{\nabla}(t) &= \frac{1}{d}\int_{t_1}^{t} p(t)\phi^{\nabla}(t)\psi(s)u(s)\Delta s + \frac{1}{d}\int_{t}^{t_n} p(t)\psi^{\nabla}(t)\phi(s)u(s)\Delta s \\
&\quad + A(u)p(t)\psi^{\nabla}(t) + B(u)p(t)\phi^{\nabla}(t) \\
(py^{\nabla})^{\Delta}(t) &= \frac{p(t)}{d}\phi^{\nabla}(\sigma(t))\psi(t)u(t)
\end{aligned}$$

$$
\begin{aligned}
&+\frac{1}{d}\int_{t_1}^{t}(p\phi^{\nabla})^{\Delta}(t)\psi(s)u(s)\Delta s\\
&-\frac{1}{d}p(t)\psi^{\nabla}(\sigma(t))\phi(t)u(t)\\
&+\frac{1}{d}\int_{t}^{t_n}(p\psi^{\nabla})^{\Delta}(t)\phi(s)u(s)\Delta s\\
&+A(u)(p\psi^{\nabla})^{\Delta}(t)+B(u)(p\phi^{\nabla})^{\Delta}(t)\\
=&\ \frac{1}{d}\int_{t_1}^{t}(p\phi^{\nabla})^{\Delta}(t)\psi(s)u(s)\Delta s\\
&+\frac{1}{d}\int_{t}^{t_n}(p\psi^{\nabla})^{\Delta}(t)\phi(s)u(s)\Delta s\\
&+A(u)(p\psi^{\nabla})^{\Delta}(t)+B(u)(p\phi^{\nabla})^{\Delta}(t), \quad t\in[\rho(t_1),t_n].
\end{aligned}
$$

Next,

$$
\begin{aligned}
q(t)y(t)-u(t) &= \frac{q(t)}{d}\int_{t_1}^{t}\phi(t)\psi(s)u(s)\Delta s\\
&\quad+\frac{q(t)}{d}\int_{t}^{t_n}\psi(t)\phi(s)u(s)\Delta s\\
&\quad+A(u)q(t)\psi(t)+B(u)q(t)\phi(t)\\
&= \frac{1}{d}\int_{t_1}^{t}(p\phi^{\nabla})^{\Delta}(t)\psi(s)u(s)\Delta s\\
&\quad+\frac{1}{d}\int_{t}^{t_n}(p\psi^{\nabla})^{\Delta}(t)\phi(s)u(s)\Delta s\\
&\quad+A(u)(p\psi^{\nabla})^{\Delta}(t)+B(u)(p\phi^{\nabla})^{\Delta}(t)\\
&= (py^{\nabla})^{\Delta}(t), \quad t\in[\rho(t_1),t_n].
\end{aligned}
$$

Thus, y satisfies the first equation of (6.12). Now,

$$
y(t_1) = \frac{1}{d}\int_{t_1}^{t_n}\psi(t_1)\phi(s)u(s)\Delta s + A(u)\psi(t_1)+B(u)\phi(t_1),
$$

$$
p(t_1)y^{\nabla}(t_1) = \frac{p(t_1)\psi^{\nabla}(t_1)}{d}\int_{t_1}^{t_n}\phi(s)u(s)\Delta s
$$

$$+A(u)p(t_1)\psi^{\nabla}(t_1)+B(u)p(t_1)\phi^{\nabla}(t_1),$$

whereupon,

$$\begin{aligned}
\alpha y(t_1)-\beta p(t_1)y^{\nabla}(t_1) &= \frac{1}{d}\int_{t_1}^{t_n}\left(\alpha\psi(t_1)-\beta p(t_1)\psi^{\nabla}(t_1)\right)\phi(s)u(s)\Delta s\\
&\quad +A(u)\left(\alpha\psi(t_1)-\beta p(t_1)\psi^{\nabla}(t_1)\right)\\
&\quad +B(u)\left(\alpha\phi(t_1)-\beta p(t_1)\phi^{\nabla}(t_1)\right)\\
&= B(u)(\alpha\phi(t_1)-\beta p(t_1)\phi^{\nabla}(t_1))
\end{aligned}$$

and

$$\begin{aligned}
y(t_j) &= \int_{t_1}^{t_n} G(t_j,s)u(s)\Delta s+A(u)\psi(t_j)+B(u)\phi(t_j),\\
a_j y(t_j) &= \int_{t_1}^{t_n} a_j G(t_j,s)u(s)\Delta s+A(u)a_j\psi(t_j)+B(u)a_j\phi(t_j),\\
b_j y(t_j) &= \int_{t_1}^{t_n} b_j G(t_j,s)u(s)\Delta s+A(u)b_j\psi(t_j)+B(u)b_j\phi(t_j), \quad j\in\{2,\dots,n-2\}.
\end{aligned}$$

Hence, using the first equation of the system (6.10), we get

$$\begin{aligned}
\sum_{j=2}^{n-1} a_j y(t_j) &= \int_{t_1}^{t_n}\sum_{j=2}^{n-1} a_j G(t_j,s)u(s)\Delta s+A(u)\sum_{j=2}^{n-1} a_j\psi(t_j)\\
&\quad +B(u)\sum_{j=2}^{n-1} a_j\phi(t_j)\\
&= B(u)\left(\alpha\phi(t_1)-\beta p(t_1)\phi^{\nabla}(t_1)\right)\\
&= \alpha y(t_1)-\beta p(t_1)y^{\nabla}(t_1).
\end{aligned}$$

Next,

$$\begin{aligned}
y(t_n) &= \int_{t_1}^{t_n} G(t_n,s)u(s)\Delta s+A(u)\psi(t_n)+B(u)\phi(t_n),\\
p(t_n)y^{\nabla}(t_n) &= \frac{p(t_n)\phi^{\nabla}(t_n)}{d}\int_{t_1}^{t_n}\psi(s)u(s)\Delta s\\
&\quad +A(u)p(t_n)\psi^{\nabla}(t_n)+B(u)p(t_n)\phi^{\nabla}(t_n),
\end{aligned}$$

and

$$\begin{aligned}\gamma y(t_n)+\delta p(t_n)y^{\nabla}(t_n) &= \frac{1}{d}\int_{t_1}^{t}\left(\gamma\phi(t_n)+\delta p(t_n)\phi^{\nabla}(t_n)\right)\psi(s)u(s)\Delta s\\ &\quad +A(u)\left(\gamma\psi(t_n)+\delta p(t_n)\psi^{\nabla}(t_n)\right)\\ &\quad +B(u)\left(\gamma\phi(t_n)+\delta p(t_n)\phi^{\nabla}(t_n)\right)\\ &= A(u)\left(\gamma\psi(t_n)+\delta p(t_n)\psi^{\nabla}(t_n)\right),\end{aligned}$$

and using the second equation of the system (6.10), we find

$$\begin{aligned}\sum_{j=2}^{n-1} b_j y(t_j) &= \int_{t_1}^{t_n}\sum_{j=2}^{n-1} b_j G(t_j,s)u(s)\Delta s\\ &\quad +A(u)\sum_{j=2}^{n-1} b_j\psi(t_j)+B(u)\sum_{j=2}^{n-1} b_j\phi(t_j)\\ &= A(u)\left(\gamma\psi(t_n)+\delta p(t_n)\psi^{\nabla}(t_n)\right)\\ &= \gamma y(t_n)+\delta p(t_n)y^{\nabla}(t_n).\end{aligned}$$

Therefore, y is a solution to the BVP (6.12). This completes the proof.

Lemma 6.10. *Assume* $(E13)$-$(E15)$ *and*

$$D<0,\quad d-\sum_{j=2}^{n-1} a_j\phi(t_j)>0,\quad d-\sum_{j=2}^{n-1} b_j\psi(t_j)>0. \tag{6.13}$$

Then, if $u\in\mathscr{C}_{rd}([\rho(t_1),t_n])$, $u\ge 0$ *on* $[\rho(t_1),t_n]$, *and* y *satisfies* (6.11), *then* $y(t)\ge 0$, $t\in[\rho(t_1),t_n]$.

Proof. By (6.13), it follows that $A(u)\ge 0$, $B(u)\ge 0$. Also,

$$G(t,s)\ge 0,\quad t,s\in[\rho(t_1),t_n].$$

Therefore, y, defined by (6.11), is a non-negative on $[\rho(t_1),t_n]$, This completes the proof.

Lemma 6.11. *Assume* $(E13)$-$(E15)$ *and* (6.13). *Fix* $\xi_1,\xi_2\in\mathbb{T}_\kappa\bigcap\mathbb{T}^\kappa$ *so that*

$$p(t_1)<\xi_1<\xi_2<t_n,$$

and set

$$\Gamma = \min\left\{ \frac{\phi(\xi_2)}{\phi(\rho(t_1))}, \frac{\psi(\xi_1)}{\psi(t_n)} \right\}.$$

If y satisfies (6.11), *then*

$$\min_{t\in[\xi_1,\xi_2]} y(t) \geq \Gamma \|y\|,$$

where

$$\|y\| = \max_{t\in[\rho(t_1),t_n]} y(t).$$

Proof. We have

$$0 \leq G(t,s) \leq G(s,s), \quad t,s \in [\rho(t_1),t_n].$$

Then

$$\begin{aligned} y(t) &= \int_{t_1}^{t_n} G(t,s)u(s)\Delta s + A(u)\psi(t) + B(u)\phi(t) \\ &\leq \int_{t_1}^{t_n} G(s,s)u(s)\Delta s + A(u)\psi(t_n) + B(u)\phi(t_n), \end{aligned}$$

$t \in [\rho(t_1),t_n]$. Next,

$$\begin{aligned} \frac{G(t,s)}{G(s,s)} &= \begin{cases} \dfrac{\phi(t)}{\phi(s)} & \rho(t_1) \leq s \leq t \leq t_n \\[2ex] \dfrac{\psi(t)}{\psi(s)} & \rho(t_1) \leq t \leq s \leq t_n \end{cases} \\ &\geq \begin{cases} \dfrac{\phi(\xi_2)}{\phi(\rho(t_1))} & \rho(t_1) \leq s \leq t \leq t_n \\[2ex] \dfrac{\psi(\xi_1)}{\psi(t_n)} & \rho(t_1) \leq t \leq s \leq t_n \end{cases} \\ &\geq \Gamma, \quad t,s \in [\rho(t_1),t_n]. \end{aligned}$$

Hence,

$$\begin{aligned} y(t) &= \int_{t_1}^{t_n} G(t,s)u(s)\Delta s + A(u)\psi(t) + B(u)\phi(t) \\ &= \int_{t_1}^{t_n} \frac{G(t,s)}{G(s,s)} G(s,s)u(s)\Delta s + A(u)\psi(t) + B(u)\phi(t) \\ &\geq \Gamma \int_{t_1}^{t_n} u(s)\Delta s + A(u)\psi(t) + B(u)\phi(t) \\ &\geq \Gamma \int_{t_1}^{t_n} u(s)\Delta s + A(u)\Gamma\psi(t_n) + B(u)\Gamma\phi(\rho(t_1)) \end{aligned}$$

$$= \Gamma \left(\int_{t_1}^{t_n} u(s)\Delta s + A(u)\psi(t_n) + B(u)\phi(\rho(t_1)) \right)$$

$$\geq \Gamma \|y\|, \quad t \in [\xi_1, \xi_2].$$

Consequently,

$$\min_{t\in[\xi_1,\xi_2]} y(t) \geq \Gamma \|y\|.$$

This completes the proof.

In addition, suppose

(E18) $\xi_1, \xi_2 \in [\rho(t_1), t_n]$ be such that

$$\xi_1 < t_* < \xi_2,$$

$$\int_{\xi_1}^{\xi_2} G(t,s)h(s)\Delta s > 0, \quad t \in [\rho(t_1), t_n],$$

$\tau \in [\rho(t_1), t_n]$ is chosen so that

$$\int_{\xi_1}^{\xi_2} G(\tau,s)h(s)\Delta s = \max_{t\in[\rho(t_1),t_n]} \int_{\xi_1}^{\xi_2} G(t,s)h(s)\Delta s.$$

Define the constant

$$K = \int_{t_1}^{t_n} G(s,s)h(s)\Delta s + A(h)\psi(t_n) + B(h)\phi(\rho(t_1)).$$

Let $B = \mathscr{C}([\rho(t_1), t_n])$ be endowed with the norm

$$\|y\| = \sup_{t\in[\rho(t_1),t_n]} |y(t)|$$

and

$$P = \{y \in B : y(t) \geq 0, \quad t \in [\rho(t_1), t_n], \quad y(t) \geq \Gamma\|y\|, \quad t \in [\xi_1, \xi_2]\}.$$

For $y \in P$, define the operator

$$Ty(t) = \lambda \left(\int_{t_1}^{t_n} G(t,s)h(s)f(y(s))\Delta s + A(hf(y))\psi(t) + B(hf(y))\phi(t) \right),$$

$t \in [\rho(t_1), t_n]$.

Lemma 6.12. *Suppose* $(E13)$-$(E18)$. *Then* $T : P \to P$.

Proof. Let $y \in P$ be arbitrarily chosen. Then

$$\begin{aligned}
Ty(t) &= \lambda\left(\int_{t_1}^{t_n} G(t,s)h(s)f(y(s))\Delta s + A(hf(y))\psi(t) + B(hf(y))\phi(t)\right)\\
&\leq \lambda\left(\int_{t_1}^{t_n} G(s,s)h(s)f(y(s))\Delta s + A(hf(y))\psi(t) + B(hf(y))\phi(t)\right)\\
&\leq \lambda\left(\int_{t_1}^{t_n} G(s,s)h(s)f(y(s))\Delta s + A(hf(y))\psi(t_n) + B(hf(y))\phi(\rho(t_1))\right),
\end{aligned}$$

$t \in [\rho(t_1), t_n]$. For $t \in [\xi_1, \xi_2]$, we have

$$\begin{aligned}
Ty(t) &= \lambda\left(\int_{t_1}^{t_n} G(t,s)h(s)f(y(s))\Delta s + A(hf(y))\psi(t) + B(hf(y))\phi(t)\right)\\
&= \lambda\left(\int_{t_1}^{t_n} \frac{G(t,s)}{G(s,s)} G(s,s)h(s)f(y(s))\Delta s + A(hf(y))\psi(t) + B(hf(y))\phi(t)\right)\\
&\geq \lambda\left(\Gamma\int_{t_1}^{t_n} G(s,s)h(s)f(y(s))\Delta s + A(hf(y))\psi(t) + B(hf(y))\phi(t)\right)\\
&\geq \lambda\left(\Gamma\int_{t_1}^{t_n} G(s,s)h(s)f(y(s))\Delta s + A(hf(y))\psi(\xi_1) + B(hf(y))\phi(\xi_2)\right)\\
&\geq \lambda\left(\Gamma\int_{t_1}^{t_n} G(s,s)h(s)f(y(s))\Delta s + \Gamma A(hf(y))\psi(t_n)\right.\\
&\quad \left. + \Gamma B(hf(y))\phi(\rho(t_1))\right)\\
&= \Gamma\|Ty\|.
\end{aligned}$$

Thus, $T : P \to P$. This completes the proof.

Exercise 6.6. Suppose $(E13)$-$(E18)$ and (6.13). Prove that $T : P \to P$ is a completely continuous operator.

Theorem 6.13. *Suppose* $(E13)$-$(E18)$ *and* (6.13). *Then, for each* λ *satisfying*

$$\frac{1}{f_\infty \Gamma \int_{\xi_1}^{\xi_2} G(\tau,s)h(s)\Delta s} < \lambda < \frac{1}{f_0 K},$$

there exists at least one positive solution of the BVP (6.8).

Proof. Take $\varepsilon > 0$ so that $\varepsilon < f_\infty$ and

$$\frac{1}{(f_\infty - \varepsilon)\Gamma \int_{\xi_1}^{\xi_2} G(\tau,s)h(s)\Delta s} \leq \lambda \leq \frac{1}{(f_0 + \varepsilon)K}.$$

By the definition of f_0, it follows that there exists a constant $R_1 > 0$ so that

$$f(y) \leq (f_0+\varepsilon)y, \quad 0<y\leq R_1.$$

Define

$$\Omega_1 = \{y \in P : \|y\| < R_1\}.$$

By the definitions of $A(\cdot)$ and $B(\cdot)$, it follows that

$$A(hf(y)) \leq A(h)\|f(y)\|,$$

$$B(hf(y)) \leq B(h)\|f(y)\|, \quad y \in \overline{\Omega}_1.$$

For $y \in P\bigcap \partial\Omega_1$, we have

$$\begin{aligned}
Ty(t) &= \lambda\left(\int_{t_1}^{t_n} G(t,s)h(s)f(y(s))\Delta s + A(hf(y))\psi(t) + B(hf(y))\phi(t)\right)\\
&\leq \lambda\Bigg(\|f(y)\|\int_{t_1}^{t_n} G(t,s)h(s)\Delta s + A(h)\|f(y)\|\psi(t)\\
&\quad +B(h)\|f(y)\|\phi(t)\Bigg)\\
&= \lambda\|f(y)\|\left(\int_{t_1}^{t_n} G(t,s)h(s)\Delta s + A(h)\psi(t) + B(h)\phi(t)\right)\\
&\leq \lambda\|f(y)\|\left(\int_{t_1}^{t_n} G(s,s)h(s)\Delta s + A(h)\psi(t) + B(h)\phi(t)\right)\\
&= \lambda K\|f(y)\|\\
&\leq \lambda K(f_0+\varepsilon)\|y\|\\
&\leq \|y\|, \quad t \in [\rho(t_1), t_n].
\end{aligned}$$

Thus,

$$\|Ty\| \leq \|y\|, \quad y \in P\bigcap \partial\Omega_1.$$

Next, by the definition of f_∞, it follows that there exists an $R_2' > R_1$ so that

$$f(y) \geq (f_\infty - \varepsilon)y, \quad y \geq R_2'.$$

Take

$$R_2 = \max\left\{2R_1, \frac{R_2'}{\Gamma}\right\}$$

and define

$$\Omega_2 = \{y \in P : \|y\| < R_2\}.$$

Let $y \in P \bigcap \partial\Omega_2$ be arbitrarily chosen. Then, for $s \in [\xi_1, \xi_2]$, we have

$$\begin{aligned} y(s) &\geq \Gamma\|y\| \\ &= \Gamma R_2 \end{aligned}$$

and

$$\begin{aligned} Ty(t) &= \lambda\left(\int_{t_1}^{t_n} G(t,s)h(s)f(y(s))\Delta s + A(hf(y))\psi(t) + B(hf(y))\phi(t)\right) \\ &\geq \lambda\int_{t_1}^{t_n} G(t,s)h(s)f(y(s))\Delta s \\ &\geq \lambda(f_\infty - \varepsilon)\int_{t_1}^{t_n} G(\tau,s)h(s)y(s)\Delta s \\ &\geq \lambda(f_\infty - \varepsilon)\int_{\xi_1}^{\xi_2} G(\tau,s)h(s)\|y\|\Delta s \\ &\geq \lambda\Gamma R_2(f_\infty - \varepsilon)\int_{\xi_1}^{\xi_2} G(\tau,s)h(s)\Delta s \\ &\geq R_2 \\ &= \|y\|, \quad t \in [\rho(t_1), t_n]. \end{aligned}$$

Therefore,

$$\|Ty\| \geq \|y\|, \quad y \in P \bigcap \partial\Omega_2.$$

Now, applying the Guo-Krasnosel'skii Theorem 1.11, it follows that the operator T has a fixed point in $P \bigcap (\overline{\Omega}_2 \backslash \Omega_1)$. This completes the proof.

Theorem 6.14. *Suppose* $(E13)$-$(E18)$ *and* (6.13). *Then, for each* λ *satisfying*

$$\frac{1}{f_0\Gamma\int_{\xi_1}^{\xi_2} G(\tau,s)h(s)\Delta s} < \lambda < \frac{1}{f_\infty K},$$

there exists at least one positive solution for the BVP (6.8).

Proof. Take $\eta > 0$ so that $\eta < f_0$ and

$$\frac{1}{(f_0 - \eta)\Gamma\int_{\xi_1}^{\xi_2} G(\tau,s)h(s)\Delta s} \leq \lambda \leq \frac{1}{(f_\infty + \eta)K}.$$

By the definition of f_0, it follows that there exists an $R_1 > 0$ so that

$$f(y) \geq (f_0 - \eta)y, \quad 0 < y \leq R_1.$$

Define

$$\Omega_1 = \{y \in P : \|y\| < R_1\}.$$

Then, for each $y \in P\bigcap \partial\Omega_1$, we have

$$y(s) \geq \Gamma \|y\|, \quad s \in [\xi_1, \xi_2],$$

and

$$\begin{aligned}
Ty(\tau) &= \lambda\left(\int_{t_1}^{t_n} G(\tau,s)h(s)f(y(s))\Delta s + A(hf(y))\psi(t) + B(hf(y))\phi(t)\right)\\
&\geq \lambda \int_{t_1}^{t_n} G(\tau,s)h(s)f(y(s))\Delta s\\
&\geq \lambda \int_{\xi_1}^{\xi_2} G(\tau,s)h(s)f(y(s))\Delta s\\
&\geq \lambda(f_0-\eta)\int_{\xi_1}^{\xi_2} G(\tau,s)h(s)y(s)\Delta s\\
&\geq \lambda\Gamma(f_0-\eta)\|y\|\int_{\xi_1}^{\xi_2} G(\tau,s)h(s)\Delta s\\
&\geq \|y\|,
\end{aligned}$$

whereupon,

$$\|Ty\| \geq \|y\|, \quad y \in P\bigcap \partial\Omega_1.$$

Now, by the definition of f_∞, it follows that there exists an $R_2' > R_2$ so that

$$f(y) \leq (f_\infty + \eta)y, \quad y \geq R_2'.$$

If f is bounded, then there exists a positive constant M so that

$$f(y) \leq M, \quad y \in (0,\infty).$$

Let

$$R = \max\left\{2R_2', M\left(\int_{t_1}^{t_n} G(s,s)h(s)\Delta s + A(h)\psi(t_n) + B(h)\phi(t_1)\right)\right\}$$

and

$$\Omega_2 = \{y \in P : \|y\| < R_2\}.$$

For $y \in P\bigcap \partial\Omega_2$, we get

$$Ty(t) = \lambda\left(\int_{t_1}^{t_n} G(t,s)h(s)f(y(s))\Delta s + A(hf(y))\psi(t) + B(hf(y))\phi(t)\right)$$

$$
\begin{aligned}
&\leq \lambda\left(M\int_{t_1}^{t_n} G(s,s)h(s)\Delta s + MA(h)\psi(t_n) + MB(h)\phi(t_1)\phi(t)\right)\\
&\leq R_2\\
&= \|y\|, \quad t\in[\rho(t_1),t_n].
\end{aligned}
$$

Thus,

$$\|Ty\| \leq \|y\|, \quad y\in P\bigcap\partial\Omega_2.$$

If f is unbounded, then take $R_2 = \max\{2R_1, R_2'\}$ such that

$$f(y) \leq f(R_2), \quad 0\leq y\leq R_2.$$

Let $y\in P\bigcap\partial\Omega_2$. Then

$$
\begin{aligned}
Ty(t) &= \lambda\left(\int_{t_1}^{t_n} G(t,s)h(s)f(y(s))\Delta s + A(hf(y))\psi(t) + B(hf(y))\phi(t)\right)\\
&\leq \lambda\left(f(R_2)\int_{t_1}^{t_n} G(t,s)h(s)\Delta s + A(h)f(R_2)\psi(t_n) + B(h)f(R_2)\phi(t_1)\right)\\
&\leq \lambda(f_\infty+\eta)R_2K\\
&\leq R_2\\
&= \|y\|, \quad t\in[\rho(t_1),t_n].
\end{aligned}
$$

Therefore,

$$\|Ty\| \leq \|y\|, \quad y\in P\bigcap\partial\Omega_2.$$

Now, applying the Guo-Krasnosel'skii Theorem 1.11, it follows that the operator T has a fixed point in $P\bigcap(\overline{\Omega}_2\backslash\Omega_1)$. This completes the proof.

Exercise 6.7. Let $\mathbb{T} = 2^{-10}\mathbb{Z}$. Find λ so that the eigenvalue problem

$$
\begin{aligned}
&y^{\nabla\Delta} + \lambda\frac{y^3}{y^2+1} = 0, \quad t\in[0,1],\\
&y(0) - y^\nabla(0) = \frac{2}{5}\left(y\left(\frac{1}{4}\right) + y\left(\frac{3}{4}\right)\right),\\
&y(1) + y^\nabla(1) = \frac{2}{5}\left(y\left(\frac{1}{4}\right) + y\left(\frac{3}{4}\right)\right)
\end{aligned}
$$

has at least one positive solution.

6.4 Advanced Practical Problems

Problem 6.1. Let $\mathbb{T} = 2\mathbb{Z}$. Prove that the eigenvalue problem

$$y^{\Delta^2} + \lambda \frac{3t+1}{4t^2+t+2} \frac{y(t+2)\left((y(t+2))^3 + (y(t+2))^2 + \frac{1}{700}\right)}{(y(t+2))^3+2} = 0, \quad t \in [0,10],$$

$$y(0) = y(12) = 0$$

has at least one positive solution.

Problem 6.2. Let $\mathbb{T} = 2^{\mathbb{N}_0}$. Prove that the eigenvalue problem

$$y^{\Delta^2} + \lambda \frac{t+1}{t^2+t+1} \frac{y(2t)\left(350(y(2t))^7 + 100(y(2t))^2 + 3\right)}{(y(2t))^7+2} = 0, \quad t \in [1,64],$$

$$y(1) = y(128) = 0$$

has at least one positive solution.

Problem 6.3. Let $\mathbb{T} = 2^{\mathbb{N}_0}$. Prove that the eigenvalue problem

$$y^{\Delta^2} + \lambda \left(u^{\frac{1}{20}} + 2u^{\frac{1}{40}} + u^{\frac{1}{8}} + \frac{1}{3}\right) = 0, \quad t \in (1,16),$$

$$y(1) = y(32) = 0,$$

has at least one positive solution for any $\lambda > 0$.

Problem 6.4. Let $\mathbb{T} = 3^{\mathbb{N}_0}$. Consider the iterative eigenvalue problem

$$u_1^{\Delta^2}(t) + \lambda_1 \frac{1+t}{7+t^4} \frac{u_2(3t)((u_2(3t))^2+1)}{(u_2(3t))^2+20} = 0,$$

$$u_2^{\Delta^2}(t) + \lambda_2 \frac{t^2+1}{t^2+t+10} \frac{u_3(3t)\left((u_3(3t))+1\right)}{(u_3(3t))+2} = 0,$$

$$u_3^{\Delta^2}(t) + \lambda_3 \frac{1+t}{1+3t^2} \frac{u_4(3t)((u_4(3t))^4+2)}{(u_4(3t))^4+u_4(3t)+1} = 0,$$

$$u_4^{\Delta^2}(t) + \lambda_4 \frac{1+t}{1+t+t^2} \frac{u_1(3t)((u_1(3t))^4+1)}{(u_1(3t))^4+u_1(3t)+1} = 0 \quad t \in [1,81],$$

$$u_i(1) = u_i(243) = 0, \quad i \in \{1,2,3,4\}.$$

Find $L_1, L_2 > 0$, so that the considered iterative eigenvalue problem to have at least one non-negative solution for any

$$L_1 < \lambda_i < L_2, \quad i \in \{1,2,3,4\}.$$

Problem 6.5. Let $\mathbb{T} = 2^{\mathbb{N}_0}$. Find λ so that the eigenvalue problem

$$\left((t^2+1)y^{\nabla}\right)^{\Delta} + \lambda \frac{y^5}{y^4+1} = 0, \quad 1 \le t \le 32,$$

$$y(1) - y^{\nabla}(1) = 3(y(2) + y(8)),$$

$$y(1) + y^{\nabla}(1) = 4(y(2) + y(8))$$

has at least one positive solution.

6.5 Notes and References

In this chapter we investigate some nonlinear second order eigenvalue problems. There are deducted and proved existence results for iterative eigenvalue problems. Some of the results in this chapter can be found in [2], [16], [40] and [46].

Chapter 7
Boundary Value Problems for Second Order Impulsive Dynamic Equations

Let $\mathbb{T}$ be a time scale with forward jump operator and delta differentiation operator σ and Δ, respectively. Let, also, $0, T \in \mathbb{T}$, $\{t_j\}_{j=1}^m \subset \mathbb{T}$, t_j are right-dense in $\mathbb{T}$, $j \in \{1, \ldots, m\}$,

$$0 = t_0 < t_1 < \ldots < t_m < t_{m+1} = T.$$

Set

$$J = [0, T], \quad J_k = [t_k, t_{k+1}], \quad k \in \{0, \ldots, m\}.$$

7.1 Existence of Solutions for Boundary Value Problems for Second Order Impulsive Dynamic Equations

In this section we will investigate the following BVP

$$\begin{aligned} -y^{\Delta^2}(t) &= f(t, y(t)), \quad t \in J \backslash \{t_j\}_{j=1}^m, \\ y(t_j^+) - y(t_j^-) &= I_j(y(t_j)), \\ y^{\Delta}(t_j^+) - y^{\Delta}(t_j^-) &= L_j(y(t_j)), \quad j \in \{1, \ldots, m\}, \\ y(0) &= y(T) = 0, \end{aligned} \tag{7.1}$$

where

(H1) $f \in \mathscr{C}([0, T] \times \mathbb{R})$, $\sigma \in \mathscr{C}(\mathbb{T})$,

$$|f(t, y)| \leq p(t), \quad (t, y) \in [0, T] \times \mathbb{R},$$

for some non-negative function $p \in \mathscr{C}([0, T])$,

(H2) $I_j, L_j \in \mathscr{C}(\mathbb{R})$,

DOI: 10.1201/9781003173557-7

$$|I_j(y)| \le c_j, \quad |L_j(y)| \le l_j, \quad j \in \{1,\dots,m\}, \quad y \in \mathbb{R},$$

c_j, l_j are non-negative constants, $j \in \{1,\dots,m\}$.

Theorem 7.1. *A function $y \in PC([0,T]) \bigcap \mathscr{C}^2([0,T]\backslash\{t_j\}_{j=1}^m)$ is a solution of the BVP (7.1) if and only if $y \in PC([0,T])$, and it satisfies the integral equation*

$$y(t) = \int_0^T G(t,s) f(s,y(s))\Delta s + \sum_{j=1}^m W_j(t,y(t_j)), \quad t \in [0,T], \tag{7.2}$$

where

$$G(t,s) = \begin{cases} \frac{T-t}{T}\sigma(s) & 0 \le s \le t \\ \frac{T-\sigma(s)}{T} t & t \le s \le T, \end{cases}$$

$$W_j(t,y(t_j)) = \begin{cases} \frac{T-t}{T} I_j(y(t_j)) - \frac{T-t}{T} t_j L_j(y(t_j)) & 0 \le t \le t_j \\ -\frac{t}{T} I_j(y(t_j)) - \frac{T-t_j}{T} t L_j(y(t_j)) & t_j < t \le T. \end{cases}$$

Proof. It follows that $y \in PC([0,T]) \bigcap \mathscr{C}^2([0,T]\backslash\{t_j\}_{j=1}^m)$ of the impulsive dynamic equation

$$\begin{aligned} -y^{\Delta^2}(t) &= f(t,y(t)), \quad t \in J\backslash\{t_j\}_{j=1}^m, \\ y(t_j^+) - y(t_j^-) &= I_j(y(t_j)), \\ y^{\Delta}(t_j^+) - y^{\Delta}(t_j^-) &= L_j(y(t_j)), \quad j \in \{1,\dots,m\}, \end{aligned} \tag{7.3}$$

if and only if $y \in PC([0,T])$ and it satisfies the integral equation

$$\begin{aligned} y(t) &= y_0 + t y_1 - \int_0^t (t-\sigma(s)) f(s,y(s))\Delta s \\ &\quad + \sum_{\{j: 0 < t_j < t\}} (I_j(y(t_j)) + (t-t_j) L_j(y(t_j))), \end{aligned} \tag{7.4}$$

where y_0 and y_1 are real constants.

1. Let $y \in PC([0,T]) \bigcap \mathscr{C}^2([0,T]\backslash\{t_j\}_{j=1}^m)$ satisfy the BVP (7.1). Then $y \in PC([0,T])$ satisfies (7.5) and

$$\begin{aligned} 0 &= y(0) \\ &= y_0. \end{aligned}$$

Hence,

$$
\begin{aligned}
0 &= y(T) \\
&= Ty_1 - \int_0^T (T-\sigma(s)) f(s,y(s)) \Delta s \\
&\quad + \sum_{j=1}^{m} (I_j(y(t_j)) + (T-t_j) L_j(y(t_j))),
\end{aligned}
$$

whereupon,

$$
Ty_1 = \int_0^T (T-\sigma(s)) f(s,y(s)) \Delta s - \sum_{j=1}^{m} (I_j(y(t_j)) + (T-t_j) L_j(y(t_j)))
$$

or

$$
y_1 = \frac{1}{T}\int_0^T (T-\sigma(s)) f(s,y(s)) \Delta s - \frac{1}{T}\sum_{j=1}^{m} (I_j(y(t_j)) + (T-t_j) L_j(y(t_j))).
$$

Then

$$
\begin{aligned}
y(t) &= \frac{t}{T}\int_0^T (T-\sigma(s)) f(s,y(s)) \Delta s \\
&\quad - \frac{t}{T}\sum_{\{j:t\le t_j<T\}} (I_j(y(t_j)) + (T-t_j) L_j(y(t_j))) \\
&\quad - \int_0^t (t-\sigma(s)) f(s,y(s)) \Delta s \\
&\quad + \sum_{\{j:0<t_j<t\}} (I_j(y(t_j)) + (t-t_j) L_j(y(t_j))),
\end{aligned}
$$

i.e.,

$$
\begin{aligned}
y(t) &= \int_0^t \frac{T-t}{T}\sigma(s) f(s,y(s)) \Delta s + \int_t^T \frac{t(T-\sigma(s))}{T} f(s,y(s)) \Delta s \\
&\quad + \sum_{\{j:0<t_j<t\}} \left(\frac{T-t}{T} I_j(y(t_j)) - \frac{T-t}{T} t_j L_j(y(t_j)) \right) \\
&\quad + \sum_{\{j:t\le t_j<T\}} \left(-\frac{t}{T} I_j(y(t_j)) - \frac{T-t_j}{T} L_j(y(t_j)) \right).
\end{aligned} \tag{7.5}
$$

2. Let $y \in PC([0,T])$ satisfy (7.2). Then $y \in PC([0,T]) \bigcap \mathscr{C}^2([0,T]\backslash\{t_j\}_{j=1}^{m})$ satisfies (7.3) and (7.5). Hence, using (7.5), we arrive at

$$y(0) = 0,$$

$$y(T) = 0.$$

This completes the proof.

Note that, if $(H2)$ holds, then

$$|W_j(t, y(t_j))| \leq c_j + Tl_j, \quad j \in \{1, \dots, m\}.$$

Let

$$\sup_{(t,s)\in[0,T]\times[0,T]} G(t,s) = A.$$

For $y \in PC([0,T])$, define the operator

$$\Phi y(t) = \int_0^T G(t,s) f(s, y(s)) \Delta s + \sum_{j=1}^m W_j(t, y(t_j)), \quad t \in [0,T].$$

Note that $\Phi : PC([0,T]) \to PC([0,T])$.

Theorem 7.2. *Assume* $(H1)$-$(H2)$. *Then the BVP* (7.1) *has at least one solution.*

Proof. 1. $\Phi : PC([0,T]) \to PC([0,T])$ is continuous. Let $\{y_n\}_{n\in\mathbb{N}}$ be a sequence of elements of $PC([0,T])$ such that $y_n \to y$, as $n \to \infty$, for some $y \in PC([0,T])$. Then

$$\begin{aligned}
|\Phi y_n(t) - \Phi y(t)| &= \left| \int_0^T G(t,s) f(s, y_n(s)) \Delta s + \sum_{j=1}^m W_j(t, y_n(t_j)) \right. \\
&\quad \left. - \int_0^T G(t,s) f(s, y(s)) \Delta s - \sum_{j=1}^m W_j(t, y(t_j)) \right| \\
&= \left| \int_0^T G(t,s)(f(s, y_n(s)) - f(s, y(s))) \Delta s \right. \\
&\quad \left. + \sum_{j=1}^m (W_j(t, y_n(t_j)) - W_j(t, y(t_j))) \right| \\
&\leq \left| \int_0^T G(t,s)(f(s, y_n(s)) - f(s, y(s))) \Delta s \right| \\
&\quad + \left| \sum_{j=1}^m (W_j(t, y_n(t_j)) - W_j(t, y(t_j))) \right| \\
&\leq \int_0^T G(t,s) |f(s, y_n(s)) - f(s, y(s))| \Delta s \\
&\quad + \sum_{j=1}^m \left| W_j(t, y_n(t_j)) - W_j(t, y(t_j)) \right|
\end{aligned}$$

$$\to 0, \quad \text{as} \quad n \to \infty, \quad t \in [0,T],$$

whereupon,

$$\|\Phi y_n - \Phi y\| \to 0, \quad \text{as} \quad n \to \infty.$$

2. $\Phi : PC([0,T]) \to PC([0,T])$ maps a bounded set into a bounded set. Let

$$B = \{y \in PC([0,t]) : \|y\| \leq C\}$$

for some positive constant C. Then, for $y \in B$, we get

$$\begin{aligned}
|\Phi y(t)| &= \left| \int_0^T G(t,s) f(s,y(s)) \Delta s + \sum_{j=1}^m W_j(t, y(t_j)) \right| \\
&\leq \left| \int_0^T G(t,s) f(s,y(s)) \Delta s \right| + \left| \sum_{j=1}^m W_j(t, y(t_j)) \right| \\
&\leq \int_0^T G(t,s) |f(s,y(s))| \Delta s + \sum_{j=1}^m |W_j(t, y(t_j))| \\
&\leq A \int_0^T p(s) \Delta s + \sum_{j=1}^m (c_j + T l_j), \quad t \in [0,T].
\end{aligned} \tag{7.6}$$

Therefore,

$$\|\Phi y\| \leq A \int_0^T p(s) \Delta s + \sum_{j=1}^m (c_j + T l_j). \tag{7.7}$$

3. $\Phi : PC([0,T]) \to PC([0,T])$ maps bounded sets into equi-continuous sets. Let $y \in B$, $t^1, t^2 \in [0,T]$. Without loss of generality, assume that $t^1 < t^2$. Then

$$\begin{aligned}
|\Phi y(t^1) - \Phi y(t^2)| &= \left| \int_0^T G(t^1,s) f(s,y(s)) \Delta s + \sum_{j=1}^m W_j(t^1, y(t_j)) \right. \\
&\quad \left. - \int_0^T G(t^2,s) f(s,y(s)) \Delta s - \sum_{j=1}^m W_j(t^2, y(t_j)) \right| \\
&= \left| \int_0^T (G(t^1,s) - G(t^2,s)) f(s,y(s)) \Delta s \right. \\
&\quad \left. + \sum_{j=1}^m (W_j(t^1, y(t_j)) - W_j(t^2, y(t_j))) \right| \\
&\leq \left| \int_0^T (G(t^1,s) - G(t^2,s)) f(s,y(s)) \Delta s \right|
\end{aligned}$$

$$
\begin{aligned}
&+\left|\sum_{j=1}^{m}(W_j(t^1,y(t_j))-W_j(t^2,y(t_j)))\right| \\
&\leq \int_0^T |G(t^1,s)-G(t^2,s)||f(s,y(s))|\Delta s \\
&\quad +\sum_{j=1}^{m}|W_j(t^1,y(t_j))-W_j(t^2,y(t_j))| \\
&\leq \int_0^T |G(t^1,s)-G(t^2,s)|p(s)\Delta s \\
&\quad +\sum_{j=1}^{m}|W_j(t^1,y(t_j))-W_j(t^2,y(t_j))| \\
&\to 0, \quad \text{as} \quad t^1\to t^2.
\end{aligned}
$$

From here, using the Arzela-Ascoli theorem, it follows that $\{\Phi y\}_{y\in B}$ is an equicontinuous family.

4. Define the set

$$E(\Phi)=\{y\in PC([0,T]): y=\lambda\Phi y, \quad 0<\lambda<1\}.$$

Let $y\in E(\Phi)$. Then, using (7.6) and (7.7), we get

$$\|y\|\leq \lambda\left(A\int_0^T p(s)\Delta s+\sum_{j=1}^{m}(c_j+Tl_j)\right).$$

Therefore, $E(\Phi)$ is a bounded set.

By Schaefer's fixed point theorem, it follows that Φ has at least one fixed point in $PC([0,T])$. This completes the proof.

Exercise 7.1. Prove that the BVP

$$
\begin{aligned}
-y^{\Delta^2}(t) &= \frac{t}{1+(y(t))^2}, \quad t\in J\backslash\{t_j\}_{j=1}^m, \\
y(t_j^+)-y(t_j^-) &= \frac{1}{1+(y(t_j^-))^2}, \\
y^{\Delta}(t_j^+)-y^{\Delta}(t_j^-) &= \frac{3}{1+4(y(t_j^-))^4}, \quad j\in\{1,\ldots,m\}, \\
y(0) &= y(T)=0,
\end{aligned}
$$

has at least one solution in $PC([0,T])$.

7.2 Triple Positive Solutions of Boundary Value Problems for p-Laplacian Impulsive Dynamic Equations

Consider the following BVP

$$\begin{aligned}
\left(\phi_p(y^\Delta(t))\right)^\Delta + a(t)f(y(t)) &= 0, \quad t \in [0,T]\backslash\{t_1\},\\
y(t_1^+) - y(t_1^-) &= I(y(t_1)),\\
y^\Delta(t_1^+) - y^\Delta(t_1^-) &= J(y(t_1)),\\
y(0) = y^\Delta(T) &= 0,
\end{aligned} \tag{7.8}$$

where $\phi_p(s) = |s|^{p-2}s$, $p > 1$, $\phi_p^{-1} = \phi_q$, $\dfrac{1}{p} + \dfrac{1}{q} = 1$,

(H1) $f \in \mathscr{C}(\mathbb{R})$ is a non-negative function,
(H2) $a \in \mathscr{C}(\mathbb{T})$ is a non-negative function,
(H3) $I \in \mathscr{C}(\mathbb{R})$ is a non-negative function, $J \in \mathscr{C}(\mathbb{R})$ is a nonpositive and non-increasing.

We will start our investigations with the following useful result.

Theorem 7.3. *The function y is a solution of the BVP* (7.8) *if and only if it satisfies the integral equation*

$$y(t) = \int_0^t \phi_q\left(\int_s^T a(\tau)f(y(\tau))\Delta\tau\right)\Delta s + W(t, y(t_1)),$$

where

$$W(t, y(t_1)) = \begin{cases} -tJ(y(t_1)) & 0 \le t \le t_1 \\ I(y(t_1)) - t_1 J(y(t_1)) & t_1 < t \le T. \end{cases}$$

Proof. Note that the solution of equation

$$\left(\phi_p(y^\Delta(t))\right)^\Delta + a(t)f(y(t)) = 0, \quad t \in [0,T],$$

is given by the representation

$$y(t) = \int_0^t \phi_q\left(\int_s^T a(\tau)f(y(\tau))\Delta\tau\right)\Delta s + c_1 t + c_2,$$

where c_1 and c_2 are real constants. Now we integrate the first equation of (7.8) from t to T and using that $y^\Delta(T) = 0$, we get

$$\phi_p(y^\Delta(t)) = \int_t^T a(s)f(y(s))\Delta s,$$

and

$$y^{\Delta}(t) = \phi_q\left(\int_t^T a(s)f(y(s))\Delta s\right).$$

In particular,

$$y^{\Delta}(t_1^+) = \phi_q\left(\int_{t_1}^T a(s)f(y(s))\Delta s\right)$$

and hence,

$$\begin{aligned} y^{\Delta}(t_1^-) &= y^{\Delta}(t_1^+) - J(y(t_1)) \\ &= \phi_q\left(\int_{t_1}^T a(s)f(y(s))\Delta s\right) - J(y(t_1)). \end{aligned}$$

1. Let $t \in [0,t_1]$. Then

$$\begin{aligned} y(0) &= c_2 \\ &= 0 \end{aligned}$$

and

$$\begin{aligned} y(t) &= \int_0^t \phi_q\left(\int_s^T a(\tau)f(y(\tau))\Delta\tau\right)\Delta s + c_1 t, \\ y^{\Delta}(t) &= \phi_q\left(\int_t^T a(\tau)f(y(\tau))\Delta\tau\right) + c_1, \\ y^{\Delta}(t_1^-) &= \phi_q\left(\int_{t_1}^T a(\tau)f(y(\tau))\Delta\tau\right) + c_1 \\ &= \phi_q\left(\int_{t_1}^T a(a)f(y(s))\delta s\right) - J(y(t_1)). \end{aligned}$$

Then

$$c_1 = -J(y(t_1))$$

and

$$y(t) = \int_0^t \phi_q\left(\int_s^T a(\tau)f(y(\tau))\Delta\tau\right)\Delta s - tJ(y(t_1)).$$

Hence,

$$y(t_1^-) = \int_0^{t_1} \phi_q\left(\int_s^T a(\tau)f(y(\tau))\Delta\tau\right)\Delta s - t_1 J(y(t_1)).$$

2. Let $t \in [t_1,T]$. Then

$$y^{\Delta}(t) = \phi_q\left(\int_s^T a(\tau)f(y(\tau))\Delta\tau\right) + c_1,$$

$$
\begin{aligned}
y^{\Delta}(T) &= c_1 \\
&= 0
\end{aligned}
$$

and

$$y(t) = \int_0^t \phi_q\left(\int_s^T a(\tau)f(y(\tau))\Delta\tau\right)\Delta s + c_2.$$

From here,

$$
\begin{aligned}
y(t_1^+) &= \int_0^{t_1} \phi_q\left(\int_s^T a(\tau)f(y(\tau))\Delta\tau\right)\Delta s + c_2, \\
y(t_1^+) - y(t_1^-) &= c_2 + t_1 J(y(t_1)) \\
&= I(y(t_1)).
\end{aligned}
$$

Therefore,

$$c_2 = I(y(t_1)) - t_1 J(y(t_1)),$$

and

$$y(t) = \int_0^t \phi_q\left(\int_s^T a(\tau)f(y(\tau))\Delta\tau\right)\Delta s + I(y(t_1)) - t_1 J(y(t_1)).$$

This completes the proof.

Let

$$E = \{y : [0,T] \to \mathbb{R}, \quad y \in \mathscr{C}([0,t_1]), \quad y \in \mathscr{C}((t_1,T]), \quad y(t_1^+) \in \mathbb{R}\}$$

equipped with the norm

$$\|y\| = \max\{\sup_{t\in[0,t_1]} |y(t)|, \quad \sup_{t\in(t_1,T]} |y(t)|\}.$$

Define the cone P as follows

$$
\begin{aligned}
P = \Big\{ & y \in E : y \quad \text{is} \quad \text{concave}, \quad \text{non-decreasing}, \quad \text{non-negative} \\
& I(y(t_1)) \geq 0, \quad J(y(t_1)) \leq 0 \Big\}.
\end{aligned}
$$

Note that for $y \in P$, we obtain

$$\|y\| = y(T).$$

Let

$$\eta = \inf\left[\frac{t_1+T}{2}, T\right) \in \mathbb{T}.$$

Fix $0 < \eta < r < T$. In addition, if $y \in P$, then

$$y(t) \geq \frac{1}{2} \sup_{s \in [\frac{t_1}{2}, t_1]} y(s)$$

$$= \frac{1}{2} y(t_1), \quad t \in \left[\frac{t_1}{2}, t_1\right],$$

and

$$y(t) \geq \frac{1}{2} \sup_{s \in [\eta, T]} y(s)$$

$$= \frac{1}{2} y(T), \quad t \in [\eta, T].$$

Define, for $y \in P$,

$$\gamma(y) = \max_{t \in [0, \eta]} y(t)$$

$$= y(\eta),$$

$$\beta(y) = \min_{t \in [\eta, r]} y(t)$$

$$= y(\eta)$$

and

$$\alpha(y) = \max_{t \in [t_1, r]} y(t)$$

$$= y(r).$$

Thus, for each $y \in P$, we obtain

$$\gamma(y) = \beta(y)$$

$$\leq \alpha(y)$$

and

$$\gamma(y) = y(\eta)$$

$$\geq \frac{1}{2} y(T)$$

$$= \frac{1}{2}\|y\|.$$

Hence,

$$\|y\| \le 2\gamma(y).$$

Denote

$$P(\gamma,d) = \{x \in P : \gamma(x) < d\}.$$

In our next result we will use the following fixed point theorem.

Theorem 7.4. *Let P be a cone in a real Banach space E and α, β and γ be increasing, non-negative, continuous functionals on P such that for some $c > 0$ and $M > 0$*

$$\gamma(x) \le \beta(x) \le \alpha(x), \quad \|x\| \le M\gamma(x), \quad x \in \overline{P(\gamma,c)}.$$

Suppose that there exist positive numbers a and b with $a < b < c$ and

$$F : \overline{P(\gamma,c)} \to P$$

is a completely continuous operator such that

1. $\gamma(Fx) < c$, $x \in \partial P(\gamma,c)$,
2. $\beta(Fx) > b$, $x \in \partial P(\beta,b)$,
3. $P(\alpha,a) \ne \emptyset$ *and* $\alpha(Fx) < 0$, $x \in \partial P(\alpha,a)$.

Then F has at least three fixed points $x_1,x_2,x_3 \in \overline{P(\gamma,c)}$ such that

$$0 \le \alpha(x_1) < a < \alpha(x_2), \quad \beta(x_2) < b < \beta(x_3), \quad \gamma(x_3) < c.$$

For $y \in E$, define the operator

$$Fy(t) = \int_0^t \phi_q\left(\int_s^T a(\tau)f(y(\tau))\Delta\tau\right)\Delta s + W(t,y(t_1)), \quad t \in [0,T].$$

Exercise 7.2. Assume $(H1)-(H3)$. Prove that $F : E \to E$ is completely continuous.

Define

$$M = \eta\left(\int_0^T a(\tau)\Delta\tau\right),$$

$$N = \eta\phi_q\left(\int_\eta^T a(\tau)\Delta\tau\right),$$

$$m = r\phi_q\left(\int_0^T a(\tau)\Delta\tau\right).$$

Theorem 7.5. *Let $(H1)$-$(H3)$ hold,*

$$0 < a < b < \frac{M}{2N}c,$$

f, I and J satisfy the following conditions

(C1)

$$f(w) \le \phi_p\left(\frac{c}{2M}\right), \quad 0 \le w \le 2c,$$

(C2)

$$f(w) > \phi_p\left(\frac{b}{N}\right), \quad b \le w \le 2b,$$

(C3)

$$f(w) < \phi_p\left(\frac{a}{2w}\right), \quad 0 \le w \le a,$$

(C4)

$$I(w(t_1)) - t_1 J(w(t_1)) \le \frac{a}{2}, \quad w: \mathbb{T} \to \mathbb{R}.$$

Then the BVP (7.8) *has at least three fixed points* $y_1, y_2, y_3 \in \overline{P(\gamma, c)}$ *such that*

$$\begin{aligned} 0 &\le \alpha(y_1) < a < \alpha(y_2), \\ \beta(y_2) &< b < \beta(y_3), \\ \gamma(y_3) &< c. \end{aligned} \tag{7.9}$$

Proof. Let $y \in \overline{P(\gamma, c)}$. By $(H1)$ and $(H3)$, it follows that $Fy(t) \ge 0, t \in [0, T]$. Next, by $(H1)$ and by the definition of the operator F, we get

$$(Fy)^{\Delta^2}(t) \le 0, \quad t \in [0, T].$$

Moreover, Fy is concave on $[0, t_1]$ and $(t_1, T]$. Also,

$$\begin{aligned} (Fy)^{\Delta}(t) &= \phi_q\left(\int_t^T a(\tau) f(y(\tau)) \Delta \tau\right) - J(y(t_1)) \chi_{[0,t_1]}(t) \\ &\ge 0, \quad t \in [0, T], \end{aligned}$$

where $\chi_{[0,t_1]}$ is the characteristic function of $[0, t_1]$. Thus, Fy is non-decreasing on $[0, t_1]$ and $(t_1, T]$. Since $Fy(0) = 0$, we conclude that

$$Fy(t) \ge 0, \quad t \in [0, t_1].$$

Also,

$$\begin{aligned} I(F(y(t_1))) &= Fy(t_1^+) - Fy(t_1^-) \\ &= W(t_1^+, y(t_1^+)) - W(t_1^-, y(t_1^-)) \end{aligned}$$

$$
\begin{aligned}
&= I(y(t_1^+)) - t_1 J(y(t_1^+)) - t_1 J(y(t_1)) \\
&= I(y(t_1^+)) - t_1 \left(J(y(t_1^+)) + J(y(t_1))\right) \\
&\geq 0, \\
J(F(y(t_1))) &= (Fy)^{\Delta}(t_1^+) - (Fy)^{\Delta}(t_1^-) \\
&= J(y(t_1)) \\
&\leq 0.
\end{aligned}
$$

Thus,

$$Fy(t_1^+) \geq Fy(t_1^-) \geq 0$$

and

$$Fy(t) \geq 0, \quad t \in (t_1, T].$$

Therefore, $Fy \in P$ and in particular,

$$F : \overline{P(\gamma, c)} \to P.$$

Let now, $y \in \partial P(\gamma, c)$. Then

$$
\begin{aligned}
\gamma(y) &= \max_{t \in [0,\eta]} y(t) \\
&= y(\eta) \\
&= c.
\end{aligned}
$$

Hence,

$$
\begin{aligned}
\|y\| &\leq 2\gamma(y) \\
&= 2c.
\end{aligned}
$$

Therefore,

$$0 \leq y(t) \leq 2c, \quad t \in [0, T].$$

Hence and by $(C1)$, we obtain

$$f(y(s)) \leq \phi_p\left(\frac{c}{2M}\right), \quad s \in [0, T].$$

Also,

$$
\begin{aligned}
\gamma(Fy) &= Fy(\eta) \\
&= \int_0^\eta \phi_q\left(\int_s^T a(\tau)f(y(\tau))\Delta\tau\right)\Delta s + W(\eta, y(t_1)) \\
&= \int_0^\eta \phi_q\left(\int_s^T a(\tau)f(y(\tau))\Delta\tau\right)\Delta s + I(y(t_1)) - t_1 J(y(t_1)) \\
&< \eta\phi_q\left(\int_0^T a(\tau)\Delta\tau\right)\frac{c}{2M} + \frac{a}{2} \\
&= \frac{c}{2} + \frac{c}{2} \\
&= c.
\end{aligned}
$$

Now, suppose that $y \in \partial P(\beta, b)$. Then

$$
\begin{aligned}
\beta(y) &= \min_{t\in[\eta,r]} y(t) \\
&= y(\eta) \\
&= b.
\end{aligned}
$$

Thus,

$$
y(t) \geq b, \quad t \in [\eta, T],
$$

and

$$
\begin{aligned}
0 &\leq y(t) \\
&\leq \|y\| \\
&= y(T), \quad t \in [\eta, T], \\
\|y\| &\leq 2\gamma(y) \\
&= 2\beta(y) \\
&= 2b, \quad y \in P.
\end{aligned}
$$

Then

$$b \leq y(t) \leq 2b, \quad t \in [\eta, T].$$

Hence and by $(C2)$, we arrive at

$$f(y(s)) > \phi_p\left(\frac{b}{N}\right), \quad s \in [\eta, T],$$

and

$$\begin{aligned}
\beta(Fy) &= Fy(\eta) \\
&= \int_0^\eta \phi_q\left(\int_s^T a(\tau) f(y(\tau)) \Delta\tau\right) \Delta s + W(\eta, y(t_1)) \\
&= \int_0^\eta \phi_q\left(\int_s^T a(\tau) f(y(\tau)) \Delta\tau\right) \Delta s + I(y(t_1)) - t_1 J(y(t_1)) \\
&\geq \eta \phi_q\left(\int_\eta^T a(\tau) \Delta\tau\right) \frac{b}{N} \\
&= b.
\end{aligned}$$

Note that $\frac{a}{2} \in P(\alpha, a)$. Then $P(\alpha, a) \neq \emptyset$. Let now, $y \in \partial P(\alpha, a)$. Then

$$\begin{aligned}
\alpha(y) &= \max_{t \in [t_1, r]} y(t) \\
&= y(r) \\
&= a, \\
0 &\leq y(t) \\
&\leq a, \quad t \in [t_1, r].
\end{aligned}$$

Since y is non-decreasing and $y(t_1^+) \geq y(t_1)$, we obtain

$$0 \leq y(t) \leq a, \quad t \in [0, r].$$

By $(C3)$, we obtain

$$f(y(s)) \leq \phi_p\left(\frac{a}{2m}\right), \quad s \in [0, r].$$

Then

$$\alpha(Fy) = Fy(r)$$

$$= \int_0^r \phi_q\left(\int_s^T a(\tau)f(y(\tau))\Delta\tau\right)\Delta s + W(r, y(t_1))$$

$$= \int_0^r \phi_q\left(\int_s^T a(\tau)f(y(\tau))\Delta\tau\right)\Delta s + I(y(t_1)) - t_1 J(y(t_1))$$

$$< r\phi_q\left(\int_0^T a(\tau)\Delta\tau\right)\frac{a}{2m} + \frac{a}{2}$$

$$= a.$$

Hence and by Theorem 7.4, we get that problem (7.8) has at least three positive solutions $y_1, y_2, y_3 \in \overline{P(\gamma, c)}$ which satisfy (7.9). This completes the proof.

Exercise 7.3. Check if the BVP

$$\left(\phi_p(y^\Delta(t))\right)^\Delta + e^{-y(t)} = 0, \quad t \in [0,T]\backslash\{t_1\},$$

$$y(t_1^+) - y(t_1^-) = 4,$$

$$y^\Delta(t_1^+) - y^\Delta(t_1^-) = -1,$$

$$y(0) = y^\Delta(T) = 0,$$

where $\phi_p(s) = |s|^{p-2}s$, $p > 1$, $\phi_p^{-1} = \phi_q$, $\dfrac{1}{p} + \dfrac{1}{q} = 1$, has at least three different solutions.

7.3 Variational Approach for Second Order Impulsive Dynamic Equations

Consider the following BVP for a class of linear impulsive dynamic equations

$$\begin{aligned} -u^{\Delta^2}(t) + \lambda u^\sigma(t) &= h(t) \quad \Delta - \text{a.e.} \quad t \in [0,T]^{\kappa^2}, \\ u^\Delta(t_j^+) - u^\Delta(t_j^-) &= d_j, \quad j \in \{1, \ldots, m\}, \\ u(0) = u(T) &= 0, \end{aligned} \tag{7.10}$$

where $h \in \mathscr{C}([0,T])$, d_j, $j \in \{1, \ldots, m\}$, are constants. Suppose that $u \in \mathscr{C}_{rd}([0,T])$, $u(0) = u(T) = 0$ and

$$u_j = u\Big|_{(t_j, t_{j+1})} \in \mathbb{H}^2(t_j, t_{j+1}), \quad j \in \{0, \ldots, m\}.$$

Definition 7.1. We say that u is a classical solution of the BVP (7.10) if the limits $u^{\Delta}(t_j^+)$, $u^{\Delta}(t_j^-)$ exist for any $j \in \{1,\ldots,m\}$, and it satisfies equation (7.10) for Δ-almost everywhere $t \in [0,T]^{\kappa^2}$.

Take $v \in \mathbb{H}_0^1(J)$. We multiply the first equation of (7.10) by v^{σ} and we integrate from 0 to T. Thus, we get

$$-\int_0^T u^{\Delta^2} v^{\sigma} \Delta + \lambda \int_0^T u^{\sigma} v^{\sigma} \Delta = \int_0^T h v^{\sigma} \Delta. \tag{7.11}$$

Since $v \in \mathbb{H}_0^1(J)$, using the integration by parts, we obtain

$$\begin{aligned}
-\int_0^T u^{\Delta^2} v^{\sigma} \Delta &= -\sum_{j=0}^{m} \int_{t_j}^{t_{j+1}} u^{\Delta^2} v^{\sigma} \Delta \\
&= \sum_{j=0}^{m} \left(-(u^{\Delta} v)\Big|_{t_j^+}^{t_{j+1}^-} + \int_{t_j}^{t_{j+1}} u^{\Delta} v^{\Delta} \Delta \right) \\
&= -\sum_{j=1}^{m} \left(u^{\Delta}(t_{j+1}^-) v(t_{j+1}^-) - u^{\Delta}(t_j^+) v(t_j^+) \right) \\
&\quad + \sum_{j=0}^{m} \int_{t_j}^{t_{j+1}} u^{\Delta} v^{\Delta} \Delta \\
&= -\sum_{j=1}^{m} \left(u^{\Delta}(t_{j+1}^+) v(t_{j+1}) - d_{j+1} v(t_{j+1}) - u^{\Delta}(t_j^+) v(t_j^+) \right) \\
&\quad + \int_0^T u^{\Delta} v^{\Delta} \Delta \\
&= \sum_{j=0}^{m} d_j v(t_j) + \int_0^T u^{\Delta} v^{\Delta} \Delta.
\end{aligned}$$

From here and from (7.11), we obtain

$$\sum_{j=0}^{m} d_j v(t_j) + \int_0^T u^{\Delta} v^{\Delta} \Delta + \lambda \int_0^T u^{\sigma} v^{\sigma} \Delta = \int_0^T h v^{\sigma} \Delta$$

or

$$\int_0^T u^{\Delta} v^{\Delta} \Delta + \lambda \int_0^T b u^{\sigma} v^{\sigma} \Delta = \int_0^T h v^{\sigma} \Delta - \sum_{j=0}^{m} d_j v(t_j).$$

Define the bilinear form $a : \mathbb{H}_0^1([0,T]) \times \mathbb{H}_0^1([0,T]) \to \mathbb{R}$ by

$$a(u,v) = \int_0^T u^{\Delta} v^{\Delta} \Delta + \lambda \int_0^T u^{\sigma} v^{\sigma} \Delta,$$

and the linear operator $l : \mathbb{H}_0^1([0,T]) \to \mathbb{R}$ by

$$l(v) = \int_0^T h v^\sigma \Delta - \sum_{j=0}^{m} d_j v(t_j).$$

Definition 7.2. We say that $u \in \mathbb{H}_0^1([0,T])$ is a weak solution of the BVP (7.10) if

$$a(u,v) = l(v), \quad v \in \mathbb{H}_0^1([0,T]).$$

Note that $a : \mathbb{H}_0^1([0,T]) \times \mathbb{H}_0^1([0,T]) \to \mathbb{R}$ and $l : \mathbb{H}_0^1([0,T]) \to \mathbb{R}$ are continuous.

Definition 7.3. Let H be a Hilbert space and let $a : H \times H \to \mathbb{R}$ be a bounded bilinear form. If there exists an $\alpha > 0$ such that

$$a(u,u) \geq \alpha \|u\|^2$$

for any $u \in H$, then we say that a is coercive.

Let λ_1 be the smallest eigenvalue of problem

$$\begin{aligned} -u^{\Delta^2}(t) &= \lambda u^\sigma(t), \quad t \in [0,T]^{\kappa^2}, \\ u(0) &= u(T) = 0. \end{aligned} \tag{7.12}$$

For our next result we need the following fixed point theorem.

Theorem 7.6 (Lax-Milgram Theorem). *Let H be a Hilbert space and let $a : H \times H \to \mathbb{R}$ be a bounded bilinear form. If a is coercive, then for any $\phi \in H'$, where H' is the conjugate space of H, there exists a unique $u \in H$ such that*

$$a(u,v) = \langle \phi, v \rangle, \quad v \in H.$$

Moreover, if a is also symmetric, then the functional $J : H \to \mathbb{R}$ defined by

$$J(v) = \frac{1}{2} a(u,v) - \langle \phi, v \rangle,$$

attains its minimum at u.

Theorem 7.7. *If $\lambda > -\lambda_1$, then $a : \mathbb{H}_0^1([0,T]) \times \mathbb{H}_0^1([0,T]) \to \mathbb{R}$ is coercive.*

Proof. We have, using Wirtinger's inequality,

$$a(u,u) = \int_0^T (u^\Delta)^2 \Delta + \lambda \int_0^T (u^\sigma)^2 \Delta$$

$$\geq \lambda_1 \int_0^T (u^\sigma)^2 \Delta + \lambda \int_0^T (u^\sigma)^2 \Delta.$$

This completes the proof.

For $u \in \mathbb{H}_0^1([0,T])$, define

$$\phi(u) = \frac{1}{2} a(u,u) - l(u).$$

We have, for $u \in \mathbb{H}_0^1([0,T])$,

$$\begin{aligned}
\phi(u) &= \frac{1}{2} a(u,u) - l(u) \\
&= \frac{1}{2} \int_0^T (u^\Delta)^2 \Delta + \frac{\lambda}{2} \int_0^T (u^\sigma)^2 \Delta \\
&\quad - \int_0^T h u^\sigma \Delta + \sum_{j=1}^m d_j u(t_j).
\end{aligned}$$

For $u \in \mathbb{H}_0^1([0,T])$, we have that there exists $\phi'(u)$ and

$$\begin{aligned}
(\phi'(u), v) &= \int_0^T u^\Delta v^\Delta \Delta + \lambda \int_0^T u^\sigma v^\sigma \Delta \\
&\quad - \int_0^T h v^\sigma \Delta + \sum_{j=1}^m d_j v(t_j) \\
&= a(u,v) - l(v),
\end{aligned}$$

and if $u \in \mathbb{H}_0^1([0,T])$ is a critical point of ϕ, then it is a solution of the BVP (7.10). Hence and by Lax-Milgram's theorem, we conclude that if $\lambda > -\lambda_1$, then the BVP (7.10) has a weak solution $u \in \mathbb{H}_0^1([0,T])$ for any $h \in \mathbb{L}^2([0,T))$. Also, if $u \in \mathbb{H}^2([0,T])$ is a classical solution of the BVP (7.10) and u minimizes ϕ, then it is a critical point of ϕ. Now we consider the nonlinear Dirichlet problem

$$\begin{aligned}
-u^{\Delta^2}(t) + \lambda u^\sigma(t) &= f(t, u^\sigma(t)) \quad \Delta - \text{a.e.} \quad t \in [0,T]^{\kappa^2}, \\
u^\Delta(t_j^+) - u^\Delta(t_j^-) &= I_j(u(t_j)), \quad j \in \{1, \dots, m\}, \qquad (7.13) \\
u(0) = u(T) &= 0,
\end{aligned}$$

where

(H1) $f : \mathbb{T} \times \mathbb{R} \to \mathbb{R}$, $I_j : \mathbb{R} \to \mathbb{R}$, $j \in \{1, \dots, m\}$, are continuous functions,
(H2) $\lambda > \lambda_1$, where λ_1 is the smallest positive eigenvalue of problem (7.12).

Definition 7.4. A weak solution of problem (7.13) is a function $u \in \mathbb{H}_0^1([0,T])$ such that

$$\int_0^T u^\Delta v^\Delta \Delta + \lambda \int_0^T u^\sigma v^\sigma \Delta = -\sum_{j=1}^m I_j(u(t_j^+))v(t_j) + \int_0^T f(t,u^\sigma(t))v^\sigma(t)\Delta t$$

for any $v \in \mathbb{H}_0^1([0,T])$.

Consider the functional

$$\begin{aligned}\phi(u) &= \frac{1}{2}a(u,u) - l(u)\\ &= \frac{1}{2}\int_0^T (u^\Delta)^2\Delta + \frac{\lambda}{2}\int_0^T (u^\sigma)^2\Delta \\ &\quad + \sum_{j=1}^m \int_0^{u(t_j)} I_j(t)dt - \int_0^T F(t,u^\sigma(t))\Delta t,\end{aligned} \tag{7.14}$$

where

$$F(t,u) = \int_0^u f(t,x)dx.$$

We have that the functional ϕ, defined by (7.14), is continuous, differentiable and weakly lower semi-continuous. Also, the critical points of ϕ are weak solutions of problem (7.13).

Theorem 7.8. *Suppose* $(H1)$, $(H2)$ *and* f, I_j, $j \in \{1,\dots,m\}$, *are bounded. Then there is a critical point of* ϕ *and problem* (7.13) *has at least one weak solution.*

Proof. Let M, M_j, $j \in \{1,\dots,m\}$, be positive constants such that

$$|f(t,u)| \le M, \quad (t,u) \in [0,T]\times\mathbb{R},$$

$$|I_j(u)| \le M_j, \quad j \in \{1,\dots,m\}, \quad u \in \mathbb{R},$$

$$N = \max\{M, M_j : j \in \{1,\dots,m\}\}.$$

By $(H2)$, it follows that there exists an $\alpha > 0$ so that

$$\int_0^T (u^\Delta)^2\Delta + \lambda\int_0^T (u^\sigma)^2\Delta \ge \alpha\|u\|^2_{\mathbb{H}_0^1}$$

for any $u \in \mathbb{H}_0^1([0,T])$. Moreover, there exists a positive constant p so that

$$\|u\|_{\mathscr{C}([0,T])} \ge p\|u\|_{\mathbb{H}_0^1([0,T])}$$

for any $u \in \mathbb{H}_0^1([0,T])$. Then, for any $u \in \mathbb{H}_0^1([0,T])$, we obtain

$$\begin{aligned}
\phi(u) &\geq \frac{\alpha}{2}\|u\|^2_{\mathbb{H}_0^1([0,T])} + \sum_{j=1}^{m}\int_0^{u(t_j)} I_j(t)dt \\
&\quad - \int_0^T F(t,u^\sigma(t))\Delta \\
&\geq \frac{\alpha}{2}\|u\|^2_{\mathbb{H}_0^1([0,T])} - \sum_{j=1}^{m} M_j|u(t_j)| - M\int_0^T |u^\sigma(t)|\Delta \\
&\geq \frac{\alpha}{2}\|u\|^2_{\mathbb{H}_0^1([0,T])} - N\left(\sum_{j=1}^{m}|u(t_j)| + \int_0^T |u^\sigma(t)|\Delta\right) \\
&\geq \frac{\alpha}{2}\|u\|^2_{\mathbb{H}_0^1([0,T])} - N\left(m\|u\|_{\mathscr{C}([0,T])} + T\|u\|_{\mathscr{C}([0,T])}\right) \\
&= \frac{\alpha}{2}\|u\|^2_{\mathbb{H}_0^1([0,T])} - N(m+T)\|u\|_{\mathscr{C}([0,T])} \\
&\geq \frac{\alpha}{2}\|u\|^2_{\mathbb{H}_0^1([0,T])} - N(m+T)p\|u\|_{\mathbb{H}_0^1([0,T])} \\
&\to \infty, \quad \text{as} \quad \|u\|_{\mathbb{H}_0^1([0,T])} \to \infty.
\end{aligned}$$

Thus, ϕ is coercive. Hence and by the Lax-Milgram theorem, we conclude that ϕ has a minimum, which is a critical point of ϕ. Consequently, problem (7.13) has a weak solution. This completes the proof.

Theorem 7.9. *Suppose* $(H1)$, $(H2)$ *and*

$$|f(t,u)| \leq a + b|u|^\gamma, \quad (t,u) \in [0,T]\times\mathbb{R},$$

$$|I_j(u)| \leq a_j + b_j|u|^{\gamma_j}, \quad j \in \{1,\ldots,m\}, \quad u \in \mathbb{R},$$

where a, b, a_j, b_j, $j \in \{1,\ldots,m\}$, *are positive constants,* γ, $\gamma_j \in [0,1)$, $j \in \{1,\ldots,m\}$. *Then there is a critical point of* ϕ *and problem* (7.13) *has at least one weak solution.*

Proof. Let

$$N_1 = \max\{a, a_j, j \in \{1,\ldots,m\}\},$$

$$N_2 = \max\{b, b_j, j \in \{1,\ldots,m\}\},$$

α and p be as in the proof of Theorem 7.8. Then for any $u \in \mathbb{H}_0^1([0,T])$ we have

$$
\begin{aligned}
\sum_{j=1}^{m}\int_0^{u(t_j)} I_j(t)dt &\geq -\sum_{j=1}^{m}\int_0^{u(t_j)} |I_j(t)|dt \\
&\geq -\sum_{j=1}^{m}\int_0^{u(t_j)} (a_j+b_j|u|^{\gamma_j})\,dt \\
&\geq -\sum_{j=1}^{m}\left(a_j|u(t_j)|+b_j|u(t_j)|\|u\|^{\gamma_j}_{\mathscr{C}([0,T])}\right) \\
&\geq -\sum_{j=1}^{m}\left(a_j\|u\|_{\mathscr{C}([0,T])}+b_j\|u\|^{1+\gamma_j}_{\mathscr{C}([0,T])}\right) \\
&\geq -\sum_{j=1}^{m}\left(N_1\|u\|_{\mathscr{C}([0,T])}+N_2\|u\|^{1+\gamma_j}_{\mathscr{C}([0,T])}\right) \\
&\geq -\sum_{j=1}^{m}\left(pN_1\|u\|_{\mathbb{H}^1_0([0,T])}+p^{1+\gamma_j}N_2\|u\|^{1+\gamma_j}_{\mathbb{H}^1_0([0,T])}\right), \\
-F(t,|u|) &\geq -\int_0^{|u|}|f(t,x)|dx \\
&\geq -\int_0^{|u|}(a+b|x|^{\gamma})\,dx \\
&\geq -\left(a|u|+b|u|^{\gamma+1}\right) \\
&\geq -\left(a\|u\|_{\mathscr{C}([0,T])}+b\|u\|^{\gamma+1}_{\mathscr{C}([0,T])}\right) \\
&\geq -\left(ap\|u\|_{\mathbb{H}^1_0([0,T])}+bp^{\gamma+1}\|u\|^{\gamma+1}_{\mathbb{H}^1_0([0,T])}\right), \\
\phi(u) &\geq \frac{\alpha}{2}\|u\|^2_{\mathbb{H}^1_0([0,T])}-\sum_{j=1}^{m}\left(pN_1\|u\|_{\mathbb{H}^1_0([0,T])}+p^{1+\gamma}N_2\|u\|^{1+\gamma_j}_{\mathbb{H}^1_0([0,T])}\right) \\
&\quad -\left(ap\|u\|_{\mathbb{H}^1_0([0,T])}+bp^{\gamma+1}\|u\|^{\gamma+1}_{\mathbb{H}^1_0([0,T])}\right) \\
&\quad \to\infty, \quad \text{as} \quad \|u\|_{\mathbb{H}^1_0([0,T])}\to\infty.
\end{aligned}
$$

Therefore, ϕ is coercive. Hence and by the Lax-Milgram theorem, we conclude that ϕ has a critical point and then problem (7.13) has at least one weak solution. This completes the proof.

Now we consider the following BVP

$$\begin{aligned}
-u^{\Delta^2}(t)-g(t)u^{\Delta}(\sigma(t))+\lambda u^{\sigma}(t) &= f(t,u^{\sigma}(t)) \quad \Delta-\text{a.e.} \quad t\in[0,T]^{\kappa^2},\\
-(u^{\Delta}(t_j^+)-u^{\Delta}(t_j^-)) &= I_j(u(t_j)), \quad j\in\{1,\dots,m\},\\
u(0)=u(T) &= 0,
\end{aligned} \tag{7.15}$$

where

(H3) $f\in\mathscr{C}(\mathbb{T}\times\mathbb{R})$, $I_j\in\mathscr{C}(\mathbb{R})$, $j\in\{1,\dots,m\}$, $g\in\mathscr{C}(\mathbb{T})$, g is non-negative,

(H4) $\lambda>-\dfrac{p\lambda_1}{P}$, where

$$p=\min_{t\in[0,T]} e_g(t,0), \quad P=\max_{t\in[0,T]} e_g(t,0),$$

λ_1 is the smallest eigenvalue of problem (7.12).

We multiply by $e_g(t,0)$ the first equation of (7.15) and we get

$$-e_g(t,0)u^{\Delta^2}(t)+g(t)e_g(t,0)u^{\Delta}(\sigma(t))+\lambda e_g(t,0)u^{\sigma}(t)=e_g(t,0)f(t,u^{\sigma}(t))$$

Δ-a.e. $t\in[0,T]$, or

$$-\left(e_g(t,0)u^{\Delta}(t)\right)^{\Delta}+\lambda e_g(t,0)u^{\sigma}(t)=e_g(t,0)f(t,u^{\sigma}(t)) \quad \Delta-\text{a.e.} \quad t\in[0,T].$$

Thus, we can rewrite the BVP (7.15) in the following equivalent form

$$\begin{aligned}
-\left(e_g(t,0)u^{\Delta}(t)\right)^{\Delta}+\lambda e_g(t,0)u^{\sigma}(t) &= e_g(t,0)f(t,u^{\sigma}(t)) \quad \Delta-\text{a.e.} \quad t\in[0,T]^{\kappa^2},\\
-(u^{\Delta}(t_j^+)-u^{\Delta}(t_j^-)) &= I_j(u(t_j)), \quad j\in\{1,\dots,m\},\\
u(0)=u(T) &= 0.
\end{aligned} \tag{7.16}$$

In $\mathbb{H}_0^1(\mathbb{T})$ we define an inner product and a norm as follows

$$(u,v)=\int_0^T e_g(t,0)u^{\Delta}(t)v^{\Delta}(t)\Delta \tag{7.17}$$

and

$$\|u\|_{\mathbb{H}_0^1(\mathbb{T})}=\left(\int_0^T e_g(t,0)|u^{\Delta}(t)|^2\Delta\right)^{\frac{1}{2}}, \tag{7.18}$$

respectively.

Exercise 7.4. Prove that (7.17) satisfies all axioms for an inner product.

Exercise 7.5. Prove that (7.18) satisfies all axioms for a norm.

Now we multiply by v^σ, $v \in \mathbb{H}_0^1(\mathbb{T})$, both sides of the first equation of (7.16), then integrate from 0 to T and using the integration by parts, we get

$$
\begin{aligned}
\int_0^T e_g(t,0) f(t,u^\sigma(t)) v^\sigma(t)\Delta &= \lambda \int_0^T e_g(t,0) u^\sigma(t) v^\sigma(t)\Delta \\
&\quad - \int_0^T \left(e_g(t,0) u^\Delta(t)\right)^\Delta v^\sigma(t)\Delta \\
&= \lambda \int_0^T e_g(t,0) u^\sigma(t) v^\sigma(t)\Delta \\
&\quad - \int_0^{t_1} \left(e_g(t,0) u^\Delta(t)\right)^\Delta v^\sigma(t)\Delta \\
&\quad - \int_{t_1}^{t_2} \left(e_g(t,0) u^\Delta(t)\right)^\Delta v^\sigma(t)\Delta \\
&\quad - \cdots \\
&\quad - \int_{t_m}^T \left(e_g(t,0) u^\Delta(t)\right)^\Delta v^\sigma(t)\Delta \\
&= \lambda \int_0^T e_g(t,0) u^\sigma(t) v^\sigma(t)\Delta \\
&\quad - e_g(t_1,0) u^\Delta(t_1^-) v(t_1) \\
&\quad + \int_0^{t_1} e_g(t,0) u^\Delta(t) v^\Delta(t)\Delta \\
&\quad + e_g(t_1,0) u^\Delta(t_1^+) v(t_1) \\
&\quad - e_g(t_2,0) u^\Delta(t_2^-) v(t_2) \\
&\quad + \int_{t_1}^{t_2} e_g(t,0) u^\Delta(t) v^\Delta(t)\Delta \\
&\quad - \cdots \\
&\quad + e_g(t_m,0) u^\Delta(t_m^+) v(t_m) \\
&\quad + \int_{t_m}^T e_g(t,0) u^\Delta(t) v^\Delta(t)\Delta
\end{aligned}
$$

$$= \lambda \int_0^T e_g(t,0) u^\sigma(t) v^\sigma(t) \Delta + \int_0^T e_g(t,0) u^\Delta(t) v^\Delta(t) \Delta$$

$$- e_g(t_1,0)\left(u^\Delta(t_1^+) - u^\Delta(t_1^-)\right) v(t_1)$$

$$- e_g(t_2,0)\left(u^\Delta(t_2^+) - u^\Delta(t_2^-)\right) v(t_2)$$

$$- \cdots$$

$$- e_g(t_m,0)\left(u^\Delta(t_m^+) - u^\Delta(t_m^-)\right) v(t_m)$$

$$= \lambda \int_0^T e_g(t,0) u^\sigma(t) v^\sigma(t) \Delta$$

$$+ \int_0^T e_g(t,0) u^\Delta(t) v^\Delta(t) \Delta$$

$$- \sum_{j=1}^m e_g(t_j,0) I_j(u(t_j)) v(t_j),$$

whereupon,

$$\lambda \int_0^T e_g(t,0) u^\sigma(t) v^\sigma(t) \Delta + \int_0^T e_g(t,0) u^\Delta(t) v^\Delta(t) \Delta = \int_0^T e_g(t,0) f(t, u^\sigma(t)) v^\sigma(t) \Delta$$
$$+ \sum_{j=1}^m e_g(t_j,0) I_j(u(t_j)) v(t_j). \tag{7.19}$$

Definition 7.5. A weak solution of the BVP (7.16) is a function $u \in \mathbb{H}_0^1(\mathbb{T})$ that satisfies (7.19) for any $v \in \mathbb{H}_0^1(\mathbb{T})$.

For $u \in \mathbb{H}_0^1(\mathbb{T})$, define

$$A(u,v) = \int_0^T e_g(t,0) u^\Delta(t) v^\Delta(t) \Delta + \lambda \int_0^T e_g(t,0) u^\sigma(t) v^\sigma(t) \Delta,$$

$$F(t,u) = \int_0^u f(t,\xi) d\xi,$$

$$\psi(u) = \frac{1}{2} A(u,u) - \sum_{j=1}^m e_g(t_j,0) \int_0^{u(t_j)} I_j(t) dt$$

$$- \int_0^T e_g(t,0) F(t, u^\sigma(t)) \Delta,$$

$$L(u,v) = \int_0^T e_g(t,0) f(t,u^\sigma(t)) v^\sigma(t)\Delta$$

$$+ \sum_{j=1}^{m} e_g(t_j,0) I_j(u(t_j)) v(t_j).$$

We have that A is a bilinear, continuous and symmetric form. Then $u \in \mathbb{H}_0^1(\mathbb{T})$ is a weak solution of (7.16) if

$$A(u,v) = L(u,v)$$

for any $v \in \mathbb{H}_0^1(\mathbb{T})$. Note that ψ is continuous, continuously-differentiable and weakly lower semi-continuous on $\mathbb{H}_0^1(\mathbb{T})$ and

$$(\psi'(u),v) = A(u,v) - L(u,v), \quad u,v \in \mathbb{H}_0^1(\mathbb{T}).$$

Hence, if $u \in \mathbb{H}_0^1(\mathbb{T})$ is a critical point for ψ, then it is a weak solution of the BVP (7.16).

Theorem 7.10. *Suppose* $(H3)$, $(H4)$. *Then there exist positive constants* $\beta > \alpha > 0$ *such that*

$$\alpha \|u\|^2_{\mathbb{H}_0^1(\mathbb{T})} \leq A(u,u) \leq \beta \|u\|^2_{\mathbb{H}_0^1(\mathbb{T})}, \quad u \in \mathbb{H}_0^1(\mathbb{T}).$$

Proof. By Wirtinger's inequality, it follows that for $N = \dfrac{P}{p\lambda_1}$ we have

$$N \int_0^T e_g(t,0)(u^\Delta)^2\Delta = \frac{P}{p\lambda_1}\int_0^T e_g(t,0)(u^\Delta(t))^2\Delta$$

$$\geq \frac{P}{\lambda_1}\int_0^T (u^\Delta(t))^2\Delta$$

$$\geq P\int_0^T (u^\sigma(t))^2\Delta$$

$$\geq \int_0^T e_g(t,0)(u^\sigma(t))^2\Delta$$

for any $u \in \mathbb{H}_0^1(\mathbb{T})$. Hence, if $\lambda > 0$, we obtain

$$\|u\|^2_{\mathbb{H}_0^1(\mathbb{T})} \leq \int_0^T e_g(t,0)(u^\Delta(t))^2\Delta + \lambda\int_0^T e_g(t,0)(u^\sigma(t))^2\Delta$$

$$= A(u,u)$$

$$\leq \int_0^T e_g(t,0)(u^\Delta(t))^2\Delta + \lambda N\int_0^T e_g(t,0)(u^\Delta(t))^2\Delta$$

$$= (1+\lambda N)\int_0^T e_g(t,0)(u^\Delta(t))^2\Delta$$

$$= (1+\lambda N)\|u\|^2_{\mathbb{H}_0^1(\mathbb{T})}.$$

If $\lambda < 0$, then

$$0 \le \left(1+\lambda\frac{P}{p\lambda_1}\right)\|u\|^2_{\mathbb{H}_0^1(\mathbb{T})}$$

$$\le (1+\lambda N)\|u\|^2_{\mathbb{H}_0^1(\mathbb{T})}$$

$$\le \int_0^T e_g(t,0)(u^\Delta(t))^2\Delta + \lambda\int_0^T (u^\sigma(t))^2\Delta$$

$$= A(u,u)$$

$$\le \int_0^T e_g(t,0)(u^\Delta(t))^2\Delta$$

$$= \|u\|^2_{\mathbb{H}_0^1(\mathbb{T})}.$$

Then, if $\lambda \ge 0$, we take

$$\alpha = 1, \quad \beta = 1+\lambda\frac{P}{p\lambda_1}.$$

If $\lambda < 0$, then we take

$$\alpha = 1+\lambda\frac{P}{p\lambda_1}, \quad \beta = 1.$$

This completes the proof.

Theorem 7.11. *If $u \in \mathbb{H}_0^1(\mathbb{T})$, then there exists a positive constant δ such that*

$$\|u\|_0 \le \delta\|u\|_{\mathbb{H}_0^1(\mathbb{T})},$$

where

$$\|u\|_0 = \max_{t\in[0,T]}|u(t)|.$$

Proof. Let $u \in \mathbb{H}_0^1(\mathbb{T})$. We have

$$|u(t)| \le \int_0^T |u^\Delta(s)|\Delta$$

$$= \int_0^T \frac{1}{\sqrt{e_g(s,0)}}\sqrt{e_g(s,0)}|u^\Delta(s)|\Delta$$

$$\leq \left(\int_0^T \frac{1}{e_g(s,0)} \Delta \right)^{\frac{1}{2}} \int_0^T e_g(s,0)|u^\Delta(s)|^2 \Delta$$

$$\leq \sqrt{\frac{T}{P}} \|u\|_{\mathbb{H}_0^1(\mathbb{T})}.$$

Hence,

$$\|u\|_0 \leq \sqrt{\frac{T}{P}} \|u\|_{\mathbb{H}_0^1(\mathbb{T})}.$$

This completes the proof.

Theorem 7.12. *Suppose* $(H3)$, $(H4)$, f *and* I_j, $j \in \{1, \ldots, m\}$, *are bounded. Then the BVP* (7.16) *has at least one weak solution.*

Proof. Let B and B_j, $j \in \{1, \ldots, m\}$, be positive constants such that

$$|f(t,u)| \leq B, \quad (t,u) \in [0,T] \times \mathbb{R},$$

$$|I_j(u)| \leq B_j, \quad j \in \{1, \ldots, m\}.$$

Let, also,

$$\alpha = \begin{cases} 1 \quad \text{if} \quad \lambda \geq 0 \\ 1 + \lambda \dfrac{P}{p\lambda_1} \quad \text{if} \quad -\dfrac{P\lambda_1}{p} < \lambda < 0, \end{cases}$$

$$\beta = \begin{cases} 1 + \lambda \dfrac{P}{p\lambda_1} \quad \text{if} \quad \lambda \geq 0 \\ 1 \quad \text{if} \quad -\dfrac{P\lambda_1}{p} < \lambda < 0, \end{cases}$$

$\delta = \sqrt{\frac{T}{P}}$. Take $u \in \mathbb{H}_0^1(\mathbb{T})$ arbitrarily. By Theorem 7.10, we have

$$A(u,u) \geq \alpha \|u\|^2_{\mathbb{H}_0^1(\mathbb{T})}.$$

By Theorem 7.11, we obtain

$$\|u\|_0 \leq \delta \|u\|_{\mathbb{H}_0^1(\mathbb{T})}.$$

Hence,

$$\psi(u) \geq \frac{\alpha}{2} \|u\|^2_{\mathbb{H}_0^1(\mathbb{T})} - \sum_{j=1}^m B_j e_g(t_j,0)|u(t_j)|$$

$$
\begin{aligned}
&-B\int_0^T e_g(t,0)|u^\sigma(t)|\Delta \\
&\geq \frac{\alpha}{2}\|u\|^2_{\mathbb{H}_0^1(\mathbb{T})} - \sum_{j=1}^m B_j P\|u\|_0 - BTP\|u\|_0 \\
&= \frac{\alpha}{2}\|u\|^2_{\mathbb{H}_0^1(\mathbb{T})} - \left(\sum_{j=1}^m B_j + BT\right) P\|u\|_0 \\
&\geq \frac{\alpha}{2}\|u\|^2_{\mathbb{H}_0^1(\mathbb{T})} - \delta\left(\sum_{j=1}^m B_j + BT\right) P\|u\|_{\mathbb{H}_0^1(\mathbb{T})} \\
&\to \infty, \quad \text{as} \quad \|u\|_{\mathbb{H}_0^1(\mathbb{T})} \to \infty.
\end{aligned}
$$

Thus, ψ is coercive. Hence and by the Lax-Milgram theorem, it follows that ψ has a minimum which is a critical point of ψ. Therefore, the BVP (7.16) has at least one weak solution. This completes the proof.

For our next result we need the following fixed point theorem.

Theorem 7.13. *For the functional $F: M \subset X \to \mathbb{R}$ with M nonempty, $\min\limits_{u\in M} F(u) = a$ has solutions in case the following hold.*

1. *X is a reflexive Banach space.*
2. *M is bounded and weak sequentially closed.*
3. *F is sequentially lower semi-continuous on M.*

Definition 7.6. Suppose that X is a Banach space and $\phi : X \to \mathbb{R}$ is differentiable and $c \in \mathbb{R}$. We say that ϕ satisfies the Palais-Small condition if every bounded sequence $\{\phi_k\}_{k\in\mathbb{N}}$ in the space X such that $\lim\limits_{k\to\infty} \phi'(u_k) = 0$ contains a convergent subsequence.

Theorem 7.14. *Suppose $(H3)$, $(H4)$ and*

(H5) there exist constants $\eta > 2$ and $\gamma > 0$ such that for all $(t,u) \in [0,T] \times \mathbb{R}$, $|u| \geq \gamma$,

$$0 < \eta F(t,u) \leq u f(t,u),$$

$$0 < \eta \int_0^u I_j(\xi)d\xi \leq u I_j(u), \quad j \in \{1,\ldots,m\},$$

(H6) there exists a positive $s \geq \eta$ such that

$$f(t,u) = o(|u|^s),$$

$$I_j(u) = o(|u|^s)$$

uniformly for $t \in [0,T]$, *as* $|u| \to \infty$, $j \in \{1,\dots,m\}$,

(H7)

$$f(t,u) = o(|u|),$$

$$I_j(u) = o(|u|), \quad j \in \{1,\dots,m\},$$

uniformly for $t \in [0,T]$, *as* $|u| \to 0$.

Then the BVP (7.16) *has at least two weak solutions.*

Proof. By $(H6)$ and $(H7)$, it follows that for any $\varepsilon > 0$ and $(t,u) \in [0,T] \times \mathbb{R}$, there exist $c_1(\varepsilon) > 0$, $c_{1j}(\varepsilon) > 0$, $j \in \{1,\dots,m\}$, such that

$$|f(t,u)| \le \varepsilon |u| + c_1(\varepsilon)|u|^s,$$

$$|I_j(u)| \le \varepsilon |u| + c_{1j}(\varepsilon)|u|^s, \quad j \in \{1,\dots,m\}.$$

Then, for any $\varepsilon > 0$ and $(t,u) \in [0,T] \times \mathbb{R}$, we have

$$\begin{aligned} F(t,u) &\le \int_0^{|u|} (\varepsilon\xi + c_1(\varepsilon)\xi^s)\, d\xi \\ &= \frac{\varepsilon}{2}|u|^2 + c_2(\varepsilon)|u|^{s+1}, \end{aligned}$$

$$\begin{aligned} \int_0^u I_j(\xi) d\xi &\le \int_0^{|u|} (\varepsilon\xi + c_{1j}(\varepsilon)\xi^s)\, d\xi \\ &\le \frac{\varepsilon}{2}|u|^2 + c_{2j}(\varepsilon)|u|^{s+1}, \quad j \in \{1,\dots,m\}, \end{aligned}$$

where

$$c_2(\varepsilon) = \frac{c_1(\varepsilon)}{s+1},$$

$$c_{2j}(\varepsilon) = \frac{c_{1j}(\varepsilon)}{s+1}, \quad j \in \{1,\dots,m\}.$$

By $(H5)$, it follows that for all $(t,u) \in [0,T] \times \mathbb{R}$ and $u \ge \gamma$, we have

$$\frac{\eta}{u} \le \frac{f(t,u)}{F(t,u)},$$

whereupon,

$$\eta \log \frac{u}{\gamma} \leq \log \frac{F(t,u)}{F(t,\gamma)},$$

and for all $(t,u) \in [0,T] \times \mathbb{R}$ and $u \leq -\gamma$

$$\frac{P\eta}{u} \geq \frac{f(t,u)}{F(t,u)},$$

whereupon,

$$\eta \log \frac{\gamma}{-u} \geq \log \frac{F(t,-\gamma)}{F(t,u)}.$$

Therefore,

$$F(t,u) \geq F(t,\gamma)\left(\frac{u}{\gamma}\right)^{\eta}, \quad (t,u) \in [0,T] \times \mathbb{R}, \quad u \geq \gamma,$$

$$F(t,u) \geq F(t,-\gamma)\left(\frac{-u}{\gamma}\right)^{\eta}, \quad (t,u) \in [0,T] \times \mathbb{R}, \quad u \geq -\gamma.$$

From here, there exists a constant $a_1 > 0$ such that

$$F(t,u) \geq a_1 |u|^{\eta}$$

for all $(t,u) \in [0,T] \times \mathbb{R}$, $|u| \geq \gamma$. Since F is continuous, there exists a constant $k > 0$ such that

$$\begin{aligned} F(t,u) &\geq -k \\ &\geq a_1 |u|^{\eta} - a_1 \gamma^{\eta} - k, \quad |u| \leq \gamma, \quad t \in [0,T]. \end{aligned}$$

Consequently,

$$F(t,u) \geq a_1 |u|^{\eta} - a_2, \quad (t,u) \in [0,T] \times \mathbb{R}.$$

As in the above, there exist constants $a_{1j}, a_{2j} > 0$, $j \in \{1,\ldots,m\}$,

$$\int_0^u I_j(\xi)d\xi \geq a_{1j}|u|^{\eta} - a_{2j}, \quad u \in \mathbb{R}, \quad j \in \{1,\ldots,m\}.$$

Hence, it follows that there exists a $\rho > 0$ such that ψ has a local minimum $u_0 \in \overline{B_\rho}$, where

$$B_\rho = \{u \in \mathbb{H}_0^1([0,T]) : \|u\| < \rho\},$$

i.e.,

$$\psi(u_0) = \min_{u \in \overline{B_\rho}} \psi(u).$$

Let α, β be as in Theorem 7.10 and δ be as in Theorem 7.11. Then

$$
\begin{aligned}
\psi(u) &\geq \frac{\alpha}{2}\|u\|^2_{\mathbb{H}_0^1([0,T])} - \sum_{j=1}^{m} e_g(t,0)\int_0^{u(t_j)} I_j(t)dt \\
&\quad - \int_0^T e_g(t,0)F(t,u^\sigma(t))\Delta \\
&\geq \frac{\alpha}{2}\|u\|^2_{\mathbb{H}_0^1([0,T])} - \sum_{j=1}^{m} e_g(t,0)\left(\frac{\varepsilon}{2}|u|^2 + c_{2j}(\varepsilon)|u|^{s+1}\right) \\
&\quad - \int_0^T e_g(t,0)\left(\frac{\varepsilon}{2}|u|^2 + c_2(\varepsilon)|u^\sigma|^{s+1}\right)\Delta \\
&\geq \frac{\alpha}{2}\|u\|^2_{\mathbb{H}_0^1([0,T])} - P\frac{\varepsilon}{2}m\delta^2\|u\|^2_{\mathbb{H}_0^1([0,T])} - \delta^{s+1}P\sum_{j=1}^{m} c_{2j}(\varepsilon)\|u\|^{s+1}_{\mathbb{H}_0^1([0,T])} \\
&\quad - \frac{P\varepsilon}{2}\int_0^T |u^\sigma(t)|^2\Delta - Pc_2(\varepsilon)\int_0^T |u^\sigma(t)|^{s+1}\Delta \\
&\geq \frac{\alpha}{2}\|u\|^2_{\mathbb{H}_0^1([0,T])} - P\frac{\varepsilon}{2}m\delta^2\|u\|^2_{\mathbb{H}_0^1([0,T])} - \delta^{s+1}P\sum_{j=1}^{m} c_{2j}(\varepsilon)\|u\|^{s+1}_{\mathbb{H}_0^1([0,T])} \\
&\quad - \frac{P\varepsilon}{2}\frac{1}{p\lambda_1}\|u\|^2_{\mathbb{H}_0^1([0,T])} - Pc_2(\varepsilon)\frac{Tk^{s+1}c^{s+1}}{(\sqrt{P})^{s+1}}\|u\|^{s+1}_{\mathbb{H}_0^1([0,T])} \\
&= \frac{\alpha - P\varepsilon\left(\frac{1}{p\lambda_1} + m\delta^2\right)}{2}\|u\|^2_{\mathbb{H}_0^1([0,T])} \\
&\quad - P\left(\delta^{s+1}\sum_{j=1}^{m} c_{2j}(\varepsilon) + \frac{c_2(\varepsilon)Tk^{s+1}c^{s+1}}{(\sqrt{p})^{s+1}}\right)\|u\|^{s+1}_{\mathbb{H}_0^1([0,T])}.
\end{aligned}
$$

Now, we take

$$
\varepsilon = \frac{\alpha}{2P\left(\frac{1}{p\lambda_1} + \delta^2 m\right)},
$$

$$
\rho_0 = \left(\frac{\alpha}{8P\left(\delta^{s+1}\sum_{j=1}^{m} c_{2j}(\varepsilon) + \frac{c_2(\varepsilon)Tk^{s+1}c^{s+1}}{(\sqrt{p})^{s+1}}\right)}\right)^{\frac{1}{s-1}}.
$$

Then, for any $u \in \partial B_{\rho_0}$, we have

$$
\|u\|_{\mathbb{H}_0^1([0,T])} = \rho_0
$$

and

$$\begin{aligned}\psi(u) &\geq \frac{\alpha-\frac{\alpha}{2}}{2}\|u\|^2_{\mathbb{H}^1_0([0,T])}\\&\quad -P\left(\delta^{s+1}\sum_{j=1}^{m}c_{2j}(\varepsilon)+\frac{c_2(\varepsilon)Tk^{s+1}c^{s+1}}{(\sqrt{p})^{s+1}}\right)\|u\|^{s-1}_{\mathbb{H}^1_0([0,T])}\|u\|_{\mathbb{H}^1_0([0,T])}\\&= \frac{\alpha}{4}\|u\|^2_{\mathbb{H}^1_0([0,T])}-M\frac{\alpha}{8M}\|u\|^2_{\mathbb{H}^1_0([0,T])}\\&= \frac{\alpha}{8}\|u\|^2_{\mathbb{H}^1_0([0,T])}\\&= \frac{\alpha}{8}\rho_0^2\\&> 0.\end{aligned}$$

Also,

$$\begin{aligned}\psi(u_0) &\leq \psi(0)\\&= 0.\end{aligned}$$

Then

$$\begin{aligned}\psi(u) &> \frac{\alpha}{8}\rho_0^2\\&> \psi(0)\\&\geq \psi(u_0)\end{aligned}$$

for any $u \in \partial B_{\rho_0}$. So,

$$\psi(u_0) < \inf\{\psi(u) : u \in \partial B_{\rho_0}\}.$$

Hence, ψ has a local minimum $u_0 \in B_{\rho_0}$. Next,

$$\begin{aligned}\psi(u) &\leq \frac{\beta}{2}\|u\|^2_{\mathbb{H}^1_0([0,T])}-\sum_{j=1}^{m}e_g(t,0)\int_0^{u(t_j)}I_j(t)dt-\int_0^T e_g(t,0)F(t,u^\sigma(t))\Delta\\&\leq \frac{\beta}{2}\|u\|^2_{\mathbb{H}^1_0([0,T])}-\sum_{j=1}^{m}e_g(t,0)\left(a_{1j}|u(t_j)|^\eta-a_{2j}\right)\\&\quad -\int_0^T e_g(t,0)\left(a_1|u^\sigma(t)|^\eta-a_2\right)\Delta\end{aligned}$$

$$\leq \frac{\beta}{2}\|u\|^2_{\mathbb{H}_0^1([0,T])} - p\sum_{j=1}^m a_{1j}|u(t_j)|^\eta + P\sum_{j=1}^m a_{2j}$$

$$-pa_1\int_0^T |u^\sigma(t)|^\eta\Delta + Pa_2T$$

and if $u \in \mathbb{H}_0^1([0,T])$, $\|u\|_{\mathbb{H}_0^1([0,T])} = 1$, we have

$$\psi(Nu) \leq \frac{\beta}{2}N^2 - pN^\eta\sum_{j=1}^m a_{1j}|u(t_j)|^\eta + P\sum_{j=1}^m a_{2j}$$

$$-pa_1N^\eta\int_0^T |u^\sigma(t)|^\eta\Delta + Pa_2T$$

$$\to -\infty, \quad \text{as} \quad N \to \infty.$$

Therefore, there exists an $N_0 > \rho_0$ such that

$$\psi(N_0u) \leq 0.$$

Hence, there exists an $u_1 \in \mathbb{H}_0^1([0,T])$ such that $\|u_1\|_{\mathbb{H}_0^1([0,T])} = N_0$ and $\psi(u_1) < 0$. Consequently,

$$\max\{\psi(u_0), \psi(u_1)\} < \inf\{\psi(u) : u \in \partial B_{\rho_0}\}.$$

Let now $\{\psi(u_k)\}_{k\in\mathbb{N}}$ be a bounded sequence such that

$$\lim_{k\to\infty}\psi'(u_k) = 0.$$

We have

$$(\psi'(u_k), u_k) = A(u_k, u_k) - \int_0^T e_g(t,0)f(t, u_k^\sigma(t))u_k^\sigma(t)\Delta$$

$$-\sum_{j=1}^m e_g(t,0)I_j(u_k(t_j))u_k(t_j)$$

and

$$\psi(u_k) - \frac{1}{\eta}(\psi'(u_k), u_k) = \left(\frac{1}{2} - \frac{1}{\eta}\right)A(u_k, u_k)$$

$$+\frac{1}{\eta}\int_0^T e_g(t,0)f(t, u_k^\sigma(t))u_k^\sigma(t)\Delta$$

$$-\int_0^T e_g(t,0)F(t,u_k^\sigma(t))\Delta$$

$$+\frac{1}{\eta}\sum_{j=1}^m e_g(t,0)I_j(u_k(t_j))-\sum_{j=1}^m e_g(t,0)\int_0^{u_k(t_j)} I_j(t)dt.$$

Let

$$\begin{aligned}\Gamma_1^k &= \frac{1}{\eta}\int_0^T e_g(t,0)f(t,u_k^\sigma(t))u_k^\sigma(t)\Delta \\ &\quad -\int_0^T e_g(t,0)F(t,u_k^\sigma(t))\Delta, \\ \Gamma_2 &= \frac{1}{\eta}\sum_{j=1}^m e_g(t,0)I_j(u_k(t_j)) \\ &\quad -\sum_{j=1}^m e_g(t,0)\int_0^{u_k(t_j)} I_j(t)dt, \\ S_1^k &= \{t\in[0,T]: |u_k(t)|<\gamma\}, \\ S_2^k &= \{t\in[0,T]: |u_k|\geq\gamma\}.\end{aligned}$$

Then

$$\psi(u_k)\geq\left(\frac{1}{2}-\frac{1}{\eta}\right)\alpha\|u_k\|^2_{\mathbb{H}_0^1([0,T])}+\Gamma_1+\Gamma_2.$$

Note that there exists a positive constant c such that

$$\begin{aligned}|F(t,u^\sigma(t))| &\leq c, \\ |f(t,u^\sigma(t))u^\sigma(t)| &\leq c, \\ \left|\int_0^{u(t_j)} I_j(t)dt\right| &\leq c, \\ |I_j(u(t_j))u(t_j)| &\leq c, \quad j\in\{1,\ldots,m\}, \quad \text{if} \quad |u|<\gamma.\end{aligned}$$

Then

$$\begin{aligned}\Gamma_1 &= \frac{1}{\eta}\int_{S_1^k} e_g(t,0)f(t,u_k^\sigma(t))u_k^\sigma(t)\Delta \\ &\quad +\frac{1}{\eta}\int_{S_2^k} e_g(t,0)f(t,u_k^\sigma(t))u_k^\sigma(t)\Delta\end{aligned}$$

$$
\begin{aligned}
&-\int_{S_1^k} e_g(t,0)F(t,u_k^\sigma(t))\Delta \\
&-\int_{S_2^k} e_g(t,0)F(t,u_k^\sigma(t))\Delta \\
\geq &-\frac{1}{\eta}\int_{S_1^k} e_g(t,0)|f(t,u_k^\sigma(t))u_k^\sigma(t)|\Delta \\
&+\frac{1}{\eta}\int_{S_2^k} e_g(t,0)f(t,u_k^\sigma(t))u_k^\sigma(t)\Delta \\
&-\int_{S_1^k} e_g(t,0)|F(t,u_k^\sigma(t))|\Delta \\
&-\int_{S_2^k} e_g(t,0)F(t,u_k^\sigma(t))\Delta \\
\geq &\ c_1.
\end{aligned}
$$

As in the above,

$$\Gamma_2 \geq c_2.$$

Here c_1 and c_2 are constants independent of $k \in \mathbb{N}$. Hence,

$$
\begin{aligned}
\psi(u_k) &\geq \left(\frac{1}{2}-\frac{1}{\eta}\right)\alpha\|u_k\|^2_{\mathbb{H}_0^1([0,T])}+\frac{1}{\eta}(\psi'(u_k),u_k)+\Gamma_1+\Gamma_2 \\
&\geq \left(\frac{1}{2}-\frac{1}{\eta}\right)\alpha\|u_k\|^2_{\mathbb{H}_0^1([0,T])}-\frac{1}{\eta}\|\psi'(u_k)\|_{\mathbb{H}_0^1([0,T])}\|u_k\|_{\mathbb{H}_0^1([0,T])}+c_1+c_2.
\end{aligned}
$$

Since $\{\psi(u_k)\}_{k\in\mathbb{N}}$ is bounded and $\lim\limits_{k\to\infty}\psi'(u_k) = 0$, we conclude that $\{\|u_k\|_{\mathbb{H}_0^1([0,T])}\}_{k\in\mathbb{N}}$ is a bounded sequence. Then there exists a subsequence of the sequence $\{u_k\}_{k\in\mathbb{N}}$, denoted again by $\{u_k\}_{k\in\mathbb{N}}$ which weakly converges to some $u\in\mathbb{H}_0^1([0,T])$. Then $\{u_k\}_{k\in\mathbb{N}}$ converges uniformly to $u\in\mathscr{C}([0,T])$. Moreover,

$$
\begin{aligned}
\|u_k\|^2_{\mathbb{H}_0^1([0,T])} = &\int_0^T e_g(t,0)\left(f(t,u_k^\sigma)u_k^\sigma-\lambda u_k^2\right)\Delta \\
&+(\psi'(u_k),u_k) \\
&+\sum_{j=1}^m e_g(t,0)I_j(u_k(t_j))u_k(t_j)
\end{aligned}
$$

and hence,

$$\lim_{k\to\infty}\|u_k\|^2_{\mathbb{H}_0^1([0,T])} = \int_0^T e_g(t,0)\left(f(t,u^\sigma)u^\sigma-\lambda u^2\right)\Delta$$

$$+(\psi'(u),u)$$

$$+\sum_{j=1}^{m} e_g(t,0)I_j(u(t_j))u(t_j).$$

Then $\{\|u_k\|_{\mathbb{H}_0^1([0,T])}\}_{k\in\mathbb{N}}$ converges. From here, $u_k \to u$, as $k \to \infty$. Now, applying the mountain pass theorem, we conclude that there exists a critical point u^* of ψ. Thus, u^* and u_0 are two critical points of ψ. Therefore, (7.16) has at least two weak solutions. This completes the proof.

Exercise 7.6. Let $\mathbb{T} = 2\mathbb{Z}$. Prove that the BVP

$$\begin{aligned}
-u^{\Delta^2}(t) - 4u^{\Delta}(\sigma(t)) + \lambda u^{\sigma}(t) &= 7u^{\sigma}(t) \quad \Delta-\text{a.e.} \quad t \in [0,12]^{\kappa^2},\\
-(u^{\Delta}(2^+) - u^{\Delta}(2^-)) &= u^{\Delta}(2)),\\
u(0) = u(12) &= 0,
\end{aligned}$$

has at least two weak solutions.

7.4 Advanced Practical Problems

Problem 7.1. Prove that the BVP

$$\begin{aligned}
-y^{\Delta^2}(t) &= \frac{t^2+t+1}{4+10(y(t))^8}, \quad t \in J\backslash\{t_j\}_{j=1}^m,\\
y(t_j^+) - y(t_j^-) &= -\frac{3}{1+3(y(t_j^-))^4},\\
y^{\Delta}(t_j^+) - y^{\Delta}(t_j^-) &= \frac{3}{7+(y(t_j^-))^2}, \quad j \in \{1,\ldots,m\},\\
y(0) &= y(T) = 0,
\end{aligned}$$

has at least one solution in $PC([0,T])$.

Problem 7.2. Prove that the BVP

$$\begin{aligned}
-y^{\Delta^2}(t) &= \frac{t^2+1}{1+t^4+3t^6+t^8}, \quad t \in J\backslash\{t_j\}_{j=1}^m,\\
y(t_j^+)-y(t_j^-) &= 1,\\
y^{\Delta}(t_j^+)-y^{\Delta}(t_j^-) &= 4, \quad j \in \{1,\dots,m\},\\
y(0) &= y(T) = 0,
\end{aligned}$$

has at least one solution in $PC([0,T])$.

Problem 7.3. Prove that the BVP

$$\begin{aligned}
-y^{\Delta^2}(t) &= \frac{y(t)}{1+(y(t))^2+4(y(t))^6}, \quad t \in J\backslash\{t_j\}_{j=1}^m,\\
y(t_j^+)-y(t_j^-) &= 1,\\
y^{\Delta}(t_j^+)-y^{\Delta}(t_j^-) &= -1, \quad j \in \{1,\dots,m\},\\
y(0) &= y(T) = 0,
\end{aligned}$$

has at least one solution in $PC([0,T])$.

Problem 7.4. Check if the BVP

$$\begin{gathered}
\left(\phi_p(y^{\Delta}(T))\right)^{\Delta} + t\frac{1}{1+e^{y(t)}} = 0, \quad t \in [0,T]\backslash\{t_1\},\\
y(t_1^+)-y(t_1^-) = 1,\\
y^{\Delta}(t_1^+)-y^{\Delta}(t_1^-) = -2,\\
y(0) = y^{\Delta}(T) = 0,
\end{gathered}$$

where $\phi_p(s) = |s|^{p-2}s$, $p > 1$, $\phi_p^{-1} = \phi_q$, $\frac{1}{p}+\frac{1}{q} = 1$, has at least three different solutions.

Problem 7.5. Let $\mathbb{T} = \mathbb{Z}$. Prove that the BVP

$$\begin{aligned}
-u^{\Delta^2}(t) - u^{\Delta}(\sigma(t)) + \lambda u^{\sigma}(t) = 7u^{\sigma}(t) \quad \Delta-\text{a.e.} \quad & t \in [0,7]^{\kappa^2},\\
-(u^{\Delta}(1^+)-u^{\Delta}(1^-)) &= u^{\Delta}(1)),\\
u(0) = u(7) &= 0,
\end{aligned}$$

has at least two weak solutions.

7.5 Notes and References

In this chapter we investigated existence of solutions to some of the main classes of nonlinear second order boundary value problems for impulsive dynamic equations. Some of the results can be found in [7], [8], [13] and [22].

Appendix A
Lebesgue Integration. $\mathbb{L}^p$ Spaces. Sobolev Spaces

Let $\mathbb{T}$ be a time scale with forward jump operator and delta differentiation operator σ and Δ, respectively. In this section, suppose $a = \inf \mathbb{T}$, $b = \sup \mathbb{T}$, $-\infty < a < \rho(b) \le b < \infty$.

A.1 Lebesgue Type Measure and Integration over $\mathbb{T}$

Define the function $E : [a,b) \to \mathbb{R}$ as follows

$$E(t) = \sup\{s \in \mathbb{T} : s \le t\}, \quad t \in [a,b].$$

Suppose that $u : \mathbb{T}^\kappa \to \mathbb{R}$ is an arbitrary function. Then the function $u \circ \mathbb{E} : [a,b) \to \mathbb{R}$ is an extension of u to $[a,b)$ and it is constant on gaps in $\mathbb{T}$, with value equal to the value of u at the left-hand end of the gap.

Definition A.1. We say that u is measurable, respectively integrable, if $u \circ E$ is measurable, respectively integrable, on the real interval $[a,b)$ in the usual Lebesgue sense. Let $\mathbb{L}^1(\mathbb{T})$ denote the set of such integrable functions on $\mathbb{T}$.

Definition A.2. For any $u \in \mathbb{L}^1(\mathbb{T})$ we define the integral of u by

$$\int_s^t u\Delta = \int_s^t u \circ E dx, \quad s,t \in \mathbb{T},$$

where the integration on the right-hand side is the standard Lebesgue integral over the real interval $[s,t] \subset [a,b]$.

Using the above construction, we can also define a measure and a class of measurable subsets of $\mathbb{T}$, and integrals over such sets. For a set $A \subseteq \mathbb{T}$, with χ_A we will denote the characteristic function of A.

Definition A.3. We say that A is measurable if χ_A is measurable in the sense of Definition A.2.

DOI: 10.1201/9781003173557-A

Definition A.4. If $A \subseteq \mathbb{T}$ is measurable, we define the integral of $u \in \mathbb{L}^1(\mathbb{T})$ over A to be

$$\int_A u\Delta = \int_a^b u\chi_A \Delta.$$

We define the measure $\mu_{\mathbb{T}}(A)$ of A to be

$$\mu_{\mathbb{T}}(A) = \int_A 1\Delta.$$

From Definition A.4, it follows

$$\mu_{\mathbb{T}}\left([s,t)\bigcap \mathbb{T}\right) = t-s, \quad s,t \in \mathbb{T}, \quad s<t,$$

and

$$\mu_{\mathbb{T}}(\{b\}) = 0, \quad \mu_{\mathbb{T}}(\{t\}) = \sigma(t)-t, \quad t \in \mathbb{T}.$$

Also, if $t \in \mathbb{T}$ is right-scattered and $u \in \mathbb{L}^1(\mathbb{T})$, then $u(t)$ is well-defined and

$$\int_t^{\sigma(t)} u\Delta = u(t)\left(\sigma(t)-t\right).$$

Remark A.1. 1. From Definition A.2, it follows that the most of the standard Lebesgue integration results on $[a,b]$ hold immediately for the above integral on $\mathbb{T}$. In particular, Minkowsky's type inequality and Hölder's type inequality, and the various formulation of the dominated convergence theorem hold.

2. A nabla integral $\int_s^t u\nabla$ could be obtained in a similar manner to the above integral, by replacing the function E with the function

$$E_1(t) = \inf\{s \in \mathbb{T} : s \geq t\}.$$

We will not consider this further here.

For $u \in \mathbb{C}_{rd}(\mathbb{T})$, we define

$$|u|_{0,\mathbb{T}} = \sup\{|u(t)| : t \in \mathbb{T}\}.$$

For $u \in \mathbb{C}^1_{rd}(\mathbb{T})$, we define

$$|u|_{1,\mathbb{T}} = |u|_{0,\mathbb{T}} + \left|u^{\Delta}\right|_{0,\mathbb{T}^{\kappa}}.$$

Theorem A.1. *If $u \in \mathbb{L}^1(\mathbb{T})$ and $s \in \mathbb{T}$, then the function*

$$U_s(t) = \int_s^t u\Delta, \quad t \in \mathbb{T},$$

is well-defined and $U_s \in \mathbb{C}(\mathbb{T})$ with

$$|U_s|_{0,\mathbb{T}} \le \int_a^b |u|\Delta.$$

Proof. By Definition A.2, we get

$$\begin{aligned} U_s(t) &= \int_s^t u\Delta \\ &= \int_s^t u \circ E dx, \quad t \in \mathbb{T}. \end{aligned}$$

Hence, using the standard Lebesgue integration theory, the integral $\int_s^t u \circ E dx$ defines an absolutely continuous function of $t \in [a,b]$. Therefore, U_s is well-defined on $\mathbb{T}$. Also,

$$\begin{aligned} |U_s(t)| &= \left| \int_s^t u \circ E dx \right| \\ &\le \int_s^t |u \circ E| dx \\ &\le \int_a^b |u \circ E| dx \\ &\le \int_a^b |u|\Delta, \quad t \in \mathbb{T}. \end{aligned}$$

Hence,

$$\begin{aligned} |U_s|_{0,\mathbb{T}} &= \sup\{|U_s(t)| : t \in \mathbb{T}\} \\ &\le \int_a^b |u|\Delta. \end{aligned}$$

This completes the proof.

Theorem A.2. *If $u \in \mathbb{C}^1(\mathbb{T})$, then*

$$u(t) - u(s) = \int_s^t u^\Delta \Delta, \quad s,t \in \mathbb{T}.$$

Proof. Let $s \in \mathbb{T}$ be fixed. We define the function

$$H(t) = u(t) - u(s) - \int_s^t u^\Delta \Delta, \quad t \in \mathbb{T}.$$

Since $u^\Delta \in \mathbb{C}(\mathbb{T})$, by Theorem A.1, it follows that $\int_s^t u^\Delta \Delta$ is a continuous function of $t \in \mathbb{T}$. Therefore, H is a continuous function on $\mathbb{T}$. Let $t \in \mathbb{T}$ be arbitrarily chosen and fixed. We take $\varepsilon > 0$ arbitrarily. Then

$$\begin{aligned} |H(\sigma(t)) - H(t)| &= \left| u(\sigma(t)) - u(s) - \int_s^{\sigma(t)} u^\Delta \Delta - u(t) + u(s) + \int_s^t u^\Delta \Delta \right| \\ &= \left| u(\sigma(t)) - u(t) - \int_t^{\sigma(t)} u^\Delta \Delta \right| \\ &= \left| u(\sigma(t)) - u(t) - u^\Delta(t)(\sigma(t) - t) \right| \\ &\le \varepsilon |\sigma(t) - t|. \end{aligned}$$

In the last inequality we have used that $u \in \mathbb{C}^1(\mathbb{T})$. Since H is a continuous function on $\mathbb{T}$, there is a $\delta > 0$ such that for any $z \in (t-\delta, t+\delta)_{\mathbb{T}}$ we have

$$|H(t) - H(z)| < \varepsilon.$$

Therefore,

$$\begin{aligned} |H(\sigma(t)) - H(z)| &= |H(\sigma(t)) - H(t) + H(t) - H(z)| \\ &\le |H(\sigma(t)) - H(t)| + |H(t) - H(z)| \\ &< \varepsilon |\sigma(t) - t| + \varepsilon \end{aligned}$$

for any $z \in (t-\delta, t+\delta)_{\mathbb{T}}$. Since $t \in \mathbb{T}$ was arbitrarily chosen, we conclude that $H^\Delta(t)$ exists and $H^\Delta(t) = 0$ for any $t \in \mathbb{T}$. Hence, using the Mean Value Theorem for time scale derivatives, we conclude that H is a constant on $\mathbb{T}$. Because $H(s) = 0$, we obtain $H \equiv 0$ on $\mathbb{T}$. This completes the proof.

A.2 Spaces $\mathbb{L}^p(\mathbb{T})$

Having defined the space $\mathbb{L}^1(\mathbb{T})$, it is usual to define $\mathbb{L}^p(\mathbb{T})$ for any $p \ge 1$ in the following way

$$\mathbb{L}^p(\mathbb{T}) = \{u \in \mathbb{L}^1(\mathbb{T}) : |u|^p \in \mathbb{L}^1(\mathbb{T})\}.$$

For any $p \ge 1$, we provide $\mathbb{L}^p(\mathbb{T})$ with a norm

$$\|u\|_{\mathbb{L}^p(\mathbb{T})} = \left(\int_a^b |u|^p \Delta \right)^{\frac{1}{p}}, \quad u \in \mathbb{L}^p(\mathbb{T}). \tag{A.1}$$

Exercise A.1. Prove that (A.1) satisfies all axioms for a norm.

Theorem A.3. *For any $p \ge 1$, we have*

$$\mathbb{C}(\mathbb{T}) \subset \mathbb{C}_{rd}(\mathbb{T}) \subset \mathbb{L}^p(\mathbb{T}).$$

Proof. Let $u \in \mathbb{C}_{rd}(\mathbb{T})$ be arbitrarily chosen. Then $|u|^p \circ E$ is bounded on $[a,b]$ and it is continuous on $[a,b]$ except possibly for right-scattered points. Since the set of such points is countable, it has Lebesgue measure zero. Thus $|u|^p \circ E$ is Lebesgue measurable and, hence, integrable. Therefore, $u \in \mathbb{L}^p(\mathbb{T})$. Because $u \in \mathscr{C}_{rd}(\mathbb{T})$ was arbitrarily chosen and we get that $u \in \mathbb{L}^p(\mathbb{T})$, we conclude that

$$\mathscr{C}_{rd}(\mathbb{T}) \subset \mathbb{L}^p(\mathbb{T}).$$

This completes the proof.

Theorem A.4. *For any $p \geq 1$ we have that $\mathbb{C}(\mathbb{T})$ is dense in $\mathbb{L}^p(\mathbb{T})$.*

Proof. Let $u \in \mathbb{L}^p(\mathbb{T})$ be arbitrarily chosen. Consider $\mathbb{T}$ as a topological space with measure $\mu_{\mathbb{T}}$. By the standard Lebesgue integration theory, it follows that there exists a sequence $\{u_n\}_{n\in\mathbb{N}}$ of elements of $\mathbb{C}(\mathbb{T})$ such that

$$||u - u_n||_{\mathbb{L}^1(\mathbb{T})} \to 0,$$

as $n \to \infty$. Thus the result is proved for $p = 1$. In addition, by the standard Lebesgue integration theory, it follows that we can suppose that there exists a constant $C > 0$ for which $|u(t)| \leq C$, $|u_n(t)| \leq C$ for any $n \in \mathbb{N}$ and for any $t \in \mathbb{T}$. Hence,

$$\begin{aligned}
||u - u_n||^p_{\mathbb{L}^p(\mathbb{T})} &= \int_a^b |u - u_n|^p \Delta \\
&= \int_a^b |u - u_n|^{p-1} |u - u_n| \Delta \\
&\leq \int_a^b (|u| + |u_n|)^{p-1} |u - u_n| \Delta \\
&\leq (2C)^{p-1} \int_a^b |u - u_n| \Delta \\
&= (2C)^{p-1} ||u - u_n||_{\mathbb{L}^1(\mathbb{T})} \\
&\to 0,
\end{aligned}$$

as $n \to \infty$. This completes the proof.

Theorem A.5. *Suppose that $\{u_n\}_{n\in\mathbb{N}}$ is a sequence in $\mathbb{L}^p(\mathbb{T})$ for some $p \geq 1$. If $||u - u_n||_{\mathbb{L}^p(\mathbb{T})} \to 0$, as $n \to \infty$, for some $u \in \mathbb{L}^p(\mathbb{T})$, and if $t \in \mathbb{T}$ is right-scattered, then $u_n(t) \to u(t)$, as $n \to \infty$.*

Proof. Let $|| \cdot ||_{\mathbb{L}^p(a,b)}$ denotes the standard $\mathbb{L}^p(a,b)$-norm on the real interval $[a,b]$. Then

$$\begin{aligned}
0 &= \lim_{n\to\infty} \|(u-u_n)\circ E\|^p_{\mathbb{L}^p(a,b)}\\
&= \lim_{n\to\infty} \int_a^b |(u-u_n)\circ E|^p\,dx\\
&\geq \lim_{n\to\infty} \int_t^{\sigma(t)} |(u-u_n)\circ E|^p\,dx\\
&= \lim_{n\to\infty} \int_t^{\sigma(t)} |u-u_n|^p \Delta\\
&= \lim_{n\to\infty} |u(t)-u_n(t)|^p (\sigma(t)-t)\\
&\geq 0.
\end{aligned}$$

Consequently, $u_n(t) \to u(t)$, as $n \to \infty$. This completes the proof.

Theorem A.6. *If $\{u_n\}_{n\in\mathbb{N}}$ is a Cauchy sequence in $\mathbb{L}^p(\mathbb{T})$, then there exists a unique $u \in \mathbb{L}^p(\mathbb{T})$ such that $\|u-u_n\|_{\mathbb{L}^p(\mathbb{T})} \to 0$, as $n \to \infty$.*

Proof. Note that the sequence $\{u_n \circ E\}_{n\in\mathbb{N}}$ is a Cauchy sequence in $\mathbb{L}^p(a,b)$. Then, by the standard Lebesgue theory, it follows that there is a unique $z \in \mathbb{L}^p(a,b)$ such that

$$\|z-u_n\circ E\|_{\mathbb{L}^p((a,b))} \to 0,$$

as $n \to \infty$. Suppose that $t \in \mathbb{T}$ is right-scattered. Then

$$\begin{aligned}
\|(u_n-u_m)\circ E\|^p_{\mathbb{L}^p(\mathbb{T})} &= \int_a^b |(u_n-u_m)\circ E|^p\,dx\\
&\geq \int_t^{\sigma(t)} |(u_n-u_m)\circ E|^p\,dx\\
&= \int_t^{\sigma(t)} |u_n-u_m|^p\,\Delta\\
&= |u_n(t)-u_m(t)|^p\,(\sigma(t)-t).
\end{aligned}$$

Therefore, the sequence $\{u_n(t)\}_{n\in\mathbb{N}}$ is a Cauchy sequence. Thus, it converges to $c(t)$ and the function z must equal to $c(t)$ almost everywhere on the interval $[t,\sigma(t))$, and so we may suppose that this equality holds everywhere on $[t,\sigma(t))$. Since the set of right-scattered points $t \in \mathbb{T}$, is at most countable, we may suppose that this is true for all right-scattered points $t \in \mathbb{T}$. Defining u to be the restriction of z to $\mathbb{T}$, it follows that $u \in \mathbb{L}^p(\mathbb{T})$, $z = u\circ E$ and $\|u-u_n\|_{\mathbb{L}^p(\mathbb{T})} \to 0$, as $n \to \infty$. This completes the proof.

Remark A.2. 1. By Theorem A.6, it follows that $\mathbb{L}^p(\mathbb{T})$, $p \geq 1$, is a Banach space with respect to the norm $\|\cdot\|_{\mathbb{L}^p(\mathbb{T})}$.

2. We can define a natural inner product on $\mathbb{L}^2(\mathbb{T})$ by

$$\langle u,v\rangle_{\mathbb{T}} = \int_a^b uv\Delta, \quad u,v \in \mathbb{L}^2(\mathbb{T}). \tag{A.2}$$

   With respect to this inner product, the space $\mathbb{L}^2(\mathbb{T})$ is a Hilbert space.
3. To simplify the notation, from now on we will use the notation $\|\cdot\|_{\mathbb{T}}$ for the norm $\|\cdot\|_{\mathbb{L}^2(\mathbb{T})}$.
4. The notations $\|\cdot\|_{\mathbb{T}}$ and $\langle\cdot,\cdot\rangle_{\mathbb{T}}$ for the above norm and inner product on $\mathbb{L}^2(\mathbb{T})$ indicate that their values depend on the entire time scale $\mathbb{T}$, even if u is not defined at b.

Exercise A.2. Prove that (A.2) satisfies all axioms for an inner product.

A.3 Sobolev Type Spaces and Generalized Derivatives

Definition A.5. For $u \in \mathbb{C}^1_{rd}(\mathbb{T})$, we define

$$\|u\|^2_{1,\mathbb{T}} = \|u\|^2_{\mathbb{T}} + \|u^{\Delta}\|^2_{\mathbb{T}}, \tag{A.3}$$

and define the space $\mathbb{H}^1(\mathbb{T}) \subset \mathbb{L}^2(\mathbb{T})$ to be the completion of $\mathbb{C}^1(\mathbb{T})$ with respect to the norm $\|u\|_{1,\mathbb{T}}$. The space $\mathbb{H}^1(\mathbb{T})$ will be called Sobolev space over $\mathbb{T}$.

Exercise A.3. Prove that (A.3) satisfies all axioms for a norm.

Theorem A.7. *$u \in \mathbb{H}^1(\mathbb{T})$ if and only if there exists a function $u^{\Delta_g} \in \mathbb{L}^2(\mathbb{T})$ such that the following condition holds: there exists a sequence $\{u_n\}_{n\in\mathbb{N}}$ in $\mathbb{C}^1(\mathbb{T})$ such that $u_n \to u$ and $u_n^{\Delta} \to u^{\Delta_g}$, as $n \to \infty$, in $\mathbb{L}^2(\mathbb{T})$. If $u \in \mathbb{H}^1(\mathbb{T})$, then u^{Δ_g} is unique in the $\mathbb{L}^2(\mathbb{T})$ sense. If $u \in \mathbb{C}^1(\mathbb{T})$, then $u^{\Delta_g} = u^{\Delta}$.*

Proof. 1. Let $u \in \mathbb{H}^1(\mathbb{T})$. By the definition of the space $\mathbb{H}^1(\mathbb{T})$, it follows that there exists a sequence $\{u_n\}_{n\in\mathbb{N}}$ of elements of $\mathbb{C}^1(\mathbb{T})$ such that $\|u_n - u\|_{1,\mathbb{T}} \to 0$, as $n \to \infty$. Hence, $\|u_n - u_m\|_{1,\mathbb{T}} \to 0$, as $m,n \to \infty$. Consequently, $\left\{u_n^{\Delta}\right\}_{n\in\mathbb{N}}$ is a Cauchy sequence in $\mathbb{L}^2(\mathbb{T})$. Because $\mathbb{L}^2(\mathbb{T})$ is a Banach space, there exists a unique $u^{\Delta_g} \in \mathbb{L}^2(\mathbb{T})$ such that $u_n^{\Delta} \to u^{\Delta_g}$, as $n \to \infty$, in $\mathbb{L}^2(\mathbb{T})$. In particular, if $u \in \mathbb{C}^1(\mathbb{T})$, then $u^{\Delta} = u^{\Delta_g}$.
2. Let $\{u_n\}_{n\in\mathbb{N}}$ be a sequence of elements of $\mathbb{C}^1(\mathbb{T})$ such that $u_n \to u$ and $u_n \to u^{\Delta_g}$, as $n \to \infty$, in $\mathbb{L}^2(\mathbb{T})$, for some $u^{\Delta_g} \in \mathbb{L}^2(\mathbb{T})$. Hence, using (A.3), we conclude that $\{u_n\}_{n\in\mathbb{N}}$ is a convergent sequence in $\mathbb{H}^1(\mathbb{T})$. By the definition of $\mathbb{H}^1(\mathbb{T})$, it follows that there exists $u_1 \in \mathbb{H}^1(\mathbb{T})$ such that $u_n \to u_1$, as $n \to \infty$, in $\mathbb{H}^1(\mathbb{T})$. Hence, $u_n \to u_1$, as $n \to \infty$, in $\mathbb{L}^2(\mathbb{T})$. Because $u_n \to u$, as $n \to \infty$, in $\mathbb{L}^2(\mathbb{T})$, we conclude that $u = u_1$ and $u \in \mathbb{H}^1(\mathbb{T})$. This completes the proof.

Definition A.6. For any $u \in \mathbb{H}^1(\mathbb{T})$, the function u^{Δ_g} in Theorem A.7 will be called generalized derivative of u.

Theorem A.8. *If $u \in \mathbb{H}^1(\mathbb{T})$, then $u \in \mathbb{C}(\mathbb{T})$. There exists a constant $C > 0$ such that*

$$|u|_{0,\mathbb{T}} \leq C\|u\|_{1,\mathbb{T}}, \quad u \in \mathbb{H}^1(\mathbb{T}). \tag{A.4}$$

Furthermore,

$$u(t) - u(s) = \int_s^t u^{\Delta_g}\Delta, \quad s,t \in \mathbb{T}. \tag{A.5}$$

Proof. Let $u \in \mathbb{C}^1(\mathbb{T})$ and $s,t \in \mathbb{T}$, $s \leq t$. By Theorem A.2, we get

$$u(t) - u(s) = \int_s^t u^{\Delta}\Delta. \tag{A.6}$$

Hence,

$$\begin{aligned}
|u(t)| &= \left|u(s) + \int_s^t u^{\Delta}\Delta\right| \\
&\leq |u(s)| + \left|\int_s^t u^{\Delta}\Delta\right| \\
&\leq |u(s)| + \int_s^t \left|u^{\Delta}\right|\Delta \\
&\leq |u(s)| + (t-s)^{\frac{1}{2}}\left(\int_s^t \left|u^{\Delta}\right|^2\Delta\right)^{\frac{1}{2}} \\
&\leq |u(s)| + (b-a)^{\frac{1}{2}}\left\|u^{\Delta}\right\|_{\mathbb{T}}.
\end{aligned}$$

Now we integrate the last inequality over $\mathbb{T}$ with respect to s and we get

$$\begin{aligned}
\int_a^b |u(t)|\Delta &\leq \int_a^b |u(s)|\Delta + (b-a)^{\frac{1}{2}}\int_a^b \|u^{\Delta}\|_{\mathbb{T}}\Delta \\
&\leq (b-a)^{\frac{1}{2}}\left(\int_a^b |u(s)|^2\Delta\right)^{\frac{1}{2}} + (b-a)^{\frac{3}{2}}\left\|u^{\Delta}\right\|_{\mathbb{T}} \\
&= (b-a)^{\frac{1}{2}}\|u\|_{\mathbb{T}} + (b-a)^{\frac{3}{2}}\|u^{\Delta}\|_{\mathbb{T}}
\end{aligned}$$

or

$$(b-a)|u(t)| \leq (b-a)^{\frac{1}{2}}\|u\|_{\mathbb{T}} + (b-a)^{\frac{3}{2}}\left\|u^{\Delta}\right\|_{\mathbb{T}}.$$

Hence,

$$|u(t)| \leq \frac{1}{(b-a)^{\frac{1}{2}}}\|u\|_{\mathbb{T}} + (b-a)^{\frac{1}{2}}\left\|u^{\Delta}\right\|_{\mathbb{T}}$$

and

$$\sup_{t\in\mathbb{T}} |u(t)| \leq \frac{1}{(b-a)^{\frac{1}{2}}} \|u\|_{\mathbb{T}} + (b-a)^{\frac{1}{2}} \left\| u^{\Delta} \right\|_{\mathbb{T}},$$

or

$$|u|_{0,\mathbb{T}} \leq \frac{1}{(b-a)^{\frac{1}{2}}} \|u\|_{\mathbb{T}} + (b-a)^{\frac{1}{2}} \left\| u^{\Delta} \right\|_{\mathbb{T}}.$$

Let

$$C = \max\left\{ \frac{1}{(b-a)^{\frac{1}{2}}}, (b-a)^{\frac{1}{2}} \right\}.$$

Then

$$\begin{aligned} |u|_{0,\mathbb{T}} &\leq C\left(\|u\|_{\mathbb{T}} + \left\|u^{\Delta}\right\|_{\mathbb{T}}\right) \\ &\leq C\|u\|_{1,\mathbb{T}}. \end{aligned}$$

By Theorem A.7, we have that $u^{\Delta} = u^{\Delta_g}$. Hence and by (A.6), we get (A.5). Let now $u \in \mathbb{H}^1(\mathbb{T})$. Then there exists a sequence $\{u_n\}_{n\in\mathbb{N}}$ of elements of $\mathbb{C}^1(\mathbb{T})$ such that $u_n \to u$, as $n \to \infty$, in $\mathbb{H}^1(\mathbb{T})$. For this sequence we have

$$|u_n - u_m|_{0,\mathbb{T}} \leq C\|u_n - u_m\|_{1,\mathbb{T}}, \quad m,n \in \mathbb{N}. \tag{A.7}$$

Note that $u_n \to u$, as $n \to \infty$, in $\mathbb{L}^2(\mathbb{T})$. Because $\|u_n - u_m\|_{1,\mathbb{T}} \to 0$, as $m,n \to \infty$, using (A.7), we get

$$|u_n - u_m|_{0,\mathbb{T}} \to 0,$$

as $m,n \to \infty$. Therefore, $\{u_n\}_{n\in\mathbb{N}}$ is a Cauchy sequence in $\mathbb{C}(\mathbb{T})$. Because $\mathbb{C}(\mathbb{T})$ is a Banach space, we conclude that the sequence $\{u_n\}_{n\in\mathbb{N}}$ is a convergent sequence in $\mathbb{C}(\mathbb{T})$. Let $u^1 \in \mathbb{C}(\mathbb{T})$ be its limit, i.e., $u_n \to u$, as $n \to \infty$, in $\mathbb{C}(\mathbb{T})$. Then

$$\begin{aligned} \|u_n - u^1\|_{\mathbb{T}}^2 &= \int_a^b \left|u_n - u^1\right|^2 \Delta \\ &\leq (b-a)\left|u_n - u^1\right|_{0,\mathbb{T}}^2 \\ &\to 0, \end{aligned}$$

as $n \to \infty$. Therefore, $u_n \to u^1$, as $n \to \infty$, in $\mathbb{L}^2(\mathbb{T})$. Because $u_n \to u$, as $n \to \infty$, in $\mathbb{L}^2(\mathbb{T})$, we conclude that $u = u^1$ and $u \in \mathbb{C}(\mathbb{T})$. From here, using that

$$|u_n|_{0,\mathbb{T}} \leq C\|u_n\|_{1,\mathbb{T}},$$

we obtain that u satisfies the inequality (A.4). Since $u \in \mathbb{H}^1(\mathbb{T})$, using Theorem A.7, there exists a unique function $u^{\Delta_g} \in \mathbb{L}^2(\mathbb{T})$ and there exists a sequence $\{v_n\}_{n\in\mathbb{N}}$ of elements of $\mathbb{C}^1(\mathbb{T})$ such that $v_n \to u$ and $v_n^{\Delta} \to u^{\Delta_g}$, as $n \to \infty$, in $\mathbb{L}^2(\mathbb{T})$. Also, we have

$$v_n(t) - v_n(s) = \int_s^t v_n^{\Delta_g} \Delta, s, t \in \mathbb{T}, \quad s \le t. \tag{A.8}$$

Note that

$$\begin{aligned}\left| \int_s^t v_n^{\Delta_g} \Delta - \int_s^t u^{\Delta_g} \Delta \right| &= \left| \int_s^t v_n^{\Delta} \Delta - \int_s^t u^{\Delta_g} \Delta \right| \\ &= \left| \int_s^t \left(v_n^{\Delta} - u^{\Delta_g} \right) \Delta \right| \\ &\le \int_s^t \left| v_n^{\Delta} - u^{\Delta_g} \right| \Delta \\ &\le (t-s)^{\frac{1}{2}} \left(\int_s^t \left| v_n^{\Delta} - u^{\Delta_g} \right|^2 \Delta \right)^{\frac{1}{2}} \\ &\le (b-a)^{\frac{1}{2}} \left\| v_n^{\Delta} - u^{\Delta_g} \right\|_{\mathbb{T}} \\ &\to 0, \quad s \le t,\end{aligned}$$

as $n \to \infty$. Hence, using that $v_n \to u$, as $n \to \infty$, in $\mathbb{C}(\mathbb{T})$, and (A.8), we conclude that u satisfies (A.5). This completes the proof.

Theorem A.9. *Let $u \in \mathbb{H}^1(\mathbb{T})$. If $u^{\Delta_g} = 0$, then u is a constant on $\mathbb{T}$.*

Proof. By (A.5), we get

$$u(t) = u(s)$$

for any $s, t \in \mathbb{T}$. This completes the proof.

Theorem A.10. *Let $u \in \mathbb{H}^1(\mathbb{T})$. If $t \in \mathbb{T}$ is a right-scattered point, then*

$$u^{\Delta_g}(t) = \frac{u(\sigma(t)) - u(t)}{\sigma(t) - t}.$$

Proof. Let $t \in \mathbb{T}$ be a right-scattered point. Since $u \in \mathbb{H}^1(\mathbb{T})$, we have that $u \in \mathbb{L}^2(\mathbb{T})$ and there exists a sequence $\{u_n\}_{n\in\mathbb{N}}$ of elements of $\mathbb{C}^1(\mathbb{T})$ and a unique $u^{\Delta_g} \in \mathbb{L}^2(\mathbb{T})$ such that $u_n \to u$ and $u_n^{\Delta} \to u^{\Delta_g}$, as $n \to \infty$, in $\mathbb{L}^2(\mathbb{T})$. Hence and by Theorem A.5, we get that $u_n(t) \to u(t)$ and $u_n^{\Delta}(t) \to u^{\Delta_g}(t)$, as $n \to \infty$. By (A.5), we have

$$u(\sigma(t)) = u(t) + \int_t^{\sigma(t)} u^{\Delta_g} \Delta$$

and

$$u_n(\sigma(t)) = u_n(t) + \int_t^{\sigma(t)} u_n^{\Delta} \Delta$$

$$= u_n(t) + u_n^{\Delta}(t)\,(\sigma(t) - t)\,.$$

Hence,

$$\begin{aligned}\lim_{n\to\infty} u_n\,(\sigma(t)) &= \lim_{n\to\infty} \left(u_n(t) + u_n^{\Delta}(t)\,(\sigma(t) - t) \right) \\ &= u(t) + \int_t^{\sigma(t)} u^{\Delta_g} \Delta \\ &= u\,(\sigma(t))\,.\end{aligned}$$

Therefore,

$$\begin{aligned} u^{\Delta_g}(t) &= \lim_{n\to\infty} u_n^{\Delta}(t) \\ &= \lim_{n\to\infty} \frac{u_n\,(\sigma(t)) - u_n(t)}{\sigma(t) - t} \\ &= \frac{u\,(\sigma(t)) - u(t)}{\sigma(t) - t}.\end{aligned}$$

This completes the proof.

Theorem A.11. *Let $u, v \in \mathbb{H}^1(\mathbb{T})$ and $\alpha, \beta \in \mathbb{R}$. Then $\alpha u + \beta v \in \mathbb{H}^1(\mathbb{T})$ and*

$$(\alpha u + \beta v)^{\Delta_g} = \alpha u^{\Delta_g} + \beta v^{\Delta_g}.$$

Proof. Since $u, v \in \mathbb{H}^1(\mathbb{T})$, there exist sequences $\{u_n\}_{n\in\mathbb{N}}$, $\{v_n\}_{n\in\mathbb{N}}$ of elements of $\mathbb{C}^1(\mathbb{T})$ and unique u^{Δ_g}, $v^{\Delta_g} \in \mathbb{L}^2(\mathbb{T})$ such that $u_n \to u$, $v_n \to v$, $u_n^{\Delta} \to u^{\Delta_g}$, $v_n^{\Delta} \to v^{\Delta_g}$, as $n \to \infty$, in $\mathbb{L}^2(\mathbb{T})$. Hence,

$$\alpha u_n + \beta v_n \to \alpha u + \beta v, \quad \alpha u_n^{\Delta} + \beta v_n^{\Delta} \to \alpha u^{\Delta_g} + \beta v^{\Delta_g},$$

as $n \to \infty$, in $\mathbb{L}^2(\mathbb{T})$. Hence and by Theorem A.7, it follows that $\alpha u + \beta v \in \mathbb{H}^1(\mathbb{T})$. Applying again Theorem A.7, we have that there exists unique $(\alpha u + \beta v)^{\Delta_g} \in \mathbb{L}^2(\mathbb{T})$ such that

$$(\alpha u_n + \beta v_n)^{\Delta} \to (\alpha u_n + \beta v_n)^{\Delta_g},$$

as $n \to \infty$, in $\mathbb{L}^2(\mathbb{T})$. Since

$$\begin{aligned}(\alpha u_n + \beta v_n)^{\Delta} &= \alpha u_n^{\Delta} + \beta v_n^{\Delta} \\ &\to \alpha u^{\Delta_g} + \beta v^{\Delta_g},\end{aligned}$$

as $n \to \infty$, in $\mathbb{L}^2(\mathbb{T})$, we conclude that

$$(\alpha u+\beta v)^{\Delta_g} = \alpha u^{\Delta_g} + \beta v^{\Delta_g}.$$

This completes the proof.

Theorem A.12. *Let $u, v \in \mathbb{H}^1(\mathbb{T})$. Then $uv \in \mathbb{H}^1(\mathbb{T})$ and*

$$\begin{aligned}(uv)^{\Delta_g} &= u^{\Delta_g} v + u^{\sigma} v^{\Delta_g}\\ &= u v^{\Delta_g} + u^{\Delta_g} v^{\sigma},\end{aligned} \tag{A.9}$$

and

$$\begin{aligned}\int_s^t u^{\Delta_g} v\Delta + \int_s^t u^{\sigma} v^{\Delta_g}\Delta &= \int_s^t u v^{\Delta_g}\Delta + \int_s^t u^{\Delta_g} v^{\sigma}\Delta\\ &= (uv)(t) - (uv)(s), \quad s,t \in \mathbb{T}.\end{aligned}$$

Proof. Since $u, v \in \mathbb{H}^1(\mathbb{T})$, there exist sequences $\{u_n\}_{n\in\mathbb{N}}$, $\{v_n\}_{n\in\mathbb{N}}$ of elements of $\mathbb{C}^1(\mathbb{T})$ and unique u^{Δ_g}, $v^{\Delta_g} \in \mathbb{L}^2(\mathbb{T})$ such that

$$u_n \to u, \quad v_n \to v, \quad u_n^{\Delta} \to u^{\Delta_g}, \quad v_n^{\Delta} \to v^{\Delta_g},$$

as $n \to \infty$, in $\mathbb{L}^2(\mathbb{T})$. Then

$$u_n v_n \to uv,$$

as $n \to \infty$, in $\mathbb{L}^2(\mathbb{T})$,

$$u^{\Delta_g} v, \quad u^{\sigma} v^{\Delta_g}, \quad u v^{\Delta_g}, \quad u^{\Delta_g} v^{\sigma} \in \mathbb{L}^2(\mathbb{T}),$$

$$u_n^{\sigma} \to u^{\sigma}, \quad v_n^{\sigma} \to v^{\sigma},$$

as $n \to \infty$, in $\mathbb{L}^2(\mathbb{T})$. Hence,

$$\begin{aligned}(u_n v_n)^{\Delta} &= u_n^{\Delta} v_n + u_n^{\sigma} v_n^{\Delta}\\ &\to u^{\Delta_g} v + u^{\sigma} v^{\Delta_g},\end{aligned} \tag{A.10}$$

$$\begin{aligned}(u_n v_n)^{\Delta} &= u_n v_n^{\Delta} + u_n^{\Delta} v_n^{\sigma}\\ &\to u v^{\Delta_g} + u^{\Delta_g} v^{\sigma},\end{aligned} \tag{A.11}$$

as $n \to \infty$, in $\mathbb{L}^2(\mathbb{T})$. Therefore, $uv \in \mathbb{H}^1(\mathbb{T})$. Hence, using Theorem A.7, there exists a unique $(uv)^{\Delta_g} \in \mathbb{L}^2(\mathbb{T})$ such that

$$(u_n v_n)^{\Delta} \to (uv)^{\Delta_g},$$

as $n \to \infty$, in $\mathbb{L}^2(\mathbb{T})$. From here and from (A.10), (A.11), we obtain (A.9). Hence and by (A.5), we get

$$\int_s^t u^{\Delta_g} v\Delta + \int_s^t u^{\sigma} v^{\Delta_g}\Delta = \int_s^t \left(u^{\Delta_g} v + u^{\sigma} v^{\Delta_g}\right)\Delta$$

$$= \int_s^t (uv)^{\Delta_g} \Delta$$

$$= (uv)(t) - (uv)(s)$$

and

$$\begin{aligned}
\int_s^t uv^{\Delta_g} \Delta + \int_s^t u^{\Delta_g} v^\sigma \Delta &= \int_s^t \left(uv^{\Delta_g} + u^{\Delta_g} v^\sigma \right) \Delta \\
&= \int_s^t (uv)^{\Delta_g} \Delta \\
&= (uv)(t) - (uv)(s), \quad s,t \in \mathbb{T}.
\end{aligned}$$

This completes the proof.

Theorem A.13. *Let $u \in \mathbb{L}^2(\mathbb{T})$ and $U(t) = \int_a^t u\Delta$, $t \in \mathbb{T}$. Then $U \in \mathbb{H}^1(\mathbb{T})$, $U^{\Delta_g} = u$, and if $V \in H^1(\mathbb{T})$ satisfies $V^{\Delta_g} = u$, then $U - V$ is a constant. In addition, there exists a constant $C > 0$ such that*

$$\|U\|_{1,\mathbb{T}} \leq C\|u\|_{\mathbb{T}}, \quad u \in \mathbb{L}^2(\mathbb{T}).$$

Proof. Let $u \in \mathbb{L}^2(\mathbb{T})$. By Theorem A.4, it follows that there exists a sequence $\{u_n\}_{n\in\mathbb{N}}$ of elements of $\mathbb{C}(\mathbb{T})$ such that $u_n \to u$, as $n \to \infty$, in $\mathbb{L}^2(\mathbb{T})$. Since $u_n \in \mathbb{C}(\mathbb{T})$, $n \in \mathbb{N}$, we have that

$$U_n(t) = \int_a^t u_n\Delta \in \mathbb{C}^1(\mathbb{T}) \quad \text{and} \quad U_n^\Delta(t) = u_n(t), \quad t \in \mathbb{T}.$$

For $n \in \mathbb{N}$, we have

$$\begin{aligned}
\|U_n - U\|_{\mathbb{T}}^2 &= \int_a^b \left| \int_a^t u_n\Delta - \int_a^t u\Delta \right|^2 \Delta \\
&= \int_a^b \left| \int_a^t (u_n - u)\Delta \right|^2 \Delta \\
&\leq \int_a^b \left(\int_a^t (u_n - u)^2\Delta \right) (t-a)\Delta \\
&\leq (b-a) \int_a^b \left(\int_a^b (u_n - u)^2\Delta \right) \Delta \\
&= (b-a)^2 \|u_n - u\|_{\mathbb{T}}^2
\end{aligned}$$

$$\to 0,$$

$$\|U_n^{\Delta} - u\|_{\mathbb{T}} = \|u_n - u\|_{\mathbb{T}}$$

$$\to 0,$$

as $n \to \infty$, i.e., $U_n \to U$, $U_n^{\Delta} \to u$, as $n \to \infty$, in $\mathbb{L}^2(\mathbb{T})$. Hence and by Theorem A.7, it follows that $U \in \mathbb{H}^1(\mathbb{T})$ and $U^{\Delta_g} = u$. Next,

$$\begin{aligned}
\|U\|_{\mathbb{T}}^2 &= \int_a^b \left| \int_a^t u\Delta \right|^2 \Delta \\
&\leq \int_a^b (t-a) \left(\int_a^t u^2 \Delta \right) \Delta \\
&\leq (b-a) \int_a^b \left(\int_a^b u^2 \Delta \right) \Delta \\
&= (b-a)^2 \|u\|_{\mathbb{T}}^2,
\end{aligned}$$

$$\|U^{\Delta_g}\|_{\mathbb{T}}^2 = \|u\|_{\mathbb{T}}^2.$$

Hence,

$$\begin{aligned}
\|U\|_{1,\mathbb{T}} &= \left(\|U\|_{\mathbb{T}}^2 + \|U^{\Delta_g}\|_{\mathbb{T}}^2 \right)^{\frac{1}{2}} \\
&\leq \left((b-a)^2 \|u\|_{\mathbb{T}}^2 + \|u\|_{\mathbb{T}}^2 \right)^{\frac{1}{2}} \\
&= \left(1 + (b-a)^2 \right)^{\frac{1}{2}} \|u\|_{\mathbb{T}}.
\end{aligned}$$

Let $V \in \mathbb{H}^1(\mathbb{T})$ satisfies $V^{\Delta_g} = u$. Then, using Theorem A.11, we obtain

$$\begin{aligned}
(U - V)^{\Delta_g} &= U^{\Delta_g} - V^{\Delta_g} \\
&= u - u \\
&= 0.
\end{aligned}$$

Hence and by Theorem A.9, we conclude that $U - V$ is a constant. This completes the proof.

Definition A.7. The function U in Theorem A.13 will be called anti-derivative of u.

Theorem A.14. *We have*

$$\mathbb{C}^1_{rd}(\mathbb{T}) \subset \mathbb{H}^1(\mathbb{T}). \tag{A.12}$$

Proof. Let $v \in \mathbb{C}^1_{rd}(\mathbb{T})$ be arbitrarily chosen and $u = v^{\Delta} \in \mathbb{C}_{rd}(\mathbb{T})$. By Theorem A.3, it follows that $u \in \mathbb{L}^2(\mathbb{T})$. Hence and by Theorem A.13, we have

$$U = \int_a^t u\Delta \in \mathbb{H}^1(\mathbb{T}) \quad \text{and} \quad U^{\Delta_g} = u.$$

Because $v^{\Delta} = v^{\Delta_g} = u$, by Theorem A.13, it follows that there exists a constant C such that $v = U + C \in \mathbb{H}^1(\mathbb{T})$. Because $v \in \mathbb{C}^1_{rd}(\mathbb{T})$ was arbitrarily chosen and for it we get that it is an element of $\mathbb{H}^1(\mathbb{T})$, we obtain the inclusion (A.12). This completes the proof.

Definition A.8. Define the space

$$\mathbb{H}^2(\mathbb{T}) = \left\{ u \in \mathbb{C}^1(\mathbb{T}) : u^{\Delta} \in \mathbb{H}^1\left(\mathbb{T}^{\kappa}\right) \right\}$$

with the norm

$$\|u\|^2_{2,\mathbb{T}} = \|u\|^2_{\mathbb{T}} + \|u^{\Delta}\|^2_{1,\mathbb{T}}.$$

We define the space

$$\mathbb{H}^2(\mathbb{T}) = \left\{ u \in \mathbb{C}^1(\mathbb{T}) : u^{\Delta} \in \mathbb{H}^1\left(\mathbb{T}^{\kappa}\right) \right\}$$

with the norm

$$\|u\|^2_{2,\mathbb{T}} = \|u\|^2_{\mathbb{T}} + \left\|u^{\Delta}\right\|^2_{1,\mathbb{T}^{\kappa}}. \tag{A.13}$$

Exercise A.4. Prove that (A.13) satisfies all axioms for a norm.

Theorem A.15. *If $\{u_n\}_{n\in\mathbb{N}}$ is a bounded sequence in $\mathbb{H}^1(\mathbb{T})$, then $\{u_n\}_{n\in\mathbb{N}}$ has a subsequence that converges in $\mathbb{C}(\mathbb{T})$.*

Proof. Since $\{u_n\}_{n\in\mathbb{N}}$ is a bounded sequence in $\mathbb{H}^1(\mathbb{T})$, there exists a constant $M > 0$ such that

$$\|u_n\|_{1,\mathbb{T}} \leq M$$

for any $n \in \mathbb{N}$. Hence and by Theorem A.8, it follows that

$$|u_n|_{0,\mathbb{T}} \leq C\|u\|_{1,\mathbb{T}}$$

$$\leq CM$$

with $C = \max\left\{ \frac{1}{(b-a)^{\frac{1}{2}}}, (b-a)^{\frac{1}{2}} \right\}$, for any $n \in \mathbb{N}$. Therefore, the sequence $\{u_n\}_{n\in\mathbb{N}}$ is a bounded sequence in $\mathbb{C}(\mathbb{T})$. Also, using Cauchy-Schwartz inequality and (A.5),

we have

$$\begin{aligned}|u_n(s)-u_n(t)| &= \left|\int_s^t u_n^{\Delta_g}\Delta\right| \\ &\le \int_s^t \left|u_n^{\Delta_g}\right|\Delta \\ &\le (t-s)^{\frac{1}{2}}\left(\int_s^t \left|u_n^{\Delta_g}\right|^2\Delta\right)^{\frac{1}{2}} \\ &\le (t-s)^{\frac{1}{2}}\|u_n\|_{1,\mathbb{T}} \\ &\le M(t-s)^{\frac{1}{2}}, \quad s,t\in\mathbb{T}, \quad s\le t.\end{aligned}$$

As in the above,

$$|u_n(t)-u_n(s)| \le M(s-t)^{\frac{1}{2}}, \quad s,t\in\mathbb{T}, \quad s\ge t.$$

Therefore, the sequence $\{u_n\}_{n\in\mathbb{N}}$ is equi-continuous on $\mathbb{T}$. Hence, using Ascoli's Lemma, there exists a subsequence $\{u_{n_m}\}_{m\in\mathbb{N}}$ of the sequence $\{u_n\}_{n\in\mathbb{N}}$ that converges in $\mathbb{C}(\mathbb{T})$. This completes the proof.

Theorem A.16. *The embedding*

$$\mathbb{H}^{i+1}(\mathbb{T}) \hookrightarrow \mathbb{C}(\mathbb{T}), \quad i=0,1,$$

are compact.

Proof. Let $i=0$. By Theorem A.8, we have

$$|u|_{0,\mathbb{T}} \le C\|u\|_{1,\mathbb{T}},$$

with $C=\max\left\{\frac{1}{(b-a)^{\frac{1}{2}}},(b-a)^{\frac{1}{2}}\right\}$. Therefore,

$$\mathbb{H}^1(\mathbb{T}) \hookrightarrow \mathbb{C}(\mathbb{T}). \tag{A.14}$$

By Theorem A.15, we have that every bounded sequence in $\mathbb{H}^1(\mathbb{T})$ has a subsequence that converges in $\mathbb{C}(\mathbb{T})$. Therefore, the embedding (A.14) is compact. Let $i=1$. Suppose that $u\in\mathbb{H}^2(\mathbb{T})$. By the definition of $\mathbb{H}^2(\mathbb{T})$, it follows that $u\in\mathbb{C}^1(\mathbb{T})$. Therefore,

$$\mathbb{H}^2(\mathbb{T}) \hookrightarrow \mathbb{C}^1(\mathbb{T}). \tag{A.15}$$

Let $\{u_n\}_{n\in\mathbb{N}}$ be a bounded sequence in $\mathbb{H}^2(\mathbb{T})$. Then $\{u_n^{\Delta}\}_{n\in\mathbb{N}}$ is a bounded sequence in $\mathbb{H}^1(\mathbb{T})$. Hence and by Theorem A.15, it follows that there exists a subsequence $\left\{u_{n_m}^{\Delta}\right\}_{m\in\mathbb{N}}$ of the sequence $\{u_n^{\Delta}\}_{n\in\mathbb{N}}$ that converges in $\mathbb{C}(\mathbb{T})$. From here, it follows that $\left\{u_{n_m}^{\Delta}\right\}_{m\in\mathbb{N}}$ is a bounded sequence in $\mathbb{C}(\mathbb{T})$ and then there exists a positive constant M such that

$$\left|u_{n_m}^{\Delta}(t)\right| \leq M$$

for any $t \in \mathbb{T}$ and for any $m \in \mathbb{N}$. Note that

$$\begin{aligned}|u_{n_m}(t) - u_{n_m}(s)| &= \left|\int_s^t u_{n_m}^{\Delta}\Delta\right| \\ &\leq \int_s^t \left|u_{n_m}^{\Delta}\right|\Delta \\ &\leq M(t-s), \quad s,t \in \mathbb{T}, \quad s \leq t.\end{aligned}$$

As in the above,

$$|u_{n_m}(t) - u_{n_m}(s)| \leq M(s-t), \quad s,t \in \mathbb{T}, \quad s \geq t.$$

Therefore, $\{u_{n_m}\}_{m\in\mathbb{N}}$ is equi-continuous on $\mathbb{T}$. Since $u_{n_m} \in \mathbb{C}^1(\mathbb{T})$, we have that $u_{n_m} \in \mathbb{H}^1(\mathbb{T})$ for any $m \in \mathbb{N}$. By Theorem A.8, it follows that

$$\begin{aligned}|u_{n_m}|_{0,\mathbb{T}} &\leq C\,\|u_{n_m}\|_{1,\mathbb{T}} \\ &\leq C\,\|u_{n_m}\|_{2,\mathbb{T}}\end{aligned}$$

with $C = \max\left\{\frac{1}{(b-a)^{\frac{1}{2}}}, (b-a)^{\frac{1}{2}}\right\}$, for any $m \in \mathbb{N}$. Since $\{u_n\}_{n\in\mathbb{N}}$ is a bounded sequence in $\mathbb{H}^2(\mathbb{T})$ and $\{u_{n_m}\}_{m\in\mathbb{N}}$ is its subsequence, we conclude that $\{u_{n_m}\}_{m\in\mathbb{N}}$ is a bounded sequence in $\mathbb{C}(\mathbb{T})$. Hence and by Ascoli's Lemma, it follows that there exists a subsequence $\left\{u_{n_{m_k}}\right\}_{k\in\mathbb{N}}$ of the sequence $\{u_{n_m}\}_{m\in\mathbb{N}}$ that converges in $\mathbb{C}(\mathbb{T})$. Because $\left\{u_{n_m}^{\Delta}\right\}_{m\in\mathbb{N}}$ converges in $\mathbb{C}(\mathbb{T})$ we have that $\left\{u_{n_{m_k}}^{\Delta}\right\}_{k\in\mathbb{N}}$ converges in $\mathbb{C}(\mathbb{T})$. Consequently, the sequence $\left\{u_{n_{m_k}}\right\}_{k\in\mathbb{N}}$ converges in $\mathbb{C}^1(\mathbb{T})$. Therefore, the embedding (A.15) is compact. This completes the proof.

Theorem A.17. *For any $\varepsilon > 0$ there exists $C(\varepsilon) > 0$ such that*

$$|u|_{0,\mathbb{T}} \leq \varepsilon\|u^{\Delta_g}\|_{\mathbb{T}} + C(\varepsilon)\left(\|u\|_{\mathbb{T}} + \|u^{\sigma}\|_{\mathbb{T}}\right), \quad u \in \mathbb{H}^1(\mathbb{T}). \tag{A.16}$$

Proof. Let $u \in \mathbb{H}^1(\mathbb{T})$. Then there exists a sequence $\{u_n\}_{n\in\mathbb{N}}$ of elements of $\mathbb{C}^1(\mathbb{T})$ such that $u_n \to u$, $u_n^{\Delta} \to u^{\Delta_g}$, as $n \to \infty$, in $\mathbb{L}^2(\mathbb{T})$. Take $n \in \mathbb{N}$ arbitrarily. Consider an arbitrary $t_0 \in \mathbb{T}$. Suppose that

$$\mathbb{T}_{0,\varepsilon} = \left[t_0 - \frac{\varepsilon^2}{2}, t_0 + \frac{\varepsilon^2}{2}\right] \bigcap \mathbb{T} \neq \{t_0\},$$

i.e., t_0 is not isolated in $\mathbb{T}$. Let

$$i_0 = \inf \mathbb{T}_{0,\varepsilon}, \quad s_0 = \sup \mathbb{T}_{0,\varepsilon}.$$

We have

$$0 < s_0 - i_0 \leq \varepsilon^2.$$

Then

$$\begin{aligned}
|u_n(t)| &= \left| u_n(s) + \int_s^t u_n^{\Delta} \Delta \right| \\
&\leq |u_n(s)| + \left| \int_s^t u_n^{\Delta} \Delta \right| \\
&\leq |u_n(s)| + \int_s^t \left| u_n^{\Delta} \right| \Delta \\
&\leq |u_n(s)| + (t-s)^{\frac{1}{2}} \left(\int_s^t \left| u_n^{\Delta} \right|^2 \Delta \right)^{\frac{1}{2}} \\
&\leq |u_n(s)| + \varepsilon \left\| u_n^{\Delta} \right\|_{\mathbb{T}}, \quad s, t \in \mathbb{T}_{0,\varepsilon}, \quad s \leq t.
\end{aligned}$$

As in the above,

$$|u_n(t)| \leq |u_n(s)| + \varepsilon \left\| u_n^{\Delta} \right\|_{\mathbb{T}}, \quad s, t \in \mathbb{T}_{0,\varepsilon}, \quad s \geq t.$$

Integrating over $\mathbb{T}_{0,\varepsilon}$ with respect to s, we get

$$\begin{aligned}
(s_0 - i_0)|u_n(t)| &\leq \int_{t_0 - \frac{\varepsilon^2}{2}}^{t_0 + \frac{\varepsilon^2}{2}} |u_n(s)| \Delta + \varepsilon (s_0 - i_0) \left\| u_n^{\Delta} \right\|_{\mathbb{T}} \\
&\leq (s_0 - i_0)^{\frac{1}{2}} \left(\int_{t_0 - \frac{\varepsilon^2}{2}}^{t_0 + \frac{\varepsilon^2}{2}} |u_n(s)|^2 \Delta \right)^{\frac{1}{2}} + \varepsilon (s_0 - i_0) \left\| u_n^{\Delta} \right\|_{\mathbb{T}} \\
&\leq (s_0 - i_0)^{\frac{1}{2}} \|u_n\|_{\mathbb{T}} + \varepsilon (s_0 - i_0) \left\| u_n^{\Delta} \right\|_{\mathbb{T}}, \quad t \in \mathbb{T}_{0,\varepsilon}.
\end{aligned}$$

Hence,

$$|u_n(t)| \leq \varepsilon (s_0 - i_0)^{-\frac{1}{2}} \|u_n\|_{\mathbb{T}} + \varepsilon \left\| u_n^{\Delta} \right\|_{\mathbb{T}}, \quad t \in \mathbb{T}_{0,\varepsilon}. \tag{A.17}$$

Now we suppose that $\mathbb{T}_{0,\varepsilon} = \{t_0\}$, i.e., t_0 is isolated in $\mathbb{T}$. If $\sigma(t_0) > t_0$, then

$$\begin{aligned}|u_n(t_0)| &= (\sigma(t_0)-t_0)^{-1}\int_{t_0}^{\sigma(t_0)}|u_n|\Delta\\ &\le (\sigma(t_0)-t_0)^{-\frac{1}{2}}\left(\int_{t_0}^{\sigma(t_0)}|u_n|^2\Delta\right)^{\frac{1}{2}}\\ &\le (\sigma(t_0)-t_0)^{-\frac{1}{2}}\,\|u_n\|_{\mathbb{T}}.\end{aligned}\tag{A.18}$$

If $t_0>\rho(t_0)$, then

$$\begin{aligned}|u_n(t_0)| &= (t_0-\rho(t_0))^{-1}\int_{\rho(t_0)}^{t_0}|u_n^{\sigma}|\Delta\\ &\le (t_0-\rho(t_0))^{-\frac{1}{2}}\left(\int_{\rho(t_0)}^{t_0}|u_n^{\sigma}|^2\Delta\right)^{\frac{1}{2}}\\ &\le (t_0-\rho(t_0))^{-\frac{1}{2}}\,\|u_n^{\sigma}\|_{\mathbb{T}}.\end{aligned}\tag{A.19}$$

Suppose that there exists $\varepsilon>0$ such that for each $m\in\mathbb{N}$ and $t_m\in\mathbb{T}$ we have

$$|u_n(t_m)|>\varepsilon\left\|u_n^{\Delta}\right\|_{\mathbb{T}}+m\left(\|u_n\|_{\mathbb{T}}+\|u_n^{\sigma}\|_{\mathbb{T}}\right),\quad t_m\to t_0.\tag{A.20}$$

If t_0 is isolated, (A.20) contradicts with (A.18) and (A.19). If t_0 is not isolated, then (A.20) contradicts with (A.17). Consequently, for each $\varepsilon>0$ there exists $C(\varepsilon)>0$ such that

$$|u_n|_{0,\mathbb{T}}\le\varepsilon\|u_n^{\Delta}\|_{\mathbb{T}}+C(\varepsilon)\left(\|u_n\|_{\mathbb{T}}+\|u_n^{\sigma}\|_{\mathbb{T}}\right).\tag{A.21}$$

Because $n\in\mathbb{N}$ was arbitrarily chosen and

$$|u_n|_{0,\mathbb{T}}\to|u|_{0,\mathbb{T}},\quad \left\|u_n^{\Delta}\right\|_{\mathbb{T}}\to\left\|u^{\Delta_g}\right\|_{\mathbb{T}},$$

$$\|u_n\|_{\mathbb{T}}\to\|u\|_{\mathbb{T}},\quad \|u_n^{\sigma}\|_{\mathbb{T}}\to\|u^{\sigma}\|_{\mathbb{T}},$$

as $n\to\infty$, by (A.21), we get (A.16). This completes the proof.

Theorem A.18. *For any $\varepsilon>0$ there exists $C(\varepsilon)>0$ such that*

$$|u|_{0,\mathbb{T}_\kappa}\le\varepsilon\|u^{\Delta_g}\|_{\mathbb{T}}+C(\varepsilon)\,\|u^{\sigma}\|_{\mathbb{T}},\quad u\in\mathbb{H}^1(\mathbb{T}).\tag{A.22}$$

Proof. Let $u\in\mathbb{H}^1(\mathbb{T})$. Then there exists a sequence $\{u_n\}_{n\in\mathbb{N}}$ of elements of $\mathbb{C}^1(\mathbb{T})$ such that $u_n\to u$, $u_n^{\Delta}\to u^{\Delta_g}$, as $n\to\infty$, in $\mathbb{L}^2(\mathbb{T})$. Take $n\in\mathbb{N}$ arbitrarily. Consider an arbitrary $t_0\in\mathbb{T}_\kappa$. Suppose that

$$\mathbb{T}_{0,\varepsilon}=\left[t_0-\frac{\varepsilon^2}{2},t_0+\frac{\varepsilon^2}{2}\right]\bigcap\mathbb{T}_\kappa\neq\{t_0\},$$

i.e., t_0 is not isolated in $\mathbb{T}$. Let

$$i_0 = \inf \mathbb{T}_{0,\varepsilon}, \quad s_0 = \sup \mathbb{T}_{0,\varepsilon}.$$

We have

$$0 < s_0 - i_0 \le \varepsilon^2.$$

Then

$$\begin{aligned}
|u_n(t)| &= \left| u_n^\sigma(s) + \int_{\sigma(s)}^t u_n^\Delta \Delta \right| \\
&\le |u_n^\sigma(s)| + \left| \int_{\sigma(s)}^t u_n^\Delta \Delta \right| \\
&\le |u_n^\sigma(s)| + \int_{\sigma(s)}^t \left| u_n^\Delta \right| \Delta \\
&\le |u_n^\sigma(s)| + (t - \sigma(s))^{\frac{1}{2}} \left(\int_{\sigma(s)}^t \left| u_n^\Delta \right|^2 \Delta \right)^{\frac{1}{2}} \\
&\le |u_n^\sigma(s)| + \varepsilon \left\| u_n^\Delta \right\|_{\mathbb{T}}, \quad s, \sigma(s), t \in \mathbb{T}_{0,\varepsilon}, \quad \sigma(s) \le t.
\end{aligned}$$

As in the above,

$$|u_n(t)| \le |u_n^\sigma(s)| + \varepsilon \left\| u_n^\Delta \right\|_{\mathbb{T}}, \quad s, \sigma(s), t \in \mathbb{T}_{0,\varepsilon}, \quad \sigma(s) \ge t.$$

Integrating over $\mathbb{T}_{0,\varepsilon}$ with respect to s, we get

$$\begin{aligned}
(s_0 - i_0)|u_n(t)| &\le \int_{t_0 - \frac{\varepsilon^2}{2}}^{t_0 + \frac{\varepsilon^2}{2}} |u_n^\sigma(s)| \Delta + \varepsilon (s_0 - i_0) \left\| u_n^\Delta \right\|_{\mathbb{T}} \\
&\le (s_0 - i_0)^{\frac{1}{2}} \left(\int_{t_0 - \frac{\varepsilon^2}{2}}^{t_0 + \frac{\varepsilon^2}{2}} |u_n^\sigma(s)|^2 \Delta \right)^{\frac{1}{2}} + \varepsilon (s_0 - i_0) \left\| u_n^\Delta \right\|_{\mathbb{T}} \\
&\le (s_0 - i_0)^{\frac{1}{2}} \|u_n\|_{\mathbb{T}} + \varepsilon (s_0 - i_0) \left\| u_n^\Delta \right\|_{\mathbb{T}}, \quad t \in \mathbb{T}_{0,\varepsilon}.
\end{aligned}$$

Hence,

$$|u_n(t)| \le \varepsilon (s_0 - i_0)^{-\frac{1}{2}} \|u_n^\sigma\|_{\mathbb{T}} + \varepsilon \left\| u_n^\Delta \right\|_{\mathbb{T}}, \quad t \in \mathbb{T}_{0,\varepsilon}. \tag{A.23}$$

Now we suppose that $\mathbb{T}_{0,\varepsilon} = \{t_0\}$, i.e., t_0 is isolated in $\mathbb{T}$. Then

$$
\begin{aligned}
|u_n(t_0)| &= (t_0-\rho(t_0))^{-1}\int_{\rho(t_0)}^{t_0}|u_n^\sigma|\Delta \\
&\le (t_0-\rho(t_0))^{-\frac{1}{2}}\left(\int_{\rho(t_0)}^{t_0}|u_n^\sigma|^2\Delta\right)^{\frac{1}{2}} \\
&\le (t_0-\rho(t_0))^{-\frac{1}{2}}\|u_n^\sigma\|_{\mathbb{T}}.
\end{aligned}
\tag{A.24}
$$

Suppose that there exists $\varepsilon>0$ such that for each $m\in\mathbb{N}$ and $t_m\in\mathbb{T}_\kappa$ we have

$$|u_n(t_m)|>\varepsilon\left\|u_n^\Delta\right\|_{\mathbb{T}}+m\|u_n^\sigma\|_{\mathbb{T}},\quad t_m\to t_0. \tag{A.25}$$

If t_0 is isolated, then (A.25) contradicts with (A.24). If t_0 is not isolated, then (A.25) contradicts with (A.23). Consequently, for each $\varepsilon>0$ there exists $C(\varepsilon)>0$ such that

$$|u_n|_{0,\mathbb{T}}\le\varepsilon\|u_n^\Delta\|_{\mathbb{T}}+C(\varepsilon)\|u_n^\sigma\|_{\mathbb{T}}. \tag{A.26}$$

Because $n\in\mathbb{N}$ was arbitrarily chosen and

$$
\begin{aligned}
&|u_n|_{0,\mathbb{T}}\to|u|_{0,\mathbb{T}},\quad \left\|u_n^\Delta\right\|_{\mathbb{T}}\to\left\|u^{\Delta_g}\right\|_{\mathbb{T}},\\
&\|u_n^\sigma\|_{\mathbb{T}}\to\|u^\sigma\|_{\mathbb{T}},
\end{aligned}
$$

as $n\to\infty$, by (A.26), we get (A.22). This completes the proof.

Appendix B
Elements of Fixed Point Index Theory

B.1 Fixed Points of Strict Set Contractions

Suppose that E is a Banach space.

Definition B.1. Let $A : E \to E$ is a mapping.

1. The mapping A is said to be compact if it maps every bounded subset of E into a relatively compact subset of E.
2. The mapping A is said to be a completely continuous mapping if it is continuous and compact.
3. The mapping A is said to be a k-set contraction if it is continuous, bounded and

$$\alpha(A(S)) \leq k\alpha(S)$$

   for any bounded set $S \subset D$. Especially, if $k < 1$, a k-set contraction A is said to be a strict set contraction.
4. The mapping A is said to be a condensing operator if it is continuous, bounded and

$$\alpha(A(S)) < \alpha(S)$$

   for any nonrelatively compact and bounded set $S \subset D$.

Theorem B.1. *Assume that Ω is a convex open set in E and $x_0 \in \Omega$. If a continuous mapping $A : \overline{\Omega} \to \overline{\Omega}$ has the following property: $C \subset \Omega$ is countable, $C \subset \overline{co}(\{x_0\} \bigcup A(C))$ implies C is relatively compact, then A has a fixed point in E.*

Proof. Define the sets

DOI: 10.1201/9781003173557-B

$$D_0 = \{x_0\},$$

$$D_n = \mathrm{co}(\{x_0\} \bigcup A(D_{n-1})), \quad n \in \mathbb{N}.$$

We have

$$D_0 \subset D_1 \subset \ldots \subset D_n \subset \ldots \subset \overline{\Omega}.$$

Each set D_n, $n \in \mathbb{N}_0$, is separable and there exists a countable set C_n, $n \in \mathbb{N}_0$, such that

$$\overline{C}_n = \overline{D}_n, \quad n \in \mathbb{N}_0.$$

Let

$$D = \bigcup_{n\in\mathbb{N}_0} D_n, \quad C = \bigcup_{n\in\mathbb{N}_0} C_n.$$

We have

$$\begin{aligned} D &= \bigcup_{n\in\mathbb{N}} \mathrm{co}(\{x_0\} \bigcup A(D_{n-1})) \\ &= \mathrm{co}(\{x_0\} \bigcup A(D)), \end{aligned}$$

$$\begin{aligned} \overline{\bigcup_{n\in\mathbb{N}_0} \overline{D}_n} &= \overline{\bigcup_{n\in\mathbb{N}_0} D_n} \\ &= \overline{D}, \end{aligned}$$

$$\begin{aligned} \overline{\bigcup_{n\in\mathbb{N}_0} \overline{D}_n} &= \overline{\bigcup_{n\in\mathbb{N}_0} \overline{C}_n} \\ &= \overline{\bigcup_{n\in\mathbb{N}_0} C_n} \\ &= \overline{C}. \end{aligned}$$

Thus

$$\begin{aligned} C &\subset \overline{C} \\ &= \overline{D} \\ &= \overline{\mathrm{co}}(\{x_0\} \bigcup A(D)) \qquad \text{(B.1)} \\ &= \overline{\mathrm{co}}(\{x_0\} \bigcup A(\overline{D})) \\ &= \overline{\mathrm{co}}(\{x_0\} \bigcup A(\overline{C})). \end{aligned}$$

Since C is countable, using (B.1), we get that $\overline{C}$ is a compact set in E and so $\overline{D}$ is a compact set in E. Because $x_0 \in D$, we have that $\overline{D}$ is nonempty. Also, we have that

$$A(D) \subset \overline{D}.$$

Hence and by Schauder's fixed point theorem, it follows that A has a fixed point in D. This completes the proof.

For the next result we will use the following extension theorem.

Theorem B.2 (Dugundji's Extension Theorem). *Let X be an arbitrary metric space, A is a closed subset of X, L is locally convex linear space, $f : A \to L$ is a continuous mapping. Then there exists an extension $F : X \to L$ of f and*

$$F(X) \subset co(f(A)).$$

Theorem B.3 (Mönch's Theorem). *Assume that D is a closed convex set in E and $x_0 \in D$. If a continuous mapping $A : D \to D$ has the following property: $C \subset D$ is countable, $C \subset \overline{co}(\{x_0\} \bigcup A(C))$ implies C is relatively compact, then the mapping A has a fixed point in D.*

Proof. By Dugundji's extension theorem, it follows that there exist a continuous mapping $A_1 : E \to D$ such that

$$A_1 x = Ax, \quad x \in D.$$

Assume that $C \subset E$ is countable, $C \subset \mathrm{co}(\{x_0\} \bigcup A_1(C))$. Since $A_1(C) \subset D$, it follows that $C \subset D$ and so $A_1(C) = A(C)$. Hence, C is a relatively compact set. From here and from Theorem B.1, it follows that the mapping A_1 has a fixed point $x^* \in E$. We have that $A_1 x^* \in D$ and then $x^* \in D$. Therefore,

$$x^* = A_1 x^*$$

$$= Ax^*.$$

This completes the proof.

Theorem B.4 (Dabor's Theorem). *Assume that D is a nonempty bounded and closed convex set. If a continuous mapping $A : D \to D$ has the following property: $C \subset D$ is countable and is not relatively compact implies $\alpha(A(C)) < \alpha(C)$, then the mapping A has a fixed point in D.*

Proof. Choose $x_0 \in D$ and a countable set $C \subset D$ with $C \subset \mathrm{co}(\{x_0\} \bigcup A(C))$. Then, using Theorem B.1 and Theorem B.3, we get

$$\alpha(C) \le \alpha(\overline{\mathrm{co}}(\{x_0\} \bigcup A(C)))$$

$$\le \alpha(A(C)).$$

Therefore, C is a relatively compact set. Hence and by Mönch's theorem, it follows that A has a fixed point theorem in D. This completes the proof.

Theorem B.5 (Sadovskii's Theorem). *Assume that D is a nonempty bounded and closed convex set. If a mapping $A : D \to D$ is condensing, then A has a fixed point in D.*

Proof. Since $A : D \to D$ is condensing, we have

$$\alpha(A(C)) < \alpha(C).$$

Hence and by Dabor's theorem, we get that A has a fixed point in D. This completes the proof.

Theorem B.6 (Darbo's Theorem). *Assume that D is a nonempty bounded and closed convex set. If a mapping $A : D \to D$ is a strict set contraction, then A has a fixed point in D.*

Proof. Since $A : D \to D$ is a strict set contraction, then it is condensing. Hence and by Sadovskii's theorem, it follows that A has a fixed point in D. This completes the proof.

Theorem B.7. *Assume that $\{S_n\}_{n\in\mathbb{N}}$ is a sequence of nonempty bounded and closed sets in E such that*

$$S_1 \supset S_2 \supset \ldots \supset S_n \supset \ldots.$$

If $\alpha(S_n) \to 0$, as $n \to \infty$, then $S = \bigcap_{n\in\mathbb{N}} S_n$ is a nonempty compact set in E.

Proof. Let $\{x_n\}_{n\in\mathbb{N}}$ be a sequence of elements of S. Then $x_n \in S_j$ for any $n, j \in \mathbb{N}$. Let

$$S_n = \bigcup_{j=1}^{j_n} S_j^n, \quad n \in \mathbb{N},$$

be such that

$$\mathrm{diam}(S_j^n) < \alpha(S_n) + \frac{1}{n}, \quad j \in \{1, \ldots, j_n\}.$$

Since $\{x_n\}_{n\in\mathbb{N}} \subset S_1$, there is a subsequence $\{x_n^1\}_{n\in\mathbb{N}}$ of $\{x_n\}_{n\in\mathbb{N}}$ so that it is included in some S_j^1 and

$$\mathrm{diam}(\{x_n^1\}_{n\in\mathbb{N}}) < \alpha(S_1) + 1.$$

Since $\{x_n^1\}_{n\in\mathbb{N}} \subset S_2$, there is a subsequence $\{x_n^2\}_{n\in\mathbb{N}}$ of $\{x_n^1\}_{n\in\mathbb{N}}$ such that

$$\mathrm{diam}(\{x_n^2\}_{n\in\mathbb{N}}) < \alpha(S_2) + \frac{1}{2}.$$

And so on. We get a sequence $\{x_n^m\}_{n\in\mathbb{N}}$ so that

$$\mathrm{diam}(\{x_n^m\}_{n\in\mathbb{N}}) < \alpha(S_m) + \frac{1}{m}.$$

In fact, for $m > n$, we get

$$\|x_n^m - x_n^n\| < \alpha(S_n) + \frac{1}{n}.$$

Thus, the sequence $\{x_n^n\}_{n\in\mathbb{N}}$ is a fundamental sequence in E. Hence, it is convergent to $x_0 \in E$. Because $x_m^n \in S_n$, $m \geq n$, $m, n \in \mathbb{N}$, and S_n is closed, $n \in \mathbb{N}$, we have that $x_0 \in S_n$, $n \in \mathbb{N}$. Then $x_0 \in S$. Consequently, S is a nonempty compact set in E. This completes the proof.

B.2 Kronecker Index

Let $n \geq 2$, $n \in \mathbb{N}$, and

$$w = dy_1 \wedge \ldots \wedge dy_n$$

be the volume n-form in $\mathbb{R}^n$, and let σ be the solid angle $(n-1)$-form

$$\sigma = \sum_{j=1}^{n} (-1)^{j-1} y_j dy_1 \wedge \ldots \wedge \widehat{dy_j} \wedge \ldots \wedge dy_n,$$

where $\widehat{dy_j}$ means that dy_j is missed. We have

$$w = d\left(\frac{1}{n}\sigma\right).$$

Let $D \subset \mathbb{R}^n$ be a bounded open set with oriented smooth boundary ∂D and $f \in \mathscr{C}^1(\overline{D})$, $f : \overline{D} \to \mathbb{R}^n$, be such that $0 \notin f(\partial D)$. Then

$$\mu = \min_{\partial D} \|f\| > 0.$$

Definition B.2. The Kronecker index $i_K(f, \partial D)$ is defined by

$$\begin{aligned} i_K(f,\partial D) &= \frac{1}{\mu_{n-1}} \int_{\partial D} f^* \left(\|f\|^{-n} \sigma \right) \\ &= \frac{1}{\mu_{n-1}} \int_{\partial D} \left(\sum_{j=1}^{n} (-1)^{j-1} \frac{f_j}{\|f\|^n} df_1 \wedge \ldots \wedge \widehat{df_j} \wedge \ldots \wedge df_n \right), \end{aligned} \tag{B.2}$$

where df_k is the 1-differential form

$$df_k = \sum_{j=1}^{n} \partial_j f_k dx_k, \quad k \in \{1, \ldots, n\},$$

and μ_{n-1} is the $(n-1)$-dimensional measure of the unit sphere S^{n-1}.

Definition B.3. If $n = 1$ and $D = (a,b)$, the Kronecker index of a continuous function $f : [a,b] \to \mathbb{R}$ such that $f(a)f(b) \neq 0$ can be defined by

$$\begin{aligned} i_K(f,\partial(a,b)) &= \frac{1}{2} \left(\frac{f(b)}{|f(b)|} - \frac{f(a)}{|f(a)|} \right) \\ &= \frac{1}{2} (\mathrm{sign} f(b) - \mathrm{sign} f(a)). \end{aligned}$$

Note that

1. if $f(a)f(b) > 0$, then
$$i_K(f,\partial(a,b)) = 0.$$
2. if $f(a) < 0 < f(b)$, then
$$i_K(f,\partial(a,b)) = 1.$$
3. if $f(b) < 0 < f(a)$, then
$$i_K(f,\partial(a,b)) = -1.$$

Definition B.4. Let $n = 2$. Then the Kronecker index is called the winding number or Poincaré index.

Now suppose that $n = 2$. Then the formula (B.2) takes the form

$$i_K(f,\partial D) = \frac{1}{2\pi} \int_{\partial D} \frac{f_1 df_2 - f_2 df_1}{f_1^2 + f_2^2}.$$

In this case, if ∂D has the parametric representation $\phi : [0, 2\pi] \to \mathbb{R}^2$ with $\phi(0) = \phi(2\pi)$, then

$$df_1 = \frac{d}{ds} f_1(\phi(s))ds,$$

$$df_2 = \frac{d}{ds} f_2(\phi(s))ds,$$

and

$$i_K(f, \partial D) = \frac{1}{2\pi} \int_0^{2\pi} \frac{f_1(\phi(s)) \frac{d}{ds} f_2(\phi(s)) - f_2(\phi(s)) \frac{d}{ds} f_1(\phi(s))}{(f_1(\phi(s)))^2 + (f_2(\phi(s)))^2} ds.$$

Example B.1. Let $D = B^2$. Then $\partial D = S^1$ and

$$\phi(s) = (\cos s, \sin s), \quad s \in [0, 2\pi].$$

Suppose that $f = I$. Then

$$f_1(\phi(s)) = \cos s,$$

$$f_2(\phi(s)) = \sin s,$$

$$\frac{d}{ds} f_1(\phi(s)) = -\sin s,$$

$$\frac{d}{ds} f_2(\phi(s)) = \cos s, \quad s \in [0, 2\pi].$$

Hence,

$$\begin{aligned} i_K(f, \partial D) &= \frac{1}{2\pi} \int_0^{2\pi} \frac{\cos s(\cos s) - \sin s(-\sin s)}{(\cos s)^2 + (\sin s)^2} ds \\ &= \frac{1}{2\pi} \int_0^{2\pi} \frac{(\cos s)^2 + (\sin s)^2}{(\cos s)^2 + (\sin s)^2} ds \\ &= 1. \end{aligned}$$

Example B.2. Let $D = B^2$,

$$f(x_1, x_2) = (x_1, -x_2).$$

Then

$$f_1(\phi(s)) = \cos s,$$

$$f_2(\phi(s)) = -\sin s,$$

$$\frac{d}{ds}f_1(\phi(s)) = -\sin s,$$

$$\frac{d}{ds}f_2(\phi(s)) = -\cos s.$$

Hence,

$$\begin{aligned} i_K(f,\partial D) &= \frac{1}{2\pi}\int_0^{2\pi} \frac{\cos s(-\cos s)+\sin s(-\sin s)}{(\cos s)^2+(\sin s)^2}ds \\ &= -\frac{1}{2\pi}\int_0^{2\pi} \frac{(\cos s)^2+(\sin s)^2}{(\cos s)^2+(\sin s)^2}ds \\ &= -1. \end{aligned}$$

Example B.3. Let $D = B^2$ and

$$f(x_1,x_2) = (x_1^2 - x_2^2, 2x_1x_2).$$

Then

$$\begin{aligned} f_1(\phi(s)) &= (\cos s)^2 - (\sin s)^2 \\ &= \cos(2s), \\ f_2(\phi(s)) &= 2\sin s\cos s \\ &= \sin(2s), \\ \frac{d}{ds}f_1(\phi(s)) &= -2\sin(2s), \\ \frac{d}{ds}f_2(\phi(s)) &= 2\cos(2s), \end{aligned}$$

and

$$\begin{aligned} i_K(f,\partial D) &= \frac{1}{2\pi}\int_0^{2\pi} \frac{\cos(2s)(2\cos(2s))-\sin(2s)(-2\sin(2s))}{(\cos(2s))^2+(\sin(2s))^2}ds \\ &= \frac{1}{\pi}\int_0^{2\pi} \frac{(\cos(2s))^2+(\sin(2s))^2}{(\cos(2s))^2+(\sin(2s))^2}ds \\ &= 2. \end{aligned}$$

Now suppose that

$$f = f_1 + i f_2.$$

Then

$$\begin{aligned}
\frac{1}{2\pi i}\int_{\partial D}\frac{df}{f} &= \frac{1}{2\pi i}\int_{\partial D}\frac{df_1 + i df_2}{f_1 + i f_2} \\
&= \frac{1}{2\pi i}\int_{\partial D}\frac{(df_1 + i df_2)(f_1 - i f_2)}{(f_1 + i f_2)(f_1 - i f_2)} \\
&= \frac{1}{2\pi i}\int_{\partial D}\frac{f_1 df_1 + i f_1 df_2 - i f_2 df_1 + f_2 df_2}{f_1^2 + f_2^2} \\
&= \frac{1}{2\pi i}\int_{\partial D}\frac{f_1 df_1 + f_2 df_2}{f_1^2 + f_2^2} + \frac{1}{2\pi}\int_{\partial D}\frac{f_1 df_2 - f_2 df_1}{f_1^2 + f_2^2} \\
&= \frac{1}{2\pi i}\int_{\partial D}\frac{d(\|f\|^2)}{2\|f\|^2} + i_K(f, \partial D) \\
&= i_K(f, \partial D).
\end{aligned}$$

Consequently,

$$\begin{aligned}
i_K(f, \partial D) &= \frac{1}{2\pi i}\int_{\partial D}\frac{\partial_1 f(\phi(s))\phi_1'(s) + \partial_2 f(\phi(s))\phi_2'(s)}{f(\phi(s))}ds \\
&= \frac{1}{2\pi i}\int_0^{2\pi}\frac{\partial_\tau f(\phi(s))}{f(\phi(s))}ds,
\end{aligned}$$

where ∂_τ denotes the tangential derivative of f along ∂D.

Theorem B.8. *Let $n = 2$ and ∂D have the parametric representation $\phi : [0, 2\pi] \to \mathbb{R}^2$ with $\phi(0) = \phi(2\pi)$. Then $i_K(f, \partial D)$ is an integer.*

Proof. Let

$$\Phi = \phi_1 + i\phi_2$$

and

$$h(t) = \int_0^t \frac{\frac{d}{ds} f(\Phi(s))}{f(\Phi(s))}ds,$$

$$g(t) = e^{-h(t)} f(\Phi(t)), \quad t \in [0, 2\pi].$$

Then

$$\begin{aligned}
g'(t) &= -\frac{\frac{d}{dt} f(\Phi(t))}{f(\Phi(t))}e^{-h(t)} f(\Phi(t)) \\
&\quad + e^{-h(t)}\frac{d}{dt} f(\Phi(t))
\end{aligned}$$

$$= 0, \quad t \in [0, 2\pi].$$

So,

$$g(0) = g(2\pi).$$

Since

$$\Phi(0) = \Phi(2\pi) \quad \text{and} \quad h(0) = 0,$$

we get

$$f(\Phi(0)) = e^{-h(2\pi)} f(\Phi(2\pi)),$$

whereupon,

$$e^{-h(2\pi)} = 1 \quad \text{and} \quad e^{h(2\pi)} = 1.$$

Thus

$$h(2\pi) = 2m\pi i$$

for some $m \in \mathbb{Z}$, and

$$\begin{aligned} i_K(f, \partial D) &= \frac{1}{2\pi i} \int_0^{2\pi} \frac{\frac{d}{ds} f(\Phi(s))}{f(\Phi(s))} ds \\ &= \frac{1}{2\pi i} 2m\pi i \\ &= m. \end{aligned}$$

This completes the proof.

Let $n \geq 2$, $n \in \mathbb{N}$, $D \subseteq \mathbb{R}^n$ be a bounded open set with oriented smooth boundary ∂D. Suppose that $g \in \mathscr{C}^2(\overline{D})$, $g : \overline{D} \to \mathbb{R}^n$, are such that

$$\|g(x)\| = 1 \quad \text{for} \quad \text{all} \quad x \in \partial D. \tag{B.3}$$

With J_g we denote the Jacobian of g.

Theorem B.9. *Let* (B.3) *hold. Then*

$$i_K(g, \partial D) = \frac{n}{\mu_{n-1}} \int_D J_g(x) dx. \tag{B.4}$$

Proof. By (B.2), we get

$$\begin{aligned} i_K(g, \partial D) &= \frac{1}{\mu_{n-1}} \int_{\partial D} g^*(\sigma) \\ &= \frac{1}{\mu_{n-1}} \int_D d(g^*(\sigma)) \end{aligned}$$

$$= \frac{1}{\mu_{n-1}} \int_D g^*(d\sigma)$$

$$= \frac{n}{\mu_{n-1}} \int_D g^* w$$

$$= \frac{n}{\mu_{n-1}} \int_D dg_1 \wedge \ldots \wedge dg_n$$

$$= \frac{n}{\mu_{n-1}} \int_D J_g(x)dx.$$

This completes the proof.

Theorem B.10. *Let $n = 2$ and* (B.3) *hold. Then*

$$i_K(g, \partial D) = \frac{1}{2\pi} \int_0^{2\pi} g(\phi(s)) \wedge \partial_\tau g(\phi(s))ds.$$

Proof. We have

$$i_K(g, \partial D) = \frac{1}{2\pi} \int_0^{2\pi} \left(g_1(\phi(s)) \frac{d}{ds} g_2(\phi(s)) - g_2(\phi(s)) \frac{d}{ds} g_1(\phi(s)) \right) ds$$

$$= \frac{1}{2\pi} \int_0^{2\pi} (g_1(\phi(s))\partial_\tau g_2(\phi(s)) - g_2(\phi(s))\partial_\tau g_1(\phi(s))) ds$$

$$= \frac{1}{2\pi} \int_0^{2\pi} g(\phi(s)) \wedge \partial_\tau g(\phi(s))ds.$$

This completes the proof.

Let $0 < \varepsilon < \mu_0$, $a \in \mathscr{C}^k([0,\infty))$, $a : [0,\infty) \to \mathbb{R}$, be such that $\operatorname{supp} a \subset [\varepsilon_0, \mu_0]$. Consider

$$w_a = a(\|y\|)dy_1 \wedge \ldots \wedge dy_n$$

$$= a(\|y\|)w.$$

Theorem B.11. *The equality*

$$w_a = d\sigma_A,$$

with

$$\sigma_A = A(\|y\|)\sigma,$$

$A \in \mathscr{C}^{k+1}([0,\infty))$, $A : [0,\infty) \to \mathbb{R}$, $\operatorname{supp} A \subset [\varepsilon_0, \infty)$, holds if and only if

$$rA'(r) + nA(r) = a(r),$$

i.e., if and only if A is given by

$$A(0) = 0, \quad A(r) = \frac{1}{r^n}\int_0^r a(s)s^{n-1}ds, \quad r > 0.$$

Proof. Let $A \in \mathscr{C}^{k+1}([0,\infty))$, $A : [0,\infty) \to \mathbb{R}$, $\operatorname{supp} A \subset [\varepsilon_0,\infty)$. Then $\sigma_A \in \mathscr{C}^{k-1}(\mathbb{R}^n)$ and $\sigma_A = 0$ on $B(\varepsilon_0)$. For $y \neq 0$, we have

$$\begin{aligned}
d\sigma_A &= d(A(\|y\|\sigma) \\
&= d(A\|y\|) \wedge \sigma + A(\|y\|)d\sigma \\
&= \sum_{k=1}^n \frac{A'(\|y\|)y_k}{\|y\|}dy_k \wedge \sigma + nA(\|y\|)w \\
&= A'(\|y\|)\frac{1}{\|y\|}\sum_{k=1}^n y_k dy_k \wedge \left(\sum_{j=1}^n (-1)^{j-1} y_j dy_1 \wedge \ldots \wedge \widehat{dy_j} \wedge \ldots \wedge dy_n\right) \\
&\quad + nA(\|y\|)w \\
&= A'(\|y\|)\frac{1}{\|y\|}\sum_{k=1}^n y_k^2 w + nA(\|y\|)w \\
&= A'(\|y\|)w + nA(\|y\|)w \\
&= \left(A'(\|y\|) + nA(\|y\|)\right) w.
\end{aligned}$$

Consequently,

$$w_a = d\sigma_A$$

if and only if

$$a(\|y\|)w = \left(A'(\|y\|) + nA(\|y\|)\right) w.$$

Hence,

$$w_a = d\sigma_A$$

if and only if for any $r \in [\varepsilon_0,\infty)$ we have

$$A'(r) + nA(r) = a(r),$$

which is equivalent to, for $r > 0$,

$$\frac{d}{dr}\left(r^n A(r)\right) = r^{n-1}a(r).$$

This completes the proof.

Corollary B.1. *If we assume that*

$$\int_0^\infty a(r)r^{n-1}dr = 0,$$

then $suppA \subset [\varepsilon_0, \mu_0]$.

Proof. Let $r > \mu_0$. Then

$$\begin{aligned} A(r) &= \frac{1}{r^n}\int_0^r a(s)s^{n-1}ds \\ &= \frac{1}{r^n}\int_0^{\mu_0} a(s)s^{n-1}ds \\ &= \frac{1}{r^n}\int_0^\infty a(s)s^{n-1}ds \\ &= 0. \end{aligned}$$

This completes the proof.

Suppose $m > \mu_0$ and define $B \in \mathscr{C}^1([0,\infty))$, $B : [0,\infty) \to \mathbb{R}$, by

$$B(r) = \begin{cases} 0 \quad \text{if} \quad r \in [0, \varepsilon_0] \\ \textit{positive} \quad \text{if} \quad r \in (\varepsilon_0, \mu_0) \\ r^{-n} \quad \text{if} \quad r \in [\mu_0, \infty). \end{cases} \tag{B.5}$$

For $r > \mu_0$, we have

$$rB'(r) + nB(r) = 0.$$

Theorem B.12. *Let B be given by* (B.5), *$b \in \mathscr{C}([0,\infty))$ defined by*

$$b(r) = rB'(r) + nB(r)$$

is such that

$$suppb \subset [\varepsilon_0, \mu_0], \quad \int_0^\infty b(r)r^{n-1}dr = 1,$$

and if $f \in \mathscr{C}^2(\overline{D})$, $\|f(x)\| \geq \mu$, then

$$\begin{aligned} i_K(f, \partial D) &= \frac{1}{\mu_{n-1}}\int_D f^* w_b \\ &= \frac{1}{\mu_{n-1}}\int_D b(\|f(x)\|)J_f(x)dx. \end{aligned}$$

Proof. We have

$$i_K(f, \partial D) = \frac{1}{\mu_{n-1}}\int_{\partial D} f^*\left(\|y\|^{-n}\sigma\right)$$

$$
\begin{aligned}
&= \frac{1}{\mu_{n-1}} \int_{\partial D} f^* \sigma_B \\
&= \frac{1}{\mu_{n-1}} \int_D d(f^* \sigma_B) \\
&= \frac{1}{\mu_{n-1}} \int_D f^* d\sigma_B \\
&= \frac{1}{\mu_{n-1}} \int_D f^* w_b \\
&= \frac{1}{\mu_{n-1}} \int_D b(\|f(x)\|) J_f(x) dx.
\end{aligned}
$$

Note that $b(r) = 0$ for $r \in [0, \varepsilon_0) \bigcup (\mu_0, \infty)$ and

$$
\begin{aligned}
\int_0^\infty b(r) r^{n-1} dr &= \int_0^\infty \left(rB'(r) + nB(r)\right) r^{n-1} dr \\
&= \int_0^\infty \frac{d}{dr} (r^n B(r)) \\
&= 1.
\end{aligned}
$$

This completes the proof.

B.3 Brouwer Degree

B.3.1 Smooth Mappings

Let $n \in \mathbb{N}$, B be open and bounded in $\mathbb{R}^n$, $f \in \mathscr{C}^2(\overline{D})$, $f : \overline{D} \to \mathbb{R}^n$ be such that $0 \notin f(\partial \overline{D})$. Then

$$
\mu = \min_{\partial D} \|f\| > 0.
$$

Let, also, $0 < \varepsilon_0 < \mu_0 < \mu$.

Definition B.5. The Brouwer degree $d_B(f, D)$ is defined by

$$
\begin{aligned}
d_B(f, D) &= \int_D f^* w_c \\
&= \int_D c(\|f\|) J_f(x) dx,
\end{aligned}
\tag{B.6}
$$

where $c \in \mathscr{C}([0,\infty))$, $c : [0,\infty) \to \mathbb{R}$, $\operatorname{supp} c \subset [\varepsilon_0, \mu_0]$ and

$$\int_{\mathbb{R}^n} c(\|f\|)dx = 1$$

or

$$\int_0^\infty c(r)r^{n-1}dr = \frac{1}{\mu_{n-1}}.$$

Suppose that $\tilde{c} \in \mathscr{C}([0,\infty))$, $\tilde{c} : [0,\infty) \to \mathbb{R}$, $\operatorname{supp}\tilde{c} \subset [\varepsilon_0, \mu_0]$ and

$$\int_0^\infty \tilde{c}(r)r^{n-1}dr = \frac{1}{\mu_{n-1}},$$

and

$$d_B(f,D) = \int_D f^* w_{\tilde{c}}.$$

Let $a = c - \tilde{c}$. Then

$$\begin{aligned}\int_0^\infty a(r)r^{n-1}dr &= \int_0^\infty c(r)r^{n-1}dr - \int_0^\infty \tilde{c}(r)r^{n-1}dr\\ &= 0.\end{aligned}$$

Hence and by Corollary B.1, it follows that $\operatorname{supp} A \subset [\varepsilon_0, \mu_0]$. Then, using Theorem B.11, we obtain

$$\begin{aligned}\int_D f^* w_c - \int_D f^* w_{\tilde{c}} &= \int_D c(\|f\|)J_f(x)dx - \int_D \tilde{c}(\|f\|)J_f(x)dx\\ &= \int_D (c(\|f\|) - \tilde{c}(\|f\|))\, J_f(x)dx\\ &= \int_D a(\|f\|)J_f(x)dx\\ &= \int_D f^* w_a\\ &= \int_D d\,(f^* \sigma_A)\\ &= 0.\end{aligned}$$

By Definition B.5, it follows that

$$d_B(f,\emptyset) = 0.$$

Also, if ∂D is smooth enough, by Theorem B.12, it follows that

$$i_K(f,\partial D) = d_B(f,D).$$

Theorem B.13. *If the open set $D' \subset D$ is such that $0 \notin f(\overline{D} \backslash D')$, then*

$$d_B(f,D) = d_B(f,D').$$

In particular, if $0 \notin f(\overline{D})$, then $d_B(f,D) = 0$.

Proof. Let

$$\mu' = \min_{\partial D'} \|f\|,$$

$$\mu'' = \min_{\overline{D} \backslash D'} \|f\|.$$

Then

$$0 < \mu'' < \mu', \quad \mu'' < \mu,$$

and take $0 < \varepsilon_0 < \mu_0 < \mu''$ in the function c defining $d_B(f,D)$. We have

$$c(\|f(x)\|) = 0, \quad x \in \overline{D} \backslash D'.$$

Consequently,

$$\begin{aligned} d_B(f,D) &= \int_D c(\|f(x)\|) J_f(x) dx \\ &= \int_{D'} c(\|f(x)\|) J_f(x) dx \\ &= d_B(f,D'). \end{aligned}$$

This completes the proof.

Definition B.6. Let $D \subset \mathbb{R}^n$ is open and bounded, $f \in \mathscr{C}^2(\overline{D})$, $f : \overline{D} \to \mathbb{R}^n$ and $z \notin f(\partial D)$. The Brouwer degree $d_B(f,D,z)$ is defined as follows:

$$d_B(f,D,z) = d_B(f(\cdot) - z, D).$$

By Definition B.6, it follows that

$$\begin{aligned} d_B(f,D,z) &= d_B(f(\cdot) - z, D) \\ &= \int_D c(\|f(x) - z\|) J_f(x) dx, \end{aligned}$$

where c is the function defining $d_B(f(\cdot) - z, D)$.

Definition B.7. $z \in \mathbb{R}^n$ is called a regular value for f if $J_f(x) \neq 0$ when $x \in f^{-1}(z)$.

Note that any z with $f^{-1}(z)$ empty is a regular value.

Theorem B.14. *If $z \notin f(\partial D)$ is a regular value for f, then*

$$\begin{aligned} d_B(f,D,z) &= \sum_{x\in f^{-1}(z)} signJ_f(x) \\ &= N_+ - N_-, \end{aligned}$$

where N_+(resp. N_-) denotes the number of elements of $f^{-1}(z)$ with positive (negative) Jacobian.

Proof. If $f^{-1}(z) = \emptyset$, then the theorem is trivial. Let $z \notin f(\partial D)$ be a regular value for f and $f^{-1}(z) \neq \emptyset$. Then $f^{-1}(z) \subset D$ is compact and discrete. Therefore, $f^{-1}(z)$ is finite and let

$$f^{-1}(z) = \{x_1, \dots, x_m\}.$$

By the inverse function theorem, for each $j \in \{1, \dots, m\}$ there exists an open neighbourhood $U_j \subset D$ of x_j such that f is a diffeomorphism on U_j and $U_j \bigcap U_k = \emptyset$, $j \neq k$, $j,k \in \{1, \dots, m\}$. Then $f(U_j)$ is an open neighbourhood of z as well as $N = \bigcap_{j=1}^{m} f(U_j)$ and $N \subset B_z(r)$ for some $r > 0$. Let

$$K = \overline{D} \backslash \bigcup_{j=1}^{m} U_j.$$

Then $f(x) \neq z$, $x \in K$. Hence and by Theorem B.13, it follows that

$$\begin{aligned} d_B(f,D,z) &= d_B\left(f, \bigcup_{j=1}^{m} U_j, z\right) \\ &= \int_{\bigcup_{j=1}^{m} U_j} c(\|f(x) - z\|) J_f(x) dx \\ &= \sum_{j=1}^{m} \int_{U_j} c(\|f(x) - z\|) J_f(x) dx \\ &= \sum_{j=1}^{m} d_B(f, U_j, z), \end{aligned}$$

where

$$\mu_0 < \min\left\{ r, \min_{\overline{D}\backslash \bigcup_{j=1}^{m} U_j} \|f(\cdot) - z\| \right\}$$

in the definition of the function c. Next,

$$\begin{aligned}
d_B(f,U_j,z) &= \int_{U_j} c\left(\|f(x)-z\|\right) J_f(x)dx \\
&= \operatorname{sign}(J_f(x_j)) \int_{U_j} c\left(\|f(x)-z\|\right) |J_f(x)|dx \\
&= \operatorname{sign}(J_f(x_j)) \int_{f(U_j)-z} c(\|y\|)dy \\
&= \operatorname{sign}(J_f(x_j)) \int_{\mathbb{R}^n} c(\|y\|)dy \\
&= \operatorname{sign}(J_f(x_j)), \quad j \in \{1,\ldots,m\}.
\end{aligned}$$

Therefore,

$$\begin{aligned}
d_B(f,D,z) &= \sum_{j=1}^{m} \operatorname{sign}(J_f(x_j)) \\
&= N_+ - N_-.
\end{aligned}$$

This completes the proof.

Let $A : \mathbb{R}^n \to \mathbb{R}^n$ be a linear isomorphism. Then

$$d_B(A,D,z) = \begin{cases} \operatorname{sign}\det A & \text{if} \quad z \in A(D) \\ 0 \quad \text{if} \quad z \notin A(\overline{D}). \end{cases}$$

If $\lambda_1,\ldots,\lambda_r$ are the real eigenvalues of A with respective multiplicities $m_1,\ldots,m_r$, and if $\mu_1,\overline{\mu_1},\ldots,\mu_s,\overline{\mu_s}$ are the non real eigenvalues of A with respective multiplicities $n_1,\ldots,n_s$, then

$$r + 2s = n$$

and

$$\det(A-\mu I) = \prod_{j=1}^{r}(\lambda_j-\mu)^{m_j}\prod_{l=1}^{s}(\mu_l-\mu)^{n_l}(\overline{\mu_l}-\mu)^{n_l}.$$

Hence,

$$\det(A) = \prod_{j=1}^{r}\lambda_j^{m_j}\prod_{l=1}^{s}|\mu_l|^{2n_l}.$$

Consequently,

$$d_B(A,D,z) = \operatorname{sign}\det A$$

$$= (-1)^{\sum_{j=1}^{p} m_j},$$

where the sum is over the negative eigenvalues $\lambda_1, \ldots, \lambda_p$, $p \in \{1, \ldots, r\}$.

B.3.2 Homotopy Invariance

Let $D \subset \mathbb{R}^n$ be open and bounded, $w \in \mathscr{C}^1(\mathbb{R}^n)$, $w : \mathbb{R}^n \to \mathbb{R}$, $F \in \mathscr{C}^2(\overline{D} \times [0,1])$, $F : \overline{D} \times [0,1] \to \mathbb{R}^n$, and

$$dF_j = \sum_{k=1}^{n} \partial_k F_j dx_k, \quad j \in \{1, \ldots, n\}.$$

Theorem B.15. *For each $\lambda \in [0,1]$, we have*

$$\partial_\lambda \left((w \circ F) dF_1 \wedge \ldots \wedge dF_n \right)$$

$$= d\left((w \circ F) \left(\sum_{j=1}^{n} (-1)^{j-1} \partial_\lambda F_j dF_1 \wedge \ldots \widehat{dF_j} \wedge \ldots \wedge dF_n \right) \right).$$

Proof. We have

$$\begin{aligned}
\partial_\lambda dF_j &= \partial_\lambda \left(\sum_{k=1}^{n} \partial_k F_j dx_k \right) \\
&= \sum_{k=1}^{n} \partial_\lambda \partial_k F_j dx_k \\
&= \sum_{k=1}^{n} \partial_k \partial_\lambda F_j dx_k \\
&= d\left(\partial_\lambda F_j \right).
\end{aligned}$$

Now, we get

$$\partial_\lambda ((w \circ F) dF_1 \wedge \ldots \wedge dF_n) = (\partial_\lambda (w \circ F)) dF_1 \wedge \ldots \wedge dF_n + (w \circ F) \partial_\lambda (dF_1 \wedge \ldots \wedge dF_n)$$

$$= \left(\sum_{j=1}^{n} (\partial_\lambda w \circ F) \partial_\lambda F_j \right) dF_1 \wedge \ldots \wedge dF_n + (w \circ F) \sum_{j=1}^{n} dF_1 \wedge \ldots \wedge \partial_\lambda (dF_j) \wedge \ldots \wedge dF_n$$

$$= \sum_{j=1}^{n} (-1)^{j-1} (\partial_j w \circ F) dF_j \wedge \partial_\lambda F_j dF_1 \wedge \ldots \wedge \widehat{dF_j} \wedge \ldots \wedge dF_n$$

$$+(w\circ F)\left(\sum_{j=1}^{n}(-1)^{j-1}d(\partial_\lambda F_j)dF_1\wedge\ldots\wedge\widehat{dF_j}\wedge\ldots\wedge dF_n\right)$$

$$=\sum_{j=1}^{n}(-1)^{j-1}\left(\sum_{k=1}^{n}\partial_k w\circ F dF_k\right)\wedge\partial_\lambda F_j dF_1\wedge\ldots\wedge\widehat{dF_j}\wedge\ldots\wedge dF_n$$

$$+(w\circ F)\left(\sum_{j=1}^{n}(-1)^{j-1}d\left(\partial_\lambda F_j dF_1\wedge\ldots\wedge\widehat{dF_j}\wedge\ldots\wedge dF_n\right)\right)$$

$$=d(w\circ F)\wedge\left(\sum_{j=1}^{n}(-1)^{j-1}\partial_\lambda F_j dF_1\wedge\ldots\wedge\widehat{dF_j}\wedge\ldots\wedge dF_n\right)$$

$$+(w\circ F)d\left(\sum_{j=1}^{n}(-1)^{j-1}\partial_\lambda F_j dF_1\wedge\ldots\wedge\widehat{dF_j}\wedge\ldots\wedge dF_n\right)$$

$$=d\left((w\circ F)\left(\sum_{j=1}^{n}(-1)^{j-1}\partial_\lambda F_j dF_1\wedge\ldots\wedge\widehat{dF_j}\wedge\ldots\wedge dF_n\right)\right).$$

This completes the proof.

Theorem B.16. *If*

$$F(\partial D\times[0,1])\bigcap suppw=\emptyset,$$

then $I_w(F(\cdot,\lambda))$ is independent of λ, where

$$I_w(F(\cdot,\lambda))=\int_D w\circ F dF_1\wedge\ldots\wedge dF_n.$$

Proof. By Theorem B.15, we get

$$\frac{d}{d\lambda}I_w(F(\cdot,\lambda))=\frac{d}{d\lambda}\left(\int_D w\circ F dF_1\wedge\ldots\wedge dF_n\right)$$

$$=\int_D\frac{d}{d\lambda}(w\circ F dF_1\wedge\ldots\wedge dF_n)$$

$$=\int_D d\left((w\circ F)\left(\sum_{j=1}^{n}(-1)^{j-1}\partial_\lambda F_j dF_1\wedge\ldots\wedge\widehat{dF_j}\wedge\ldots\wedge dF_n\right)\right)$$

$$=\int_{\partial D}(w\circ F)\left(\sum_{j=1}^{n}(-1)^{j-1}\partial_\lambda F_j dF_1\wedge\ldots\wedge\widehat{dF_j}\wedge\ldots\wedge dF_n\right)$$

$$=0.$$

This completes the proof.

Theorem B.17. *If* $0 \notin F(\partial D \times [0,1])$, *then* $d_B(F(\cdot,\lambda),D)$ *is independent of* $\lambda \in [0,1]$.

Proof. We have

$$\mu = \inf_{\partial D \times [0,1]} \|F\| > 0.$$

Take $0 < \varepsilon_0 < \mu_0 < \mu$ in the definition of the function c defining the Brouwer degree. Then

$$d_B(F(\cdot,\lambda),D) = I_w(F(\cdot,\lambda)).$$

Hence and by Theorem B.16, it follows that $d_B(F(\cdot,\lambda),D)$ is independent of $\lambda \in [0,1]$. This completes the proof.

Theorem B.18. *Let* $f,g \in \mathscr{C}^2(\overline{D})$, $f,g : \overline{D} \to \mathbb{R}^n$ *and*

$$\nu = \max_{\partial D} \|f - g\| < \mu = \min_{\partial D} \|f\|.$$

Then

$$d_B(f,D) = d_B(g,D).$$

Proof. Define $F : \overline{D} \times [0,1] \to \mathbb{R}^n$ as follows.

$$F(x,\lambda) = (1-\lambda)f(x) + \lambda g(x).$$

Then

$$\begin{aligned}
|F(x,\lambda)| &= \|(1-\lambda)f(x) + \lambda g(x)\| \\
&= \|f(x) + \lambda(g(x) - f(x))\| \\
&\geq \|f(x)\| - \lambda\|g(x) - f(x)\| \\
&\geq \|f(x)\| - \|g(x) - f(x)\| \\
&> 0
\end{aligned}$$

for any $(x,\lambda) \in \partial D \times [0,1]$. Hence and by Theorem B.17, we have that $d_B(F(\cdot,\lambda),D)$ is independent of $\lambda \in [0,1]$. Then

$$\begin{aligned}
d_B(f,D) &= d_B(F(\cdot,0),D) \\
&= d_B(F(\cdot,1),D) \\
&= d_B(g,D).
\end{aligned}$$

This completes the proof.

For our next result we will use the Sard theorem.

Theorem B.19 (Sard's Theorem). *The set of critical values (that is, the image of the set of critical points) of a smooth function f from one Euclidean space or manifold to another is a null set, i.e., it has Lebesgue measure* 0.

Theorem B.20. *For each* $f \in \mathscr{C}^2(\overline{D})$, $f : \overline{D} \to \mathbb{R}^n$, *and* $z \notin f(\partial D)$, $d_B(f,D,z)$ *is an integer.*

Proof. By Sard's theorem, it follows that there is a regular value u such that

$$\|u - z\| < \mu = \inf_{\partial D} \|f(\cdot) - z\|.$$

Hence,

$$\max_{\overline{D}} \|z - u\| = \max_{\overline{D}} \|f(\cdot) - z - (f(\cdot) - u)\|$$

$$< \mu.$$

From here and from Theorem B.18, we get

$$d_B(f,D,z) = d_B(f,D,u).$$

By Theorem B.14, we have that $d_B(f,D,u)$ is an integer. Therefore, $d_B(f,D,z)$ is an integer. This completes the proof.

B.3.3 Continuous Mappings: Basic Properties

Let $D \subset \mathbb{R}^n$ be open and bounded, $f \in \mathscr{C}(\overline{D})$, $f : \overline{D} \to \mathbb{R}^n$, $z \notin f(\partial D)$ so that

$$\mu = \min_{\partial D} \|f(\cdot) - z\| > 0.$$

Then there exist $g, h \in \mathscr{C}^2(\overline{D})$ such that

$$\max_{\overline{D}} \|g - f\| < \frac{\mu}{4}, \quad \max_{\overline{D}} \|h - f\| < \frac{\mu}{4}.$$

Hence,

$$\|g(\cdot) - z\| = \|g(\cdot) - f(\cdot) + f(\cdot) - z\|$$

$$\geq \|f(\cdot) - z\| - \|g - f\|$$

$$\begin{aligned} &\geq \mu - \frac{\mu}{4} \\ &= \frac{3\mu}{4}, \end{aligned}$$

$$\begin{aligned} \|g-h\| &= \|g-f+f-h\| \\ &\leq \|g-f\| + \|f-h\| \\ &\leq \frac{\mu}{4} + \frac{\mu}{4} \\ &= \frac{\mu}{2}. \end{aligned}$$

Hence and by Theorem B.18, it follows that

$$d_B(g,D,z) = d_B(h,D,z).$$

Definition B.8. The Brouwer degree $d_B(f,D,z)$ is defined by

$$d_B(f,D,z) = d_B(g,D,z),$$

where $g \in \mathscr{C}^2(\overline{D})$, $\max_{\overline{D}} \|g-f\| < \frac{\mu}{4}$.

Theorem B.21. *If the open subset $D' \subset D$ is such that $z \notin f(\overline{D} \backslash D')$, then*

$$d_B(f,D,z) = d_B(f,D',z).$$

Proof. Let $g \in \mathscr{C}^2(\overline{D})$ be such that

$$\max_{\overline{D}} \|g-f\| < \frac{\mu}{4} \quad \text{and} \quad \max_{\overline{D}} \|g-f\| < \min_{\overline{D} \backslash D'} \|f(\cdot)-z\|.$$

Then, by Definition B.8, we get

$$d_B(f,D,z) = d_B(g,D,z),$$

$$d_B(f,D',z) = d_B(g,D',z),$$

and by Theorem B.13, we obtain

$$d_B(g,D',z) = d_B(g,D,z).$$

Hence,

$$d_B(f,D,z) = d_B(f,D',z).$$

This completes the proof.

Theorem B.22 (Existence Property). *If $z \notin f(\overline{D})$, then $d_B(f,D,z) = 0$. Equivalently, if $d_B(f,D,z) \neq 0$, there exists at least one $x \in D$ such that $f(x) = z$.*

Proof. Take $g \in \mathscr{C}^2(\overline{D})$ such that

$$\max_{\overline{D}} \|g - f\| < \frac{\mu}{4} \quad \text{and} \quad \max_{\overline{D}} \|g - f\| < \min_{\overline{D}} \|f(\cdot) - z\|.$$

Then, using Theorem B.21, it follows that

$$\begin{aligned} d_B(f,D,z) &= d_B(f,\emptyset,z) \\ &= d_B(g,\emptyset,z) \\ &= 0. \end{aligned}$$

This completes the proof.

Theorem B.23. *If $F \in \mathscr{C}(\overline{D} \times [0,1])$, $F : \overline{D} \times [0,1] \to \mathbb{R}^n$ and $z \notin F(\partial D \times [0,1])$, then $d_B(F(\cdot,\lambda),D,z)$ is independent of $\lambda \in [0,1]$.*

Proof. Let

$$\mu = \min_{\partial D \times [0,1]} \|F(\cdot,\cdot) - z\| > 0.$$

Then there exists a $G \in \mathscr{C}^2(\overline{D} \times [0,1])$ such that

$$\max_{\overline{D} \times [0,1]} \|G - F\| < \frac{\mu}{4}.$$

Consequently,

$$d_B(F(\cdot,\lambda),D,z) = d_B(G(\cdot,\lambda),D,z)$$

for any $\lambda \in [0,1]$. By Theorem B.16, it follows that $d_B(G(\cdot,\lambda),D,z)$ is independent of $\lambda \in [0,1]$. Therefore, $d_B(F(\cdot,\lambda),D,z)$ is independent of $\lambda \in [0,1]$. This completes the proof.

Theorem B.24. *Let $f,g \in \mathscr{C}(\overline{D})$, $f,g : \overline{D} \to \mathbb{R}^n$, $z \notin f(\partial D)$,*

$$\max_{\partial D} \|f - g\| < \mu = \min_{\partial D} \|f(\cdot) - z\|.$$

Then

$$d_B(f,D,z) = d_B(g,D,z).$$

Proof. Take

$$F(\cdot,\lambda) = (1-\lambda)f(\cdot) + \lambda g(\cdot), \quad \lambda \in [0,1].$$

We have

$$\begin{aligned}
\|F(\cdot,\lambda) - z\| &= \|f(\cdot) - z - \lambda(f(\cdot) - g(\cdot))\| \\
&\geq \|f(\cdot) - z\| - \lambda\|f - g\| \\
&\geq \|f(\cdot) - z\| - \|f - g\| \\
&> 0.
\end{aligned}$$

Hence and by Theorem B.23, we obtain

$$\begin{aligned}
d_B(f,D,z) &= d_B(F(\cdot,0),D,z) \\
&= d_B(F(\cdot,1),D,z) \\
&= d_B(g,D,z).
\end{aligned}$$

This completes the proof.

Theorem B.25. *If $f \in \mathscr{C}(\overline{D})$, $f : \overline{D} \to \mathbb{R}^n$, $z \notin f(\partial D)$, then for any*

$$y \in B_z\left(\min_{\partial D} \|f(\cdot) - z\|\right),$$

we have

$$d_B(f,D,y) = d_B(f,D,z).$$

Proof. Note that

$$\begin{aligned}
\|f(\cdot) - z - (f(\cdot) - y)\| &= \|z - y\| \\
&< \min_{\partial D} \|f(\cdot) - z\|.
\end{aligned}$$

Hence and by Theorem B.24, it follows that

$$d_B(f(\cdot) - y, D) = d_B(f(\cdot) - z, D),$$

whereupon,

$$d_B(f,D,y) = d_B(f,D,z).$$

This completes the proof.

Theorem B.26. *If $f \in \mathscr{C}(\overline{D})$, $f : \overline{D} \to \mathbb{R}^n$, is such that $0 \notin f(\partial D)$ and $f(\overline{D}) \subset Y$, where Y is a proper vector subspace of $\mathbb{R}^n$, then*

$$d_B(f, D, 0) = 0.$$

Proof. Let $y \notin Y$ be such that

$$\|y\| < \min_{\partial D} \|f\|.$$

Then, using Theorem B.25, we get

$$d_B(f, D, 0) = d_B(f, D, y).$$

Since $y \notin Y$, we have $y \notin f(\overline{D})$. Therefore, using Theorem B.21, it follows that

$$\begin{aligned} d_B(f, D, y) &= d_B(f, \emptyset, y) \\ &= 0. \end{aligned}$$

Hence,

$$d_B(f, D, 0) = 0.$$

This completes the proof.

Theorem B.27. *Let $f \in \mathscr{C}(\overline{D})$. Then $d_B(f, D, z)$ is a constant on each connected component of $\mathbb{R}^n \backslash f(\partial D)$.*

Proof. Let Ω be a connected component of $\mathbb{R}^n \backslash f(\partial D)$. Then Ω is open and connected and, hence, it is path-connected. Let $y, z \in \Omega$. Then there is a continuous path $\theta : [0,1] \to \Omega$ such that

$$\theta(0) = y, \quad \theta(1) = z.$$

For any $(x, \lambda) \in \partial D \times [0,1]$, one has $f(x) \neq \theta(\lambda)$. Then, using Theorem B.23, it follows that

$$\begin{aligned} d_B(f, D, y) &= d_B(f, D, \theta(0)) \\ &= d_B(f(\cdot) - \theta(0), D, 0) \\ &= d_B(f(\cdot) - \theta(1), D, 0) \\ &= d_B(f, D, \theta(1)) \\ &= d_B(f, D, z). \end{aligned}$$

This completes the proof.

Theorem B.28. *Suppose that* $\{D_j\}_{j\in\mathbb{N}}$ *is a sequence of open and mutually disjoint subsets of* D*. If* $z \notin f\left(\overline{D}\backslash \bigcup\limits_{k\in\mathbb{N}} D_k\right)$*, then* $d_B(f,D_j,z) = 0$ *for all but finitely many* $j \in \mathbb{N}$*, and*

$$d_B(f,D,z) = \sum_{j=1}^{\infty} d_B(f,D_j,z).$$

Proof. By the assumptions, we have

$$0 < \mu_0 = \min_{\overline{D}\backslash \bigcup_{j\in\mathbb{N}} D_j} \|f(\cdot) - z\|$$

and

$$\mu_0 \leq \mu = \min_{\partial D} \|f(\cdot) - z\|,$$

$$\mu_0 \leq \mu_j = \min_{\partial D_j} \|f(\cdot) - z\|, \quad j \in \mathbb{N}.$$

Let $g \in \mathscr{C}^2(\overline{D})$, $g : \overline{D} \to \mathbb{R}^n$, be such that

$$\max_{\overline{D}} \|f - g\| < \frac{\mu_0}{4},$$

$$\frac{3\mu_0}{4} \leq \tilde{\mu} = \min_{\partial D} \|g(\cdot) - z\|,$$

$$\frac{3\mu_0}{4} \leq \tilde{\mu}_j = \min_{\partial D_J} \|g(\cdot) - z\|.$$

Then

$$d_B(f,D,z) = d_B(g,D,z),$$

$$d_B(f,D_j,z) = d_B(g,D_j,z), \quad j \in \mathbb{N}.$$

Take $0 < r < \tilde{\mu}$. By Sard's theorem, it follows that there exists a regular value $u \in B_z(r) \subset D$ of f. Hence,

$$d_B(g,D,z) = d_B(g,D,u),$$

$$d_B(g,D_j,z) = d_B(g,D_j,u), \quad j \in \mathbb{N}.$$

Since u is regular, $g^{-1}(u)$ is finite and hence, it is contained in a finite number D_j's, namely $D_{j_1}, \ldots, D_{j_m}$. Then

$$d_B(f,D,z) = d_B(g,D,z)$$

$$
\begin{aligned}
&= d_B(g,D,u) \\
&= d_B\left(g,\bigcup_{l=1}^{m} D_{j_l},u\right) \\
&= \sum_{l=1}^{m} d_B(g,D_{j_l},u) \\
&= \sum_{l=1}^{m} d_B(g,D_{j_l},z) \\
&= \sum_{l=1}^{m} d_B(f,D_{j_l},z).
\end{aligned}
$$

This completes the proof.

Example B.4. Let $a < b$ and $f : [a,b] \to \mathbb{R}$ is continuous. Then

$$d_B(f,(a,b),0) = \frac{1}{2}\left(\frac{f(b)}{|f(b)|} - \frac{f(a)}{|f(a)|}\right).$$

Example B.5. Let

$$p(x) = \sum_{k=1}^{n} a_k x^k$$

be a real polynomial with $a_n \neq 0$. Let $R > 0$ be sufficiently large. Then

$$
\begin{aligned}
d_B(p,(-R,R),0) &= \frac{1}{2}\left(\frac{f(R)}{|f(R)|} - \frac{f(-R)}{|f(-R)|}\right) \\
&= \begin{cases} 0 \quad \text{if} \quad n \quad \text{is} \quad \text{even} \\ \dfrac{a_n}{|a_n|} \quad \text{if} \quad n \quad \text{is} \quad \text{odd}. \end{cases}
\end{aligned}
$$

B.3.4 The Case $f : \partial D \subset \mathbb{R}^n \to S^{n-1}$

Let $D \subset \mathbb{R}^n$ be open and bounded and $f : \partial D \to S^{n-1}$ be continuous. Let, also, $g,h : \overline{D} \to \mathbb{R}^n$ be two continuous extensions of f to $\overline{D}$. Then, by Theorem B.24, it follows that

$$d_B(g,D) = d_B(h,D).$$

Definition B.9. The Brouwer degree $d_B(f,\partial D)$ is defined as follows.

$$d_B(f,\partial D) = d_B(g,D),$$

where $g : \overline{D} \to \mathbb{R}^n$ is any continuous extension of f to $\overline{D}$.

Theorem B.29. *If* $f : \partial D \to S^{n-1}$ *is not onto, then*

$$d_B(f, \partial D) = 0.$$

Proof. Let $y \in S^{n-1} \backslash f(\partial D)$ and $\tilde{f} : \mathbb{R}^n \to \overline{B}(1)$ be a continuous extension of f. Define the homotopy $F : \overline{B}(1) \times [0,1] \to \mathbb{R}^n$ as follows.

$$F(x, \lambda) = (1-\lambda)\tilde{f}(x) - \lambda y.$$

Suppose that there exists a $(x, \lambda) \in \partial D \times [0,1]$ so that

$$F(x, \lambda) = 0.$$

Then

$$\begin{aligned} 0 &= (1-\lambda)\tilde{f}(x) - \lambda y \\ &= (1-\lambda) f(x) - \lambda y \end{aligned}$$

or

$$(1-\lambda) f(x) = \lambda y.$$

Hence,

$$\begin{aligned} 1-\lambda &= (1-\lambda)\|f(x)\| \\ &= \lambda \|y\| \\ &= \lambda. \end{aligned}$$

Therefore, $\lambda = \frac{1}{2}$ and $f(x) = y$, which is a contradiction. By Theorem B.23, we get

$$\begin{aligned} d_B(f, \partial D) &= d_B(\tilde{f}, D) \\ &= d_B(F(\cdot, 0), D) \\ &= d_B(F(\cdot, 1), D) \\ &= d_B(-y, D) \\ &= 0. \end{aligned}$$

This completes the proof.

Corollary B.2. *If $d_B(f,\partial D) \neq 0$, then f is onto.*

Theorem B.30. *Let $g \in \mathscr{C}(S^{n-1})$, $g : S^{n-1} \to S^{n-1}$ have no fixed points. Then*

$$d_B(g, S^{n-1}) = (-1)^n.$$

Proof. Let $\tilde{g} : \mathbb{R}^n \to \overline{B}(1)$ be a continuous extension of f. Consider the homotopy $F : \overline{B}(1) \times [0,1] \to \mathbb{R}^n$ defined by

$$F(x,\lambda) = (1-\lambda)\tilde{g}(x) - \lambda x.$$

Assume that there is a $(x,\lambda) \in S^{n-1} \times [0,1]$ such that

$$F(x,\lambda) = 0.$$

Hence,

$$0 = (1-\lambda)g(x) - \lambda x$$

or

$$(1-\lambda)g(x) = \lambda x.$$

Then

$$\begin{aligned} 1-\lambda &= (1-\lambda)\|g(x)\| \\ &= \lambda\|x\| \\ &= \lambda, \end{aligned}$$

whereupon, $\lambda = \frac{1}{2}$ and $g(x) = x$. This is a contradiction. Therefore,

$$\begin{aligned} d_B(g, S^{n-1}) &= d_B(\tilde{g}, B(1)) \\ &= d_B(F(\cdot,0), B(1)) \\ &= d_B(F(\cdot,1), B(1)) \\ &= d_B(-I, B(1)) \\ &= (-1)^n. \end{aligned}$$

This completes the proof.

Corollary B.3. *Any mapping $g \in \mathscr{C}(S^{n-1})$, $g : S^{n-1} \to S^{n-1}$ such that $d_B(g, S^{n-1}) \neq (-1)^n$, has at least one fixed point.*

Proof. Assume that g has no any fixed points. By Theorem B.30, it follows that $d_B(g, S^{n-1}) = (-1)^n$. This is a contradiction. This completes the proof.

Definition B.10. Let $D \subset \mathbb{R}^n$ is open and bounded, ∂D is smooth and $f \in \mathscr{C}^1(\partial D)$, $f : \partial D \to S^{n-1}$. The Kronecker index $i_K(f, \partial D)$ is defined by

$$i_K(f, \partial D) = i_K(g, \partial D),$$

where g is any $\mathscr{C}^1$ extension of f to $\overline{D}$.

Remark B.1. Suppose that all conditions of Definition B.9 hold. Then

$$\begin{aligned} d_B(f, \partial D) &= d_B(g, D) \\ &= i_K(g, \partial D) \\ &= i_K(f, \partial D). \end{aligned}$$

B.4 Leray-Schauder Degree

In this section we construct the Leray-Schauder degree. We need the following result on the approximation of a compact mapping by finite dimensional mappings.

Theorem B.31. *Let E be a real Banach space, $\Omega \subset E$ be an open and bounded subset and $T : \overline{\Omega} \to E$ be a completely continuous mapping. Then for any $\varepsilon > 0$ there exist a finite dimensional space F and a continuous mapping $T_\varepsilon : \overline{\Omega} \to F$ such that*

$$\|T_\varepsilon x - Tx\| < \varepsilon, \quad x \in \overline{\Omega}.$$

Proof. Since $T : \overline{\Omega} \to E$ is a completely continuous mapping, we have that $T(\overline{\Omega})$ is relatively compact in E. Then for any $\varepsilon > 0$ there exist a finite number of elements of $\overline{\Omega}$, $\{x_1, \ldots, x_m\} \subset \overline{\Omega}$, such that

$$T(\overline{\Omega}) \subset \bigcup_{j=1}^{m} B_{Tx_j}(\varepsilon).$$

Define

$$F = \bigcup_{j=1}^{m} B_{Tx_j}(\varepsilon),$$

$$\phi_j(x) = \max\{0, \varepsilon - \|Tx - Tx_j\|\}, \quad j \in \{1, \ldots, m\},$$

$$\Gamma(x) = \sum_{j=1}^{m} \phi_j(x),$$

$$T_\varepsilon x = \sum_{j=1}^{m} \frac{\phi_j(x)}{\Gamma(x)} Tx_j.$$

Let $A \subseteq \{1, \ldots, m\}$ so that for $j \in A$ we have

$$\|Tx_j - Tx\| < \varepsilon.$$

Then, for $j \in \{1, \ldots, m\} \backslash A$, we have $\phi_j(x) = 0$ and

$$\begin{aligned}
\|T_\varepsilon x - Tx\| &= \left\| \sum_{j=1}^{m} \frac{\phi_j(x)}{\Gamma(x)} Tx_j - Tx \right\| \\
&= \left\| \sum_{j=1}^{m} \frac{\phi_j(x)}{\Gamma(x)} (Tx_j - Tx) \right\| \\
&\le \sum_{j=1}^{m} \frac{\phi_j(x)}{\Gamma(x)} \|Tx_j - Tx\| \\
&= \sum_{j \in A} \frac{\phi_j(x)}{\Gamma(x)} \|Tx_j - Tx\| \\
&< \varepsilon \sum_{j \in A} \frac{\phi_j(x)}{\Gamma(x)} \\
&\le \varepsilon \sum_{j=1}^{m} \frac{\phi_j(x)}{\Gamma(x)} \\
&= \varepsilon, \quad x \in \overline{\Omega}.
\end{aligned}$$

This completes the proof.

Theorem B.32. *Let E be a real Banach space, $B \subset E$ be a closed bounded subset and $T : B \to E$ be a completely continuous mapping. Suppose that $Tx \neq x$, $x \in B$. Then there exists an $\varepsilon_0 > 0$ such that*

$$x \neq tT_{\varepsilon_1} x + (1-t) T_{\varepsilon_2} x, \quad t \in [0,1], \quad x \in B,$$

where $\varepsilon_1, \varepsilon_2 \in (0, \varepsilon_0)$, and $T_{\varepsilon_i} : B \to F_{\varepsilon_i}$, $i \in \{1, 2\}$, are as in Theorem B.31.

Proof. Suppose the contrary. There exist $\varepsilon_1^j, \varepsilon_2^j \to 0$, $t_j \to t_0$, as $j \to \infty$, $t_0 \in [0,1]$, such that

$$t_jT_{\varepsilon_1^j}x_j+(1-t_j)T_{\varepsilon_2^j}x_j=x_j,\quad x_j\in B,\quad j\in\mathbb{N},\tag{B.7}$$

and $T^j_{\varepsilon_1}$, $T^j_{\varepsilon_2}$ are as in Theorem B.31, $j\in\mathbb{N}$. By the compactness of T, the sequence $\{Tx_j\}_{j\in\mathbb{N}}$ has a subsequence $\{Tx_{j_n}\}_{n\in\mathbb{N}}$ converging to $y\in E$ and by Theorem B.31, it follows that

$$T_{\varepsilon_i^{j_k}}x_{j_k}\to y\quad\text{as}\quad k\to\infty,\quad i\in\{1,2\}.$$

Hence and by (B.7), we conclude that

$$x_{j_k}\to y\quad\text{as}\quad k\to\infty.$$

Therefore, $Ty=y$, $y\in B$. This is a contradiction. This completes the proof.

Definition B.11. Let E be a real Banach space, $\Omega\subset E$ be an open and bounded set and $T:\overline{\Omega}\to E$ be a completely continuous mapping. By Theorem B.32, it follows that there exists an $\varepsilon_0>0$ such that

$$x\neq tT_{\varepsilon_1}x+(1-t)T_{\varepsilon_2}x,\quad t\in[0,1],\quad x\in\partial\Omega,$$

where $\varepsilon_i\in(0,\varepsilon_0)$, $T_{\varepsilon_i}:\overline{\Omega}\to F_{\varepsilon_i}$, $i\in\{1,2\}$, as in Theorem B.32. Then

$$d_B(I-T_{\varepsilon_i},\Omega\bigcap F_{\varepsilon_i},0),\quad i\in\{1,2\},$$

is well defined, and

$$d_B(I-T_{\varepsilon_1},\Omega\bigcap F_{\varepsilon_1},0)=d_B(I-T_{\varepsilon_2},\Omega\bigcap F_{\varepsilon_2},0).$$

Consequently,, we define the Leray-Schauder degree as follows.

$$d_{LS}(I-T,\Omega,0)=d_B(I-T_{\varepsilon},\Omega\bigcap F_{\varepsilon},0)$$

for any $\varepsilon\in(0,\varepsilon_0)$. For the general case, if $p\notin(I-T)(\partial\Omega)$, we define

$$d_{LS}(I-T,\Omega,p)=d_{LS}(I-T-p,\Omega,0).$$

Below we will list some of the properties of the Leray-Schauder degree.

Theorem B.33 (Normality). *We have*

$$d_{LS}(I,\Omega,0)=1$$

if and only if $0\in\Omega$.

Proof. Let $\varepsilon_0>0$, $\varepsilon\in(0,\varepsilon_0)$, O_{ε}, F_{ε} be as in Theorem B.31 and Theorem B.32. Then

$$d_{LS}(I,\Omega,0) = d_B(I - O_\varepsilon, \Omega \bigcap F_\varepsilon, 0)$$

$$= 1$$

if and only if $0 \in \Omega \bigcap F_\varepsilon$. This completes the proof.

Theorem B.34 (Solvability). *If*

$$d_{LS}(I - T, \Omega, 0) \neq 0,$$

then equation $Tx = x$ has a solution in Ω.

Proof. Let $\varepsilon_0 > 0$, $\varepsilon \in (0, \varepsilon_0)$, T_ε, F_ε be as in Theorem B.31 and Theorem B.32. Then

$$d_{LS}(I - T, \Omega, 0) = d_B(I - T_\varepsilon, \Omega \bigcap F_\varepsilon, 0).$$

Since

$$d_B(I - T_\varepsilon, \Omega \bigcap F_\varepsilon, 0) \neq 0,$$

by Theorem B.22, it follows that there exists an $x \in \Omega \bigcap F_\varepsilon$ such that

$$T_\varepsilon x = x.$$

Hence and

$$\varepsilon > \|T_\varepsilon x - Tx\|$$

$$= \|(T_\varepsilon - I)x + (I - T)x\|$$

$$= \|(I - T)x\|,$$

we conclude that $Tx = x$. This completes the proof.

Theorem B.35 (Homotopy). *Let $T_t : [0,1] \times \overline{\Omega} \to E$ be a completely continuous mapping and $T_t x \neq x$ for any $(t,x) \in [0,1] \times \partial\Omega$. Then $d_{LS}(I - T_t, \Omega, 0)$ does not depend on $t \in [0,1]$.*

Proof. Let $\varepsilon_0 > 0$, $\varepsilon \in (0, \varepsilon_0)$, $T_{t\varepsilon}$, $F_{t\varepsilon}$ be as in Theorem B.31 and Theorem B.32. Then

$$d_B(I - T_{t\varepsilon}, \Omega \bigcup F_{t\varepsilon}, 0)$$

is independent of $t \in [0,1]$. Hence and

$$d_{LS}(I - T_t, \Omega, 0) = d_B(I - T_{t\varepsilon}, \Omega \bigcap F_{t\varepsilon}, 0),$$

we conclude that $d_{LS}(I - T_t, \Omega, 0)$ is independent of $t \in [0,1]$. This completes the proof.

Theorem B.36 (Additivity). *Let Ω_1, Ω_2 be disjoint open subsets of Ω and $0 \notin \overline{\Omega} \setminus (\Omega_1 \bigcap \Omega_2)$. Then*

$$d_{LS}(I-T,\Omega,0) = d_{LS}(I-T,\Omega_1,0) + d_{LS}(I-T,\Omega_2,0).$$

Proof. Let $\varepsilon_0 > 0$, $\varepsilon \in (0,\varepsilon_0)$, T_ε, F_ε be as in Theorem B.31 and Theorem B.32. Then

$$\begin{aligned} d_{LS}(I-T,\Omega,0) &= d_B(I-T_\varepsilon,\Omega \bigcap F_\varepsilon,0) \\ &= d_B(I-T_\varepsilon,\Omega_1,0) + d_B(I-T_\varepsilon,\Omega_2,0) \\ &= d_{LS}(I-T,\Omega_1,0) + d_{LS}(I-T,\Omega_2,0). \end{aligned}$$

This completes the proof.

B.5 Fixed Point Index for Completely Continuous Mappings

Let X be a Banach space, K be a closed convex subset of X and $\Omega \subset K$ be open and bounded.

Definition B.12. If $F : \overline{\Omega} \to K$ is a completely continuous mapping such that $x \neq Fx$, $x \in \partial\Omega$, $r_1, r_2 : X \to K$ are retractions, then

$$d_{LS}(I-Fr_1, r_1^{-1}(\Omega), 0) \quad \text{and} \quad d_{LS}(I-Fr_2, r_2^{-1}(\Omega), 0)$$

are well defined and

$$d_{LS}(I-Fr_1, r_1^{-1}(\Omega), 0) = d_{LS}(I-Fr_2, r_2^{-1}(\Omega), 0).$$

Define the fixed point index $i(F,\Omega,K)$ as follows.

$$i(F,\Omega,K) = d_{LS}(I-Fr, r^{-1}(\Omega), 0)$$

for any retraction $r : X \to K$.

Below suppose that $F : \overline{\Omega} \to K$ is a completely continuous operator such that $x \neq Fx$ for $x \in \partial\Omega$.

Theorem B.37 (Normality). *If F is a constant mapping, then*

$$i(F,\Omega,K) = 1.$$

Proof. Let $r : X \to K$ be any retraction. Then

$$d_{LS}(I - Fr, r^{-1}(\Omega), 0) = 1.$$

Hence,

$$i(F, \Omega, K) = d_{LS}(I - Fr, r^{-1}(\Omega), 0)$$

$$= 1.$$

This completes the proof.

Theorem B.38 (Solvability). *If $i(F, \Omega, K) \neq 0$, then $Fx = x$ has a solution in Ω.*

Proof. Let $r : X \to K$ be any retraction. Then

$$i(F, \Omega, K) = d_{LS}(I - Fr, r^{-1}(\Omega), 0)$$

$$\neq 0.$$

Hence and by Theorem B.34, it follows that there exists an $x \in r^{-1}(\Omega) \subseteq \Omega$ such that $Frx = x$. Since $r : X \to K$ is a retraction and $\Omega \subset K$, we get $Fx = x$. This completes the proof.

Theorem B.39 (Homotopy). *Let $T_t : [0,1] \times \overline{\Omega} \to E$ be a completely continuous mapping and $T_t x \neq x$ for any $(t,x) \in [0,1] \times \partial\Omega$. Then $i(T_t, \Omega, K)$ is independent of $t \in [0,1]$.*

Proof. Let $r : X \to Y$ be any retraction. Then, by Theorem B.35, it follows that $d_{LS}(I - T_t r, r^{-1}(\Omega), 0)$ is independent of $t \in [0,1]$. Hence and

$$i(T_t, \Omega, K) = d_{LS}(I - T_t r, r^{-1}(\Omega), 0),$$

it follows that $i(T_t, \Omega, K)$ is independent of $t \in [0,1]$. This completes the proof.

Theorem B.40 (Additivity). *Let Ω_1, Ω_2 be two disjoint open subsets of Ω and $0 \notin \overline{\Omega} \backslash (\Omega_1 \bigcup \Omega_2)$. Then*

$$i(F, \Omega, K) = i(F, \Omega_1, K) + i(F, \Omega_2, K).$$

Proof. Let $r : X \to K$ be any retraction. Then

$$i(F, \Omega, K) = d_{LS}(I - Fr, r^{-1}(\Omega), 0),$$

$$i(F, \Omega_1, K) = d_{LS}(I - Fr, r^{-1}(\Omega_1), 0),$$

$$i(F, \Omega_2, K) = d_{LS}(I - Fr, r^{-1}(\Omega_2), 0).$$

Hence and by Theorem B.36, it follows that

$$
\begin{aligned}
i(F,\Omega,K) &= d_{LS}(I - Fr, r^{-1}(\Omega), 0) \\
&= d_{LS}(I - Fr, r^{-1}(\Omega_1), 0) + d_{LS}(I - Fr, r^{-1}(\Omega_2), 0) \\
&= i(F,\Omega_1,K) + i(F,\Omega_2,K).
\end{aligned}
$$

This completes the proof.

References

1. D. Anderson. Existence of solutions for a first-order p-Laplacian BVP on time scales, Nonlinear Analysis, 69(2008), 4521-4525.
2. D. Anderson and R. Ma. Second-order n-point eigenvalue problems on time scales, Advances in Difference Equations, Vol. 2006, pp. 1-17.
3. F. M. Atici and S. G. Topal. The generalized quasilinearization method and three point boundary value problems on time scales, Applied Mathematics Letters, Vol. 18, no. 5, pp. 577-585, 2005.
4. F. M. Atici and G. Sh. Guseinov. On Green's functions and positive solutions for boundary value problems on time scales, Journal of Computational and Applied Mathematics, 141(2002), pp. 75-99.
5. R. I. Avery and J. Henderson. Two positive fixed points of nonlinear operators on ordered Banach spaces, Communications on Applied Nonlinear Analysis, vol. 8, no. 1, pp. 27-36, 2001.
6. M. Benchohra, J. Henderson, S.K. Ntouyas and A. Quahab. On first order impulsive dynamic equations on time scales, J. Differ. Equ. Appl., 6, pp. 541-548(2004).
7. M. Benchohra, S. K. Ntouyas and A. Ouahab. Extremal solutions of second order impulsive dynamic equations on time scales, J. Math. Anal. Appl., 324(2006), pp. 425-434.
8. M. Benchohra, S. K. Ntouyas and A. Ouahab. Existence results for second order boundary value problem of impulsive dynamic equations on time scales, J. Math. Anal. Appl., 296(2004), pp. 65-73.
9. M. Bohner and A. Peterson. Dynamic Equations on Time Scales: An Introduction with Applications, Birkhäuser, Boston, 2003.
10. A. Cabada and D. R. Vivero. Existence of solutions of first-order dynamic equations with nonlinear functional boundary value conditions, Nonlinear Analysis: Theory, Methods and Applications, vol. 63, no. 5-7, pp. e697-e706, 2005.
11. A. Cabada. Extremal solutions and Green's functions of higher order periodic boundary value problems in time scales, Journal of Mathematical Analysis and Applications, vol. 290, no. 1, pp. 35-54, 2004.
12. Y. Chang and W-T. Li. Existence results for impulsive dynamic equations on time scales with nonlocal initial conditions, Mathematical and Computer Modeling, 43(2006), pp. 377-384.
13. H. Chen and H. Wang. Triple Positive Solutions of Boundary Value Problems for p-Laplacian Impulsive Dynamic Equations on Time Scales, Mathematical and Computer Modeling, 47(2008), pp. 917-924.
14. Q. Dai and C. C. Tisdell. Existence of solutions to first-order dynamic boundary value problems, International Journal of Difference Equations, vol. 1, no. 1, pp. 1-17, 2006.
15. F. Davidson and B. Rynne. Eigenfunction Expansions in L^2 spaces for Boundary Value Problems on Time Scales. J. Math. Anal. and Appl., Vol. 335, Issue 2, 2007, pp. 1038-1051.
16. J. Davis, J. Henderson, K. Rajendra Prasad and W. Yin. Solvability of a nonlinear second order conjugate eigenvalue problem on a time scale, Abstract and Applied Analysis, 5:2, 2000, pp. 91-99.
17. A. Dogan. Existence of multiple positive solutions for p-Laplacian multipoint boundary value problems on time scales, Advances in Difference Equations, 2013.
18. A. Dogan, J. Graef and L. Kong. Higher-order singular multi-point boundary value problems on time scales. Proc. Edinb. Math. Soc. 54, 345-361 (2011).
19. L. Erbe and A. Peterson. Green's functions and comparison theorems for differential equations on measure chains. Dynam. Contin. Discrete Impuls. Systems, 6(1), 121-137, 1999.
20. L. Erbe and A. Peterson. Positive solutions for a nonlinear differential equation on a measure chain. Math. Comput. Modelling, 32(5-6), 571-585, 2000.
21. L. Erbe, A. Peterson and C. Tisdell. Existence of solutions to second-order BVPs on time scales, preprint.
22. V. Otero-Espinar and T. Pernas-Castano. Variational Approach to Second-Order Impulsive Dynamic Equations on Time Scales, Boundary Value Problems, 2013, pp. 1-15.

23. S. Georgiev and Z. Khaled. Multiple fixed-point theorems with applications in ODE, FDE and PDE, CRC Press, 2020.
24. S. Georgiev. Integral equations on time scales, Atlantis Press, 2016.
25. C. Goodrich. The existence of a positive solution to a seciond-order delta-nabla p-Laplacian BVP on time scale, Applied Mathematics Letter, 25(2012), pp. 157-162.
26. W. Guan. Positive solutions for nonlinear first-order impulsive dynamic equations on time scales. Advances in Difference Equations, (2015), pp. 1-9.
27. W. Guan, D-G. Li and Sh-H. Ma. Nonlinear first-order periodic boundary-value problems of impulsive dynamic equations on time scales, EJDE, Vol. 2012(2012), No. 198, pp. 1-8.
28. D. Guo and V. Lakshmikantham. Nonlinear Problems in Abstract Cones, Academic Press, New York, 1988.
29. Z. He. Existence of two solutions of m-point boundary value problem for second order dynamic equations on time scales, J. Math. Anal. Appl., 296(2004), pp. 97-109.
30. J. Henderson, A. Peterson and C. Tisdel. On the existence and uniqueness of solutions to boundary value problems on time scales, Advances in Difference Equations, 2004:2(2004), pp. 93-109.
31. G. Sh. Guseinov. Eigenfunction expansions for a Sturm-Liouville problem on time scales. Int. J. of Diff. Eq., Vol. 2, No. 1, 2007, pp. 93-104.
32. A. Huseynov and E. Bairamov. On expansions in eigenfunctions for second order dynamic equations on time scales. Nononlinear Dynamics and Systems Theory, Vol. 9, No. 1, 2009, pp. 77-88.
33. E. Kaufmann, N. Kosmatov and Y. Raffoul. Impulsive dynamic equations on a time scale, EJDE, Vo. 2008(2008), pp. 1-9.
34. E. Kaufmann. Impulsive periodic boundary value problems for dynamic equations on time scales, Advances in Difference Equations, Vol. 2009, 2009, pp. 1-10.
35. C. Kunkel. Positive solutions to singular second-order boundary value problems on time scales, Advances in Dynamical Systems and Applications, Vol. 14, No. 2, pp. 201-211, 2019.
36. R. W. Leggett and L. R. Williams. Multiple positive fixed points of nonlinear operators on ordered Banach spaces, Indian Univ. Math. J., 28(1979), pp. 673-688.
37. Y. Li and J. Shu. Multiple positive solutions for first-order impulsive integral boundary value problems on time scales, Boundary Value Problems, 2011, pp. 1-19.
38. Y. Li and H. Zhang. Extremal solutions for periodic boundary value problems for first order impulsive integro-differential equations of mixed type on time scales, Boundary Value Problems, Vol. 2007, 2007, pp. 1-16.
39. Y. Li and T. Zhang. Multiple positive solutions for second-order p-Laplacian dynamic equations with integral boundary conditions, Boundary Value Problems, Vol. 2011, 17 pages.
40. W-T. Li and X-L. Liu. Eigenvalue problem for second-order nonlinear dynamic equations on time scales, J. Math. Anal. Appl., 318(2005), pp. 578-592.
41. S. Liang and J. Zhang. The existence of countably many positive solutions for nonlinear singular m-point boundary value problems on time scales, Journal of Computational and Applied Mathematics, Vo. 223, no. 1, pp. 291-303, 2009.
42. S. Liang, J. Zhang and Z. Wang. The existence of three positive solutions of m-point boundary value problems for some dynamic equations on time scales, Mathematical and Computer Modeling, Vol. 49, no. 7-8, pp. 1386-1393, 2009.
43. X. Lin and Z. Du. Positive solutions of m-point boundary value problem for second-order dynamic equations on time scales, Journal of Difference Equations and Applications, Vol. 14, no. 8, pp. 851-864.
44. H. Luo. Bifurcation from interval and positive solutions of a nonlinear second-order dynamic boundary value problem on time scales, Abstract and Applied Analysis, Vol. 2102, 15 pages.
45. A. Oguz and F. Topal. Symmetric positive solutions for second order boundary value problems with integral boundary conditions on time scales, Journal of Applied Analysis and Computation, Vol. 6, No. 2, 2016, pp. 531-542.
46. A. Kameswara Rao and S. Negaswara Rao. Eigenvalues for iterative systems of nonlinear second order boundary value problems on time scales, Bulletin of International Mathematical Virtual Institute, Vol. 2, (2012), pp. 47-57.

47. B. Rynne. L^2 spaces and Boundary Value Problems on Time Scales. J. Math. Anal. and Appl., 328, 2007, pp. 1217-1236.
48. Y. Sang and H. Su. Several existence theorems of nonlinear m-point boundary value problem for p-Laplacian dynamic equations on time scales. J. Math. Anal. Appl. 340, 1012-1026 (2008).
49. A. Sinanoglu, I. Karaca. F. Tokmak and T. Senlik. Existence of three symmetric positive solutions for a second-order multi-point boundary value problem on time scales, Advances in Difference Equations, 2014.
50. J.-P. Sun and W.-T. Li. Existence of solutions to nonlinear first-order PBVPs on time scales, Nonlinear Analysis: Theory, Methods & Applications, vol. 67, no. 3, pp. 883-888, 2007.
51. J.-P. Sun and W.-T. Li. Existence and multiplicity of positive solutions to nonlinear first-order PBVPs on time scales, Computers & Mathematics with Applications, vol. 54, no. 6, pp. 861-71, 2007.
52. J.-P. Sun. Twin positive solutions of nonlinear first-order boundary value problems on time scales, Nonlinear Analysis: Theory, Methods & Applications, vol. 68, no. 6, pp. 1754-1758, 2008.
53. J.-P. Sun and W.-T. Li. Positive solution for system of nonlinear first-order PBVPs on time scales, Nonlinear Analysis: Theory, Methods & Applications, vol. 62, no. 1, pp. 131-139, 2005.
54. J.-P. Sun and W.-T. Li. Positive solutions to nonlinear first-order PBVPs with parameter on time scales, Nonlinear Analysis: Theory, Methods & Applications, vol. 70, no. 3, pp. 1133-1145, 2009.
55. H.-R. Sun and W.-T. Li. Positive solutions for nonlinear three-point boundary value problems on time scales, Journal of Mathematical Analaysis and Applications, Vol. 299, no. 2, pp. 508-524, 2004.
56. H. -R. Sun. Triple positive solutions for p-Laplacian m-point boundary value problem on time scales. Comput. Math. Appl. 58, 1736-1741 (2009).
57. Y. Tian and W. Ge. Existence and uniqueness results for nonlinear first-order three-point boundary value problems on time scales, Nonlinear Analysis: Theory, Methods & Applications, vol. 69, no. 9, pp. 2833–2842, 2008.
58. S. G. Topal and A. Yantir. Positive solutions of a second order m-point BVP on time scales, Nonlinear Dynamics and Systems Theory, Vol. 9, no. 2, pp. 185-197, 2009.
59. D-B. Wang. Positive solutions for nonlinear first-order periodic boundary value problems of impulsive dynamic equations on time scales. Computers and Mathematics with Applications, 56(2008), pp. 1496-1504.
60. Z. Wang and P. Weng. Existence of solutions for first order PBVPs with impulses on time scales. Computers and Mathematics with Applications, 56(2008), pp. 2010-2018.
61. Y. Yang and D. Xiao. On existence of multiple positive solutions for -Laplacian multipoint boundary value. Nonlinear Anal. 71, 4158-4166 (2009).
62. I. Yaslan. Positive solutions for nonlinear first-order m-point boundary value problem on time scales, Abstract and Applied Analysis, Vol. 2012, 12 pages.
63. I. Yaslan. Multiple positive solutions for nonlinear three-point boundary value problems on time scales, Compures & Mathematics with Applications, Vol. 55, no. 8, pp. 1861-1869, 2008.
64. C. Yuan and Y. Liu. Multiple positive solutions of a second order nonlinear semipositone m-point boundary value problem on time scales, Abstract and Applied Analaysis, Vol. 2010, 19 pages.
65. Q. Zhang, X. He and H. Sun. Positive Solutions for Sturm-Liouville BVPs on Time Scales via ub-Supersolution and Variational Methods. BVP, 2013.
66. Y.-H. Zhao and J.-P. Sun, Monotone iterative technique for first-order nonlinear periodic boundary value problems on time scales, Advances in Difference Equations, vol. 2010, Article ID 620459, 10 pages, 2010.
67. Ya-Hong Zhao. First-order boundary value problem with nonlinear boundary condition on time scales, Discrete Dynamics in Nature and Society, Vol. 2011, Article ID 845107, 8 pages.

Index